MODERN DEVELOPMENTS IN POWDER METALLURGY

Volume 20

MODERN DEVELOPMENTS IN POWDER METALLURGY

Edited by Henry H. Hausner

1966: Proceedings of the 1965 International Powder Metallurgy Conference, New York, NY USA
- *Volume 1: Fundamentals and Methods*
- *Volume 2: Applications*
- *Volume 3: Development and Future Prospects*

1971: Proceedings of the 1970 International Powder Metallurgy Conference, New York, NY USA
- *Volume 4: Processes*
- *Volume 5: Materials and Properties*

Edited by Henry H. Hausner and Walter E. Smith

1974: Proceedings of the 1973 International Powder Metallurgy Conference, Toronto, Canada
- *Volume 6: Applications and Processes*
- *Volume 7: P/M Forging, Copper P/M*
- *Volume 8: Ferrous P/M, Special Materials*

Edited by Henry H. Hausner and Pierre W. Taubenblat

1977: Proceedings of the 1976 International Powder Metallurgy Conference, Chicago, Illinois USA
- *Volume 9: P/M Principles and Production Processes*
- *Volume 10: Ferrous and Non Ferrous P/M Materials*
- *Volume 11: P/M Special Materials and Applications*

Edited by Henry H. Hausner, Harry W. Antes and Gaylor D. Smith

1981: Proceedings of the 1980 International Powder Metallurgy Conference, Washington, DC USA
- *Volume 12: Principles and Processes*
- *Volume 13: Ferrous and Nonferrous Materials*
- *Volume 14: Special Materials*

Edited by Edward N. Aqua and Charles I. Whitman

1985: Proceedings of the 1984 International Powder Metallurgy Conference, Toronto, Canada
- *Volume 15: Principles and Processes*
- *Volume 16: Ferrous and Nonferrous Materials*
- *Volume 17: Special Materials*

Compiled by P. Ulf Gummeson and Donald A. Gustafson

1988: Proceedings of the 1988 International Powder Metallurgy Conference, Orlando, Florida USA
- *Volume 18*
- *Volume 19*
- *Volume 20*
- *Volume 21*

MODERN DEVELOPMENTS IN POWDER METALLURGY

Proceedings of the 1988 International Powder Metallurgy Conference, sponsored by the Metal Powder Industries Federation and the American Powder Metallurgy Institute, June 5–10, 1988, Orlando, Florida.

Compilers

P. ULF GUMMESON

Vice President, Research and Development
Hoeganaes Corporation
Riverton, New Jersey

DONALD A. GUSTAFSON

Account Executive
Ferraloy, A Pullman Company
Salem, Indiana

Volume 20

METAL POWDER INDUSTRIES FEDERATION
AMERICAN POWDER METALLURGY INSTITUTE
105 College Road East, Princeton, New Jersey 08540 • (609) 452-7700

These four-volume Proceedings of the 1988 International Powder Metallurgy Conference, published under the title Modern Developments in Powder Metallurgy, also comprise Volume 44 of the series Progress in Powder Metallurgy, published by the Metal Powder Industries Federation.

The papers contained in this publication were prepared by the author(s) and reviewed by the session chairmen and submitted for subsequent publication by direct reproduction. The publisher is not responsible for content and/or any deviations from oral presentations from which the papers may have been based.

ISSN No. 0074-7513
ISBN Number 0-918404-79-7
ISBN Number 0-918404-81-9 (for set of volumes 18,19,20,21)
Library of Congress Catalog Card Number 66-5483

105 College Road East, Princeton, N.J. 08540

Printed in the United States of America

FOREWORD
P/M '88 Proceedings

The 1988 International Powder Metallurgy Conference, sponsored by the Metal Powder Industries Federation and the American Powder Metallurgy Institute, was held June 5-10, 1988 at the Hyatt Regency Grand Cypress Hotel and the Buena Vista Palace Hotel, in Orlando, Florida. The proceedings of this Conference, Volumes 18–21 of "Modern Developments in Powder Metallurgy," contain the reports of original research and the status of current technologies as presented in technical sessions, seminars, and poster sessions. These papers are supplemented by plenary presentations on the status of the P/M industries in North America, Europe, the United Kingdom, West Germany, Italy, Scandinavia, Benelux, France, Korea, Japan, the People's Republic of China, India, the U.S.S.R., and Australia.

The theme of the Conference, "World Showcase of Powder Metallurgy" was appropriately supported by an International Coordinating Council, comprised of worldwide association and society leaders, and by the participation of more than 2,200 delegates from thirty-eight countries attending the various technical presentations and the large exhibition. The gratifyingly large number of abstracts submitted for inclusion in the 1988 program were evaluated by a Program Committee consisting of some thirty-nine dedicated individuals and the necessary difficult judgments made. The challenge to and responsibility of the Program Committee members differed somewhat from previous years in that once the abstracts were chosen and approved and divided into sessions, the members assumed complete responsibility for one or two sessions. This responsibility included working with the authors to assure that all deadlines were met, instructing the authors on the routine to be followed during the presentations, in most instances, chairing the particular sessions, and assuring the quality of the written presentation which is included in these proceedings.

The Technical Program Committee Co-Chairmen express their thanks for the very excellent job done by this year's Program Committee. Their very professional work contributed immensely to the success of P/M '88.

A number of seminars were included in this conference. The topics were: Secondary Operations; Metal Injection Molding; Advanced Sintering Technology; Properties, Evaluation and Testing of P/M Materials; and Prevention and Detection of Cracks in Ferrous P/M Parts. Seminar pre-prints were provided to the participants of the various sessions during the Conference and have subsequently been available through MPIF.

A very special event at this year's Conference was the awarding of the prestigious Pioneer Award to Sven Hulthen of Sweden in recognition of his many contributions to the growth of the technology worldwide. Mr. Hulthen is officially retired but continues to exhibit an active interest in P/M and can frequently be found at industry meetings.

The four volumes which comprise the P/M '88 proceedings are organized in a manner that allows the papers on related topics to appear together. The chapter titles correspond to the sessions in which they were presented at the Conference.

P. Ulf Gummeson
Program Co-Chairman

Donald A. Gustafson
Program Co-Chairman

PREFACE

1988 International Powder Metallurgy Conference

P/M '88 offered an unique opportunity to its participants to communicate with one another, t exchange ideas and to establish an international relationship that makes an industry grow and technology move forward. The recording of the formal presentations in these four volumes preserves fo posterity those contributions and makes them available to a still wider audience in perpetuity.

The sponsors of P/M '88, the Metal Powder Industries Federation and the American Powde Metallurgy Institute, as well as all those who were present, owe a debt of gratitude to the many individual who served on committees; to the professional staff of the Federation and the Institute at Princeton wh were responsible for all of the details and their coordination; to those companies whose contribution made the event a pleasant and memorable one; to the exhibitors who displayed the latest examples o advances in powder metallurgy technology; and, certainly, to the authors and moderators and thos associated with the technical aspects of the conference. As are all Federation sponsored technica conferences, P/M '88 was conducted under the auspices of the Federation's Technical Board. W acknowledge and thank all.

Below are the names of those who expended such efforts, who devoted the time and that of thei company, and gave of themselves to make P/M '88 a memorable, worthwhile and pleasant event for a who were privileged to be there:

Technical Program Committee

P. Ulf Gummeson, Co-Chairman
Hoeganaes Corporation

Donald A. Gustafson, Co-Chairman
Ferraloy

Eugene Andreotti
New Industrial Techniques

Harry W. Antes
SPS Technologies

Raymond C. Benn
INCO Alloys International Inc.

John W. Burlingame
U.S. Army Armament R&D Command

Joseph M. Capus
Quebec Metal Powders Limited

Earl A. Carlson
Ipsen Commercial Heat Treating Co.

Manek R. Dustoor
Ferraloy

Mark R. Eisenmann
Newmet Krebsoge

Donald L. Erich
Battelle Columbus Laboratories

Gary Gaines
Linde Div. Union Carbide Corp.

Carl J. Garcia
Honeywell Inc.

Peter W. Lee
The Timken Company

Paul Loewenstein
Nuclear Metals, Inc.

Patrick J. McGeehan
Hoeganaes Corporation

Henry Meeks
Ti Industries Corp.

Angela Moran
David Taylor Research Center

Karl D. Morrison
Detroit Diesel Allison-GMC

Kenneth H. Moyer
Magna-Tech P/M Labs

Aspi N. Patel
Battelle Columbus Division

Burton R. Patterson
University of Alabama at Birmingham

Leander F. Pease III
Powder-Tech Associates Inc.

Frank Rizzo
Crucible Compaction Metals

Reynold Sansoucy
The Presmet Corporation

Haskell Sheinberg
Los Alamos National Laboratory

Randall M. German
Rensselaer Polytechnic Institute

William Greim
Form Physics Corporation

Frank W. Heck
Novamet Specialty Products Corp.

W. Brian James
Hoeganaes Corporation

William F. Jandeska
General Motors Research Labs

Erhard Klar
SCM Metal Products, Inc.

Michael J. Koczak
Drexel University

Alan Lawley
Drexel University

Dennis E. Shewell
Ferraloy

Joseph Skurka
Federal-Mogul Corp.

Darrell W. Smith
Michigan Technological University

Victor C. Straub
Keystone Carbon Company

Krishna Vedula
Case Western Reserve University

Craig Waldo
IBM Corporation

J. Charles Webster
Ford Motor Company

International Coordinating Council

Europe
European Powder Metallurgy Federation
Dr. Ivor Jenkins

West Germany
Fachverband Pulvermetallurgie
Dr. Lothar Albano-Muller
Dr. Hermann Hassel

Ausschuss fur Pulvermetallurgie
Dr. Fritz Aldinger
Dr. Hermann Hassel

United Kingdom
The Institute of Metals
V. A. Tracey

British Powder Metal Federation
Peter Brewin

British Hard Metal Association
H. King

Austria
Fachausschuss fur Pulvermetallurgie
O. Prof. Dr. B. Lux

Italy
Associazone Italiana di Metallurgia
Dr. Gian Filippo Bocchini

Assinter
Dr. Ricardo Tosi

France
Societe Francais de Metallurgie
Prof. Andre Accary

Australia
Powder Metal Industries Association of Australia
David Whitelaw

India
Powder Metallurgy Association of India
N. T. George

Japan
Japan Powder Metallurgy Association
Kunizo Imanishi

The Japan Society for Powder and Powder Metallurgy
Dr. Tosio Takada

Korea
Korea Institute of Machinery and Metals
Prof. Sang-Kee Suh

People's Republic of China
The Powder Metallurgy Institute
Chinese Society of Metals
Prof. Lai Ho-yi

Chinese Society of Non-Ferrous Metals
Huang Peiyum

The Institute of Powder Metallurgy
China Mechanical Engineering Society
Shi Feng

U.S.S.R.
Byelorussian Powder Metallurgy Association
Prof. Dr. Pyotr A. Vityaz

Belgium, Netherlands, Luxembourg
Benelux Metallurgie
Yvan Du Bois

Scandanavia
Jernkontoret, The Powder Metallurgy Society
Bjoern Widell

South Africa
Powder Metallurgy Association of South Africa
Dr. Steven Parsons

A special thanks to the Session Chairmen who reviewed the papers published in the proceedings. Also acknowledged with appreciation is Teresa F. Stillman for coordinating the efforts of the Program Committee, organizing the sessions and subsequently, in conjunction with Elizabeth E. Carey, for preparing the proceedings for publication.

Princeton, New Jersey

Donald G. White
Conference Chairman

Contents

Volume 20

CHAPTER ONE - INFORMATION AND APPLICATION TRENDS

Award-Winning Parts Show Broad Spectrum of Powder Metallurgy Applications - P. K. Johnson 3

Information and Education Problems in Powder Metallurgy - R. A. Andrievski 13

New Product and Process Development in the P/M Parts Industry - C. I. Whitman 27

How Highly Innovative Components Developed for Business Machines May Foster Applications in Other Industrial Fields - O. Morandi and E. Mosca 35

CHAPTER TWO - POWDER PRODUCTION: MECHANICAL PROCESSES

Novel Process for Safe, Contamination-Free Size Reduction of Ferroalloys - D. J. Eddington 53

Preparation of Bulk Amorphous Ni60Nb40 - F. Petzoldt, B. Scholz, H-D. Kunze and M. Held 63

CHAPTER THREE - POWDER PRODUCTION: GAS ATOMIZING

Atomized Powders with Reduced Microstructure Variations - W. A. Kaysser, K. Rzesnitzek and G. Petzow 79

Surface Reactions During Atomization, Handling and Hot Consolidation of Fe- and Ni-Base Alloys - I. Olefjord and L. Nyborg 93

Structure and Properties of Atomized Beryllium Powder - J. M. Marder and D. H. Hashiguchi 109

Homogeneous Copper—Lead Powders - T. E. Fine 125

Atomized Material Metallurgy: Problems and Prospects - L. K. Druzhinin 137

Metal Powder Production by Ultrasonic Capillary or Standing Wave Technique - M. Hohmann, S. Jönsson and E. Lierke 159

The Tandem Atomization Process for Ultrafine Metal Powder Production - C. P. Ashdown, J. G. Bewley and G. B. Kenney 169

A New Technique for the Production of Gas Atomized Powder - C. Aslund and T. Tingskog 181

Internal Porosity of Gas Atomized Powders - L. L'Estrade, H. Hallen and R. Ljunggren 187

Observations of Gas Atomization Process Dynamics -
I. E. Anderson and R. S. Figliola 205

The Influence of Process Variables and Equipment Characteristics on the Purity of Atomized Powders - H. W. Meinhardt and P. Kunert 225

CHAPTER FOUR - POWDER PRODUCTION: IRON POWDERS

The Diverse Effects of Fe, Cu, Ni, Ag, Sio_2 Addition to Carbon on the Reduction Rate of Ferric Oxide - J. Jin 243

Morphological Changes During Reduction of Magnetite into Sponge Iron Powder - R. K. Dube and B. Deo 265

Perspectives in Iron Powder Manufacture: A Resource Approach -
K. C. Sahoo, D. K. Lahiri, S. Barpanda, G. S. Bhattacharjee and T. P. Bagchi 279

P/M Dual Phase Steel Processed from Ball Bearing Steel Swarf -
P. Ramakrishnan, A. N. Tiwari, V. S. Gorantiwar and J. K. Beddow 293

CHAPTER FIVE - POWDER PRODUCTION: METHODS AND PROPERTIES

Nickel Coated Powders Via the Carbonyl Nickel Gas Process -
E. L. LL. Rees, F. W. Heck and G. A. DiBari 311

Oil Atomization—A Method for the Production of Rapidly Solidified Iron-Carbon Alloys - K. U. Kainer and B. L. Mordike 323

Characterization of Carbonyl Iron Powder—Particle Size Distribution by Image Analysis and Light Scattering Techniques - J. E. Japka 339

CHAPTER SIX - P/M COMPOSITE MATERIALS

The Development of a New Dispersion Strengthened Vitallium® Alloy for Medical Implants - K. K. Wang, L. J. Gustavson and J. H. Dumbleton 361

Mechanical and Physical Properties of Iron-Based Cermets - N. Petersen ... 377

Duplex 4600 Powder Forged Structures Processing, Properties and Mechanics Analysis - M. N. H. Ahmed, M. J. Koczak and M. I. Negm .. 391

Wear Resistance of Al_2O_3-Reinforced High Speed Tool Steel -
R. A. Queeney, R. E. Masters and R. J. Beltz and J. D. Dankoff 409

Mechanical Behaviour of Cold Sintered High Speed Steel-Carbides Composites - E. Y. Gutmanas and D. Zak 421

Sintered Steel with High Carbide Content -
F. Thümmler, R. Oberacker and R. Klausmann 431

Interface Bonding in a Diamond/Metal Matrix Composite -
Z. Lin and R. A. Queeney 443

Titanium Diboride and Silicon Carbide Particulate Reinforced Aluminium Alloy Metal Matrix Composites – B. Roebuck and A. E. J. Forno 451

Design and P/M Processing of New Hierarchical Materials Structures – E. Y. Gutmanas and A. Lawley 467

Metal-Polymer Composite Material – M. S. Stoichev, Y. T. Simeonov, Zh. H. Zhelazkov and S. B. Milkov 481

CHAPTER SEVEN – INTERMETALLIC COMPOUNDS

Oxide Dispersion Strengthened Aluminides – K. Vedula, G. M. Michal and A. M. Figueredo 491

Production of a Precipitation Strengthening Cu-0.6wt%Zr Alloy by Hot Extrusion of a Rapidly Solidified Powder – A. M. Hallén, L. Arnberg, N. Bäckström and U. Backmark 503

Processing Effects on Densification in Reactive Sintering of Nickel-Aluminum Powder Mixtures – B. H. Rabin, A. Bose and R. M. German 511

Development of Intermetallic Matrix Composites for Elevated Temperature Service – G. E. Fuchs and W. H. Kao 531

Powder Processing of Ni_3Al – V. K. Sikka 543

The Effect of Order on the Tensile Properties of as Sintered and HIP'ped Ni_3Fe Intermetallic Compound – B. Lograsso, A. Bose, G. Camus, N. S. Stoloff, D. J. Duquette and R. M. German 559

CHAPTER EIGHT – HIGH-TEMPERATURE NICKEL-BASE MATERIALS

Process Development for Low-Cost, High-Strength P/M Ni-Base Superalloy Turbine Disk – H. Takigawa, N. Kawai, K. Iwai, S. Furuta and N. Nagata 575

Rapidly Solidified Ni-Mo-Fe-B Alloy for High Temperature Wear Resistance Applications – Y. J. Park, S. K. Suh, W. W. Park and C. J. Choi 587

Superplasticity of HIPped P/M Superalloys Made from Attrited Prealloy Powder – N. Uenishi, Y. Takeda and K. Kuroishi 599

CHAPTER NINE – RAPIDLY SOLIDIFIED MATERIALS

Consolidation of Rapidly Solidified Magnesium Alloy Powders – S. Krishnamurthy and Y-W. Kim 615

Fabrication of Shaped Charge Liners by Rapid Solidification Techniques and P/M Processes – C. Gélinas, R. Angers and B. Voyzelle 629

CHAPTER TEN - STRUCTURE AND PROPERTIES OF FERROUS ALLOYS

Measurement of Compressive Yield Strength and Poisson's Ratio of FN-0208, Heat Treated, at 6.85 g/cm³ - J. C. Webster, J. J. Lengyel, L-L. Tu and L. F. Pease III 643

Mechanical Properties and Microstructure of Phosphorus Alloyed Sintered Steel - P. Engdahl 655

CHAPTER ELEVEN - TECHNICAL BRIEFS— POSTER PRESENTATIONS

Properties of Densified Steel Grit Powders with Carbides or Iron Additions - A. Kobylanski, J. Leroux, P. VuChat and D. Huet 669

The Effects of Microstructure of Gas-Atomized Zn-22%Al Alloy Powder on Densification Process During Hot Pressing - M. Tokizane, K. Isonishi, M. Hashii and K. Ameyama 675

Manufacturing and Processing of HIPped and/or Forged Steel and Stellite Compounds - B. Hribernik, J. Stamberger, H. Jäger and O. Pacher 681

Mechanical Alloying of Aluminum-Lithium-Magnesium Powders - A. A. Layyous, S. Nadiv and I. J. Lin 691

New Applications of Sinter Technology for Production of Machine Elements - Z. Marcela, R. Moszumański, Z. Polański, A. Sulkowski and W. Tabor 701

Preparation of Amorphous FeTa-Powders by Mechanical Alloying (MA) - G. Veltl, B. Scholz and H-D. Kunze 707

Aerodynamic Study of Gas Flow from a Molten Metal Atomizing Nozzle - M. S. Stoichev and I. S. Antonov 713

A Modern Powder Metal Press for Universal Applications - W. Beisel 723

Development of Multiphase Alloys of Refractory Metals by Using Sinteractive Composite Powders - S. Haerdtle and R. Schmidberger 729

Novel and Alternative Materials for Production of Harrow Tines - O. Kraemer and L. Roesen 733

HIP—A Way to Increase Life Time of Cast Inconel 713 C - C. Persson, P. O. Persson and M. Ostensson 739

Material Defects in a P/M-Nickel-Base Superalloy - B. Nowak, H. Lahodny, E. Arzt, G. König and E. Affeldt 745

Pressureless Sintering and HIPping of β_{ss}-Si_3N_4 + SiC for Cutting Tools - A. A. Layyous, P. Greil and G. Petzow 751

Compression Molding of Fine Ceramic Powder by Using Water Binder - T. Nakagawa, L. Zhang, H. Noguchi, N. Takahashi and K. Suzuki 763

Variation of the Microstructure of P/M Astroloy LC and Its Effect on Static Strength at Room Temperature - W. Track, H. Floege, and H. Pillhoefer 773

Studies on the Sintering of Parts of Austenitic Stainless Steel - M. Rosso, G. Scavino, E. Galetto and G. Pesce 779

High Resolution Auger Analysis of 12Cr-Steel Powder Annealed at Consolidation Temperatures - L. Nyborg and I. Olefjord 787

Author Index 793

Subject Index 799

Contents

Volume 18

The Growing Internationalization of Powder Metallurgy - D. G. White 1

CHAPTER ONE - HOT ISOSTATIC PRESSING

Prediction of the HIP Behaviour of Complex Materials - A. Frisch, W. A. Kaysser, M. Aslan, and G. Petzow 11

Shape Development of Powder Bodies Modelled as a Result of the Densification Process in Hot Isostatic Pressing - B. G. Arén and E. Navara 23

The Application of HIP/PM Technique for Gas and Steam Turbines - R. Ekbom 37

Microstructure of Hot Isostatic Pressed P/M Medium-Carbon 12% Chromium Steels as a Function of the Initial Powder Particle Size and Composition - R. L. Brown 57

Near Net Gear Preforms From Prealloyed Gas Atomized Powders - W. L. Wentland, V. K. Chandhok and G. R. McIndoe 71

Inhomogeneities in Materials HIPped at Lower Pressures - W. A. Kaysser, M. Aslan, A. Frisch and G. Petzow 77

Sinter-HIP Furnaces—Sintering and Compacting in a Combined Cycle - R. E. Bauer 91

CHAPTER TWO - POWDER FORGING

Manufacturing of Clean Steel Powders for P/F Applications - M. Gagné, C. Ciloglu and Y. Trudel 107

Microcleanliness Studies of Low Alloy and Carbon Steel Powders Intended for Powder Forging Applications - W. B. James, R. J. Causton, J. M. Castelli, T. F. Murphy and H. S. Shaw 119

Fatigue Life as a Function of the Mean Free Path Between Inclusions - T. Prucher 143

Mechanical Properties of P/F Connecting Rod Materials - A. Bhambri and L. F. Pease III 155

The Fatigue Properties of Heat-Treatable Low Alloy, Powder Metallurgy Steels - I. J. Mellanby and J. R. Moon 183

Properties of Powder-Forged Type 46XX Steels - J. W. Burlingame, S. R. Buzolits, T. Leister and H. W. Antes 197

Optimization of the Properties by Processing and Heat Treatment of Powder Forged 4600 Alloys - M. N. H. Ahmed and M. J. Koczak 213

Application of Powder Forging (P/F) to Ring Shaped Ordnance Components - J. R. Senick, Jr., P. A. Thornton and P. S. Raghupathi ... 231

Powder Forging of Helical Gears for Passenger Car Manual Gear Boxes—Concept and Properties - W. König, G. Röber, K. Vossen and M. Strömgren ... 253

Powder Forging of Cylindrical Gears with Multiphase Structure - W. König, H. Saljé, G. Rober, M. Strömgren, M. Weber, H-J. Dorweiler, K-H. Lindner, H. Schiller and H. Schiefer ... 263

Application of Finite Element Modeling to Powder Forging - B. L. Ferguson ... 269

CHAPTER THREE - METAL INJECTION MOLDING

Complete Densification of Copper Sintered Compact by Addition of Iron Powder - K. Hayashi, T. W. Lim and M. Itabashi ... 287

Injection Molding of an Intermetallic Matrix Composite - A. Bose and R. M. German ... 299

Nitrogen/Hydrogen Atmospheres for Continuous Sintering of Metal Injection Molded Components - C. W. Parten and T. V. Hermes .. 315

Evaluation of MCM Alloy in the Production of Injection Molded Steel Components - H. E. Amaya and J. St. Pierre ... 323

Mechanical Properties of Injection Molded Iron-Nickel Alloys - P. Bhave, W. Dormon and D. Teel ... 333

Microstructural Design of Metal Injection Molded Parts by Enhanced Sintering Techniques - J. Beuers, W. Achterath and J. H. Hauβelt ... 351

High Performance Soft Magnetic Components by Powder Metallurgy and Metal Injection Molding - C. Lall and L. W. Baum ... 363

The Effect of Powder Characteristics on Binder Incorporation for Injection Molding Feedstock - J. Warren and R. M. German ... 391

Influence of Process Variables on Debinding by Melt Wicking - C. S. Aria and B. R. Patterson ... 403

Modeling and Testing of Powder Injection Molding Processes - D. Lee, K. F. Hens, B. O. Rhee and C. M. Sierra ... 417

CHAPTER FOUR - MODELING OF CONSOLIDATION PROCESSES

Stress Analysis in a P/M Tool During Compacting Operation - M. B. Santos, M. T. Pellegrini and J. E. L. Bicudo ... 433

A Simple Method for Predicting Density Heterogeneity and Tool-Powder Interactions in P/M Consolidation Processes Involving Plastic Deformation of the Particles - P. Lasne and C. Levaillant ... 449

A Constitutive Model for the Compaction of Metal Powders - S. B. Brown and G. G. A. Weber ... 465

CHAPTER FIVE - CONSOLIDATION AND COMPACTION: ALTERNATIVE APPROACHES

Unconventional Methods of Pulsatory Pressing of Crushed Materials - R. Moszumański 479

Rotary Cold Repressing—The Key to Restructurization of the Forging Industry? - J. Pregowski and W. Tabor 493

Advantages and Limitations of Sizing and Coining of Sintered Parts - G. F. Bocchini 507

CHAPTER SIX - LIQUID PHASE SINTERING

New Alloy Systems for High Strength Applications - K. H. Moyer and J. B. Ryan 535

Optimization of the Processing of Iron-Phosphorus Materials - J. D. Ruhkamp and R. Kubicki 551

Enhancement of Alloying in Mixed Powder Ferrous Alloy Compacts - J. A. Lund and X. Qu 569

Effect of Ternary Additions on the Sintered Properties of 316L Stainless Steel and Its Yttria Containing Composites - S. Lal and G. S. Upadhyaya 581

CHAPTER SEVEN - THEORY OF LIQUID PHASE SINTERING

Effect of Metalloid Additions on Sintering of Rapidly Solidified Powders - S. T. Lin, R. M. German, S. Farooq, L. L. Bourguignon and D. P. Agarwal 597

Effect of Initial Particle Size Distribution on Ostwald Ripening During Liquid Phase Sintering - Z. Fang and B. R. Patterson 615

CHAPTER EIGHT - NOVEL P/M TECHNOLOGY

Properties of Beryllium Consolidated by Several Near-Net-Shape Processes - D. H. Hashiguchi, T. P. Clement and J. M. Marder 627

Explosive Compaction of Copper and Tungsten Powders - D. Chaiat, G. Raz and S. Nadiv (deceased) 645

Ballistic Particle Manufacturing: A New Processing Technology - D. Hauber 661

Joining of Al_2O_3 to P/M Iron and Nickel Alloys by Cold Sintering - I. Gotman and E. Y. Gutmanas 675

CHAPTER NINE - INNOVATION IN TOOLING

The Strategic Role of Tools in P/M Parts Production - G. F. Bocchini 691

Update of Preweighing Press Charges - L. L. Hierath and C. R. Riblett 711

Lubrication of Compacting Dies - J. Planas and C. B. Molins 723

CHAPTER TEN - MAGNETICS: SOFT MAGNETIC MATERIALS

Soft Magnetic Materials in P/M: Current Applications and State-of-the-Art - J. A. Bas, J. Puig and C. B. Molins 745

Emerging P/M Alloys for Magnetic Applications - K. H. Moyer and J. B. Ryan 757

CHAPTER ELEVEN - MAGNETICS: SUPERCONDUCTIVITY AND SHIELDING

Manufacturing Rapidly Solidified FE-ND-B Powders - J. Dickson 775

The Research of Superconducting Thermal Spray Powder - C-M. Liao, X-L. Chen, D-M. Du and X-W. Qi 789

Fabrication of a Magnetic Composite Material by Use of Zn-22A1 Superplastic Powder - K. Okimoto, T. Satoh and N. Horiishi 799

Author Index 815

Subject Index 821

Contents

Volume 19

CHAPTER ONE - CARBIDES: PROCESSING

Effects of Cobalt Level on Attritor Milling Behavior of WC-Co Powder Mixtures - J. A. Payne, D. W. Smith and T. E. Hale ... 3

Investigation on Spinodal Decomposition of WC-TiC-ZrC-Co Cemented Carbide - R. Xu, P. Huang and H. Yang ... 15

Investigation of Heat Treatment of WC-Co Hardmetals and Transformation Temperature of Cobalt Binder Phase - H-Y. Lai and J. Yang ... 23

CHAPTER TWO - CARBIDES: PROPERTIES

Palmqvist Indentation Fracture Toughness of WC-Co Cemented Carbide Alloys - S. H. Zhang and Y. X. Liu ... 33

Nondestructive Evaluation of Mechanical Properties of WC-Co Alloys - A. Haerian ... 43

Investigation on Repeated Impact Property of WC-Co Cemented Carbides - R. Xu, C. Wang, X. Ling, C. Xiong and X. Qu ... 53

CHAPTER THREE - TUNGSTEN MULTI-PHASE MATERIALS

Liquid Phase Sintering Under Microgravity in Space - L. B. Ekbom and A. Eliasson ... 63

Production and Sintering of Ag-W Composites Containing Ni - F. R. Sale and J. N. Albiston ... 75

Solid Phase Sintering of W Base Alloy with Al_2O_3 Additions by Hot Isostatic Pressing (HIP) - O. Botstein and B. Shpigler ... 91

CHAPTER FOUR - NEW TUNGSTEN ALLOYS

Effect of Silicon Addition in Rapidly Solidified Tungsten Alloys - S. H. Whang, S. U. Kim and D. Kapoor ... 105

Microstructural Characteristics and Mechanical Behavior in Rapidly Solidified W-Ge Alloys - D. Vujic, S. H. Whang and D. Kapoor ... 121

Properties of Swaged and Aged Molybdenum Doped Heavy Alloys - A. Bose and R. M. German ... 139

Heavy Alloys from Rapidly Solidified Powders - R. F. Cheney, R. L. Daga, R. M. German, A. Bose and J. W . Burlingame ... 155

CHAPTER FIVE - REFRACTORY ALLOYS

High Temperature High Strength Molybdenum and Molybdenum-Tungsten Ti-Zr-Hf-C-Alloys - R. Eck and J. Tinzl 173

Evaluation of Powder Metallurgy Processed Nb-30Hf-9W (WC-3009) - C. C. Wojcik 187

ODS-Niobium Alloy Its Properties and Applications - U. Gennari, E. Kny, P. Warbichler and H. Hofer 201

CHAPTER SIX - ACTIVATED SINTERING OF TUNGSTEN ALLOYS

Effects of Ni-Additive on the Grain Growth of Tungsten Wire - I-H. Moon, S. T. Oh and Y. L. Kim 223

Rate Controlled Sintering of Ni Doped W - W. A. Kaysser and I. S. Ahn .. 235

The Effect of Particle Size Distribution on Grain Growth in Hot Isostatically Pressed Tungsten - C. French and B. R. Patterson 249

High Temperature Properties of Ni-Activated Sintered W-Powder Compact - I-H. Moon and Y-H. Kim 259

CHAPTER SEVEN - TOOL STEELS AND CUTTING MATERIALS

Drilling Performance of AISI T1 and M2 High Speed Steel Drill Fabricated by Normal P/M Process - S-C. Yoo, H-S. Kim and I-H. Moon 271

Toughness Comparison Between Vacuum Sintered Plus Forged PMS T15 and Conventional Tool Steels - P. Beiss and R. Wähling 281

The Effect of Structural Development on Wear Properties of Sintered High Speed Steels - P. D. Nurthen, P. Brewin, B. Toloui, J. V. Wood, M. Igharo, D. S. Coleman and Q. Shaikh 305

Combined Effects of Sintering, HIP'ing and Heat Treating Parameters on the Cutting Properties of T15 High-Speed Steel - I. M. Martins, M. Santos, M. M. Oliveira and H. Carvalhinhos 319

Effect of Particle Size and Shape on Sintering of High Speed Steels - K. M. Kulkarni 329

New Cutting Materials Based on Titanium Carbonitride - R. Porat and A. Ber 345

Wear Behaviour of Sintered T15 High Speed Steel Inserts - R. M. D. Mesquita and M. J. M. Barata Marques 365

Cutting Performance of PCBN and PCD for P/M Parts - T. Nakai, M. Goto, Y. Saegusa and S. Nakatani 379

CHAPTER EIGHT - SPRAY FORMED ALUMINUM ALLOYS

Fundamentals of Droplet Consolidation During Spray Deposition - D. Apelian, A. Lawley, P. Mathur and X. J. Luo 397

Comparative Evaluation of the Osprey Process with Powder Metallurgy and Ingot Metallurgy for a High Strength Aluminium Alloy - J. F. Faure and L. Ackermann 425

The Osprey Preform Process and Its Application to Aluminium Alloys Including Wear and High Temperature Resistance Al-20Si-Cu Alloys - J. Duszczyk, J. L. Estrada, B. M. Korevarr, T. L. J. de Haan, D. Bialo, A. G. Leatham and A. J. W. Ogilvy 441

CHAPTER NINE - SPRAY FORMING: PROCESSING, STRUCTURE AND PROPERTIES

Heat Transfer and Solidification Behavior During Spray Atomization and Deposition of Low Carbon Steels - E. Lavernia, E. Gutierrez-Miravete and G. Trapaga 457

The Production of Advanced Materials by Means of The Osprey Process - A. G. Leatham, A. J. W. Oglivy, P. F. Chesney and O. Metelmann 475

Evolution of Microstructure in Spray Cast Cu-Zr - R. P. Singh and A. Lawley 489

Alloy 625 Piping Produced Via Near-Net-Shape Processing - A. L. Moran and W. A. Palko 503

CHAPTER TEN - STRUCTURE AND PROPERTIES OF ALUMINUM BASE ALLOYS

Characteristics of Powders and Mechanical Properties of Extrusions for Rapidly Cooled Al-8Mass% Alloy Powders Produced by Flame Spraying Technique - M. Yokota and K. Shoji 519

Hot Forgeability of Hot Extruded P/M Al-Si-Fe-X Alloys - Y. Takeda, T. Hayashi, Y. Odani, N. Amano and N. Kuroishi 533

Extrusion of High-Strength Aluminium P/M Alloys and Composites - A. Jokinen 547

Forging of a P/M Aluminium Alloy Connecting Rod: Experimental Results and Computational Preform Design - C. Laugée, J. L. Querbes, H. Grazzini, P. Lasne and C. Levaillant 565

Boron/Magnesium-Aluminum Sintered Composite Alloy - E. Horikoshi, T. Iikawa and T. Sato 579

Interfaces of Electro-Discharged Compacts - H-R. Pak, D. K. Kim and K. Okazaki 591

Sinter-Forged Al 2214: Processing, Structure and Property Relations - J. F. Faure and G. Brunel 603

CHAPTER ELEVEN - COPPER BASE AND BEARING ALLOYS

Superior Quality Bronze Bearings - K. B. M. Patel 621

Changes in Dimensional Change for Bronze Premixes as Premix Components Change - A. B. Backensto 641

Copper-Nickel-Tin Spinodal Alloys for Powder Metallurgy Applications - J. H. Reinshagen and A. J. Neupaver 655

The Effects of Oxygen on Alloy Formation, Grain Growth and Dimensional Change in the Alpha Microstructure of 90/10 Bronze - W . D. Marks 671

Structure, Property and Performance in Fe-Bi-Sn Bearing Alloy - Y. T. Chen 681

The Use of Hot Isostatic Pressing (HIP) for the Development of New High Efficiency Materials - A. Havalda, S. Emmer, V. Pulc and A. Salák 697

CHAPTER TWELVE - P/M TITANIUM

Review and Status of Titanium Materials Produced from Spherical Prealloyed Powder - J. H. Moll and C. F. Yolton 709

Tensile Properties and Corrosion Behavior of Hot Isostatically Pressed Ti-Fe Alloy - K. Majima, K. Amafuji and K. Shoji 727

CHAPTER THIRTEEN - PLASMA PROCESSING

Microlaminate Composite Structures by Low Pressure Plasma Spray Deposition - R. G. Castro and P. W. Stanek 743

Thin Beryllium Structure by Powder Metallurgy - W. G. Northcutt, Jr. and V. M. Hovis, Jr. 753

Author Index 769

Subject Index 775

Contents

Volume 21

CHAPTER ONE - FATIGUE PROPERTIES OF P/M MATERIALS

Methods for Predicting the Fatigue Strength of P/M-Materials - B. Weiss and R. Stickler 3

Statistical Analysis of Density and Particle Size Influences on Microstructural and Fatigue Properties of a Ferrous Alloy - K. D. Christian, R. M. German and A. S. Paulson 23

The Influence of Pore Morphology on the Monotonic and Cyclic Properties of Sintered Iron - L. Ledoux and C. Prioul 41

Influence of Mechanical Surface Treatments on the Fatigue Properties of Sintered Steels Under Constant and Variable Amplitude Loading - C. M. Sonsino, F. Müller, V. Arnhold and G. Schlieper 55

Influence of Microstructure and Porosity on Fatigue Properties of Sintered Steels - B. Lindqvist 67

CHAPTER TWO - IMPROVED MATERIAL PROPERTIES

Microalloying: A Suitable Method for the Production of Low Alloyed High Strength P/M Steels? - R. Oberacker and F. Thümmler 85

Properties of Parts Made from Ancorbond® Processed F-0008/Carbon Steel Mix - F. J. Semel 101

High Impact Strength P/M Steel by Copper Infiltration: Theory and Practice - D. F. Berry, M. Svilar and E. Klar 123

Sintering of 70 Ni/30 Cu Powder Compacts Through Prealloy and Premix Routes - G. S. Upadhyaya and R. Hulyal 139

CHAPTER THREE - ENGINEERING PROPERTIES VIA SINTERING

New Sintered Valve Seats for Internal Combustion LPG Engines - K. Suzuki, Y. Ikenoue, H. Endoh and M. Uchino 157

Sintering and Properties of Cr, Mo and W Alloyed P/M Steels - H. Danninger and O. Z. Zengin 171

High Performance Components Produced Via Vacuum Sintering - V. Arnhold and R. Wähling 183

High Strength Sintered Steel Without Heat Treatment - T. Saito and M. Obayashi 197

CHAPTER FOUR - GENERATION AND EFFECTS OF SINTERING ATMOSPHERES

Alloy Corrosion Behavior in the Various Zones of a Computer Optimized Nitrogen-Based Atmosphere Sintering Furnace - G. Smith and J. McGeever 215

Producing a Sintering Atmosphere Through a Hollow Fiber Membrane - K. S. McGinn and Mark S. Pfitzinger 229

Productivity and Quality Improvements Via Design and Process Changes in the Conventional Sintering Equipment - H. S. Nayar 243

CHAPTER FIVE - ATMOSPHERE CONTROLLED SURFACE TREATMENTS

Influence of Pretreatment Cycles on Case Characteristics of Nitrocarburized Sintered Steel Parts - M. L. P. Rosso and G. Scavino ... 259

Mechanical Properties of Steam Blackened P/M Materials - L. F. Pease III, J. P. Collette and D. A. Pease 275

Heat Treatments of Sintered Steels Alloyed with Nickel. Hardenability - J. M. Ruiz, J. M. Torralba, J. M. Luna and C. Ranninger 301

CHAPTER SIX - IMPROVED DYNAMIC PROPERTIES IN P/M PARTS

Microstructural and Electrochemical Study of Copper Alloyed AISI 316L Stainless Steel - A. Molinari, L. Fedrizzi, P. L. Bonora, A. Tiziani and A. Tomasi 313

CHAPTER SEVEN - STEPS TO IMPROVE MACHINABILITY

Relationship Between Machinability and Strength in a Prealloyed Manganese Sulfide Sintered Material - L. G. Roy, A. F. de Rege and L. F. Pease III 327

Manganese Sulfide in Machining Grade Ferrous P/M Alloys - K. S. Chopra 361

Different Methods to Improve Machinability of High Strength Sintered Steels - R. Johansson 381

CHAPTER EIGHT - NONEQUILIBRIUM P/M PROCESSING

The Mechanical Alloying Process - J. S. Benjamin 397

Amorphous Powders of Al-Hf Prepared by Mechanical Alloying - R. B. Schwarz, J. W. Hannigan, H. Sheinberg and T. Tiainen 415

Application of Mechanical Alloying to Titanium Systems - R. Sundaresan and F. H. Froes 429

Phase Redistribution by Mechanical Alloying in an RS Immiscible Copper Automotive Bearing Alloy - S. Diamond and A. N. Patel 445

Commercial Status of Mechanically Alloyed Materials – J. J. Fischer and R. M. Haeberle 461

Intermetallic Systems Produced by Mechanical Alloying – R. C. Benn, P. K. Mirchandani and A. S. Watwe 479

Elevated Temperature Al-Ti-X Alloys: The Effect of Ce and V Additions on Microstructural Thermal Stability W. E. Frazier and J. J. Thompson 495

The Production of Mechanically Alloyed Aluminum – D. O. Gothard 511

CHAPTER NINE – JAPAN'S PROGRESS IN FERROUS P/M

Recent Developments in Low-Alloy Steel Powders – Y. Morioka and S. Takajo 521

The Effect of Alloying Additions on the Properties of Partially Prealloyed Powders – H. Sakuma, H. Manto, K. Homma, J. Takata and T. Shimmura 535

Development of Ferrous Sintered Parts in Japan – T. Kimura 551

Properties of Sintered Wear-Resistant Alloys Having High Volume Fraction of Carbides – T. Tanase, O. Mayama and H. Matsunaga 563

Mechanical Properties of High Density Sintered Compacts Obtained by Cold-Working – Y. Sakamoto and Y. Hirai 575

The Recent Trends of Sintered Components for Automobiles – I. Nagare 591

Ni-Cu-Mo Alloyed Steel Powder for High Strength Sintered Components Produced by Repressing – S. Takajo, I. Sakurada, J. Ohta, K. Ogura and M. Kawano 605

CHAPTER TEN – CERAMICS

Enhancing Thermo-Mechanical Properties of Si_3N_4 with Non-Toxic Non-Oxide Additive – C-C. Ge, Y-L. Xia and L-M. Chen 619

CHAPTER ELEVEN – HIGH POROSITY MATERIALS

P/M Composite Structures for Porous Applications – M. Eisenmann, A. Fischer, H. Leismann and R. Sicken 637

Fiber Metallurgy and Its New Applications – T. Nakagawa 653

Densification Behavior of Early-Stage Sintering with Simultaneous Surface and Volume Diffusions – J. C. Wang 669

CHAPTER TWELVE – PROCESS CONTROL AND SIMULATION

Integrated Design – B. L. Ferguson 685

Ferrous Powder Metallurgy Mechanical Properties Prediction System – A. C. Dornisch and J. T. Dornisch 699

Spreadsheet Based Cost Estimation for the P/M Fabrication Process - Y. C. Han, J. C. Park, J. P. Clark and F. R. Field 723

Determination of Forward Slip and Friction-Coefficient in Cold Rolling of Sintered Billets - D. Zeng and W. Zhou 735

A Critical Assessment of Strength Development in Porous Materials - V. A. Tracey 745

Experimentally Determined and Calculated Temperature Profiles in P/M Compacts - A. M. Hallén and W. Neumann 755

Author Index 765

Subject Index 771

Chapter One

Information and Application Trends

AWARD-WINNING PARTS SHOW BROAD SPECTRUM OF POWDER METALLURGY APPLICATIONS

Peter K. Johnson
Metal Powder Industries Federation

Winners of the 1988 Powder Metallurgy (P/M) Part-of-the-Year Design Competition show the growing international diversity of P/M applications. Markets for the winning parts include: computers, automobiles, high-security locks, firearms, chemical processing and biomedical prostheses. The competition is sponsored by the Metal Powder Industries Federation.

Computer Printer Platen Adjust Handle Wins Ferrous Grand Prize

A nickel-steel (MPIF FN-0408-45HT), platen adjust handle used to adjust the main high-speed printer platen in an office computer won the Grand Prize in the Ferrous category of the competition (Fig.1). The multilevel steel part is made by Pacific Sintered Metals Company, Los Angeles, California, for Hewlett-Packard Co., Boise division, Boise, Idaho.

Powder metallurgy replaced an investment casting, offering a 40% cost savings. The part is heat treated to HRC 35-42 and offers close tolerances, +/-.002" (.0508 mm) on the ID of the hub. Secondary operations include, drilling, tapping, pressing a screw machined pin into the handle, and a black oxide coating.

Mortise Lock Brass Dead Bolt Wins Nonferrous Grand Prize

A brass dead bolt assembly, used in a commercial mortise lock in industrial and institutional buildings won the Nonferrous Grand Prize award (Fig.2). The part is made by the Presmet Corporation, Worcester, Massachusetts for Best Lock Corporation, Indianapolis, Indiana.

Figure 1. Computer platen adjust handle.

Figure 2. Brass dead bolt.

The multi-level P/M dead bolt edged out a forging that required secondary machining and had assembly problems.

The P/M bolt passed an ANSI security test of 150 foot pounds of impact force, in addition to a shear test of 1,200 pounds of force when seated in a door strike plate. The part is also 100% tested by Presmet.

Critical tolerances include the parallelism of the flange section to the body controlled to .010" (.254 mm). Secondary operations on this design are limited to coining to hold flatness and critical dimensions, and pressing a pin into the end section.

Key Segments Win Stainless Steel Grand Prize

P/M 316L (MPIF SS-316L-15) segments which form a mechanical recording key for a high-security lock box won the Grand Prize in the Stainless Steel category of the competition (Fig.3). The segments are made by ASCO Sintering Company, Commerce, California for Supra Products, Inc., Salem, Oregon.

ASCO makes seven stainless steel ductile P/M segments that are staked to a wrought hex pin making up the key. Secondary operations include a corrosion-resistant seal, double disc grinding and deburring.

The segments have a minimum yield strength of 15,000 psi (103 N/mm^2) and a minimum elongation of 12%. Each segment has a different angular orientation to the hex ID hole. The hex ID hole is .136" +/- .001" (3.454 +/- .025 mm). The overall length must be held to .306" +/- .001" (7.766 +/- .025 mm).

The multiple segment key design could not be made economically by any other manufacturing method, including die casting and investment casting. The key is used with four other P/M parts.

Roll Compacted PC Board Contact Wins Fully Dense Grand Prize

A roll compacted contact made from a copper-nickel-tin alloy won the Grand Prize in the Fully Dense P/M products category (Fig.4). The P/M strip is made by Ametek, Inc., Wallingford, Connecticut for the Bristol Spring Manufacturing Company, Plainville, Connecticut.

The contact, stamped from the P/M strip, is locked into a PC board in a lapstyle computer. Aged, cleaned and nickel-plated, the part has an ultimate tensile strength of 170,000 psi (1172 N/mm^2) and a yield strength of 160,000 psi (1103 N/mm^2).

P/M replaced a beryllium-copper 172 alloy cast strip that needed several qualifying operations to avoid fracturing and to control distortion after heat treating. The P/M material is more consistent, had no distortion and provided a cost savings of 30% compared to the cast strip.

Pistol Sear Housing Wins Metal Injection Molding Grand Prize

A nickel-steel sear housing, used in a 10 mm police pistol won the Grand Prize in the Metal Injection Molding products category (Fig.5). The part is made by Multimaterial Molding, a div. of Fine Particle Technology Corp., San Diego, California for a firearms manufacturer.

The MIM sear housing has a tensile strength of 200,000 psi (1379 N/mm^2), a yield strength of 150,000 psi (1034 N/mm^2) and a HRC 56 hardness. P/M replaced a machined part at a cost savings of more than 60%.

Computer Core Wins Overseas Grand Prize

A low carbon steel magnetic core that operates in the support system for the read and write head of a computer disk drive won the Grand Prize in the Overseas category (Fig.6). The part is made by AMES, S.A., Barcelona, Spain, for IBM Corporation, United Kingdom.

Made from a phosphorous-iron material, the bushing must have good magnetic characteristics and high precision. Its density is 7.35 - 7.45 g/cm^3.

Secondary operations include rough and high precision turning, drilling, tapping, shot blasting, and plastic impregnation prior to nickel plating. The P/M part was made at half the cost of investment casting, an alternative manufacturing technique.

Ferrous Awards of Distinction

A copper infiltrated steel (MPIF FX-1005-40) control lever for a hydrostatic transaxle on a lawn and garden tractor, won a Ferrous Award of Distinction (Fig.7). The lever is made by Advance Products Corp., Beaver Dam, Wisconsin for John Deere Horicon Works, Horicon, Wisconsin. The customer considered a weldment of four individual pieces at an estimated cost of four times the P/M lever.

The part has a minimum yield strength of 40,000 psi (276 N/mm^2) and a hardness of 82 HRB. Tolerance control is important and the cross drilled hole must withstand a 45 foot pound bolt torque test. Advance Products designed a unique shell die to allow compaction and ejection of the complex shape.

Another Ferrous Award of Distinction goes to a copper steel (MPIF FC-0208-50) weight carrier that holds four flyweights in a fuel system in Caterpillar diesel engines (Fig.8). The part is made by DuPont/Remington Powder Metal Products, Wilmington, Delaware, for Caterpillar, Inc., Peoria, Illinois.

Figure 3. Key segments.

Figure 4. Roll compacted contact.

Figure 5. Sear housing.

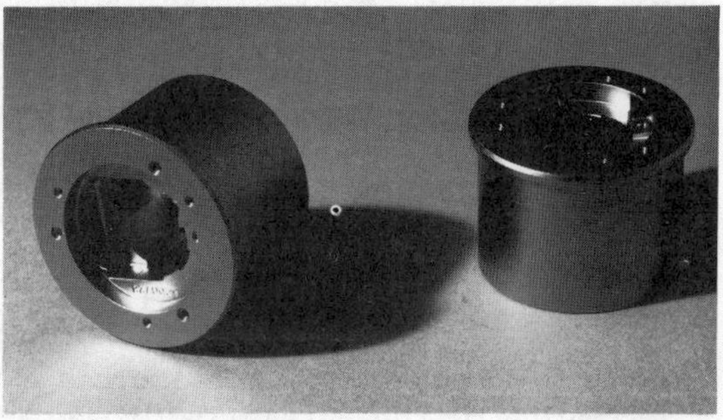

Figure 6. Steel core.

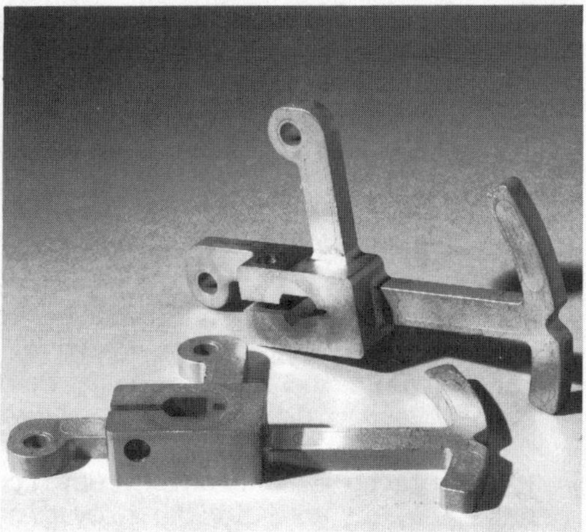

Figure 7. Control lever.

The weight carrier has a minimum yield strength of 50,000 psi (345 N/mm^2). Remington makes the complex part as a net-shape with no secondary operations and ready for assembly into a Caterpillar fuel system. A counterbore tolerance of +/-.003" (0.08 mm) must be met along with parallelism, +/-.005" (0.13 mm), of the opposite face of the part. Tooling for this part required 15 punches operating at five different levels, as well as the full utilization of all available press motions.

Nonferrous Award of Distinction

IBM Corp., Endicott, New York, won the Nonferrous Award of Distinction for a bronze hammer pivot pin used in medium- and high-speed impact printers for IBM computers and other electronic equipment (Fig.9).

The oil-impregnated pin serves as the bearing for the hammers and must operate for years without failure. A diameter of +/- .0002" (.0051 mm), must be held along its entire length. The ability to hold this tolerance on a swager is quite a feat.

P/M eliminated seven production steps. Secondary operations include swaging, cutting the pin to length, and oil impregnation.

Stainless Steel Award of Distinction

A 316L controlled porosity ball valve seat used in chemical processing won the Stainless Steel Award of Distinction (Fig.10). The part is made by Newmet Krebsoge, Terryville, Connecticut for Worcester Controls Corp., Marlboro, Massachusetts.

Important design features include a uniform density of 70% on the total concavity of the part. The diameter of the part must be held to 3.73" +/- .015" (947 mm).

The complex valve seat is impregnated with a Teflon-type material or graphite for abrasion resistance and lubrication. It is then coined to seal in the Teflon or graphite. P/M replaced a machined valve seat, while providing a 75% cost savings and increasing the operational life of the ball valve seat in an abrasive environment from one day to 30 days.

Fully Dense Award of Distinction

A P/M extruded diamond rod used to make teeth for lumber industry band saws won the Fully Dense Award of Distinction (Fig.11). The part is made by Deloro Stellite Inc., Belleville, Ontario, Canada, from a P/M cobalt base nickel chromium alloy.

P/M replaced a continuous cast rod, improving production speeds and handling, in addition to providing higher strength saw teeth. The P/M material has a room temperature hardness of HRC 44.

Figure 8. Weight carrier.

Figure 10. Ball valve seat.

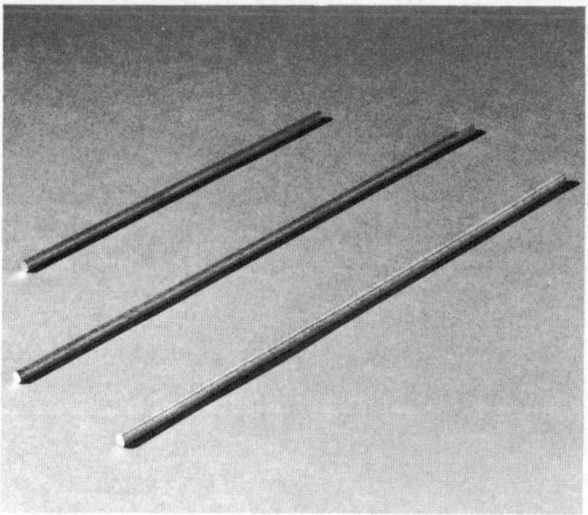

Figure 9. Hammer pivot pin.

Figure 11. Extruded rod.

Metal Injection Molding Awards of Distinction

A 17-4PH stainless steel actuator used in the clamp device of a fighter engine shroud won an Award of Distinction in the Metal Injection Molding (MIM) products category (Fig.12). The part is made by Form Physics Corp., San Diego, California.

The P/M design meets stringent military specifications and eliminated the need for machining the cross holes.

Although investment casting was originally considered, the P/M actuator provided a minimum yield strength of 130,000 psi (896 N/mm^2) and a tensile strength of 150,000 psi (1034 N/mm^2), in addition to a 50% cost savings.

A nickel-steel rear sight base and slide assembly for the mechanical sighting system on Remington centerfire and rimfire rifles also won an Award of Distinction in the Metal Injection Molding products category (Fig.13). The assembly is made by DuPont Company, Wilmington, Delaware for Remington Arms Co., Ilion, New York.

Mechanical properties are: tensile strength 50,000 psi (345 N/mm^2), yield strength 25,000 psi (172 N/mm^2), elongation 20% and a minimum hardness of HRB 50.

Metal injection molding enabled Remington to eliminate 15 manufacturing operations between the two parts, which allowed a significant reduction in manufacturing costs. Another important MIM benefit is the very low interconnected porosity of less than 0.2%, which allowed black oxide coating without the need for resin impregnation.

Overseas Awards of Distinction

A P/M timer and support plate assembly used in an automobile electronic ignition system won an Award of Distinction in the Overseas category (Fig.14). The assembly is made by Tecsinter, S.p.A., Ivrea, Italy for Industrie Magneti Marelli, Milano, Italy.

The assembly consists of a timer, made from iron powder with an ultimate tensile strength of 28,000 psi (193 N/mm^2) and a steel support plate, with an ultimate tensile strength of 49,000 psi (338 N/mm^2).

The P/M design requires no secondary operations except electrical projection welding and hot riveting of the studs. The complex part offers cost savings of more than 30%.

The other Award of Distinction in this category goes to a copper steel auto camshaft pulley made by Hitachi Powdered Metals, Ltd., Chiba-ken, Japan (Fig. 15). The part has a minimum ultimate tensile strength of

40,000 psi (276 N/mm^2) and a minimum hardness of HRB 45. An original P/M design, Hitachi reports a 20% cost advantage, compared to making the part by conventional machining. Secondary operations include turning, drilling and caulking of a wrought steel plate.

Hip Prosthesis Wins Award of Achievement

A high-strength orthopaedic hip joint replacement won the Award of Achievement in the Other P/M Materials category (Fig.16). The prosthesis, consisting of a femoral head and stem and an acetabular cup, are made by the Howmedica Division of Pfizer Hospital Products Group, Inc., Rutherford, New Jersey.

The acetabular cast cup is coated with a cobalt-chromium-molybdenum porous P/M alloy via vacuum sintering for improved adhesion and fatigue properties in the hip joint. The stem is made from the same P/M alloy by hot extrusion from gas atomized powder, followed by forging and coating of the alloy in selected areas. A uniform fine oxide dispersion stabilizes the alloy's grain structure during the porous coating sintering process.

The P/M dispersion-strengthened material provides fatigue properties almost double those of investment casting or conventional forging. The porous P/M coating can extend the functional life of hip prostheses by helping to solve the chronic problem of loosening implants due to loss of fixation in the body.

The P/M material has a tensile strength of 180,000 psi (1241 N/mm^2), a yield strength of 90,000 psi (620 N/mm^2) and an elongation of 30% and a fatigue strength of 90,000 psi (620 N/mm^2).

Figure 12. Actuator.

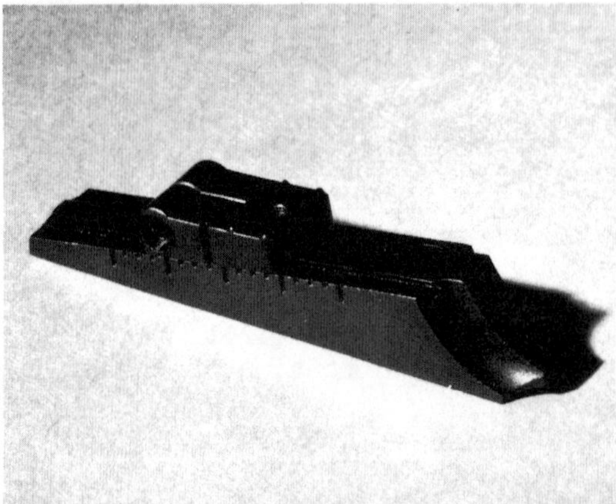

Figure 13. Rear sight base and slide assembly.

Figure 14. Timer and support plate.

Figure 15. Camshaft pulley.

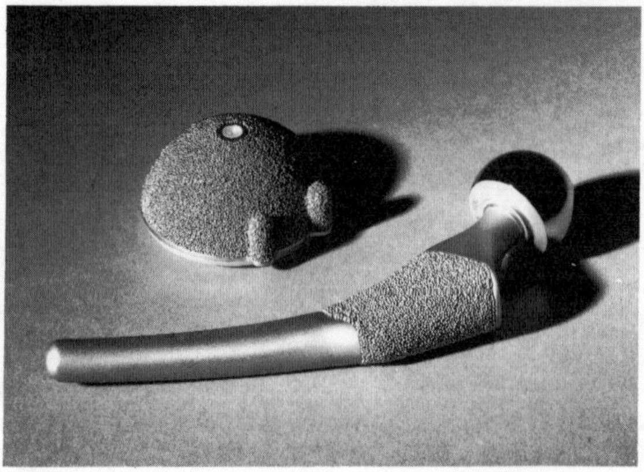

Figure 16. Acetabular cup and femoral stem and head.

INFORMATION AND EDUCATION PROBLEMS IN POWDER METALLURGY

R.A.Andrievski

Institute of Physics, Kirghizian Academy of Sciences, Frunze, USSR, 720071

Abstract

The present paper describes the current state of PM information. Analysis of three abstract editions (Soviet, American and English) revealed duplicate and general information flow (IF). In the structure of IF three main components have been distinguished: journal papers, unjournal publications, patent documents. At present all over the world there are published 13 specialized PM journals in which about 800 papers are annualy published. General IF may be of 6000-7000 units. This estimation is discussed in the present paper.

Interbranch character of PM and scattered information make training in this field not simple. Absence of modern PM textbooks is observed throughout of the world and it can become the object of international cooperation.

Introduction

Successful development of science and technology is defined in many respects by information and specialists providing. It is especially important for new branches which are arised at the joint of traditional directions. Powder metallurgy (PM) as a field of engineering is distinguished by variety of technological processes which belong to the metallurgy, chemical technology and machine-building. As a scientific direction PM is characterized by a multidisciplinary peculiarity due to its close collaboration with solid state physics, physical chemistry of dispersed objects, mechanics of deformed solid, cybernetical material science and others. At last PM is traditionally connected with ceramics forming zones of overlapping in general technological processes and development of tooling, magnetic, atomic and other materials. For the last 10-15 years close connections have been appeared with technologies of coatings and high energy influence (explosion, plasma, high pressure etc). All this leaves traces on the peculiarities of information and education in the field of PM.

Information flow

It is known that growth rate of PM as a scientific and industrial direction has been very high for a long time. This results in the growth of information flow (IF) which, taking into account abstract journals data, follows an exponential rule and considerably outstrips traditional fields of metallurgy in the rate of growth /1/. Hence, while beginning from 1960 up to 1980 IF upon metallurgy increased by factor of 1,5, in the field of PM it increased by factor of 2,5 and as for PM patent information it increased approximately by factor of 4. The last 8-10 years have also been characterized by essential enlargement of information due to development of such directions as application of refractory compounds in engineering, tooling and electronic materials (advanced ceramics), rapidly solidified and amorphous alloys, vapour-deposited coatings, ultrafine powders, high temperature supperconductors etc. Increasing of IF volume and its scattering obliged us to take special attention on abstract journals (secondary sources of information).

Table 1 shows the results of the analysis of three most representative abstract editions: Soviet abstract journal "Metallurgy" (section "Powder Metallurgy"; from 1986 it takes the title "Powder Metallurgy. Coatings and films obtaining by physical and metallurgical methods"; RJM), English journal "Metal Powder Report" (MPR) and American abstract journal "Powder Metallurgy. Science and Technology" (PMST). In the structure of IF three main components have been distinguished: journal papers, unjournal publications (papers in the collections, proceedings; reports, books etc.), patent documents. As we can see from Table 1, correlation between various components fluctuate a little bit, nevertheless it is evident some advantage of journal papers and patents. Last time beginning from 1985 in MPR patent information has not been reviewed due to publication of "International Powder Metallurgy Patent Bulletin" (IPMPB).

It was interesting, as it was fulfilled for publications of 1965 /3/, to reveal duplicate and integral IF in PM. Table 2 shows these results.

It is important to pay attention on changing of IF structure as a whole for 17 years: considerable growth of unjournal papers and patent documents. Firstly it reflects the growth of communications in modern science (increasing of number of symposiums, conferences, seminars); on the other hand, it is connected with the growth of technical applications. In integral IF all three components are approximately equal that is mostly characteristic for publications in RJM and MPR in 1979-1982. The part of duplicated flow in integral quantity of abstracts in 1965 and 1982 is relatively small - 15,5 and 13,8%, respectively. It means that over 4/5 documents are reviewed only once and each abstract journal issue reflects mainly exceptional IF. A little bit of duplication (11%) is also noted when analysing four leading abstract journals on electronics /4/.

Figure (a,b) shows schemes which illustrate duplication in pairs and that common for three abstract journals. Journal papers are mainly duplicated and those are from leading specialized journals: "Powder Metallurgy" (USSR), "International Journal of Powder Metallurgy" (USA), "Powder Metallurgy" (UK), "Metal Powder Report" (UK), "Powder Metal International" (FRG) and others. Common small level of IF duplication in abstract journals is connected with differences of thematic direction (the greatest wideness of thematic direction is presented in RJM), specific of accessibility, selectiveness and operativeness of reviewing. Particularly it is related to unjournal papers and patent documents.

Figure (c) shows the scheme of duplication of journals reviewed in analysed issues. In 1982 the IF of journals was 209 units, in 1979-1982 487 (including 88 Soviet journals) journals were reviewed in all abstract journals. It should be noted that 171 from all over the journals are reviewed only once, that is, scattering zone of journal publications on PM (zone of small productive journals) is very large.

For last 4-5 years IF on PM is going on considerably increasing (Table 1). A number of specialized PM journals increased up to 13 (Table 3); these journals are published in 11 countries. Comparing with 1982 a number of papers which are published in specialized journals has been increased as well (in 1982-600-640 /2/, in 1987 - about 800). Using in the first approximation a well-known zone model (the Bradford rule) a common number of papers in periodic journals will be about 2400. Accordingly to Table 1 journal publications make up 0,35-0,4 of general IF, thus the latter may be about 6000-7000 units. This number may be thought to be quite real, although it may be considered only as estimated prognostically and this estimation would likely be minimum. For example, it may be doubtful that patent information is reviewed so exhaustively. In 1987 1404 titles of PM patents were published in IPMPB; those belong to USSR, USA, UK, FRG and Japan. This number of patents essentially exceeds the results given in Table 1. On the other hand, in reviewed IF on PM information on ceramic materials is probably reflected not enough. So, Soviet abstract journal "Chemistry" (section "Technology of silicate and refractory nonmetallic compounds") and American journal "Ceramic Abstracts" in 1986-1987 every year published about 2500-3000 abstracts which belong to the problems of above-mentioned zone of PM and ceramic interest overlapping.

It should be noted as well that there is the fourth significant component of IF-firm documentation, publicity editions,but it is impossible to evaluate this information quantitatively.

Close results of determination of all PM publication by means of analysis of primary and secondary information sources (4500-4800 units in 1980-1982) testifies about the reasonability of estimation of IF in accordance with zone model. Using zone model mentioned above and taking into account 13 specialized journals it should be noted that PM papers have been published in about 400 journals /1/. As it has already been mentioned above during 1979-1982 487 periodical issues have been reviewed, that is enough close to the estimated value.

When considering PM information significant growth and scattering automatic information retrieval systems (AIRS) become very important for effective information service. This system - AIRS "Document" - has been created and exploited in the USSR /5,6/. This system contains information on the problem of "PM, composite materials and coatings". The input flow of documents by AIRS, including firm documentation, in 1984 was about 8000 units.

Education problem

Interbranch character of PM and scattered information make training in this field not simple. PM engineers must naturally combine by themselves chemical technologists, metallurgy specialists, mechanic engineers and specialists on material science. This problem has been discussed in literature /7,8/; during the International PM Conference (1986, Dusseldorf) the day of students has been carried out. In the USSR training of PM specialists is realized in a number of higher colleges - Institute of Steel and Alloys, Institute of Fine Chemical Technology, Avia-Technological Institute (Moscow); Polytechnical Institutes in Kiev, Minsk, Perm, Tomsk, Novocherkassk et al . In the USSR monography of educational program, edited by prof.W.Schatt (GDR) /9/, has been translated into Russian; textbook "Powder Metallurgy and Vapour-Deposited Coatings"(edited by prof.B.S.Mitin) has been published /10/; recently a number of generalized monographs and reference books has been issued (for instance, /11-15/).

In general, PM book literature is published quite widely. During 1985-1987 in the USSR every year about 20 monographs and about 20 collections were published in this field. Most of collections reflecting symposium, conference, seminar proceedings published abroad is predominated (about 25-30 units annualy). Certainly, conferences, seminars, symposiums, schools themselves promote increasing of educational level and communications. At present about 30 such steps of international range take place throughout of the world, not saying about special regional steps.

However the absence of modern PM textbooks is observed all over the world; the well-known monographs by Balshin, Goetzel, Eisencolb, Jones, Fedorchenko and the author of the present report /16-20/ became out of date both as with fact matter as methodologically.

Creation of a monograph on modern PM fundamentals can become the object of international cooperation. It is important to give an account of principal regularities of powder production, their pressing (shaping), sintering and property formation of powder materials at the modern scientific level and united principles.

It is also important to take into consideration alternative and variable nature of PM processes (producing of powders by several methods, various ways of consolidation), variety of structural components and their morphology in powder materials. This is the reason for more active application of computers in PM that will permit to save it from empirical approach.

References

1. R.A.Andrievski. Powder Metallurgy (USSR), 1982, N 8 , 96-101 .
2. R.A.Andrievski, G.M.Zeer, ibid, 1984, N 6, 102-105.
3. G.V.Gedrimovich, NTV, series 1, 1969, N 9, 16-22.
4. G.V.Gedrimovich, ibid, 1980, N 3, 15-21.
5. E.L.Shvedkov, I.I.Kovenski, E.T.Denisenko et al, Powder Metallurgy (USSR), 1982, N 1, 93-98.
6. E.L.Shvedkov, ibid, 1985, N 6, 96-102.
7. R.A.Andrievski, ibid, 1982, N 10, 95-100.
8. E.Mosca, Powder Metallurgy International, 1986, v. 18, N 3, 170-171.
9. Pulvermetallurgie Sinter-und Verbundwerkstoffe, herausgegeben von W.Schatt, Leipzig VEB Deutscher Verlag für Grondstoffindustrie, 1979 (second edition in 1985); soviet translated edition in 1983.
10. Powder Metallurgy and Vapour-Deposited Coatings. Ed. by B.S.Mitin, Moscow, Metallurgy, 1987.-792 p.
11. V.V.Skorochod, S.M.Solonin. Physical-Metallurgical

Fundamentals of Powder Sintering, Moscow, Metallurgy, 1984.-199 p.
12.J.E.Geguzin. Physic of Sintering, Moscow, Nauka, 1984.-312 p.
13.I.D.Morochov, L.I.Trusov, V.N.Lapovok. Physical Phenomenon in Ultrafine Objects, Moscow, Energoizdat, 1984.-224 p.
14.Powder Metallurgy Materials.Technology. Properties. Applications. Reference book, ed. by I.M.Fedorchenko, Kiev, Naukova dumka, 1985.-624 p.
15.Powder Metallurgy in USSR. History. Present status. Outlooks. Ed. by V.I.Trefilov and I.N.Franzevitsch, Moscow, 1986.-294 p.
16.M.J.Balshin. Powder Metal Science, Moscow, Metallurgizdat, 1948.
17.C.Goetzel. Treatize on Powder Metallurgy, vol. 1-3, N.Y., Interscience Publishers, 1949.
18.W.D.Jones. Fundamental Principles of Powder Metallurgy, London, Eduard Arnold, 1960; soviet edition in 1964-1965.
19.I.M.Fedorchenko, R.A.Andrievski. Fundamentals of Powder Metallurgy, Kiev, Edition AS UkrSSR, 1961.-420 p.
20.F.Eisenkolb. Fortschritte der Pulvermetallurgie. Bd I und II, Berlin, Akademie Verlag, 1963; soviet translated edition in 1969.

TABLE 1 : ANALYSIS OF PUBLICATIONS IN ABSTRACT JOURNALS

Abstract Journal	Year	Journal papers	Unjournal publications	Patents	Common number of abstracts on PM
RJM	1979	536(0,33)	340(0,21)	761(0,46)	1637
	1980	537(0,33)	494(0,3)	593(0,37)	1624
	1981	669(0,4)	397(0,23)	621(0,37)	1687
	1982	728(0,34)	708(0,33)	728(0,33)	2164
	1986	1381(0,45)	935(0,32)	753(0,24)	3069
	1987	1403(0,5)	618(0,22)	787(0,28)	2808
MPR	1979	609(0,39)	375(0,24)	589(0,37)	1573
	1980	597(0,36)	500(0,31)	527(0,33)	1624
	1981	517(0,34)	429(0,29)	555(0,37)	1501
	1982	514(0,31)	472(0,28)	694(0,41)	1680
	1986	495	865	-	1360
	1987	592	250	-	842
PMST	1979	535(0,54)	241(0,25)	205(0,21)	981
	1980	681(0,61)	160(0,14)	270(0,25)	1111
	1981	575(0,39)	621(0,42)	289(0,19)	1485
	1982	704(0,5)	461(0,33)	227(0,17)	1392
	1985	795(0,5)	264(0,16)	551(0,34)	1610

TABLE 2 : COMMON NUMBER OF ABSTRACTS, DUPLICATE AND INTEGRAL IF ON PM IN 1965 /3/ AND 1982

Type of flow	Journal publications		Unjournal publications		Patents		All over	
	1965	1982	1965	1982	1965	1982	1965	1982
Sum of abstracts in RJM, MPR, PMST (ASM review*)	611	1946	228	1641	219	1649	1158	5236
Duplicate (coincided) flow	130	520	48	121	1	83	179	724
Integral flow	581	1426	180	1520	218	1566	979	4512

* In 1965 journal PMST has not been published and information on PM has been reviewed in ASM review of metal literature.

TABLE 3 : SPECIALIZED JOURNALS ON PM

NN	Journal	Country	Annualy numbers
1.	**Порошковая металлургия**	USSR	12
2.	Metal Powder Report	UK	12
3.	Journal of the Japan Society of Powder and Powder Metallurgy	Japan	8
4.	Powder Metal International	FRG	6
5.	Powder Technology	Switzerland	6
6.	International Journal of Powder Metallurgy	USA	4
7.	Powder Metallurgy	UK	4
8.	International Journal of Hard and Refractory Metals	UK	4
9,	Pokroky Praskove Metallurgie	Czechoslovakia	4
10.	Powder Metallurgy Association of India. Newsletter	India	4
11.	Powder Metallurgy Technology	China	4
12.	Science of Sintering	Yugoslavia	3
13.	Metallurgia Proszkow	Poland	3

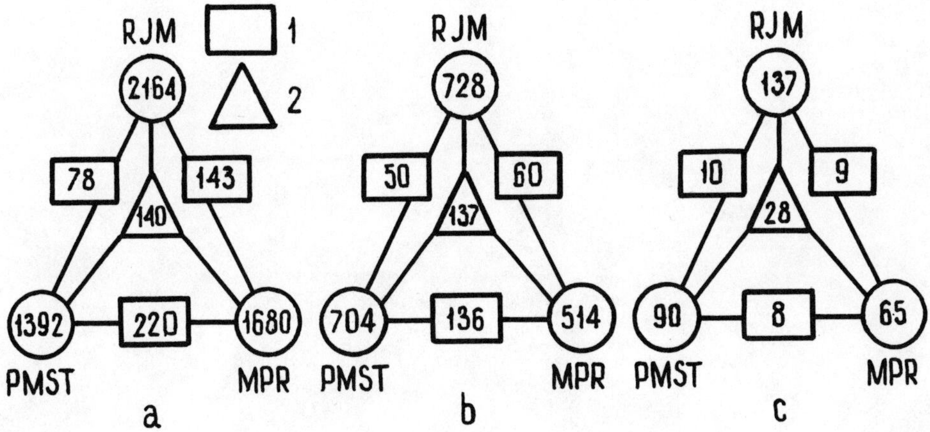

Figure: Scheme, illustrating duplications in pairs (1) and common duplication (2) of the whole abstracts (a), abstracts of journal papers (b) and periodical journals, reviewed in abstract editions (c).

Common number of abstracts and journals are given in circles.

NEW PRODUCT AND PROCESS DEVELOPMENT IN THE P/M PARTS INDUSTRY

Charles I. Whitman

Abstract

The P/M parts industry needs a constant stream of new and improved products and processes to maintain and enlarge its market position vis à vis other parts making technologies in this global economy.

Unfortunately, the industry's splintered and highly competitive structure has inhibited marshalling the necessary patient resources for such advances while global innovation overall is increasing in breadth and speed.

The fundamentals of the innovation cycle and systems considerations explain why this structure prevents competitive response, with specific examples cited from the development of corrosion resistant stainless steel parts, Metal Injection Molding (MIM), and the new high performance infiltration technology (HPIT).

Increased technical interchange and collaboration of parts makers with suppliers and customers is recommended as a first step to improving industry response.

Introduction

Radical change is required! True, the P/M Parts Industry can look back with pride on forty years of growth into a major industry. True, this great conference is testimony to the potential ahead for further growth. Yet, given the increasing pace of global innovation, this industry risks a slide to mediocrity if it does not change radically its approach to new product and process development.

Too bold a statement perhaps? Why can it be said at all? After all, this industry's technical capability has increased vastly over the years. For example, atomized iron powder, with its greater

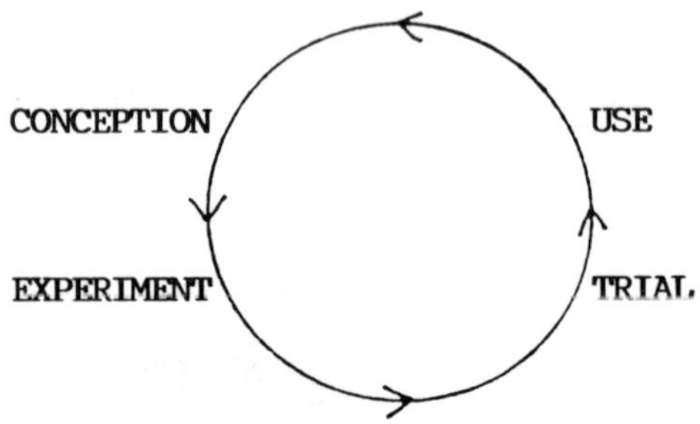

FIGURE 1: THE INNOVATION CYCLE

compressibility, apparent density, and purity compared to sponge iron allowed stronger, tougher, and more precise sintered parts. And P/M forging is finally in use for engine parts. Such technical advances will continue.

But the key questions for this industry are, "At what rate? Will that rate be sufficient to maintain and strengthen powder metallurgy's competitive position, and if not, what needs to be done?" The answers lie in the fundamentals of innovation, and how these fundamentals apply today.

The Innovation Cycle

An Innovation Cycle (Figure 1) with successive steps of Conception, Experiment, Trial, and Use describes these advances in P/M, and in fact all innovation. Through repeated cycles advances accumulate, much like compound interest.

Stripped to its basics, the innovation cycle is simply the development of new information pertinent to an agreed goal, and often hard won. The cycle rarely yields the best information easily; many defeats disrupt the path to success. Key also then, is selecting what information to develop and the speed of its development to complete the cycle.

And the returns can be extraordinary. In semiconductor electronics, the innovation cycle has increased the number of transistors on a chip by a factor of a thousand over the past twenty years, and this trend is expected to continue (Figure 2). Mead projects "a 10,000 fold increase in cost effectiveness of computing" over the next ten years from custom chip design and manufacturing advances (1). These continuing advances are the base for profound changes, not only in other technologies (including powder metallurgy), but in our business, social, and political systems as well.

Now conventional wisdom tells us such innovation cycles can reach a point of diminishing returns, a point where the old technology is vulnerable to replacement by a new one, such as jet engines replacing propeller types. This may be the conventional wisdom, and it does happen, but incremental innovation can give handsome returns indeed. All of us have seen the radical changes in automobile manufacturing stemming from sustained innovation cycles to improve quality while also reducing cost. Today, General Motors and Ford single source about 98% of their auto parts purchased from outside suppliers as a result of such efforts (2). The key to such success has been the use of statistical tools to select the best information to develop and thereby accelerating the cycle. All entities from the supplier to the customer participate in the cycle, with risks and gains shared.

The Pace of Innovation

We are seeing a rapid increase in the pace of innovation, fueled partly by resources made available under competitive stress or in a national interest, partly by the availability of inexpensive computer power which allows tackling <u>systems</u> problems far more complex than could be considered only a few short years ago (e.g. CAD-CAM), partly by availability of high speed communication links, and partly by integration of efforts among scientific disciplines and business functions as noted above. This last is essentially forced by the prior three.

To illustrate, a few years ago product development efforts at the Aluminum Company of America would stress replacement of individual automobile parts with aluminum. Today this company has a vigorous interdisciplinary effort on engineered material structures, using combinations of metals, ceramics and polymers , all in collaboration with potential customers (3). Further, in response to severe competition many U.S. companies are cutting product development cycle times in half by a combination of such measures (4). Chrysler recently announced an expanded effort to "reduce the development time needed for a new vehicle from five to six years currently to three to three and a half years..." (5).

In contrast, the P/M parts industry, splintered and highly competitive, has been accustomed to technology being brought to it from without, by suppliers of powder, furnaces, atmospheres and so on, and purchasing such when proven. This position often impedes the information development and flow required to complete a product innovation cycle for what is really a system, not a collection of independent items. To be sure, the end user, the parts industry

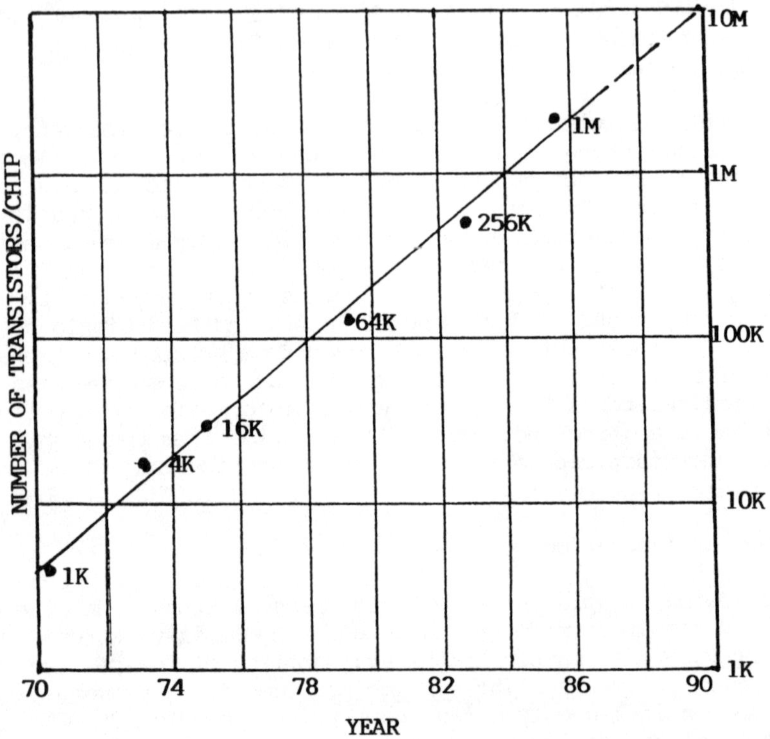

FIGURE 2: INCREASE IN NUMBER OF TRANSISTORS/CHIP

customer, can contribute his share by demanding guarantees of performance that stifle risk taking. Today more than ever, a P/M part is manufactured in a powder, tooling, and equipment system, and used in a material system.

Examples

Stainless Steel Parts

As one case history consider the long standing belief in the industry that a substantial market for stainless steel parts could develop but for the inferior corrosion resistance of sintered parts compared to wrought stainless steel. Much effort was devoted to this question at the SCM Metal Products Laboratory by Klar and his co-workers (6). This work was the base for a special stainless steel powder first offered to the market in 1980, and for a definitive understanding of the role of nitrides, carbides, oxygen, and ferrous contaminants in corrosion performance. Part performance here is strongly dependent on both powder and process, i.e. a system. Notwithstanding this, and work by others, development of this market has lagged.

I lay this partly to the traditional expectation by the parts industry that the technical push to the sintering process complexities should come from the powder supplier and that the market pull should come from parts user. However, if the prospective parts user demands guarantees i.e. no risk for him, the chance for an effective innovation cycle becomes slim.

P/M Infiltration

As a result of the above, SCM has chosen licensing to exploit its recent development of High Performance Infiltration Technology (HPIT) (7). This technology uses a special infiltrant and process conditions to obtain mechanical properties in infiltrated parts that rival those obtained by P/M forging. The license provides a basis for information flow under circumstances that protect each party's commercial interest and further encourages the licensee to contribute his resources to customize the technology for the specific part desired. Again, we have a system. The need remains to develop participation with end users to close the cycle for maximum progress.

Metal Injection Molding

This technology, some sixteen years in incubation (mostly outside the traditional parts industry), seems close to breaking out from the small, high value added, specialty parts that have characterized its beginning years to applications with market potential of $200,000,000 (8,9). Much credit is due those persistent pioneers.

Still, the P/M parts industry is presently more waiting for this technology to come to it, than pushing to make it happen. Development effort in this large interdisciplinary opportunity is fragmented; almost without exception those in the field are tackling only a part of this complex system owing to limits on technical and financial resources. Powder maker and furnace builders are involved more as suppliers than as collaborators. Haynes International, a licensor of the Rivers patent, has found a demonstration manufacturing line necessary to entice slow to come licensees. In this situation a real risk exists that the maturing of this technology will be unduly delayed. Don't expect your competition to wait for you!

And much is yet to be done. For instance, powder, binder, and processing variables intertwine in complex ways which limit the size and tolerance of parts. Tooling is critical. The drive for lower costs seeks ways to replace fine powders with coarser ones, prealloyed powders with elemental ones, and long binder removal cycles with short ones. Furnaces, batch and continuous, vacuum and atmosphere, are part of this system.

To speed advance for MIM a coming together of those with the skills and knowledge who can contribute to the many facets of this system is in order, from powder maker and equipment maker to end user, from fundamental scientist to engineer, all in a climate that allows sharing of risks and gains.

What Can Be Done?

The need is clear to set aside the traditional competitive posture of the P/M parts industry towards innovation. Satisfactory for its time, it is inadequate to deal with the current breadth, pace, and systems orientation of technological innovation. The industry needs much stronger technical interaction with its suppliers, with its customers, and with itself to foster the continuing innovation cycles needed for its long term competitive strength.

Given this cannot happen overnight, here are some suggestions to get it started.

-Strengthen your company's technical depth. It will be needed for the task ahead.

-Build on the current collaboration on quality and productivity improvement issues with customer and supplier to increase technical interchange. Consider single sourcing.

-Seek joint efforts with common interests between competitor, supplier, and customer. Expand to include university efforts. For instance, the National Science Foundation is supporting a number of industry- university multidisciplinary centers.

-Consider what can be accomplished through your trade associations to strengthen interactions.

These suggestions are just a beginning. Much time and patience will be required. But without a start, nothing will ever happen. Good luck!

References

1. Forbes, v.141, no.7, p.88ff, April 4, 1988.

2. Industry Week, p.34, February 1, 1988.

3. P. R. Bridenbaugh, Zay Jeffries Lecture, Cleveland, Ohio, February 1, 1988.

4. Wall Street Journal, p.1, February 23, 1988.

5. Wall Street Journal, April 1, 1988.

6. E. Klar, Proceedings of 1987 International Symposium on Powder Metallurgy, October 23-24 1987, Taiwan, ROC. pp.17-50.

7. D. F. Berry, M. Svilar, and E. Klar, Modern Developments in Powder Metallurgy, v.18-21, 1988.

8. L. Pease, Progress in Powder Metallurgy, v.43, 1987.

9. L. Pease, International Journal of Powder Metallurgy, pp.129-133, April 1988.

HOW HIGHLY INNOVATIVE COMPONENTS DEVELOPED FOR BUSINESS MACHINES MAY FOSTER APPLICATIONS IN OTHER INDUSTRIAL FIELDS

O. Morandi and E. Mosca

Tecsinter S.p.A. and Ing. C. OLIVETTI & C. , S.p.A

Ivrea - Italy

Abstract

In the past the business machine industry has primarily used P/M components in mechanical devices because very complex shapes could be produced at low cost by means of sophisticated techniques (split-die concept in pressing, projection welding to assemble multifunctional components, etc.) and improved physical and mechanical characteristics could be obtained by adopting special alloy systems and vacuum sintering.

The aim of this paper is to indicate, through some significant case histories, not only the evolution of the business machine parts through the design of unusual P/M components and the combination of different processing techniques, but also how it is possible to transfer these innovative concepts into other industrial fields.

Introduction

From the beginning of the Fifties, the business machine industry has increasingly used P/M technology by adopting highly complex design criteria; office products designers, working closely together with P/M experts, succeeded in exploiting the tremendous potential of powder technology, showing how original and innovative solutions could help to produce new devices at low cost.

Over more than 30 years, an exceptional technological know how was generated first at Olivetti and then at Tecsinter. Several aspects of both materials and processing were especially developed and succesfully applied in office-machine products; among them we should particularly note the split-die pressing principle, the assembling of complex parts by electrical projection welding, the use of sophisticated P/M components of suitable composition in magnetic devices, and the combination of proper processing steps, in order to be able to make parts otherwise difficult to produce economically by conventional compacting technologies.

During the Eighties that know-how has been placed at the disposal of the market and new, interesting and successful applications have been developed. In the following sections, several case histories are discussed which show how some of these ideas can be applied universally to the manufacture of items intended for use in mechanical industries.

Split-die pressing principle

This technique which uses an advanced tool motion, is not really new; it has been exploited very profitably for years in the manufacture of office-machine components and is quoted in P/M textbooks as a process capable of producing parts with complex multilevel shapes, for instance helical gears with hub, components with tangential flanges or undercuts, etc. Fig. 1 shows the split-die compaction sequence; it requires special presses, developed and constructed to work reliably on this principle.

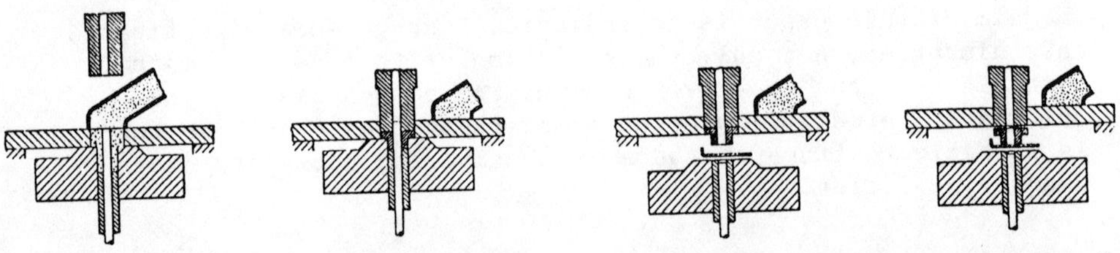

Figure 1: The Tecsinter split-die compacting system

The following group of case histories illustrates how this technique can be used effectively not only to produce business machine parts (Fig. 2, 3 and 4), but also parts for cars, power tools and appliances (Fig. 5, 6, 7, 8 and 9).

Fig. 2 demonstrates the ability to press shapes with undercuts with a group of cams used in typewriters and calculating machines, while Fig. 3 shows a group of gears, some of them with a hub; the hub was formed in the lower die and the gear in the upper die. The helix angle can reach 45°.

Fig. 2: Cams used in typewriters and calculating machines.

Fig. 3: Gears used in typewriters and calculating machines.

Fig. 4 illustrates parts with several hubs, formed at a remarkable distance. This is feasible because in the split-die tool it is possible to eject the hubs from the lower die before ejecting the flange. In the single-die system, hubs will break during the ejection step, owing to the spring back of the flange, while hubs are still held in the die.

These concepts are now exploited in similar applications but in quite different industrial sectors.
Fig. 5 shows in the upper row two double gears used in hand power tools; they are made of Cu Ni Mo sintered steel, gas carburized and hardened, density 7 g/cm^3, UTS> 740 N/mm^2, and hardness HRA>70.

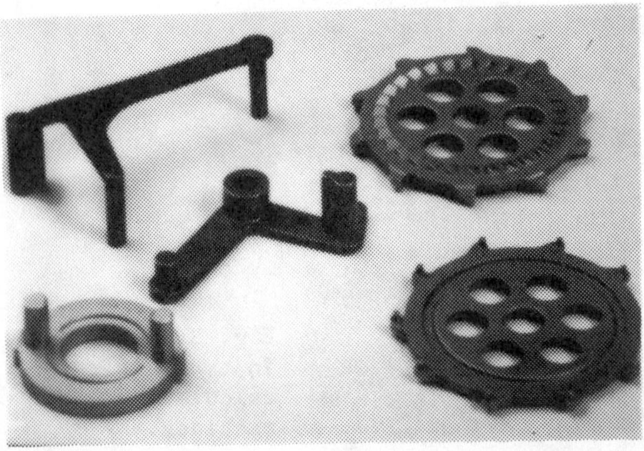

Figure 4: Parts with several hubs used in business-machines.

Figure 5: Double gears and parts with undercuts used in different industrial sectors.

The double ratchet wheel with opposed teeth is made of Cu Ni sintered steel, gas carburised and hardened, density 7 g/cm^3, UTS>530 N/mm^3 and HRA>53. The lower row, starting from the left, includes together with a typewriter part (composition and properties similar to the double ratchet wheel), a gear for a door window regulator assembly (automotive), a part of a syncrhonizing gear (automotive, manual transmission) and finally a part of a knitting machine.
The composition and properties of these three components are: Cu Ni Mo sintered steel, gas carburised and hardened, density 6,8 g/cm^3, UTS>650 N/mm^2, HRA>64.

Fig. 6 shows at the left a double gear with undercut used as an actuator in a centralized door-locking device. The idea of this component was originated by a similarity with gears for calculating machines (automotive, Cu Ni Mo sintered steel, gas carburized and hardened, density 7 g/cm^3, UTS>710 N/mm^2, HRA>68). At the right a cam, with an undercut which eliminates a secondary machining operation; this component is utilised in appliances,(Cu Ni Mo sintered steel, density 6,8 g/cm^3, UTS>360 N/mm^2 and HRB>54)

Figure 6: Double gear (weight 12 g) and cam (weight 62g)

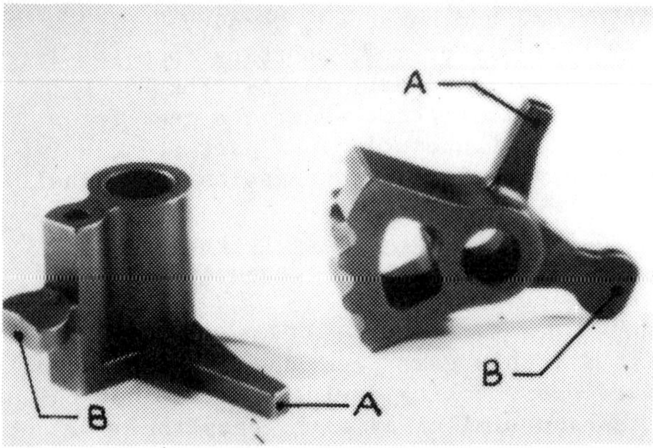

Figure 7: Speed-selector levers for a manual transmission (lever on the left: weight 103 g) (lever on the right: weight 108 g)

Figure 8: Helical gears for hand power tools (weight 20 g and 105 g)

Fig. 7 illustrates two speed-selector levers used in a manual transmission. Split die technique in this case represents the right solution because standard compacting tooling, apart from its complexity and cost, would not guarantee the integrity of the green compact and require secondary operations to obtain undercut shapes (automotive, Cu sintered steel, combined carbon 0,8%, density 7 g/cm^3, UTS>470 N/mm^2, HRB>80 while zones indicated by A and B are induction-hardened and hardness is HRA>70)
The two helical gears pictured in Fig. 8 are used in power hand tools; also in these cases the split die system makes it very easy to combine in a monolithic piece a helical with a straigh-tooth gear or a helical gear with a grooved bore. In both cases, part specifications are: Cu Ni Mo sintered steel, carburised and hardened, density 7 g/cm^3, UTS>740 N/mm^2, HRA>70.

Fig. 9 is a photograph of two parts having small pins at a distance which does not allow pressing in a standard tool without problems. The bigger part is a lever used in a powered hedge cutter; this component was designed directly for P/M technology in cooperation with P/M experts. Material is Cu Ni Mo sintered steel, carburised and hardened, density 7 g/cm^3, UTS>740 N/mm^2, HRA>70. The smaller part is used in a refrigerator compressor; formerly the part was made of solid steel (stamped plate) but this P/M solution allowed a better design and saving in costs. Also in this case, Cu Ni Mo sintered steel is used (density 6,8 g/cm^3, UTS >650 N/mm^2, HRA>64 after carburising and hardening).

Figure 9: On the left, component for refrigerator compressor (weight 11 g); on the right, lever for Powered hedge cutter. (weight 57g)

The advantages of electrical projection welding

Projection welding has been largely adopted by the business machine industry because in the old-generation products, based on very complex mechanical devices, there was a need to produce shapes much more complicated than those possible using the split die technique. In the last few years Tecsinter has conducted a considerable research effort to expand the range of applications of projection welding in the P/M field, in terms not only of material combinations, but also of reliability and profitability.

Projection shapes were studied and optimized to establish rules for a correct design (projections are moulded) and at the same time all factors like alloy composition, density, metallurgical structure together with the settings of the welding machine and the design of welding fixtures were considered.

In conclusion, it was found that projection welding gave very good technical results at a relatively low cost: as an example, sintered steels containing up to 0,8% of combined carbon can be joined reliably.

Fig. 10 gives some typical examples of parts used in business machines, each of them has been produced by joining two or more P/M components; multiple welds are often obtained in a single welding cycle. Owing to the great number of cases studied a remarkable know-how was put together and then successfully applied to innovative parts in other industrial fields, as shown in the following examples.

Figure 10: parts obtained by projection welding of two or more components

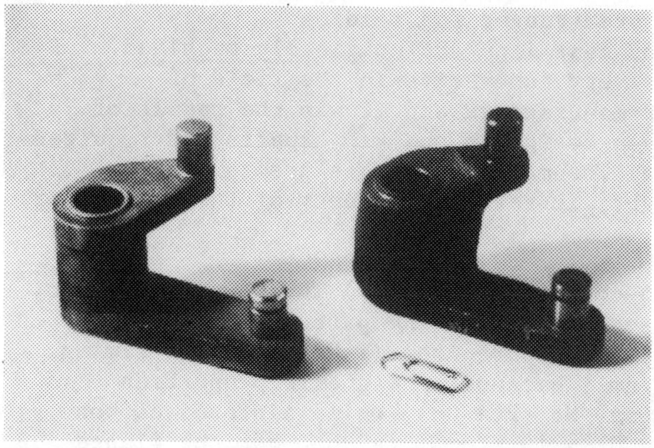

Figure 11: Lever for lifting attachment of farm tractors (weight 150 g)

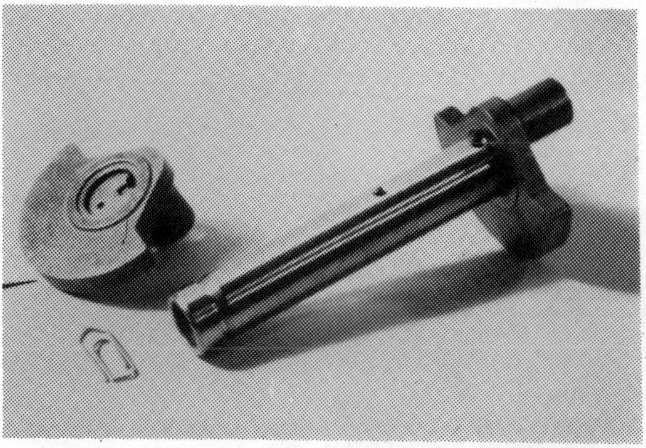

Figure 12: Eccentric shaft for a refrigerator compressor (weight of sintered component 78 g)

Fig. 11 shows a lever obtained by joining two P/M components and used in the lifting attachment of farm tractors. Formerly the part was forged, as shown on the right of the picture, and then finished by machining. Now the P/M solution gives a cost saving of 50%. The two single components are made of Cu Ni Mo sintered steel, gas carburised and hardened to HRA>64, density 6,8 g/cm^3, UTS>650 N/mm^2.

In Fig. 12 is shown an eccentric shaft for a refrigerator compressor; the shaft is made of solid steel whereas the flange is sintered (Cu Ni steel, density 6,8 g/cm^3, UTS>330 N/mm^2, HRB>49) and the two components are joined.

It is possible to see, on the sintered component (on the left of the picture), the zone prearranged for welding.
Formerly the part was of cast iron and completely machined.
The present solution is very innovative in the field of refrigerator compressor manufacture and it was the result of close cooperation between designers and P/M experts and required intense experimentation. The main advantages are:
cost saving, lower weight, improved efficiency of the compressor.

A timer for an ignition coil (car industry) is shown in fig. 13. On the left is the present solution. The part is made of two sintered components; sintered pure iron is used for the timing gear with density >7 g/cm^3, because of magnetic properties required (Bmax >15 kgauss, Hc <2,4 Oe), while sintered carbon steel (density 7 g/cm^3, UTS>210 N/mm^2) is used for the flange with the central hub and small pins.

On the right of the figure is the previous solution made of five components welded together, four of them made of solid steel. The main advantages of the present solution are: cost saving, lower weight, improved efficiency of the ignition coil.

Figure 13: Timer for ignition coil (weight 55 g)

Fig. 14 shows a camshaft with its distribution gear used in a small engine (farm sector). It is obtained by joining, three different P/M components; they are made of Cu Ni sintered steel, density 6,8 g/cm^3, UTS>330 N/mm^2, hardness HRB>49.

The last case history (see Fig. 15) shows an example of the industrial application of Laser welding; the figure shows a stator of a rotating electromagnet used in an electronic typewriter and is produced by Laser joining of two identical pure-iron sintered components. Laser joining was chosen for magnetic reasons because it eliminates the air gap between the two components (very narrow, but still present when projection welding is used).

Figure 14: Camshaft with distribution gear (weight 400 g)

Figure 15: Stator of a rotating electromagnet (weight 35 g)

PM components for sophisticated magnetic circuits

The transition from the old mechanical concept of the business machine to the modern one has brought an extended use of electromechanical devices which have forced the design of components, often with a shape so complex that it was desirable to produce them by P/M technology; in this case, standard materials (pure iron, iron-nickel, iron-phosphorus, silicon-iron, etc.) and processing cycles (for instance H2 purification combined with vacuum sintering) were adopted.

But applications became less conventional, because of the capability to produce, by P/M route, shapes economically not feasible by competing technologies.

Figure 16: Components for magnetic circuits

Fig. 16 shows several self-explanatory examples of what has been realised in the business-machine field; there are magnetic cores (some of them also have mechanical functions), cups of electromagnets, pole pieces, rotors for hybrid stepping motors, etc. Obviously the know-how accumulated in these developments was effectively used to design and produce components for other industrial sectors.

Figure 17: On the left, components for textile machine (weight of the group 15 g); on the right components for a relay (weight of the group 32 g)

Figure 18: On the left, stator for a small DC motor (weight 44 g); on the right, pole-pieces for magnetic ignition (weight 22 g and 35g)

Fig. 17 shows at the left three parts of an actuator for textile machinery; two pole pieces and a keeper (sintered iron-phosphorus, density 7,3 g/cm^3, Bmax> 16 kgauss, Hc < 1,2 Oe) and at right, three parts of a relay for industrial applications.

Fig. 18 shows at the left a stator for a small DC motor and at the right two pole-pieces used in the magnetic ignition of a small engine. In these last examples material used is sintered pure iron, density 7 g/cm^3, Bmax > 15 kgauss, Hc < 2,4 Oe. All the applications pictured above were discussed with the designers of the respective customers and P/M was chosen mostly for economic reasons.

Combination of proper processing steps

The case histories shown in Fig. 19 are a good example of a suitable choice of processing parameters; in this case vacuum sintering of stainless steel components which must be coined to control dimensions. The common practice of sintering in cracked ammonia,was not possible because of problems with nitrogen pick-up (especially by chromium) that rendered the material hard and brittle, making coining impossible.

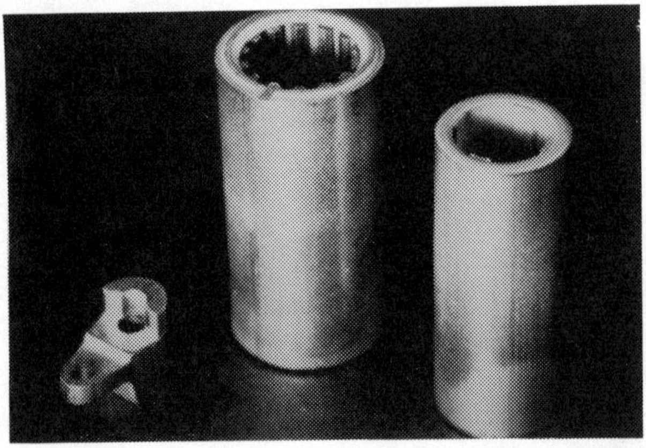

Figure 19: On the left, component for a coin box (weight 6 g); on the right two couplings for water pumps (weight 72 g and 32 g)

The small part on the left is utilised in a coin box for a public telephone. The cylindrical parts shown on the right are couplings for water pumps. In all cases, material is 316L, stainless steel, density 6,6 g/cm^3 and UTS > 340 N/mm^2.

Conclusions

It is interesting to note that usually, for historical reasons, every industrial sector has developed some technologies more actively than others and this has led to a certain industrial culture; P/M in business machines or in cars is a good example. Nevertheless, the knowledge of the full range of opportunities offered by a certain process, in terms of performance or cost saving, is sometimes lacking because standard communication routes do not give the information to the designers and the engineering departments.

Starting from a series of case histories, the present paper tries to develop the message that good achievements in one industrial activity can stimulate applications in other fields. Actually, P/M offers a lot of interesting perspectives, provided that the technical information can reach the right people.

References

1. E. Mosca, G. Ferlito, "Il metodo di compattazione Olivetti", La Metallurgia Italiana n. 4 - 1973.

2. Donald R. Dreger, "Split-dies make impossible P/M parts", Machine Design, July 21 1977.

3. "Powder Metallurgy in Italy", Metal Powder Report, Vol. 33, n. 8, August 1978.

4. V. Actis Dato, "Les avantage de l'utilisation des presses a double matrice pour la production de pièces sans defaut", proceeding of the "Colloque sur la compression des poudres metalliques" - Paris - March 15th 1979.

5. A. Marchetti, E. Mosca, O. Morandi, "Automatic Control of Pressing in Presses Using the Split Die Principle". Proceedings of the 1986 international Powder Metallurgy Conference. - P. 429.

Chapter Two

Powder Production: Mechanical Processes

NOVEL PROCESS FOR SAFE, CONTAMINATION-FREE SIZE REDUCTION OF FERROALLOYS

D. J. Eddington

Alpine American Corporation
5 Michigan Drive
Natick, MA 01760

Abstract

The paper describes design principles and engineering features for a novel fluidized bed jet milling system as it is applied to produce powdered materials with particle sizes finer than those obtainable from spray atomization. In order to prevent reaction of the finely divided powder with oxygen in air, the dry milling process is inerted with nitrogen gas which is recirculated in a closed loop to improve process economics. Case studies show the results of test runs with ferrophosphorus, ferrochrome, ferrosilicon, ferromanganese, and ferromolybdenum materials.

Introduction

Ferroalloys are used in a wide variety of metallurgical applications. Ferromanganese is an essential component in steel manufacturing. Ferrosilicon deoxidizes steel and improves the castability of iron and steel. Ferrochrome is used for making stainless steel and high temperature alloys. Continuing research focuses on developing new ferroalloy applications which will improve process efficiency, lower costs, and improve product strength or performance characteristics.

This research relates to alloys containing many different metals. Among these are nickel, copper, molybdenum, and titanium.

Nickel, for instance, improves ferrous powder metallurgy alloys by solid-solution strengthening, and increases their hardenability. The presence of nickel may aggravate the problem of shrinkage during sintering; additions of copper will help to maintain dimensions. Molybdenum may also be added to iron powders to improve hardenability. Elements whose oxides are stable in typical sintering atmospheres are used sparingly, if at all, in ferrous powder metallurgy materials. While manganese, phosphorus, and chromium are sometimes used in very small amounts, vanadium and titanium are rarely encountered.

Ferroalloys are used as additives to enhance the sintering response and the sintered properties of P/M alloys. Also, some ferroalloys are used in pure form to fabricate high performance parts for defined applications. Metals companies have discovered that powder metallurgy techniques can be used to make components for demanding jobs calling for combinations of high temperature resistance/high impact strength/corrosion resistance. In some cases, this has led to a demand for making very fine powders from ferroalloy materials. This is because the finer additive powder particles are more homogeneously distributed and thus provide a larger number of additive to base powder particle contacts, resulting in a more homogeneous final product. In general, the fine powders needed must be produced without introducing any appreciable contamination from the pulverizing process.

There are two basic potential sources of contamination from conventional solid state dry size reduction processes. The mill may introduce impurities caused by the abrasion of the process material against the mill components. Or, the process may allow for oxygen contamination of the feed material by exposing newly created surface areas on the powder to air. The possibility of oxygen contamination calls for adoption of an inert processing environment. Inerting also protects against burning or explosion of the finely divided powders.

Disadvantages of conventional metal grinding processes are summarized in Figure 1.

- Very difficult to inert (if not impossible)

- "Flat" granulometry with excessive fines and/or oversize particles

- Mill wear contaminates product

- Excessive space required

- Frequently a batch process

- (Wet Milling) requires drying product, waste water disposal

Figure 1: Disadvantages of conventional processes

A new process has been developed which eliminates many of the disadvantages of conventional milling and thereby permits production-scale pulverizing of brittle materials without contamination. This process has been extensively used for making fine powders from lanthanide materials, primarily neodymium-iron-boron and samarium-cobalt. In the form described here, it has also been used to perform tests and custom processing for ferroalloy products manufacturers.

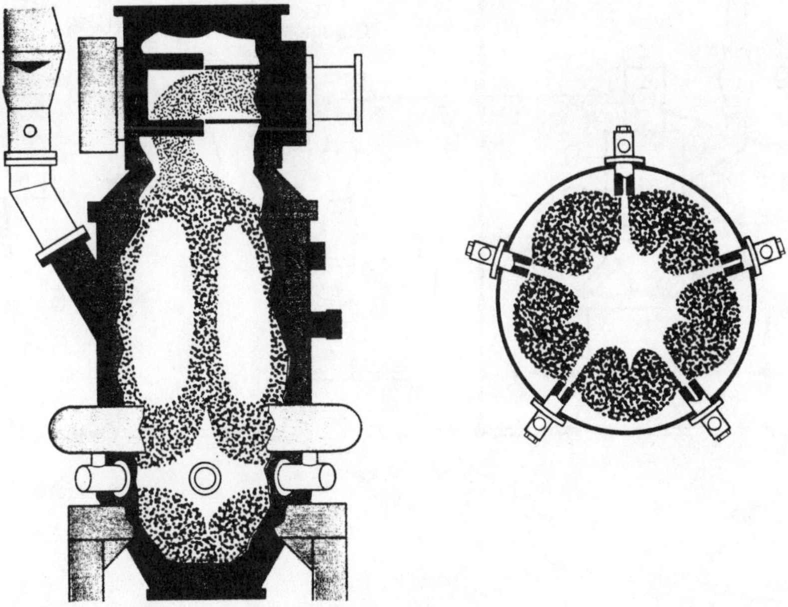

Figure 2: Schematic of fluidized bed jet mill

New Process Prevents Contamination

While the most frequent application of the system we discuss is making fine powders from the lanthanides, it should be noted that fluidized bed jet milling can be considered whenever a contamination problem occurs during dry processing of a reactive material. Other applications for this type of system are abrasive powders for advanced ceramics, engineering plastics powders, and high-purity specialty chemicals.

The heart of the system is a fluidized bed opposed jet mill (Figure 2) which consists of two sections: a bottom milling section with a fluidized bed, and a top section with a radial vane air classifier that controls the fineness of the product exiting the mill.

The entire system, as illustrated in Figure 3, operates under inert gas, typically using nitrogen gas as a fluid medium for grinding and transport of the product. The gas is recirculated to minimize loss and concomitant expense.

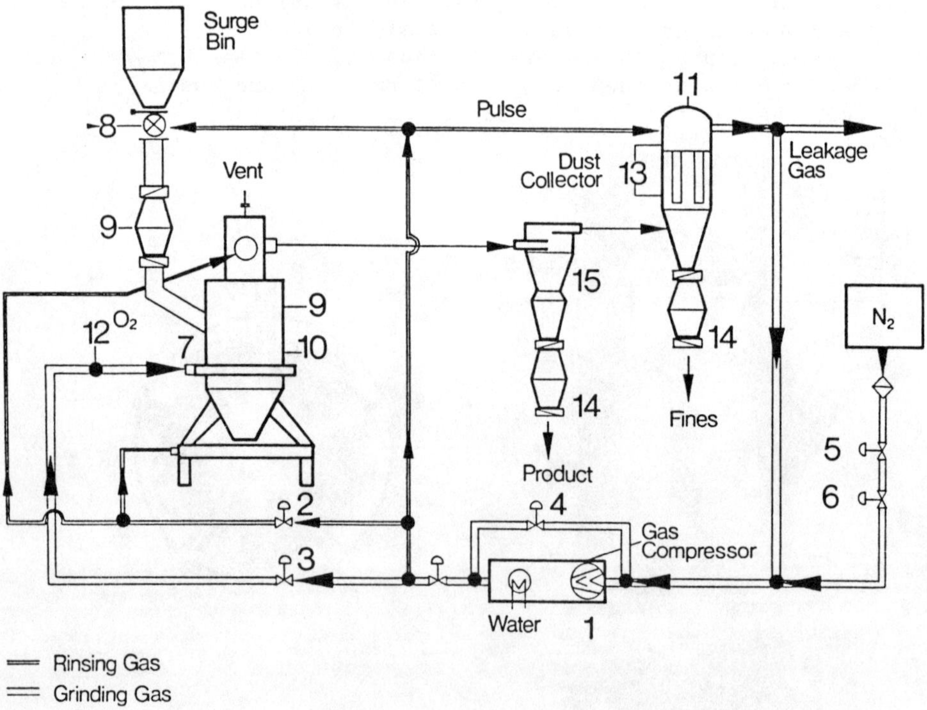

Figure 3: Schematic of a fluidized bed jet mill operating under an inert atmosphere

System Operation

An easy way to understand how the system works is to follow the start-up routine. First, we start the compressor (1). Then we open the purge gas valve (2) followed by the grinding gas valve (3). Then, after closing the bypass around the compressor (4), we open the main nitrogen gas line (5) and set the control for the nitrogen gas (6).

This enables purging the system of air until the oxygen level is brought down to below the safe level, which in turn permits opening the rotary valve for the feed (8). The material level control in the milling chamber (9) serves to start and stop the feed and discharge of the system and thus maintain, as closely as possible, a constant amount of material throughout the process. This control function of the level indicator is shown by the number (9) next to the feed double flap valve.

Sensors are placed to monitor the system in operation and, if necessary, to actuate the overpressure alarm for the grinding chamber (10), the overpressure alarm for the dust collector (11), the oxygen level safety alarm (12), and the differential pressure alarm for the dust collector (13).

The figure schematically shows raw feed material ($<$20 mesh) entering the grinding system (8) through a feeding apparatus which assures maintenance of constant gas pressure within the system. The surge bin is also sealed, because it is necessary to prevent exposure of the process material to air.

The material passes from the feed area into the milling chamber via a gravity feed chute. This design feature prevents wear of iron from a screw feed and, more importantly, facilitates the "trigger" effect of the load level indicator on the feed. The true significance of this automatic control over the feed and discharge is that it permits unattended operation of the milling process, allowing the operator to set reliable process parameters which can be monitored and analyzed by statistical process control routines.

Nitrogen gas is introduced (7) via nozzles arranged symmetrically around the milling chamber. The gas can be heated or chilled, or the diameter of the nozzle orifices can be modified, in order to influence the mechanics of particle fracture within the fluidized bed. (The expansion of gas as it emerges from the nozzles also exerts an adiabatic cooling effect on the milling process.) Size reduction of the particles takes place in the focal point of the jets. It is only here that the individual particles exhibit high kinetic energy levels.

Particle speed progressively decreases as the particles approach the interior wall of the milling chamber. At the wall itself, the velocity approaches zero. This means that minimum abrasion or wear is experienced. Published cases confirm that this mill introduces iron contamination of $<$ 20 ppm even with very abrasive materials over 9 Mohs hardness.

The oxygen level of the gas is monitored (12) and controlled within acceptable limits. This monitor serves as an emergency alarm, causing system shutdown and flooding the entire system with inert gas from an auxiliary source, if the oxygen exceeds a predetermined safe level.

The pressure within the grinding chamber is maintained at 1.0 psig (10). The expanded gas in the fluidized bed conveys the finer particles upward into the classification zone, where the product fraction is separated. Product is collected in the downstream cyclone (15) and dust collector (13). This equipment and connecting ductwork are sealed and pressurized to approx. 0.3-0.5 psig. The dedusted gas is then returned to the compressor (1) and reintroduced into the milling process. In a recently published case, the gas loss in continuous operation was reported at 1.3%.

The system as illustrated produces fine powders of neodymium-iron-boron to a top particle size (d99) $<$ 3.8 microns, (d99 means that 99% of the product particles are smaller than the designated size) and has been adapted to a wide range of ceramic and powder metallurgy applications. The primary modifications for grinding more abrasive materials are the installation of a polyurethane lining in the milling chamber and aluminum oxide ceramic vanes in the classifier wheel. Typical system throughput for the lanthanide product fineness mentioned is about 35 lb (16 kg) per hour.

By enlarging the nozzle orifices, this milling system can also be adapted to reprocess precrushed scrap material. These may be coarser than the $<$20 mesh of a typical feed. The overall adaptability of this system is further enhanced because -- unlike other jet mill designs -- this mill provides separate controls for milling and classifying. This allows the processor to "tailor" the particle size distribution of the powder with great precision.

Case Studies

The following case studies are offered to illustrate some of the process parameters and capabilities for the milling and classification system described.

Case 1: Ferrophosphorus

A feed material 99% < 50 microns was tested on a lab/pilot scale jet milling system to determine if it was feasible to obtain a product 61.4%< .2 microns and 99.99%< 8.3 microns. The throughput rate of the test was approximately 2kg/hr, although the targeted production rate was much greater. The test machine (100 AFG) was equipped with a ceramic classifier rotor and three 2mm gas nozzles; rotor speed was set at 16,000 rpm. Product was collected in a filter bag without cyclone. Test results at two different pressure levels were recorded as follows:

%<	1u	1.5u	2u	3u	4u	6u
Run #1 = 6-7 bar	17.7	25.6	45.7	73.3	88.7	trace
Run #2 = 5 bar	19.7	29.1	56.0	87.3	98.8	trace

Run #2 achieved the required fineness at 8.3 microns but could not attain a level of 61.4%< 1.2 microns.

Case 2: Ferrochrome

A feed material 98.4% < 3.15mm was tested on a small production scale system to determine feasibility of obtaining several products: 1) max. 2% over 15 microns; 2) 100% < 20 microns; 3) 100% < 15 microns. The material was processed at 7bar with 5mm nozzles (Run #1), and at 10bar with 4mm nozzles (Run #2). The runs utilized the same mill (200 AFG) with classifier rotor speed at 4500-5000 rpm (the slower speed results in a coarser product). Test results were recorded as follows:

%	10	50	90	RPMS
Run #1	2.53	7.06	15.53	4500
Run #2	2.33	7.35	15.25	4700
Run #3	2.84	6.76	13.93	5000
Run #4	2.46	6.56	14.56	4800

These results suggest that the required particle size distribution can be obtained after some changes in operating parameters (such as a faster classifier speed).

Case 3: Ferrosilicon

Two feed materials were tested in this case. The first was analyzed at 6% <100 microns with traces over 3mm; the second was very coarse (over 8mm) and was preground in the laboratory to less than 3mm (this is about the practical limit to feed size for the fluidized bed jet mill). The objective was to grind both materials to a level 99%< 75 microns. Using a lab/pilot system (100 AFG) equipped with three 2mm nozzles, operating under 7bar overpressure, with classifier rotor speed set at 2200 rpm, the material was successfully processed to the required level of fineness. The classifier rotor was made from aluminum oxide ceramic to protect against excessive wear and the associated contamination of the product.

In these cases air was used as the grinding medium because there was little likelihood of explosion with the fine powder product. The introduction of inert gas, as has been mentioned, also prevents oxidation of the process material. Where oxidation poses a problem in terms of contamination of the fine powder, it is necessary to operate the system with an inert processing gas such as nitrogen or, less frequently, argon. Cases 4 and 5 describe size reduction in this process using nitrogen.

Case 4: Ferromanganese

A 20 mesh feed was tested on a lab/pilot scale jet milling system operated under nitrogen gas. The goal of the test was to determine if it would be possible to get a d90=15micron product at a projected throughput of 40 kg/hr. The feed was milled with 2mm nozzles and classified at a rotor speed of 4,000 rpm. Product particle size analysis (Microtrac) yielded the following data: d10=2.98microns; d50=6.85microns; d90=14.03microns. It was predicted that the required throughput could easily be obtained with a small production scale system (200 AFG).

Case 5: Ferromolybdenum

This material is described as an easily inflammable and spontaneously combustible metal powder. A coarse feed 99.7%< 1.6mm was to be ground to a powder 100%< 20microns with a maximum 0.5% residue over 15 microns. With classifier rotor speed set at 4,000 rpm and utilizing three 5mm nozzles at 5.5bar overpressure, the feed was milled under nitrogen. The product was 99.98%< 20microns, 99.4%< 15microns. The summary of all test runs indicated a possible throughput of about 20kg/hr could be projected on a full-scale production system (400 AFG). This was the most difficult ferroalloy material to grind of those cited here.

If all other process parameters remain the same, the

substitution of nitrogen for air as the fluid energy medium in this process will result in an overall reduction in specific energy levels of about 20%. This will appear as a reduction in

throughput which cannot be accurately estimated without testing on full-scale production equipment.

Summary

New processing capabilities incorporate closed-loop inert gas circulation into fluidized bed opposed jet milling. This new technology, now used for production of lanthanide powders, is shown to generate ferroalloy powders finer than those typically produced with spray atomization. Ferroalloy materials tested include ferrophosphorus, ferrochrome, ferrosilicon, ferromanganese and ferromolybdenum.

# # #

PREPARATION OF BULK AMORPHOUS Ni60Nb40

F. Petzoldt, B. Scholz, H.-D. Kunze
Fraunhofer-Institut für angewandte Materialforschung, Bremen, FRG

M. Held, MBB, Schrobenhausen, FRG

Abstract

Recently the P/M method of mechanical alloying (MA) has been firstly applied to produce amorphous metal powders by a high energy milling process of at least two elemental polycrystalline powders.

The present paper presents results on amorphization by MA of elemental Ni- and Nb-powders and the production of an amorphous semifinished part.

The amorphization process with proceeding milling time is characterized by x-ray diffraction and hardness measurements. The completely amorphous Ni60Nb40 powder is homogeneous in chemical composition and shows a relatively narrow size distribution of globular shaped particles. The crystallization temperature and the activation energy for crystallization have been determined by differential scanning calorimetry. The crystallization temperature is similar to that of rapidly quenched Ni60Nb40 samples prepared by melt spinning. Therefore one might expect that melt-spun ribbons and MA-powders of Ni60Nb40 have similar properties. But one important difference between both methods is the geometry of the amorphous product. The powder geometry is advantageous for compaction and production of bulk amorphous material. The MA-Ni60Nb40 powder has therefore been produced in larger powder charges and was compacted. The compacted cylinder is approximately 8 mm in diameter and 100 mm long. This sample is still amorphous as examined by x-ray. The properties of the compacted part such as hardness and crystallization behaviour are compared of the MA-Ni60Nb40 powder.

Introduction

Mechanical alloying (MA) is a powder metallurgical method applied since 20 years to produce compound powders. On an industrial scale for example nickel superalloys like MA 6000 are fabricated by attrition milling [1]. These alloys are superior to both the cast nickel superalloy and ODS nickel because the two strengthening mechanisms (oxide dispersion hardening and γ'-precipitation hardening) supplement one another.

In 1983 Koch and co-workers [2] published a paper describing the possibility of amorphizing $Ni_{60}Nb_{40}$ by MA. In the following years also some other groups of scientists prepared amorphous powders by this method. A successful amorphization was achieved for a lot of binary TM_1-TM_2-alloys (TM_1 = early transition metal, TM_2 = late transition metal), see e.g. [3].

Of major scientific interest is to understand the mechanism of amorphization by MA. One could imagine that local melting of powder particle surfaces occurs during milling followed by rapid quenching. If the cooling rate is high enough an amorphous layer could be formed by this process. Combined with the continuous fracturing and rewelding of powder particles during MA this might result in completely amorphous powders. Another possible mechanism

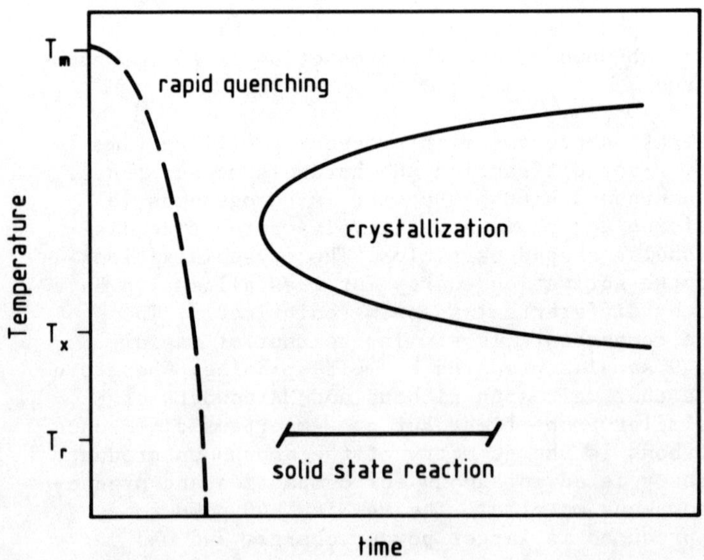

Figure 1: Schematic time-temperature-transformation diagram

of amorphization during MA is a solid state interdiffusion reaction similar to that reported by Schwarz and Johnson [4]. They prepared a thin amorphous film by an isothermal annealing of alternating polycrystalline Au- and La-layers. Necessary conditions for this solid state reaction are a large negative heat of mixing and the fast diffusion of one of the elements.

In Fig. 1 the two proposed mechanisms of amorphization are depicted schematically in a TTT-diagram. The dashed line represents the amorphization by rapid quenching from the melt. A solid state reaction must occur at temperatures below the crystallization temperature T_x of the amorphous phase.

While a rapid quenching process takes only microseconds, the time scale for amorphization by solid state reaction takes about 10^5 s.

Mechanical alloying is a well established process in P/M-industries. Therefore it suggests itself to use this technique for an industrial production of amorphous alloy powders. Furthermore manufacturing of semifinished products out of MA amorphous powder seems to be relatively easy. In the following sections the production steps from polycrystalline elemental powders to bulk amorphous material are described.

Powder Production

Elemental polycrystalline Ni (FSSS 4.6 µm) powder and Nb (FSSS 63 µm) powder were mixed in the desired starting composition. This powder mixture was milled in a high energy planetary mill. The MA was carried out under argon atmosphere. Milling parameters were a steel ball (100 Cr 6) to powder weight ratio of 15:1. In our milling equipment the powder charge in a single milling vessel is therefore limited up to about 100 g.

The progress in MA was controlled and characterized by different analytical methods. Optical microscopy and hardness measurements (HV0.025) were carried out to describe structural refining. Scanning electron microscopy (CAMSCAN 4) was used to investigate changing powder morphology. X-ray analysis was performed using Cu-K_α radiation (λ = 1,5418 Å) in reflection geometry to detect the proceeding amorphization. Additionally the crystallization temperature of the completely amorphous powders was measured by differential scanning calorimetry (DSC 7, Perkin Elmer).

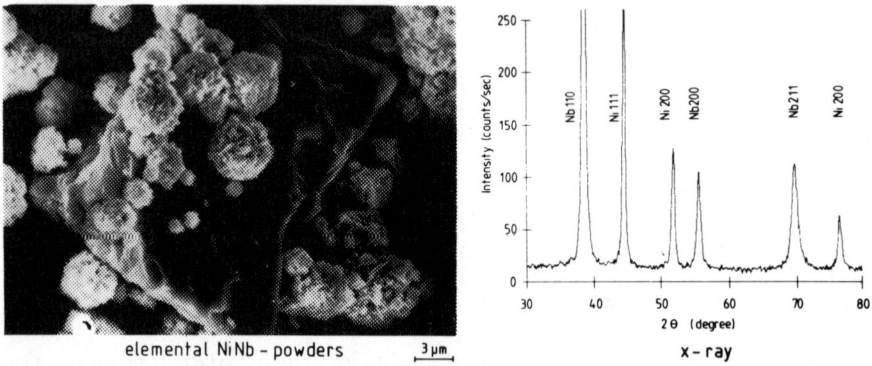

Figure 2: SEM micrograph and x-ray diffraction patterns of crystalline powders

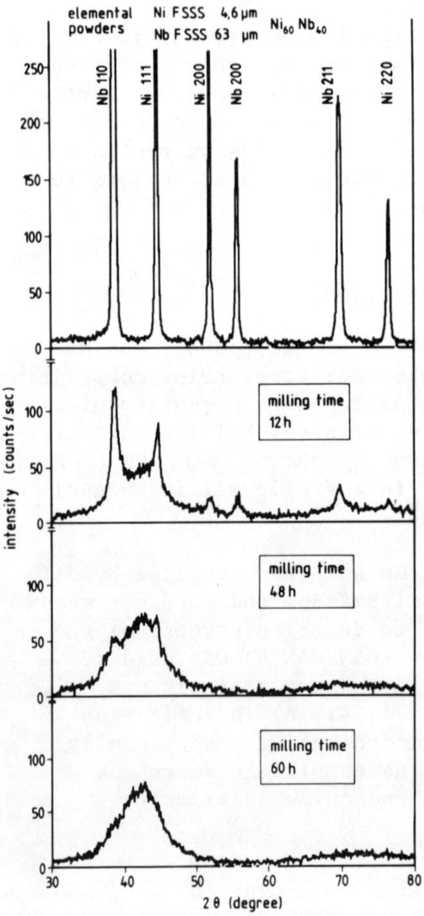

Figure 3: Comparison of x-ray diffraction patterns of $Ni_{60}Nb_{40}$ before milling and after different MA-processing times

Fig. 2 shows a SEM-micrograph of the starting materials. The small particles have the typical shape of general-purpose nickel powder produced by carbonyl decomposition. Furthermore a blocky coarse Nb-particle can be seen. The corresponding x-ray diffraction patterns of the crystalline powder mixture is plotted in Fig. 2 and the top of Fig. 3. After 12 h of MA the x-ray diagram shows decreasing intensities of crystalline Ni and Nb. Simultaneously a broad diffraction maximum forms being characteristic for an amorphous phase. With proceeding milling time the amount of this new phase increases. After 60 h of MA no crystalline reflection can be observed in the diffraction pattern (bottom of Fig. 3). This amorphization process is accompanied by a structural refinement. The optical micrographs of Fig. 4 render clearly the microstructure typical for any MA-process [1]. With proceeding milling time the lamellar structure (5 h) changes to an irregular looking microstructure (28 h). After longer milling times powders were analysed by transmission electron microscopy showing crystalline Ni and Nb lamellae (20-30 nm after 40 h milling) in an amorphous

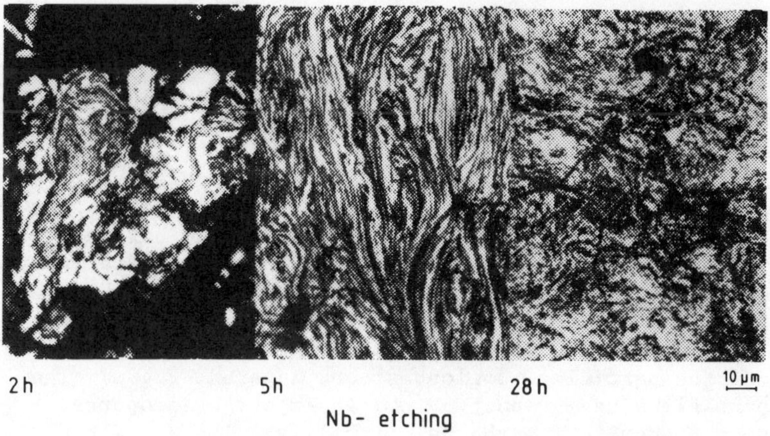

Figure 4: Optical micrographs of MA-$Ni_{60}Nb_{40}$ powders after different milling times

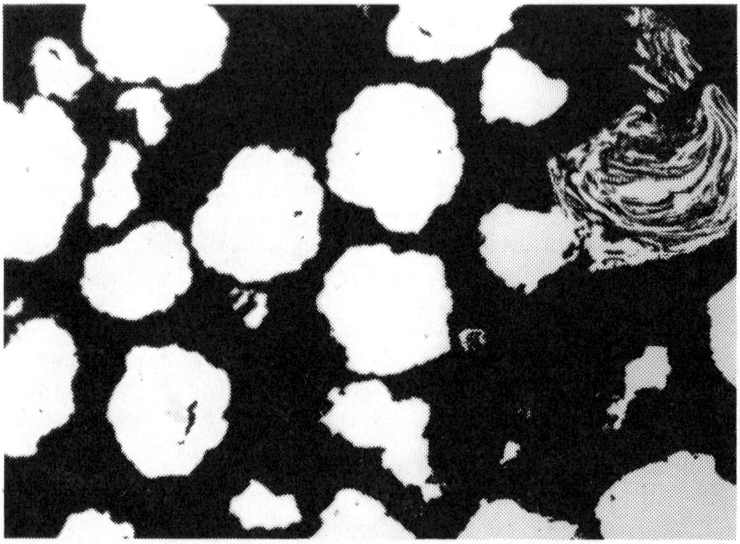

Figure 5: Optical micrograph of amorphous $Ni_{60}Nb_{40}$ powder

matrix [5]. After 60 h of MA the powder is amorphous as detected by x-ray. This powder is hardly etchable, Fig. 5. Another interesting property is the microhardness of powder particles after different milling times giving some hints on the actual state of the alloying process. The increasing microhardness is a linear function of milling time for $4\ h < t < 40\ h$, Fig. 6. The hardness of the amorphous powder is about 850 HV0,025. This is a relatively high value being typical for amorphous metals. A too high hardness of powder particles may render a compaction process more difficult. During MA also the morphology has changed completely. The shape of the particles is globular and layered, Fig. 7. The average particle size is about 50 μm. An important property of the amorphous powder is its thermal stability. Therefore the crystallization temperature was measured by DSC at a heating rate of 40 K/min. In Fig. 8 the exothermal heat flow during crystallization is drawn. The crystallization temperature is 678 °C.

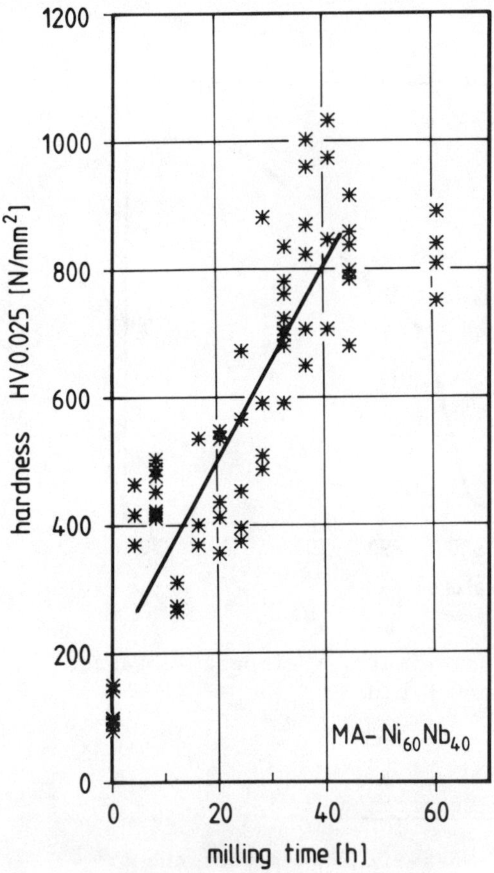

Figure 6: Microhardness (HV0,025) of $Ni_{60}Nb_{40}$ versus milling time

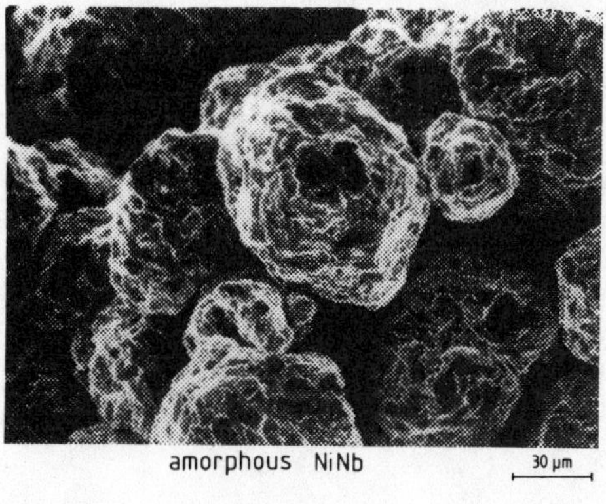

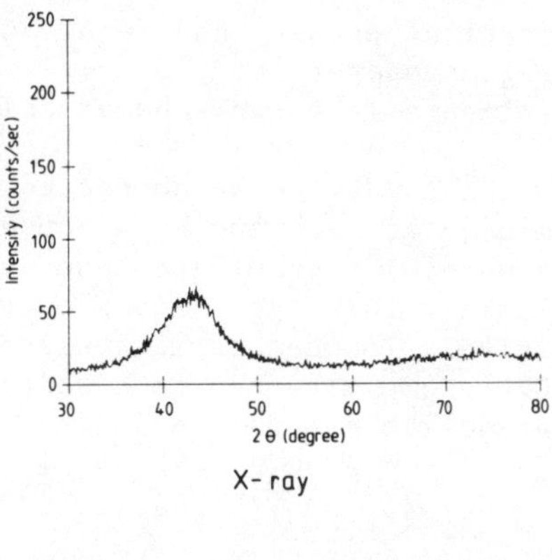

Figure 7: SEM micrograph and x-ray diffraction patterns of amorphous powder

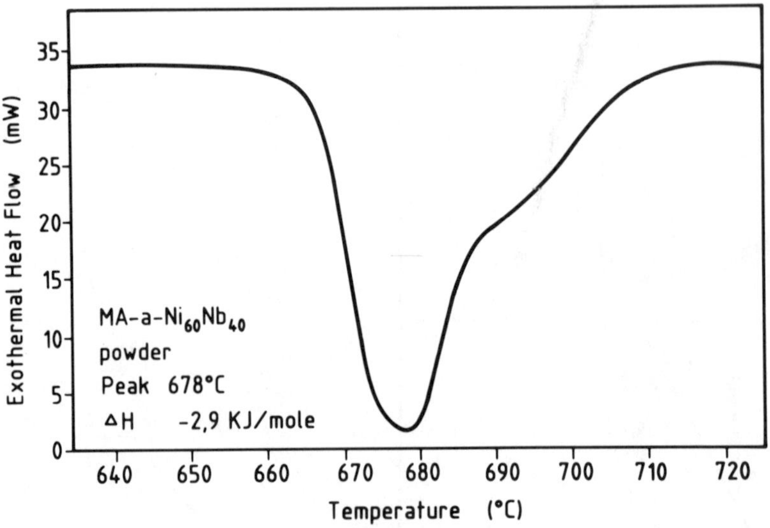

Figure 8: Differential scanning calorimetry thermogram obtained with a heating rate of 40 K/min

Explosive Compaction

During consolidation of the MA-a-$Ni_{60}Nb_{40}$ powder the crystallization temperature should not be exceeded. Explosive compaction is a dynamic compaction method leading to nearly 100 % theoretical density and being accompanied by a relatively low average temperature rise [6,7]. The arrangement used in our experiments is shown schematically in Fig. 9.

The powder to be compacted is placed in a copper tube which is closed at each end by a copper plug. This tube was evacuated by a water-jet pump. The tube is surrounded by the explosive (Volumex). This type of explosive was chosen because of its relatively low detonation velocity (2200 m/s) and its density of 0,6 g/cm³. Thereby no high pressures of the shock waves are created. The detonation is initiated by an electric detonator

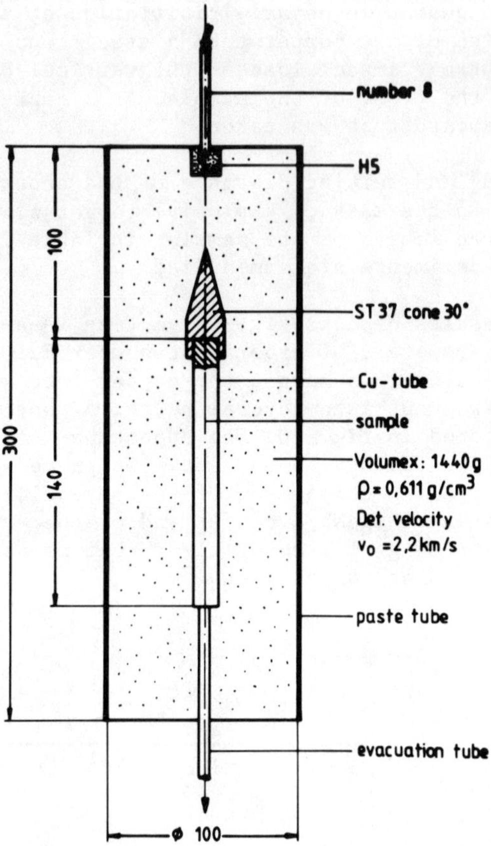

Figure 9: Scheme of a cylindrical compaction device

TABLE 1: PARAMETERS OF EXPLOSIVE COMPACTION

Experiment number	diameter of cardboard (mm)	length of cardboard (mm)	starting/setting distance (mm)	mass of explosive (g)
E1	50	240	50/50	200
E2	100	300	100/60	1440
E3	200	440	200/100	8000

KX 1/20 (similar to detonator No. 8) attached to a small booster charge chosen to guarantee an axial initiation of the shock wave in Volumex. On top of the copper tube a steel cone was mounted to avoid a too strong impact load of the vertical propagating shock wave onto the front of the sample. As a containment of the explosion a cardboard tube was taken.

The starting and setting distance of the shock wave were varied adequate to the mass of explosive to get always a constant pressure along the length of the sample. In Table 1 parameters of the different experiments are summarized.

The best results of explosive compacting MA-a-$Ni_{60}Nb_{40}$ were achieved using parameters of experiment number E2. For the following investigations of this sample the copper tube was turned off and the compact was cut into several parts. A photograph of such a sample is depicted in Fig. 10. The appearance of the 100 mm long sample is satisfactory. X-ray analysis proves the amorphicity of the bulk material. A comparison of the crystallization behaviour of the amorphous MA-a-$Ni_{60}Nb_{40}$ powder and of the explosively compacted cylinder, Fig. 11, makes clear that no essential changes have taken place during compaction.

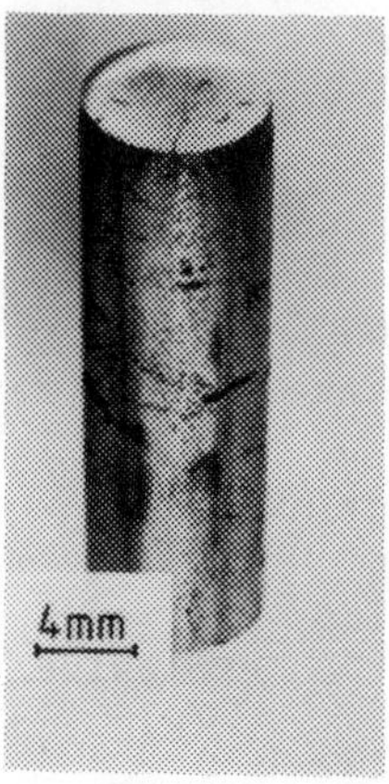

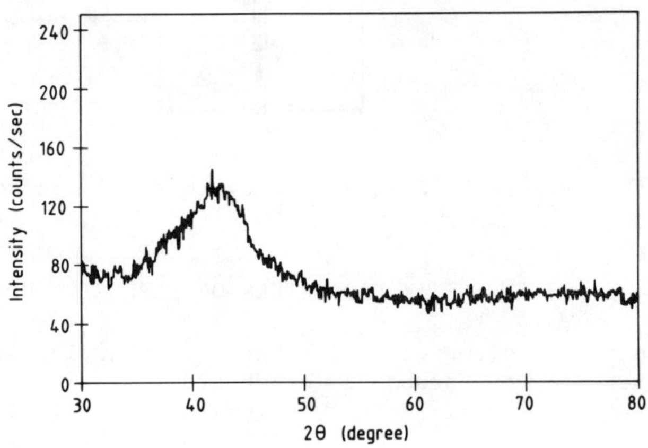

Figure 10: Explosively compacted cylinder and x-ray diffraction pattern

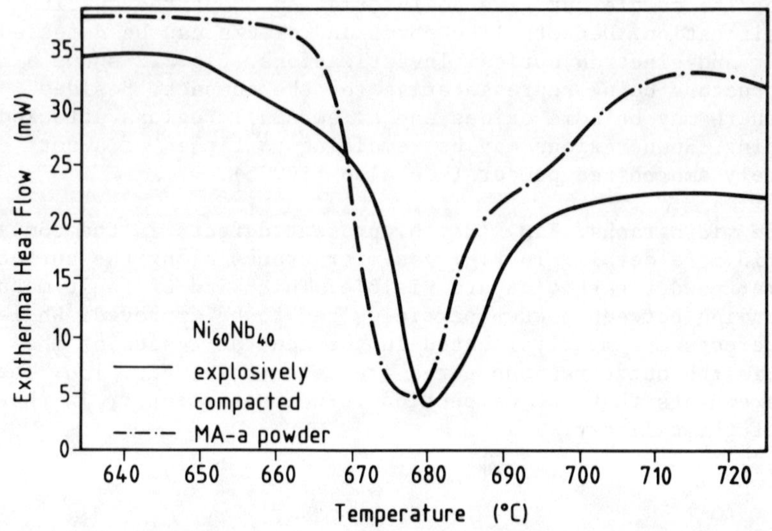

Figure 11: Comparison of DSC-scans of amorphous $Ni_{60}Nb_{40}$ before and after compaction

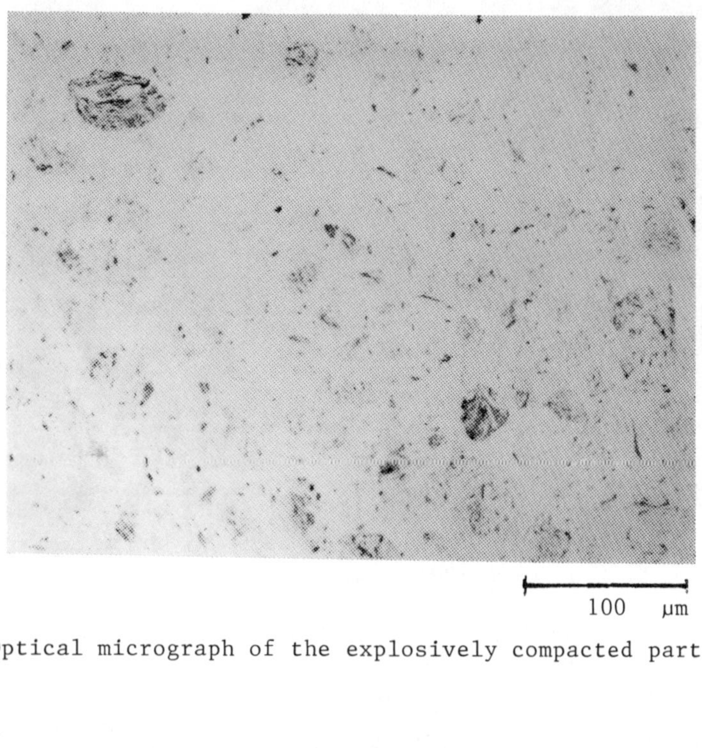

Figure 12: Optical micrograph of the explosively compacted part

These examinations are necessary to prove the amorphous state of the sample but give no information about the quality of densification. Defects like pores and cracks can be detected by light and electron optical investigations. Fig. 12 shows a microstructure being representative for the compact. Besides pores there may be some oxides and a few small regions attacked by etching. Such regions may be remaining particles of a not completely amorphized powder (see also Fig. 5).

The micrographs, Fig. 13a, b, present defects in the compacted sample in more detail. In Fig. 13a microcracks along the surfaces of former powder particles are visible. This kind of fault means that bonding between powder particles has to be improved. While these defects are mostly located in the central region of the sample in the outer regions pores are detectable, Fig. 13b. The pores are hints that the compaction is not satisfactory in these parts of the cylinder.

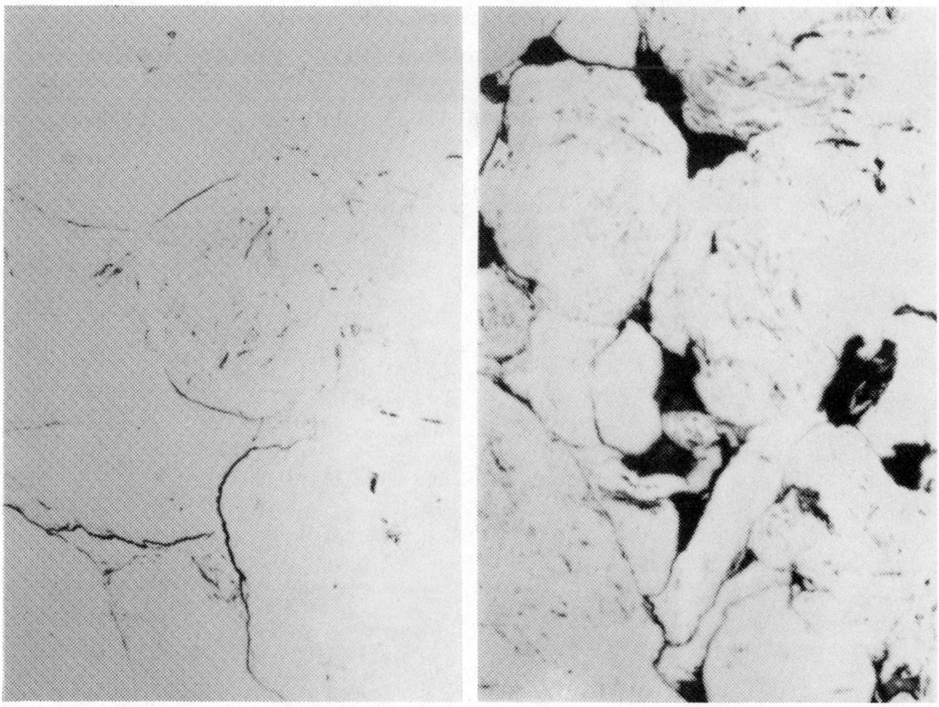

a) 10 μm b) 10 μm

Figure 13: Optical micrographs showing microcracks along former powder surfaces (left) and pores in the outer region of the sample (right)

The sample was broken to investigate the fracture behaviour. Fig. 14 depicts the fracture surface and details in higher magnification. The fracture occurs almost along the surfaces of original powder particles. No transgranular separations can be observed. This is also a hint that bonding between powder particles has to be further optimized to get good mechanical properties.

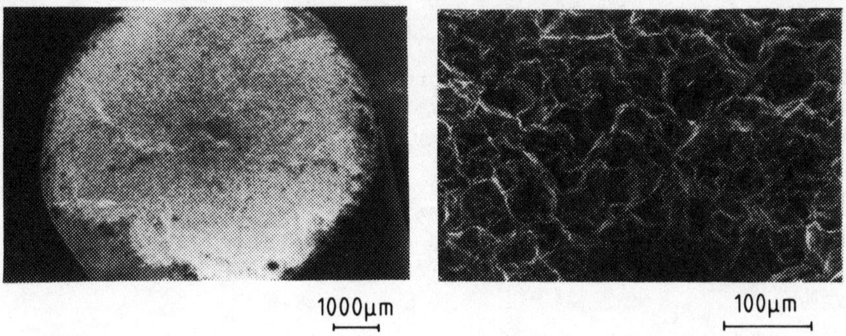

Figure 14: SEM micrographs of a fracture surface

There are different methods to achieve a machinable semi-finished product. First the parameters of explosive compaction have to be further optimized. An additional thermomechanical treatment could also improve bonding between particles and microcracks might simultaneously be annihilated. A different compaction method to get an amorphous sheet is hot rolling of the powder at temperatures just below the crystallization temperature [8]. At the temperature of the so called glass point the amorphous material behaves like a metastable liquid. That means the viscosity is lower than 10^{13} poise, the amorphous material becomes quite ductile and can flow freely. Finally it shall be mentioned that recently performed experiments point out the chance to produce amorphous coatings by thermal spraying [9].

Final Remarks

The presented experiments demonstrate that amorphous Ni-Nb-powders can be fabricated by MA. Furthermore these powders can be manufactured by means of P/M-technologies. One process to get a semifinished amorphous product is to compact the powder explosively.

The properties of the MA-amorphous powder - e.g. amorphicity, hardness and crystallization temperature - were conserved during the compaction process. Compared to rapidly quenched amorphous ribbons the physical properties of the powders are similar. An advantage of the newly developed amorphous MA-powder is firstly its easy performance and the possibility to produce large powder charges on an industrial scale. Secondly the preparation of bulk amorphous material seems to be more easy via powder technology because of the appropriate morphology of the mechanically alloyed powder.

The amorphous metals might perhaps find new interest and new applications because restrictions in the dimensions existing for application of melt-spun ribbons of 50 µm thickness are now overcome by MA.

References

[1] P.S. Gilman and J.S. Benjamin
Ann. Rev. Mat. Sci. 13 (1983) 279

[2] C.C. Koch, O.B. Cavin, C.G. McKamey and J.O. Scarbrough
Appl. Phys. Letters 43 (1983) 1017

[3] L. Schultz and E. Hellstern, in:
Science and technology of rapidly quenched alloys, eds.
M. Tenhover et al. (MRS 1987) paper G1.1

[4] R.B. Schwarz, W.L. Johnson
Phys. Rev. Letters 51 (1983) 415

[5] F. Petzoldt
Proceedings Conf. on Solid State Amorphizing Transformation,
Los Alamos 1987, J. Less Common Metals (in print)

[6] R. Prümmer
Explosivverdichtung pulvriger Substanzen
Springer 1987

[7] R. Prümmer
Proc. of the 6th Int. Conf. on Rapidly Quenched Metals
RQ6, Montreal 1987, J. Mat. Sci and Engineering 1988 (in print)

[8] P.H. Shingu
Proc. of the 6th Int. Conf. on Rapidly Quenched Metals
RQ6, Montreal 1987, J. Mat. Sci and Engineering 1988 (in print)

[9] F. Petzoldt, H. Grützner, et al.
to be published

Chapter Three

Powder Production: Gas Atomizing

ATOMIZED POWDERS WITH REDUCED MICROSTRUCTURE VARIATIONS

W.A.Kaysser, K.Rzesnitzek, G.Petzow

Max-Planck-Institut für Metallforschung
Institut für Werkstoffwissenschaften
Heisenbergstr.5, D-7000 Stuttgart 80, West Germany

ABSTRACT

Fine melt droplets solidify during gas and water atomization at high cooling rates, up to 10^6 K/s, after extensive undercooling of up to some hundred degrees centigrade. Uncontrolled undercooling provides a range of unexpected phases. Theoretically the fine dendritic, cellular and featureless microstructures formed under virtually equal atomization conditions can be understood as a result of metastable local interfacial equilibria and interfacial non-equilibria. In practice the uncontrolled variety of microstructures in atomized powders provides consolidated materials which are prone to local property variations and hence are of reduced value.

In the present work, a distinct undercooling was achieved by adding or eliminating heterogeneous nuclei in the melt. The microstructures of fine particles of Cu, Sn and Zn alloys which were solidified after adding a large number of artificial nuclei with calculated cooling rates between 10^3 and 10^6 K/s showed very fine and homogeneous dendritic microstructures. The elimination of heterogeneous nuclei in the melt before atomization resulted in extended undercooling and the formation of microstructures with featureless areas. A homogeneous dispersion of fine inert oxide particulates present in the melt, was still maintained in the solidified featureless areas.

INTRODUCTION

The influence of cooling rate, undercooling and melt conditioning on the homogeneity of the solidified microstructure of Cu-Sn alloys is described. Melt confining includes seeding and refining in respect to heterogeneous nuclei.

Compared to the frequent use of gas atomization and despite a considerable number of investigations, the microstructural development during rapid solidification of the small droplets is still understood unadequately. Major problems in interpretation and prediction of rapidly solidified microstructures arise from the combined effects of cooling rate and undercooling. It was the purpose of this investigation to modify the atomization conditions in a broad range in one binary system. Cu-Sn was chosen, since many data required for quantitative estimations are known.

EXPERIMENTAL EQUIPMENT AND PROCEDURE

In a laboratory scale atomizer, shown in fig.1, gas atomization with Ar was performed at a maximum gas pressure at the nozzle of 120 bar. Different types of nozzles, with varying jet angles and diameters were used. The alloys were molten in crucibles with a induction heating system. Two high frequency generators with 30 kW and 4 kW, respectively, provided temperatures in the main crucible up to 2000 °C and allowed the "opening" of a thermo valve starting atomization.

When temperature was reached treatments of the melts in order to adjust the nucleation conditions of the alloy were performed. Metals or fluxes were dipped into the melt by a mechanical system. The atomized powders were collected in the lower parts of the atomizing device or in a cyclone. They were sieved and metallographically embedded. The samples were polished and examined by optical and scanning electron microscopy.

COOLING RATE DURING ATOMIZATION

It is well accepted that the cooling of droplets during atomization is mainly controlled by the convective heat transfer between the droplet and the atomizing gas. The radiative heat transfer is only of minor importance and can be neglected in the calculation. On the other side, the velocity difference between gas and particles is an important factor that determinates the cooling by a large value. Often the velocity difference at the beginning of the atomization, when the gas stream hits the melt, is taken to calculate the Nusselt number which gives the heat transfer coefficient. But the droplets are accelerated by the gas and so the velocity difference decreases and therewith the heat transfer coefficient. The particle acceleration depends on the particle diameter, small ones will very soon attain the velocity of the surrounding gas and solidify in that state. Larger particles are slowly accelerated and leave the zone of the rapid gas stream before the onset of solidification, the last happens during the free flight of the droplet in the atomizer.

In both cases, the starting gas velocity plays no dominant role for the heat transfer during the solidification. So the cooling rate is mainly

determined by the thermal properties of the used gas and the particle diameter. The velocity difference between gas and particle during the particle flight and the onset of solidification is shown in fig. 2a. The heat transfer coefficient until the end of solidification for different particle diameters is shown in fig. 2b.

EXPERIMENTAL RESULTS

Untreated melts

Different cooling conditions were realized in Cu-7wt.%Sn by slow furnace cooling, water quenching, gas and water atomization and splat cooling of untreated melts. Microstructures of Cu-7wt.%Sn, solidified at different cooling rates, are shown in fig.3 a-c. In the atomized samples larger particles sometimes show a microstructure that resembles directional solidification, fig.4a, starting from a discrete location. The majority of larger particles, fig.4b, shows several nucleation sites. Atomization of Cu-20wt.%Sn, fig.5 a-b, resulted in the dendritic α-phase. Smaller powders also showed segregation-free solidified β-phase.

Preconditioned melts

When a treatment of the melt prior to atomization was carried out, different microstructures were found. After adding Al shortly before atomization particles with areas of segregation-free solidification, or cellular microstructures that appeared only after the melt treatment occured. The addition of iron and zirconium to the molten Cu-7wt.%Sn alloy lead to very fine dendritic microstructures.

DISCUSSION OF THE SOLIDIFICATION BEHAVIOUR

The extremes of solidification of atomized droplets are solidification by external heat flow only or solidification after large initial undercooling.

In practice always a mixture of both extremes will occur. In this investigation a certain separation of these extremes would be achieved by melt conditioning.

Solidification by external heat flow

In this case remarkable undercooling is avoided by the presence of a sufficient number of heterogeneous nuclei.

Solidification after seeding of the melt

The heterogeneous nuclei in the melt are of different nucleation potential. The nuclei that are almost present in the melt, especially the metaloxides, are of less potential than some other nuclei that are also used in the foundry practice as seeding materials. Cu-7Sn ingots show for example a significant grain refinement and a smaller degree of undercooling when seeded with Fe and Zr. The addition of 0.1% Fe and 0.1% Zr results in ingots solidified, after nearly zero-undercooling,

with grain sizes about 30 times smaller than those of the untreated material.

We used this effect for atomization. No remarkable undercooling occured and the solidification was only controlled by the external heat flow. This method resulted in powders that showed the same microstructure in particles of different sizes. The secondary dendrite arm spacing was determined by the cooling rate, that means by the particle diameter.

The velocity of the solidification front with no initial undercooling can easily be calculated using a simple model. A spherical powder particle is approximated by a cylinder of diameter d and height d. Solidification will occur in the isothermal sample with a planar interface moving at velocity v_E from one cylinder face to the other. If we assume that the particle is isothermal at any instant, the solidification velocity V_E is given by a simple heat balance. In this case the solidification is only controlled by the external heat flow /5/.

$$V_E = \frac{6\, h_i\, \Delta T}{L}$$

In this formula h_i means the heat transfer coefficient, L the latent heat of fusion and ΔT is the temperature difference between the cooling gas and the alloy droplet.

Solidification after weak undercooling

Atomization was done with materials of no special purity. The atmosphere in the atomization chamber contained enough oxygen to form oxides in the melt that were able to act as nuclei.

Under usual conditions, the larger particles always showed multiple nucleation. This is indicated by the different growth directions of the dendrites in Fig. 4b. After a certain undercooling a number of nuclei provides the formation of several crystals in one particle. If nucleation occurs closer to the solidus temperature, by a single more potent nucleus, the whole particle forms a single crystal. This case is shown in Fig. 4a. Here the liquid droplet was hit during flight by an already solidified small particle. This particle, a so called "satellite", caused at once nucleation and subsequent solidification of the bigger particle. The undercooling of the big particle was smaller than in the case of multiple nucleation.

In a technical Cu-7wt.%Sn alloy the resulting microstructures were almost equal in most of the particles yielded. When the Cu-7wt.%Sn alloy was produced from pure Cu and Sn precursors, microstructures as in the technical alloy, but also, especially in the small size fractions, particles that exhibited other kinds of microstructural features were obtained.

The difference in the chemical composition is too small to explain the severe change in the microstructures. Hence the nucleation conditions are considered to be the dominant factor. The technical alloy contains more oxides that may act as heterogeneous nuclei. Undercooling is suppressed and the main part of solidification is governed by external

heat flow. Therefore nearly all particles show a similar microstructure.

In a given melt the density of heterogeneous nuclei is approximately constant in the whole volume. Subdividing the volume in very small droplets increases the probability for the small droplets to contain no heterogeneous nucleus.So the production of highly undercooled powders is connected with the production of very fine powder sizes.

Using convenient alloy compositions the nucleant densities can be estimated easily. For example the Cu-20wt.%Sn alloy, fig. 5, shows a typical change in the formed phases in the large and the small particle size fractions. The nucleant density follows the Poisson distribution $X=\exp(-Mv)$. Herein X is the nucleant free droplet fraction, M is average number of nucleants per volume element and v is the droplet volume. Quantifying the existent phases in the size fractions the density of the nuclei can be calculated /6/.

The most potent nucleus is that which leads to the formation of the α-phase. Particles containing the β-phase only exist in the size fraction smaller 50 µm. The nucleant density for formation of the α-phase is $5*10^{15}/m^3$. The development of the β-phase requires an undercooling of at least 50 K.

The presence of heterogeneous nuclei determines the possible undercooling. The number of nucleant free particles can be increased by producing very fine powders as shown above, or by reducing the number of potent nuclei in the bulk melt prior to atomization.

Solidification after Enhanced Undercooling

In the melt always oxides exist which act as heterogeneous nuclei. If these oxide nuclei can be eliminated, large undercoolings and therefore high solidification velocities do result. One method to achieve melt purification is, in the case of the Cu-7wt.%Sn alloy, the addition of metallic aluminum to the bulk melt. The high oxygen affinity of the aluminum will cause a reduction of other metal oxides. Fine dispersed alumimumoxide will be formed. Aluminumoxides, however, do not act as heterogeneous nuclei in the Cu-7wt.%Sn alloys. Therefore the number of nuclei decreases and the possibility of large undercoolings increases. With an initial undercooling large enough to allow the particle to be its own heat sink, very rapid solidification velocities are realized. If the particle reaches, due to the release of the latent heat of fusion, the solidus temperature again before solidification is completed, the remaining liquid solidifies by the normal external heat flow. The particle in fig. 6 was strongly undercooled before the onset of nucleation. Until the solidus temperature was reached again, a part of the liquid solidified in a segregation-free manner. The rest of the melt solidified with the normal external heat flow, as can be seen in the usual dendritic structure.

The change in microstructural development is caused by the decreasing solidification front velocity. A mathematical model describes the growth velocities as a function of recalescence /7/. It was shown that the growth velocity may differ by a factor of 25 between the start and the end of solidification of a strongly undercooled droplet.

Boettinger gave a hierarchy of equilibrium (fig. 7) /8/. In the normal metallurgical practice condition III is required. Condition IV shows a significant loss of interfacial equilibrium. This is important for simple metallic phases when the crystal growth rate exceeds the diffusive speed of solute atoms in the liquid phase. In this case solute trapping occurs and leads to a segregation free solidification. Segregation free solidification is possible if the growth velocity V_E of the solidification front is larger than the diffusional speed of the solute atoms /9/. In the above example a solidification front velocity of about 4 m/s is required. The usual solidification front velocity in our particles by cooling by convective heat transfer without initial undercooling was calculated before:

$V_E = 2.4$ cm/s

The growth velocity during the segregation-free solidification must exceed the required 4 m/s.

Another effect of this high growth velocities is the entrappment of the aluminumoxide particles in the segregation-free solidified alloy regions. Figure 8 shows an example.The white dots in the matrix are the aluminumoxide particles. The velocity of the growth front determines whether these particles are entrapped or not. In the particles without segregation-free solidification the aluminumoxide ist always found in the interdendritic space.

The entrappment of those particles in the solidification front gives some information about the growth velocity. The velocity necessary to entrap the particles in the solidification front can be calculated /10/.Unfortunately a lack of data makes calculations impossible until now.

CONCLUSIONS

The coarsening of the secondary dendrite arms occurs during the time interval when the initially formed dendrites and the melt are in contact. A model of Kattamis describes the dendrite ripening based on thermodynamical data. Our present calculations of the cooling rate from thermal data and the measured secondary dendrite arm spacing are in good agreement with the model.

The gas velocity at the nozzle determines the droplet size during fragmentation. The cooling rate during gas atomization is dependend on the reciprocal square of the diameter and increases with decreasing droplet diameter. During solidification of the particles velocity difference between the gas and the droplets was found to be small, independend from the initial gas velocity. Hence the cooling rate during solidification cannot be increased by an increased gas velocity at the nozzle outlet.

The growth velocities of particles with small initial undercooling are in the range of some cm/s. They result in a refinement of the microstructures. The production of segregation-free solidified powders, however, requires growth velocities in the order of some m/s. These high velocities can only be attained if solidification does not dependend on the external heat flow. Undercooling is mainly controlled by

the nucleation conditions within the melt droplet. Usually a large number of heterogeneous nuclei exist in a particle and multiple nucleation occurs at medium undercooling.

The nucleant densities can be decreased by a treatment of the melt. The metaloxides in the melt that act as heterogeneous nuclei can be transformed into inert oxide-particles. In the case of Cu-7wt.%Sn aluminum was used to reduce these metaloxides. The generated Al_2O_3-particles did not act as nuclei and larger undercooling was observed.

ACKNOWLEDGEMENT

This work was supported by Deutsche Forschungsgemeinschaft

REFERENCES

/1/ P.Duwez, R.H.Willens, W.Klement, J. Appl. Phys. ,31, pp.1136,1137, (1960)

/2/ T.W.Clyne, R.A.Ricks, P.J.Goodhew, "The Production of Rapidly-Solidified Aluminum Powder by Ultrsonic Gas Atomization. Part I: Heat and Fluid Flow", Int. Jour. Rapid Solidification, 1, pp. 59-80(1984)

/3/ T.Z.Kattamis, J.Coughlin, M.C.Flemings, Trans. Met. Soc. AIME, 239: 1504(1967)

/4/ M.Flemings, "Solidification Processing", McGraw-Hill Series in Materials Science and Engineering, New York, 1974

/5/ W.J.Boettinger, J.H.Perepezko "Fundamentals of Rapid Solidification", in "Rapidly Solidified Crystalline Alloys", Proceedings of the TMS-AIME, New Jersey, May 1-3, 1985 eds. S.K.Das, B.H.Kear, C.M.Adam, pp. 21-58(1985)

/6/ J.H.Perepezko, I.E.Anderson, "Synthesis and Properties of Metastable Phases", (Warrendale, PA: TMS AIME, 1980), pp. 31-63

/7/ C.G.Levi, R.Mehrabian "Heat Flow During Rapid Solidification of Undercooled Metal Droplets", Met. Trans. A, 13 A, 221-234(1982)

/8/ W.J.Boettinger, S.r.Coriell, "Microstructure Formation in Rapidly Solidified Alloys" in Science and Technology of the Undercooled Melt, eds. P.R.Sahm, H.Jones, C.M.Adam. NATO ASI Series E; no. 114, pp. 81-110,(1986)

/9/ M.J.Azis, "Model for Solute Redistribution During Rapid Solidification", J. Appl. Phys.; Vol. 53(2), pp. 1158-1168 (1982)

/10/ J.Poetschke, Spacelab-Nutzung, Status-Seminar 1980 des BMFT, DFVLR-Bericht 80-02, pp. 141-168, (KFI-UB 025/80)

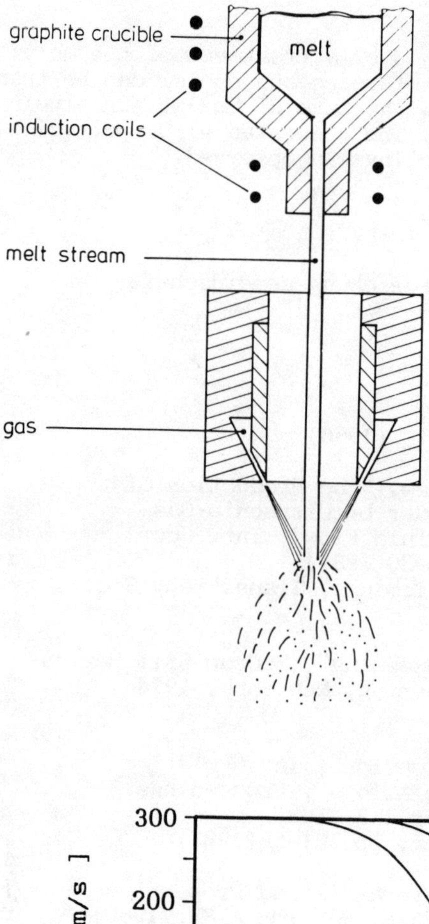

Fig.1 Principle arrangement of the atomizer

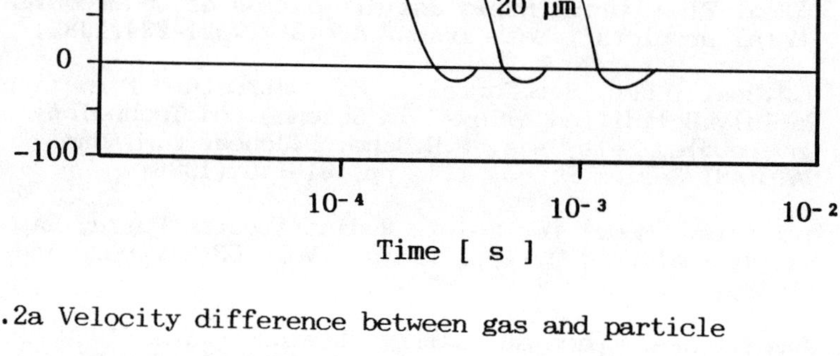

Fig.2a Velocity difference between gas and particle

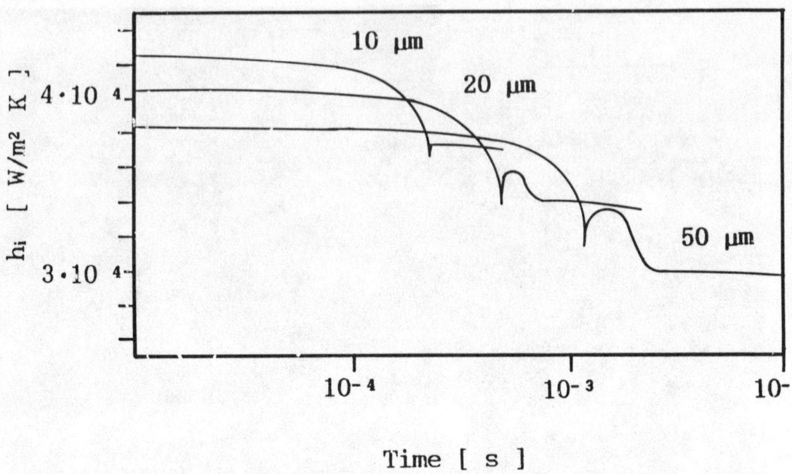

Fig.2b Heat transfer coefficient until the end of solidification

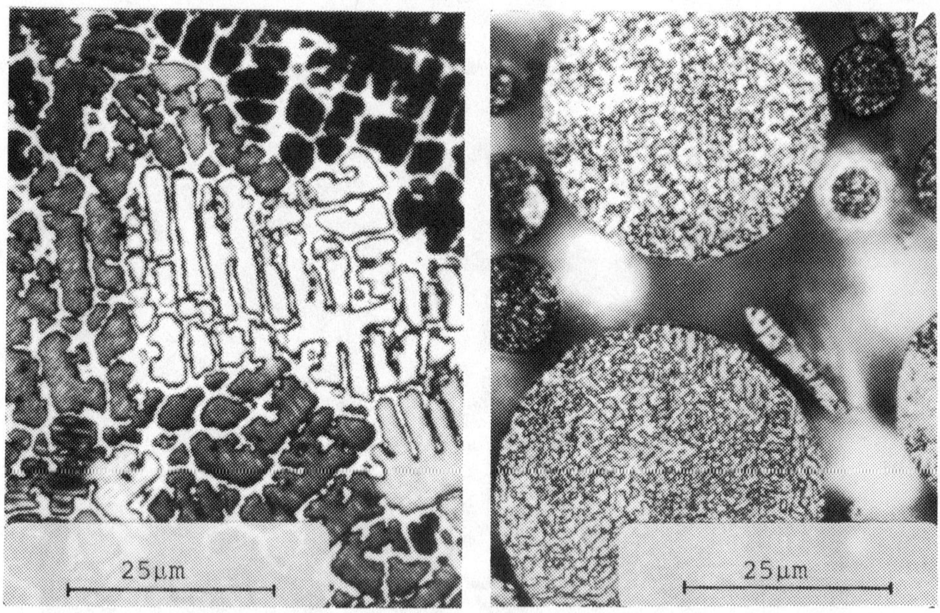

Fig.3a-c Microstructures of Cu-7Sn

a. Water quenched droplet b. Gas atomized powder

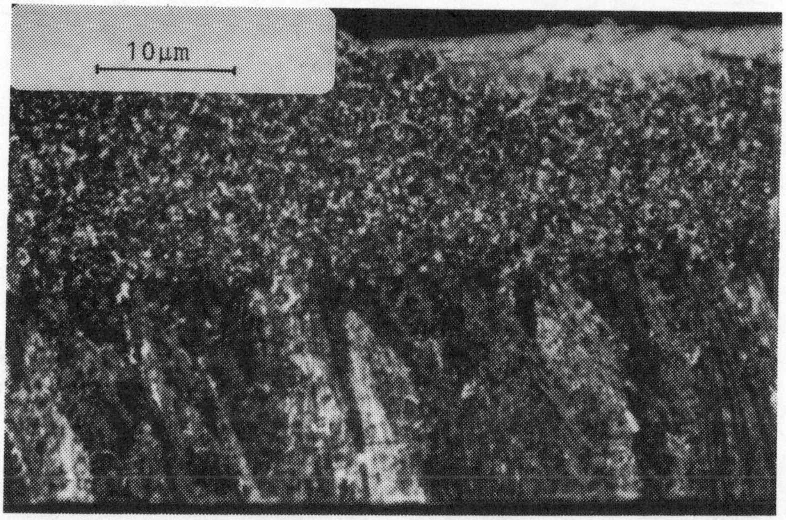

c. Splat cooled material

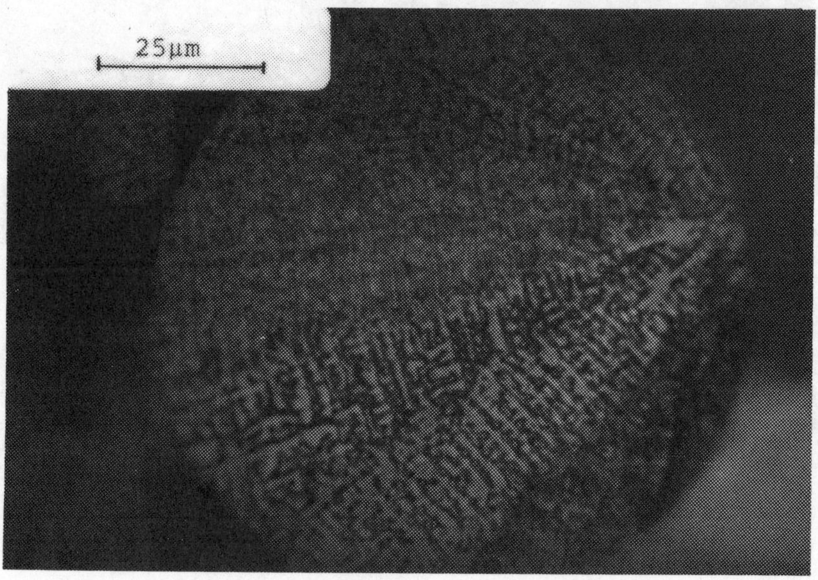

Fig.4a Single nucleation event in Cu-7Sn

Fig.4b Multiple heterogeneous nucleation in Cu-7Sn

Fig.5 Microstructure of atomized Cu-20Sn. α-dendritic, ß-light phase

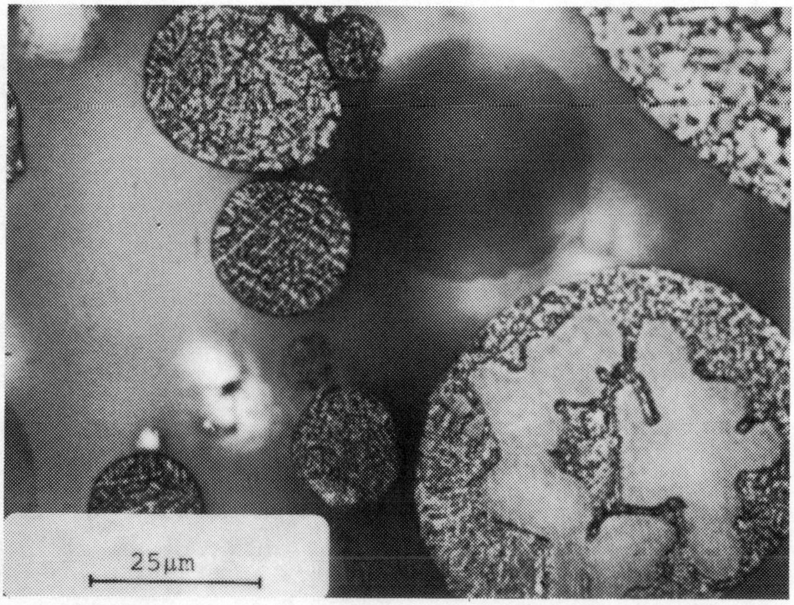

Fig.6 Cu-7Sn with segregation free solidified areas

Increasing Solidification Rate (↓)

I. Full Diffusional (Global) Equilibrium

- A. No chemical potential gradients (composition of phases are uniform)
- B. No temperature gradients
- C. Lever rule applicable

II. Local Interfacial Equilibrium

- A. Phase diagram gives compositions and temperatures only at liquid-solid interface
- B. Corrections made for interface curvature (Gibbs-Thomson Effect)

III. Metastable Local Interfacial Equilibrium

- A. Stable phase can not nucleate or grow sufficiently fast
- B. Metastable phase diagram (a true thermodynamic phase diagram missing the stable phase or phases) gives the interface conditions

IV. Interfacial Non-Equilibrium

- A. Phase diagram fails to give temperature and compositions at interface
- B. Chemical potentials are not equal at interface
- C. Free energy functions of phases still lead to criteria for the impossible

Fig.7 Hierarchy of equilibrium

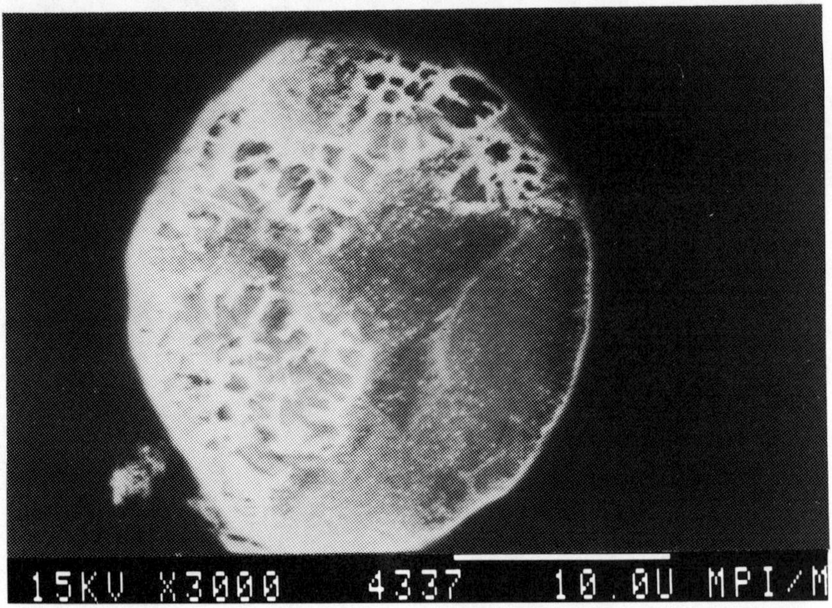

Fig.8 SEM micrograph of Cu-7Sn, partial segregation free. The white dots are entrapped aluminumoxide particles

SURFACE REACTIONS DURING ATOMIZATION, HANDLING AND HOT CONSOLIDATION OF Fe- AND Ni-BASE ALLOYS

I. Olefjord and L. Nyborg

Department of Engineering Metals,
Chalmers University of Technology
S-412 96 Gothenburg, Sweden

INTRODUCTION

The utilization of hot isostatic pressing (hipping) for the consolidation of inert-gas-atomized metal powders allows for the production of high quality materials. The P/M processing route has two main stages: the production of metal powder (atomization) and the compaction of powder into metal components. The competitive strength of the hipping technique is the opportunity to produce advanced alloys with properties superior or equivalent to those which are achieved by casting and/or forging.

Reaction products are formed on the surface during atomization and handling of the metal powder. The reactants are either present in the gas atmosphere or in the metal phase. In spite of the fact that the oxygen content in the gas atomizer is low surface oxides are formed (1-12). Other compounds formed are carbides, nitrides and sulphides (1-4, 7-11). Handling of the metal powder causes further oxidation of the powder and adsorption of water vapour (4,13).

To obtain the enhanced chemical and mechanical properties, possible with hot isostatically pressed alloys, the bonding between the metal particles must be strong. Thus, the reaction products formed on the particle surface during atomization and handling has to be reduced during consolidation so that metal bonding between the particles is established.

The chemical composition of the metal powder and fracture surfaces has been analysed by utilizing surface sensitive techniques such as ESCA (Electron Spectroscopy for Chemical Analysis) and Auger spectroscopy (1-11, 14-18). High resolution electron microscopy has also provided information about precipitation of compounds on the prior particle boundaries (PPBs) during hipping.

Nickel-base superalloys are the class of materials which have been studied most extensively. One of the reasons why these materials are produced by P/M techniques is that macrosegregation of alloying elements can be avoided. On the other hand, alloying elements such as Cr, Al, Ti and Mo form oxides on the surface during atomization (6-10). During consolidation MC and $M_{23}C_6$ carbide particles have been found on the PPBs (7-10, 16, 19).

The aim of this paper is to describe the chemical reactions which take place on the powder surfaces during the atomization, handling and hot isostatic pressing of Fe-base and Ni-base alloys. The reactions are illustrated by results obtained from studies of Fe-base alloys.

EXPERIMENTAL PROCEDURE

The powders were atomized in nitrogen at Avesta-Nyby and ABB-Powder Met in Sweden. Table 1 shows the composition of the steels. The martensitic steel was hipped at different temperatures in order to study the surface reactions during the hipping cycle. The hipping parameters and the post heat treatments are given in Table 2.

TABLE 1. CHEMICAL COMPOSITION OF THE POWDERS, WT-%

	Fe	C	Si	Mn	P	S	Cr	V	Nb	N	O
Ferritic	73.5	0.10	0.55	0.65	0.02	0.01	24.6	-	-	0.20	0.0150
Martensitic	84.9	0.20	0.30	0.60	0.02	0.01	12.0	0.3	0.1	0.06	0.0135

TABLE 2. THE HIPPING PARAMETERS AND HEAT TREATMENTS

Steel	Hipping parameters			Heat treatment	
	T(°C)	P(MPa)	t(h)	T(°C)	t(h)
	770	150	8	no	
Martensitic	1000	50	2	760	5
	1150	150	2	760	5

Surface analyses of the powders and the fracture surfaces were performed by utilizing ESCA (HP 5950A) and Auger microprobe (PHI 600).

RESULTS

Powder

Figure 1 illustrates the ESCA spectra recorded from the atomized ferritic powder in the as-received condition and after ion etching to 5

nm below the original surface. The upper row shows that both the metallic and the oxide states of Fe and Cr are present in the spectrum. The ratio between the metallic and the oxide states increases after 5 nm ion etching due to thinning of the oxide. However it appears that the intensities of Cr^{3+} and Mn^{2+} do not change at this ion etch depth. This fact implies that the surface oxide is not uniform in thickness, and that instead discrete oxide islands are present on the surface. The islands are formed at high temperatures during cooling of the metal particles. The area between the oxide islands is oxidized at low temperature during cooling and later during handling of the powder. The thickness of this oxide is less than 5 nm.

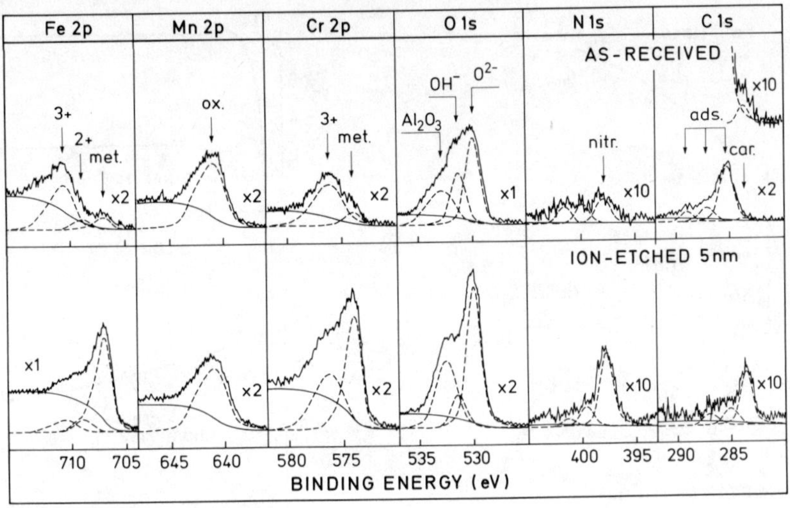

Figure 1. ESCA spectra recorded from gas-atomized ferritic stainless steel powder (1).

The oxygen signal "OH^-" suggests that the oxide is covered by a thin layer of hydroxide which is formed during handling of the powder in air. The oxygen peak denoted "Al_2O_3" is the oxygen signal from the support plate into which the metal powder was pressed. The main peak of the carbon signal represents a hydrocarbon species which was present on the surface. The other carbon peaks denoted "ads" originate from carbon bonded to nitrogen and oxygen. The nitrogen and the carbon signals marked "nitr" and "car" represent nitride and carbide formed on the surface during cooling of the metal powder. After ion etching the intensities of these signals were increased due to removal of the contamination layer. The metal components of the carbide and the nitride compounds cannot be identified directly from the spectra because the chemical shifts for the covalently bonded metal elements are small.

The composition of the surface layer was calculated from the measured spectral intensities. The quantitative technique has been given elsewhere (1). The etch depth was calibrated for Fe_2O_3. Figure 2a shows the variation of the composition with depth. The profile shows that the

thickness of the oxide products is very thin; after removing 30 nm the concentration of the cations and O^{2-} is low. At the etch depth of 15 nm Fe-oxide cannot easily be determined while Mn- and Cr-oxides are detected. The dashed curve in the figure shows the theoretical intensity for the metallic state of Fe which should have been obtained if the metal particle had been covered with a 10 nm thick oxide. The shape of the curve has been experimentally verified (18). The fact that the measured Fe^{met}-content is higher at low etch depths and lower at higher etch depths than the theoretical curve supports the suggestion above from the ESCA spectra that the surface of the metal particles is covered with oxide whose thickness varies laterally over the surface.

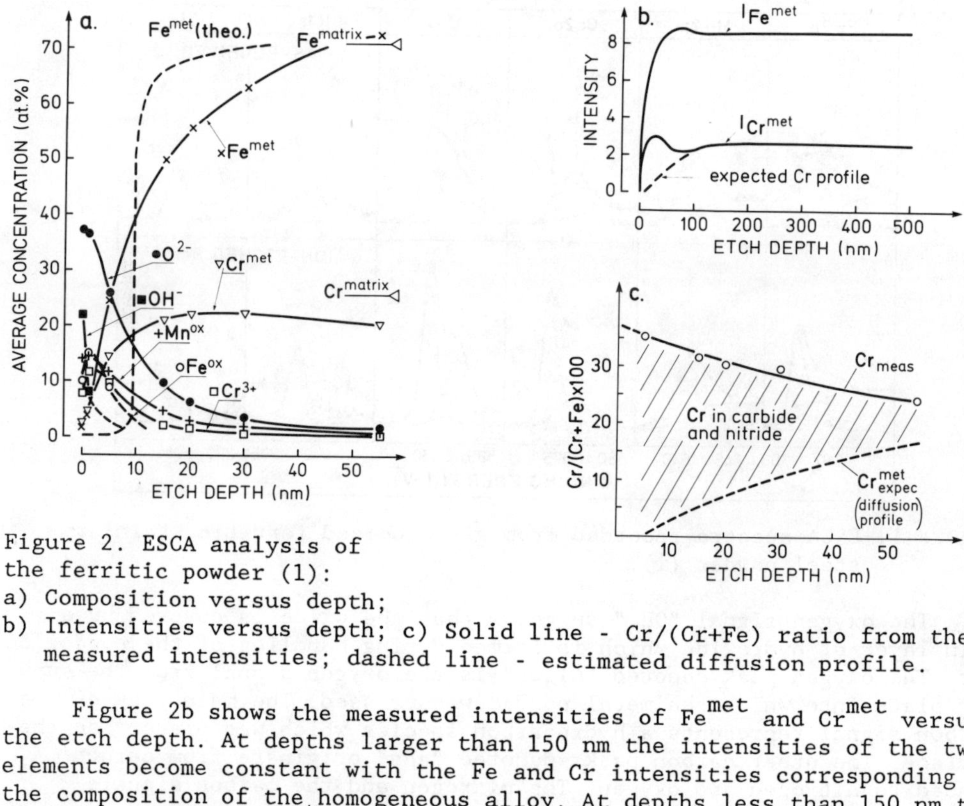

Figure 2. ESCA analysis of the ferritic powder (1):
a) Composition versus depth;
b) Intensities versus depth; c) Solid line - Cr/(Cr+Fe) ratio from the measured intensities; dashed line - estimated diffusion profile.

Figure 2b shows the measured intensities of Fe^{met} and Cr^{met} versus the etch depth. At depths larger than 150 nm the intensities of the two elements become constant with the Fe and Cr intensities corresponding to the composition of the homogeneous alloy. At depths less than 150 nm the intensities vary; Cr^{met} shows a maximum value at 30 nm and a minimum value at 80 nm where the intensity of Fe^{met} is maximum. The depletion of Cr^{met} in the surface zone is due to the selective oxidation of Cr. The high Cr content close to the surface is caused by the formation of Cr-containing carbides and nitrides. The positions of the Cr spectra for these compounds are almost the same as for the metallic state so that no chemical shift can be noted. The dashed curves in Figs. 2b-c are the estimated concentration profiles of Cr^{met} in the metal phase (1). The shadowed area in Fig. 4b corresponds to Cr bound in carbide and nitride, $Cr^{C,N}$. The atomic ratio $Cr^{C,N} / (C^{carbide} + N^{nitride})$ varies between 2.5

and 3.3 in the depth region 5 to 60 nm (1). Thus, the constant ratio indicates that Cr-containing carbonitride is present on the surface.

Auger analysis (1) of the surface shows that Si and S segregate to the surface during atomization. Silicon is present in its oxidized state to the same depth as Cr-oxide. The enrichment of Si is about 8 times larger than the enrichment of Cr. Sulphur is also present to the same depth. It is suggested that sulphur is bonded to Mn as MnS particles on the surface. By using the depth at which the oxygen signal approaches the half intensity value of its maximum as a measure of the average oxide thickness, it appears from Fig. 2a that the oxide products are 10 nm thick. Figure 2b shows that carbon- and nitrogen-containing particles thicker than 50 nm exist on the surface. These coarse particles appear in the scanning electron micrograph (x46000) in Fig. 3. The lateral dimension of the particles is in the range 30-100 nm which is of the same order of magnitude as the depth dimension obtained from the etch profile. Thus, the carbonitride particles are present as spherical particles on the surface. The oxide particles are too small to be detected on the surface (the resolution of the instrument is about 30 nm).

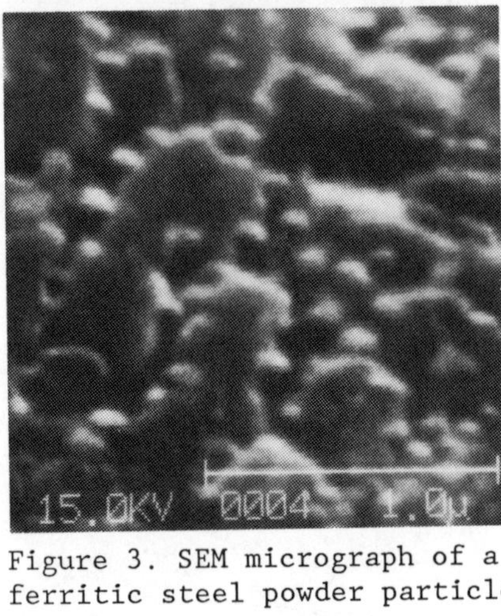

Figure 3. SEM micrograph of a ferritic steel powder particle (1).

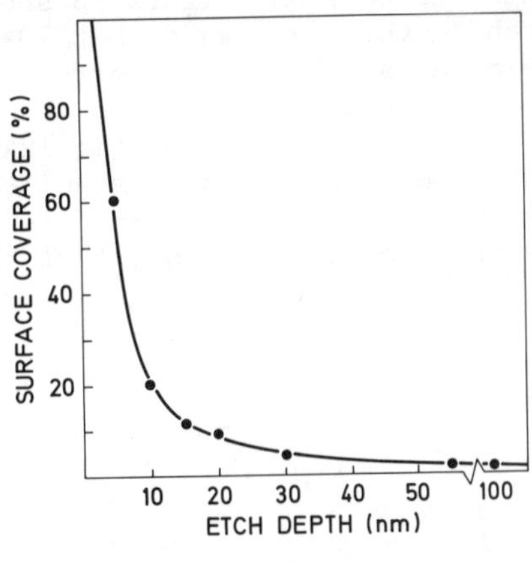

Figure 4. The oxide surface coverage of the ferritic steel (1).

From the recorded intensities of the ESCA signals the surface coverage of the oxide particles can be calculated (1). Figure 4 shows the surface oxide coverage versus the etch depth. It appears that 60% of the surface is covered with oxide particles thicker than 5 nm. The amounts of oxygen present in the surface oxide is calculated from the composition of the surface oxide and the distributions of the oxide and the metal particles (1). Table 3 shows the coverage, the average thickness, the density and the weight fraction of oxygen in the oxides present on the surface of the ferritic powder. The surface content of oxygen is calculated from the expression:

$$C_o = \sum_{r_j} \sum_{i} 3t_i A_i \rho_i X_i^0/(\rho_m r_j) \qquad (1)$$

where summations are done over the metal particle distribution, r_i, and the kind of oxides present on the surface. The density of the metal phase is ρ_m. The rest of the symbols are given in Table 3. It was found in the case of the ferritic steel (Table 1) that 60% of the total amount of oxygen (150 ppm) was present in the surface oxide.

TABLE 3. OXIDES PRESENT ON THE SURFACE OF THE FERRITIC POWDER

Oxide	A_i	t_i (nm)	ρ_i (g/cm^3)	X_i^O (wt%)
MnO	0.56	15	5.4	22
Cr_2O_3	0.16	15	5.2	32
Fe_2O_3	0.20	3	5.2	30
SiO_2	0.08	3	2.6	53

A_i = coverage; t_i = average thickness; ρ_i = density; X_i^O = oxygen content

A chemical composition profile obtained by ESCA of the martensitic stainless steel is shown in Fig. 5a. The average oxide thickness is estimated to be 7 nm. Figure 5b shows the cation concentration versus the etch depth. It appears that 70% of the cations in the outer region of the oxide are Fe^{ox}. The corresponding values for Cr^{3+} and Mn^{2+} are 10% and 20% respectively. Thus the relative Fe^{ox} content in the surface oxide formed on the martensitic 12%Cr-steel is much higher than that found on the ferritic 25%Cr-steel. For etch depths of 5 nm and larger the ratio between Cr^{3+} and Mn^{2+} is about 2:1. This indicates that the coarse oxide particles consist of the mixed oxide $MnCr_2O_4$.

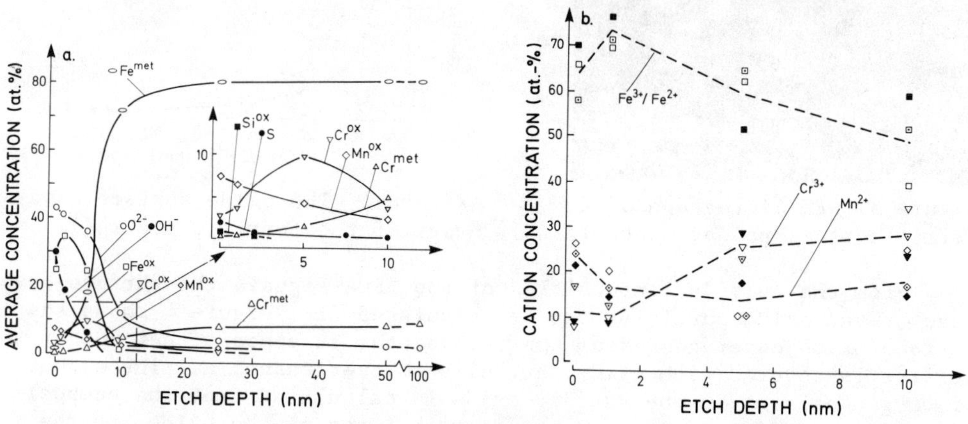

Figure 5. ESCA analysis of the martensitic steel powder (3): a) Composition versus depth; b) Cation concentration in the surface oxide.

The distribution of the oxide particle size is calculated (3) from the etch profiles in the same way as in Fig. 4. The oxide formed at low temperature and during handling of the powder is thinner than 5 nm and

covers about 30% of the surface. The rest of the surface is covered with oxide particles containing the cations Fe^{2+}, Fe^{3+}, Cr^{3+} and Mn^{2+}.

Knowing the composition of the surface oxide the thickness of the oxide can be estimated from the analysis of the total oxygen content of the powder. Figure 6 (3) shows the measured oxygen content versus the radius, r, of the metal particles. The crosses are the measured oxygen contents from six different sieved size fractions. The dashed line represents the theoretical total weight fraction of oxygen, $X_{O,tot}$ (3):

$$X_{O,tot} = X_{O,b} + 3t\ X_{O,ox}\ \rho_{ox}/(\rho_m\ r) \tag{2}$$

where $X_{O,b}$ is the bulk composition, t the average thickness of the oxide, ρ_{ox} and ρ_m are the densities of the oxide and the metal phases respectively. Figure 6 shows that the bulk content of this powder is 80 ppm. The average content of surface bound oxygen is found to be 55 ppm and the corresponding oxide thickness is 7 nm. The values were calculated from the slope of the line in Fig. 6 and the metal particle size distribution. Thus, the oxide thickness obtained from the chemical analysis of the oxygen content is in good agreement with the value obtained from the surface analysis.

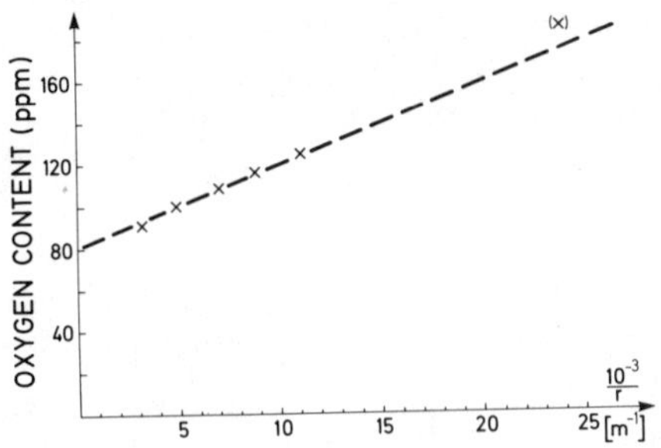

Figure 6. Chemical analysis of the oxygen content versus the reciprocal of the metal particle radius (3).

Compacted Material

Figure 7 shows SEM micrographs, magnifications 70 and 2000 times, of fracture surfaces of the hipped steels. It appears that the samples consolidated at 770°C and at 1000°C fracture along the PPBs. The steel compacted at 1150°C shows only transparticulate ductile failure. The material compacted at 1000°C shows that the fracture path follows the particle boundaries. The surface of the steel particles compacted at 770°C are smooth showing that the bonding between the particles is very weak. The measured impact energies (Charpy V) of the steels hipped at 770°C, 1000°C and 1150°C are 3J, 12J and > 50J respectively. Thus, it is obvious that the particle boundaries are embrittled during consolidation at the lowest temperatures.

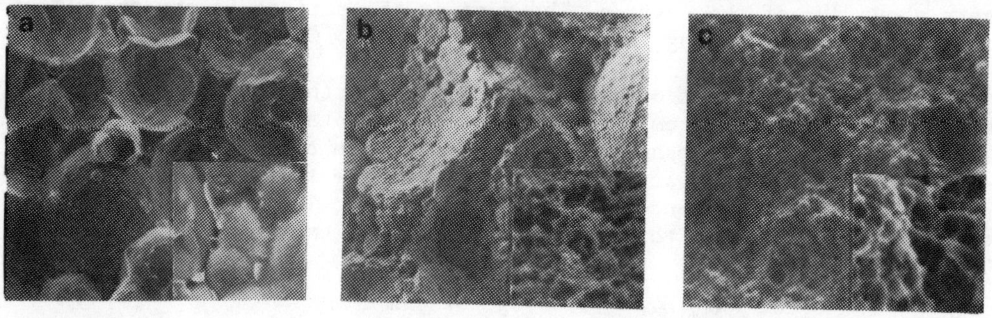

Figure 7. SEM micrographs (70x and 2000x) of fracture surfaces of martensitic steel specimens hipped at: a) 770°C; b) 1000°C; c) 1150°C (14).

Auger spectra recorded from the fracture surfaces of the hipped materials are shown in Fig. 8. The spectral intensities from the elements O, P, S, C, N, V and Si are highest for the samples compacted at the lowest temperature. Signals from the elements N, V and Si were not observed on the atomized powder. Thus, these elements have segregated to the surface during consolidation, Figure 8a shows that a Mn- and Cr-rich oxide is present on the surface during hipping at 770°C. The low kinetic energy peaks of Fe shows that the metallic state of Fe dominates on the surface at 770°C and higher temperatures. Thus, the Fe-oxide formed on the surface at low temperatures and during handling of the powder is reduced during heating from room temperature to 770°C.

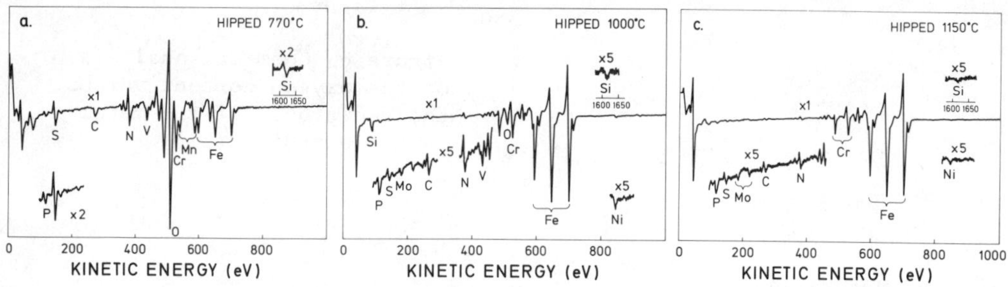

Figure 8. Auger electron spectra recorded from fracture surfaces of martensitic steel hipped at: a) 770°C; b) 1000°C; c) 1150°C (14).

The sample consolidated at 1000°C shows lower oxygen and Cr-signals and hardly any detectable Mn-signal compared with the material hipped at 770°C. The lower intensities are to some extent due to the fact that the surface is only partly covered with the reaction products because the plastic flow of the material creates clean metal surfaces.

The Auger spectrum, Fig. 8c, from the fracture surface of the 1150°C hipped material shows no detectable signals from O, C and Mn. The measured Cr/Fe ratio is the same as for the matrix. The intensities of

the P, S, C and N signals are lower than those obtained from the materials hipped at lower temperatures. The N-signal is slightly higher than the C-signal. The fact that the carbon content in the alloy is 4.5 times higher than the N-content and that the sensitivities of N and C are about the same suggests that nitride inclusions are present on the fracture surface and initiate preferentially the fracture.

Depth profiles recorded from the material compacted 770°C and 1000°C are shown in Fig. 9. The oxygen profile in Fig. 9a shows that the oxide products are significantly coarsened during hipping at 770°C. The half height of the oxygen profile shows that the average height of the oxide particles is about 50 nm. The oxygen profile in Fig. 9b shows a very low oxygen concentration at 100 nm indicating that the Mn- and Cr-oxide is at least partly reduced at 1000°C. Hence, the oxide particles present on the surface after hipping at 1000°C are smaller than after hipping at 770°C. The slope of the oxygen profile in Fig. 9a is steeper than those of carbon and nitrogen indicating that the carbide and nitride particles are coarser than the oxide particles. The etch profiles of the hipped materials show that the obtained Cr/Fe ratios are about twice as high as for the bulk alloy. Thus, Cr is enriched on the fracture surface in coarse carbide particles.

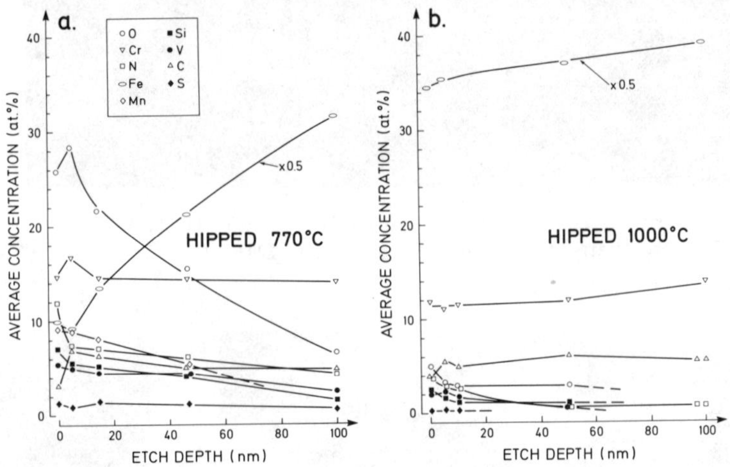

Figure 9. Auger depth profiles of the martensitic steel hipped at: a) 770°C; b) 1000°C (14).

It appears from Fig. 9 that the carbon profiles are independent of the etch depth while the N-profiles decrease showing that the nitride particles are smaller than the carbide particles. The shape of the N-profiles are about the same as those of the V- and Si-profiles indicating that the nitride contains these elements. However, the possibility that V and Si are also present in the oxide phase cannot be excluded. Indeed we have shown that at least SiO_2 is formed at high temperatures (20).

SEM and TEM micrographs (17) of the steel compacted at 1000°C are shown in Figs. 10 and 11 respectively. The former is taken from the sample used for the TEM analysis. It clearly appears from the figures that precipitates occur on the PPBs and that these particles are coarser than those present in the matrix. The particles on the boundaries were identified (17) with EDX and selected area electron diffraction to be V(C,N) and $M_{23}C_6$. The TEM micrograph in Fig. 11a shows contrast in the V(C,N) particle and from the EDX analysis it was found that Si is present in the particle. It is suggested that the Si-compound is Si-oxide and/or Si-nitride on which V(C,N) precipitates. The micrographs show that the coarsest particles are about 200 nm and that a large fraction of much smaller particles exist on the PPBs. Qualitatively the particle size distribution agrees well with the size distribution obtained from the Auger analysis.

Figure 10. SEM micrograph (700x) of the martensitic steel hipped at 1000°C (17).

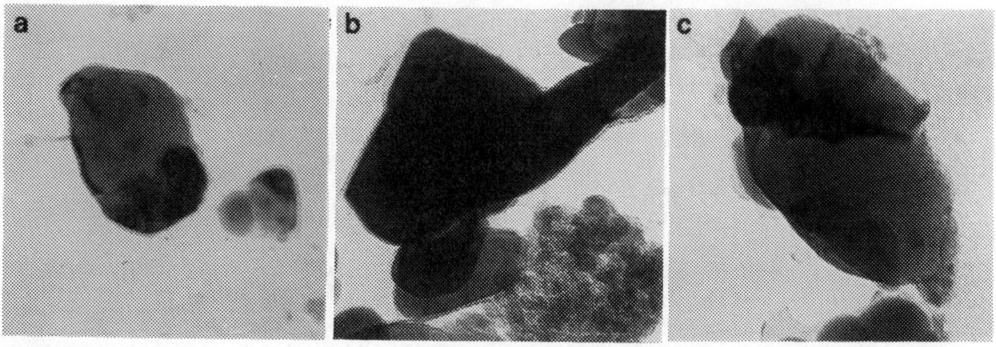

Figure 11. TEM micrographs (140.000x) of PPB-precipitates in the martensitic steel hipped at 1000°C: a) V(C,N) co-precipitated with Si-compound; b) $M_{23}C_6$-carbide; c) V(C,N) precipitated on $FeCr_2O_4$ (17).

The co-precipitation of V(C,N) on cubic oxide has also been demonstrated (17). The electron diffraction pattern, Fig. 12, shows epitaxial growth in the <112> direction of V(C,N) on the cubic oxides $FeCr_2O_4$ or Fe_3O_4. Thus the surface of the oxide formed during atomization and handling of the powder acts as low energy surface for nucleation and growth of the V(C,N) compound.

Figure 12. Composite selected area electron diffraction pattern of V(C,N) and $FeCr_2O_4$ from <112> direction of both phases (17).

DISCUSSION

The experiments described in this paper show that during atomization of Fe-base alloys oxide particles are formed on the surface at high temperatures. The oxide particles consist of the strongest oxide forming elements Cr, Mn and Si. During cooling at low temperatures and during handling of the powder a thin uniform layer of Fe-oxide, less than 5 nm in thickness, is formed between the oxide particles. The diameter of the oxide particles formed at high temperatures is in the range 5 - 50 nm. The average oxide thickness was found to be about 10 nm.

The surface analyses of the powders show that carbonitride is formed on the ferritic steel but not on the martensitic steel. The difference in behaviour is due to the lower solubility of C and N in ferrite than in austenite. The carbonitride compounds occur as particles on the surface. They are coarser than the oxide particles and observable in SEM. The ion etching profile and the SEM micrograph indicate that the carbonitride particles are spherically shaped and occur on the dendrite boundaries. Precipitation of carbide particles on the interdendritic regions of superalloy powders were observed by Ingesten et al. (8) and Aubin et al. (9).

The reactivity of the alloying elements can be understood from the free energy diagram in Fig. 13. The main diagram (21) is the standard free energy diagram for oxide formation, while the detailed diagram (3) represents the martensitic steel (12Cr, 0.2C, 0.3Si, 0.6Mn). It appears that the maximum oxidation temperature for Mn and Si is $1400^{\circ}C$. Above that temperature dissolved C reacts with oxygen and forms CO. The maximum oxidation temperature of Cr in the martensitic steel is 1300°C. For the ferritic steel Cr is oxidized at higher temperatures due to the higher Cr-content and lower C-content. The higher oxidation temperature of Cr contributes to the formation of more Cr_2O_3 on the ferritic steel than on the martensitic. It was observed that the amount of Cr_2O_3 on the former is twice as high as that formed on the latter. However, the oxygen content in the atmosphere of the atomizers was not known, so in these particular cases the amount of Cr_2O_3 could be controlled by the availability of oxygen.

The oxidation behaviour of the Ni base superalloys from which data have been reported in the literature (5-9) can also be understood from the thermodynamic diagram in Fig. 13. Nickel is a weak oxide forming element, like Fe, while the alloying elements, Ti, Al, Cr, etc, have a large oxygen affinity. It is therefore suggested that oxides such as TiO_2, Al_2O_3 and Cr_2O_3 form oxide islands on the surface at high temperatures. The oxides Al_2O_3 and TiO_2 can be formed at higher temperatures than Cr_2O_3. This prediction is in good agreement with the observation of Warren et al. (7). They have shown that the oxide formed on Ni-base superalloys consists of more than 3 times more Al_2O_3 than Cr_2O_3 in spite of the fact that the Cr-content is four times higher than the Al-content. Surface enrichment of Ti-carbide, oxides and amorphous carbon have been reported by Aubin et al. (9). Ross et al. (11) found by using Auger spectroscopy that the elements C, Cr, O, S and Ti are enriched on the surface. They state that the correlation between O and S strongly indicates the formation of oxy-sulphide. Further, it was suggested that a monolayer of NiS is formed on a Ni-base alloy with high C-content. To avoid carbide formation on the surface of the powder during atomization the carbon content of some superalloys is kept low (22). However, the low carbon content will result in an increased formation of Cr_2O_3 and NiO. On the other hand oxidation of Al and other strong oxide forming elements will not be influenced by the carbon content due to the fact that their oxides are formed at very high temperatures. In some cases even on the surface of the melt.

The amounts of the compounds formed on the surface is controlled by the availability of the reactants in the metal and gas phases. Estimation of the kinetics oxidation of the martensitic powder indicates that the availability of oxygen is the rate determining step. The amount of oxygen transported from the gas phase to the metal during cooling from 1400°C to 1000°C is found to be $6.9\ 10^{-4}$ mole/m^2 (3-4) which assumes that the oxygen content in the gas phase is 50 ppm; the particle size is 200 μm; and the mass transfer resistance is localized to a thin gas boundary layer. The corresponding value indicated from surface analysis is $6\ 10^{-4}$ mole/m^2. Thus, the agreement between the calculated and the observed take up of oxygen is relatively good. The amount of Cr transported to the surface in the solid state is $2\ 10^{-4}$ mole/m^2 which shows that significant parts of the oxide formed may be Cr_2O_3.

The observed amount of Mn on the surface is too large compared to the amount which can diffuse in the solid state to the surface at 1400°C even if it is assumed that the metal particle solidifies at this temperature. The solidification time is 18 ms (3). The required diffusion time is 100 ms. Manganese has a high vapour pressure. Thus, it is suggested that Mn evaporates from the melted metal droplet and that the Mn vapour condensates on the metal particles in a later stage.

The hipping temperature has to be chosen within a restricted temperature range in order to optimize the properties of the consolidated products. For the martensitic steel, studied in this paper, the hipping temperature is about 1150°C. During the rise in temperature chemical reactions take place on the surface of the powder. The hipping temperature has to be high enough so that the oxide and carbide products formed are dissolved or coarsened at the maximum temperature. The bonding between the metal particles can therefore be achieved. If the temperature is too high there is a risk that coarse sulphides are formed on the PPBs (20) and the material becomes embrittled.

The hipping at 770°C and 1000°C was performed in order to clarify whether solid products are formed at these temperatures. The Auger analyses of the fracture surfaces show that the Fe-oxide is not detectable. Thus, the Fe-oxide formed during atomization and handling of the powder is reduced during heating from room temperature up to 770°C. The reduction of Fe-oxide is supposed to be caused by oxidation of Cr and Mn.

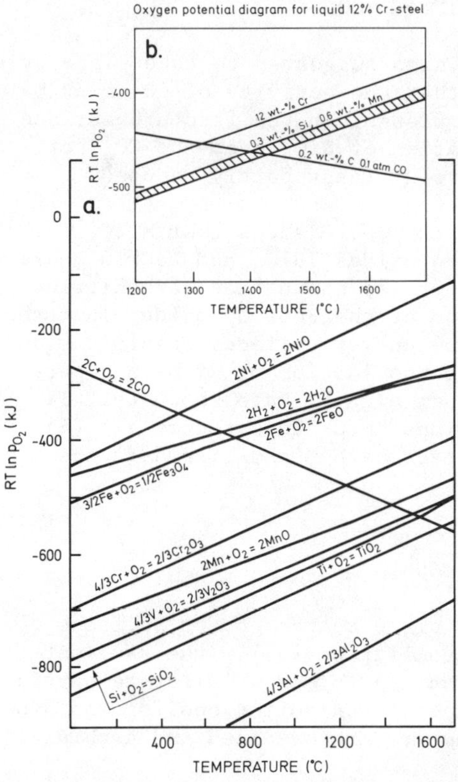

Figure 13. Free energy diagram for martensitic stainless steel powder (21).

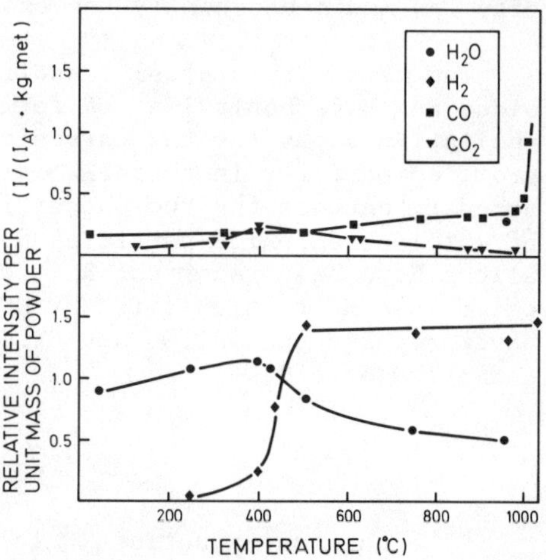

Figure 14. Gas analysis of the martensitic stainless steel powder (23).

Gas reactions are involved in the red-ox reactions which take place during the hipping cycle. By utilizing mass spectroscopic techniques the composition of the gas phase can be analysed. Figure 14 shows the relative partial pressure versus the temperature for the martensitic powder. The powder was exposed for a short time in air before being heated in a closed canister connected to the mass-analyser. It appears that the maximum H_2O content decreases as the hydrogen content increases. Further the CO content increases slowly with temperature up to 700°C where a very strong increase in the CO content occurs. The carbon dioxide contents show a maximum at 400°C. At higher temperature CO_2 is hardly detectable.

It is obvious that during the rise in temperature up to 400°C H_2O is desorbed from the powder. At 400°C the alloying elements are oxidized and hydrogen is released, for example

$$2Cr + 3H_2O \rightarrow Cr_2O_3 + 3H_2$$

The increase in CO above 500°C is partly caused by the reaction:

$$H_2O + C_{met} \rightarrow H_2 + CO$$

The reduction of Fe-oxide is suggested to occur above 700°C and is controlled by the reaction of the type:

$$Fe_3O_4 + 4C_{met} \rightarrow 3Fe + 4CO$$

The reduction of Cr_2O_3 and MnO is also suggested to take place by a gas reaction. From Fig. 13 it appears that the activity of CO_2 must be very low. It was observed that SiO_2 is present on the fracture surface of the 1050°C hipped steel. Thus, formation of SiO_2 keep the CO_2 activity low and consequently the oxides are at least partly reduced.

The martensitic steel consolidated at 1000°C shows that $M_{23}C_6$ carbides and V-carbonitride are formed. The oxides, SiO_2 and $FeCr_2O_4$, are nucleation sites for the carbonitride. It was shown that carbonitride grows epitaxially in the <112> direction of the $FeCr_2O_4$ oxide. Thus the boundary between the two phases is a low energy surface. Precipitation of carbides on oxide particles during consolidation of Ni-base superalloys have been reported: R.E. Waters et al. (8) have shown that Ti-oxide is a nucleation site for MC carbides; N.G. Ingesten et al. (6) have proposed that Al_2O_3 and Cr_2O_3 attract carbon and Ti and causes formation of TiC.

CONCLUSIONS

The chemical reactions which take place on the powder surfaces of Febase and Ni-base alloys during gas atomization follow the same pattern: thermodynamically stable compounds are formed as discrete particles at high temperatures; at lower temperatures during cooling and handling of the powder mainly the base elements are oxidized and a thin continuous oxide layer is formed between the previously formed oxide islands. The superior mechanical properties of the hipped materials obtained is a result of the reduction of the continuous oxide film at high temperatures in combination with the stable compounds appearing as particles on the metal particle boundaries.

REFERENCES

1. I. Olefjord and L. Nyborg, Powder Metall., 1985, 28 (4), p. 234.
2. L. Nyborg and I. Olefjord, Horizons of Powder Metallurgy (eds. W.A. Kaysser and W.J. Huppmann), p. 755, 1986, Verlag Schmid, Freiburg.
3. L. Nyborg and I. Olefjord, Powder Metall., 1988, 31 (1), p. 33.
4. L. Nyborg and I. Olefjord, Powder Metall. Int., 1988, 2, p. 11.
5. L. Nyborg and I. Olefjord, Proceedings 1987 Int. P/M Conf., 7-10 Dec. 1987, Bombay, India.
6. D.P. Ferriss, Int. J. Powd. Met. Powd. Technol., 1983, 19 (2) p. 11.
7. R. Warren, N.G. Ingesten, L. Winberg and T. Rönnhult, Powder Metall., 1984, 27 (3), p. 141.
8. N.G. Ingesten, R. Warren and L. Winberg, in High Temperature Alloy for Gas Turbines 1982 (eds. R. Brunetaud et al.), p. 1013, 1982, Dordrecht, Reidel.
9. C. Aubin, J.H. Davidson and J.P. Trottier, Proc. of the 4th Int. Symp. on Superalloys (ed. J.K. Tien), p. 345, 1980, American Society for Metals, Metals Park, Ohio.
10. R.E. Waters, J.A. Charles and C. Lea, Metals Technology, 1981, 194.
11. P.N. Ross and B.H. Kear, Rapidly Quenched Metals III, vol. 3 (ed. B. Cantor), p. 102, 1978, Chameleon Press Limited, London.
12. J.J. Dunkley, Met. Powd. Rep., 1981, 36 (5), p. 226.
13. K. Hajmrle and R. Angers, Int. J. Powd. Met. Powd Technol., 1979, 15 (4), p. 319.
14. L. Nyborg and I. Olefjord, Powder Metall., 1988, 31 (1), p. 40.
15. R.M. German and J.E. Smugeresky, Mat. Sci. Eng., 1978, 36, p. 223.
16. J.H. Davidson and C. Aubin, in "High Temperature Alloys for Gas Turbines 1982" (eds. R. Brunetaud et al.), p. 853, 1982, Dordrecht, Reidel.
17. L. Nyborg and G. Wahlberg, presented at the Int. Conf. on Hot Isostatic Pressing, 15-17 June 1987, Luleå, Sweden, Centek Publ., Luleå.
18. L. Nyborg, A. Nylund and I. Olefjord, Surface and Interface Analysis in press.
19. T.C. Lu, Y. Bienvenu, M. Jeudin, O. Farul and J. H. Davidson, AIME TMS, Technical report no. F86-9, The Metallurgical Society of Aime, Commonwealth Drive, Warrendale.
20. L. Nyborg and I. Olefjord, presented at this conf.
21. F.D. Richardson and J.H.E. Jeffes, J Iron Steel Inst. 1948, 160, p.261.
22. M.T. Pobod, Modern Developments in Powder Metallurgy, 1977, 11, p. 25.
23. L. Nyborg, PhD thesis, Chalmers University of Technology, Gothenburg, 1987.

STRUCTURE AND PROPERTIES OF ATOMIZED BERYLLIUM POWDER

J. M. Marder and D. H. Hashiguchi

Brush Wellman Inc.
17876 St. Clair Avenue
Cleveland, Ohio 44110

Abstract

Beryllium has the close-packed-hexagonal crystal structure, which can result in mechanical and physical property anisotropy of consolidated powder metallurgy billets. The anisotropy, is in large part, a result of the shape of the powder particles which are the product of conventional mechanical comminution processes, such as attritioning, ball milling, and impact grinding. The use of atomization as a powder-making technique was postulated to be a means of reducing or eliminating anisotropy of beryllium billets. It was the purpose of this work to investigate methods of atomizing beryllium, and to determine the degree to which anisotropy could be eliminated. In addition, the mechanical properties of billets made from spherical powder were determined.

Introduction

Conventional beryllium powder is produced by one of several mechanical comminution processes. The resultant powder reflects the crystal structure and deformation characteristics of beryllium. In mechanical comminution, fracture of machined beryllium chips, derived from vacuum melted ingot, occurs upon {0002} basal planes. This produces flat particles with faces corresponding to the basal planes (Figure 1). During handling and vibration of the powder prior to consolidation, the faces tend to align and produce extended areas of non-random orientation. Ball milled powder shows a flake-like morphology as a result of fracture predominantly upon the {0002} basal planes. Impact ground powder, which represents the vast majority of the powder produced, shows a more blocky structure, because under impact conditions, fracture planes other than the basal planes are activated.

Spherical powder should show no crystallographic relationship between the surface features and crystallographic planes. This is the essential element which could serve to reduce the anisotropy of powder metallurgy (P/M) beryllium consolidated billets. If the flat faces, representing {0002} planes could be eliminated, then large regions would not align during powder die loading and vibratory packing. This alignment is largely a result of the vibration modes used in loading the vacuum-hot-pressing dies. During consolidation by uniaxial vacuum hot pressing, these areas align with the basal planes parallel to the pressing axis, resulting in a substantial degree of preferred orientation in the consolidated billet[1]. In view of these considerations, it becomes obvious that spherical particles should reduce or eliminate anisotropy in both vacuum-hot-pressed (VHP) and hot-isostatically-pressed(HIP) material.

50 μm 880614-1 50 μm 880614-2

Figure 1 - Scanning Electron Photomicrograph of Beryllium Powder. Left - Attritioned (Flake-like structure) Right -Impact Ground (Blocky shape).

In critical components such as precision optics and inertial guidance systems, crystallographic texture can be a major design obstacle. The directionality of properties such as the coefficient of thermal expansion and elastic modulus are factors which must be accounted for in these sensitive applications. The objective of this program was to determine the feasibility of producing clean, spherical beryllium powder, as well as to evaluate the properties and isotropy of the billets made using this powder.

Experimental Design

Process Selection

The first task of this effort was to identify a suitable atomizing technique for beryllium. The technique would have to display the potential for producing powder with a fine grain size to yield the required strength in a consolidated billet, be mechanically simple and prevent excessive oxidation. Beryllium, shows a very strong relationship between particle size and grain size. Therefore, it was expected that a significant fraction of the particles would have to be smaller than 74 microns (-200 mesh powder). The Rotating Electrode Process (REP) and the Rapid Solidification Process (RSP) powder techniques were considered. These processes both use rotating components to impart energy to molten metal. Beryllium, with its low density and high liquid surface tension, would require excessive rotational speed to produce the required particle size. The current state of the art limits the processes to rotational speeds of 25,000 to 35,000 rpm[2]. In order to produce usable beryllium powder by these techniques, a rotational speed of about 125,000 rpm would be required. Therefore, these two processes were eliminated from consideration. Soluble Gas Atomization[3] was considered, but the potential lack of control of the process of applying the requisite gas pressure to achieve the desired powder characteristics appeared to make it a poor choice for beryllium. Excessive soluble gas pressure would result in explosive decompression of the powder melt which could pose a hazard. Inert gas atomization[4] is mechanically simple, and has been successfully used as a large-scale production technique for other reactive metals such as magnesium and aluminum. It had been investigated during the 1960's for use with beryllium with no real success, the primary drawbacks found being the excessive oxide content of the powder and freezing of the liquid metal stream in the atomization nozzle. Equipment improvements in the ensuing decades gave reason to believe that these potential drawbacks could be overcome. Therefore, inert gas atomization was selected as the process for making spherical beryllium powder.

Atomization

The second task of this investigation was to produce spherical beryllium powder using the selected process.

Atomization was carried out in a five pound melt capacity unit, Figure 2, constructed for this program at the Elmore, Ohio plant site. An induction melting furnace, utilizing a 30 KW power supply was used in this laboratory scale unit. The atomizer was designed with several unique features specific to beryllium processing. The atomization fluid was helium, required for its high heat capacity. The use of helium maximized the probability of achieving fine grain sizes in the powder particles, and the production of non-equilibrium, rapidly solidified structures. The helium flow path was a leak tight closed loop to prevent the escape of beryllium powder into the environment. The closed loop recirculation system employed cyclones and filters to prevent fine particles of beryllium from entering the compressor and causing damage when the helium was re-cycled. The powder was collected in special transfer cans which could be detached from the atomizer without exposing the environment to uncontained powder.

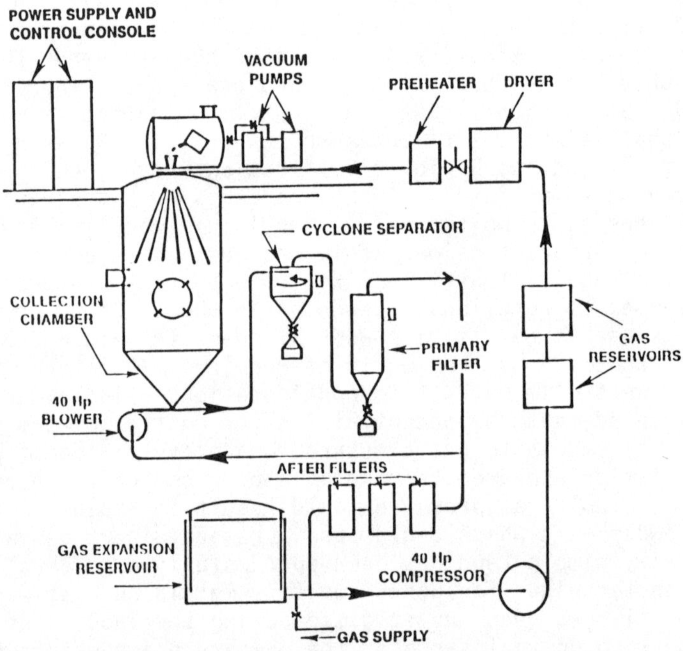

Figure 2 - Schematic Diagram of Beryllium Inert Gas Atomizer

Consolidation

Consolidation of the powder was carried out by two techniques, VHP and HIP. VHP was carried out using 2.54 cm (1 inch) diameter graphite dies and punches. HIP was carried out in 10 cm (4 inch) diameter mild steel cans. The VHP process was conducted at 1100C (2012F) and a pressure of 6.9 MPa (1 ksi), whereas the HIP process used 1000 C - 103.4 MPa (1832 F - 15 ksi). The 1100 C (2012 F) used for VHP was required to achieve an acceptable density. In both cases, the -200 mesh (-74 micron) sieve fraction was used. It should be noted that an attempt was made to Cold Isostatically Press (CIP) the spherical powder, with no success at pressures up to 275.8 MPa (40 ksi).

Results

The final task was to characterize the products derived from inert gas atomization, including powder and consolidated billets. The results of these evaluations are presented in the following sections.

In general it was noted that as the atomization gas pressure increased and liquid metal diameter decreased, the particle size distribution tended to decrease. The atomization unit was sound, and the precautions taken to preclude the escape of beryllium powder into the atmosphere were effective.

Powder Characterization

The as-atomized powder was characterized in terms of its tap density, flow rate, chemistry, and thermal stability. Optical and scanning electron metallography (SEM) were used to determine particle shape and microstructural characteristics.

As expected, the powder resulting from the atomization process was spherical, as shown in Figure 3. The difference in particle shape between atomized and mechanically comminuted powders is striking.

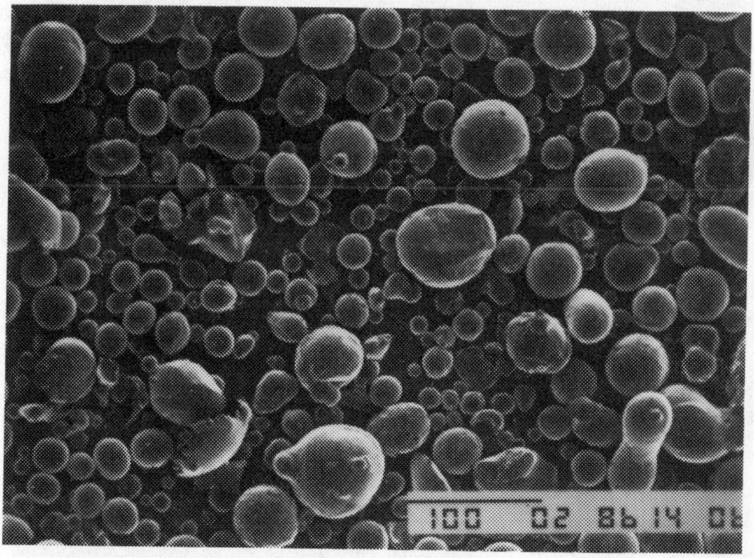

Figure 3 - Scanning Electron Micrograph of -200 Mesh Atomized Beryllium Powder

The tap density of the -200 mesh spherical powder was measured and compared to conventional powders. The spherical powder value of 64% of theoretical was significantly greater than the 50-55% of mechanically comminuted material. Flow characteristics were also significantly better with a Hall flow rate of 213 seconds as compared to mechanically comminuted powder which showed no flow. These results are summarized in

Table 1 - Characteristics of Mechanically Comminuted and Atomized Powder

	Impact Ground	Atomized
Apparent Density (% theoretical)	34	54
Tap Density (% theoretical)	53-55	64
Hall Flow Rate (seconds/50g)	Not Measurable	213

The microstructure of the powder was generally polycrystalline, with only occasional examples of monocrystalline particles. It was noted, although not quantitatively demonstrated that the smaller powder particles generally showed a greater abundance of monocrystalline particles. This is the expected behavior in view of the nucleation model advanced by Turnbull[4]. An optical photomicrograph of the cross sections of several powder particles is presented in Figure 4.

Figure 4 - Polarized Light Optical Micrograph of Spherical Powder Particles in the As-Atomized Condition.

The stability of the powder microstructure is as important as the as-atomized structure because significant grain growth during consolidation and subsequent heat treatment could compromise mechanical properties. Figure 5 shows the microstructure of the powder after 3 hours at 1000 C (1832 F), a typical thermal exposure used in HIP consolidation. It is apparent that the grain size is largely unaffected, although the grain boundaries do show signs of being straighter than those of the as-atomized powder. This finding was particularly encouraging in view of the relatively low oxide content of the atomized powder, shown in comparison with the input melt stock and impact ground powder chemistries presented in Table 2. Of particular importance is that the oxide content is approximately the same level as a typical structural grade, S-65. This data confirmed that the equipment did embody the technology necessary to produce a low oxide content, fine spherical powder.

Table 2 - Composition of Atomization Feed, Atomized Powder and Typical Impact Ground Grade Be Powders

Element Powder	Atomizer Feed (vacuum melted)	Atomized Powder (-270 mesh)	IG (S-65)
BeO (%)	0.09	0.70	0.7 max
Al (ppm)	410	635	600 max
Fe	540	2820	800 max
Mg	615	1155	600 max
Si	145	345	600 max
Ca	<20	<3	100 max
Cr	110	250	100 max
Co	5	11	10 max
Cu	10	70	150 max
Ni	85	185	300 max
Mo	<20	<20	20 max
N	10	120	300 max
Pb	<20	<20	20 max

Consolidated Billet Characterization

The billets made by HIP and VHP were evaluated in terms of microstructure, tensile properties, and preferred orientation as determined by X-Ray pole figure analysis. The results of this characterization are presented below.

The microstructure of the -200 mesh powder HIP'd billet, Figure 6, exhibits a grain size of approximately 12 microns. This grain size is similar to that of a typical structural grade, S-65. The grains are equiaxed, with little evidence of beryllium oxide delineating prior particle boundaries. The derivation of this microstructure from spherical powder is in evidence, however. This microstructure is similar to the VHP microstructure, with the exception of the grain size, which varies from 17 to 28 microns in the VHP material, Figure 7. Some directionality is in evidence in the VHP microstructure as a result of uniaxial compaction. Two additional HIP billets using -150/+200 mesh and -100/+150 mesh powders exhibited grain sizes of 16.8 and 17.1 microns respectively. It is interesting to note that the HIP'd -200 mesh spherical powder yields the same grain size as -325 mesh impact ground VHP powder (S-65).

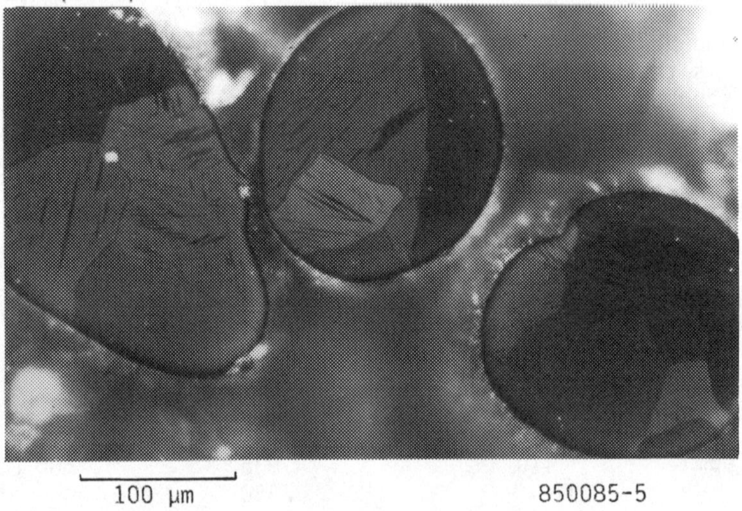

Figure 5 - Polarized Light Optical Photomicrograph of Spherical Powder Particles after 1000 C - 3 Hrs (1832 F). Note That Grain Size is Similar to That Shown in Figure 4.

The isotropy of the VHP and HIP billets were evaluated using X-Ray Pole figure analysis. The {0002} plane poles and the {1010} plane pole figures for both types of billets are shown in Figures 8-11. The HIP billet results show no preferred orientation, as expected. The VHP results also reveal little to no texture, sustaining the assumption that most preferred orientation results from powder morphology, rather than from uniaxial pressing. In addition to the pole figure results, coefficient of thermal expansion (CTE) testing was also performed. The HIP'd -100/+150 mesh powder billet was subject to CTE measurement in both the longitudinal and transverse directions. The CTE values were identical, 11.38μin./in./C in both directions. In typical vacuum hot pressed billet, longitudinal values of 11.29μin./in./C and transverse values of 11.37μin./in./C are typical. This property is very sensitive to preferred crystallographic orientation, and demonstrates the excellent isotropy possible with spherical beryllium powder.

100 µm 850218-4

Figure 6 - Polarized Light Optical Photomicrograph of HIP'd Spherical Powder. Grain Size is 12.3 Microns.

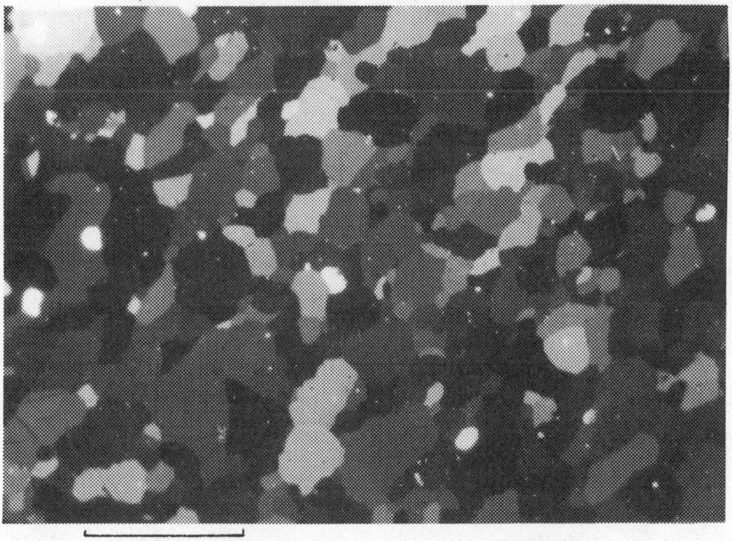

100 µm 850216-3

Figure 7 - Polarized Light Optical Photomicrograph of VHP Consolidated Spherical Powder. Grain Size is 17.1 Microns.

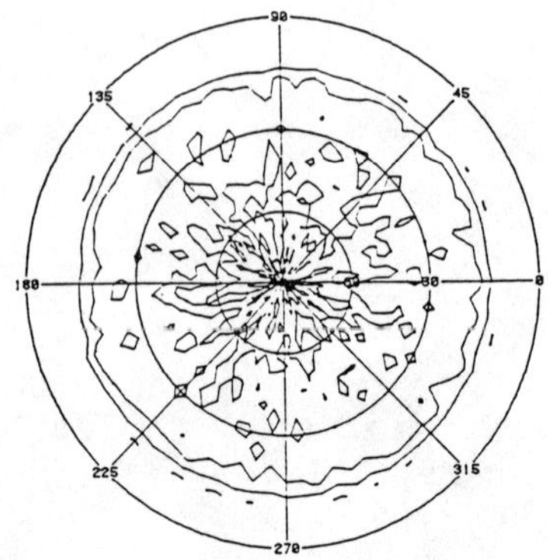

Figure 8 - X-Ray Pole Figure for [0002] Poles, HIP'd Spherical Powder. Note the randomness of C-Axis Orientation.

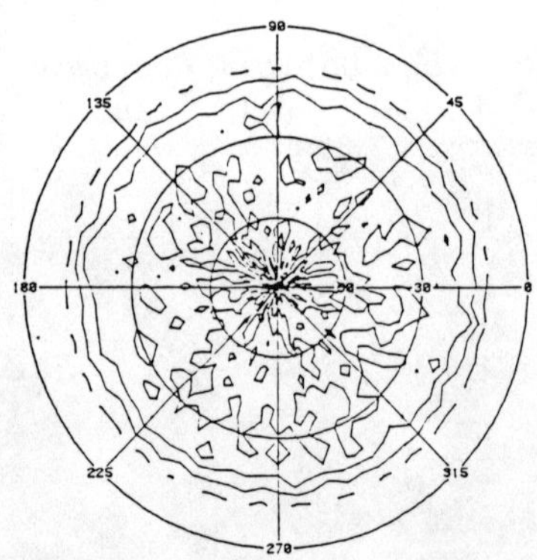

Figure 9 - X-Ray Pole Figure for [$10\bar{1}0$] Poles, HIP'd Spherical Powder. Note the Lack of Preferred Orientation.

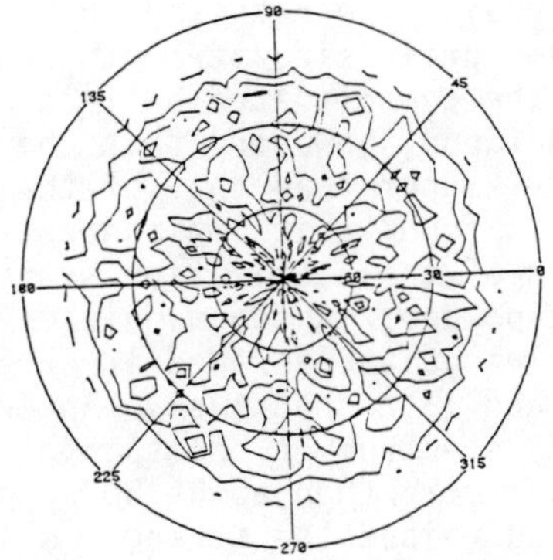

Figure 10 - X-Ray Pole Figure for [0002] Poles, VHP'd Spherical Powder. Note the Lack of Preferred Orientation.

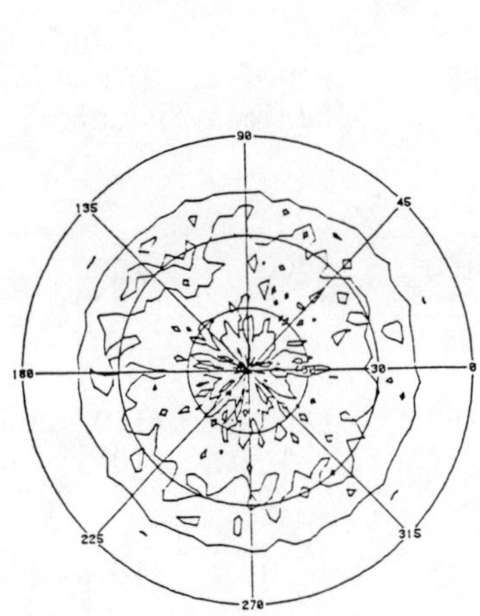

Figure 11 - X-Ray Pole Figure for $[10\bar{1}0]$ Poles, VHP'd Spherical Powder. Note the Lack of Preferred Orientation.

The tensile properties of the consolidated billets made by VHP and HIP are presented in Table 3. It is apparent that the strengths are comparable to the values obtained for impact ground powder of similar grain size. A 12 micron grain size resulted in a yield stress of 257 MPa (37 ksi). This is the same grain size/strength characteristic as seen in S-65. The ductility of the consolidated billets was found to be lower than anticipated. It was found, however, that the iron content of the billets was substantially higher than that of the conventional material. Upon investigation, it was found that impingement of atomized powder particles upon components of the atomization system had abraded stainless steel into the powder. A scanning electron micrograph and iron dot map of a typical angular, abraded iron-rich particle are presented in Figures 12 and 13,respectively. An Energy Dispersive Analysis of X-Rays (EDAX) spectrum of one such particle is presented in Figure 14. The presence of more than about 800 ppm iron results in the formation of $FeBe_{11}$, which raises the strength slightly, and severely reduces ductility. The presence of an iron-rich inclusion will also cause premature failure. The low ductility, therefore, is not inherent in the atomization process, but is the result of a detail of this particular atomizing unit.

Table 3 - Consolidated Be Powder Properties

Process Powder	VHP Spherical	HIP Spherical	HIP Spherical	HIP Spherical	VHP ImpactGround (S-65)
Temperature - C (F)	1100 (2012)	1000 (1832)	1000 (1832)	1000 (1832)	-
ASTM Sieve Micron	-200 -74	-200 -74	-150/+200 -105 - 74	-100/+150 -149 - 105	-325 -44
Longitudinal Grain Size - µm	17.1	12.8	16.8	17.1	12 Typ
0.2% YS - MPa (ksi)	200.3 (29.0)	256.9 (37.3)	249.1 (36.1)	266.8 (38.7)	206.8 min (30 min)
UTS - MPa (ksi)	300.6 (43.6)	352.1 (51.1)	316.7 (45.9)	298.9 (43.3)	289.6 min (42 min)
Elongation - %	2.6	2.4	1.7	0.8	3% min
BeO - % Fe - ppm	- -	0.45 900	0.18 925	0.23 1560	1% max 800 max

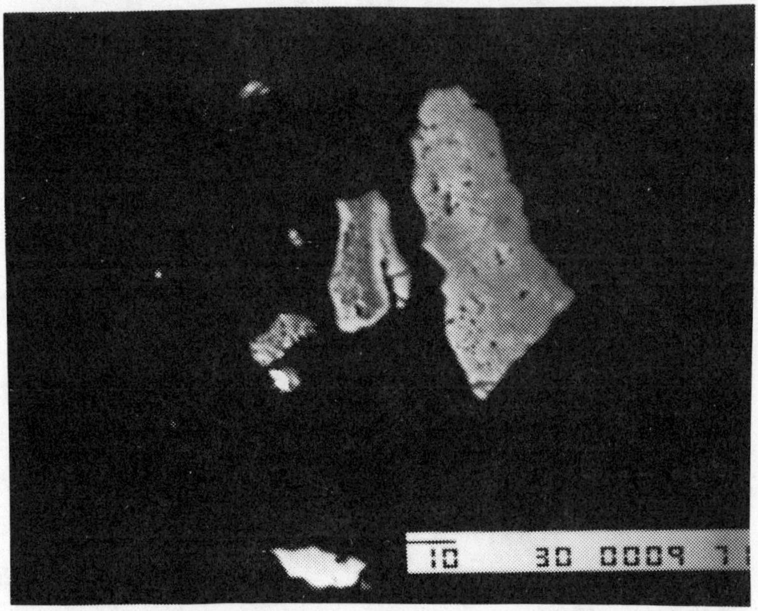

Figure 12 - Back Scatter Electron Image of a Typical Angular, Steel Contaminated Atomized Beryllium Powder Particle.

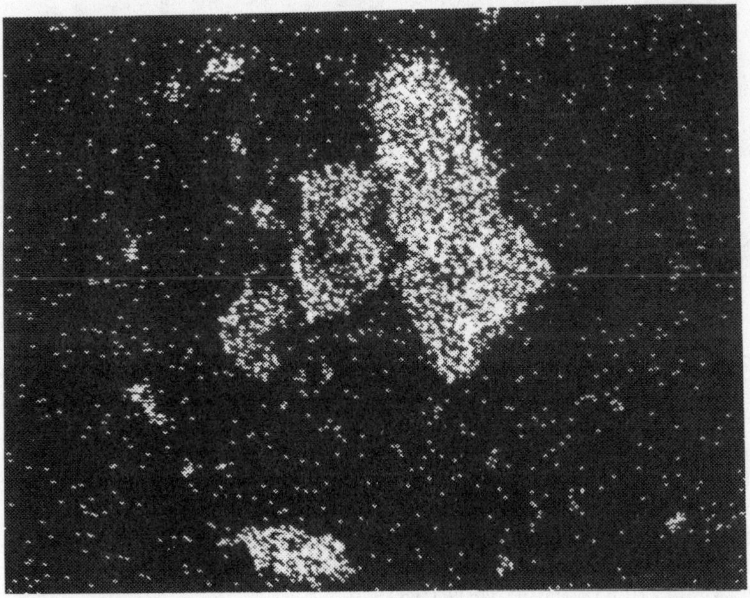

Figure 13 - Iron Dot Map of the Angular Particle Seen in Figure 12.

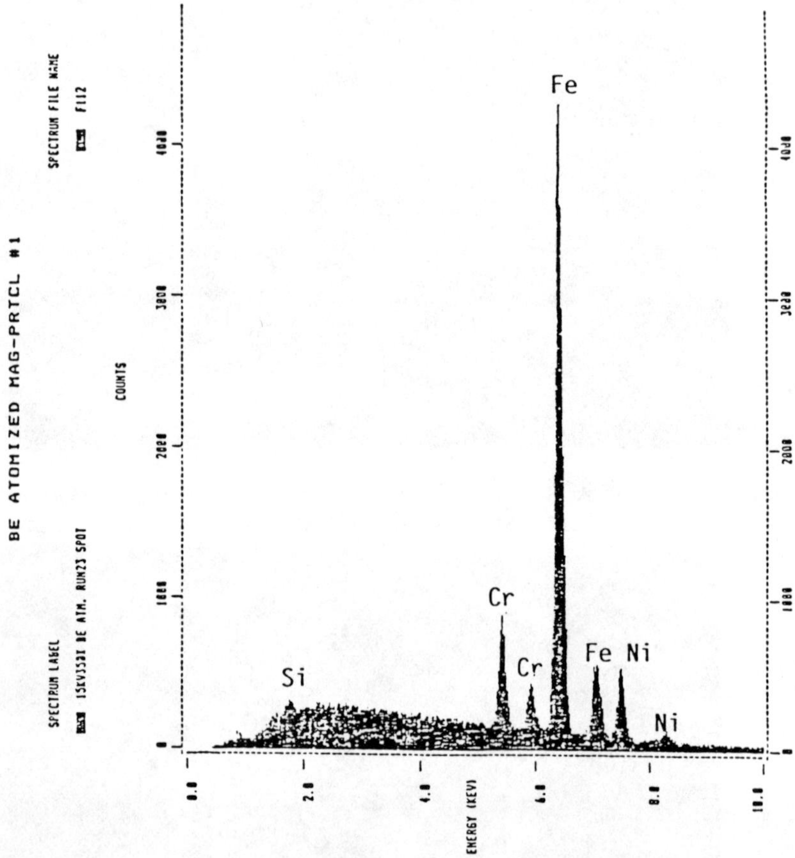

Figure 14 - EDAX Spectrum of Iron-Rich Contamination Found in Spherical Powder. The Spectrum Corresponds to 304 Stainless Steel.

In view of these findings, Brush Wellman has embarked upon an atomizer modification program. The aims of this effort are to change the design to preclude iron contamination, and to increase the atomization pressure to improve the yield of the fine particles for use in structural grade materials.

This work was conducted under the sponsorship of the Department of Energy, Albuquerque Operations Office under Contract DE-AC04-82AL18071. Dr. Richard Jiacoletti was the DOE program sponsor.

Conclusions

1. Inert gas atomization is a viable process for producing spherical beryllium powder with an acceptable oxide content.

2. Spherical powder exhibits the predicted advantages of flowability and packing density over conventional, mechanically comminuted powders. Powder particles were generally polycrystalline.

3. Billets consolidated from spherical powder were isotropic. In addition, the grain size of the powder was stable upon consolidation. The strength of billets reflected the grain size.

4. A modification of the atomizer to prevent iron contamination will permit the full potential of atomized powder to be realized.

References

1. Paine R.M. and A.J. Stonehouse, "Investigation of the Effects of Micro-Alloying and Thermal Treatment on the Properties of Beryllium," Report BW-TR-549, Naval Ordnance Laboratory Contract N609021-72-C-0284, August, 1974.

2. Lowenstein,P., "Specialty Powders by the Rotating Electrode Process," Progress in Powder Metallurgy, v. 37, p.9. MPIF, Princeton, N.J. 1982.

3. AGARD CP-200, "Advanced Fabrication Techniques in Powder Metallurgy and Their Economic Implications," NATO Conference Proceedings, Ottawa April 1976.

4. Gummeson, P.U., " Powder Metallurgy," v. 15, no. 29, p. 67, 1972.

5. Hollomon, J.H. and Turnbull, D., "Nucleation", Progress in Metal Physics, v.4, ed. B. Chalmers, Interscience, New York, 1953.

HOMOGENEOUS COPPER - LEAD POWDERS

T.E.Fine

Radian Corporation
Milwaukee, Wisconsin 53214

ABSTRACT

Conventionally processed high leaded copper alloys demonstrate significant degrees of lead segregation. This investigation was performed to study the effects of various patented and practical methods of treating a molten copper-lead mixture such that the resulting powder or ingot product retained a homogeneous dispersion of lead within the copper matrix. A further goal was to investigate the effect of a uniform lead distribution on friction and wear. Three 40% lead-60% copper samples were melted, heated to 2250/2300°F, and inert gas atomized. One heat was treated with a homogeneity promoter, one had been pretreated, and the third had no treatment and served as a control. The resulting powders were characterized as to size distribution, chemistry, morphology, and microstructure. Portions of each heat were used in a remelting experiment to determine the effect of repeated remelting on the lead distribution. Other samples were incorporated into a grease formulation and tested for effects on reduction of friction. It was found that all samples of the atomized copper-lead mixture formed very fine, spherical particles with the lead phase well distributed within the copper; the sample treated in the melt showed the greatest uniformity. Upon remelting, however, the untreated sample reverted totally to a segregated microstructure, while the sample treated with the homogeneity promoter retained a significant degree of uniformity. The treated samples incorporated into the grease formulation demonstrated a significant reduction in friction and wear behavior.

INTRODUCTION

Lead films have been employed to provide an effective lubricant for metallic components for many years. Highly leaded (15% to 22% Pb is normal) copper bearings have become the standard for use in applications involving heavily loaded rotating machinery or where the use of liquid lubricants is prohibited (1). The main problem in these situations lies in the fact that most traditional techniques employed to create highly leaded mixtures (greater than 22% Pb) with copper, aluminum and tin, result in a grossly segregated lead distribution and resultant non-uniformities in lubricating ability, preferential regions of wear, and localized failures.

Copper and lead do not form alloys, but only mixtures (2). The only apparent method to retain the homogeneous structure presumed to exist in the liquid state above the miscibility gap is through extremely rapid cooling. This method is practiced, but the resultant products revert to their elemental state upon low temperature heating, such as could be experienced in normal bearing use. Many investigators have tried, with varying degrees of success, to reduce the natural tendency of lead segregation through the use of various alloying elements such as nickel, antimony, silver, etc. (1). The best current bearing alloys, such as CDA 937/SAE 64 (80% Cu, 10% Pb, 10% Sn), are the results of these investigations. However, there are side effects detrimental to bearing or lubricating performance which accompany alloying, including increased hardness, decreased conductivity and occasionally, lowered operating temperature. It has been speculated for many years that the best lubricating situation would occur if the lead content could be increased to above 22% and was uniformly distributed as particles in a non-dendritic manner throughout a pure copper matrix.

In the early 1960's, two independent groups of investigators apparently determined methods of processing a copper-lead melt such that upon solidification, the lead was uniformly distributed. Within the patents issued to these groups, claims were made that suggested that this uniform distribution was attained through the use of a "homogeneity promoter" and carefully specified thermal processing (3-8). The patents claimed that the process could achieve the uniform lead distribution regardless of whether the treated liquid mixture was poured into sand or graphite ingot molds, chilled steel molds, centrifugally cast, or atomized to metal powder. During the next two decades, several attempts were made at commercializing various forms of this material including ingot, tubulars and powder. All of these attempts failed due to a variety of causes, mainly non-technical in nature.

In the mid-1980's, a new investigation into the nature of this "phenomenon" was sponsored, with the intent of systematically determining the underlying mechanisms, the nature of the "homogeneity promoter", the properties of carefully prepared samples and ultimately, whether the patented technology (the patents are still active and in force) could be turned into useful commercial products. Several series of melts were prepared, cast and examined to determine the relationship between the composition of the "homogeneity promoter" and the resulting microstructural characteristics. In total, twelve 120 pound heats were prepared and poured as splits into both sand and graphite molds to effect a change in cooling rate. The ingots were sectioned and examined at five internal locations, to determine an additional effect of cooling rate and to observe the progress of liquation. Standard high leaded material from commercially prepared bearing stock was also examined as a base line measurement. From these experiments, an optimum composition for the "homogeneity promoter" was chosen and subsequently employed in an atomization trial.

This paper details the procedures and results of an attempt to produce a homogeneous copper-lead mixture through the use of inert gas atomization. A theoretical explanation for the observed behavior is presented along with the results of some attempts at application of the technology to useable products.

EXPERIMENTAL PROCEDURES AND RESULTS

Atomization

In early 1987, three 40 pound heats of a nominal 60% copper, 40% lead composition were melted and atomized at the facilities of Alloy Metals, Inc. in Troy, Michigan. The Alloy Metals, Inc. atomization process is proprietary, but in general terms, they use an inert gas atomization technique with ceramic lined tundishes and a multi-port gas nozzle. Cooling rates resulting from this process have been estimated at $10^5/10^6$ °C/s and they are capable of achieving a very fine powder size distribution in many alloys.

The specifics of the three heats are as follows:

Heat 1: 24 lbs of Cu, 16 lbs of Pb; no additives; melt/pour temperature = 2280°F; atomization run time = 2 minutes. The copper was heated to the melting point in the furnace first, followed by the lead addition. After a totally molten condition had been achieved, the temperature was raised to about 2300°F, the melt poured into the tundish and the atomization occurred. A total of 36 pounds of product was recovered from the original 40 pound charge.

Heat 2: 40 lbs of a 60%Cu/40%Pb product, Mixture "T"; no other additives; melt/pour temperature = 2300°F; atomization run time = 2 minutes. This feed stock, termed mixture "T" was produced at an off-shore plant by a semicommercial enterprise (9). Examination of the microstructures found in the ingot product showed a good lead distribution, reportedly caused by a proprietary melt treatment process. The entire 40 pounds was charged cold into the furnace and heated directly to 2280°F. The atomization went smoothly and 37 pounds of powder was recovered from the 40 pound charge.

Heat 3: 24 lbs of Cu, 16 lbs of Pb; 334 grams of Na_2CO_3 and 124 grams of graphite encased in copper foil; melt/pour temperature = 2250°F, atomization run time = 2 minutes. The "homogeneity promoter" consisted of measured amounts of graphite and sodium carbonate. This mixture was preblended and wrapped with copper foil into one pound packets. The lead and copper additions were charged cold into the furnace and heated to around 2250°F. At this point, the packets were plunged below the surface of the melt using a steel "fork"; the melt was also stirred with the fork. At the melt temperature, the addition of each packet caused an evolution of copious quantities of green/yellow flames and yellow smoke, plus a violent roiling of the surface. The graphite seemed to all rise and cover the melt surface like a blanket. This condition continued for about 10 to 15 minutes, then the graphite was skimmed from the surface and the atomization began. It was anticipated that due to the length of time at reaction temperature, a lot of the lead would have been lost and that the powder would contain a high carbon content. Subsequent analysis showed that neither situation occurred; 37 pounds of powder was recovered from the initial 40 pound charge, about the same as in the other two heats.

The powders from the three heats were examined for chemical composition, size distribution and microstructure. In view of the natural segregation tendency of lead and copper, the results were surprisingly good. Chemical analyses were performed by a combination of Leco analyzers and traditional wet chemical analysis.

TABLE I - CHEMICAL ANALYSIS (Weight %)

	Carbon	Oxygen	Nitrogen	Copper	Lead
Heat 1	0.008	0.041	0.006	60.0	39.2
Heat 2	0.005	0.080	0.0009	60.1	39.5
Heat 3	0.005	0.094	0.001	60.3	39.3

The powder size distribution was determined using an L&N Microtrac. This analysis calculates the cumulative amounts of powder less than preselected percentages, in this case, 10%, 50% and 90%.

TABLE 2 - POWDER DIAMETERS (microns)

	10%	50%	90%
Heat 1	5.0	17.6	52.0
Heat 2	3.7	12.5	49.4
Heat 3	3.7	14.7	55.3

This analysis was subsequently checked through both Scanning Electron Microscope (SEM) and optical microscopy of raw powder and found to be essentially correct. In the metallographic examination, it was determined that there was a very small fraction of large particles.

Small amounts of powder from each heat were mounted in the standard metallographic manner and then polished to expose interior cross sections. The powder from each heat showed similar structures, the lead was distributed somewhat uniformly throughout the particles of copper (Figures 1, 2, & 3). There were variations in lead content from particle to particle, but a mixture was present in all particles. A careful examination of hundreds of particles from each heat revealed that the distribution of lead in copper was most uniform in Heat 3, less so in Heat 2 and worst in Heat 1.

The powder samples were further examined using an SEM. While the uniformity of lead distribution was still evident, particles from Heats 1 & 3 appeared to have primarily an exterior sheath of copper while most of those examined from Heat 2 had a mixed sheath of lead and copper (Figures 4, 5 & 6). This explained the visual color differences noted in the product of the three heats wherein Heat 2 had a much darker, grayer appearance.

Heat Treatment

One of the main features of the patented treatment process is that the homogenized product should be capable of being remelted several times and always return to the homogenized state (3,8). No restrictions were given in the patents as to the method of remelting or to any need to protect the surface of the melt. To test this claim, green compacts of the three powders were made by pressing loose powder in a hydraulic press at 7,000 psi pressure. The compacts were heated in clay crucibles under a nitrogen atmosphere to 1900°F, held for 5 minutes and naturally cooled. The powders melted and resolidified into a mini-ingot which was subsequently cut in half lengthwise and polished for metallographic examination.

Samples were prepared by remelting once, by remelting, cooling and remelting a second time prior to examination and by melting and cooling a total of three times. The samples from all three heats were processed at the same time in the same furnace to insure an equivalence of melt temperature and cooling rates.

In general, all nine samples showed some degree of lead segregation from top to bottom of each mini-ingot. The lead distribution became coarser from top to bottom and there was a lump of apparently pure lead at the bottom of several ingots. However, there were obvious differences in the behavior of the material from each of the three heats. The plain copper-lead material from Heat 1 showed gross segregation in all three remelt samples with large lumps of lead at the bottom of each ingot. This is just what would be expected from an examination of the Cu-Pb phase diagram (2) and from previous studies (10). The remelted material from Heat 3 showed the least segregation of the three materials, with almost no lead puddling and a relatively uniform distribution throughout the mini-ingot. The microstructure was coarser than that found in the original powder, but the fact that a significant degree of uniformity persisted has profound implication for industrial application. The behavior of the material from Heat 2, mixture "T", was intermediate between the plain copper-lead and the freshly treated copper-lead.

Application

Powder treated to produce the homogeneous structure was formulated into an industrial grease base and tested in both a Shell 4-Ball wear tester and a Timken loading test (11). Both tests were run according to ASTM specifications and in both cases, the tests were performed by independent testing laboratories. The grease base was formulated by a company considered to be expert in the field of lubrication and wear and associated with a major Midwestern university. In both tests, the addition of 25% by volume of the treated copper-lead powder, with an average diameter less than 25 microns, resulted in approximately a 30% decrease in wear indicators when compared to the base. The copper-lead addition level was varied from 0% to 60%; however, the performance was the best at an addition level of 25%. The likely mechanism for this improvement is that the lead fills any wear scars as they form and then provides a high degree of extra lubrication to augment the grease base. It has been shown that a lead layer approximately 0.004 inches thick provides an excellent lubricating surface (12). It is also likely that the uniform distribution of the lead in the copper allowed an even deposition of the lead on the test bearings and provided a readily available source of lead to renew any that was lost during the tests.

DISCUSSION AND CONCLUSIONS

The results of these simple experiments seem to corroborate some of the claims made in the original patents and leave room for doubt on others. That the "homogenizing" treatment creates a more desirable microstructure and improves some physical properties, including reduction of friction, appears correct. It is still unclear whether this treatment truly induces a material with "memory", i.e., one which survives the total atomic rearrangement that accompanies the melting process.

The original patents claimed that a variety of "homogeneity promoters" could be used to create the effects. These included the use of alkali or alkali earth carbonates, copper phosphate, rare earth fluorocarbonates plus finely divided carbon powder in the form of graphite. The patent postulated that when the alkali carbonates and graphite were added to the molten copper-lead mixture at 2300°F, the carbonate decomposes and forms CO plus alkali metal oxide. The CO forms a reducing atmosphere above the bath and additionally reduces the alkali metal oxides to alkali and additional CO gas. The graphite and the CO gas aid in excluding oxygen from entering the bath.

The combined action of the CO evolution and the boiling of the alkali metal gas below the surface of the melt causes a vigorous agitation of the melt which persists during cooling to the solidification point.

The agitation does two things: it breaks up the lead particles into very small droplets and disperses them throughout the melt, and it breaks up the copper dendrites as they form with decreasing temperature, resulting in many more nucleation sites and a very fine grained structure.

This mechanism, while probably correct, cannot explain the "memory" feature of the treated system. This feature may be explained if the creation of an emulsion is postulated. In a true emulsion, such as homogenized milk, the second phase droplets will exist below the critical size for normal nucleation and growth phenomenon to occur. A true emulsion can survive repeated freeze/thaw cycles with the particle distribution relatively intact. In the case of the treated copper-lead, a true emulsion is not formed, and there is sufficient time for some agglomeration to occur during the cooling. If the particle size existing after the initial solidification is greater than the critical diameter then on subsequent remelts, some fraction of the particles will grow with a greater fraction growing upon each subsequent remelting. It is also apparent that the homogeneity promoter creates favorable surface tension effects since the lead assumes a spherical (rounded) shape without the addition of any other known surface tension altering elements.

The technical literature has some interesting points to contribute to the understanding of this phenomenon. The copper-lead phase diagram shows that this is a totally immiscible system with a two liquid region at lead contents greater than 36% and at temperatures between 954°C and 1325°C. At the 40% lead content and temperatures greater than 1325°C, a single liquid exists with the lead in the form of an emulsion (13). The exact boundaries of the two liquid region is in doubt, since a second reference indicates the existence of a single liquid region above 990°C (2). In either case, slow cooling of an untreated mixture will result in massive segregation (10). In general, in immiscible systems, if the freezing rate is sufficiently rapid, the minor phase is trapped in the spaces between the growing major phase and its shape is determined solely by surface tension considerations (14).

The influence of additives on lead distribution in copper-lead mixtures has been studied by various investigators. It has been claimed that the addition of nickel, sulphur and lithium function to retard lead segregation. It was also observed that the alkali metals may have a favorable effect on the primary crystallization of copper (15). Residual phosphorus raises the surface tension of the lead, but causes coagulation of the particles, while the addition of mischmetal (rare earth) lowers the surface tension and aids in preventing segregation (15). Silver has been shown to blunt the copper dendrite arms, while excessive oxygen leads to the formation of a second copper oxide dendritic network (16).

If the information from the literature is combined with the observations in this experiment, it seems reasonable to conclude that the operative mechanisms in the homogenization of copper-lead mixtures includes the following: 1) alteration of the surface tension relationships to create small, stable globules; 2) strong agitation of the melt in a temperature region where a two liquid emulsion already exists; 3) continued agitation of the melt as temperature decreased to help reduce the crystallite size; 4) exclusion of oxygen; 5) accelerated cooling rates through the solidification range to promote retention of the shape relationships existing in the liquid state; and 6) possibly the retention of some elements around the lead particles to promote continued stability.

It was also obvious that this process is fragile and does not reoccur under all melting/refreezing conditions. The eventual development of commercially acceptable bearing alloys will require additional systematic study of the effects hypothesized in this work.

REFERENCES

1. ASM Handbook, 8th Edition, 1972, Vol. 5 (pp416-432), Vol.7 (pp 231-240, 273-296) & Vol.8 (pp130-132).

2. Constitution of Binary Alloys, M. Hansen and K. Anderko, Mc-Graw-Hill Book Company, 1958.

3. U.S. Patent 3,566,779, Copper-Lead Alloys, R. Turkisher and C.E. Lundin, 1971.

4. U.S. Patent 3,719,477, Copper-Lead Alloys, R. Turkisher and C.E. Lundin, 1973.

5. U.S. Patent 3,720,507, Copper-Lead Alloys, C. E. Lundin, 1973.

6. U.S. Patent 3,894,957, Copper-Lead Alloys for Lubricants and Bearings, R. Turkisher and C.E. Lundin, 1975.

7. U.S. Patent 3,948,651, Alloys of Aluminum-Lead-Copper, C.E. Lundin, 1976.

8. U.S. Patent 3,544,314, Homogeneous Copper-Lead Metal and Method of Making, C.E. Bazley, 1970.

9. Confederation Suisse Brevet d'invention CH 658 671, O. Sultan, 1986.

10. Interpretation of Metallographic Structures, W. Rostoker and J.R. Dvorak, Academic Press, 1965, pp 76-80.

11. Private Communication, Preventative Maintenance Design Co.

12. The Friction and Lubrication of Solids, P. Bowden and D. Tabor, Oxford Press, 1950, pp122-146.

13. Applied Science in the Casting of Metals, K. Strauss, Pergamon Press, 1970, (UWM TS 230, A7).

14. Physical Metallurgy Principles, R.E. Reed-Hill, D.Van Nostrand Company, Inc., 1964, pp 367-398.

15. J.S. Hirshhorn, Journal of Metals, October, 1970, pp. 40-43.

16. Casting Copper-Base Alloys, American Foundryman Society, M.T. Rowley, Contributing Ed., 1984.

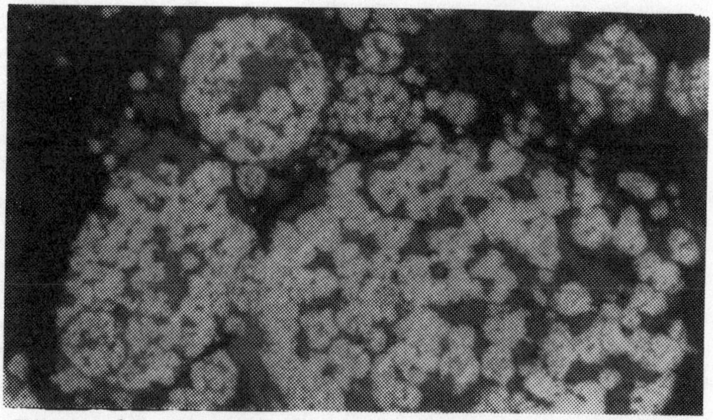

Figure 1: Optical Microstructure; Heat 1; 800X

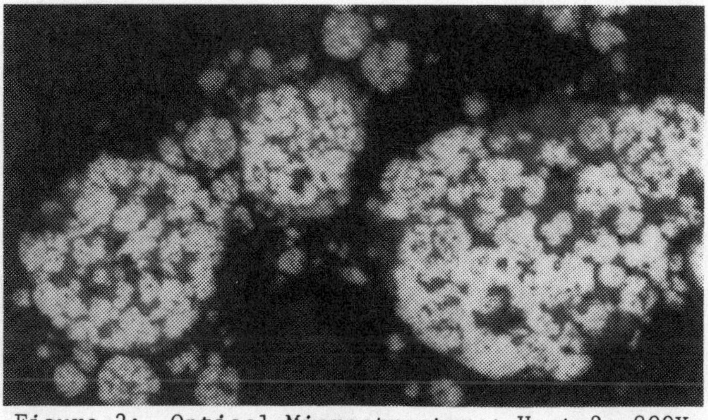

Figure 2: Optical Microstructure; Heat 2; 800X

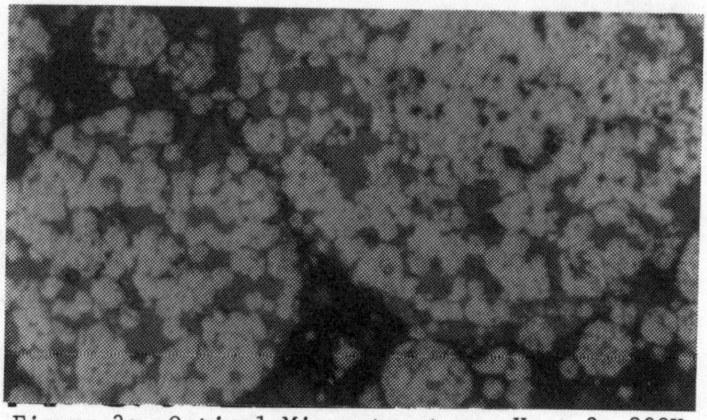

Figure 3: Optical Microstructure; Heat 3; 800X

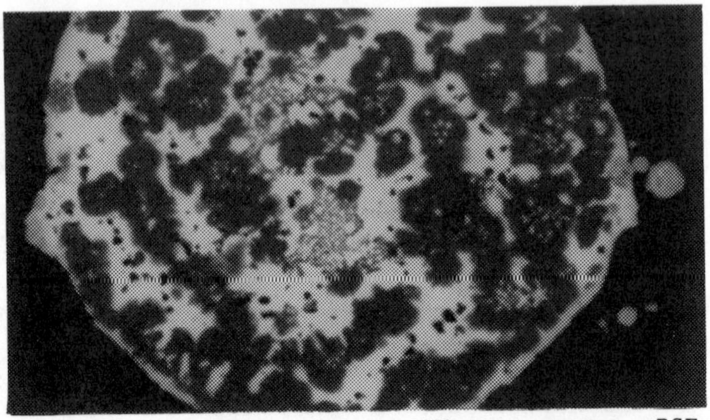

Figure 4: Scanning Electron Microstructure, BSE; Heat 1; 1000X

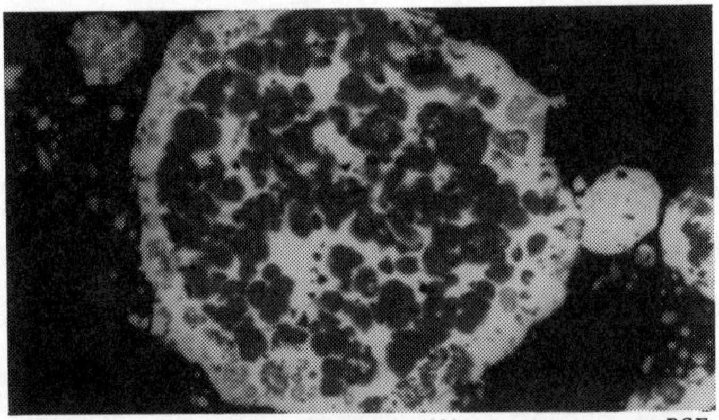

Figure 5: Scanning Electron Microstructure, BSE; Heat 2; 1000X

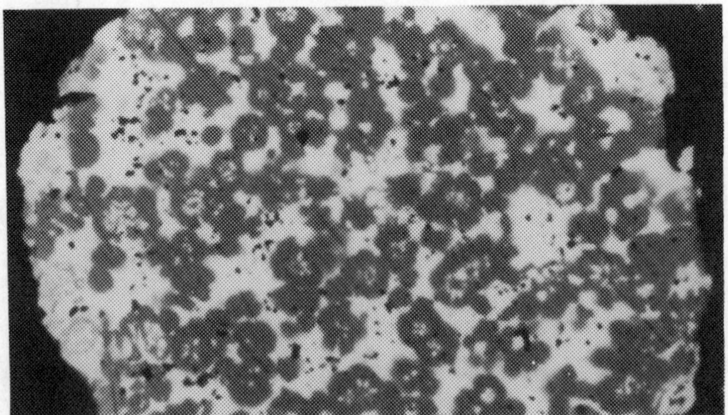

Figure 6: Scanning Electron Microstructure, BSE; Heat 3; 1000X

ATOMIZED MATERIAL METALLURGY: PROBLEMS AND PROSPECTS

d.t.s., prof. Druzhinin L.K.

All Union Scientific and Research Institute
of Inorganic Materials
Moscow 123060, USSR

Abstract

Special features of traditional powder and granule metallurgy are discussed. As compared to the classical metallurgical methods (melting-casting-mechanical working) these methods have a relatively limited usage for mass metal product manufacture; the reasons are discussed.

An alternative scheme of product manufacture, viz., ARS process (atomization and rapid solidification) is suggested; it combines processes of powder production by melt atomization and hot vacuum pressing with the traditional powder metallurgy methods.

The potentialities of the RIBS process are considered.

Beginning from the time of P.G.Sobolevsky work (1826) powder metallurgy took its prominent place in stores of technical means and methods of the effective working of metals and alloys. Having unique process, materials science and economic features powder metallurgy provided for the manufacture of products from refractory, wear resistant and frictional materials, a wide range of electrotechnical products as well as many mass products from structural materials.

Nevertheless, in the total balance of metallurgical production the fraction of powder metallurgy is negligibly small. Of 650-750 mln tons of metals and alloys annually produced in the world only 600-650 thousand tons are produced as a powder of which no more than 350-400 thousand tons are used for the powder metallurgy production. The fraction of powder metallurgy in the total balance of metal production is about 0.05%.

What is the cause? The "classical" metallurgical scheme - casting + plastic working - has disadvantages not less "classical": non-uniform phase, structural and chemical compositions of large ingots due to the inevitable segregation of elements on ingot crystallization; presence of large mainly intergranular refractory inclusions; a great number of operations accompanied by the repeated air heatings of ingots and pressings; a low metal utilization factor (MUF); for main products it is no more than 5-30% when calculated using the charge; low automation; arduous working conditions in hot shops.

The principal cause is in the fact that "traditional" powder metallurgy while having significant technological advantages, substanially yields to "classical" metallurgy in the properties and density of a metal with its inevitable porosity. It is this cause that substantiated the widely spread thesis that powder metallurgy is successfully employed only in cases where classical metallurgy (casting + plastic working) are helpless. It seemed one time that this thesis would be shaken due to the success of the new development stage of powder metallurgy - granule metallurgy.

"Granule metallurgy" through the use of spherically shaped powders produced by inert gas atomization of a melt at high rates of their crystallization and compaction by hot isostatic pressing in sealed capsules made it possible to drastically improve the properties of a blank metal and make them significantly higher than those produced by the "classical" metallurgy methods. At the same time to-day according to publications the bulk of products produced using granule metallurgy methods did not significantly alter the powder metallurgy fraction in the total balance of metal production. The vast majority of granule metallurgy products belong to special fields of engineering: cosmic, military and aviation engineering. There is no grounds to expect this situation to change in future.

In our opinion, the main causes of a limited usage of granule metallurgy are the following.

1. Both in this country and abroad granule metallurgy uses powders of spherically shaped particles that form in a free flight on their crystallization in argon or helium. The technological advantages of spherical particles are evident: proper mixing, simple transportation and metering, the adequate filling of a capsule, minimum specific surface, high bulk density etc. However, spherically shaped powders have a significant disadvantage, viz., zero mouldability which practically fully excludes the possibility of their compaction with traditional and cheap cold pressing methods.

That is why, the shaping of atomized spherical powders into the form close to a final product by hot plastic working is accomplished through the use of special containers - capsules approfimating the shape of a final product. The use of capsules -containers is the basis of powder compaction by hot isostatic pressing (HIP). The HIP process requires subsidiary production of capsules - containers upon which stringent requirements are placed as to the welding quality, air-tightness, compliance with the product design etc. For each product to be fabricated from atomized powders by the HIP process a metal capsule - container must be prepared that after a once-through usage must be removed mechanically or chemically which entails negative consequences in waste reprocessing.

With all its advantages the HIP process is expensive, of low productivity, has a low equipment utilization factor and is difficult to automate. The inert gas consumption is significant, its reuse is difficult etc. The HIP process is a large technical achievement, but it is not possible to supplant the old classical metallurgy on its basis.

2. Granule metallurgy is based on the use of an inert gas as a protective medium for centrifugal methods or as an alloy atomizer for gas-jet methods. At the same time on an inert gas atomization the inert gas filled pores in atomized particles are inevitable. The gas porosity is present even in particles ~ 50 mcm dia where it constitutes ~ 1% and increases to 2.0-6.0% in particles 300 mcm dia depending on the relative rates of a gas-melt motion. The gas porosity of powders is also stably retained on thermocentrifugal melt atomization in an immobile atmosphere.

The presence of helium in atomized powders and their products is reliably revealed on their electron beam melting in high vacuum with the evolved helium determined by a helium leak detector. The residual porosity of a compact metal can show up on polished surfaces on provocative heating to high temperatures (1500°C for zirconium). Heating induced expansion of an inert gas results in numerous microblisters appearing at the polished surface (fig.1).

3. The most developed method of producing atomized powders with a rotating electrode that is realized commer-

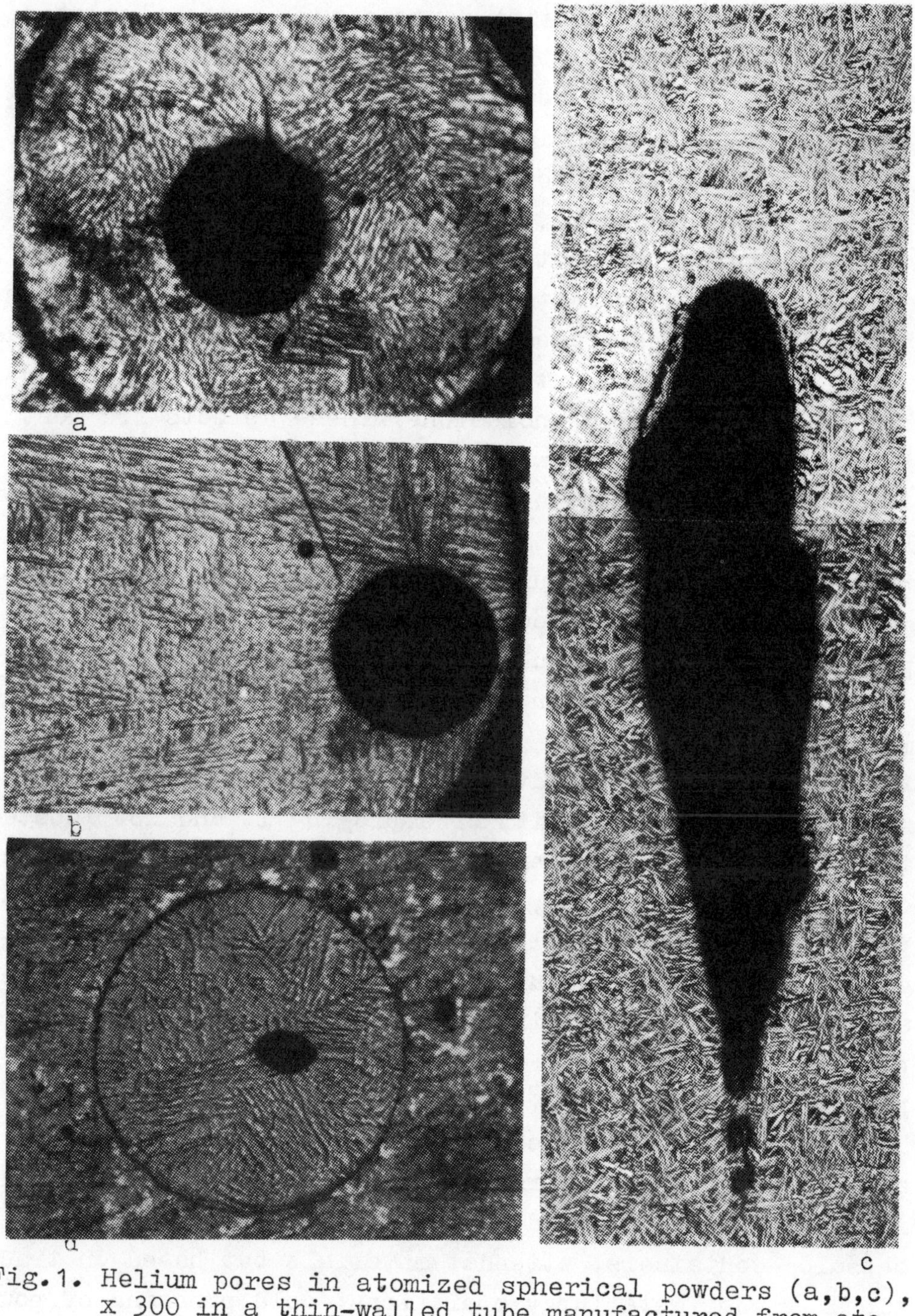

Fig. 1. Helium pores in atomized spherical powders (a,b,c), x 300 in a thin-walled tube manufactured from atomized powders after provocative anneal 1500°C (d) x 100

cially, uses plasma heating sources and, hence, tungsten electrodes. As a rule, the gas-jet atomization requires crucibles. Thus, most commercially realized techniques employ attendant materials: tungsten, copper, ceramics etc. Therefore, it is possible that foreign metal particles should uncontrollably escape to atomized powders and, hence, to future products (fig.2). It is difficult to expect a positive effect on a product of scattered pores filled with an inert gas at a pressure of several hundreds atmospheres (following the HIP process) or foreign uncontrollably distributed inclusions. These defects are most dangerous for products working under alternating loads.

4. Due to the whole complex of causes granule metallurgy cannot ensure the high MUF and lower recycles compared to classical metallurgy. When calculating from a charge especially taking account of the capsule - container production it is difficult to expect a higher MUF for comparable products than in case of classical metallurgical methods.

The indicated disadvantages of granule metallurgy do not cancel the possibility of employing it and hot isostatic pressing to solve many problems in the special fields of engineering. However, in the people's economy for mass production one can hardly expect its wide usage. In principle, "granule" metallurgy rejected the main technological advantages of the "traditional" powder metallurgy: cold pressing, high economy, 100% automation, high MUF, low power expenses etc.

In our opinion, one can expect a wide employment of atomized powder metallurgy, the people's economy included, if new flow sheets are developed not using inert gases, capsule -containers, attendat materials but based on traditional powder metallurgy operations: cold pressing of powders, thermal treatment of cold pressed blanks with simulta-

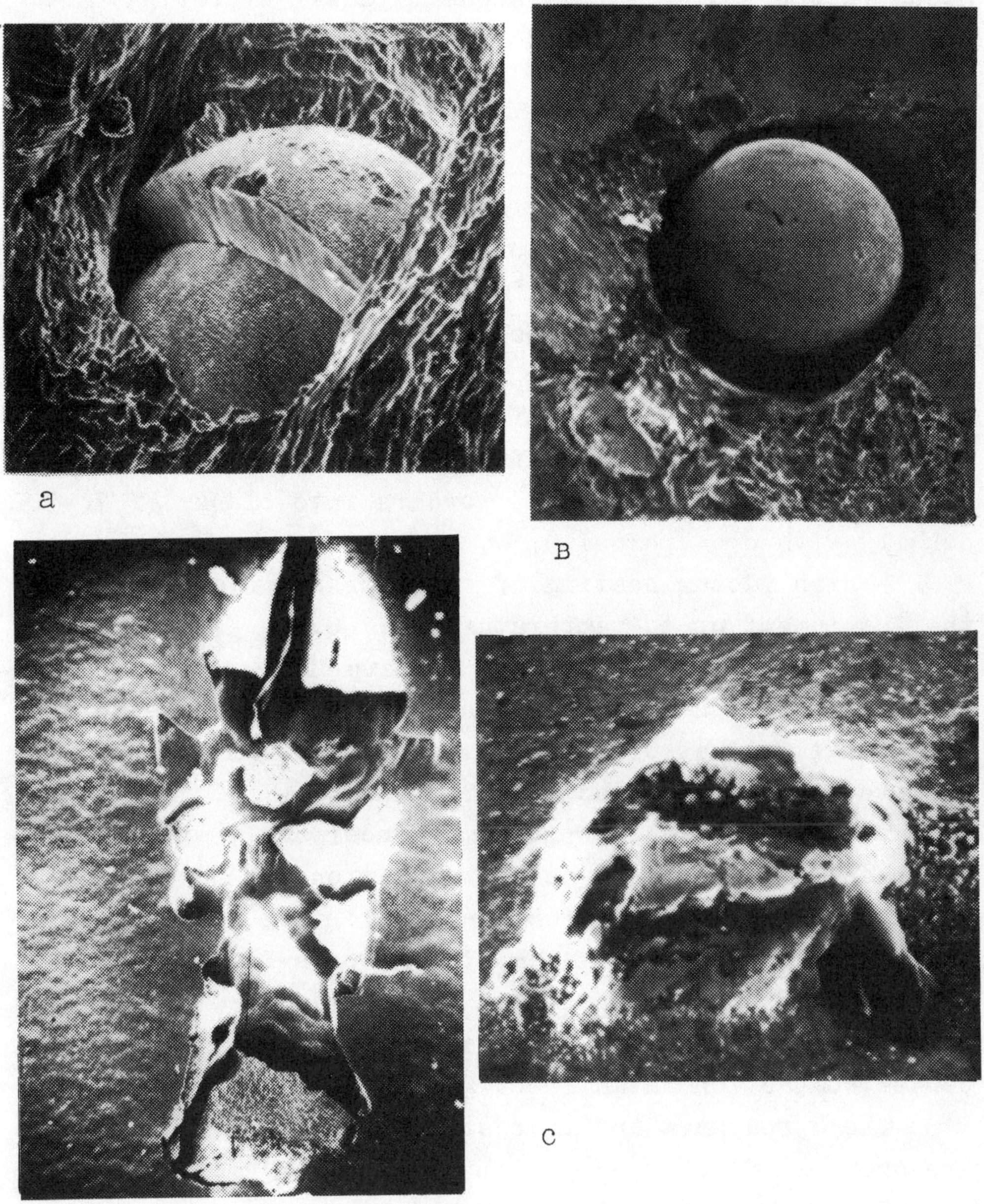

Fig.2. Appearance of inclusions: a) tungsten in a tube fracture, x 250; b) tungsten in a billet fracture, x 50; c) tungsten in a tube, x 250; d) copper at a tube surface, x 70

neous (hot pressing) or subsequent (after sintering) hot plastic working (forging, forming).

At the AUSRIIM a new flow sheet is being developed that meets requirements. It is called a "RIBS-process" (atomization and fast solidification). It incorporates the following principal operations (fig.3):

- high vacuum melting of a raw material with the melt atomization by centrifugal forces during melt particle crystallization on water cooled copper screens; powder particles acquire irregular shapes and uniform fine-grained metastable structure - RIBS powders;
- powder sieving and if necessary other processing;
- cold pressing of RIBS powders into blanks of 70-90% density with open porosity;
- high vacuum heating of cold pressed blanks outside the instrument up to temperatures 0.7-0.8 $T_{melt.}$ followed by vacuum pressing in the warm instrument to the theoretical density and shaping the blank into a form close to that of a final product or by extrusion of a metal to produce billets of the tube, rod or wire types.

To solve the above problems concurrently with the development of methods of melt atomization and powder compaction at the AUSRI of Inorganic Materials with the participation of AUSRI Met Mach, E.O.Paton's IEW and oth. a series of laboratory and pilot facilities were established which realize different schemes of melt atomization and powder compaction in an inert gas in high vacuum (fig.4).

The appearance and microstructure of Zr-1% Nb powders are shown in fig.5.

RIBS powders are liable to blending, are easily batched and transported by gravity etc. The contents of gas impurities -oxygen, nitrogen, hydrogen - of flake-like powders are at a minimum which is fully determined and ensured (in addition to the required composition of the initial

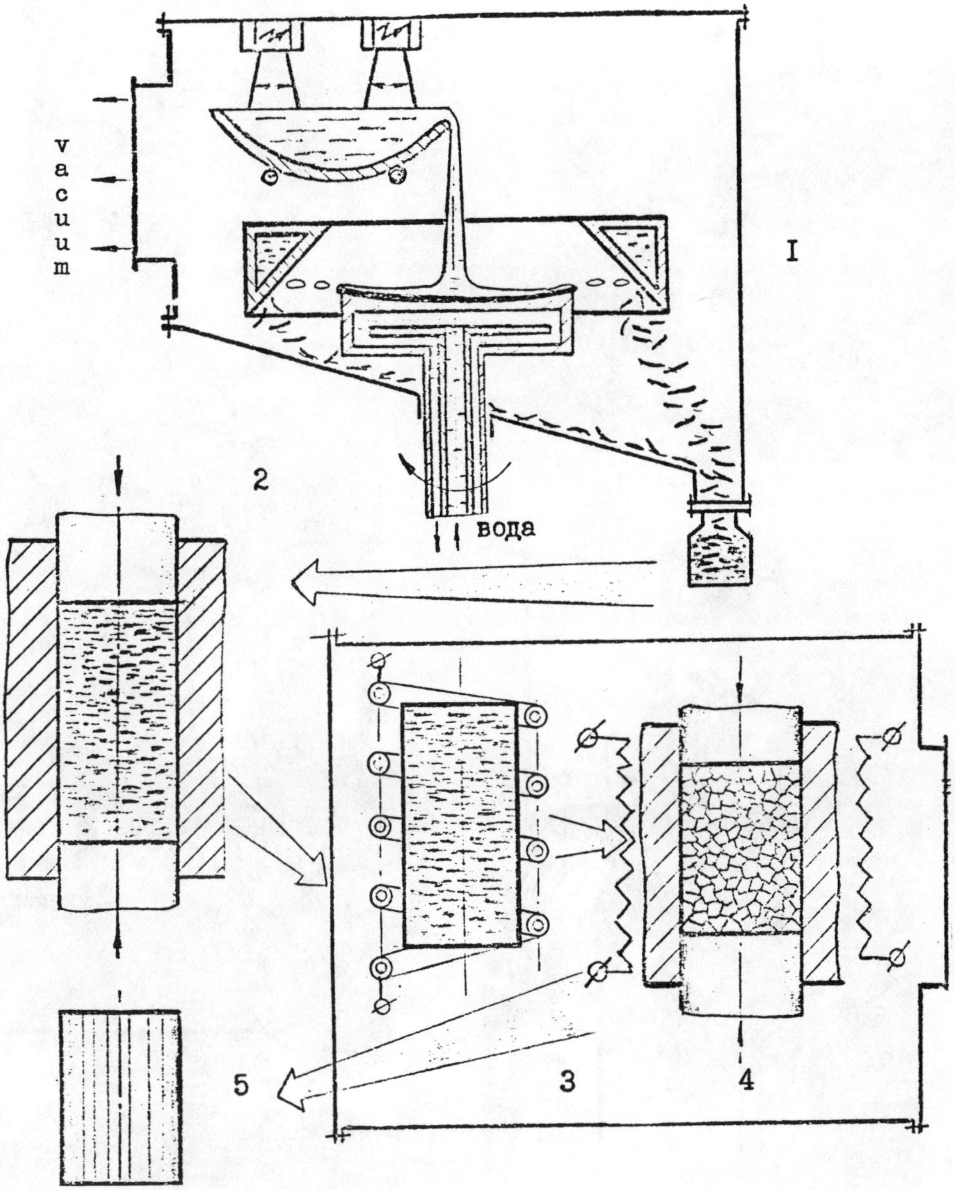

Fig.3. Principal RIBS flow shart:
1 - yacuum melting of metal, preparation of RIBS powders by melt atomization and fast solidification; 2 - cold compaction; 3 - vacuum heating outside the istrument; 4 - hot compaction (forming, extrusion) without lubricant; 5- final product-billet

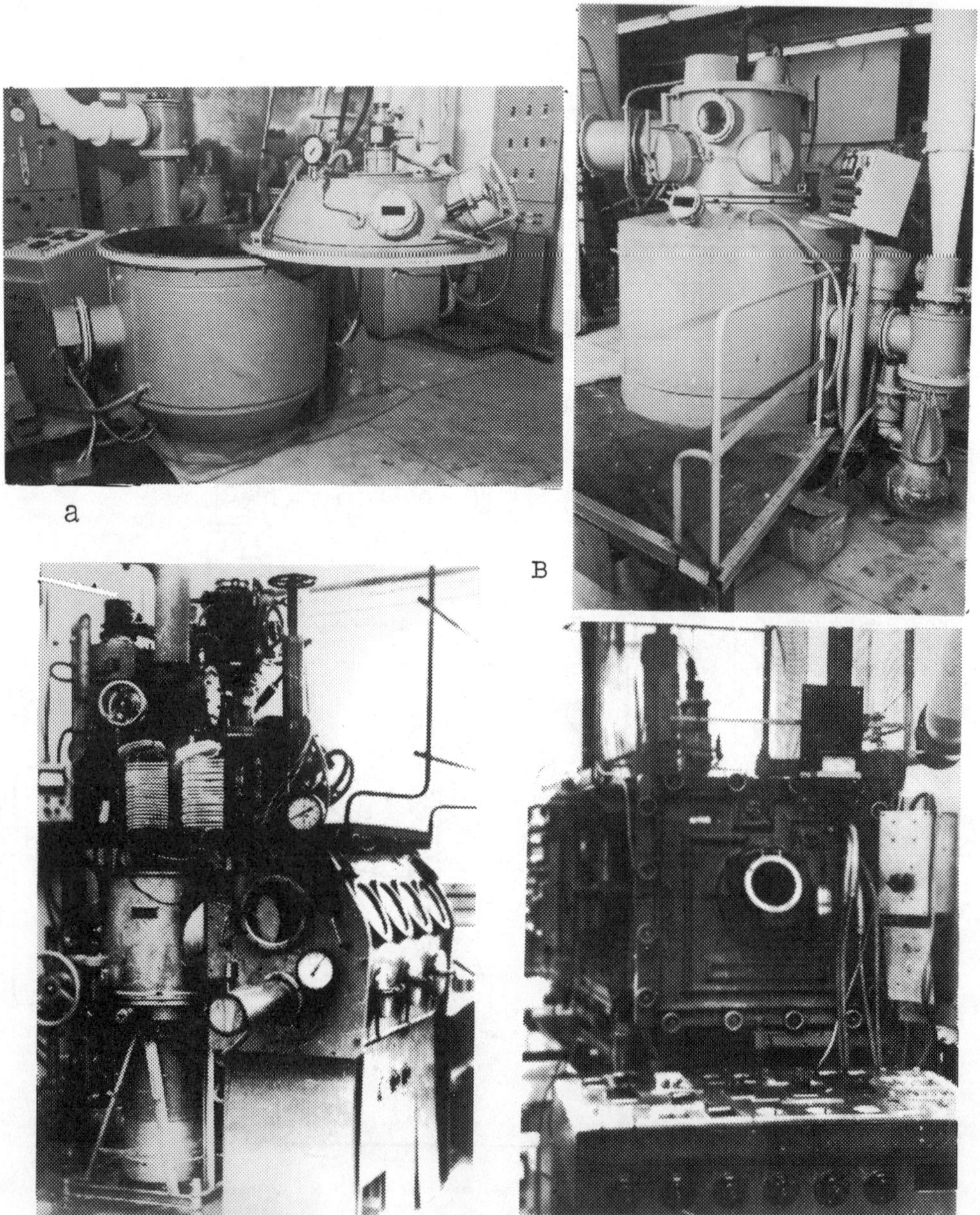

Fig.4. External view of facilities producing powders from different metals and ceramic materials bu inert gas and vacuum atomization of melts (a,b,c,d, e,f,g,h,i,e)as well as hot compaction and extrusion in vacuum

f

e

g

h

i

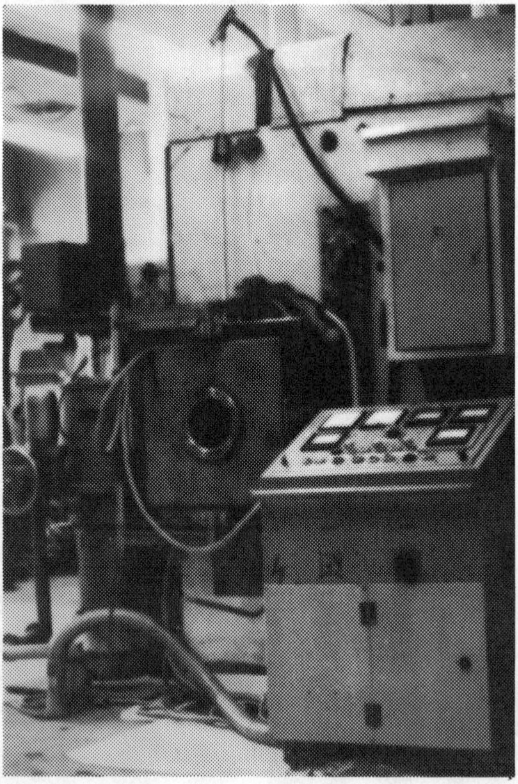

l

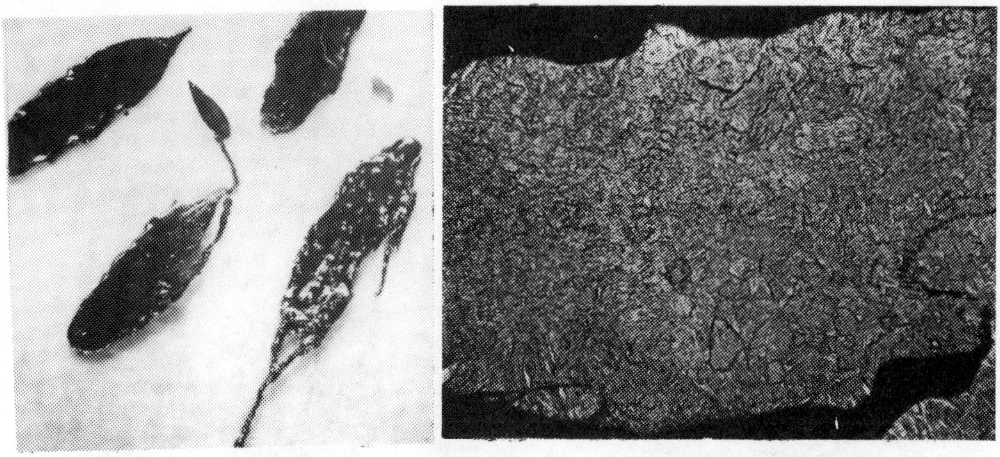

a b

Fig.5. Appearance (a) and microstructure (b) of Zr 1% Nb RIBS powders

charge) by melting and atomizing conditions, viz., high vacuum and electron beam heating.

As compared to spherically shaped powders, RIBS powders have a more stable composition, a more fine-grained microstructure due to a significantly higher cooling rate on crystallization, a high metal density without gas pores as well as provide for substantial technological advantages - the possibility of a marked reduction of sizes and metal content of units producing powder by atomization of melts.

Flake-like RIBS powders are readily cold pressed in a wide range of pressures from 100 to 1000 MPa to provide blanks of a required shape (fig.6).

A superplastic compact material can be produced from RIBS powder. This possibility became feasible as a result of the revealed peculiar features of the structure and properties of atomized powders. Due to the particle crystallization by a spontaneous and simultaneous growth of a large number of nuclei in the whole volume of a flying liquid particle at superhigh cooling rates atomized powders have a structure complying to a considerable extent with

Fig.6. Appearance of billets compacted from RIBS Zr-alloy powders.
a - compaction pressure of 1000 MPa;
b - compaction pressure of 100 MPa

academician A.A.Bochvar's conditions [1,2] which must be met by a metal structure potentially admitting superplastic straining. The mechanism of atomized spherical particle crystallization in a flight with numerous growing crystallites appearing in each particle volume is confirmed by us in the studies of powder surfaces. The surface of a single particle that originated from the impingement with the screen clearly shows crystal nuclei revealed on the impingement but formed within the body of a flying liquid particle prior to the impingement (fig.7). The high degree of the dispersity of a metal particle structure caused by a simultaneous growth of a great number of crystal nuclei and the high degree of microdefects in the crystal structure of particles solidified under the conditions of intensive heat transfer at the cooling rates of 10^4-10^7 °C/s document the possible superplastin straining of the metal of atomized powders as well as materials produced from atomized powders 3,4 .

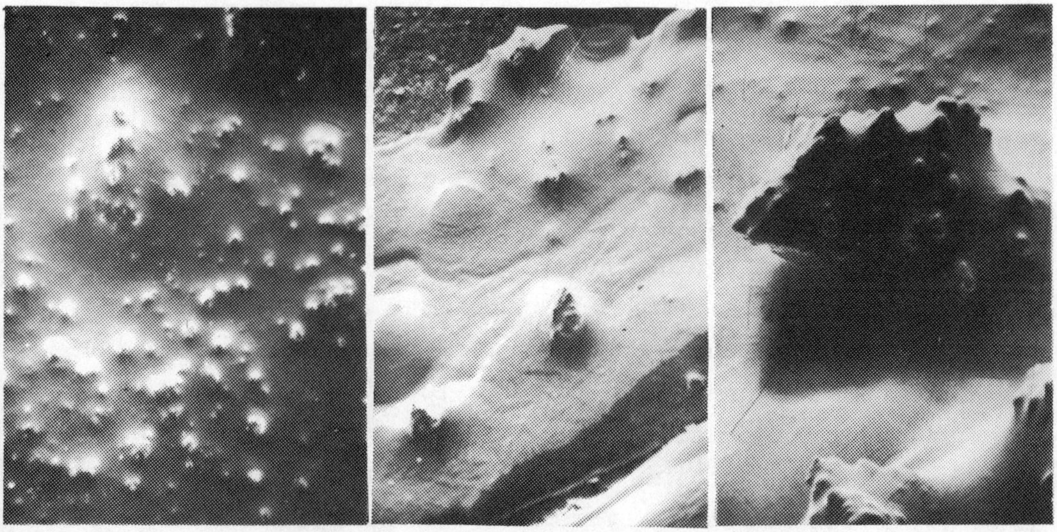

Fig.7. Surface appearance of RIBS powders prepared bu vacuum atomization

Based on the indicated backgrounds the process was developed for hot vacuum compaction of round billets using atomized Zr-1% Nb alloy (fig.8).

Fig.8. Appearance of round billets

The compaction pressure is 10-20 MPa. The density is 99.8%. As to the density, the macrostructure of a billet metal fully conforms to a compact material manufactured according to the traditional scheme: casting, plastic working. When compacting at the low pressure the metal of atomized particles deforms and flows to the die-punch gaps 0.1-0.2 mm wide rising to a height of 25-40 mm.

The possibility of producing a compact metal through superplastic deformation of atomized powders is confirmed using a brittle material such as white cast iron (1.5% Cr, 0.7% Si, 0.5% Mn, 3.5% C) hot compacted in vacuum (fig.9). The voids between the spherical particles were filled due to the intensive travel of the particle surface layers accompanied by a drastic change of their structure. However, the structure of the internal volumes of atomized cast iron particles practically remained unchanged.

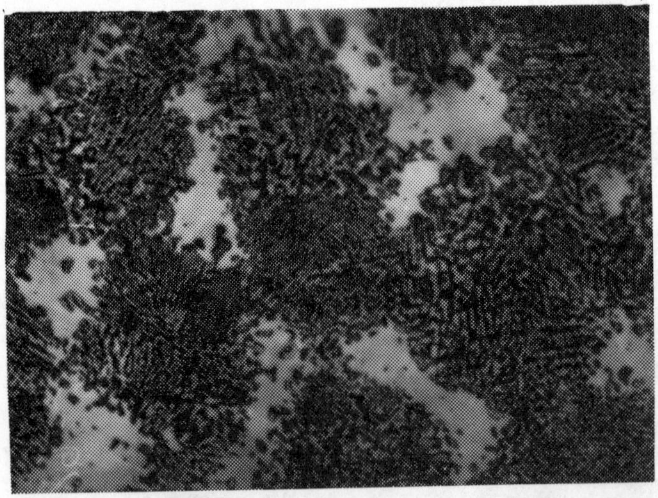

Fig.9. Microstructure of cast iron produced by hot compaction of atomized spherical powders , x

The indicated features of the atomized particle structure show up both in compact brittle cast iron having

3.5% C and in low plasticity eutectoid steel if they are produced from atomized powders by hot compaction (the work on superplasticity was carried on at AUSRIIM under the scientific direction and with the participation of academician A.A.Bochvar). As the experiments show these materials are capable of superplastic straining and at 700°C and the strain rate of 0.25-0.20%/min^{-1} the elongation of eutectoid steel and cast iron specimens is 550% and 150% [2], respectively, fig.10.

Commercial high vacuum hot compaction of blanks prepared from atomized powders is actually impossible in mass production until the problem of the instrument lubrication is solved. We have obtained the first favourable results in this field. Fig.11 shows round billets which were compacted using atomized powders without the instrument lubrication. It is supposed that due to a decreased friction at the instrument surfaces protected with coats the powder compaction was accompanied by considerably less metal flowing into the gaps than in case of lubricant application. The instrument disassembly after the end of compaction did not reveal any traces of setting. The blank surface is light grey with metallic lustre and reproduces sharply the relief of the instrument surface.

The possibility of powder extrusion in vacuum without a lubricant according to a waste -free process is confirmed by producing zirconium alloy bars (fig.12).

While developing processes for the production of tubes and items from atomized zirconium alloy and stainless steel powders numerous investigations confirmed the high properties of the tube metal that were often higher than those of traditionally worked metals. The high properties of a metal produced from an atomized powder are also confirmed by its microstructure, density, corrosion resistance, microhardness and chemical composition, including

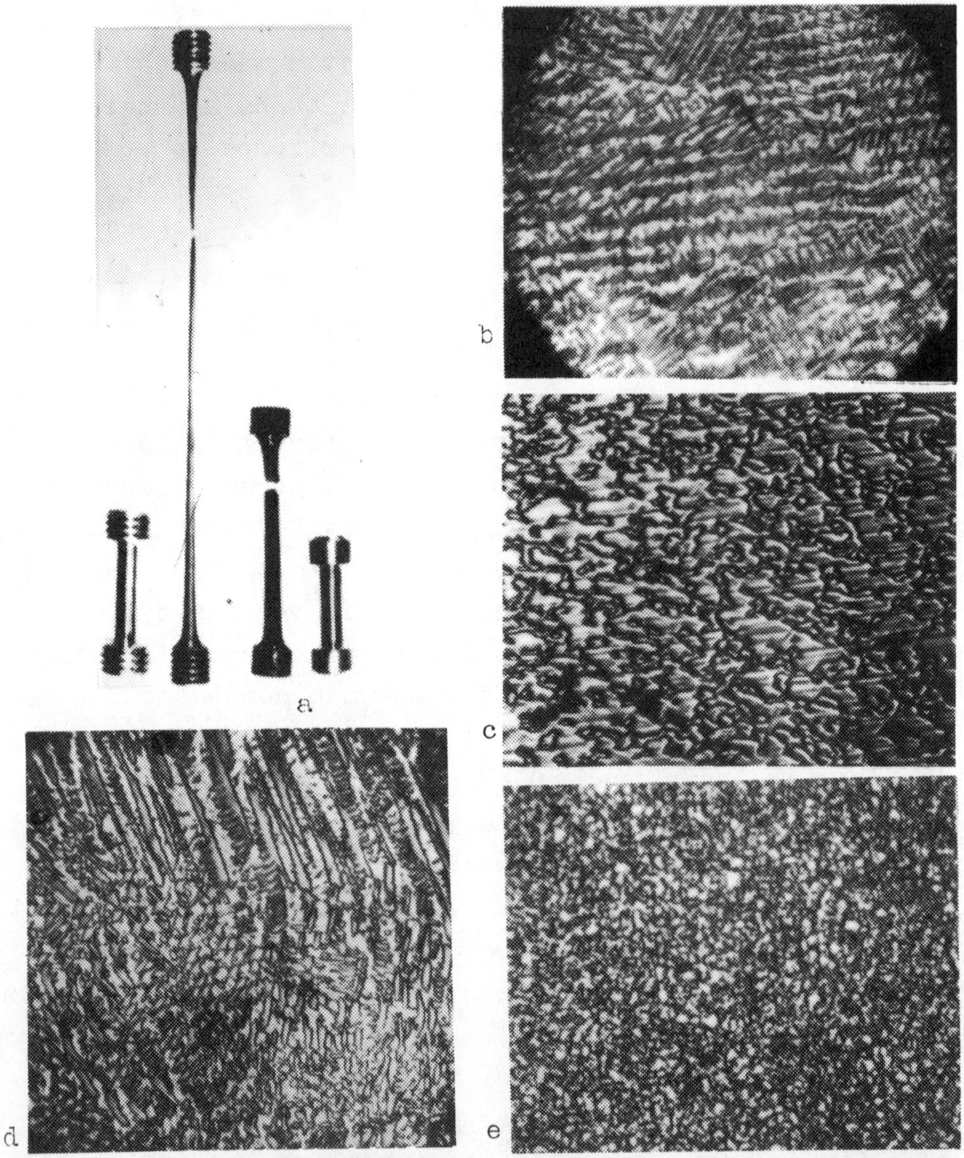

Fig. 10. Appearance and microstructure of eutectoid steel (0.8% C) and cast iron (3.5% C) before and after superplastic tension tests: eutectoid steel (δ = 500%, T=700°C, $\dot{\varepsilon}$ =0.1 mm/min) a) specimen appearance; b) atomized powder microstructure, x 1000; c) compact metal, x 1000; cast iron (δ =150%, T=700°C, $\dot{\varepsilon}$ =0.1 mm/min); d) specimen appearance; e) atomized powder microstructure, x 1000; f) compact metal, x 1000

x 0,5

Fig.11. Appearance of round billet hot compacted in vacuum without instrument lubrication (the temperature of the instrument and powders is 900°C)

gaseous impurity content: oxygen, nitrogen, hydrogen.

The developed pilot method for producing atomized zirconium alloy and stainless steel powders and the processes of their compaction and working made it possible to manufacture tubes that fully comply with the current specifications. The microstructures of atomized powder manufactured tubes are illustrated in fig.13.

Atomized powders open up interesting possibilities of producing composites having unusual properties. We have checked this possibility in experiments with powders of two different zirconium alloys. Mixtures of such atomized powders taken in different ratios were hot compacted in vacuum into round billets from which thin-walled tubes were manufactured. Fig.14 shows the metal microstructure

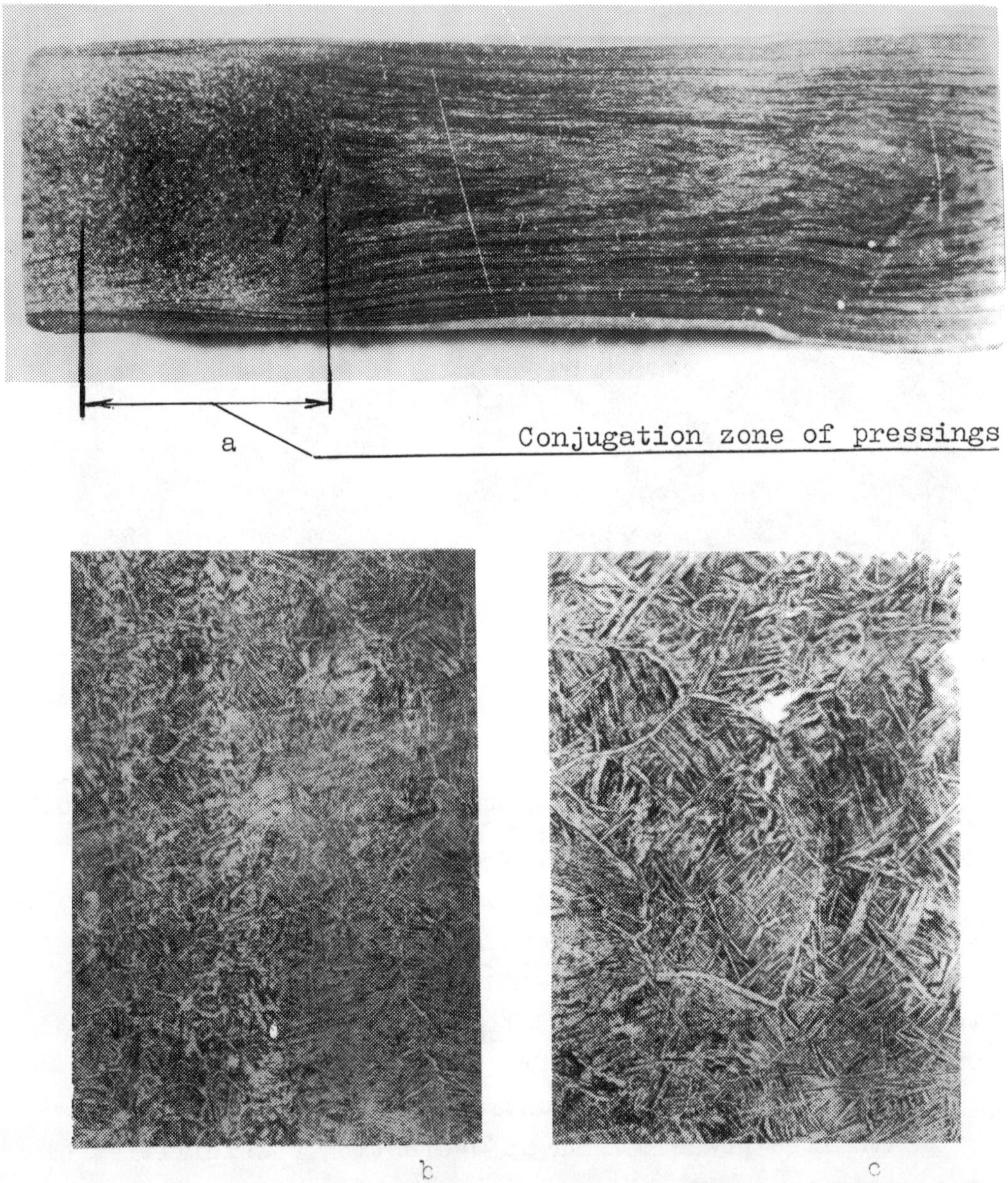

Fig. 12. Micro- and macrostructure of zirconium alloy bars extruded from RIBS powders in vacuum without instrument lubrication; waste free flow chart.
a - macrostructure of conjugation zone of two pressings, x 3;
b - the same, microstructure, x 300;
c - microstructure of bar metal

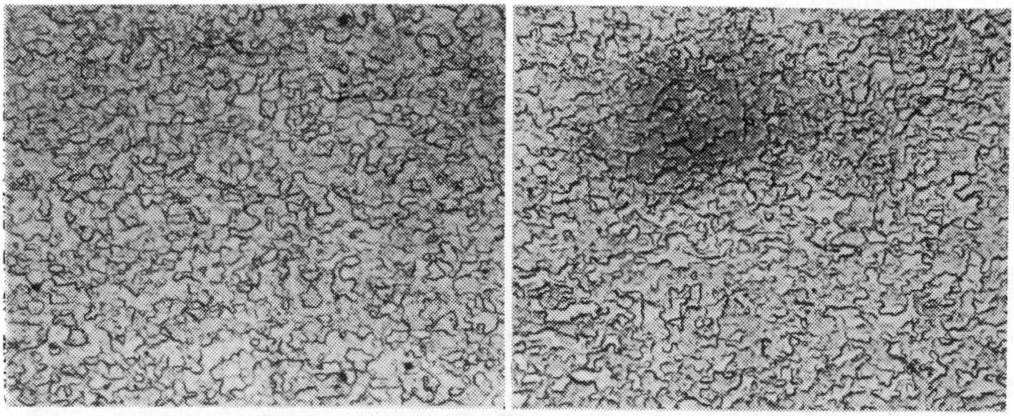

Fig.13. Microstructure of tubes produced from RIBS powders (a) and a commercial one (b)

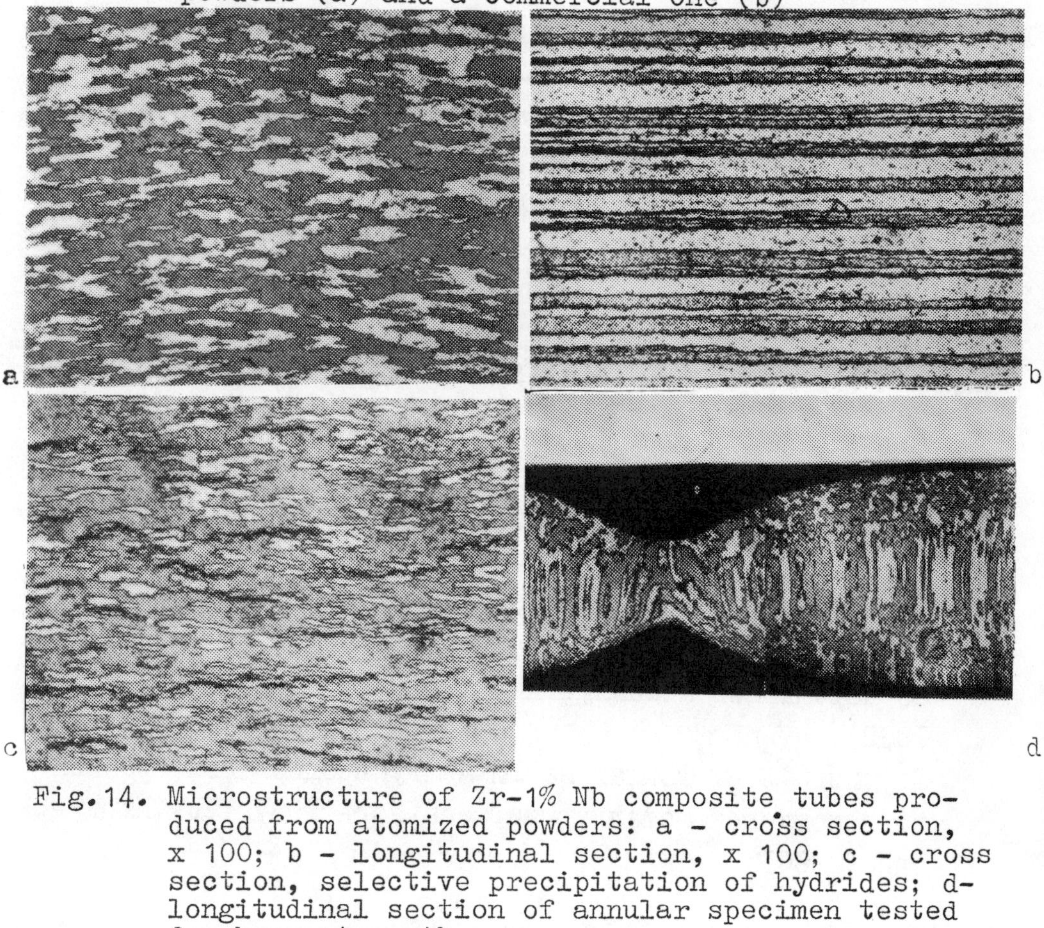

Fig.14. Microstructure of Zr-1% Nb composite tubes produced from atomized powders: a - cross section, x 100; b - longitudinal section, x 100; c - cross section, selective precipitation of hydrides; d-longitudinal section of annular specimen tested for hoop strength

in the longitudinal and cross sections of these composite tubes. In addition to the confirmation of the additive regularity in mechanical property variations at different ratios of the two compositions of those atomized powders, adequate corrosion resistance and weldability it was established that they have an interesting ability of selective hydride precipitation. On hydrogen saturation in homogeneous alloys hydrides precipitate uniformly throughout the tube body while in a composite metal they precipitated in a strictly selective and oriented mode and only in the metal of one of the components.

The RIBS flow chart based on the atomized powder metallurgy is a third stage of powder metallurgy development (after "classical" and "granule"). It is being developed commercially to produce zirconium alloy and stainless steel tubes, end pieces etc. If it is a success, this flow chart can also sharply broaden the scopes of atomized powder metallurgy application to mass production in people's economy.

References

I. Бочвар А.А. Физика и химия обработки материалов, I982, № 6, с.85-94.
2. Бочвар А.А., Давыдов В.А., Дружинин Л.К. ДАН АН СССР, I976, т 230, № 2, с. 3I8-32I.
3. Druzhinin L.K.,Safronov B.W.-Metal Powder Report,1983, № 8, p 447-449 (Доклад на международной конференции по порошковой металлургии, ПМ-82, Италия, Флоренция, I982).
4. Дружинин Л.К. Доклад на УП международной конференции по порошковой металлургии, сентябрь I987г.,г. Пардубице, ЧССР.

METAL POWDER PRODUCTION BY ULTRASONIC CAPILLARY OR STANDING WAVE TECHNIQUE

M. Hohmann and S. Jönsson
Leybold AG, Hanau, F.R.G.

E. Lierke
Battelle Institut, Frankfurt, F.R.G.

Abstract

The ecomonic potential of the P/M-technology for parts production can only be utilized if the required metal powder is available at a reasonable price. For specific applications, ultrasonic atomization has to be regarded as an alternative powder production technique with a considerable potential for cost reduction.

There are two possibile ultrasonic techniques for the atomization of liquid metal: the capillary and the standing wave technique. The ultrasonic capillary wave atomization is based on the effect that a thin layer of liquid metal covers the surface of an active solid resonator. In the ultrasonic standing wave system, the atomization as such takes place in the velocity knot of an ultrasonic standing wave between a 20 kHz transmitter and a reflector.

The first test results prove that ultrasonic atomization can be successfully applied to metal alloy powder production with the advantage of low production costs.

Introduction

The leading process for the production of high-grade metal alloy powder with spherical particle shape is inert gas atomization. The dimension of the atomization chambers must be designed big enough to ensure that the fast flying droplets are solidified before they touch the chamber wall. Therefore, the characteristic design of such equipment is dominated by huge atomization towers of up to 8 m in height and 3 m in diameter. These big units are very expensive and difficult to process. Another cost factor is the high gas consumption of about 30 Nm^3/min during the atomization of the melt.

This paper indicates that for specific applications ultrasonic atomization has to be regarded as an alternative powder production technique with considerable potential in cost reduction because this type of process does not require large atomization chambers and has hardly any gas consumption during the atomization process. Two possibilities of ultrasonic techniques for the atomization of liquid metal will be described: the capillary and the standing wave technique.

Capillary Wave Atomization

Process Principles

A thin layer of liquid metal evenly covering the surface of a solid resonator which vibrates vertically to its surface plane at a frequency f, which forms a chessboard-like pattern of stationary capillary waves as soon as the elongation-amplitude exceeds a certain limiting value. The lenghth of these capillary-waves is given by

$$\lambda_K = 2\sqrt[3]{\frac{\sigma}{\rho} \cdot \frac{\pi}{f^2}}$$

with σ = surface tension
ρ = density of the liquid

The mechanism of the capillary-wave atomization process is schematically shown in Fig. 1.

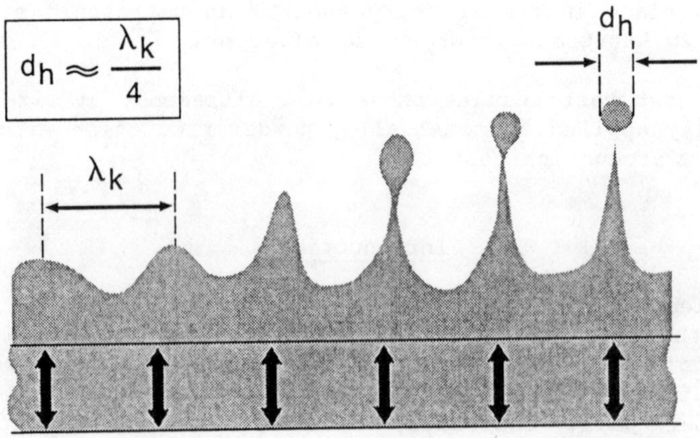

Figure 1: Priciple of capillary-wave atomization

On further increase of the velocity amplitude small droplets are expelled from the elevated cones of the capillary-wave pattern as soon as the characteristic value V_N is exceeded.

$$V_N \approx 30 \cdot \frac{\eta}{\rho} \cdot \sqrt[3]{\frac{\rho}{\sigma} \cdot \frac{f^2}{\pi}} = \frac{60}{\lambda_K} \cdot \frac{\eta}{\rho}$$

with η = viscosity of the liquid

The most frequent diameter of the droplets which form spherical metal powder particles approximately corresponds to one fourth of the capillary-wave-length Fig. 1 and it decreases with increasing frequency /1/. This is shown in Fig. 2 for water as model liquid.

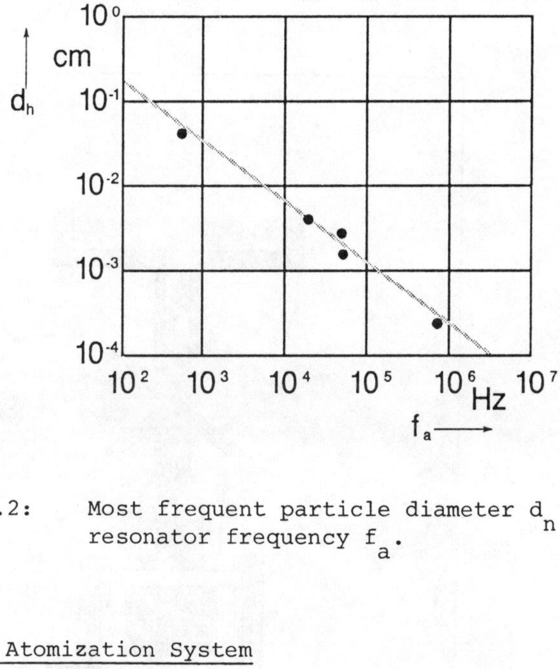

Fig.2: Most frequent particle diameter d_n versus resonator frequency f_a.

Design of the Atomization System

The design of the ultrasonic capillary wave system is shown in Fig. 3. The liquid melt is lead through a nozzle system with defined outlet diameter and covers the surface of the resonator with a thin film. When the correct amplitude of oscillation is reached, small droplets begin to fly off from the antinodes, solidifying to spherical metal powder particles as they cool down. The generator is most conveniently of phase-locked loop-design /2/. The half wavelength compound transducer consists of two piezo ceramic discs with common electrodes. To this base, the stepped horn amplitude transformer is attached on top to which the bending resonator is mounted. Compound transducers can be build for frequencies of up to 150 kHz whereby an area-specific throughput of up to 4 l/h x cm^2 can be achieved.

The total atomization capacity is largely depending on the shape of the bending resonator. The best results have been attained with a hollow cone driven by a cylindrical stem and vibrating in bending resonance. This looks like an umbrella on top of the vibrator to the center of which the stream of molten metal from the furnace is directed. The radial node-lines on the upper surface of the resonator represent the "umbrella rips" in the molten metal. Of course during atomization the bending resonator must be heated to the melting temperature of the metal to be atomized in order to avoid solidification of the melt on top of the resonator.

Because of the high specific efficiency, the consumption of energy in this atomization process is extremely low. The fact that the micro droplets leave the surface of the resonator at a relatively low speed allows the use of a compact atomization chamber.

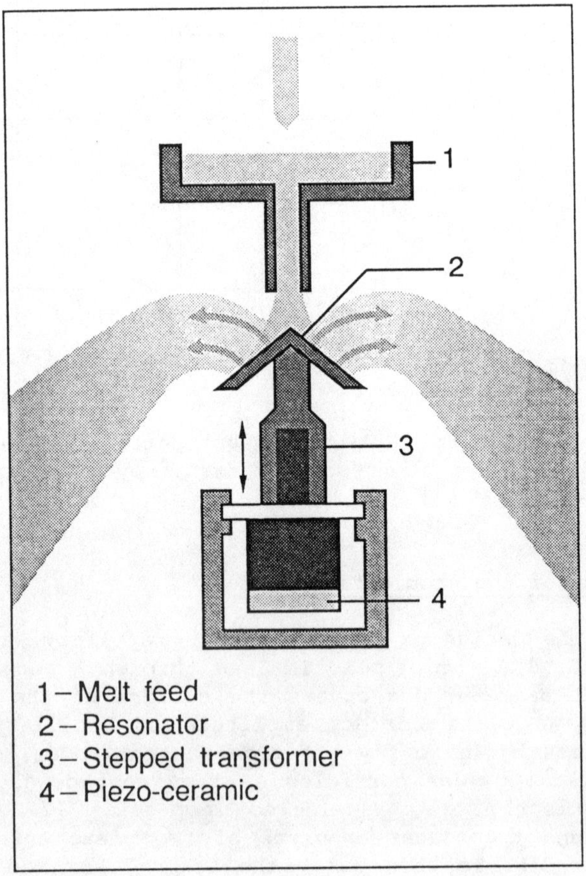

Fig. 3: Schematic design of the ultrasonic capillary wave atomization

Results of Atomization Tests

The atomization tests were done with a AgSnCu-alloy. With respect to the particle size distribution, a considerable decrease can be obtained by ultrasonic atomization, as shown in Fig. 4.

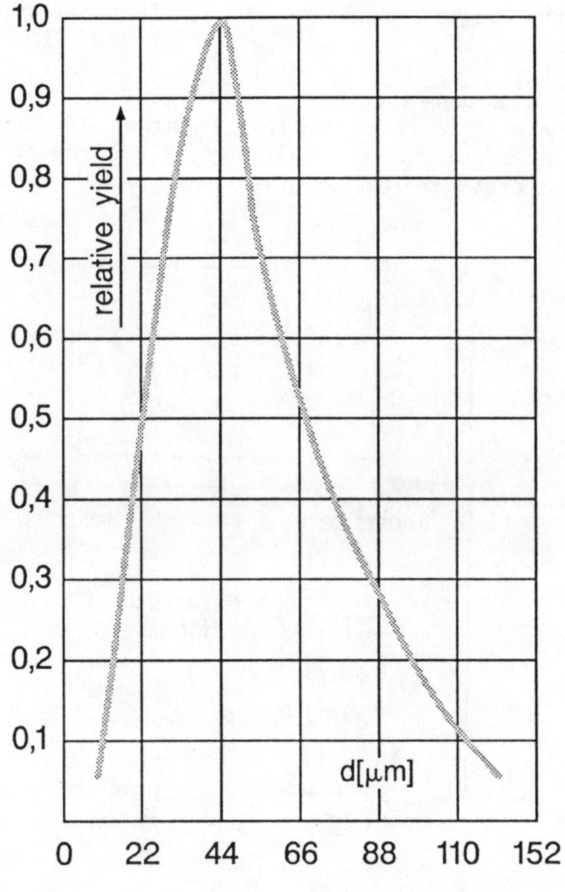

Figure 4: Particle size distribution for capillary-wave atomization (AgSnCu-alloy at 20 kHz)

The Standing Wave Atomization

Process Principles

The standing wave atomization takes place in the velocity antinode of the standing wave whose forces at the boundary of a liquid act in a similar way as a gas jet except that no gas is consumed. As shown in Fig. 5, Bernoulli forces form a cone with a central diaphragm of the liquid flow near the atomization axis. The shear forces of the dynamic gas oscillation disrupts the diaphragm. The viscosity of the liquid is a secondary factor while the surface tension is dominant.

Water requires a sound intensity of approx. 170 dB, while molten metal with very high surface tension requires 180 dB. Because Bernoulli forces depend on the gas velocity as well as on the gas density, an increased pressure can improve the atomization effect /3/.

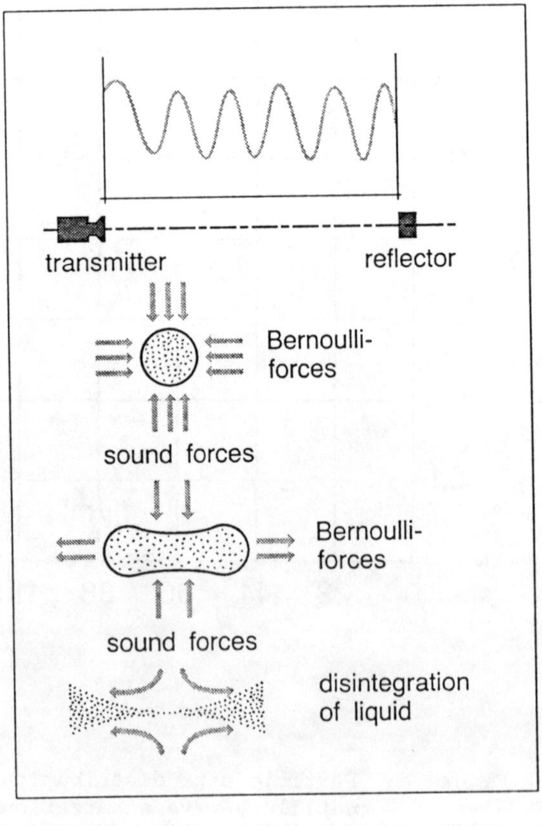

Fig.: 5 Principle of the standing wave atomization

The Design of the Atomization System

This technique is based on static and dynamic forces and gas currents in an intensive standing ultrasonic wave between a sound-radiating transducer and a coaxially aligned reflector at resonance distance (n /2). At a typical frequency of 20 kHz, several nodes and antinodes of the standing wave are observed. They are each spaced at /2 distance. The pressure amplitude and the velocity amplitude vary between minimal values (nodes) and maximal values (antinodes). The design of a standing wave system is given in Fig. 6. Liquid metal is fed through a nozzle into the pressure nodes of a horizontally alingned standing wave. The shear force of the dynamic gas oscillation is responsible for the atomization.

The sound intensity in the standing wave is responsible for the droplet size. The droplet size of the atomized melt decreases with increased gas pressure in the knot because of the increased sound intensity. The flight distance of the molten droplets is drastically reduced when the gas pressure in the chamber is increased. Since the fluid flight distance is short, the dimension of the atomization chamber can be minimized.

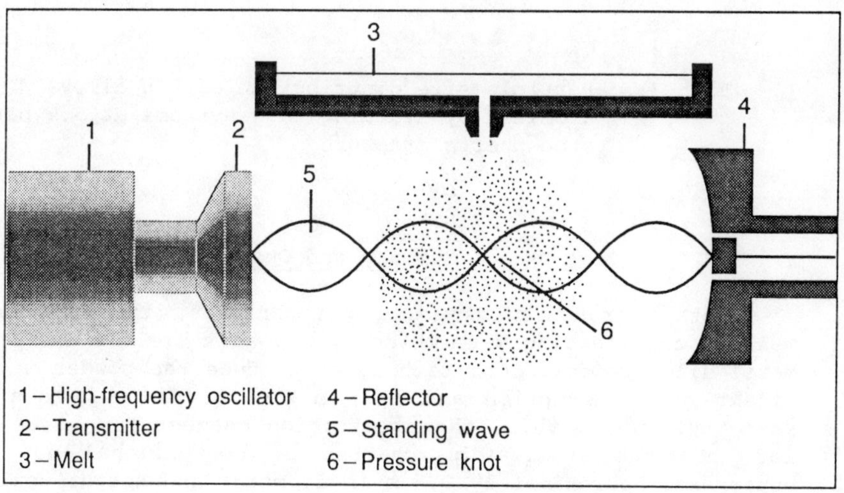

Fig. 6: Schematic design of the ultrasonic standing wave atomization

Results of Atomization Tests

Several tests with tin-alloys were performed in a laboratory atomization system. These atomization trials show clearly that the particle size distribution depends on the pressure in the atomization chamber and the amplitude of the ultrasonic standing wave. Higher pressure and amplitude result in smaller size droplets. Figure 7 shows the particle size distribution of an atomized tin alloy with the standing wave system.

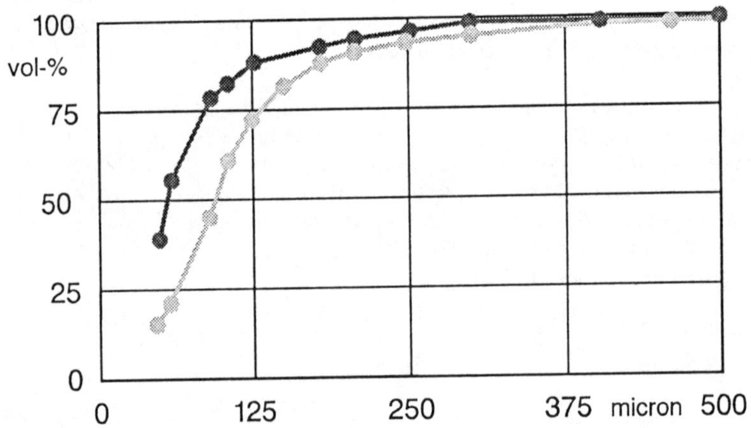

Fig. 7: Particle size distribution of tin alloy atomized by standing wave systems at 5.5 bar atmospheric pressure

Summary and Outlook

Both ultrasonic atomization techniques, the capillary wave as well as the standing wave methods, are suitable for the production of metal alloy powder. Both techniques produce the powder without significant gas consumption and the emitted droplets have a low flight velocity. Due to this, the atomization chamber can be designed smaller than the atomization chamber of conventional gas atomization equipment. Approximately 1,5 m in diameter and height would be sufficient. Therefore the investment costs as well as the actual production costs would be lower for the ultrasonic atomization. On the other side, both techniques have some typical disadvantages which to date prevent their application on a large commercial production scale.

With respect to the capillary wave atomization technique, it has been mentioned that the bending resonator has to be heated to the melting temperature of the metal alloy in order to avoid solidification of the melt on the surface of the resonator. Additionally, it has to be ensured that the resonator surface will be well wettet by the melt but not disolved. Therefore, in any special case it must be checked if the resonator material is consistent with the metal alloy melt. This problem increases with higher melting temperatures.

The ultrasonic standing wave technique allows a contact free atomization of the melt. In order to have a good efficiency of the melt disintegration, the melt stream must be accurately fed into the standing wave velocity antinode. This needs a melt stream guiding system which has to be carefully designed to prevent freezing of the melt in the outlet nozzle system. Furthermore, the spray cone of the produced droplets causes accumulation of material at the melt nozzle and the bottom section of the furnace. These problems have to be solved by further development of the process technique and design. Nevertheless, the first results prove that ultrasonic atomization can be successfully applied to metal alloy powder production with the advantage of low production costs.

References

/1/ Boguslavskii, Y., Physical Mechanism of the Acoustic Atomization of Liquid, Soviet Physics Acoustic Vol. 15, No. 1, 1968, pg. 14 - 21

/2/ Ruthart R., Lierke E., A new Ultrasonic Atomization Technique for the Production of Metal Powder. Modern Developments In Powder Metallurgy Vol. 12 - 14, ed., H. H. Hausner, H.W. Antes, G.D. Smith, Metal Powder Industries Federation Princeton, 1984, pp. 105 - 111

/3/ Lierke E., Heide M., Großbach R., Berührungslose Flüssigkeitszerstäubung im intensiven Ultraschallfeld. Deutsche Arbeitsgemeinschaft für Akustik 1980, München, VDE-Verlags GmbH, Berlin

THE TANDEM ATOMIZATION PROCESS FOR ULTRAFINE METAL POWDER PRODUCTION

C. P. Ashdown, J.G. Bewley, and
G. B. Kenney

UFP Technology
P.O. Box 444
Ayer, MA 01432

ABSTRACT

A number of interesting applications require metal powders less than 20 um in size. Such powders cannot be produced economically by conventional gas atomization. The Tandem Atomization Process (TAP) has been developed and patented (U.S. Patent #4,626,278) by UltraFine Powder Technology, Inc. to address this identified need. TAP dissolves gas in a metal melt, and then gas atomizes the melt in a conventional manner. Substantial quantities of micron sized material are produced by the process. A model for TAP is proposed, with theoretical calculations of the expected minimum particle sizes for various gas-metal combinations. The effect of varying gas overpressures on TAP will be discussed.

INTRODUCTION

The term ultrafine powders usually is taken to mean powders less than 20 microns in diameter. Of course atomized metal powder producers have been making very fine or ultrafine metal powders for many years. However, the yield of minus 20 micron powder typically obtained from conventional gas atomization ranges from a high of about 15 percent to as low as 2 percent. With these low yields, the cost of producing ultrafine powders by classifying out the coarse fraction is reflected in a relatively high price. The Tandem Atomization Process (TAP) (1) which will be described below potentially offers an alternative method of producing gas atomized ultrafine powders at substantially higher yields and lower costs than possible with existing processes.

Ultrafine powders have a number of physical properties which make them attractive candidate materials for some extremely interesting applications. First of all, ultrafine powders have, by definition, an enormous surface area. Since surface area can be considered a measure of the energy state of the material, fine powders are in a high energy state, and therefore tend to react more readily in most thermal processes. The same high energy state which leads to rapid surface oxidation of fine metal powders, serves to promote accelerated solid state diffusion and sintering of compacted bodies of these powders to high densities.

Ultrafine powders which are made by atomization from a liquid metal bath undergo extremely rapid cooling at rates on the order of one million degrees C per second or more. Such rapid cooling from the molten state largely avoids the chemical segregation which can occur under slower cooling rates, and thereby can produce promising homogeneous, single phase structures of alloy compositions which normally exist only as multiphase structures.

Rapid cooling of ultrafine metal powders during atomization can also produce structures which exhibit unique electrical and magnetic properties. Work carried out at Allied Chemical (2) and other laboratories has shown that certain alloy compositions can be cooled by melt spinning from the molten state at a rate sufficiently fast to produce a non-crystalline or amorphous structure. Atomized ultrafine metal powders are expected to show the same type of structure and because of their near spherical shape and higher packing density, should readily lend themselves to consolidation into useful shapes. Other alloy compositions which do not form amorphous structures upon rapid cooling may be induced to nevertheless form unique and in some cases very desirable structures. For example, we anticipate that the combination of extremely rapid cooling and ultrafine particle size may produce particles made up of single crystals and perhaps of single magnetic domains.

In addition to new potential applications suggested by these unique properties of fine metal powders, there are a number of existing applications which depend primarily on the fine size and near spherical shape of ultrafine powders. The most outstanding application is powders for metal injection molding. The high yield which we expect to obtain of minus 15 micron alloy powders, combined with the nearly spherical shape, high flow rate, high packing density, and high purity characteristics suggest that powders produced by the Tandem Atomization Process (TAP) will offer attractive feed stock for the metal injection molding industry.

Other existing or potential applications for ultrafine metal powders include: spray powders for producing oxidation and corrosion resistant surface coatings, superalloy powders free of harmful non-metallic inclusions, conductor paths for hybrid circuitry and multilayer capacitors, and controlled structure powders for permanent magnets.

Development of the TAP process grew out of some work with another patented process (3), the analytical probe. The analytical probe is a device (see Figure 1) designed to generate a representative sample of fine powder from a molten metal bath. This sample is then delivered pneumatically to an analytical instrument where it is chemically analyzed. The probe operates as a small atomizing chamber containing an integral refractory atomizing die which is submerged beneath the surface of a molten bath. Liquid metal is introduced into the chamber via an orifice connecting to the bath. Pressurized argon is delivered through an annular gas passage encircling the orifice so that the molten metal stream is atomized into fine droplets as it exits the die and enters the atomizing chamber. Solidified powder particles are carried by the exiting argon atomizing gas through pneumatic lines, separators, and classifiers to the analytical instrument. This process provides for on-line, realtime analysis of molten metal baths and has many potential applications in the steel industry as well as in the nonferrous metal industries.

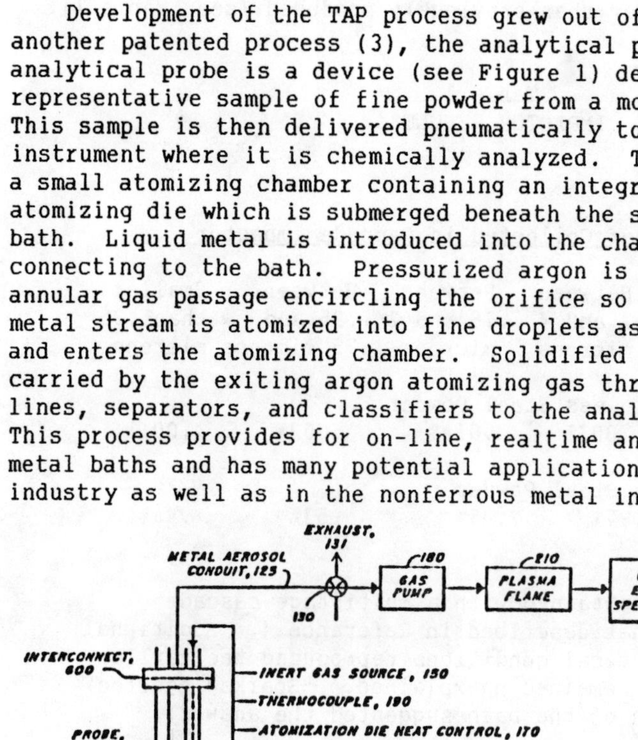

Figure 1 Analytical Probe

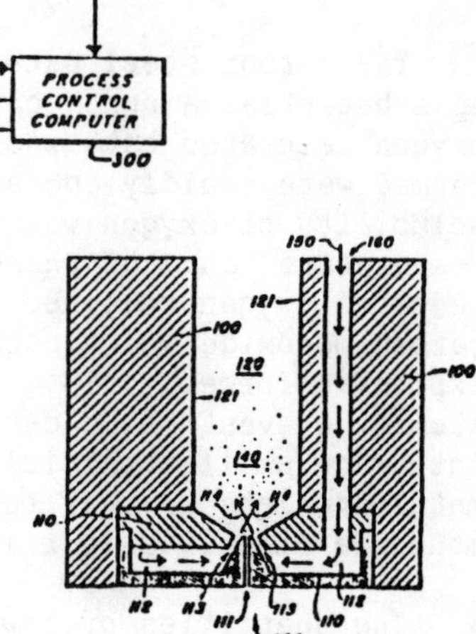

DEVELOPMENT TRIALS AND RESULTS

In early trials of the analytical probe device, a molten bath of carbon steel was held in the molten state for an extended period of time while adjustments were made to the probe apparatus. When a small portion of this molten bath was later atomized into powder by the analytical probe, the particle size of the powder was much finer than previously produced (see Table 1).

Table 1
IMPACTOR RESULTS

Powder Retained In Cyclone	Powder Collected in Cascade Impactor				
Larger than 10 microns	Between 4 and 10 microns	Between 2 and 4 microns	Between .75 and 2 microns	Between .35 and .75 microns	Smaller than .35 microns
Argon Atomized Stainless Steel Powder					
>99%	.977%	.007%	.014%	.002%	<.001%
TAP Atomized Carbon Steel Powder					
<1%	92.52%	2.97%	2.49%	.53%	.48%

These results were obtained with a multistage cascade impacter, similar to that described in Reference 4. Additional tests under nearly identical conditions reproduced the earlier results, but the cause remained unexplained. "Sparks" ejected from a pin sample taken of the bath suggested the answer.

The carbon steel bath, while held in the molten state, picked up substantial amounts of oxygen from the atmosphere. When this oxygen saturated bath was atomized in the probe, the droplets so formed were rapidly cooled from the molten state and the solubility of oxygen was dramatically reduced by both the falling temperature and the phase transition from liquid to solid. The rejected oxygen combined with the carbon in the steel to form carbon monoxide which escaped from the solidifying particles with explosive force. This gas evolution produced a secondary atomizing event which caused the original droplets to disintegrate into many smaller particles. Similarily, the pins effervesced metal droplets as they solidified, due to the evolution of carbon monoxide (the classic rimming action) (5).

The quantities of powder produced were very small, approximately milligrams per minute, but the principle was firmly established. The Tandem Atomization Process (TAP), so called because of the two distinct atomization events, had first been observed and sufficient data was generated to support a patent application covering the process.

As discovered in the analytical probe experiments and described in Patent No. 4,626,278, the TAP process works as follows (see the schematic of the process in Figure 2). A soluble gas is introduced into a molten metal bath by maintaining a pressurized atmosphere of the gas above the bath and optionally bubbling additional gas directly into the bath. The saturated bath is then atomized into droplets by fairly conventional inert gas atomization. As these droplets begin to cool gas is rejected from the solidifying metal dendrites (gas solubility in the solid state is far less than in the liquid state - Figure 3). The remaining liquid metal becomes more and more supersaturated until a gas bubble nucleates. This bubble then grows explosively to trigger a secondary atomization event. The partially solidified droplets shatter into many much finer droplets which subsquently solidify into ultrafine powder particles.

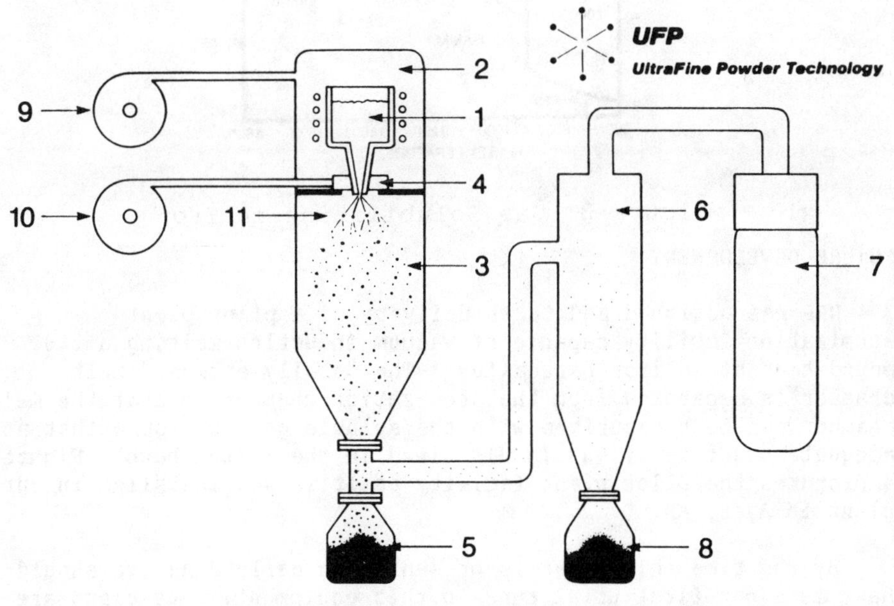

Figure 2. Schematic of TAP Process

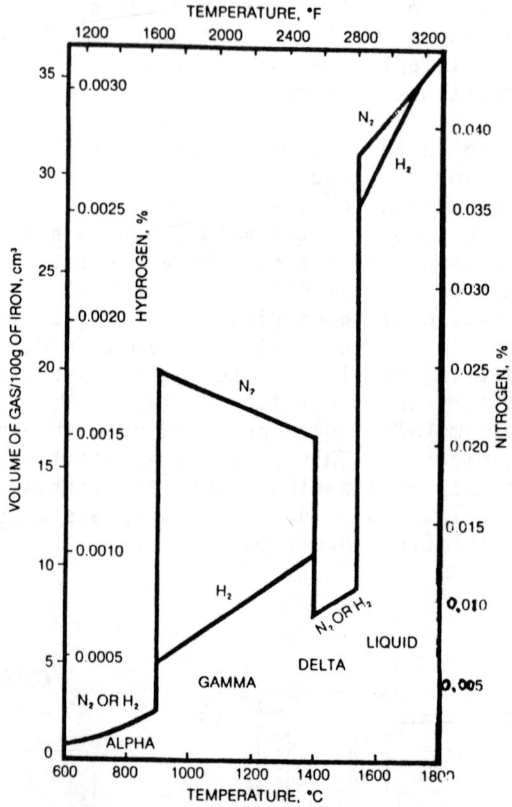

Figure 3 Gas Solubilities in Iron

FUTURE DEVELOPMENT

UFP has designed and taken delivery of a pilot plant atomization facility capable of vacuum induction melting a 200 pound heat of an iron base alloy. The totally enclosed melt chamber is separated from the atomization chamber so that the melt chamber can be pressurized with the soluble gas to insure that an adequate quantity of gas is dissolved in the molten bath. Figure 4 pictures the pilot plant facility as it is now installed in our plant in Ayer, MA.

By the time this paper is presented in early June, we should have made our first trial runs in this equipment. Our plans are to continue the scale-up development of the TAP atomization process in this pilot plant facility over the next year in order to apply the process to a number of different materials systems, targeting the markets and applications described in the beginning of the paper.

TAP THEORETICAL MODEL

All the solutes considered diffuse sufficiently quickly in solid metal at the melting point, so that equilibrium solidification takes place. The lever rule therefore holds, and:

$$\frac{\%Xl}{\%Xi} = \frac{1}{Fs(Kx-1)+1}$$

where Kx = segregation coefficient of solute element
Fs = weght fraction solid
%Xi, %Xl = weight fraction solute element in alloy, liquid

The pressure corresponding to this increased concentration of solute follows from the various equilibria:

(Fe) $\underline{C} + \underline{O} \rightarrow CO(g)$ $\quad \frac{P_{CO}}{P_{CO_i}} = \left[\frac{\%Ol}{\%Oi}\right]\left[\frac{\%Cl}{\%Ci}\right]$

(Ag) $2\underline{O} \rightarrow O_2\ (g)$ $\quad \frac{P_O}{P_{O_i}} = \left[\frac{\%Ol}{\%Oi}\right]^2$

(Cu) $2\underline{O} + \underline{S} \rightarrow SO_2(g)$ $\quad \frac{P_{SO_2}}{P_{SO_{2i}}} = \left[\frac{\%Ol}{\%Oi}\right]^2\left[\frac{\%Sl}{\%Si}\right]$

The approximate solubility products for the compound gas equilibria at the melting points and one atmosphere are:

$$(Fe):[\%Ci]\ [\%Oi] = .002$$
$$(Ag):[Oi]^2 = .09$$
$$(Cu):[\%Si]\ [\%Oi]^2 = .012$$

Using the above, the ratio of evolving gas pressure to ambient pressure when a bubble first nucleates can be calculated:

Table 2
GAS PRESSURE RATIO

	Fe/CO	Ag/O_2	Cu/SO_2
Fs	Kc = .20 Ko = .07	Ko = .02	Ks = Ko = .02
.25	1.63	1.75	2.35
.50	3.12	3.84	7.53
.75	8.28	14.20	54.90

If it is assumed that the gas is ideal and expands isothermally, then the work available per unit mass of gas is:

$$W = \frac{RT}{M} \ln Pr$$

and per unit mass of molten metal:

$$W = \frac{RT}{M} \frac{(w/o\ gas)}{100} \ln Pr$$

where M = gas molecular weight
R = universal gas constant = 8.314 X 10 erg/mole K
w/o gas = weight per cent of gas released
Pr = ratio of gas pressure at release to ambient pressure

The surface energy per unit mass of particles of radius r is:

$$\sigma A_m = \left(\frac{4\pi r^2}{\frac{4}{3}\pi \rho r^3}\right) = \frac{3\sigma}{r\rho}$$

where σ = liquid metal surface tension
ρ = liquid metal density
Am = area per unit metal mass

If one makes the assumption that all the energy available from the expanding gas bubble is consumed in creating new metal surface area:

$$r = \frac{300M\sigma}{(w/o\ gas)\rho RT \ln Pr}$$

The thermochemical data required are as follows:

Table 3

	Fe/CO	Ag/O_2	Cu/SO_2
σ (erg/cm^2)	1800	900	1300
M(g/mole)	28	32	64
ρ (g/cm^3)	8	10.5	9
Xi(w/o)	Ci=.07	Oi=0.3	Oi=.23
	Oi=.03		Si=.23
T(K)	1800	1200	1300

Calculations based on the above data agree with the limited experimental results (one atmosphere of CO over iron), as seen in Table 4.

Table 4

ESTIMATES OF PARTICLE SIZE
GENERATED BY TAP
(Diameters in um)

		1 atm		10 atm		50 atm	
System	Fs	w/o gas	D	w/o gas	D	w/o gas	D
Fe/CO	.25	.02	22.4	.08	6.4	.19	2.8
	.50	.06	3.6	.22	1.0	.50	0.4
	.75	.14	0.8	.54	0.2	1.21	0.1
Ag/O_2	.25	.10	3.0	.32	1.0	.71	0.4
	.50	.29	0.6	.92	0.2	2.05	0.08
	.75	.83	0.07	2.62	0.02	5.87	0.01
Cu/SO_2	.25	.15	4.0	.07	1.8	.55	1.0
	.50	.44	0.6	.20	0.3	1.62	0.2
	.75	1.28	0.1	2.76	0.04	4.71	0.03

When these results are extended to the higher pressures and other gas/metal combinations, then the great potential of the TAP process as a method for ultrafine powder production is apparent.

It is desirable to find a gas with more universal utility than those shown in table 4. All three gases will oxidize most alloying elements at the concentrations considered, and even oxidize the base metals at sufficiently high pressures. A reducing gas would therefore be very useful. Hydrogen is a good candidate, as it has high solubility in many liquid metal systems. Particularizing the preceding equation for dissolved hydrogen:

$$r = \frac{600\sigma}{\sqrt{Pamb}(w/o\ diss)(\sqrt{Pr}-1)\rho RT\ln Pr}$$

where w/o diss = weight percent of hydrogen dissolved at one atm pressure

Pamb = ambient hydrogen pressure over melt

The results of such calculations are shown in Table 5.

Table 5

ESTIMATE OF PARTICLE SIZE ARISING FROM HYDROGEN TAP
(particle diameters in um)

	P(atm)		
	Fe	Cu	Ni
Fs	Kh = .44	Kh = .28	Kh = .40
.25	1.35	1.49	1.38
.50	1.93	2.44	2.04
.75	2.97	4.73	3.31
σ (erg/cm^2)	1800	1300	1800
M(g/mole)	2	2	2
ρ(g/cm^3)	8	9	9
Hi(w/o)	.0023	.00041	.0036
T(K)	1800	1300	1700

	D(um)		
	Fe	Cu	Ni
Fs			
.25	164	446	80
.50	30	78	16
.75	10	22	5

It appears that atmospheric pressures of hydrogen are insufficient to produce ultrafine powder, as observed experimentally. Elevated pressures will be required, as seen in Table 6. Other process modifications, of a proprietary nature, promise to lower particle sizes even further, perhaps to the 1 um size range.

Table 6

ESTIMATE OF PARTICLE SIZE FOR Fe-H AS A FUNCTION OF PRESSURE
(particle diameters in um)

	P(atm)		
	1	10	50
Fs			
.25	164	52	24
.50	30	10	4.4
.75	10	3.2	1.4

Figure 4 TAP Pilot Plant Facility

REFERENCES

1 Kenney, G. B., and Ashdown, C.P., U.S. Patent 4,626,278
2. Ray, R., Metals Handbook, Vol. 7, A.S.M., Metals Park, 1984
3. Kenney, G. B., U.S. Patent 4,578,022
4. Newton,J.G., Raake,O.G.,and Mokler, B.V., J. Aerosol Sci., Vol. 8, pp.339-347, 1977
5. The Making, Shaping and Treating of Steel, 10th ed., A.I.S.E., Pittsburgh, 1985
6. Jackson, W. J., Steelmaking for Steel Founders, Monograph 4, SCRATA, Sheffield, 1973

A NEW TECHNIQUE FOR THE PRODUCTION OF GAS ATOMIZED POWDER

Christer Åslund and Torbjörn Tingskog

Anval Nyby Powder AB (ANPAB)
P O Box 45
S-644 00 TORSHÄLLA, SWEDEN

Abstract

A new principle of atomization, the VH atomization technique, is described. The advantage of this technique is a low gas consumption and low gas pressures combined with a fine screen cut ideal for applications such as spray powder and consolidated products where max grain sizes of about 100 um are required.

Background

Anval Nyby Powder AB (ANPAB), a subsidiary of Valinox S A, Paris, is one of the prime producers of gas atomized powders in stainless steel and nickel base alloys. The main volume is used for the production of seamless tubes to a proprietary process, the so-called ANVAL™ process. The ANVAL process is today used by the most important tube producers including Sumitomo Metal Industries, Sanyo Special Steel, Valinox and ANPAB. More than 25,000 tons of consolidated products have been produced by the ANVAL process including all types of demanding applications in the chemical industry, off-shore industry and power generating industry. The ANVAL process has proved to be a reliable process to supply the market with consolidated products like tubes, bars and profiles and through its flexibility also gives an excellent customer service.

Discussion

For many applications it is of utmost importance to be able to produce powders with a fine grain size. This applies specifically to applications like spray powder and metal injection moulding powder (MIM).

Spray powders have different screen cuts, e g 5-33, 5-44, 44-88, 44-105 micron etc. For many applications the trend is, especially in the aircraft industry, that the screen cuts get finer.

The MIM-technique is now under strong growth especially in the USA. Two of the most common screen cuts used are minus 15 and minus 22 micron. The most common grades are AISI 316 and 304 but also grades like 17-4 PH and tool steels are frequently used.

It has also shown that when atomizing alloys with reactive elements like titanium and aluminium which must be atomized with argon, it is of utmost importance that the produced powder is virtually free from hollow particles. Due to the fact that the atomizing gas is argon the hollow particle contains argon. Entrapped argon acts as a slag and causes detrimental effects on the mechanical properties of the consolidated products.

The percentage of hollow particles is strongly depending on the grain size, see fig 1. Therefore it is essential for powders alloyed with titanium and aluminium, used for consolidated products, that the produced powders have a fine grain size, preferably 100 micron and finer. The present state of art generally includes different types of gas atomization nozzles. The nozzles can be discrete sets of nozzles in different angles or it can be so-called integrated or semi-integrated nozzles. To reach a fine grain size distribution the nozzles are generally super-sonic. The total result with these types of nozzles is that high gas pressures and high flow rates are used which gives a high total production cost of the powders produced.

New Atomization Technique

ANPAB has developed a new atomization process which eliminates many of the disadvantages with existing processes, the so-called VH process (vertical horisontal process). The principle of the process is shown in figure 2. There are three options to prepare the melt, either air melting, melting under gas protection or vacuum. This means that the strictest demands on powders can be met.

The atomization process is controlled by three gas pressures, namely the tank P1, the purging gas P2 and the atomization gas P3. The pressures are regulated by valves through a microprocessor. The dynamic behaviour of the process makes this necessary.

The process is started by increasing the furnace tank pressure P1. When the furnace on its platform (4) is raised the ceramic tube (7) plunges into the steel bath and the steel raises up in the ceramic tube. Bubbling gas is added in the ceramic tube (22) which creates a froth when the steel enters the horisontal part of the nozzle (8). Entering cold gas into the ceramic tube where it meets the molten steel creates a violent reaction which helps to disintegrate the steel.

The froth enters the horisontal atomizing space where the atomizing gas is added. The exit of the atomizer is equipped with a converging insert (9), venturi shaped, which accelerates the gas to high velocities. The discrete metal particles are not accelerated in the same degree as the gas which causes a "drag force" on the metal. This drag force causes the metal to further disintegrate. The formulas are given below showing the parameters controlling the process.

As mentioned, the process is determined by the furnace pressure P1, the froth gas P2 in the vertical tube and the atomizing gas P3. Together with the diameter of the vertical raiser tube the furnace tank pressure P1 regulates the total metal flow. The froth gas and the atomization gas causes a back pressure on the metal which means that all three gas pressures must be in balance.

When looking more in detail at the theory of the nozzle design we can observe the following. By redesigning the horisontal nozzle into a convergent section and installing properly sized jets aft of the metal input port and maintaining an elevated upstream pressure, sufficient energy can be injected into the nozzle chamber to atomize the metal into fine droplets. The convergent nozzle can be sized so that the acceleration of the droplets through the nozzle will cause further atomization. The determination of the nozzle parameters can be calculated using the following derived formula:

$$F = ma$$
$$m = 4/3\pi r^3 \rho$$
$$W = Pdv = \gamma dA$$
$$Pdv = \gamma dA$$
$$P4\pi r^2 dr = \gamma\ 8\pi r dr$$
$$P = \frac{2\gamma}{r}$$
$$P = \frac{F}{A} = \frac{F}{\pi r^2} = \frac{2\gamma}{r}$$
$$F\ 2\pi r\gamma \text{ or } \gamma = \frac{F}{C}$$

where F = force
m = mass
a = acceleration
W = work
P = pressure
r = particle radius
γ = surface tension
C = particle circumference
ρ = metal density
A = area
V = volume

$$\therefore\ \gamma = \frac{F}{C} = \frac{ma}{C}$$

$$\gamma = \frac{4}{3}\frac{\pi r^3 \rho a}{2\pi r}$$

$$\gamma = 2/3\ r^2 a \rho$$

Using the above derivation and knowing the surface tension of the atomized material, which is available in published literature, the nozzle can be sized to give the proper particle acceleration for the desired particle size. Of course, the particle will not move at the same rate as the carrier gas and therefore the force on the particle will be the drag force caused by the gas and can be calculated from the formula

$$Fd = \frac{C \ Ap \ \rho \ U^2}{2 \ g_c}$$

where C = drag coefficient
Ap = area of particle
ρ = density of gas
U = velocity
g_c = 32,17

Thus the above described nozzle configuration can be designed by an expert in the field to give virtually any particle size desired based on properly sizing the primary high pressure ports to give the mass flow velocity rates necessary for the primary atomization and a secondary affluent port to accelerate the particle further for the secondary atomization.

Characteristic for the process is that much lower gas consumption is needed, both Nm^3/hour and gas pressure. Due to the fact that low pressures are needed, this means that the gas can be recycled in an economical way over a compressor. This means that, if recycled or not, the gas costs can be reduced substantially. The principle of melting, where also pouring of the steel into a tundish is avoided, gives the possibility to produce high alloyed steels with a high quality level.

Production Examples

A nickel base superalloy, Nimonic 80A, was melted under argon protection. Nimonic 80A is a standard material used among other things in high performance exhaust valves and in demanding applications in the chemical industry. It is highly desirable to deliver a powder with a fine grain size and ANPAB delivers these types of powder with a grain size of max 105 um (<140 mesh). This gives good properties when consolidating products e g by extrusion or hot isostatic pressing.

The 3000 lbs (1500 kgs) melt was atomized under the following conditions:

Max gas pressures	7 bar
Gas consumption	250 Nm^3/hour
Metal flow	200 lbs/min

The powder was collected in a receiver tank under argon protection and was then cooled under argon to room temperature. The analysis of the powder was as follows:

C	Si	Mn	P	S	Cr	Ni	Ti	Al	N	O ppm
.059	.09	.10	.003	.001	19.12	75.4	2.59	1.56	.004	95

The powder has an average grain size (median 50%) of 54 um. The powder showed a spherical grain size with a dense structure.

It can be noted that despite the low gas pressures and the low gas consumption, it is possible to achieve such a fine grained material of high quality with a median size of around 50 um. The powder tested as extruded also showed a clean structure with an ASTM E45-D value of 0.9.

Conclusion

The paper has described a new type of atomization, the so-called VH-atomizer, which is now in full scale operation at ANPAB. The process offers big advantages through low gas consumption, high cleanliness and its possibility to produce fine grain size powder. As low gas pressures are used, the investment costs are low and it is easy to recycle the gas to further reduce the processing costs.

FIGURE 1

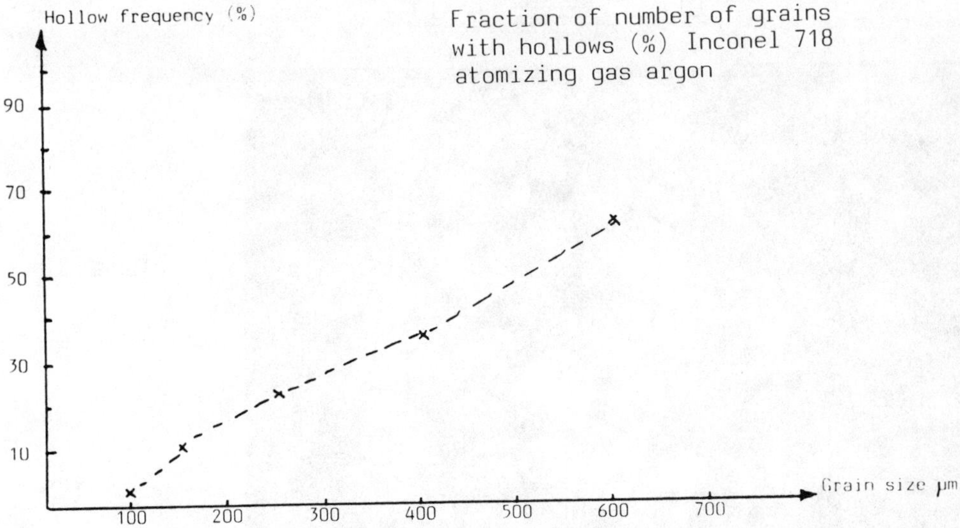

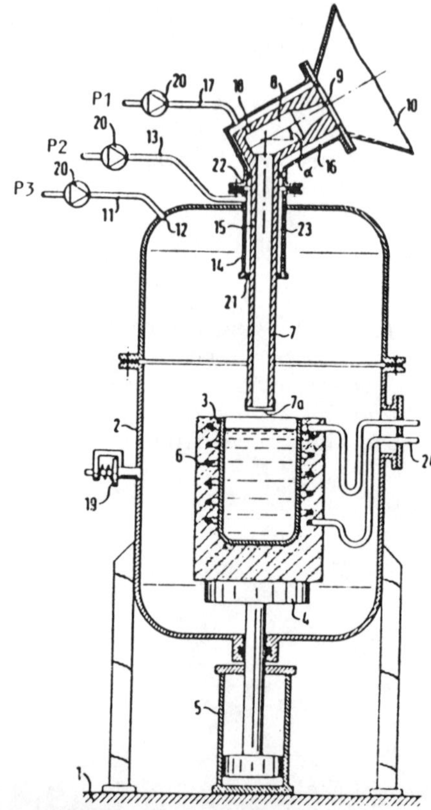

FIGURE 2

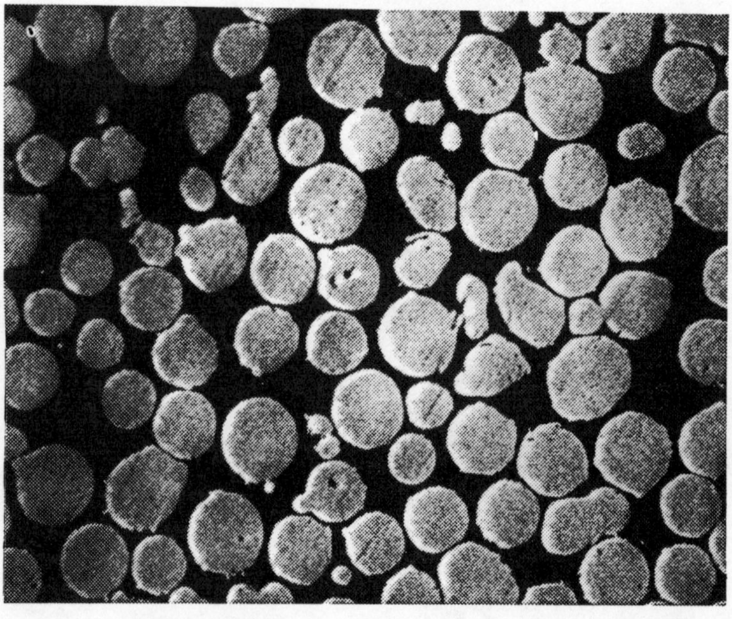

FIGURE 3

Inconel 718,
45-90 μm

200x

INTERNAL POROSITY OF GAS ATOMIZED POWDERS

L. L'Estrade, H. Hallén and R. Ljunggren

HÖGANÄS AB

S-263 83 HÖGANÄS, Sweden

Abstract

All atomized powders contain more or less pores. Internal pores are closed, i.e. they have no contact with the particle surface. Porosity is generally considered to be detrimental. This is because it decreases the tap density, causes gas bubbles in thermal spray deposits, results in uncontrolled dimensional change during sintering (and hot consolidation) and it may give residual micro-pores in the finished product.

Internal porosity is a result of many factors: viscosity and surface tension of the liquid metal, melt temperature, alloy composition and reactivity, atomizing gas and surrounding atmosphere, gas pressure and flow, jet geometry, metal flow rate and quenching rate of the particles. The porosity always increases with the particle size. This paper presents results from an investigation where some of these parameters were studied.

The porosity was measured using a new automatic image analyzing technique, in which 2000-10000 particles were scanned in each case. A mathematical model was developed in an attempt to explain the pore formation. This considers the time for nucleation and growth of a bubble compared to solidification time of a particle and diffusion of argon, nitrogen, oxygen, carbon and hydrogen. The solubility of a gas in the melt is important. Particle collision, gas entrapment and solidification shrinkage are other phenomena affecting pore formation.

1. Introduction

Atomized metal powder particles are always containing porosity, regardless of atomizing media or melting process. Figure 1 illustrates a typical case. Some particles contain such large pores that they become hollow spheres. Other particles have small pores whereas some particles seem to be completely dense (but with enough magnification, micro-pores can be observed in them). Porosity measurements of a large number of powders from different manufacturers have shown 0.5-5 volume percent pores, occasionally more or less than that interval. It is easy to see that the porosity increases with the particle size. Comparisons must therefore be made on identical size fractions. Another obvious state is that some alloys are more prone to pore formation than others.

The general opinion today seems to explain this so-called internal porosity by entrapment of atomizing gas during particle formation and collisions in the liquid state. Our early experience has pointed to other more important factors, connected to the melt properties prior to the atomization. There is not much information in the literature. Argon and helium give higher porosity than nitrogen as atomizing gas (1). If argon reacts with the melt before atomization it causes some porosity (2). The carbon-oxygen reaction may also cause severe porosity (3). The particle porosity is often detrimental:

- If the gas is insoluble in the solid metal, the pore does not close during hot consolidation and can act as a fatigue nucleation site (4).
- It causes uncontrolled density decrease during hot consolidation (1).
- It decreases apparent and tap density and "disturbs" the important relationship between flow rate and apparent density in P/M.
- It may give bubbles and residual porosity when the powder is used in thermal spraying and welding applications.

There are many different ways to study internal porosity. Most common is metallography where a number of particles are studied using various magnifications. Another way is to count floating particles when a powder is spread on a liquid surface. It is important to consider the particle size and the number of particles being studied.

2. Experimental

The alloy chosen is a self fluxing nickel base material with nominal composition and properties as in the following table.

TABLE I

Composition (weight-%)							
C	B	Cr	Fe	Si	S	Al	Ni
0.75	3.1	15	3.5	4.3	0.005	0.01	Bal

Hardness: 62 Rockwell C
Liquidus: 1300°C
Solidus: 940°C

This alloy is produced regularly by Höganäs, thus it was possible to sell or recycle the powder produced in the study. The alloy is also quite sensitive to gas pick-up and finally we know the techniques of chemical and metallographical examination of this alloy.

Twelve heats were manufactured in the study, four of them in a one ton air melting induction furnace by nitrogen atomization and water quenching, the rest in the GAPDRY process (200 kg inert gas melting induction furnace) by argon or nitrogen with complete dry collection. All heats were blended before splitting and sampling. The samples were sieved into different size fractions and evaluated with regard to porosity, physical properties and chemical analysis. The parameters used as variables were gas content of the melt, pouring temperature and atomizing gas. Gas pressure, jet geometry, gas and metal flow and base composition were kept constant. The two plants differ concerning melt atmosphere, crucible material, charge weight and quenching media of the particles. Some samples were also extracted at different levels from the jets during atomizing. Table II gives the heats made, with size, tundish temperature and main parameter in question. Note that half of nominal charge weight was always used as to minimize the amount of possible rejects. All heats, except D 1214, were nitrogen atomized. Heats D 1209-11 and D 1214-17 were inert gas melted.

TABLE II

Heat #	Size (kg)	Temp. (°C)	Main parameter investigated
D 1209	100	1600	Lower gas content
D 1210	100	1400	Reference 1400°
D 1211	100	1600	Reference 1600°
D 1213	100	1400	Air melting
D 1214	100	1400	Ar atomization
D 1215	100	1400	Sampling different levels
D 1216	100	1400	Lower gas content
D 1217	100	1400	Sampling different levels
B 3762	500	1600	High H_2 content
B 3793	500	1400	Reference 1400°
B 3794	500	1600	Reference 1600°
B 3884	500	1400	High H_2 content

The sieve analyses used from each heat were: the full fraction, 250-500 µm, 150-250 µm, 106-150 µm, 80-106 µm, 63-80 µm, 45-63 µm, 36-45 µm, 20-36 µm and finally -20 µm. A Ro-Tap machine with well defined sieve sizes was employed. Carbon and sulphur were analyzed on a LECO CS-244 apparatus. Two measurements were made on each sample. Oxygen and nitrogen were analyzed on a LECO TC-136. Again, two measurements were made on each sample. Hydrogen was analyzed on a LECO RH-402. Argon was analyzed by Kloster Speedsteel, Sweden, on a proprietory equipment using thermoconductivity. All heats, but not always all fractions, have been analyzed with respect to gas contents. Porosity was measured by an image analyzer, Leitz TAS-plus. The equipment consists of a light optical microscope with a remote controlled sample table and a camera. The picture is shown on a monitor beside and is computer analyzed according to a specially made programme. The table can take 18 samples of Ø 25 mm cylinders and is controlled by the computer. It can thereby work during the night without supervision. If something remarkable is observed, it will tell the operator exactly what and where the next day. Before the study, the necessary number of particles and the magnification, was optimized for each size fraction. It was decided to study 10,000 particles in the small fractions and 1500 in the large fractions. Some closed pores will obviously look like open and vice versa in the microscope, depending on how and where the particle was cut. The chosen alloy is very hard so no pores are believed to have been changed during polishing.

3. Results

The porosity, given as area-%, is shown by Table III for the 12 experimental heats. For each heat, six size fractions are included. The table also includes the total porosity for each heat, i.e. for the fraction 0-500 µm. The mean value of each column is found at the bottom of the table.

The apparent density, flow rate and sieve analysis are given by Table IV and V for D- and B-heats separately. The total values are shown as well.

TABLE III

Heat #	250/500	150/250	80/106	45/63	20/36	0/20	Total
D 1209	1.46	1.27	0.60	0.19	0.13	0.18	0.56
D 1210	1.74	1.37	0.74	0.20	0.25	0.08	0.67
D 1211	0.95	1.21	0.36	0.15	0.18	0.05	0.45
D 1213	0.78	0.89	0.26	0.24	0.24	0.07	0.46
D 1214	2.79	1.53	0.58	0.21	0.15	0.04	0.64
D 1215	1.38	1.05	0.67	0.25	0.22	0.06	0.57
D 1216	1.78	1.38	0.60	0.24	0.33	0.06	0.68
D 1217	2.57	1.24	0.70	0.43	0.24	0.12	0.74
B 3762	7.11	2.49	0.98	0.75	0.60	0.19	1.69
B 3793	1.82	0.63	0.74	0.27	0.27	0.09	0.44
B 3794	3.00	2.03	0.62	0.32	0.21	0.08	0.81
B 3884	5.74	3.40	0.90	0.57	0.34	0.07	0.90
$\bar{x}$	2.59	1.54	0.65	0.32	0.26	0.09	0.72

TABLE IV

Heats	Apparent density		Flow rate	
	$\bar{x}$	σ	$\bar{x}$	σ
D	4.73	0.056	14.42	1.526
B	4.59	0.143	14.01	0.604
Total	4.68	0.11	14.29	1.274

TABLE V

Fraction	D-heats	B-heats	Total
250-500	5.8	2.9	4.8
150-250	13.6	10.2	12.5
106-150	13.3	19.8	15.5
80-106	15.3	17.1	15.9
63 - 80	10.2	9.3	9.9
45 - 63	15.5	17.5	16.1
36 - 45	5.7	6.6	6.0
20 - 36	15.1	12.3	14.2
0 - 20	5.5	4.3	5.1

Note
Table III: Porosity in area-%.
Table IV: Apparent density in g/cc. Flow in sec/50 g.
Table V: Sieve analysis in weight-%.

The carbon and sulphur values are also quite constant, only differing somewhat between B- and D-heats. See Table VI, which shows the coarsest and the finest fractions. All values are given in weight-%.

TABLE VI

Heats	Carbon (%)		Sulphur (%)	
	250-500	0-20	250-500	0-20
D	0.685	0.711	0.002	0.003
B	0.758	0.745	0.003	0.002
Total	0.709	0.724	0.002	0.003

The oxygen is variating a lot more, so extra data are presented in Table VII from different heats and fractions. Values in ppm.

TABLE VII

Heat #	250/500	150/250	80/106	45/63	20/36	0/20
D 1209	40	120	80			250
D 1210	20	130	120	130	120	260
D 1211	70	120	150			240
D 1213	60	120	160	120	170	300
D 1214	40	50	130	130	140	260
D 1215	30	110	120			270
D 1216	30	90	130			300
D 1217	20					
B 3762	510	380	370	380	390	500
B 3793	200	200	250	210	280	390
B 3794	400	330	340			350
B 3884	700	290	250			390
All D	39	106	127	127	143	269
All B	452	300	302	295	335	408
Total	177	176	191	194	220	319

The nitrogen values are not differing much, which is shown in Table VIII for three size fractions and a number of heats.

TABLE VIII

Heats #	Nitrogen (ppm) 250/500	80/106	0/20
D 1210	10	30	40
D 1211	30	40	50
D 1213	30	40	50
D 1214	10	20	30
B 3793	10	20	30
B 3794	20	30	40
All D	20	33	44
All B	20	28	32
Total	20	31	40

The argon analysis was made on D-heats only. All results are found in Table IX, data in ppm.

TABLE IX

Heats #	Argon (ppm) 250-500	150-250	80-106	0-20
D 1210	<0.05		<0.05	
D 1213	<0.05		<0.05	
D 1214	1.99	1.47	1.02	0.08

At last it remains to present the hydrogen analysis. Here we have relatively large differences between the heats and the size fractions, see Table X. At the bottom of the table are the D-and B-heats separately plus the total value for all heats. All data are in ppm.

TABLE X

Heats #	Hydrogen (ppm) 250/500	150/250	80/106	45/63	20/36	0/20
D 1209			3.25			
D 1210	2.50	1.95	2.65	3.77	5.80	11.1
D 1211			2.45			
D 1213	2.75	1.85	2.95	4.25	6.60	13.6
D 1214	2.53	1.50	2.60	3.20	5.30	9.9
D 1216			2.45			
D 1217	2.37	1.65	2.30	2.70	4.15	9.0
B 3762	18.0	9.45	9.25	9.20	11.6	15.3
B 3793	6.40	4.70	5.00	4.95	6.37	8.35
B 3794			6.40			
B 3884			5.35			
All D	2.54	1.74	2.66	3.48	5.46	10.9
All B	12.2	7.08	6.50	7.08	8.96	11.8
Total	5.76	3.52	4.06	4.68	6.63	11.2

A great many photos were taken on particles sampled from various parts. Figure 2 illustrates two different kinds of porosity. We will call the left pore primary and the one to the right secondary. Figure 3 shows particles extracted 20 cm below the atomizing focus, so they have just been formed. Figure 4 a and b exhibit large (250-500 μm) and small (0-20 μm) particles respectively. The appearance of argon and nitrogen atomized is very similar (not shown here). The same applies to the powder with increased hydrogen content.

4. Discussion

We will split the discussion in parts according to the theory of pore formation, i.e. particle collision, gas entrapment and gases dissolved in the melt.

4.1 Particle collisions

A widespread opinion is that this is the main reason for porous particles. However, we have not been able to find a single evidence of this in our study. There are plenty of photos showing a small particle penetrated in a bigger one, but in no case that was connected with porosity. It cannot be ruled out that particle collisons may cause pores in certain cases; perhaps in larger particles and other alloys.

4.2 Gas entrapment

In atomization, the molten stream is broken up by gas impingement to filaments which are subsequently forming powder particles. For alloys with a low surface tension, parachute like membranes are formed as an intermediary stage, especially when the solidification rate is low. In such case, gas is easily entrapped when the fragments are closing themselves to particles. Some of the gas is dissolved in the metal, some diffuses away but enough remains as a pore in the solid particle. Since argon is insoluble in metals, has a lower quenching effect than nitrogen and does not disappear by diffusion it will almost certainly form porosity by this mechanism. The porosity as a function of the argon content for heat D 1214 is plotted in Figure 5, where the particle size is also indicated. The porosity of this heat is not only determined by argon; the reference heats do also contain pores.

How much porosity can a gas develop? If we assume 1 ppm concentration, formation at 1000°C and an internal pressure of 1 bar (we do not assume overpressure in large pores), the data in Table XI are calculated.

TABLE XI

	Ar	H_2	N_2	CO
Area % porosity when 1 ppm gas is entrapped	2	40	3	3

In heat D 1214 there is 2.8% pores and 2 ppm Ar in the large fraction. Some microporosity is not detected, but there should have been more porosity than this, because argon is not the only source. The formation temperature may be wrong, but more plausible is that we have an overpressure (about 1 bar) in the pores. That would cut the numbers of Table XI in half. Note that only argon can be used for calculation here, since the other elements are soluble. But it proves how a very small gas concentration can cause severe porosity. We cannot find that other gases than argon (as would helium) cause porosity by gas entrapment. From the gas contents, Tables VI-IX, it is evident that oxygen, nitrogen and hydrogen (except in the doped heats) decrease with particle size, whereas argon increases. But certainly we have gas entrapment pores also from nitrogen and hydrogen (water quenching in B-heats). The so-called primary pores, see Figure 2, are formed by gas entrapment. Due to a different solidification pattern, when the entrapped cold gas is cooling the metal, a skin is formed around the pore. Since we can observe primary pores in all heats, it must be concluded that other gases than argon may be entrapped. But argon is always contributing very much to porosity by entrapment.

4.3 Gas formation in the melt

The melt, prior to atomization, contains elements (O, C, H and N) which tend to form gases during solidification, when their solubility decreases. We have carried out calculations regarding diffusion, nucleation and growth. Hydrogen is the only element which can move fast enough to form pores during the short times available. Solidification times were estimated using data from a previous study (5). In particles larger than one mm, it is quite possible to get pores of N_2 and CO in certain materials. We will therefore discuss hydrogen here.

Looking back at Table X, it is obvious that hydrogen is higher in the fine fractions due to gas molecules adsorbed on the particle surface. We have to account for that before relating porosity with hydrogen. Using the equation

$$H\ (\text{total}) = H\ (\text{bulk}) + k/d$$

where k = a constant and d is the particle diameter. We can then construct a new sheet, where the surface hydrogen has been removed, see Table XII. However, two difficulties must be pointed out here. The analysis depends heavily on the sampling procedure and it is hard to put a correct value of d in the equation.

TABLE XII

Heats	Corrected H_2-values (ppm)					
#	250/500	150/250	80/106	45/63	20/36	0/20
D 1209			(1.6)			
D 1210	2.0	1.2	1.1	1.0	0.5	(0.5)
D 1211			(0.9)			
D 1213	2.2	0.9	0.9	0.8	0.4	(0.3)
D 1214	2.1	0.8	1.2	0.8	0.6	(0.5)
D 1216			(0.9)			
D 1217	2.0	1.1	1.1	0.6	0.4	(0.3)
B 3762	17.8	9.1	8.5	7.9	9.1	10.3
B 3793	6.2	4.4	4.3	3.7	4.0	3.7
B 3794			(5.7)			
B 3884			(4.7)			
All D	2.1	1.0	1.1	0.8	0.5	0.4
All B	12.0	6.8	5.8	5.8	6.6	7.0
Total	13.1	3.9	3.5	3.3	3.5	3.7

The B-heats are water quenched and would have more hydrogen in the fines in spite of the thorough drying process. In addition, heats

B 3762 and B 3884 are doped with hydrogen. In the D-heats hydrogen is lower in the fines, because it escapes by diffusion during atomization. Diffusion takes place in large particles as well, but the distances are much longer.

Let us now plot porosity versus hydrogen (excluding heat D 1214) using the two largest fractions only, see Figure 6. For the B-heats, there is a clear relation, but that cannot be said for the D-heats. Hydrogen is certainly playing a role, but there must also be another explanation. What about oxygen (as gas it must be as CO)? Again we have to account for the surface. Using correction factors, calculated from Table VII, and plotting porosity versus oxygen (excluding D 1214) for the two largest fractions, we obtain Figure 7. Again, there is no correlation for the D-heats, whereas the B-heats show a more clear relationship. An attempt to plot porosity versus nitrogen does not show any relationship at all. We may combine oxygen and hydrogen (multiplied by a suitable factor, say 25) and plot this versus porosity, but there is still no relation for the D-heats. So we end up with just a small indication of relation between porosity and hydrogen for the D-heats.

When a pore is formed by dissolved gas atoms it becomes very smooth and spherical, a secondary pore as in Figure 2. The bubble cannot be homogeneously nucleated, it would require a tremendous pressure. So it must form at some kind of heterogeneity, e.g. a slag inclusion. The growth requires less energy in a large particle. That, together with a slow cooling rate explains why porosity increases with particle size as shown for some heats in Figure 8.

If we look at Figure 4, it is obvious that some particles contain one or more pores, while others are fully dense. This can probably be explained by availability of nucleation sites.

The lowest porosity in the large fractions is found in heats D 1211 and 1213. It is impossible to explain why. The lowest total porosity is found in heats B 3793, D 1211 and D 1213. The worst porosity in the large fractions is in heats B 3762 and 3884. The reason is doping with hydrogen. The highest porosity in the D-heats has the argon atomized powder, but only for the coarse fractions. The experiment to lower the gas content (heats D 1209 and 1216) did not work well. A reasonable explanation is that electrolytic raw materials were used. However, clean as they are, the hydrogen content is higher.

No attempt was made to estimate the size and shape of the pores in relation to particle size, gas content and production parameters. In one case (6) particles up to 2500 μm were studied. It was observed that the porosity increased almost linearly. Some pores looked like solidification shrinkage, but we can now explain many of them as primary pores (gas entrapment).

5. Conclusions

From the results and discussion in the present study, we have made the following conclusions:

- Internal (closed) porosity exists in all powders regardless of manufacturing process.

- The porosity (as area- or volume-%) increases with the particle size of the powder.

- Gas entrapment is one reason for porosity, especially when the atomizing gas is argon (or helium), which is insoluble in the melt. The pores are surrounded by a shell like layer and have an irregular shape. We name this kind of porosity "primary", since it is formed very early in the atomization.

- Bubble formation of dissolved gases (hydrogen, and perhaps carbon/oxygen) is a large source of porosity. These pores are very smooth and spherical. We call them "secondary", because they are formed later, when the particles start to solidify.

- Bubble formation requires that the gas in question has a high solubility in the melt and a high diffusion rate compared with the solidification time of the particle.

- Nucleation of a bubble must be heterogeneous (since homogeneous nucleation is impossible). It must take place on a slag inclusion or some other irregularity. This requires much less energy in a large particle.

- Nitrogen does not cause porosity in particles less than about 500 μm.

- Comparison of porosity must be made on particles within a narrow size fraction. The number of particles measured has to be large enough to assure a good reproducibility. This is because the pores are unevenly distributed in the particles; some are dense while others contain several pores. The pore size and shape are also very different between the particles.

- When analyzing gas contents, sample procedures must be strictly adhered, too. It has to be remembered, that surface concentration of gases becomes important with decreasing particle size.

- Argon content is very low (<0.05 ppm) in all powders except the Ar-atomized powder, where the Ar-content is high, especially in particles >100 μm.

- Hydrogen content decreases with particle size, except in the cases with H_2 doping.

- Oxygen decreases with particle size, except when the powder is water quenched.

- All gases (Ar, H_2, N_2, CO) can cause severe porosity, even when only a few ppm are present.

- The sample preparation is important. A soft alloy may have most of its pores welded together during cutting and polishing.

Acknowledgement

The authors wish to thank Kanthal Höganäs AB for allowing us to present this paper, Mr Carl-Axel Blände for developing the metallographic technique and many people of the laboratories for performing the powder analyses.

References

1. R.H. Van Stone, F.J. Rizzo, J.F. Radavich: "Microstructure of a gas atomized Rene '95 powder", Proc. of the 2nd Internat. Conf. on Rapid Solidification, March 1980, Virginia, USA.

2. Yu F. Ternovoi, A.G. Tsipunov, S.B. Kuratchenko, O.M. Kuimova, K.V. Kondakova: "Pore formation in atomized powders", Sov. Powder Met Ceram, 24, 1985.

3. N. Ch. Pioro, E. Ch. Pioro, A.G. Zherdin, V.P. Smirnov: "Structural and morphological investigation of an atomized powder", Sov. Powder Met Ceram, 24, 1985.

4. E.N. Bamberger, J.S. Mosier, R.W. Harrison: "Materials for Advanced Turbine Engines", General Electric Aircraft Engine Group Report, R 78 AEG 265, 1978.

5. J. Liu, L. Arnberg, N. Bäckström, H. Klang, S. Savage: "Fluid flow behaviour, solidification studies and some practical experience from USGA", Swedish Institute of Metals Research, 1987.

6. Internal R&D report, Höganäs AB, 1987.

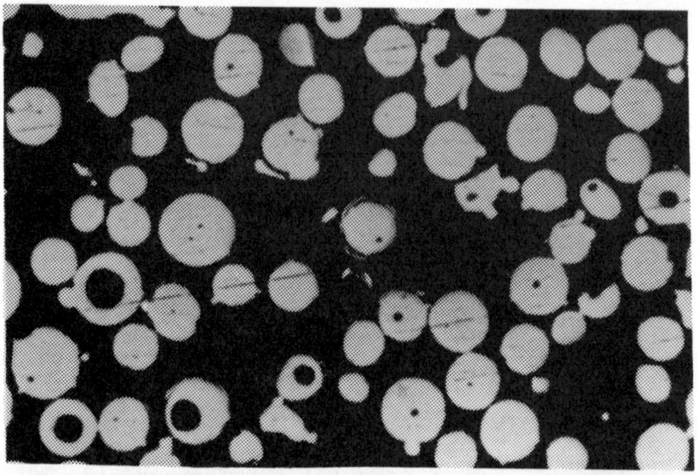

Fig. 1

x50

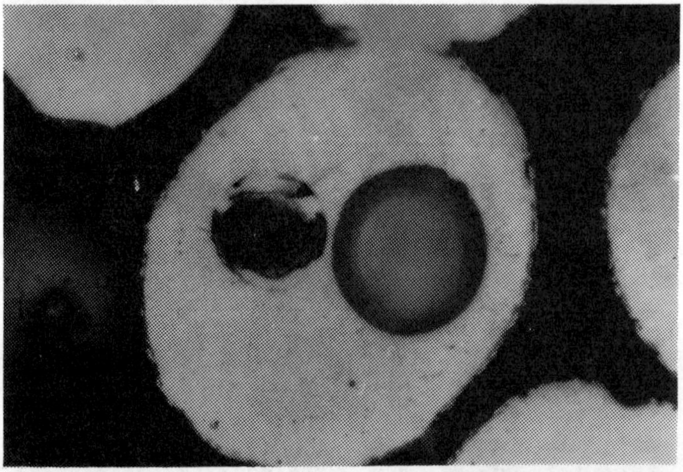

Fig. 2

x1000

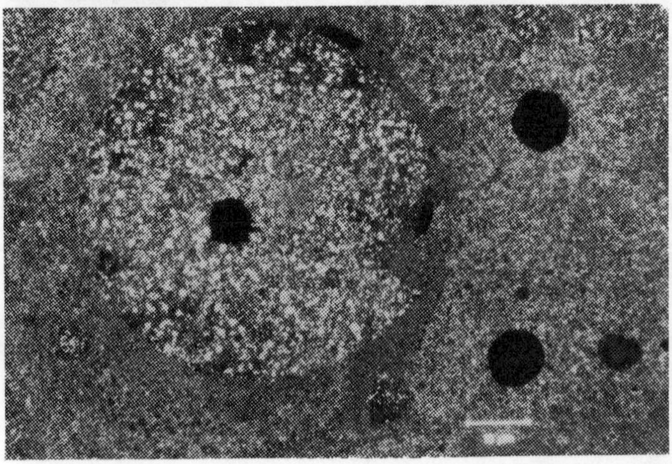

Fig. 3

x1000

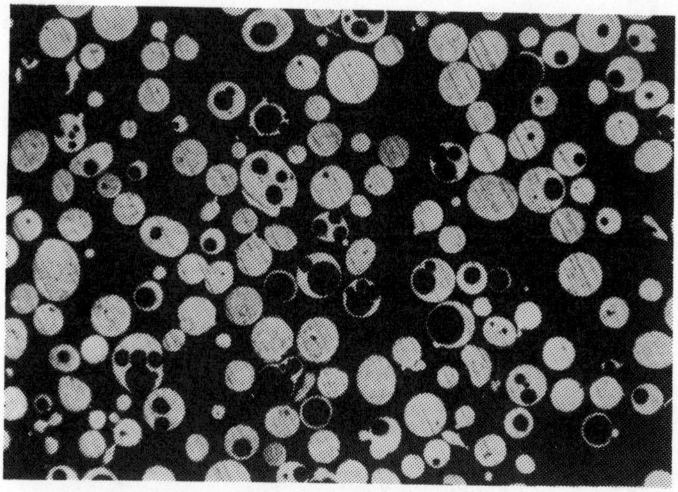

Fig. 4 a

x20

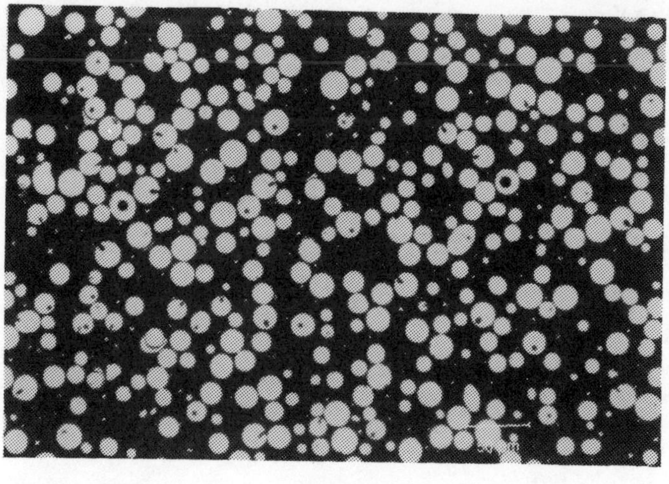

Fig. 4 b

x200

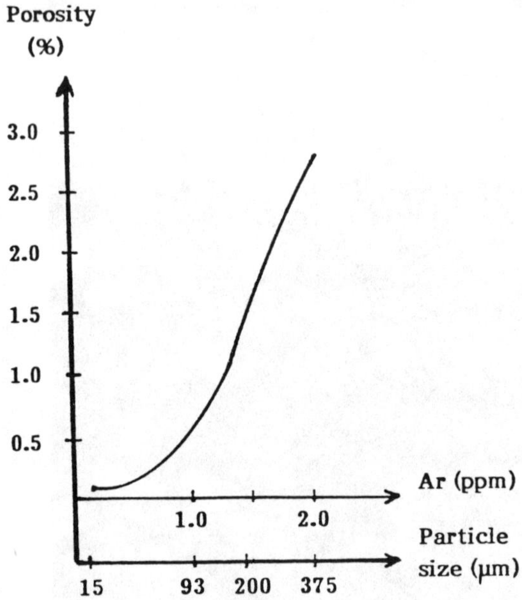

Figure 5

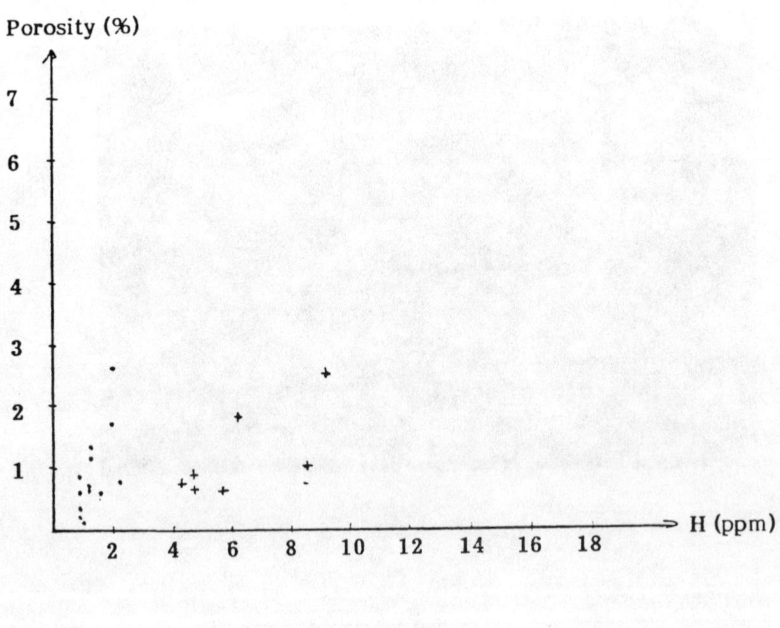

Figure 6

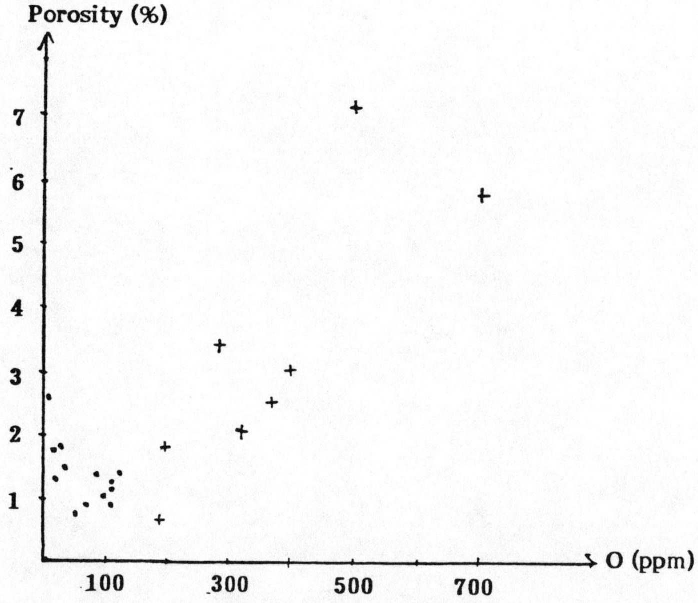

Figure 7

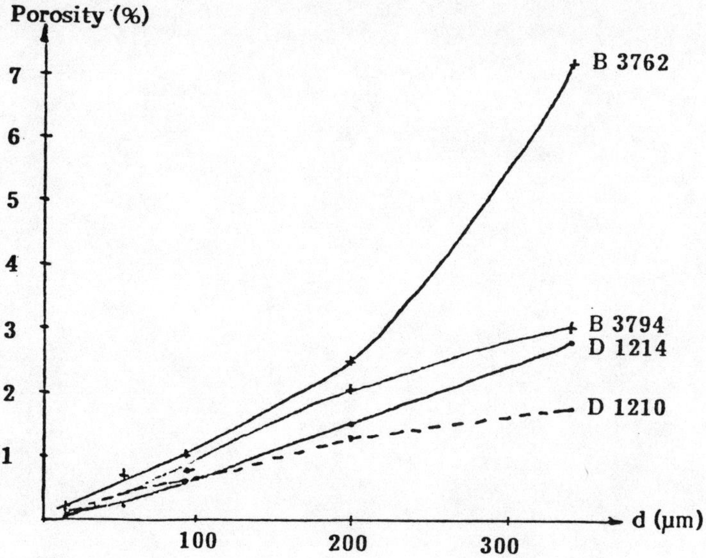

Figure 8

OBSERVATIONS OF GAS ATOMIZATION PROCESS DYNAMICS

I. E. Anderson* and R. S. Figliola**

*Ames Laboratory, Metals and Ceramics Div.,
Ames, IA 50011
**Clemson University, Mechanical Eng. Dept.,
Clemson, SC 29634

ABSTRACT

A critical understanding of the process dynamics of gas atomization is an essential step toward achievement of full control over powder particle size and size distribution. This report summarizes the results of an investigation of the gas and melt flow fields in a multiple discrete jet gas atomization nozzle with a confined melt feed arrangement operated at elevated gas pressures. Several types of measurement and visualization techniques, including static gas pressure probes, gas flow imaging, and high speed photography were used for the study of gas only flow (argon) and of atomization process dynamics. These results indicate that gas flow effects, primarily the action of a normal shock front and a strong recirculation region attached to the melt orifice tip, appear to significantly influence the atomization mechanism. A hydraulic analogy has also been used as a surrogate system to study these gas dynamics. Direct comparisons were made between the flow patterns from schlieren photographs of gas only flow from the atomization nozzle and the flow patterns from shadow photographs of water flow from two dimensional nozzle analogs. Another comparison was made between the static gas pressure measurements and the analogous water depth measurements on the water table. The water table appears to provide a convenient tool for study of a broad range of configurations for atomization nozzle jets and melt feed tube tips. In general, the results of this investigation enhance the general knowledge base for the design and control of metal powder production by gas atomization processes.

INTRODUCTION

High pressure gas atomization (HPGA) shows promise as a method for efficient production of very fine (<10um dia.) metal and alloy powders

(1). Such powders have attractive properties for rapid solidification processing, injection moulding, thermal spraying, and other emerging powder processing approaches. Since previous work showed that finer powders resulted from gas atomization at increased gas nozzle inlet pressures (2), very high pressure gas atomization was investigated to exploit the size refinement effect to its maximum extent. This investigation was undertaken to explore gas atomization process dynamics, especially in the high pressure regime, to allow more informed selection of process parameters and to improve further the performance of HPGA. A variety of diagnostic techniques were utilized in this study, probing primarily the gas flow effects, to understand the distribution of gas stream energy available for atomization. Additionally, a surrogate flow system was studied to physically model the gas flow dynamics in a more easily observed manner. Finally, a high speed movie of the atomizer in full operation in an HPGA mode was taken to see how our understanding of the pieces of the atomization process fit together in a process operational model.

Aspiration effects

Stable atomization with such a highly energetic gas flow was reasoned to occur only under conditions where a subambient pressure exists at the melt orifice, or melt feed tube exit point, as in Figure 1. This "aspiration" condition would be induced by the flow of gas past the exterior edge of the feed tube, like the wake of a boat. The gas flow would create a partial vacuum in the "wake" region, permitting the melt to be drawn into contact with the gas flow for initiation of the atomization process. Unstable atomization conditions, where a positive pressure exists at the melt orifice, can be maintained only when the melt bath surface is pressurized or when a high metallostatic head exists to counteract the positive pressure. The consequences of unstable atomization conditions, a tendency for gas to backstream through the melt feed tube and to erupt from the crucible, were observed in early experiments since melt overpressure provisions were not incorporated in our atomizer.

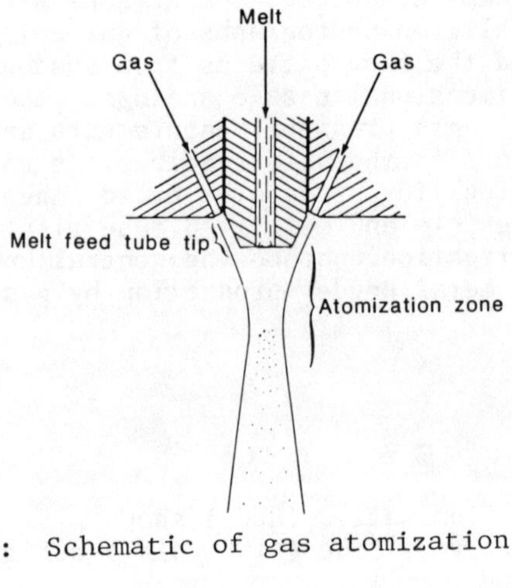

Figure 1: Schematic of gas atomization nozzle.

Atomization diagnostics

Achievement of stable atomization required development of a diagnostic procedure involving melt orifice pressure measurements with gas only flow to test various melt feeding configurations over a range of gas inlet pressures. The procedure and its results will be described in this report. Another report (3) contains the results of a series of atomization experiments to test the inlet pressure dependence of gas atomized particle size in the high pressure regime of 10-14MPa (1500-2000psig). A tin alloy, Sn-5w/oPb, was atomized with Ar at three inlet pressures. The maximum size refinement effect for Ar atomization was found at a pressure of about 12MPa. This result (3) will be compared to the results of the orifice pressure measurements as a function of inlet pressure over a similar range.

Gas flow visualization techniques and high speed cinematography also were incorporated in an analysis of the operation of the atomization nozzle, especially in the high pressure regime. The atomization nozzle gas flow patterns for argon gas were observed using schlieren and high speed shadowgraph techniques. These techniques have been applied commonly for the study of rocket nozzle exhaust patterns (4) which operate in a similar pressure ratio condition, i.e. at similar supersonic gas velocities. These studies of the HPGA nozzle gas flow were supplemented by high speed cinematographic observations of the actual atomization process to investigate the interaction of a melt with the gas flow.

Hydraulic model for gas flow effects

A method which probes the high velocity flow field in a surrogate (water) flow system was also used to enhance this exploration of the HPGA process. The hydraulic analogy (5,6) was used to study the fluid flow region encountered within the HPGA melt/gas coupling zone and in the adjacent region downstream of the zone. The intent of the use of the analogy was to test its applicability to the physical modeling of the HPGA gas flow and, thereby, to study conveniently the effects of two-dimensional changes in design of the nozzle.

The hydraulic analogy between acoustic waves in a two dimensional, compressible gas flow and long gravity waves on the surface of an incompressible open channel flow was implemented through the use of water table experiments. The analogy provides a convenient technique for the study of model geometry effects in complex, high speed flow systems. The analogy stems from the similarity in the equations of motion for two dimensional perfect gas flow and open channel flow.

For a two-dimensional gas flow the energy equation can be written between a reservoir at stagnation temperature, pressure, and density of T_o, P_o, ρ_o, respectively, and a downstream point of velocity, V, and temperature, T, pressure, P, and density, ρ, as:

$$C_pT + V^2/2 = C_pT_o \qquad [1]$$

where C_p is the specific heat of the gas. The Mach number, M, of the gas at any position, which compares the local velocity at that position to the speed of propagation of acoustic waves through the gas, is given by:

$$M = V/a = V/\sqrt{\gamma RT} \qquad [2]$$

where a = the speed of sound in the gas and $\gamma = C_p/C_v$. For an isentropic process, the relationship between the stagnation properties and any downstream position on terms of temperature is given by:

$$P_o/P = (T_o/T)^{\gamma/(\gamma-1)} \qquad [3a]$$

$$\rho_o/\rho = (T_o/T)^{\gamma/(\gamma-1)} \qquad [3b]$$

where the temperature ratio must be:

$$T_o/T = 1 + [(\gamma-1)/2]M^2 \qquad [4]$$

A similar energy equation for free surface water flow, assuming the flow to be two dimensional, can be written between a reservoir at depth, h_o, and a local channel depth downstream, h, as:

$$h + V^2/2g = h_o \qquad [5]$$

where g is the gravitational acceleration constant. The ratio between the local velocity, V, and the speed of propagation of long gravity waves, c, is given by the Froude number, Fr, as:

$$Fr = V/c = V/\sqrt{gh} \qquad [6]$$

The relation between the depth ratio and the Froude number is written as:

$$h_o/h = 1 + Fr^2/2 \qquad [7]$$

For a perfect gas having a specific heat ratio of $\gamma = 2$, the analogy between equations [4] and [7] is exact. As such, all of the isentropic relationships for a perfect gas become directly related to the hydraulic analogy. The Mach number is directly analogous to the Froude number. The density ratio and the temperature ratio are directly analogous to the depth ratio. The pressure ratio is directly analogous to the square of the depth ratio. This similarity between the two flows permits the simulation of complex gas flows with relative ease on a water table while maintaining the essential gas dynamic features.

Although the analogy holds in the strictest sense for a gas of γ = 2, the error involved in extrapolating the results to other gases having γ less than 2 is not serious enough to preclude the use of the analogy for supersonic flows. Methods to correct the analogy for other gases have been proposed (7,8,9). Harlemann (10) has demonstrated that the relationship between aerodynamic pressure and hydraulic depth within an expansion fan, such as in the underexpanded region of the atomizer gas jet, can be better approximated for N_2 or air, where $\gamma = 1.40$, by:

$$P/P_o = (h/h_o)^{1.4} \qquad [8]$$

Since argon has a γ of 1.67, a proportionally smaller correction is likely to be necessary.

EXPERIMENTAL PROCEDURE

Our current HPGA system utilizes a gas atomization nozzle consisting of 18 cylindrical gas jets that are fed from an annular manifold and are positioned around a central hole at an apex angle of 45°, as illustrated in Figure 1. This "ultrasonic" nozzle (11) may be classified as a gas atomization nozzle with multiple discrete jets and a confined melt feed arrangement. Figure 1 illustrates that a melt guide tube is positioned within the nozzle central hole to direct the melt from the crucible outlet (not shown) to the gas interaction zone where atomization occurs. The design of the guide tube portion that protrudes out of the central hole turned out to be crucial to stable operation and some critical aspects of its design are described below.

Pressure probe measurements

A static pressure probe, fixed in a melt tube at the orifice position shown in Figure 2, was used to explore aerodynamic pressure effects which appeared to control melt feeding into the atomization zone. The probe consisted of a 3 mm tube which was sealed to an electronic pressure transducer and had an open end to sample the static pressure environment of the melt orifice when high velocity gas was blown through the nozzle. The voltage output from the pressure transducer was detected and analysed with a digital storage oscilloscope

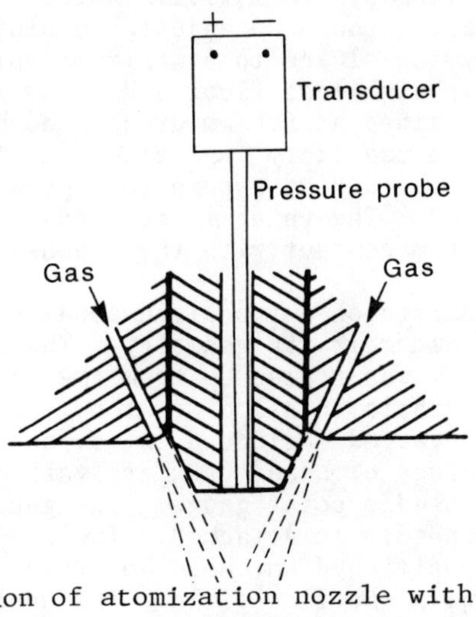

Figure 2. Configuration of atomization nozzle with static pressure probe and transducer for measuring orifice pressure with gas only flow.

where a reported value refers to a midpoint reading of the steady state transducer signal. This system permitted testing of different melt guide tube tip configurations for aspiration potential without requiring multiple atomization attempts.

Gas flow imaging

Schlieren and shadowgraphy techniques were applied to gain an understanding of the gas only fluid dynamics of the atomization nozzle. For the schlieren experiments most of the components of the gas handling system were transferred from the atomizer to an optical bench. The nozzle was positioned in the center of a parallel light field between two concave mirrors. A model of the preferred melt guide tube tip design was also inserted in the nozzle central hole during the schlieren imaging to simulate the gas flow conditions of the nozzle in an atomization experiment. The light source was a high intensity stroboscopic flash lamp operated at frequencies in the range of 330Hz. A beam splitting mirror permitted continuous viewing of the gas flow image with a video system and permanent recording of the image with 35mm camera when the flow conditions appeared to reach a steady state. High speed shadowgraph images of the nozzle gas flow were recorded with an optical system mounted on the HPGA chamber where the light path entered and exited through glass sight ports on opposite sides at the level of the atomization nozzle. Two plano-convex lenses directed the light from the light pipe exit, through the gas flow field, and to the photographic plate. A ruby laser (694.3 nm wavelength) equipped with a pockel cell served as the light source and permitted a 35 ns exposure time for each shadowgraph image.

Water table measurements

The water table used in this study, depicted in Figure 3, has a working channel 1.6 m long by 0.8 m wide. Water enters the working channel from a reservoir through an adjustable sluice. The reservoir depth is controlled by the sluice to a maximum depth of 0.4 m. The working channel has a plate glass floor and 60 mm tall sides. Back pressure depth is maintained at 6.5 mm using a downstream weir, a depth found suitable to reduce capillary wave effects. The water is recirculated from a catch tank by a pump through a flow control valve and a venturi flow meter. The water is treated with a surfactant to reduce flow friction from contact with the channel floor and sides.

The test model consists of a two dimensional cross section replica of the actual HPGA nozzle/melt tip geometry. The nozzle model, illustrated in Figure 4, was scaled up from the actual size of the gas nozzle/tip by a 6.8:1 ratio in all dimensions. The model dimensions which could be readily varied include d, s, α, and the ratio P/P_0. In this paper only the values of s and P/P_0 are varied. Depth measurements are made using a fine needle point gauge. The gauge consists of a micrometer to which a needle is attached. The gauge is mounted on a trolley which can be positioned anywhere above the channel surface without interferring with the flow field. The uncertainty in the depth

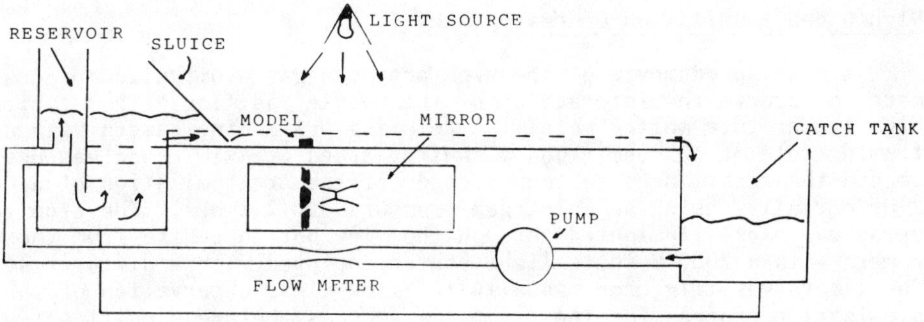

Figure 3: Schematic of water table test facility flow scheme.

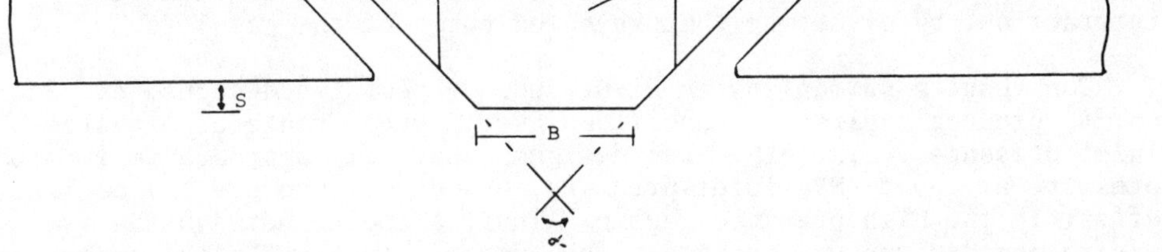

Figure 4: Nomenclature and view of water table nozzle model. For this study, D = 114 mm, d = 5.7 mm, B = 67 mm, and α = 45°.

measurements based on the use of this gauge are estimated at $\pm$ 0.5 mm (95%). Shadowgraph photography was used to record the wave patterns generated by the model at each given value of control parameter, P/P_o, for each model geometry. This was accomplished by passing a beam of unfiltered white light through the model flow field and recording the image on film.

High Speed Atomization Cinematography

A high speed movie of the high pressure gas atomization process was made to observe the interaction of the nozzle gas flow with a typical high temperature molten metal. A rotating prism cine camera with a framing rate of $10^4 sec^{-1}$ and a shutter speed of $4x10^{-5}$ sec was used to record the first three to four seconds of the Ar atomization of a Cu-2.0w/oZr alloy using an inlet gas pressure of 12.1 MPa. The atomizer spray was back-illuminated through the view port opposite from the camera with a 2000 W white light source equipped with a diffuser screen. The camera was triggered manually to allow clear observation of the atomization process for the first few seconds, before significant clouding of the viewports occurs.

RESULTS AND DISCUSSION

Pressure transducer results

The influence of the tip taper angle on aspiration was explored through a comparison of the melt tip designs illustrated in Fig. 5. These designs included two taper angles, 45° and 63°, and a straight wall cylinder. The 45° taper, no. 4 in Fig. 5, matches the apex angle of the array of gas jets, effectively selecting the projected core of each jet. The 63° taper, no. 3 in Fig. 5, allows for off-axis spreading of each gas jet (12) to minimize interaction with the gas flow. The straight wall tip, in common usage, was tested in a retracted position, no. 1 in Fig. 5, and in an extended position, no. 2 in Fig. 5. The retracted position would have no possible contact with the flowing gas jets, while the extended position would have the maximum gas interaction, by diverting the projected cores of the gas jets.

The results summarized in Figure 6 show that two designs, no. 2 and no. 4, produce aspiration conditions over a broad range of elevated inlet pressures. The other two designs apparently approach ambient base pressure at about 3MPa inlet pressure, but do not produce any aspiration effect in the high pressure regime. Thus, a choice between the two aspirating designs was necessary for further study. The 45° taper was selected over the extended cylinder because of an obvious advantage in potential utilization of gas stream energy for atomization, ie., no diversion of the gas jet core flow. Also, problems were anticipated with gas induced erosion of the straight wall tip at high gas velocities.

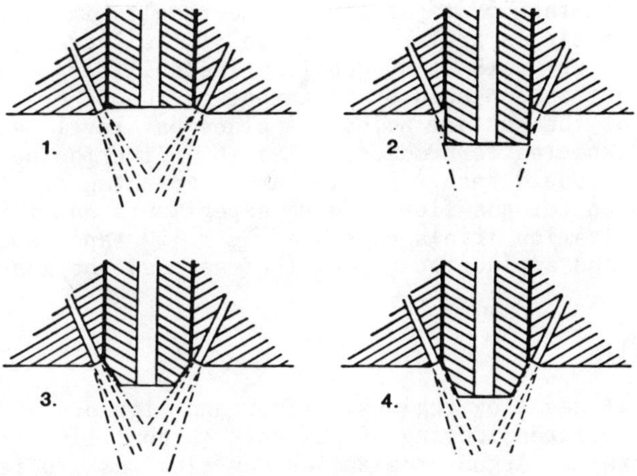

Figure 5: Four melt feed tube tip designs. Design features include: 1.) straight wall cylinder, retracted, 2.) cylinder, extended, 3.) 63° apex angle, 4.) 45° apex angle.

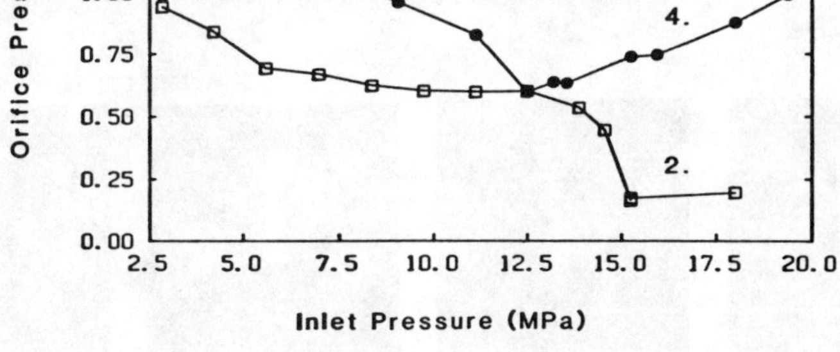

Figure 6: Summary of melt orifice pressure results for four tip designs.

Two other tip design parameters, the length of the tapered section, s, and the placement level (relative to the gas jet holes) were investigated with the 45^0 taper tip. The results indicate that a relatively short tip is preferred because of its increased range of aspiration behavior, without sudden flow instabilities. Also, the aspiration effect is enhanced by placement of the taper onset at the precise level of the gas jet holes. A placement level below this level also would be expected to produce aspiration, like the no. 2 design of Figure 5, but it would result in similar disadvantages. Therefore, the tip design chosen for gas flow imaging experiments and for all subsequent atomization trials consisted of a 45^0 taper angle, a length of s = 1.93mm, and a placement level that matched the gas jet hole circle.

Schlieren results

Two distinct gas flow regimes, at low and high pressures, were indicated by schlieren imaging of gas only (argon) flow from the atomization nozzle. Argon atomization gas flow was studied by the schlieren method for inlet gas pressures from 2 to 17MPa. The low pressure regime, from 2 to about 5MPa, consists of a symmetrical array of individual gas flow patterns with a pressure dependent "diamond" shock pattern for each of the 18 jets, as shown in Figure 7. In the

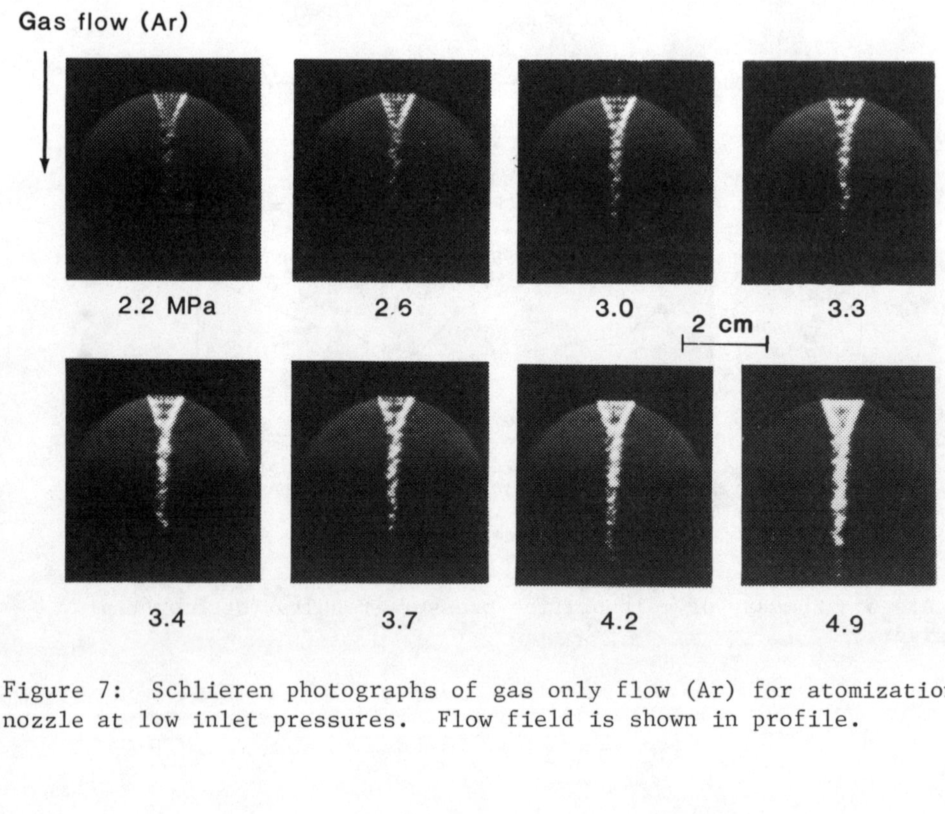

Figure 7: Schlieren photographs of gas only flow (Ar) for atomization nozzle at low inlet pressures. Flow field is shown in profile.

high pressure regime, from approximately 6 to above 17MPa, the individual flow patterns merge to form a unified gas flow field consisting of an initial normal or Mach shock disk and a large diamond pattern extending downstream. As Figure 8 shows, the most prominent feature of these high pressure flow images seems to be the initial normal shock of the unified flow. As the inlet pressure is increased to 12.4MPa, the intensity and localization of this Mach shock disk also is progressively increased. At higher inlet pressures the Mach disk appears to split and, again, to become diffuse. Eventually, the interior of the unified flow seems to develop a low density "core", as indicated by the 16.3 and 17.3 images of Figure 8. A schemmatic summary

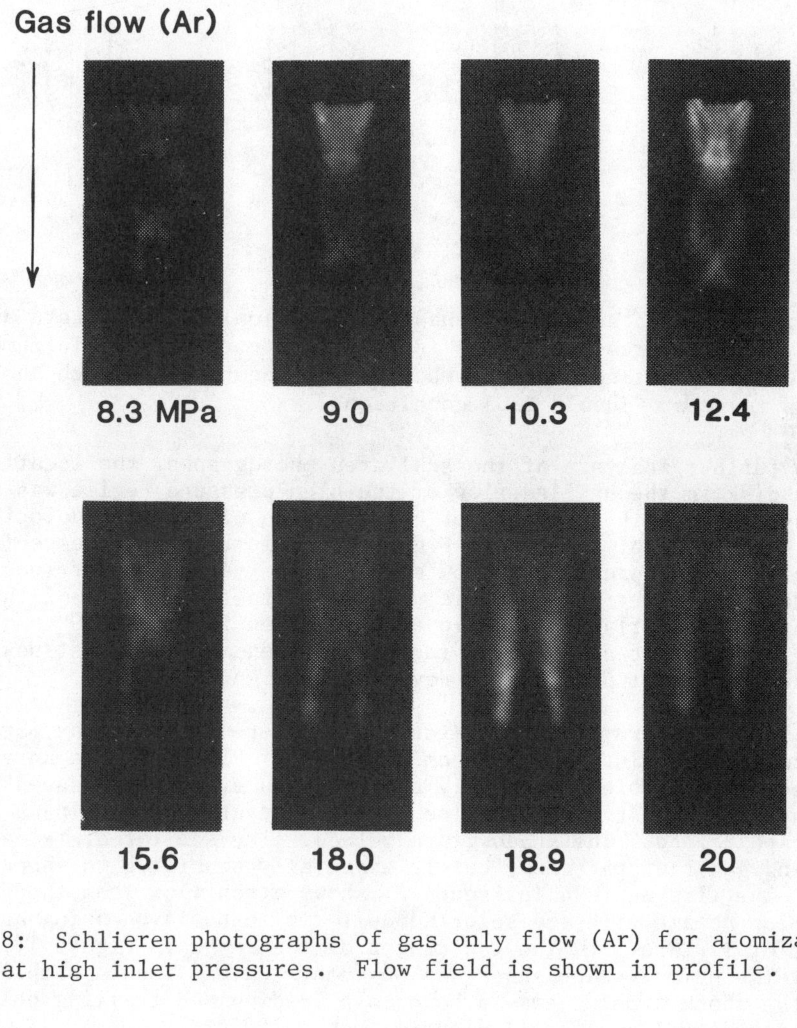

Figure 8: Schlieren photographs of gas only flow (Ar) for atomization nozzle at high inlet pressures. Flow field is shown in profile.

of the evolution of the gas flow patterns from low to high pressures is given in Figure 9, where, for clarity, only two gas jet flows are illustrated.

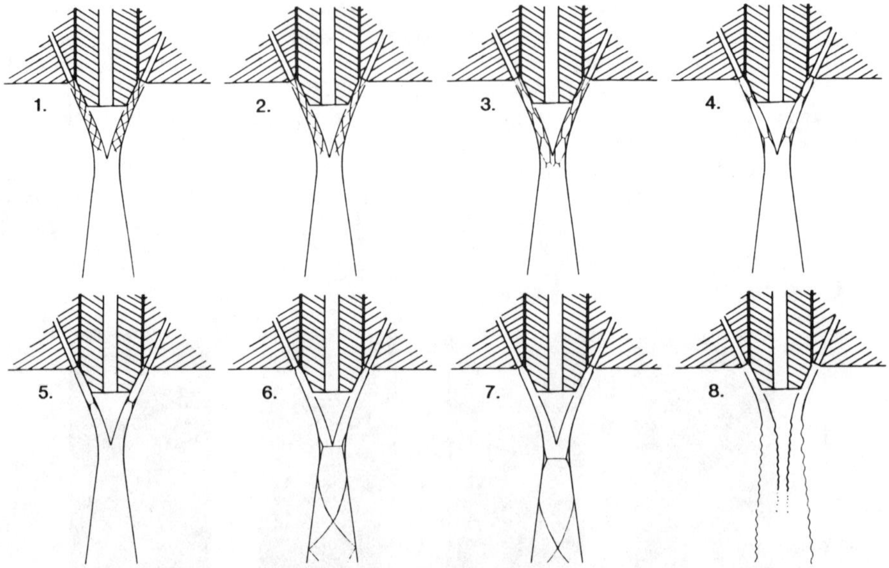

Figure 9: Schematic summary of changes in gas only flow pattern as a function of inlet pressure, where increasing pressure level increments are indicated by an increasing number in the sequence. Sketch number 7. indicates image of "ideal" HPGA conditions.

In a further analysis of the schlieren photographs, the location of the Mach disk in the unified flow of the high pressure regime was found to be associated with a "waist" or constriction of the high velocity flow pattern profile, as seen in Figure 8. This effect was especially clear in the inlet pressures below about 18MPa. A summary of the measurements from the photographic negatives (in true scale) of the waist position relative to the tip edge is given in Figure 10 as a function of inlet pressure. The increasing trend of the waist position with increased inlet pressure is revealed by Figure 10.

The long exposure times used for the schlieren photographs may have caused some image blur in the shock patterns of Figure 8. In an effort to correct this problem, extremely short exposure, or "high speed", photographs of the Ar flow from the nozzle were attempted using a shadowgraph method. The shadowgraph method, like schlieren, is capable of imaging gas flow patterns, but is especially sensitive to shock fronts. The flow pattern in Figure 11 shows argon flow from the atomization nozzle with the selected melt tip in position using an inlet pressure of 12.1MPa. In the top center of the image in Figure 11, an image of the primary Mach shock can be seen. Rather than a simple disk shape, the shock front seems to have both leading and trailing oblique shocks extending from a small diameter central disk.

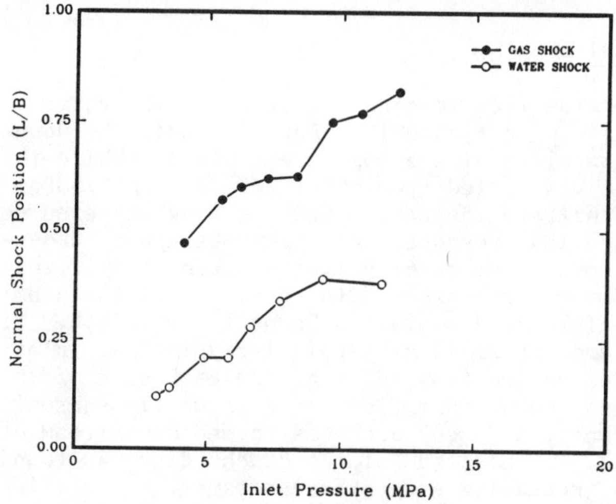

Figure 10: Analysis of schlieren results and water table shadowgraphs. Reveals inlet pressure dependence of downstream displacement, L, of normal shock front or "Mach" disk from tip, where B is tip diameter or width. The water table results are for s = 6mm.

Figure 11: Laser shadowgraph of Mach disk with 35ns exposure time.

Water table results

A summary of the flow imaging results from the water table experiments is given in Figure 12. For all tests the downstream ambient pressure was maintained at 1 atmosphere (101kPa), while the stagnation (inlet) pressure was varied from about 2MPa through 15MPa. Figures 12a through 12d demonstrate the progression of flow patterns observed with increasing stagnation pressure. All flows behave as underexpanded high velocity base flows. The exact values and ranges over which each pattern existed was a function of the geometry of the atomizer model, here manifested through a variation in melt tip length. Figure 12a is typical of patterns observed at a relatively low gas inlet pressure. A diamond shaped expansion flow issuing from each side of the tip can be seen. The general downstream flow field is mostly subsonic. With increasing pressure, a "Mach" disk was formed downstream of the melt tip. In general, the position of the "Mach" disk was found to move downstream with increasing stagnation pressure. This effect has been documented in Figure 12c for one value of melt tip length. At very high water pressures, Figure 12d, the "Mach" disk could no longer be discerned.

Figure 12: Flow patterns downstream of water table nozzle model for s = 6mm at stagnation pressures (MPa) of a) 4, b) 9, c) 11, and d) 14.

Figure 13 documents the results of the base (melt orifice) pressure measurements with stagnation (inlet) pressure for three values of melt tip length, s, with a scale factor of 6.8. The base pressure is seen to decrease continuously with increased stagnation pressure to a minimum followed by a slow increase for all configurations. The curve minimums display some sensitivity to melt tip length, but the two longest tips have very similar behavior. Since the longest tip, 13.0mm, is the scaled equivalent of the 1.93mm atomizer tip, the extended length found to produce gas flow instabilities was not studied as of yet with this water model.

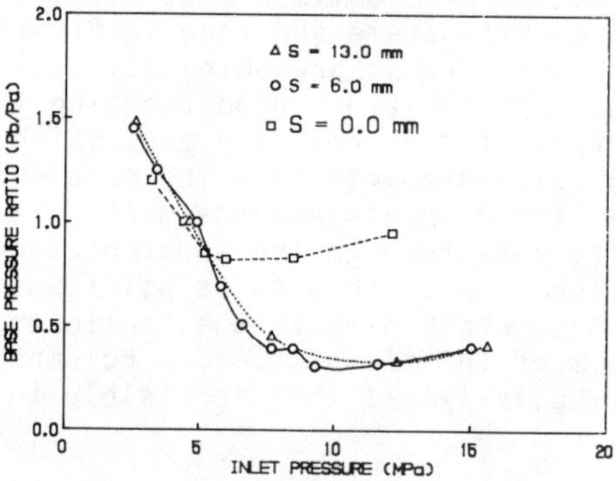

Figure 13: Base pressure ratio vs. inlet stagnation pressure with three values of tip length, s, measured on water table.

It is interesting to compare Figures 13 and 10. The region of a local maxima in Figure 13 corresponds directly to a region of minima in the water table results of Figure 10. This may reveal the effect of the "Mach" disk location on the base (orifice) pressure. The flow patterns observed at stagnation (inlet) pressures below the minima in Figure 13 were typical of Figure 12b, while those noted at stagnation pressures above the minima in Figure 13 were typical of Figure 12c.

In comparison to gas atomizer tests, the base (orifice) pressure observed on the water table follows the same tendency as those observed for the gas only atomizer flow field. The variation in base pressures measured on the water table exhibit a greater range of values than observed in the actual atomizer tests. However, the minima discussed in Figure 12 are in very reasonable agreement to those measured in the atomizer and the flow patterns observed are in good agreement with those observed using schleiren photography in the atomizer. In addition, a comparison of the gas flow image results and the water table results for the normal shock position dependence of inlet pressure in Figure 10 show another of good agreement of the gas flow and the water flow analogy, albiet with some displacement of the inlet pressure scale. Efforts to improve the uncertainty in depth measurements made on the water table are expected to provide more quantitative base pressure data from the water table.

High speed cinematography results

A preliminary viewing of the resulting movie revealed that the initial atomization event was interrupted by a brief melt freeze up, ie. only gas was flowing into the atomization zone. This was followed by resumption of the melt flow and further atomization. The brief freeze up period enabled us to observe the onset of melt flow into a fully established gas flow pattern. The sequence of movie frames in Figure 14 show the path of the melt stream in stop action from the time of initial exit from the melt tube orifice until a full atomized spray pattern was reestablished. As the still frame sequence in Figure 14 reveals, the initial melt stream trajectory is predominantly horizontal, toward the external edge of the feed tube tip. Upon reaching the edge, the melt mass encounters the flow of high velocity gas, which causes some immediate shearing of fine droplets from the mass and a filming of the remaining melt. The fine droplets and the melt film are directed by the flowing gas curtain to pass through the apparent focus point of the gas jets. The measured location of this focus point seems to coincide with the location of the Mach shock disk in the schlieren results for gas only flow. Downstream of the focus point, a columnated spray of very fine particles is essentially all that is visible in the movie still frames of Figure 14.

CONCLUSIONS

The HPGA flow field is similar in many respects to a high Mach number, underexpanded plug nozzle flow in rocket nozzles, a fluid dynamic system which was studied with great intensity in the 1960's. The results of this study and other results (3) suggest a strong correlation between the features of the HPGA gas only flow field and the resulting powder size and size distribution. Based on an analysis of the rocket nozzle findings and these results, the operational features of the gas atomization nozzle in the HPGA regime can be described, with reference to Figure 15. As the results in Figure 14 suggest, no approximation of an intact melt stream projects into the atomization zone, as has been reported (13). Instead, the melt forms a complete or partial film immediately upon exit from the orifice and the film flows in a radial outward direction. The melt behavior is probably influenced by two effects in this base flow region. First, a local subambient pressure could cause initial splitting and fanning of the stream. Second, a strong recirculation flow (5), which is also found in plug nozzle flow, could accelerate the melt film in a horizontal direction toward the external edge of the feed tube. A significant portion of the melt disintegration probably occurs when the melt film makes initial contact with the supersonic gas flow at the tube edge. Further disintegration and refinement of the droplets can also occur in the jet focus zone due to the action of an intense shock structure that appears to persist in this zone. Downstream of the focus zone, the gas flow entrains the very fine droplet spray in a narrow spray cone which gradually broadens as the velocity decays. Solidification of the droplets occurs in this spray to produce very fine powder particles.

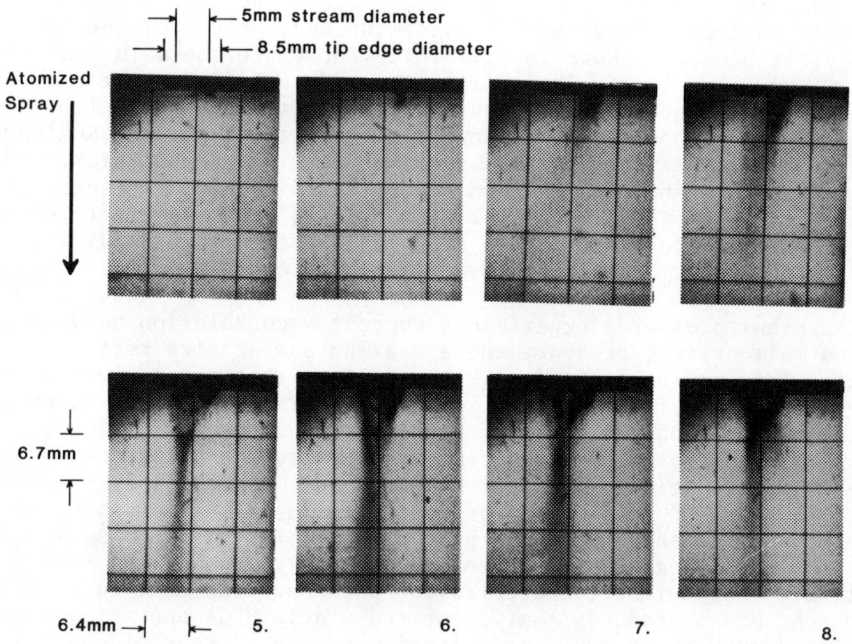

Figure 14: Sequence of still frames from high speed movie of argon atomization of Cu-2wt.%Zr alloy. The frame interval is 231μs and the exposure time for each frame is 92μs.

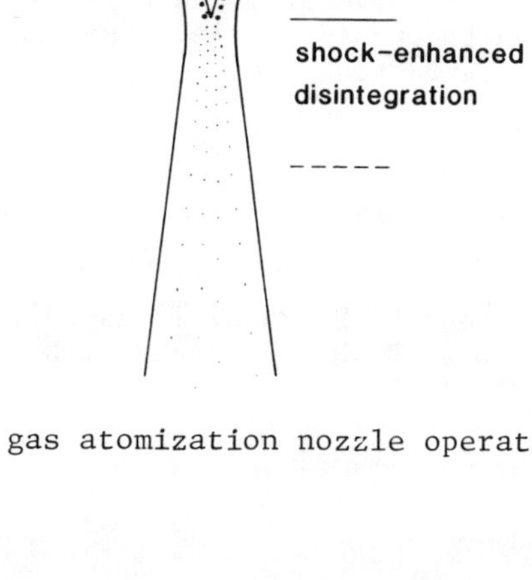

Figure 15: Schematic of gas atomization nozzle operating under HPGA conditions.

The results presented here also demonstrate the similarities between the actual three dimensional atomizer flow and the two dimensional hydraulic analogy. Despite the two dimensional nature of the hydraulic analogy, it has been shown to faithfully reproduce the main features of the HPGA gas flow field. The ability to study modified geometries and to obtain detailed, qualitative visualization of the resulting flow fields with ease and without atomizer down time make the hydraulic analogy a valuable additional tool for the study of this gas flow system. The use of the hydraulic analogy has been demonstrated to be quite successful in the study of other complex flow fields of base flow systems (5,6) and will be utilized for an expanded study of the effects of modified atomizer configurations on the overall flow field.

The results of HPGA experiments suggest a correlation between a minimum melt orifice pressure and a maximum powder size refinement effect. This may be coupled with the finding that the maximum intensity of the Mach shock front also is reached at this same nozzle operating condition in gas only flow. If the atomization process at these conditions takes place as suggested above, a sensor and process control scheme may also be suggested for the atomizer. The Mach disk region and the low base pressure zone may be preferred locations for an optical image comparator and a pressure probe, respectively. Such sensor approaches could enable real time process analysis and lead to intelligent control of the atomization process. The melt tube design philosophy and procedures that permitted stable HPGA operation may also be applied to high pressure modification of other types of gas atomization nozzles.

ACKNOWLEDGEMENTS

The authors wish to express appreciation to J. D. Ayers, W. P. Robey, P. Slebodnick, and R. G. Hughes for valuable technical participation in this investigation. One of the authors (I.E.A.) wishes to acknowledge the support received at the US Naval Research Laboratory, where most of the gas atomization analysis was performed, and the current support of the Ames Laboratory (US DoE).

REFERENCES

1. I. E. Anderson, K. M. Rusin, and B. B. Rath: in Rapid Solidification Processing; Principles and Technologies, IV, R. Mehrabian and P. Parrish, eds., Claitor's, Baton Rouge, LA, 1988, p. 114.

2. S. Small and T. J. Bruce, Int. J. of Powder Metallurgy, 4, no. 3, p. 7 (1968).

3. I. E. Anderson and B. B. Rath: in Rapidly Solidified Crystalline Alloys, S. K. Das, B. H. Kear, and C. M. Adam, eds., TMS-AIME, Warrendale, PA, 1985, p. 219.

4. M. Van Dyke: An Album of Fluid Motion, Parabolic Press, 1982.

5. R. H. Page and A. P. Meyer, A. R. S. Journal, March issue (1961).

6. H. S. Ningappa and S. Narasimhan, A. I. A. A. Journal, 14, no. 5 (1973).

7. D. M. Adams, Journal of Spacecraft, 4 (1967).

8. C. J. Schorr, Journal of Spacecraft, 5 (1968).

9. J. S. Rao, V. V. Ramana Rao, and V. Seshadri, Journal of Spacecraft, 8 (1970).

10. D. R. F. Harlemann and A. T. Ippen: in Anniversary Volume for M. D. Riabouchinsky, Science and Technology, Paris, 1954.

11. V. Anand, A. J. Kaufman, and N. J. Grant: in Rapid Solidification Processing, Principles and Technologies, II, R. Mehrabian, B. H. Kear, and M. Cohen, eds., Claitor's, Baton Rouge, LA, 1978, p. 273.

12. G. H. Geiger and D. R. Poirier: Transport Phenomena in Metallurgy, Addison-Wesley, Reading, MA, 1973, p. 163.

13. J. S. Thompson, J. Inst. Metals, 74, p. 101 (1948).

THE INFLUENCE OF PROCESS VARIABLES AND EQUIPMENT CHARACTERISTICS ON THE PURITY OF ATOMIZED POWDERS

H.W. Meinhardt and P. Kunert

Hermann C. Starck Berlin
GmbH & Co. KG
Niederlassung Laufenburg

Abstract

The most economical method of manufacturing ultra-pure, high-grade alloyed powders for P/M parts such as turbines is inert gas atomization. This process is characterised by the fact that the alloy melt comes into contact with ceramic parts during melting, pouring, and atomizing.

This process is economical only if different alloys can be produced in the same apparatus. However, this can contaminate the alloy powders with ceramic, metallic, and organic materials. Consequently it is very important to be able to identify the possible sources of contamination and to identify foreign inclusions such as oxides and other alloy particles in a special alloy like Udimet 700*.

This paper describes the influence of process variables and equipment characteristics on the physical and chemical properties of atomized powders. Special attention is devoted to the sources of oxygen impurities. In addition the authors have investigated possibilities of metallic contamination and demonstrated that foreign alloy particles, even those very similar in chemical composition to Udimet 700, can be easily identified in HIPed parts.

*) Trade mark of Special Metals Corp.

Introduction

Manufacturers of sintered parts evaluate atomized powders largely on the basis of quality - in particular, purity. However, even manufacturers of aircraft engine parts are increasingly cost-conscious, requiring the economical production and processing of atomized powders. The most demanding requirements are placed on powders which are HIPed to form rotating parts for turbines.

The quality of these powders depends on the physical and chemical properties of the individual particles an on bulk properties like the shape, chemical composition, surface impurities, flowability, bulk density, and contamination by foreign particles. For good flowability the individual particles should be spherical and have a packing density of about 62-67 % of the theoretical value. There should be no particles with gas inclusions. The influence of different atomization processes on the particle shape is well known. Figure 1 depicts particles atomized in inert gas, fluid gas, and ultrasonic gas.

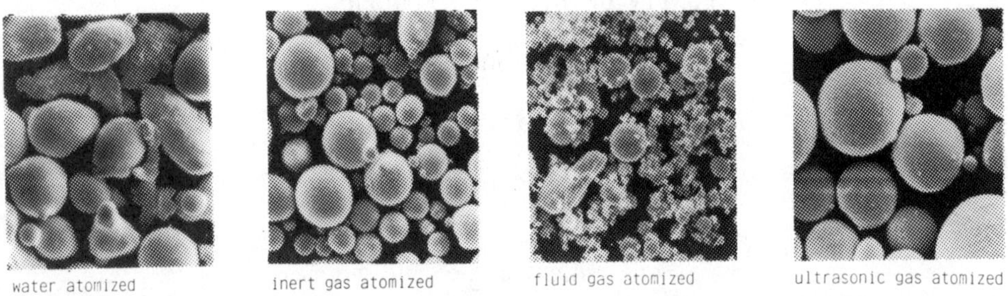

Figure 1: Alloy particles atomized with different media

Figure 2 demonstrates that different conditions during the production process yield various particle shapes and qualities. Particles with many satellites cause bad flowability and a low packing density.

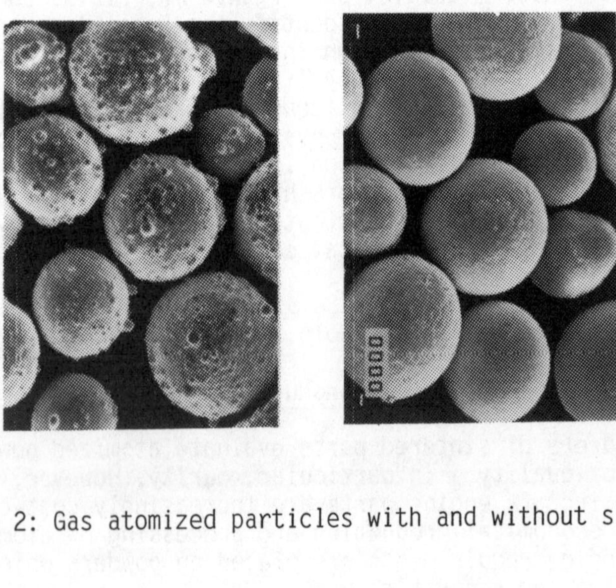

Figure 2: Gas atomized particles with and without satellites

If one uses a gas atomization equipment of sufficient height to provide all the melt droplets adequate time to solidify, the atomization and heat removal parameters can be varied so that satellite formation is almost eliminated. The possibility of varying the solidification process in an apparatus with a short atomizing tower is limited because a faster removal of heat is required; consequently a short tower leads to increased satellite formation.

Not only the physical properties but the chemical composition and the homogeneity of the individual particles and of the powder have a decisive influence on the properties of the finished sintered parts.

Three types of inclusions or foreign particles which produce chemical inhomogeneity are usually mentioned in the literature:

organic, ceramic and metallic foreign particles.

Organic particles and their residues (fig. 3) can be eliminated by avoiding contact with organic materials like plastic, rubber etc. during the production, handling, and processing of the powder.

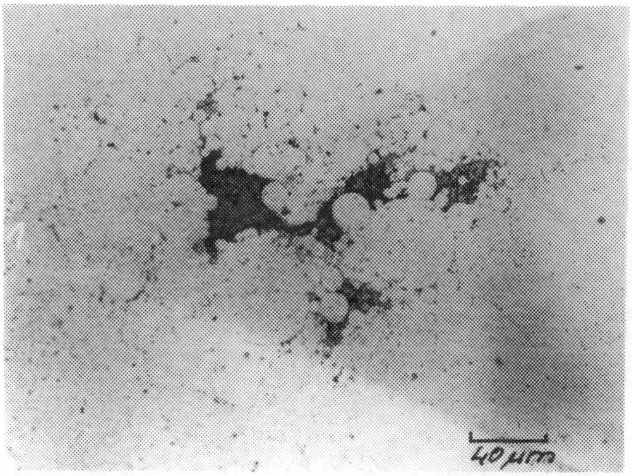

Figure 3: Residue of an organic particle after HIPing in Udimet 700

Oxidic foreign particles and the oxygen content

Oxidic contaminations are usually attributed to contact between the alloy melt and the melting crucible, the tundish and the filter ceramics. They result from chemical reactions between the alloying elements and ceramic crucibles or are broken up by microcracks in the ceramic material. Slag particles are also oxidic in nature. The quantity of slag particles present depends on the metallic starting material - in this case vacuum induction melted Udimet 700 or master alloys. In addition oxygen is interstitially combined in the metallic lattice or adsorbed at the surface of the powder particles. This portion of the oxygen content can be regarded as non oxidic. All of these source are responsible for the oxygen content of the powder.

Figure 4 depicts the origin of oxygen and the type of binding of each variety.

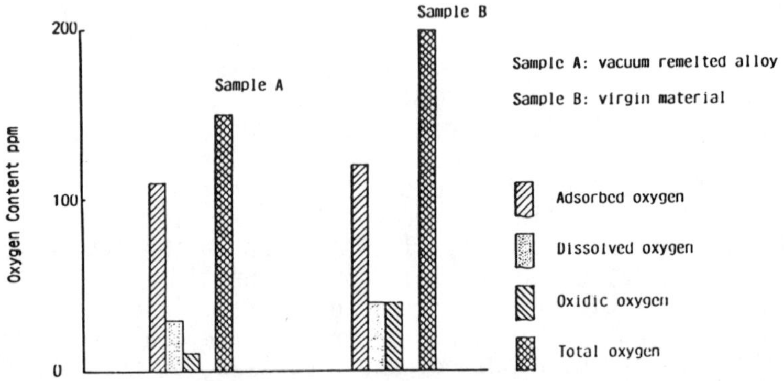

Figure 4: Origin and type of binding of all varieties of oxygen in a powder

By analysing the powders at different temperatures one can determine the quantities of each sources of oxygen present. Figure 5 shows the results of this chemical analysis.

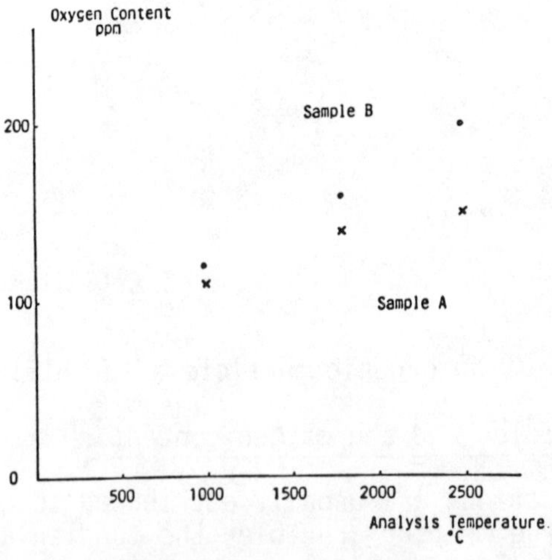

Figure 5: Oxygen content as a function of the analysis temperature

By adjusting the ceramic materials to the reactivity of the melt and optimizing the smelting method the amount of oxidic, endogeneous inclusions and interstitially combined oxygen can be diminished significantly. The result is an acceptable oxygen content of about 100 ppm in the < 45 microns fraction. This adjustment and optimization is necessary for each alloy, because the critical amount of ceramic particles is different for different alloys. Oxygen adsorbed at the surface can be largely eliminated by treating the powder in an inert gas. The following results were obtained with Udimet 700 powders or HIPed parts made from them. These powders are produced using our equipment depicted in figure 6.

Figure 6: HCST`s atomizing equipment

Altogether about 90 batches have been evaluated. On the basis of the data gathered we have determined:

- the most suitable ceramic materials for the melting crucible, the tundish, and the tundish outlet;
- the influence of ceramic filter on the quantity of oxidic particles in the powder and on the oxygen content;
- the optimal pouring temperature of Udimet 700 in this equipment as far as the oxygen content, particle shape, and the yield in the desired grain size spectrum are concerned;
- the influence of the inert gas atmosphere on the oxygen content of the powders.

Our tests have shown that the application of a suitable ceramic filter can decrease the oxygen content of the powder by about 25-30 %.

HCST`s atomizing equipment has been built - as shown in fig. 6 -

in such a manner that powders can be collected in two containers, marked by A und Z. The arrangement of the second container under a cyclone leads to a separator effect, which can be seen clearly by the particle size distribution. Powder A is coarser than powder Z. While the percentage of material in A < 45 microns is about 45 % by weight, the percentage in Z is about 85 % by weight, that is twice as high as in A. An analysis of the oxygen content of both powders reveals that the finer powders have a higher oxygen content than the coarser ones (fig. 7).

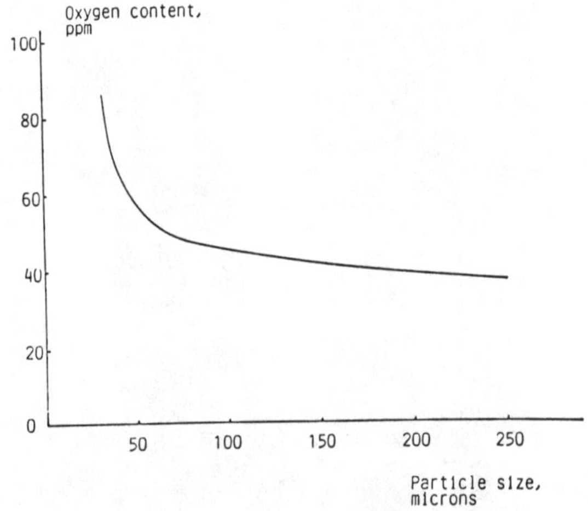

Figure 7: Oxygen content of Udimet 700 as a function of grain size

The determination of oxygen at different temperatures demonstrates that the higher oxygen content is caused by superficially adsorbed oxygen, e. g. moisture. In fact, the oxide particles are subject to the same separator effect caused by the two containers. Thus the quantity of coarser (endogeneous) oxide particles in the finer powder is significantly reduced before sieving. The quantity and the size of the oxide inclusions determine, together with other factors, the mechanical strength of P/M parts. Figure 8 shows an oxide inclusion in Udimet 700.

In order to characterize the purity of different "A" and "Z" powders Thyssen Guß, Bochum (Germany) manufactured samples from them by HIPing at 1,150°C and 1,400 bar for 4 hrs. We analysed metallographic cross-sections of these samples using a special method provided by MTU, Munich (Germany). The cross-sections have been etched by a reagent which contains primarily FeCl3 and CuCl2. Thereafter an area of about 30 cm² was visually (microcopically) investigated and all ceramic particles classified. Normally organic and metallic foreign particles are only present in small quantities, if at all; most of the inclusions are oxides. Their quantity and size can be represented in a log-log-diagram depicted in figure 9, according to MTU specifications. The straight line delineates the border between usable and unusable powders (also according to MTU specifications).

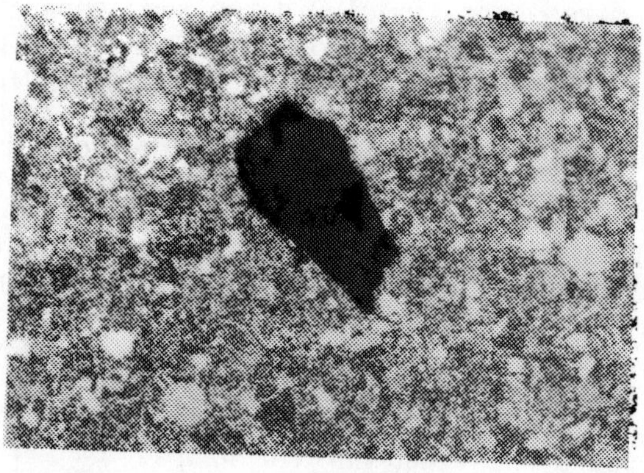

Figure 8: Oxide inclusion in Udimet 700

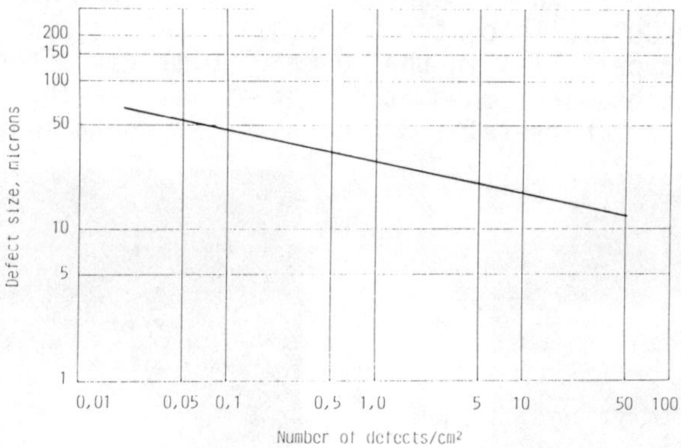

Figure 9: Classification of oxide inclusions in P/M parts according to quantity and size

This method of testing showed that all the powders produced from vacuum induction melted Udimet 700 contain oxidic particles < 20 microns in the HIPed part smaller than maximum tolerance. Given a batch of about 100 kg, the total number of ceramic inclusions in all size ranges is within the range of tolerance after a few batches of saturation.

Metallic foreign particles

Atomized powders can only be produced economically if different alloys can be atomized successively in the same apparatus. Therefore in the equipment shown earlier various Ni, Co, and Fe base alloys are atomized in an appropriate sequence. However, high-purity alloy powders may not contain cross contaminations, that is particles of other alloys. This requires special cleaning methods and preparatory batches of the desired alloy. These precautions can effectively eliminate almost all cross contaminations.

A suitable method of testing had to be developed to determine the presence of foreign metallic particles especially in small quantities. Tests have shown that the etching procedure described above provides reliable results. For these tests Udimet 700 powders, < 45 microns have been intentionally contaminated with 0,1-0,3 % by weight of various Fe, Co, and Ni base alloys. These contaminated powders were then HIPed at 1,150°C and 1,400 bar for 4 hrs.

Cross contaminations by Fe base alloy powders

Figure 10a shows an Udimet 700 structure in the HIPed condition, etched by FeCl3/CuCl2. In contrast to this flawless state figure 10b reveals metallic particles in the Udimet 700 matrix after the same HIP procedure, cross-sectioned at different levels. This sample contains steel particles (AISI 316 L) as cross contamination.

a)

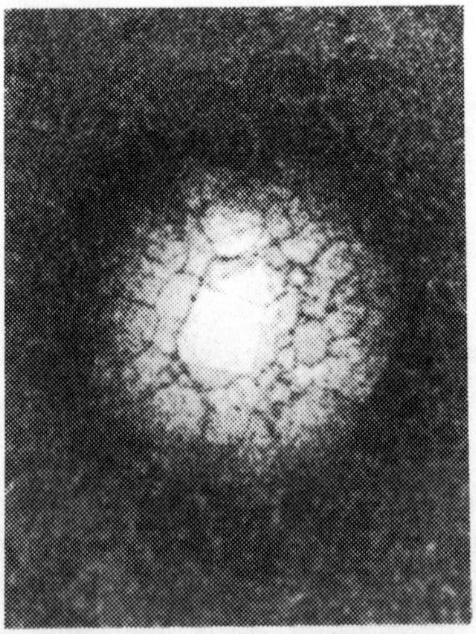

b)

Figure 10: a) Udimet 700 structure, HIPed, 500 : 1, etched
b) Udimet 700, doped with AISI 316 L-powder, 500 : 1, etched

Figure 11 shows the SEM micrograph of such a particle with the qualitative Fe distribution along the straight line indicated.

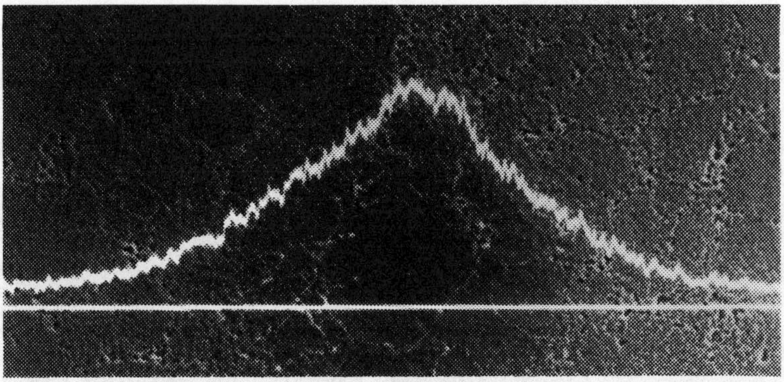

Figure 11: SEM micrograph with Fe line scan of an AISI 316 L inclusion in Udimet 700, 500 : 1, etched

Table I: Concentrations of the essential alloying elements in the matrix and the inclusion zone, AISI 316 L in Udimet 700.

Element	Matrix U 700	Reaction zone periphery (% by weight)	Reaction zone center (% by weight)	AISI 316 L
Ni	53,5	51,3	47,1	12,0
Co	17,6	16,5	18,4	-
Cr	17,0	17,3	18,4	17,5
Mo	4,6	4,4	4,7	3,0
Al	4,0	3,9	2,9	-
Ti	3,3	3,2	2,6	-
Fe	-	3,5	5,9	65,5

In table I the determined values for the main constituents in different areas are compiled:
- for the unaffected matrix
- for the dark diffusion zone (periphery) between Udimet 700 and 316 L and
- for the center of the steel particle

The Fe content of the center has decreased by diffusion compared with the chemical composition of AISI 316 L from 65 to about 6 % by weight. The Cr content is essentially unchanged, whereas the Co and Ni content have increased almost to the level in the matrix due to their inward diffusion and to the outward diffusion of Fe. Al and Ti do not quite attain the matrix level. This test demonstrates that steel particles can be identified in an Udimet 700 matrix despite the fast concentration compensation by diffusion. This is also true for defects which are cut only close to their periphery.

Cross contamination by Co base alloy powders

Co base alloy inclusions in Udimet 700 can also be clearly identified in the matrix with the etching method. This is demonstrated using CoCrAlY and F 75 particles depicted in figures 12 and 13.

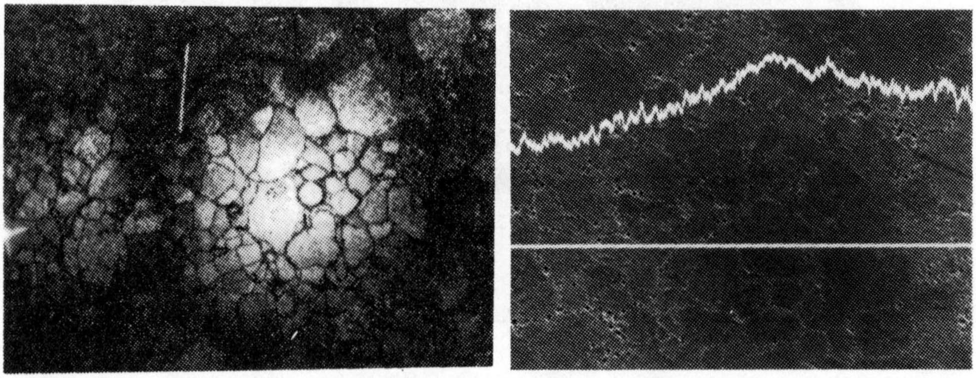

a) b)

Figure 12: CoCrAlY particles in Udimet 700, HIPed
a) metallographic cross section, 500 : 1, etched
b) SEM micrograph, 500 : 1, Co line scan

Table II: Concentrations of the essential alloying elements in the matrix and the inclusion zone, CoCrAlY in Udimet 700.

Element	Matrix U 700	Reaction zone (% by weight)	CoCrAlY
Ni	52,6	43,8	-
Co	17,7	25,0	64,0
Cr	16,6	20,1	29,0
Mo	5,4	5,3	-
Al	3,9	3,0	6,0
Ti	3,7	2,8	-

Between the center of the CoCrAlY particles and the matrix one can see a diffusion zone without a sharp border line in the cross section (etched condition). The SEM micrograph shows that an essential equalization by diffusion has occurred here too. This is confirmed by the chemical analyses listed in table II and the qualitative Co distribution.

The quantity of Co within the reaction zone of 64 wt.-% is diminished to 25 % by weight after HIPing. Similarly the Al content has decreased in all samples compared to the original CoCrAlY composition. On the other hand Ni, Mo, and Ti have penetrated into the CoCrAlY particle. These foreign particles can be clearly recognized with the simple etching method even though the concentration differences are very low. This is especially true for F 75 particles as shown in figure 13.

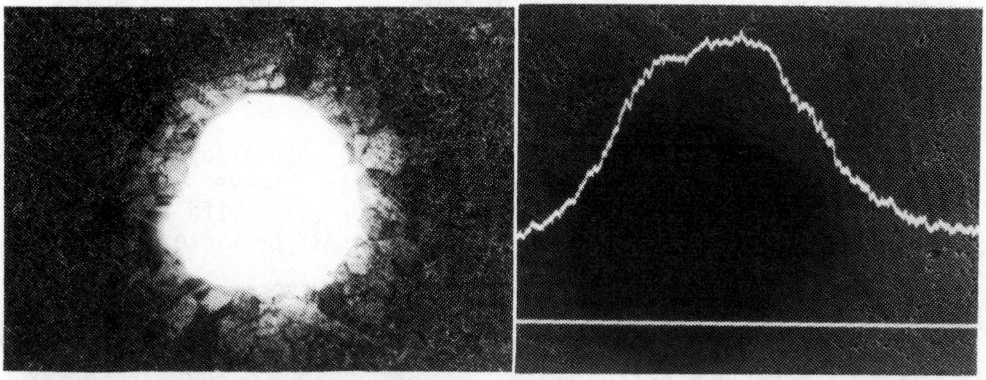

a) b)

Figure 13: F 75 particles in Udimet 700, HIPed
a) metallographic cross section, 500 : 1, etched
b) SEM micrograph, 500 : 1, Co line scan

In this case the alternations caused by diffusion are less clear if one compares the nominal composition of F 75 with the values determined for the center of the foreign particle.

Table III: Concentration values of the essential alloying elements in the matrix and in the inclusion zone, F 75 in Udimet 700.

Element	Matrix U 700	Reaction zone periphery (% by weight)	Reaction zone center	F 75
Ni	50,4	46,6	2,0	-
Co	19,5	23,7	61,0	65,0
Cr	17,6	18,2	30,7	28,5
Mo	5,4	4,5	6,1	6,0
Al	3,4	3,3	-	-
Ti	3,4	3,6	0,3	-

The measurements reveal that the Co content is decreased by about 4 % by weight. Ni has diffused into the outer zone of the original particle and to a lesser degree into the center. The Cr content has increased slightly. Al and Ti have diffused into the outer zone but not into the center. These results may indicate that a very large F 75 particle has been cut. However it is also possible that the diffusional compensation between F 75 and Udimet 700 takes places more slowly than between CoCrAlY and Udimet 700. Of all the cross-contaminationed particles studied F 75 could be recognized most clearly in the Udimet 700 matrix. The distinct differences compared to CoCrAlY are probably due to the Mo content (6 % in F 75).

Cross contamination by Ni base alloy powders

In our experience foreign particles of Ni base alloys are most difficult to recognize in the Udimet 700 matrix. However they are also visible in the etched cross section, especially with low magnification as shown in figure 14 using NiCrAlY particles.

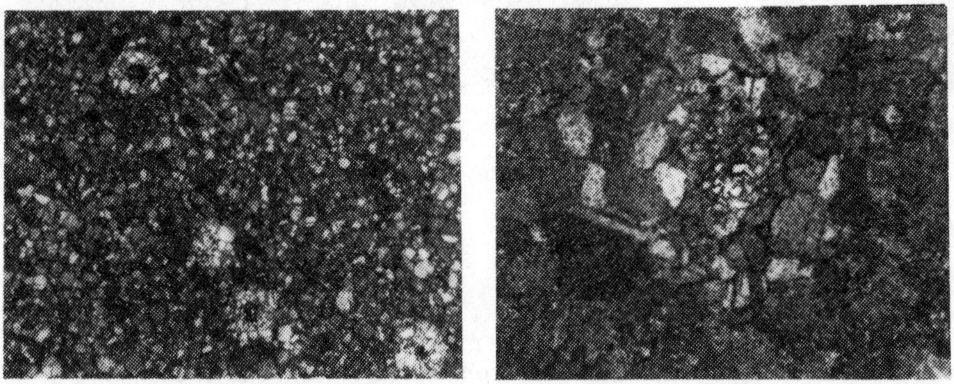

a) b)

Figure 14: NiCrAlY particles in Udimet 700, HIPed metallographic cross section, etched
a) 100 : 1 b) 500 : 1

In contrast to Fe and Co base alloys, Ni containing particles can be identified by their dark central areas surrounded by a bright diffusion zone. However, these characteristics depend on the level at which the particle was cut. The SEM micrograph (fig. 15) shows clearly only the dark region which consists of many precipitations. Concentration values are given in table IV.

The precipitations are probably finely dispersed γ'-particles which are coarser in the region of the original foreign particle. Further research will attempt to prove this. The smallest difference between the initial composition of Udimet 700 and the foreign alloy occurred in our tests using Udimet 700 and IN 100.

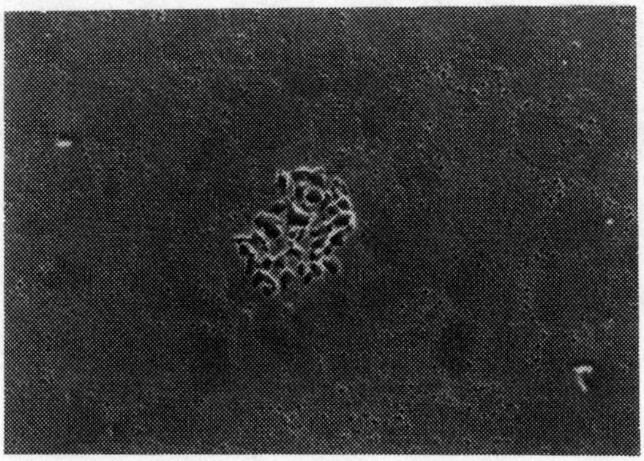

Figure 15: NiCrAlY particles in Udimet 700, HIPed, SEM micrograph, 500 : 1

Table IV: Concentration values of the essential alloying elements in the matrix and in the inclusion zone, NiCrAlY in Udimet 700.

Element	Matrix U 700	Reaction zone (% by weight)	NiCrAlY
Ni	53,9	56,6	66,7
Co	17,6	15,5	-
Cr	16,4	16,0	22,0
Mo	4,7	3,6	-
Al	3,7	4,1	6,0
Ti	3,0	4,2	-

However even these foreign alloy particles are clearly identifiable by their dark nuclei after the consolidation procedure (fig. 16). Here too the dark particles are presumably coarse γ'-precipitations formed by diffusion conditioned concentration changes. Figure 17 shows an SEM micrograph of this region and table V lists the determined compositions of the matrix, the nucleus, and the nominal composition of IN 100.

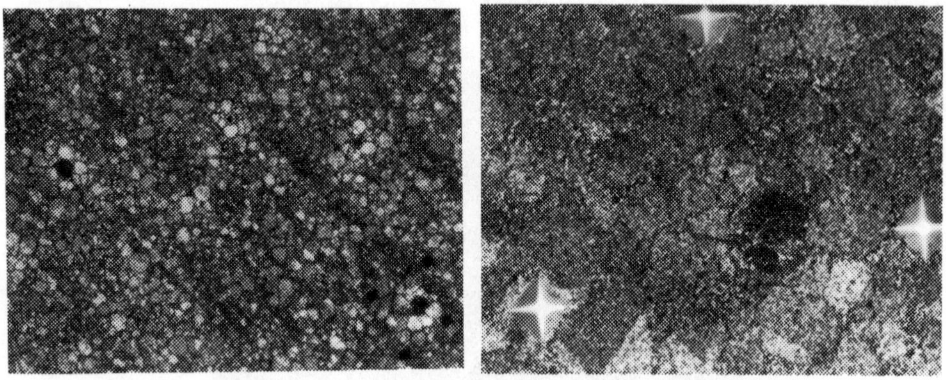

a) b)

Figure 16: IN 100 particles in Udimet 700, HIPed
a) 100 : 1 b) 500 : 1

Table V: Concentration values of the essential alloying elements in the matrix and in the inclusion zone, IN 100 in Udimet 700.

Element	Matrix U 700	Reaction zone	IN 100
		(% by weight)	
Ni	54,9	57,9	60,6
Co	16,6	15,2	15,0
Cr	15,8	13,8	10,0
Mo	4,6	3,8	3,0
Al	4,6	3,7	5,5
Ti	3,5	5,6	4,7

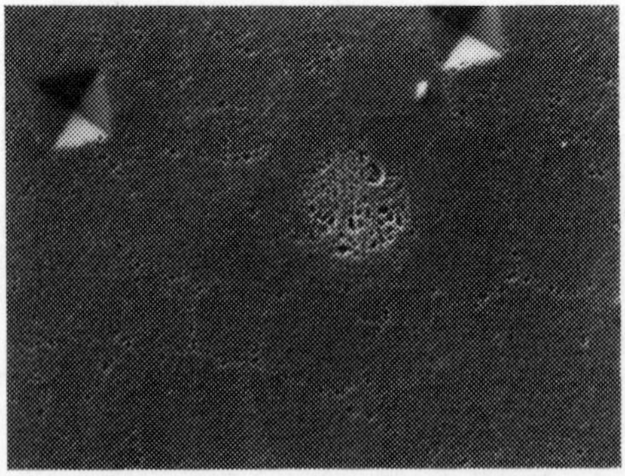

Figure 17: IN 100 particles in Udimet 700, HIPed
SEM micrograph, 500 : 1

Summary

The research described above has helped HCST to improve the production of vacuum melted powders atomized in inert gas - in particular the nickel base alloy Udimet 700. The improvements in quality extend from the shape of the particles to the amount of fine powder produced, the purity of the powder and other properties. By using suitable ceramic materials for the crucible, the tundish, the tundish outlet and the filter, the oxygen content of the atomized powders was lowered to less than 100 ppm. Many samples of different batches were HIPed, and metallographic cross-sections proved that the quantity of oxide particles in the < 45 micron material could be lowered to an acceptable level. The adsorption of oxygen is largely eliminated by handling the powders in an inert gas and by avoiding contact with organic materials during the entire production and handling process. The evaluation of the Udimet 700 samples doped with foreign particles shows that metallic cross contaminations with a chemical analysis which vary insignificantly from Udimet 700 can be identified as faults in the HIPed part. On the assumption that alloy particles which are similar in chemical composition will either be not tolerated at all or tolerated only in small quantities we have studied how such particles can be identified in HIPed Udimet 700.

The fact that they are recognizable as faults after HIPing at 1,150°C, 1,400 bar, for 4 hrs proves that the contamination alloy particles used produces chemical inhomogeneity which exceeds the specification. Therefore the entire atomizing apparatus, including all auxiliary devices, must be thoroughly cleaned before high purity powders are produced.

Recent tests with Udimet 700 produced in our equipment prove that the measures described above are very effective for attaining high purity. This means that high purity powders for aircraft engine parts can be produced economically in atomizing equipment which is used for the production of other alloys.

ACKNOWLEDGEMENT

The authors would like to thank Dr. H. Kaiser, Mr. E. Kress (Thyssen Guß Bochum, West Germany) and Mr. W. Track (MTU München, West Germany) for their critical discussion and support of this paper.

Chapter Four

Powder Production: Iron Powders

THE DIVERSE EFFECTS OF Fe, Cu, Ni, Ag, SiO_2 ADDITION TO CARBON ON THE REDUCTION RATE OF FERRIC OXIDE

Jin Jiaming

Shanghai Research Institute of Materials
China

Abstract

The catalyzing and poisoning effects of Fe, Cu, Ni, Ag, SiO_2 addition to the carbon on the reduction rate of ferric oxides with carbon were investigated. An Electron Cyclic Donor-Acceptor Catalyzing Theory (ECDAT) was proposed by author. Upon this theory, the catalytic effects of various addition on the carbon gasification reaction, which is a determinative step in the reduction processes, are attributed to the electron transfer. It is the mechanism that facilitates the restoration of the distorted carbon molecular orbits to its original symmetric state and therefore the formation of the carbon monoxide. The relative catalytic activity depends on the electronegativity of the catalyst involved. such as elements or compounds with lower electronegativity than that of carbon possess the catalytic activity. The lower the electronegativity, the higher the catalytic. In contrast with the former, such as elements or compounds with higher electronegativity than that of carbon possess poisoning effects. The higher the electronegativity, the more serious the poisoning effects. The electronegativity of element or compound depends on the chemical state under different reaction condition. A number of data are identical with this theory.

Introduction

Iron powder has become an important industrial raw material. The total capacity in the world was estimated over 600Kt, of which about 50% is the reduced sponge iron powder, and the another 50% is the atomized iron and steel powder. The sponge iron powder was mostly produced by carbothermic reduction of iron oxide, which was called also as Höganas method. In which the fine concentrated iron ore powder or mill-scale powder was used as a raw material. The reducing agent is mostly to use the metallurgical coke powder. Both iron oxide and coke powder were filled into the reaction sagger separately. At high temperature 1150-1250 °C, the iron oxide is reduced to metallic state by a two-step reaction sequence as now widely accepted:

$$FeO(s)+CO(g)=Fe(s)+CO_2(g) \quad (1)$$
$$CO_2(g)+C(s)=2CO(g) \quad (2)$$

of which, the reaction(1) should take place in the iron oxide powder layer, and the reaction(2) in the carbon layer. Although these reduction processes are very complicate, a numerous investigations showed that reaction(2) or Boudouard reaction($C+CO_2=2CO$) constituted the rate-controlling step in the uncatalytic reduction system, especially at the low temperature. Hence, it is reasonable to enhance Boudouard reaction rate may increase the whole reduction process rate. The key to obtain rapid reduction rate lies in how to increase Boudouard reaction rate.

As carbon gasification reaction is certainly of great important in the worlds energy sources, in the extractive metallurgy and in the human livelihood. It is useful, therefore, to examine in more detail the diverse effects of various mineral impurities on the carbon gasification.

Research publications and patents appeared so far were summarized competently by P.K.Walker(1) and D.W.Mckee(2). According to a number of experimental data presented in literature and the practice in the industrial production for a long time, we should arrived at two conclusions, namely,first, a number of compounds (for instance, K_2CO_3, Na_2CO_3, Li_2CO_3, $BaCO_3$, $SrCO_3$, KCl, NaCl, $BaCl_2$, K_2SO_4 etc.) and metals (for instance, Fe, Co, Cu, Ag, Pt etc.) are well-known to catalyze carbon gasification involving CO_2, O_2,H_2O and SO_2. Second, the catalytic activity of various catalysts is quite different. For example, the influence of various catalysts on the carbonthermic reduction of Fe_2O_3 had been tested by Y.K.Rao(3). The test results showed that the $BaCO_3$, $CaCO_3$, $SrCO_3$ are inferior catalysts. Whereas K_2CO_3, Na_2CO_3, Rb_2CO_3 and Cs_2CO_3 exhibit vastly superior catalying activity. Uncatalyzed Fe_2O_3+3C mixture reduces by about 12% in a 5-hours span, the K_2CO_3 catalyzed charge is completely reduced in the same duration under exactly similar conditions. Na_2CO_3 has some what less activity than K_2CO_3. The subject on the catalysis of carbon gasification has been studied for many years. A numerous efforts to find an appropriate explanation for the mechanism ofcatalysis on the reaction in terms of experimental results have been made. At present there are mainly two catalytic mechanism proposed in the literature, namely, Oxygen Transfer Theory(OTT) and Electron Transfer Theory(ETT). In the literature, relating to the carbon combustion, the surface carbonization of steel parts and the iron making process, the OTT has been extensively used. Unfortunately, these theories could not well interpret a series of experimental results, especially the relative activity of various compounds or metals. The purpose of this research is to obtain experimental data, so that we can demonstrate satisfactorily the Electron Cyclic Donor-Acceptor Catalyzing Theory proposed by author.

Experiment

Mill-scale (-60+80mesh) and metallurgical coke powder (-40mesh) used in the sponge iron production were used as samples for this study. Coke powder was treated with aqueous solution of 50% hydrochloric acide for about 10 hours to avoid the interference of alkali compounds in the coke. Both the powder were filled separately into the iron box in sequence.The mill-scale powder was arranged in the middle of the coke powder layer. A refractory sieve (200mesh) was placed between the powder layers to avoid the falling down of iron oxide powder into the coke powder layer. The box was made of steel plate 4mm thick. Its size is 65x65x100mm and is shown in Fig. 1.

The different catalysts were separately added to the coke powder. Except the addition of different catalysts, the samples are all the same. After the samples were filled into the iron box, whole things were sealed by weld. There is a Φ 8 vent hole on the box wall and a Φ 8mm stainless tube was fitted so that gas mixture produced during the reduction process can be collected to determine the composition and to measure the flow rate.

To compare the relative activity of two catalysts, a couple box were used at each time. For example, None-Fe, None-Cu, Fe-Cu, Fe-Ni, Cu-Ag etc. The boxes in which they contain different catalysts entered into furnace at the same time and heated zone. The end of thermocouple was seated between both boxes.

The stainless tubing was joined with gas flow meter, from which the flow rate of the produced gas can be measured and the composition of mixture gas was determined by gas analysor. When the boxes entered to furnace were being heated up to 400-500°C, the air in the boxes had been evacuated by vacuum pump before the reduction reaction took place. When the reduction process has completed, both boxes were taken out from the furnace at the same time, and cooled with water. Then the boxes were cut and the sponge iron was taken out for chemical analysis.

Results and Discussion

The compositions and the flow rates of the produced mixture gas were given in Fig.2 to Fig.10 for Fe-None, Cu-None, Fe-Cu, Ni-Cu, Ni-Co, Cu-Ag, None-SiO_2 of the couple experiments.

The compositions of gases evolved from box and gas compositions of FeO+CO=Fe+CO_2, CuO+CO=Cu+CO_2 equilibrium reactions were given in the Table 1. The carbon contents of sponge iron for Fe-None, Cu-None, Fe-Cu couple experiment were given in the Table 2.

The higher concentration of CO in the reducing gas, the quicker flow rate of the gas and the higher content of carbon in the sponge iron all shown that catalysts possess higher catalytic activity. These experimental results confidently show that:

1) Fe, Cu, Ni, Ag, Co, all possess catalytic activity on the carbothermic reduction of iron oxide. But the SiO_2 is a poisonor.
2) The relative activities of some metallic elements are as follows: Fe $>$ Cu, Fe $>$ Ni, Cu $>$ Ni, Ag $>$ Cu, Co $>$ Ni.
3) The comparison between the gas compositions measured and the equilibrium reaction gas phase compositions can prove that the Fe, Cu, Ni, Co, Ag, as a catalyst appears to be in a metallic state, and can not form the oxide, carbide or carbonate.
4) The comparision between the gas phase CO_2% determined and the equilibrium gas phase CO_2% of the reaction(1) as well as (2) demonstrates still that the reaction (2) is the rate-controlling step of whole reduction process.

As previously mentioned, the OTT was extensively adopted. In the theory a catalyst on the surface of carbon is regarded as an oxygen carrier by an oxidation-reduction cycle reaction. Thus the catalyst would repeat the cycle between oxidation and reduction, such as metal-oxide or lower oxide higher oxide. For metallic catalyst, the mechanism is given by the following equation:

$$YMe(s) + XCO_2(g) = Me_yO_x(s) + XCO(g) \quad (3)$$

$$Me_yO_x(s) + XC(s) = YMe(s) + XCO(g) \quad (4)$$

For alkali carbonate, the vapor cycle mechanism envisages the role of an active intermediate for the alkali metal vapor(m) produced by direct interaction between the catalyst and the carbon substrate. The mechanism may be represented by (4), (5), (6):

$$M_2CO_3(l\ , s) + 2C(s) = 2M(g) + 3CO(g) \quad (5)$$

$$2M(g) + CO_2(g) = M_2O(s,1) + CO(g) \quad (6)$$

$$M_2O(s,1) + CO_2(g) = M_2CO_3(s,1) \quad (7)$$

Upon summation, one obtains $2C(s) + 2CO_2(g) = 4CO(g)$.

A modification of the foregoing mechanism involves alkali-carbide intercalates, i.e. $C_{2n}M$, as the active intermediates(7). This consists of the following step:

$$M_2CO_3(s,1) + 2C(s) = 2M(g) + 3CO(g)$$

$$2M(g) + 4nC(s) = 2C_{2n}M(s) \quad (8)$$

$$2C_{2n}M(s) + CO_2(g) = (4nC).\ M_2O(s) + CO(g) \quad (9)$$

$$(4nC).\ M_2O(s) + CO_2(g) = 4nC(s) + M_2CO_3(s,1) \quad (10)$$

Here M is alkali metal K, Na, Cs, Li etc.

For the alkaline earth carbonates, as the group 11A oxides are more stable at high temperatures than are the monoxides of the group 1A alkali metals, reaction of the carbonates with carbon proceeds only as far as the oxides in the former case. A two-step cycle of the type(8) is feasible on thermodynamic ground.

$$BaCO_3(s) + C(s) = BaO(s) + 2CO(g) \quad (11)$$

$$BaO(s) + CO_2(g) = BaCO_3(s) \quad (12)$$

To summarize, this mechanism rests on the hypothesis that the catalysts of metals and carbonates undergo the repeat cycle of oxidation and reduction, of which the active intermediate may be the metal vapors(K, Na, Li, Rb), carbides (C_2K, C_2Na etc.), metals (Fe, Co, Ni), and oxides (BaO, SrO).

Unfortunately the OTT can not explains a series of experimental results and it is contrary to thermodynamic ground, for example:

1. The OTT can not explain the catalytic behaviour of Fe, Cu, Ag, Ni, Co, in this research as previously described.
2. According to author's experiemtal results on the carburization of steel part surface(9), the ambient partial pressure of CO in the Carburizing box is higher than the equilibrium pressure of reaction(11) as shown in Fig. 11. Hence, such as the carbonates of Barium, Sodium and Potassium

can not undergo such repeat cycle. As a matter of fact, they exhibit vastly superior catalytic activity.

3. The OTT can not explain that the various carbonates with very large difference in stability exhibit catalytic activity at the same temperature. In as much as the thermodynamic requisite of the repeated cycle reaction must be equivalance between the reaction equilibrium pressure and the ambient pressure of CO or CO_2 at the same temperature.

   Upon the same reason, the OTT can not explain also that one carbonate(for example K_2CO_3) exhibits always good catalytic activity over the wide temperature range (700-1100^oC). Because the dissociation pressure for one carbonate has a very great difference at various temperatures.
4. For some chlorides, for example KCl, NaCl, BaCl etc., which can not undergo absolutely such as repeat cycle reaction on the thermodynamic restriction. However, these chlorides are well-known active catalysts.
5. The OTT can not explain the relative activity of various compounds and metals.

A series of facts show just that Oxygen Transfer Theory is too hard for us to believe.

Long and Sykes(10) made the first and as yet the most detailed attempt to formulate an electron transfer theory of catalyzed carbon gasification. In this theory, only those substances which can accept electron are catalysts, while those substances which can not accept electron are not catalysts. It excludes the metal as active catalyst. Rakszawski et al. (11) extents the apparent validity of ETT to metals having catalytic activity. His argument is that transition metals with unfilled d bands do satisfy the basic requirements in serving as a electron acceptor.

However, the ETT can not also explain many experimental data. Especially that ETT is contrary to the electronegativity theory because the electronegativity of Fe, Cu, Ni, Co, Ag, in this study are all lower than that of the carbon. Therefore, the ETT can also hardly believe.

The Electron Cycle Donor-Acceptor Catalyzing Theory(ECDAT)

As we mentioned previously, it is well known that Boudouard reaction constitutes the rate-controlling step in the process of the carbothermic reduction of Fe_2O_3. This fact was also established by author(12). And therefore an increase of reduction reaction rate should be ascribed to increase of Boudouard reaction rate. According to experimental results in this study and data presented in the literature, as for Boudouard reaction an ELECTRON CYCLE DONOR-ACCEPTOR CATALYZING THEORY(ECDAT) was proposed by author. To introduce this mechanism, first of all we should have to discuss the mechanism of Boudouard reaction. The believable mechanism is as follows(13):

1. The first step of interaction between CO_2 and solid carbon is that CO_2 molecule in the gas phase diffuses to the surface of carbon, and then is absored actively by carbon atoms. For thermodynamic reasons possible, the carbonoxygen bonds of CO_2 molecule are weakened and in consequence of the bonds being broken up, then the Ketonegroup(➤C=O) and the ethenonegroup(➤C+C=O) are formed, as in Fig.12.
2. The next step is the thermal dissociation of ethenonegroup and the subsequent formation of CO molecule as in Fig.13. Owing to the thermal dissociation of ethenonegroup does not break the C-C bond of carbon crystal,

the energy required is smaller and hence the rate is great.

3. The next step of reaction is the dissociation of Ketonegroup, one form for it may be thermal dissociation, such as in Fig. 14. And another form may be collision mechanism, as in Fig. 15.
   Whatever the forms of dissociation of Ketonegroup are, the C-C bond of carbon crystal is always broken up. Now Let Ea be the activation energy for dissociation of ethenonegroup, Eb be the activation energy for dissociation of Ketonegroup. Tabel 3 gives the values of Ea ans Eb(14). The table indicates the kind of carbon employed, and kind and amount of impurities added.

From Table 3, it is very clear that:

1. In general the Eb values are greater than the Ea.
2. The Ea values are not related to the kind of carbon employed as well as kind of impurities, whereas the Eb values are related closely to the kind of carbon employed and kind of impurities.

According to above discussion, we should reach the following conclusions:

1. The dissociation of Ketone-group is the rate-determining step of Boudouard reaction.
2. The effect of catalysts or impurities should ascribed to decrease of energy for breaking the C-C bond.

Because the electronegativity of oxygen is greater than that of carbon(X_o = 3.5, X_c=2.5 from Pauling's data), once the Ketone-group was formed, the π and/or σ electrons of the graphite crystal must move toward the atom of oxygen. As a result of the motion, the symmetrical molecular orbits of carbon crystal are distorted. Such distortion signifies an increase in internal energy of system, thus the system becomes more unstable. To maintain the original symmetrical stable molecular orbits, the carbon molecule(graphite crystal) must bind the π and/or σ electron which always move toward oxygen atom. As a result, the attraction of interaction electrons between carbon and oxygen atoms must be occurred. At this time any chemical or physical factor that can provide electron for graphite crystal will facilitate the restoration of orbits distorted, and consequently can enhance the reaction rate of carbon with CO_2, O_2 and H_2O. From this view point, any substance which can provide electron for carbon must be a catalyst. Conversely, those substance which can accept electron from carbon will be a poisonor. The catalyst is thought to be able to provide electron.

Once the Ketone-group has been dissociated or the C-C bond has been broken, an excess electron in carbon appears, resulting in an decrease of electronegativity of carbon, its value may be lower than the catalyst, thus, the electron transferred from catalyst to carbon must be returned back to catalyst. This is the essence of Electron Cycle Donor-Acceptor Catalyzing theory(ECDAT).

Careful investigation of the conception of ECDAT and other experimental data, which will be seen later, has led author to arrive at following four conclusions:

A. There is a boundary between catalyst and poisonor.
B. The relative activities depend upon the electronegativity value of compounds or metals.
C. The catalytic activity is related to chemical state of catalyst.
D. Inhibitor of gasification reaction.

For the case of gasification reaction, the boundary is the electronegativity value of carbon.

All substances with electronegativity value less than that of carbon should be a catalyst which can donate electron towards carbon. Conversely, all those substances with electronegativity value greater than that of carbon should be an inhibitor which can accept electron from carbon.

For the sake of contrast in the periodical table of elements, the element electronegativity in a given period progressively increases from left to right and decreases in a group from top to bottom. All of the elements which are situated on the left and lower side of carbon should be catalysts, while those elements which are on the right and upper side of carbon should be poisonors. In this study, the Fe, Co, Ni, Cu, Ag, all are on the left and lower side of carbon, thus, these elements exhibit the catalytic activity. Although the relative activity is different.

As for the relative activity, a number of workers have investigated, some qualitatively and some quantitatively, the relative activities of various catalysts for enhancing the gasification of carbon. As previously decribed, the OTT or ETT can not explain the numerious experimental results. According to the ECDAT view point, the relative activity of various elements or compounds depends upon the electronegativity value of catalyst concered. Elements or compounds with lower electronegativity than that of carbon must be a catalyst. The lower the electronegativity, the higher the catalytic activity. For poisonor, the higher the electronegativity, the severer the poisoning effect.

From the changing tendency of element electronegativity in the periodical table, we have a good reason to believe the order of catalytic activity should be Cs > Rb > K > Na > Li. Their electronegativity values are $X_{Cs}=0.7$, $X_{Rb}=0.8$, $X_K=0.8$, $X_{Na}=0.9$, $X_{Li}=1.0$. The activity of alkali metals should be greater than that of alkline earths. While the order of catalytic activity on the transition metals should be Fe > Co > Ni > Cu. Such results are almost consistent with experimental data in this study as well as other data.

For the inhibitor, the order of poisoning ability should be F > Cl > Br > I and F > O > N. Their electronegativity values are $X_F=3.9$, $X_{Cl}=3.0$, $X_{Br}=2.8$, $X_I=2.5$, $X_O=3.5$, $X_N=3.0$.

Fig.16(15) shows the Arrhenius plots for gasification of pure graphite and samples containing Fe, Co, and Ni in CO_2. This result is consistent with ECDAT.

The question of how the relative activities depend on the chemical state of catalysts is of great complex and importance. Due to different reaction condition, the chemical state of element or compound is always different, as a result the activity is also different. As a matter of fact, the variation of activity should be primarily attributed to change of electronegativity. When some atoms of initially different electronegativity combine chemically, their electronegativity becomes equalized in the molecule and the intermediate electronegativity of the molecule is crudely taken as geometric mean of electronegativity of all atoms before combination.

As in this study, the iron is an active catalyst. In the Turkdogen and Vinter's study(16), the addition of less than 0.01% iron increased the gasification rate of graphite in CO_2 by a factor of 2000 in the temperature range of 700-1000°C, where as 2% iron increased the rate by a factor of about a million at 800°C. But the iron once combined with oxygen, its electronegativity should be increased and would become an inactive catalyst or decrease its catalytic activity. This conclusion had been demonstrated by Walker et al(17). Fig.17 shows reactivation of iron in H_2 atmosphere at 950°C. Following reactivation, the reaction rate in CO_2 steadily decreases to a limiting value as metallic Fe was reoxidized to an inactive stage. When subsequent heating of the deactivated iron in hydrogen at $800^{o}C$ for 3hrs reduced the oxide to the metallic state, the catalytic activity was restored again. Gulbransen and coworkers(18) studied the catalytic activity of Fe on Boudouard reaction, they found that Fe can increase the rate to 530 times(at $700^{o}C$), while its oxide was a poor catalyst. It is concluded that metallic iron is an active catalyst in gasification of carbon. But it is not an active catalyst in the combustion of carbon. As for those compounds, the effects of cation and anion in compound on the catalytic activity would be as follows:

The relative activity or electronegativity of compounds depends upon the cation and anion by which the compound was formed.

If the anion is all the same in compounds, then its electronegativity decreases with the decrease of electronegativity of cation.

If the cation is all the same in the compounds, its electronegativity increases with increase of electronegativity of anion in the compounds.

For example chloride salt, $X_{LiCl}=2.13$, $X_{NaCl}=1.93$, $X_{KCl}=1.77$, $X_{RbCl}=1.74$(19), their relative activities should increase in the order LiCl < NaCl < RbCl. Kayembe and Pulsifer(20) studied the catalytic activity of various salts on the Kinetics of gasification of coal chars by steam between 600-$850^{o}C$ at atmospheric pressure. The observed order of activity was $K_2CO_3 > Na_2CO_3 > Li_2CO_3 >$ KCL > NaCl. Fig.18(21) summarizes weight-loss measurements for graphite-alkali-carbonate mixture(1:1 by weight) on heating in flowing CO_2 in a thermobalance at a linearly increasing temperature of $10^{o}C\ min^{-1}$. In addition to the salt of lithium, the relative activity is arranged in order $Cs_2CO_3 > Rb_2CO_3 > K_2CO_3 > Na_2CO_3$. Fig.19 shows the catalytic effects of VB group oxides on the oxidation of graphite in air, the order of activity was $Bi_2O_3 > Sb_2O_3 > As_2O_3$(22) . fig.20 shows the catalytic effect of alkaline earth carbonates on the reaction of graphite with water vapor(23). Table 4 shows the gasification rates and apparent activation energies of graphite with water vapor for the various carbonate additions. The above account shows the order of activity was $BaCO_3 > SrCO_3 > CaCO_3 > MgCO_3$. In the study by Mckee(24), the pure graphite powder was doped with alkaline earth carbonates and the kinetics of gasification in CO_2 measured between $700^{o}C$ and $1000^{o}C$, the order of catalytic activity was found to be $BaCO_3 > CaCO_3 > MgCO_3$. fig.22 shows the catalytic activity of $BaCO_3$, Na_2CO_3 on the carburizing process of steel surface(25), the relative activity was $Na_2CO_3 > BaCO_3$. According to Amariglio and Duval(26), when graphite was impregnated with an aqueous solution of soluble salts, followed by freeze drying, the catalytic activity order for graphite oxidation was Na > Ba > Sr > Mg > Ca. As previously mentioned, in the study by Rao, the Na_2CO_3, K_2CO_3, Rb_2CO_3, Cs_2CO_3 exhibit vastly superior catalytic properties. The order of activities was arranged to be $Cs_2CO_3 > Rb_2CO_3 > K_2CO_3 > Na_2CO_3$.

In the study by Mckee(27), SnO showed a slight catalytic activity for graphite oxidation, but the more stable SnO_2 was inactive. K_2CO_3 has certainly more activity as compared with KCl. For potassium salts, once it is combined with F, Cl, Br, I. The molecular electronegativity is decreased in the order: $X_{KF}=2.02$, $X_{KCl}=1.77$, $X_{KBr}=1.70$, $X_{KI}=1.62$(19). On the basis of ECDAT, the relative catalytic activity of various compounds or metals may be judged on the electronegativity of compounds or metals. However, the electronegativity of compound does not only depend on the cation and anion by which the compound was formed, but, also depends on the molecular construction or induced effect of the anion on the cation, i.e. the electropolarity of unit bond distance. The calculation of group or molecular electronegativity is extremely complicated. Besides simple compounds, the available data about complex compounds are still absent so far, and await elucidation by chemists. A long series of proofs mentioned previously demonstrated the ECDAT is extremely consistent with numerous experimental data.

Author has considered that there are some discrepancies in various experiments, for example Li_2CO_3, $CaCO_3$ etc., in addition to different chemical states at the various reaction conditions, another reason is possible differnet physical states and temperatures. The temperature is also an effective factor on electronegativity and the effect is different for various compounds or metals. Tamal et al(28) examined the reactivity toward CO_2 of an activated carbon that had been impregated with group VIII METALS. They found that Ni, Ru, Rn, Os, Ir, Fe, Pd showed difference changing with temperature.

It is well-known that the catalyzing phenomenon is very complicated. The ECDAT is only refers to the chemical properties of substance, which, of cause, is most essential, but sometimes the physical properties of catalyst appear to have great effect exceeding the effect of chemical property. For example, Al_2O_3 in synthetic ammonia catalyst is an active substance.

The prevention of gasification reaction sometimes is very important, as is the case in CO_2-cooled graphite-moderated nuclear reaction, new space technology, high-temperature application of carbon, composite carbon-metal materials, and electrodes for electric arc furnace. As previously mentioned, according to ECDAT, those elements or compounds made of elements with higher electronegativity than that of carbon must be a inhibitor. It may be certain that SiO_2 is a poisonor such as in this study. It would be expected that inhibitory effects of Chlorine and its compounds such as CCl_4 and CCl_2F_2 have been well documented and $POCl_3$, which inhibit the combustion of carbon in O_2, would be effective in the case also.

CONCLUSION

1. These experimental results have demonstrated that Fe, Cu, Ni, Co, Ag exhibit all a catalytic effect on the carbothermic reduction of iron oxide, the relative activity of them is as follows: Fe > Cu, Fe > Ni, Cu > Ni, Ag > Cu, Co > Ni. SiO_2 is a poisonor.
2. The effect enhancing the carbothermic reduction of iron oxide should be attributed to the increase of Boudouard reaction rate.
3. The ECDAT proposed by the author have been conspicuously successful in interpreting the relative activity of transition metals and oxides, alkali-metal salt, alkaline-earth metal salts in the various type of gasification reaction.
4. The ECDAT involves the boundary between catalyst and poisonor impurities, relative activity of various catalysts and the effects of chemical state of catalyst, and anion as well as cation in compound on the catalytic activity.

REFERENCE

1. P.L.Walker, Jr.,M.Shelef, and R.A.Anderson, chem. and Phy. of Carbon, Vol.4 (P.L.Walker, Jr., Ed), Marcel Dekker, NewYork, 287, 1968
2. D.W.Mckee, Chem. and Phy. of Carbon, Vol.16(P.L.Walker Jr., Ed), Marcel Dekker, New York, 1. 1981
3. Y.K.Rao, J of Metals, July, 46,(1983)
4. D.W.Mckee and D.Chatterji, Carbon, 13, 381(1975)
5. D.A.Fox, and A.H.White, Ind. Eng. Chem. 23, 259(1931)

6. R.A.Ragatz and T.C.Fong, Trans. ASM, 32, 162(1974)
7. W.Y.When, Catal. Rev. Sci. eng. 22(1), 1(1980)
8. D.W.Mckee, Fuel, 59, 308(1980)
9. Jin Jiaming, In 1st. Conf. of Heat Treatment of Steel, in Chinese, 1965
10. F.T.Long and R.W.Sykes, Proc. Roy. Soc., A215, 100(1952)
11. J.K.Rakszawski, Phy. D thesis, Pennsylvania State University, University park(1960)
12. Jin Jiaming, "The reduction processes of the mill-scall in the tunnel Kiln" to be published
13. S.T.Rostofcef, The theory of metallurgical processes, Metallurgy Press,(1956)
14. O.A.Yesin and P.B.Geld, Phy. Chem. High Met., part 1, Reaction between Gas Phase and Solid Phase. Svedlofsk Press, 193, 1962
15. H.Harker, Proc. of the 4th Conf. on Carbon. Buffals, 1959. Pergamon Press, New York, 125, (1962)
16. E.T.Turkdogen, T.V.Vinters, Carbon 10, 97(1972)
17.
18. E.A.Gulbranson, K.F.Andrew, Ind. Eng. Chem., 44, 1048 (1952)
19. G.Y.Chang, "The relationship of electronegativity of bonds with force constant and breaking energy of bond" unpublish research, in Chinese, 1978
20. N.Kayember, A.H.Pulsifer, Fuel 55, 211 (1976)
21. D.V.Mckee and D.Chatterji, Carbon 13, 381 (1975)
22. D.W.Mckee, Chem. and Phy. of Carbon (P.L.Walker Jr., ed), Marcel Dekker, New York, 16 (1981)
23. D.W.Mckee, Carbon, 17,419 (1979)
24. D.W.Mckee, Fuel 59, 308 (1980)
25. T.C.Long and R.A.Ragatz, trans. ASM, 32, 162 (1944)
26. H.Amariglio and X.Dural, Carbon 4, 323 (1966)
27. D.W.Mckee, Ref. 2
28. Y.T.amai, H.Watanable and A.Tomita, Carbon 15, 103 (1977)

Table 1 Composition of Gas Evolved from Box and Equilibrium Gas Phase of $RO+CO=R+CO_2$ Reaction

Addition or Equilibrium Reaction	Gas Phase Composition	
	CO_2%	CO%
None catalyst	20%	balance
8% Fe catalyst	15	"
10% Cu catalyst	17.5	"
$FeO+CO=Fe+CO_2$(9)	26	74(at 1000°C)
$CuO+CO=Cu+CO_2$(9)	100	10^{-7}(at 1000°C)

This experimental work has carried out mainly at the Chemical Dept. of East China Normal University, Shanghai and Shanghai Powder Metallurgy Works. Author wishes to thank Prof. Ren JingYing and her graduate student.

Table 2 Carbon Content of Sponge Iron

Sample Couple	Fe-None		Cu-None		Fe-Cu	
Temperature(°C)	1000		990		990	
Time(hr.)	14		7		6.5	
Additive Catalyst	Fe	None	Cu	None	Fe	Cu
Carbon Content of Sponge Iron	0.59	0.35	0.41	0.35	0.51	0.34
	0.52	0.32				

Table 3 Values of Ea and Eb

Carbon characteristics	Ea, KCal	Eb, KCal
Nonash charcoal	26.7	62.5
" "	27.3	59.6
Sugar C+9% Al_2O_3	27.0	55.0
Sugar C+7%Fe	25.7	22.8
Coke,P_{CO2}=10-100mmw	26.8	46.6
Coke,P_{CO2}=100-1000mmw	26.8	51.6

Table 4 Effect of Catalysts on the Steam Gasification of Catalyst

Catalyst(10% by weight)	Rate Constnat K(hr^{-1}
CaO	$2.438X10^{-3}$
Fe_2O_3	$3.640X10^{-3}$
None	$4.250X10^{-3}$
CuO	$1.014X10^{-2}$
NaCl	$4,849X10^{-2}$
KCl	$1.116X10^{-1}$
Li_2CO_3	$1.670X10^{-1}$
Na_2CO_3	$3.154X10^{-1}$
K_2CO_3	$5.388X10^{-1}$

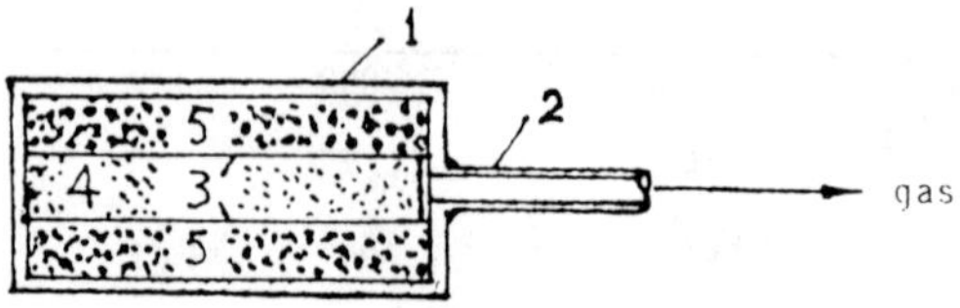

Figure 1 Reaction Box

1. Iron box 2. 8mm stainless steel tube

3. Sieve 4. Mill-scale powder

5. Coke powder

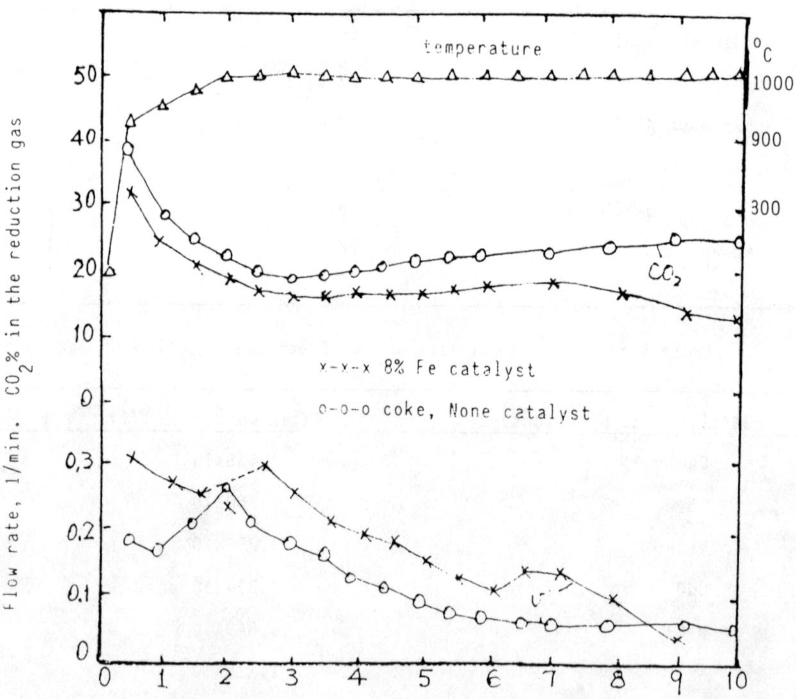

Figure 2 Catalytic effects of Fe on the reduction rate of iron oxide by carbon

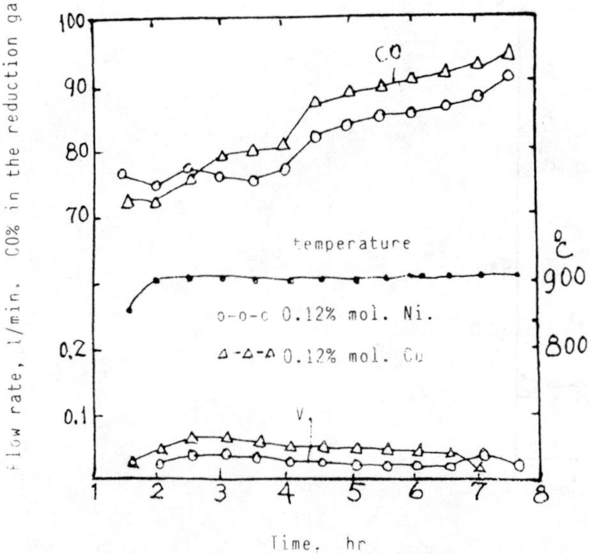

Figure 3 Relative activity of Cu, Ni

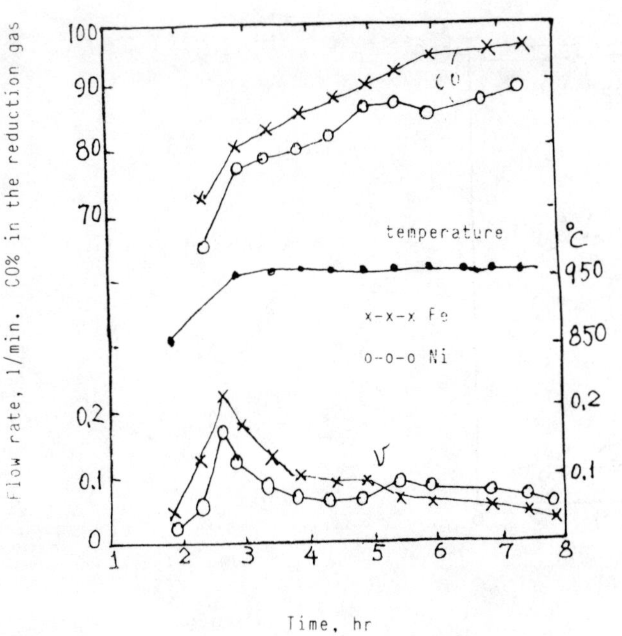

Figure 4 Relative activity of Fe, Ni

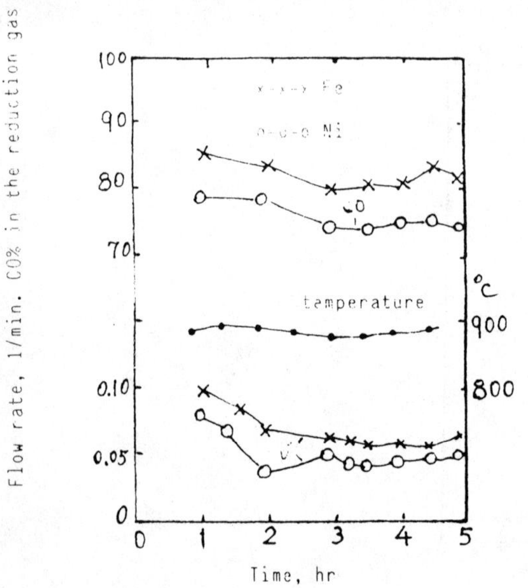

Figure 5 Relative activity of Fe, Ni

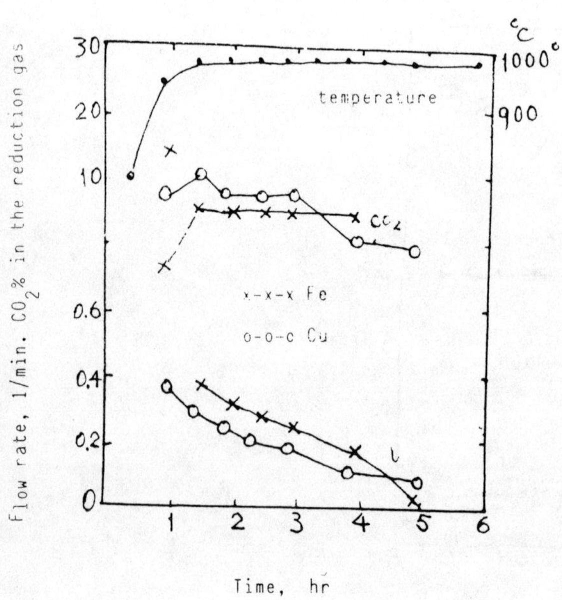

Figure 6 Relative activity of Fe, Cu

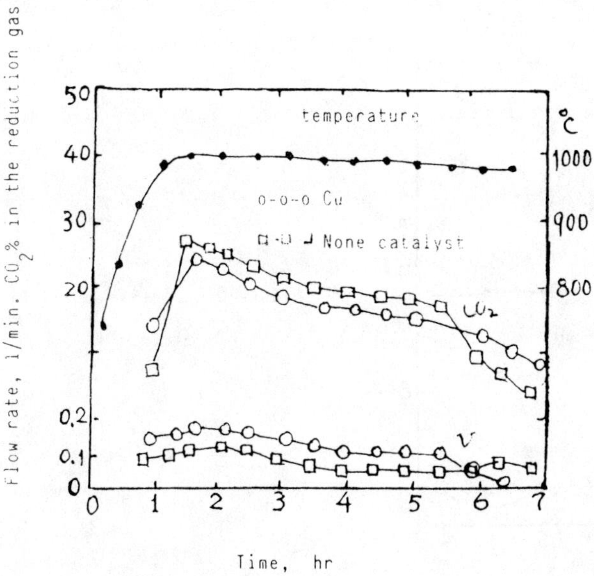

Figure 7 Relative activity of Cu

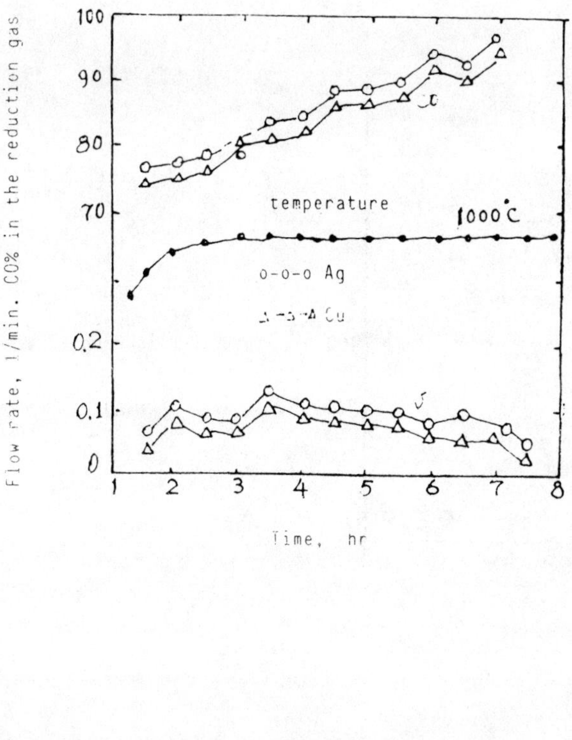

Figure 8 Relative activity of Ag, Cu

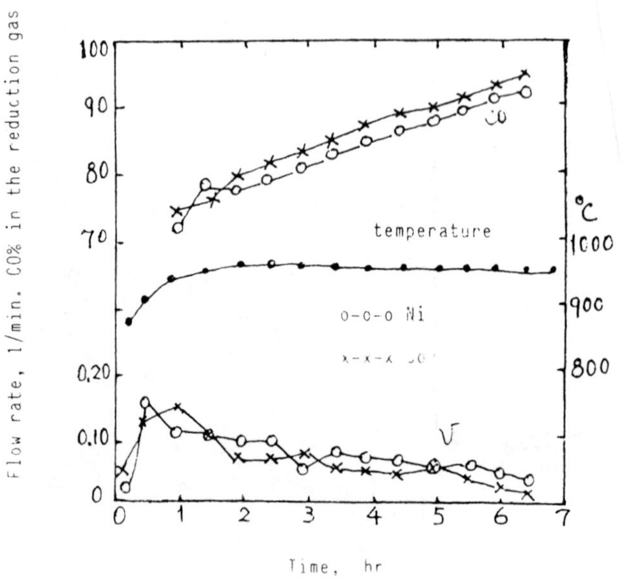

Figure 9 Relative activity of Co, Ni

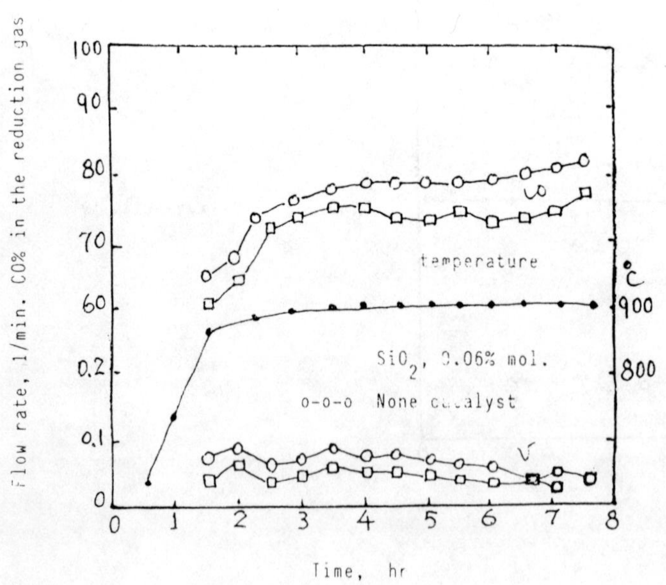

Figure 10 Poisoning effect of SiO_2

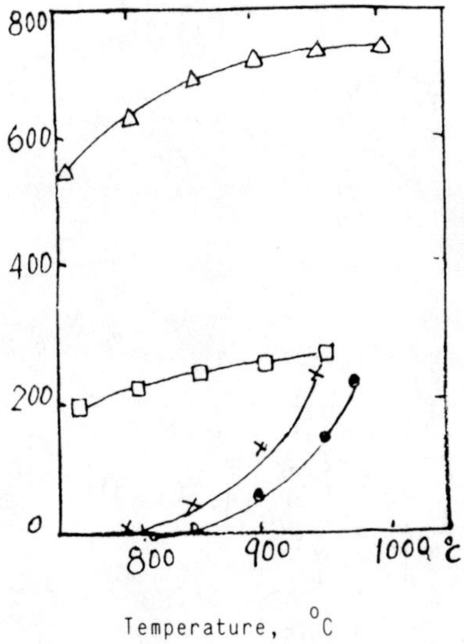

Figure 11 Partial pressure of CO in carburizing box and equilibrium pressure of reduction

x-x-x $BaCO_3$+C(charcoal)=BaO+2CO

o-o-o $BaCO_3$+C(graphite)=BaO+2CO

$P_{co}+P_{co2}+P_{N2}$=1atm.

$P_{co}+P_{co2}$=1atm.

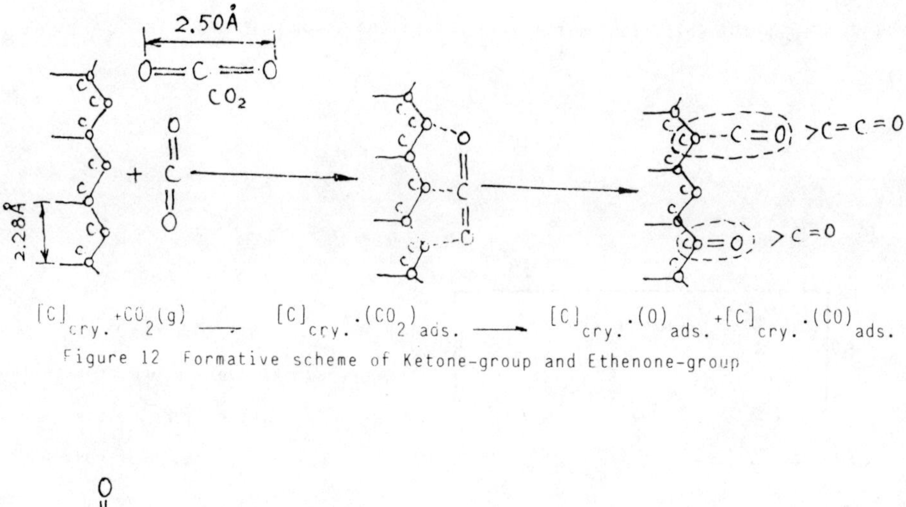

$[C]_{cry.}+CO_2(g) \longrightarrow [C]_{cry.}\cdot(CO_2)_{ads.} \longrightarrow [C]_{cry.}\cdot(O)_{ads.}+[C]_{cry.}\cdot(CO)_{ads.}$

Figure 12 Formative scheme of Ketone-group and Ethenone-group

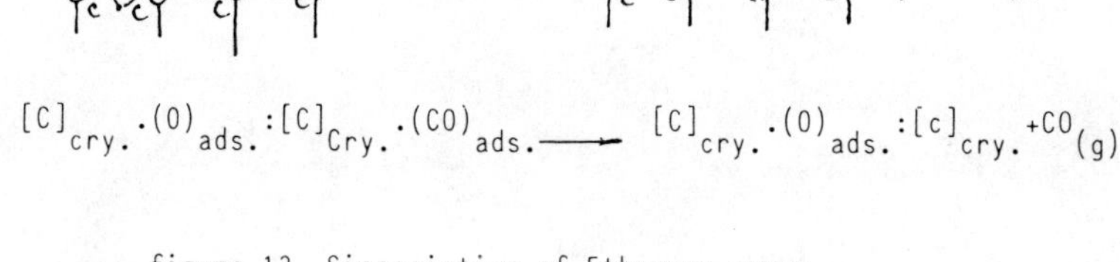

$[C]_{cry.}\cdot(O)_{ads.}:[C]_{Cry.}\cdot(CO)_{ads.} \longrightarrow [C]_{cry.}\cdot(O)_{ads.}:[c]_{cry.}+CO_{(g)}$

figure 13 Sissociation of Ethenone-group

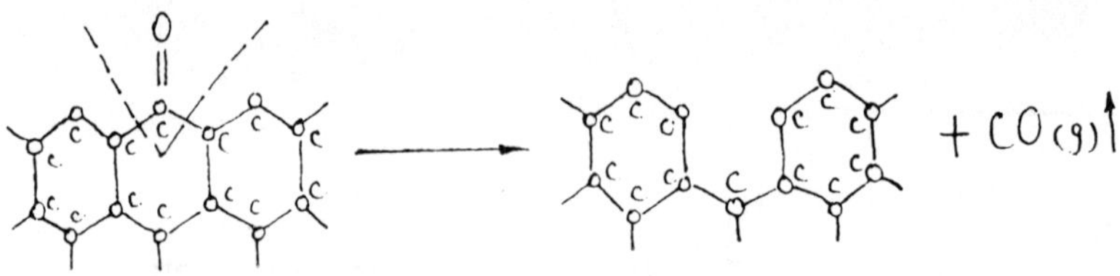

$[C_n]_{cry.} \cdot (O)_{ads.} \longrightarrow [C_{n-1}]_{cry.} + CO_{(g)}$

Figure 14 Dissociation of Ketone-group

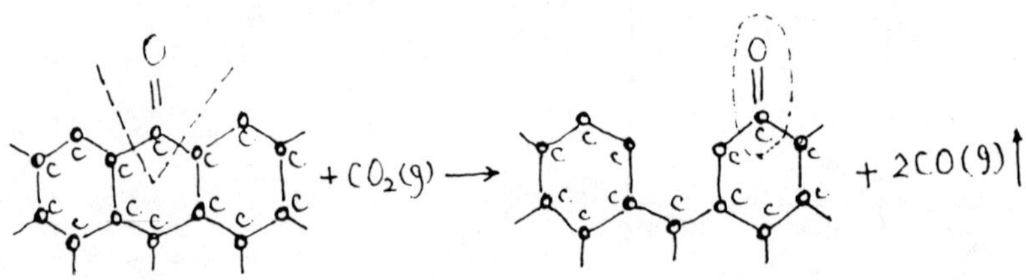

$[C_n]_{cry.} \cdot (O)_{ads.} + CO_2^*(g) \longrightarrow [C_{n-1}]_{cry.} \cdot (O)_{ads.} + 2CO(g)$

Figure 15 Collision dissociation of Ketone-group

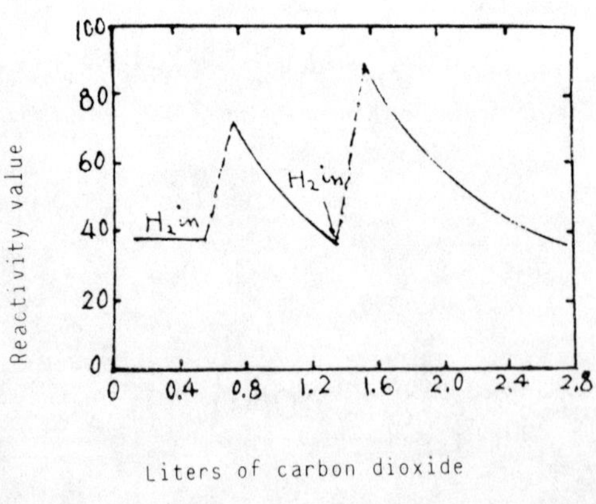

Figure 17 Activation and deactivation cycles of Fe catalyst during the reaction of coke with CO_2, activation is achieved using H_2.(17)

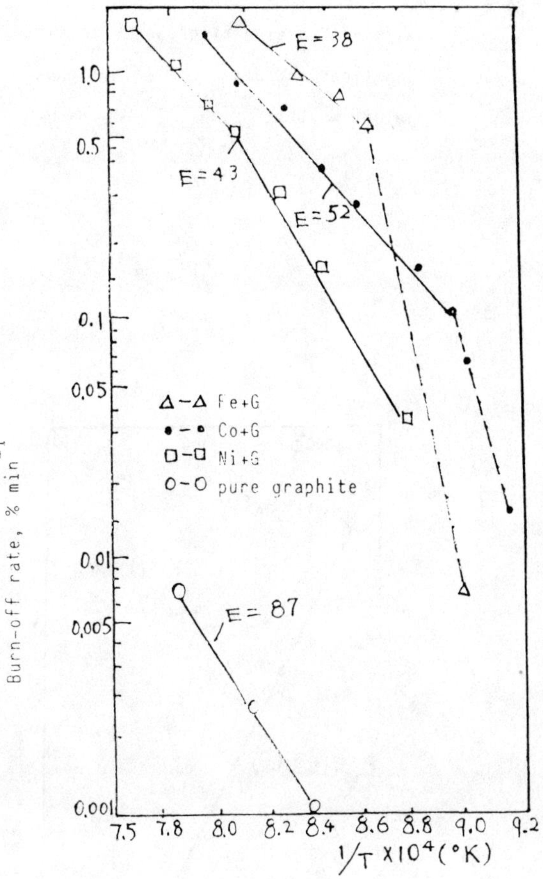

Figure 16 Arrhenius plots for gasification of pure graphite and sample containing Fe, Co, Ni, in CO_2 (15).

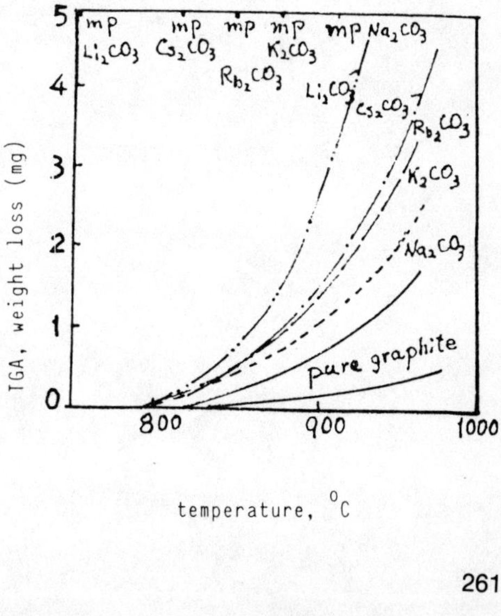

Figure 18 Effect of âlkali metal carbonates on the gasification of graphite powder in CO_2 (21)

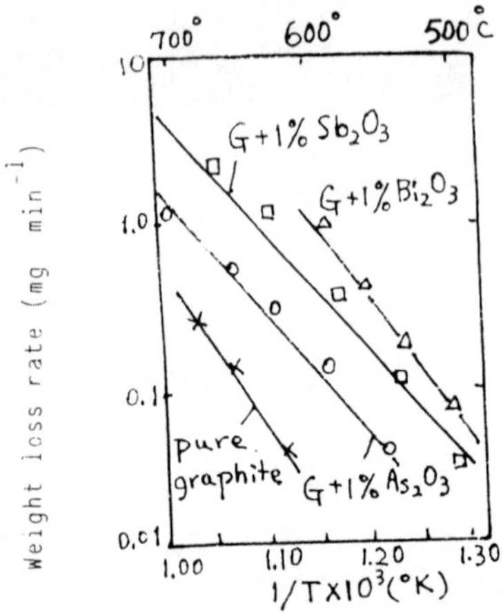

Figure 19 Catalytic effects of group VB oxides on the oxidation of graphite in air. Gasification rates vs. $1/t^{\circ}K$. Sample weight = 200mg[22].

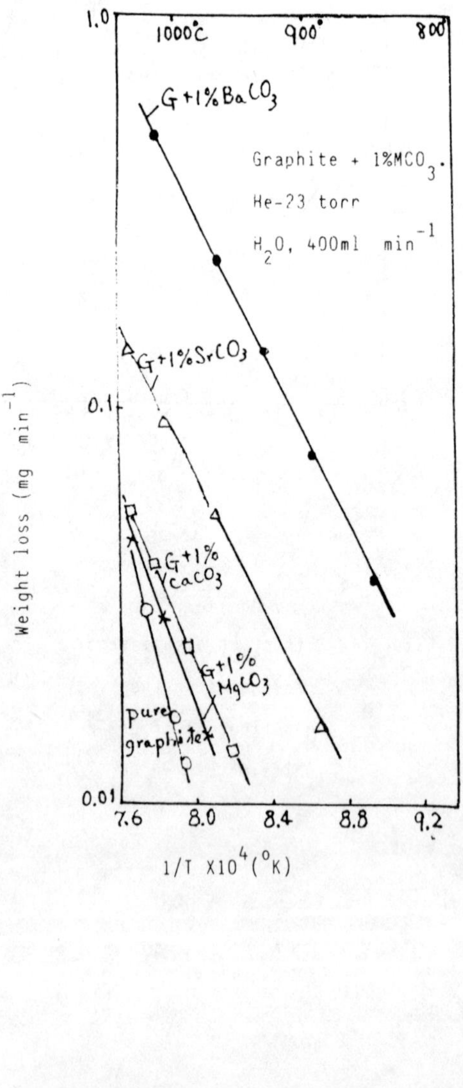

Figure 20 Catalytic effect of alkaline earth carbonates on the reaction of graphite with water vapor. Weight loss rates in wet He vs. 1/T K[23].

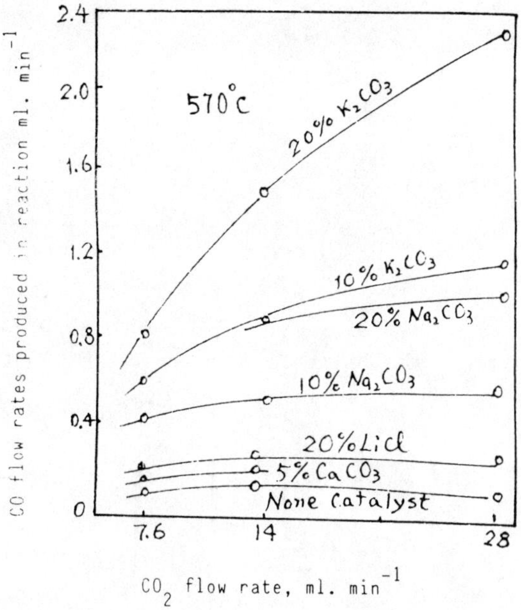

Figure 21 Catalytic effects of carbonates on the reaction rates of graphite with carbon dioxide.

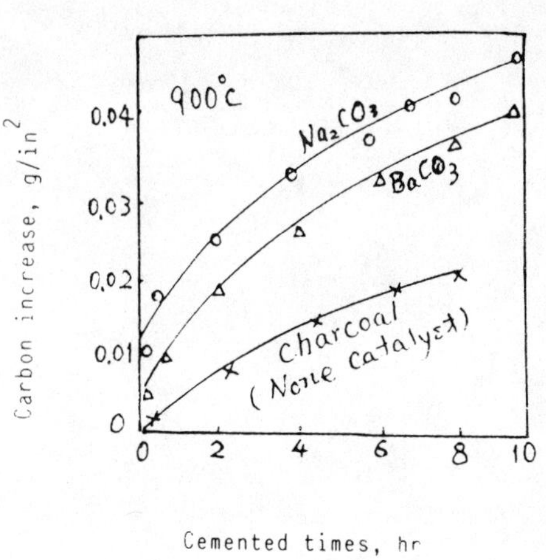

Figure 22 Catalytic activity of $BaCO_3$, Na_2CO_3, at the solid carburizing of steel[25]

MORPHOLOGICAL CHANGES DURING REDUCTION OF MAGNETITE INTO SPONGE IRON POWDER

R.K.Dube and B.Deo

Department of Metallurgical Engineering
Indian Institute of Technology
Kanpur 208016 (India)

Abstract

The morphological changes occurring at the various stages of the reduction of magnetite compacts were observed under the scanning electron microscope. It has been shown that considerable fragmentation of the parent magnetite particles occurs during solid state reduction to wustite by iron. It is expected that such fragmentation would also occur during gaseous reduction of magnetite to wustite. During reduction of wustite the iron nuclei formed have a tetrahedral shape. At low reduction temperatures (1173 K) the nuclei distribution density is large which eventually leads to the formation of dense iron layer on wustite. At high temperatures (1423 K), the growth rate of nuclei in the vertical direction is fast, and the nuclei distribution density is comparatively lower. It has been shown that the genesis of porosity in the reduced iron is also due to the fragmentation of parent magnetite particle during reduction and the growth and subsequent sintering of iron nuclei.

Introduction

Sponge iron powder is produced by reduction of iron ore concentrates followed by grinding and magnetic separation, if needed. In the well-known Höganas process for making sponge iron powder, the ore powder is placed in the centre of cylindrical ceramic containers surrounded on the outside by a concentric layer of a mixture of coke and limestone. On heating, the carbon monoxide produced from the coke reduces the ore to iron. The reaction is essentially a gaseous reduction process, although coke is used as a reducing agent. In the H-iron and Pyron processes for making iron powder, hydrogen gas is used as a reducing agent (1). The powder produced from the above processes are in general porous; the extent and nature of the porosity are not same in all the cases because of the difference in times and temperatures of the reduction process.

The reduction of magnetite above 843K takes place in stages, viz.

$$Fe_3O_4 \longrightarrow FeO \longrightarrow Fe \qquad (1)$$

An enormous amount of research work has been carried out on the gaseous direct reduction of iron ores. Most of the studies were concerned with the measurement of reduction kinetics and evolving mathematical models for the reduction rate. Very few studies (2-5) have been carried out to investigate the structural and morphological changes occurring at each stage of the reduction step. It is important to note that these changes, both on macroscopic and microscopic scales, also have a direct bearing on the overall reduction kinetics, and the final porosity generated in the sponge iron powder. So far, in the published literature, very little information is available on the role of solid state diffusion of iron in the oxides formed at the intermediate stages of reduction and its effect on the porosity created.

The present paper describes the morphological changes occurring at the various stages of reduction of magnetite superconcentrate with hydrogen and carbon monoxide, as well as metallic iron powder. It must be noted that the solid state reduction of magnetite with metallic iron powder will produce only FeO.

Theoretical Background

The crystal structure of magnetite is known to be a cubic lattice with an inverse spinel structure. There are 8 formula units (Fe_3O_4) per unit cell in which 8 ferric ions are tetrahedrally coordinated and 8 ferric and 8 ferrous ions are octahedrally coordinated. Wustite has a simple sodium chloride type face centered cubic lattice. The transformation of Fe_3O_4 to FeO occurs by incorporation of iron, as follows:

$$Fe_3O_4 + Fe^{2+}|Fe^{2+}| = 4\ FeO + |Fe^{2+}|'' + 2|\dot{e}|$$

$$Fe + |Fe^{2+}|'' + 2|\dot{e}| = Fe^{2+}|Fe^{2+}|$$

$$Fe_3O_4 + Fe = 4\ FeO \qquad (2)$$

where $Fe^{2+}|Fe^{2+}|$ is a divalent iron in normal lattice position, $|\dot{e}|$ is hole electron deficiency, and $|Fe^{2+}|''$ is an iron ion vacancy. During the reduction of FeO to Fe, slight surface roughness on newly formed wustite, which is inevitably present, is progressively intensified and eventually leads to hole-formation and slitting, due to migration of iron ion vacancies preferentially to surface depressions, (6) according to the reaction:

$$FeO + |Fe^{2+}|'' + 2|\dot{e}| + H_2 = Fe^{2+}|Fe^{2+}| + H_2O \qquad (3)$$

where $|Fe^{2+}|''$ is an iron ion vacancy, $Fe^{2+}|Fe^{2+}|$ is a divalent iron in normal lattice position, and $|\dot{e}|$ is an electron hole. Structurally, the vacancies $|Fe^{2+}|''$ are created at the Fe_3O_4/FeO interface according to the reaction:

$$Fe_3O_4 + Fe^{2+}|Fe^{2+}| = 4\ FeO + |Fe^{2+}|'' + 2|\dot{e}| \qquad (4)$$

These vacancies migrate to FeO/Fe interface and are eliminated by reaction (3). In spite of the similarity (cubic lattices in either oxides) the transformation of Fe_3O_4 to FeO is shown in this paper to be accompanied by a volume increase due to large scale slitting and fragmentation resulting from solid state diffusion of iron (equation 2) (under the experimental conditions employed in the present work).

Experimental

The detailed experimental conditions have been described elsewhere (7). Magnetite superconcentrate powder supplied by L.K.A.B. of Sweden had the following chemical analysis (by weight)

Fe : 71.7-71.8%, 0.001%, S : 0.003%, CaO : 0.05-0.07%,

MgO : 0.15-0.20%, Al_2O_3 : 0.15-0.20% and SiO_2:0.05-0.10%.

The sieve analysis of the superconcentrate powder was as follows:

417 μm : 100%
208 μm : 99.9%
104 μm : 96.3%
62 μm : 85.8%

There was no porosity in the starting magnetite particles, as verified from the scanning electron microscopy.

Two series of experiments were carried out. In the first series, cylindrical compacts of magnetite superconcentrate having 25 mm diameter, 5 mm thickness, and approximately 40% porosity were prepared at a pressure of 930 MN/m^2 in a die using single action punch. These compacts were partially reduced for 300-600 s in hydrogen atmosphere at 1173K and 1423 K at a linear gas velocity of 0.021 m/s. For the second series of experiments magnetite superconcentrate powder was intimately mixed with pure iron powder (-100 mesh size, and 150% in excess of the stoichiometric requirements) to produce wustite according to reaction (2). The cylindrical compacts having 25 mm diameter, 5 mm thickness, and approximately 30% porosity were prepared from these mixtures using a pressure of 930 MN/m^2. These compacts were heated in argon atmosphere at 1373 K for 3600 s to ensure complete reduction of magnetite to wustite which was confirmed qualitatively by X-ray diffraction patterns.

The details of reduction chamber used for the reduction of the magnetite superconcentrate are described elsewhere (7). All the samples from both the series were fractured across the thickness direction and were then observed under the scanning electron microscope from surface to centre. This revealed the morphology of the products at the various stages of reduction.

A few compacts were also reduced by carbon monoxide. This was accomplished by packing the compacts in a coal bed and heating the entire mass in the same reduction chamber which was employed for hydrogen reduction. During heating coal generated CO which eventually carried out the reduction. A flow of nitrogen was maintained inside the chamber to avoid any re-oxidation of the samples.

Results and Discussion

Solid State Reduction of Magnetite with Metallic Iron Powder:

Figure 1 shows the fractured surface of wustite prepared by solid state reduction of magnetite with 150% excess of the stoichiometric amount of iron powder. It can be clearly seen that the wustite formed is progressively breaking into small fragments.

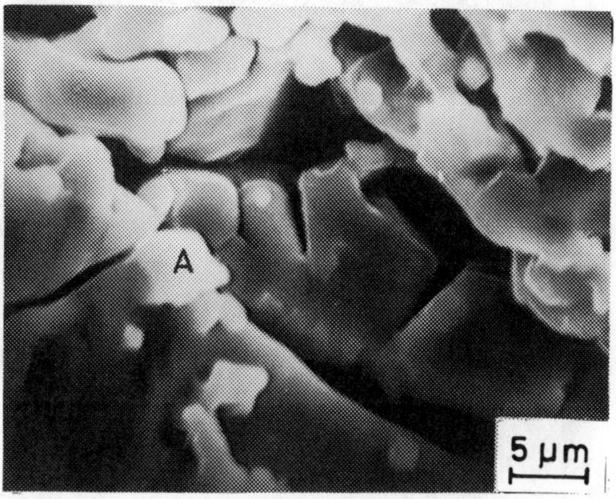

Figure 1 : Fractured surface of wustite prepared by solid state reduction of magnetite

There is a large scale migration of iron through wustite to the Fe_3O_4/FeO interface, resulting into hole formation on the surface of the parent magnetite particles. Migration of iron was facilitated by the vacancies existing in the wustite (reaction 2). These holes progressively grew into slits and cracks, due to further migration of iron leading eventually to the fragmentation of the particle. Slits are formed in the bulk of the particles due to enlargement of the holes and coalescence of neighbouring holes. Slits may turn into cracks when they reach the edges of the particles, as shown by point A in Fig.1.

It is therefore reasonable to believe that during reduction of magnetite with hydrogen, such a fragmentation, slitting, and cracking would be observed on the parent magnetite particles.

Reduction of Magnetite with Hydrogen:

Figures 2 and 3 show the iron nuclei formed on the parent magnetite surface subsequently reduced to wustite (reduction temperature: 1423 K). It can be seen that the nuclei have a tetrahedral shape, and slitting and cracking of the particles has taken place.

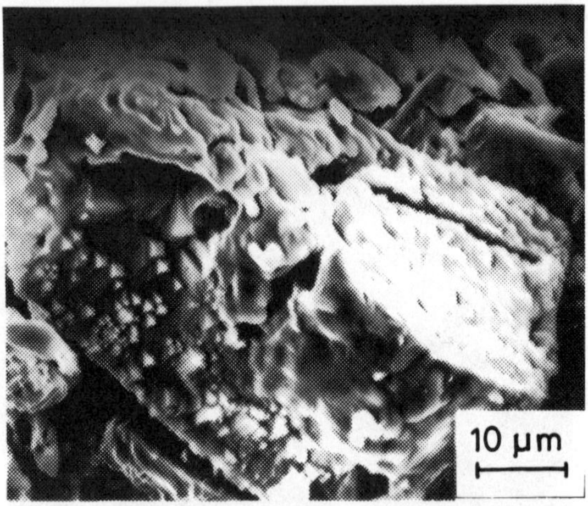

Figure 2 : Iron nuclei just after formation on the parent magnetite surface subsequently reduced to wustite by hydrogen at 1423 K

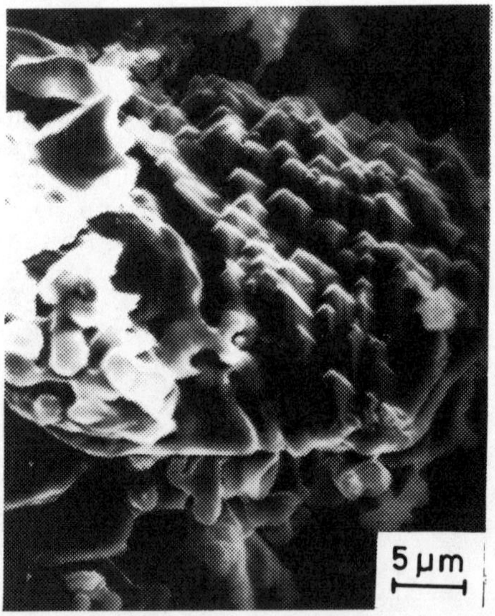

Figure 3 : Grown iron nuclei on the parent magnetite surface subsequently reduced to wustite by hydrogen at 1423 K

Figure 4 shows the general view of the fragmentation, cracking and slitting of the particles during reduction of magnetite with hydrogen at 1173 K. The observation of the same area at a higher magnification shows small holes formed on wustite surface (Fig.5). During the reduction of magnetite

Figure 4 : General view of the fragmentation, cracking and slitting of the parent magnetite particles during reduction with hydrogen at 1173K

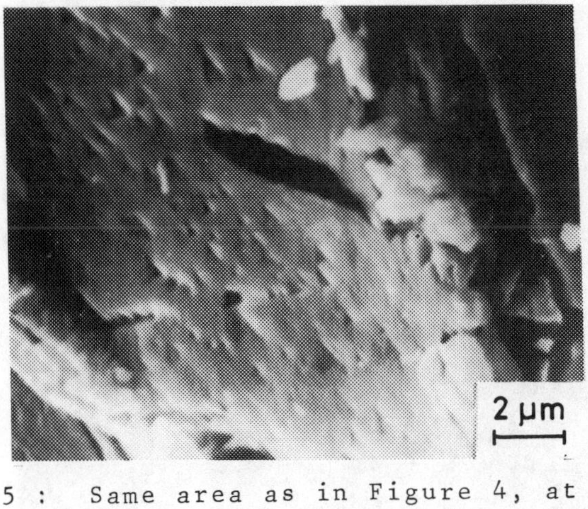

Figure 5 : Same area as in Figure 4, at higher magnification showing small holes formed on the wustite surface

particle, the formation of a covering layer of wustite first takes place. The liberated iron diffuses to the Fe_3O_4/FeO interface and further formation of wustite according to reaction (2) takes place. There exists a permanent flow of vacancies through the wustite layer and therefore the wustite/gas interface must continuously grow rougher which eventually leads to pore formation on the wustite surface. No nucleation of iron has yet started on the particles shown in Fig. 4. Similar to the reduction temperature of 1423 K, the nuclei formed on the wustite surface at the reduction temperature of 1173 K were also of tetrahedral shape (Fig.6).

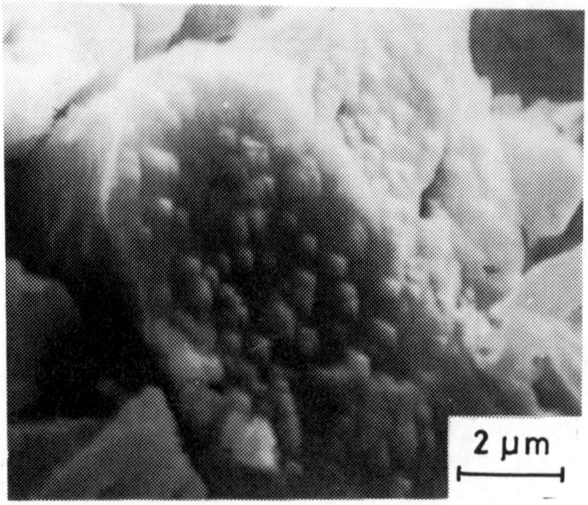

Figure 6 : Iron nuclei formed on the wustite surface during reduction of magnetite with hydrogen at 1173 K,

It is interesting to note that in case of reduction temperature of 1173 K, the nuclei distribution density was large and the nuclei have not grown in vertical direction (Fig. 7). Such a phenomenon coupled with the sintering

Figure 7 : Iron nuclei formed on the wustite surface during reduction of magnetite with hydrogen at 1173 K, showing large nuclei distribution density

between the neighbouring nuclei that would occur, eventually leads to formation of a dense metal layer on wustite surface. This restricts the further reduction of particles, since the gas no longer has direct access to the unreduced core of the parent oxide particle. On the other hand, at reduction temperature of 1423 K, the nuclei distribution density is relatively low, and the tendency for the vertical growth of the nuclei is relatively high.

The cause for the slitting and cracking of the particles during the reduction has been explained earlier.

Because of the removal of oxygen from the surface of wustite by hydrogen, a local variation in the iron/oxygen ratio takes place. After a certain supersaturation of iron has been reached, iron phase separates out on the surface of wustite. The first separation takes place in the form of nuclei. If the nuclei are to grow, there must be transport of iron to the nuclei from the points, where oxygen removal has occurred. The transport of iron takes place by counter diffusion of iron ions and vacancies through the oxide lattice or along the surface. It is the rate of oxygen removal from wustite surface on one hand and of the bulk diffusion of iron for the growth of nuclei on the other hand which will determine the nuclei distribution density in a given area. If the bulk diffusion of iron predominates, which is the case at high temperatures of 1423 K, it is expected that the nuclei formed will grow and the supersaturation in the regions between the originally formed nuclei will not be built up which is needed for additional nuclei to form.

On the other hand, if the rate of bulk diffusion of iron in wustite is not fast enough, which is the case at lower temperatures of 1173 K, the vertical growth of nuclei would be slow and the supersaturation in iron due to surface reduction of wustite in the neighbouring regions would build up leading to further formation of iron nuclei. In such cases, there would be a large number of small nuclei formed on the surface of wustite (Fig. 7).

It is interesting to note that the rate of oxygen removal upto, say 30-40% of reduction, is similar for the reduction temperatures of 1173 K and 1423 K (8). It is therefore reasonable to expect that the supersaturation build-up (at surface) would also be similar for both reduction temperatures. The bulk diffusion of iron seems to be the predominating factor in controlling the vertical growth of nuclei vs. nuclei distribution density; higher the temperature larger the vertical growth rate and lesser the number of nuclei and vice versa,

Reduction with Carbon Monoxide:

Figure 8 shows the iron nuclei formed on the parent magnetite surface subsequently reduced to wustite (reduction temperature: 1423 K). Similar to the case of hydrogen reduction, the nuclei have a tetrahedral shape. Slitting and cracking of the particle was also observed. One of the interesting feature observed was that the growth of the nuclei was more towards the formation of long thin stringers, as shown in Fig. 9. There was no other observable difference in the morphology of CO reduced samples when compared with H_2 reduced samples.

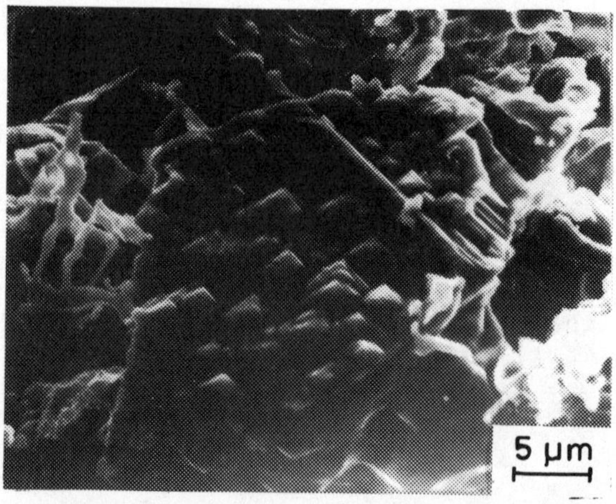

Figure 8 : Iron nuclei formed on the parent magnetite surface subsequently reduced to wustite by CO at 1423 K

General Discussion:

It has been shown that large scale pitting, slitting, cracking and fragmentation essentially occur in the stage of magnetite to wustite reduction. These are further enhanced during the reduction of wustite to iron due to counter migration of iron ions and vacancies in wustite. Further, the nuclei were randomly distributed. The pattern of nucleation and growth of iron on the wustite surface governs the reduction behaviour. If the vertical growth rate is slow, as is the case at the reduction temperature of 1273 K, a dense metal layer is soon formed which restricts further

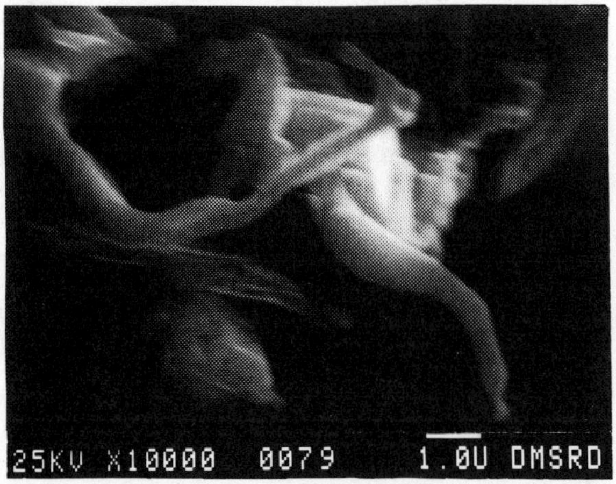

Figure 9 : Grown iron nuclei during reduction of magnetite with CO at 1423 K, showing formation of long thin stringers

the reduction. On the other hand, if the vertical growth rate is relatively fast, as is the case at the reduction temperature of 1423 K, relatively lower number of islands of iron on the wustite are formed and the region between the originally formed nuclei islands are still exposed to reducing gases. By the time the sintering between the grown iron islands starts, a large proportion of wustite has already been reduced to iron. It must be realised that the following occur simultaneously during the reduction process:

(a) fragmentation, slitting and cracking of magnetite (step $Fe_3O_4 \rightarrow FeO$)

(b) formation and growth of nuclei (step $FeO \rightarrow Fe$)

(c) sintering of the grown nuclei.

It is a well-known fact that the reduced iron powder contains porosity. This has been attributed in part to the difference in density values of iron oxides and iron, while maintaining the apparent volume of the particle. The present investigation throws some fresh light on the genesis of porosity formation in the reduced iron. In addition to the above factor, the porosity is generated by two further mechanisms. The reduced iron cake is an agglomerate of iron formed according to the mechanism described earlier. Firstly, as is evident from Fig. 5, holes are formed in the wustite phase caused perhaps by large scale counter migration of iron ions and vacancies in wustite (reaction 2). It is clear from Fig. 1 that these holes progressively enlarge into slits and cracks. Eventually the parent magnetite particles break into several fractions, also having pores, during the reduction stage FeO $\longrightarrow$ Fe. However, these fractions of parent magnetite get joined together due to sintering of resulting iron particles.

Secondly, iron nucleates on the wustite particles which grow in different directions. When a nucleus impinges upon the neighbouring one, having a different orientation, a void may be formed which can lead to porosity in the reduced iron. The size and shape of the void may change during sintering. Sintering of iron islands (a cluster of sintered nuclei) progressively increases with time and eventually the tetrahedral appearance of the iron nuclei islands disappears leading to the formation of iron having sinuous porosity. Such a cake after grinding results into powder having porosity as shown in Fig. 10.

Figure 10 : A view of typical porosity formed in the sponge iron powder.

Thus, in addition to inherent porosity created due to differences in densities of Fe_3O_4, FeO and Fe, the degree of porosity generated is also a result of interplay between the degree of fragmentation of magnetite particles due to solid state diffusion of iron, nucleation and growth rate of iron nuclei on wustite, and the extent of sintering between the product iron and islands formed.

Conclusions

1. During the reduction of magnetite to wustite by solid iron, considerable fragmentation of particles take place. It is expected that similar behaviour occurs during reduction of magnetite by hydrogen and carbon monoxide.

2. The iron nuclei formed on the wustite surface grow in a typical tetrahedral pattern till they impinge or coalesce with the neighbouring nuclei.

3. At high temperatures (1423 K), the bulk diffusion of iron in wustite predominates, and the nuclei formed grow in the vertical direction (perpendicular to the surface), leading to lower nuclei distribution density.

4. At low temperatures (1173 K), the bulk diffusion of iron in wustite is not fast enough, the vertical growth of nuclei would be slow, leading to higher nuclei distribution density. This eventually covers the entire wustite surface with a dense iron layer, which slows down the reduction.

5. In addition to the porosity created in the reduced iron, due to density differences in Fe_3O_4, FeO and Fe, the porosity is also a function of interplay between (a) fragmentation of magnetite during the reduction to wustite, (b) growth rate of iron nuclei on wustite and its impingement with the neighbouring nuclei, and (c) sintering between the product iron islands formed.

References

1. F.V. Lenel, "Powder Metallurgy: Principles and Applications", MPIF, New Jersey, 1980, 23.

2. J.O. Edstrom, J. Iron and Steel Institute, 1953, 175, 289.

3. N.J. Themelis and W.H. Gauvin, Trans. ASME, 1962, 65, 229.

4. D.H. St. John, and P.C. Hayes, Met. Trans., 1982, 13B, 117.

5. D.H. St. John, S.P. Matthew, and P.C. Hayes, Met. Trans., 1984, 15B, 701, and 709.

6. L. Von Bogdandy, and H.J. Engell, "The Reduction of Iron Ores", Springer Verlag, Berlin, 1971, 95.

7. R.K. Dube and B. Deo, Steel Research, 1987, 58, (9), 395.

8. R.E. Pocovi, T.F. Talbot, and A.R.E. Singer, Metals Technology, 1975, 2, (11), 505.

PERSPECTIVES IN IRON POWDER MANUFACTURE: A RESOURCE APPROACH

K.C. Sahoo, D.K. Lahiri, S. Barpanda
National Mineral Development Corporation Ltd.
and
G.S. Bhattacharjee, T.P. Bagchi
Defence Metallurgical Research Laboratory

ABSTRACT

India has a large deposit of high quality blue dust in fine form. Blue dust is essentially haematite (Fe_2O_3) with small percentage of silica and alumina and is singularly free from carbon, sulphur and phosphorus. Based on these resources, premium grade sponge iron powder was produced by hydrogen and carbothermic reduction. By suitable post reduction treatment, a wide spectrum of properties as demanded by various industries were prepared. Prealloyed powder of different commercial requirement were produced by coreduction of blue dust with other oxides. The powders were characterised and their green and sintered properties were studied. Sintered compacts were infiltered with copper yielding interesting results.

The commercial application of these powders in different areas like sintered friction material, high strength low density structural parts, chemical industry, addition as alloying element in critical alloy systems, soft magnetic strips etc. are discussed in brief.

This resource as converted into iron powder with attractive combination of properties, is considered to have bright commercial potentiality and cost effectiveness.

INTRODUCTION

The cost and quality of the iron powder are two of the important factors which limit the success in replacement of several engineering components, normally produced by other techniques, by sintered parts. Today, it is possible to produce high strength (upto 1200 MPa) P/M components with requisite toughness through sinter forge or injection moulding followed by post sintering treatments. In sinter forge route, the processing cost is higher. There lies a possibility of upsetting this cost by making the input raw material cheaper. Injection moulding, on the other hand, demands powder of much finer grade (10 micron) which is several times costlier than usual -100 mesh (i.e. -150 micron) powder. Production of such fine grade of powder by electrolysis, atomisation or sponge grinding is expensive. Mill scale reduction by hydrogen could yield fine powder, but the availability of consistent high purity mill scale is a constraint. An alternative source of cheap but high purity iron oxide fines, amenable to hydrogen reduction, could be an answer to this problem.

Many Indian iron ore mines, particularly the Bailadila mines located in Central India, have substantial reserves of blue dust in fine form. This blue dust is much pure than the same available in other parts of the country. The total reserve of the material is about 22 million tonnes (with 1 to 4 % silica) of which more than about 2.2 million tonnes are of high grade with less than 1 % silica. With conventional physical beneficiation, the silica content can be reduced to less than 0.2 %. This paper contains part of the results of the programme undertaken to convert the blue dust concentrate to iron powder by hydrogen and carbothermic reduction. The results of investigation have been highly encouraging. The paper further highlights the future prospects of these investigations.

CONCENTRATE PREPARATION

The R.O.M. blue dust (with 0.5 - 1.5 % silica) was subjected to beneficiation to bring down silica content to below 0.2 %. The physical and chemical properties of the concentrate are shown in Table 1.

HYDROGEN REDUCTION

The blue dust concentrate was reduced directly by hydrogen in multitube pusher-type furnace. Under optimised conditions, about 10 tonnes of powder was produced. The reduced sponge cake was subjected to mechanical comminution and post reduction treatment in addition to blending in order to impart a large spectrum of powder characteristics. With varied conditions of reductions and post reduction treatments, it was shown that powders with different characteristics could be produced

Table 1 Typical characteristics of blue dust concentrate

Physical proerties	
Average particle size, μm	17
Apparent density, g/cm^3	1.87
Tap density, g/cm^3	2.27
Flow rate, s/50 g	38
Screen analysis (μm), %	
+147	
-147, +104	4.8
-104, + 74	18.9
- 74, + 63	10.2
- 63, + 53	19.0
- 53, + 44	13.8
- 44	33.3
Chemical analysis, %	
Iron	69.80
Silica	0.10
Alumina	0.10
Sulphur	0.005
Phosphorous	0.005

(Table 2). SEM photographs of the powder particles are shown in Fig. 1 (as reduced) and Fig. 2 (after milling). It could be noted that the degree of sponginess of hydrogen reduced haematite (Fe_2O_3) powder is generally more than that normally obtained by carbothermic reduction of magnetite (Fe_3O_4). It is due to the higher oxygen content of haematite which when removed during low temperature gaseous reduction leaves higher and finer particle porosity. After prolonged milling, the surface irregularities are smoothened which causes reduced interparticle friction and hence increased level of apparent density.

CARBOTHERMIC REDUCTION

Attempts were also made to reduce the blue dust concentrate with coke breeze at temperature around 1000° C. To avoid impurity pick-up from coke breeze, blue dust and coke breeze were arranged in layers with intermediate layers of porous refractory felt. Refractory felt permits passage of gas freely while preventing admixture of ash and metallic iron. Coke and limestone mixture was taken with only 20 % in excess of stoichiometrical requirement. Complete reduction could

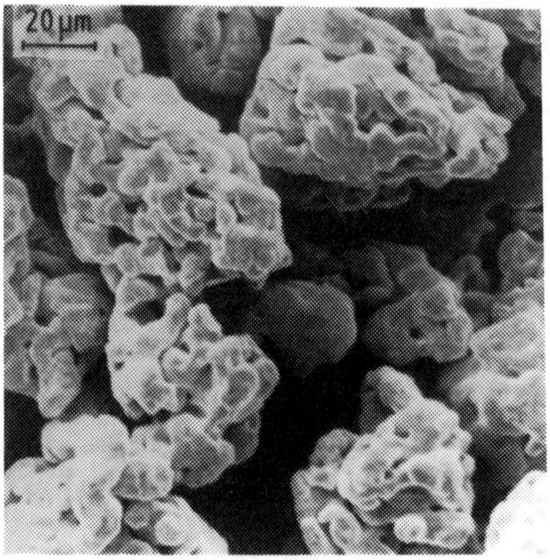

Fig. 1 : SEM Photograph of the Sponge Iron Powder (As reduced)

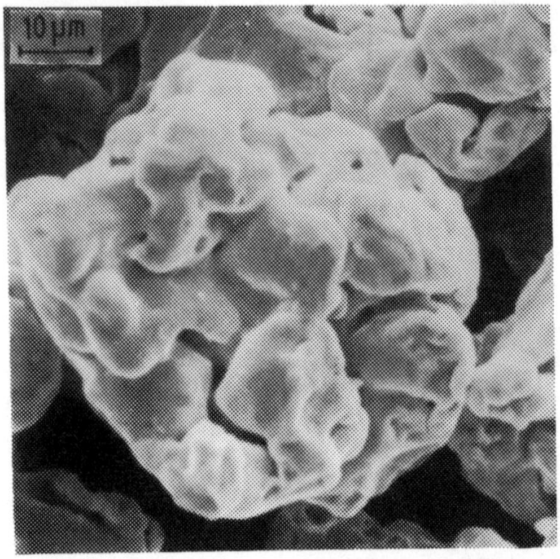

Fig. 2: SEM Photographs of the Sponge Iron Powder (After milling)

Table 2 Properties of different grades of sponge iron powder obtained from blue dust concentrate by different reduction and treatments

Property	Powder type A	B	C	D
Physical properties				
Average particle size, µm	6-10	10-12	10-12	10-12
Surface area, m^3 /kg	98	110	114	95
Apparent density, g/cm^3	1.4	2.3	2.5	2.7
Flow rate, s/50g	Poor	35	30	30
Mechanical properties				
Green density, g/cm^3	6.2	6.3	6.4	6.5
Green strength, MPa	385-405	370-400	360-385	317
Chemical analysis, %				
Acid insolubles	0.2	0.2	0.2	0.2
Hydrogen loss	0.18	0.10	0.10	0.10
Carbon	0.005	0.005	0.005	0.005
Sulphur	0.005	0.005	0.005	0.005
Phosphorous	0.005	0.005	0.005	0.005
Metallisation	99.6	99.7	99.7	99.7
Oxygen free basis				

be achieved in multilayered charge at 1000°C within 12 hours. Some studies were also carried out at 1200°C for 30 hours for achieving better properties. The physical and chemical characteristics of as produced sponge iron powder is shown in Table 3. It is seen that reduction at higher temperature yields powder with higher density. Further work is in progress to produce sufficient quantity of this type of powder in pilot scale for post reduction treatment and characterisation studies.

ALLOY POWDER PRODUCTION

Blue dust mixed homogeneously with stoichiometrical proportions of nickel oxide, copper oxide and molybdenum oxide was subjected to coreduction in dry hydrogen (dew point - 40°C) at 1000°C. Reduced prealloyed cake was ground. The chemical and physical properties of the coreduced powder (designated as AP - 1) are shown in Table 4.

FERRO ALLOY PRODUCTION

Blue dust mixed with proportional amount of tungsten oxide was coreduced in hydrogen at 1100°C for 2 hours. The reduced sponge (50 % Fe) was ground and the powder was compacted at 200 MPa. The compacts were found to have excellent green strength. Ferrotungsten compacts are found to be fully alloyed in X-ray diffraction studies. Such pellets could be used directly in high-speed steel making lending practical control on precise macroalloy additions.

Table 3 Properties of sponge iron powder obtained by carbothermic reduction

Property	Reduced at 1000°C, 12 hrs	Reduced at 1200°C, 30 hrs
Physical properties		
Average particle size, μm	12.3	18
Apparent density, g/cm^3	1.69	2.31
Tap density, g/cm^3	2.12	2.9
Flow rate, s/50g	45	28
Green density g/cm^3	6.2	6.38
Chemical analysis, %		
Total Iron	99.00	98.00
Metallic iron	98.50	98.50
Silica	00.30	00.30
Alumina	00.10	00.10
Carbon	00.05	00.05
Sulphur	00.004	00.004
Phosphorous	00.010	00.010
Hydrogen loss	0.15	0.10

Table 4 Typical properties of alloy grade powders (coreduced)

Physical properties	
Average particle size, μm	10
Apparent density, g/cm^3	2.6
Tap density, g/cm^3	3.26
Flow rate, s/50g	Poor
Compressibility g/cm^3	6.6
Screen analysis (μm), %	
+147	0.9
-147, +104	1.9
-104, + 74	7.7
- 74, + 63	4.4
- 63, + 44	36.3
- 44	48.8
Chemical analysis, %	
Iron	92.70
Nickel	4.75
Copper	1.46
Molybdenum	0.54
Acid insolubles	0.35
Hydrogen loss	0.20

COMPACTION AND SINTERING STUDIES

Sponge iron powder of Type B and Type C (vide Table 2) were pressed at various loads. Fig. 3 (a-e) shows the plots of compaction pressure with green density, green strength, sintered hardness, sintered strength and dimensional change. Alloy powder AP-1 was compacted at various pressures and sintered at 1100°C for 1 hour, Table 5 shows the sintered density, hardness dimensional change and sintered strength of the alloy powder. To assess the suitability of Type C powder for self-lubricating bearing application, some experiments were carried out in collaboration with sintered parts manufacturers. Table 6 shows the results of the study. It would be noticed that strength factor for the bushes made from this powder is about 1.5 times higher than standard acceptable values while adopting existing pressing parameters. Dimensional changes with reference to green size is higher. This is because of higher compact relaxation due to higher spongy nature of the powder. However, dimensional change with (reference) to die dimension is less than 0.06 %.

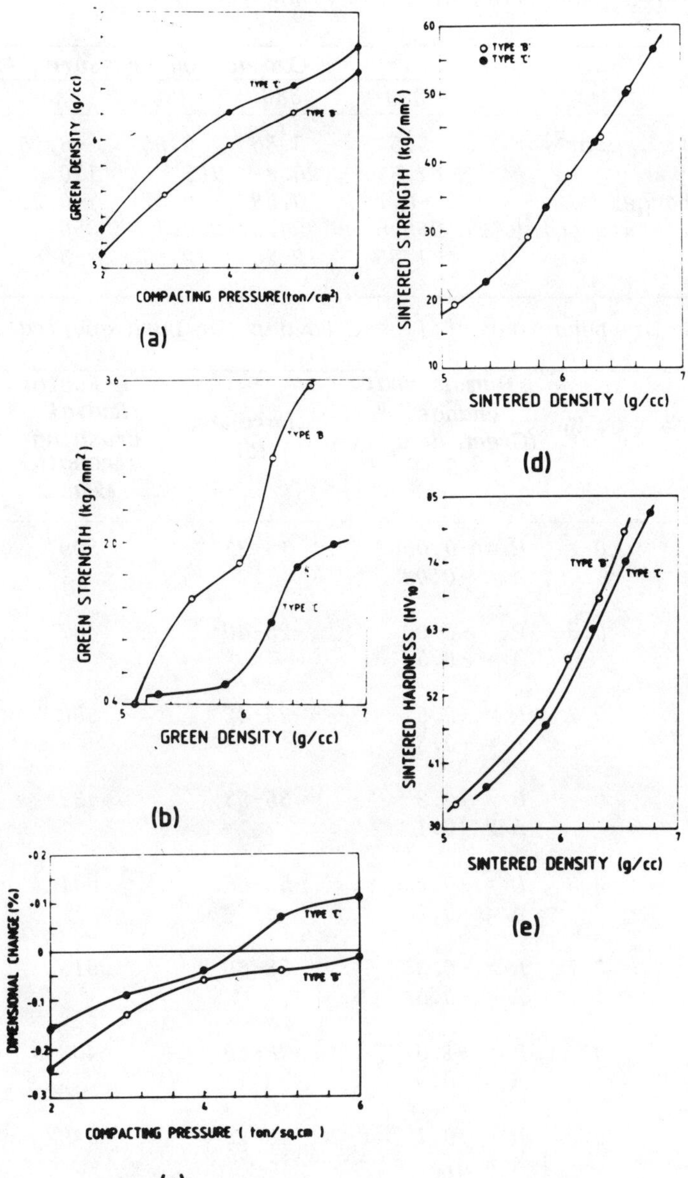

Fig. 3 (a-e) : Plots showing relationship between
(a) compaction pressure vs green density
(b) green density vs green strength
(c) compacting pressure vs dimensional change
(d) sinter strength vs sintered density
(e) sintered density vs hardness

Table 5 Sintered properties of alloy grade Powder

Property	Compaction Pressure, MPa				
	300	380	455	530	600
Sintered density, g/cm^3	5.54	5.86	6.65	-6.75	6.92
Hardness, VPN	62.8	96.3	105	120	135
Dimensional change, %	-0.74	-0.68	-0.37	-0.22	-0.17
Ultimate tensile strength, MPa	259.6	304.6	324.1	350	410
Elongation, %	1.85	2.20	2.64	3.8	4.0

Table 6 Sintering behaviour of Type C Powder for bush application

Sl.No.	Cu, %	C, %	Dimensional change, % (Green density = 5.8 g/cm^3)	Hardness RB	K-Factor (Radial crushing strength) MPa
1	0	0	R = -0.05 A = -0.03	15-25	309
2	0	0.3	R = -0.06 A = -0.5	25-40	487
3	0	0.7	R = -0.6 A = -0.3	35-42	556
4	3	0	R = +0.3 A = +0.1	35-53	432
5	3	0.3	R = -0.2 A = -0.2	58-66	641
6	3	0.7	R = -0.35 A = -0.6	58-69	618
7	5	0	R = +1.6 A = +0.9	48-55	463
8	5	0.3	R = +0.1 A = 0.3	58-70	649
9	5	0.7	R =-0.38 A = -0.4	61-76	641

MIXED GRADES

Besides the production of Iron Powder with varied apparent densities and green densities through post reduction treatments, some studies were undertaken to blend as produced sponge iron powder with commercially available atomised and electrolytic powders to improve their existing powder characteristics. Results of this study is shown in Table 7. It could be seen that addition of 10 % of fine hydrogen reduced sponge iron powder in atomised or electrolytic iron powder improves their compressibility at 455 MPa with only marginal effect on apparent density. This blend with fine sponge iron powder not only enables optimal packing pattern, but also acts as a lubricant. A short diffusion treatment of the blend could eliminate gravitational segregation, if any. Blending is conspicuously seen to improve the green strength of atomised and electrolytic powder.

Table 7 Effect of addition of hydrogen reduced sponge iron powder on the properties of electrolytic and atomised iron powder

Blending ratio		Apparent density g/cm^3	Green density g/cm^3 (P = 455 MPa)
Electrolytic :	**Sponge**		
100	0	2.86	6.0
90	10	2.8	6.2
Atomised :	**Sponge**		
100	0	2.8	6.2
90	10	2.75	6.4

INFILTRATION STUDIES

The sponge iron powder produced by hydrogen reduction of blue dust are porous the pore size is fine; the pores are distributed uniformly with high degree of surface irregularity. Further, very fine particles, obviously, resulted in less compressibility and the particle porosities could not be eliminated even at higher compacting pressure. These aspects are considered conducive to infiltration. Accordingly, compacts made at various pressure levels (ranging from 230 MPa to 460 MPa) and sintered at 1150°C for 3 minutes were infiltered with iron-copper alloy at 1100°C for 15 minutes. Both sintering and infiltration were carried out in dry hydrogen (dew point -40°C). Iron-copper alloy was chosen mainly to limit dissolution of iron from the surface layer of the compact. It was possible to carry out infiltration in finished products with major dimensional changes. Infiltered compacts were subjected to metallographic studies (Fig. 4). It has been noticed that apart from having

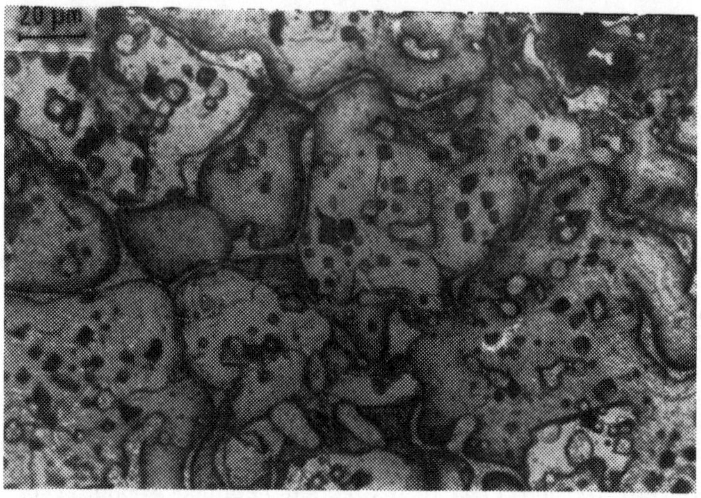

Fig. 4 Microstructure of copper infiltered iron specimen

complete infiltration over entire cross section uniformly, fine porosity within the iron particle also got filled up with copper. Due to high surface area the meanfree path of the infiltrant is considerably smaller. The authors' experience with copper brazing of steel revealed that smaller manfree path results in higher braze joint strength approaching the strength of the steel. Also higher amount of copper in solution resulted due to higher surface area for diffusion; would give higher strength. This is also supported by higher particle microhardness.

APPLICATIONS OF SPONGE IRON POWDER

From the present exercise, it is evident that iron powders with wide range of properties suitable for various applications can be produced by direct reduction of blue dust. However, the process of reduction and reduction parameters play a vital role in determining powder characteristics. Sponge iron powder with low to medium AD were successfully used in various applications, viz., sintered friction materials, low density structural parts, chemical industry, heating bag, alloying element, soft magnetic strip etc.

SINTERED FRICTION MATERIAL

A 40 kg batch of iron base friction material was produced under production environment. The powder (Type A) was found to be eminently suited for this application. Due to its low AD, the powder produces very uniform mix with graphite and refractory powder which are the major constituents of friction material

Fig. 5 Intricate shape prepared from sponge iron powder by cold Isostatic Pressing

formulation. Because of large volume of non-metallic constituents, the green strength of friction components is generally low. Friction components are very thin and handlability of green compacts is an important technical requirement. Type A powder which yields compacts with higher green strength is found best suited for this application.

LOW DENSITY STRUCTURAL PARTS

Low AD iron powder produced by hydrogen reduction has been found to be highly suitable for low density high strength structural parts. This powder is considerably finer and also more spongy than carbon reduced sponge iron powder produced at higher reduction temperatures. However, high temperature reduction with larger reduction time cycle could produce coarser powders with reduced particle porosity. Low temperature hydrogen reduced powder sinters faster than many commercial iron powder because of their fine sponge structure. As this powder has highly irregular surface with internal porosity, their compressibility (i.e. green density as 420 MPa compacting pressure) is slightly lower than Hoganas sponge iron powder NC 100.24. Highly irregular surface results higher mechanical interlocking and cold welding yielding high green strength. Such type of compacts also result in higher sintered strength. Hydrogen reduced sponge iron powder parts are often stronger than high density powder parts made out of commercially available powders and have better strength to weight ratio - a distinct resource conservation approach with better economy. The reproducibility of intricate shape in cold isostatic pressing was remarkable with this powder. Fig. 5 distinctly shows faithful reproduction of intricacies of the component. This component could not be reproduced under similar conditions using other types of iron powders.

CHEMICAL INDUSTRY

Iron powder finds diversified application in chemical industries, eg. food processing industries (as additives), pharmaceutical industries (as catalysts), paint industries and reprographic industries (as developer). In view of high purity and high surface area, this powder would be ideal for chemical applications. However, actual user's trial could be conducted only with pharmaceutical industries where large amount of iron powder is needed as catalysts in paracetamol production.

Another interesting area where Type A powder could be used with much advantage is in the manufacture of heating bags used as body and food warmers in cold weather. The heating efficiency of bag both in terms of maximum temperature attainable and sustenance time was higher (by about 30 to 50 %) than that obtainable with electrolytic iron powder. This is due to higher surface area of the sponge iron powder available for chemical reaction.

ALLOYING ELEMENT

Sponge iron powder produced by hydrogen reduction of blue dust is practically free from sulphur, phosphorus and carbon. Free silica and alumina to the extent of 0.3 % is the only impurity present in the powder. This could be broughtdown further by high intensity magnetic separation. Work on this area is still in progress.

Experiments were carried out to use this high purity sponge iron powder in tungsten-base heavy alloy (eg. W-Ni-Fe etc.) production. Mechanical properties of heavy alloy system are known to be highly susceptible to even ppm level impurities (particularly, S, P and alkaline metals). Use of this sponge iron in heavy alloy production resulted in about 3 - 5 % improvement in ductility (i.e., percent elongation) over the heavy alloy produced with carbonyl/electrolytic iron powder under identical processing parameters.

SOFT MAGNETIC STRIP

Sponge cake as obtained from the reduction furnace was subjected to hot rolling in controlled atmosphere followed by cold roll finishing. The resultant strips were fully dense. Magnetic properties of the strips were measured and were compared with those normally obtained in high purity soft magnetic strips produced by vacuum melting of Armco or sponge iron.

Magnetic induction of sponge roll product is similar to that of samples produced by melting route. But coercive force is higher than the acceptable limit. This could be attributed to higher oxide inclusions and smaller grain size of P/M strips. Attempts are being made to improve these aspects through better control of inclusion and heat treatments. The cost of production through P/M route could be only about 50 % of that of vacuum melted rolled strips.

COST EFFECTIVENESS

Pilot scale production studies carried out in multitube pusher furnace showed that the requirement of hydrogen gas is only 0.6 normal cubic meters per kg of iron powder. Hence, power consumption from the use of hydrogen gas (obtained by electrolysis of water) is only about 2.5 kW per kg of iron powder. The power requirement for reduction is 2.1 kW. This, probably, compares favourably with the power requirement of carbothermic reduction process. This level of power requirement coupled with negligible cost of raw material (blue duest), could emerge cheaper than the powder produced by other conventional processes. Use of continuous strip belt furnace could further improve productivity and power consumption factor.

CONCLUSIONS

1. *Sponge iron with wide range of characteristics can be produced by hydrogen reduction of blue dust. This powder can find wide applications in manufacture of friction materials, low density high strength P/M parts and infiltrated P/M components.*

2. *Fineness of this powder with high surface irregularities can make this powder best suited for injection moulding and slip casting. Slip consistency and settling rate in slip solution could be much favourable with fine powder as has been observed by some preliminary study; detailed studies are in progress.*

3. *Because of high purity, the powder is best suited for chemical industries, alloy production etc.*

4. *High surface area makes this powder as potential candidate for catalysts and heating bag applications.*

5. *Above all, the powder could be highly cost-effective and amenable to small to medium scale production.*

6. *The physical properties of these powders, associated with their very favourable behaviour on mechanical working, should enable to consider continuous strip making from blue dust as a distinct possibility.*

ACKNOWLEDGEMENT

The authors would hereby thank Shri PC Gupta, Chairman cum Managing Director, National Mineral Development Corporation Limited

(NMDC), and Dr. P. Rama Rao, Director, Defence Metallurgical Research Laboratory (DMRL), for providing all out organisational support during this work. Mere words are definitely not enough to express our indebtedness to Shri SLN Acharyulu, Director II, DMRL, who has guided us by providing most useful suggestions during the course of this study. Thanks are also due to Shri YP Bhalla Director (Planning) and Shri K. Sivaji, Chief Mineral Dressing Engineer - both of NMDC. Shri KP Rao, DMRL, helped in conducting some experiments and powder characterisation during one phase of this study. Our sincere thanks also go to different organisations like M/s. Hindustan Aeronautics Limited, (Hyderabad Division), M/s. Mahindra Sintered Products, Pune, M/s. Sundaram Fasteners, Bangalore, M/s. Andhra Sintered Products, Hyderabad, for characterisation of the powder and testing its applicability towards manufacture of different commercial components.

P/M DUAL PHASE STEEL PROCESSED FROM BALL BEARING STEEL SWARF

P. Ramakrishnan, A.N. Tiwari, V.S. Gorantiwar

Department of Metallurgical Engineering
I.I.T.,BOMBAY-400076, INDIA.

J.K. Beddow
University of Iowa, IOWA-52242, USA.

Abstract

In this work, the feasibility of recycling ball bearing steel swarf generated by various industries through the powder metallurgy process has been investigated. The AISI-52100 grade steel swarf has been embrittled by hydrogen charging and the size reduced by milling in a high energy attritor. The steel powders of the required particle size distribution, after annealing have been blended with pure iron powders in different proportions so as to achieve homogeneous mixed powders containing 30, 45 and 60 weight percent of steel powders. In one batch 2 wt % Cu was also added to facilitate sintering. The mixed powders have been compacted and sintered in the range of 1150°C to 1300°C in controlled atmosphere. The sintered and densified products have been subjected to heat treatment consisting of heating to the intercritical temperature of 800°C or 820°C and water quenching, to develop the dual phase microstructure. The tensile behaviour studies showed the absence of yield point and high value of strain hardening exponent (0.74-0.89). Tensile strength and ductility have also been correlated with microstructure.

Introduction

During the last decade extensive research efforts have been undertaken for the development of dual-phase steels. The microstructure of dual-phase steels is usually characterized by 15-25% martensite islands distributed in a 75-85% fine-grained ferrite martrix. Perhaps the most striking property of such steels is the remarkably large work hardening which gives rise to a high tensile strength with large elongation. In order to exploit the benefits of dual-phase microstructure in P/M components, Anand et al.(1) investigated the possibility of obtaining dual-phase microstructure in P/M components. They showed that by blending low alloy steel powders prepared from ball bearing steel swarf with iron powder and subjecting the green compacts to sintering and heat treatment, can result in samples with dual-phase microstructure. These samples exhibited good strength and high strain hardening, however, their ductility was low due to high porosities ($\sim$ 10 %). This paper deals with the development of dual-phase steels processed by P/M route utilizing iron powder and attritor milled ball bearing steel (AISI-52100) swarf. The influence of microstructure on the mechanical properties has also been reported.

Experimental Procedure

Ball bearing steel swarf (AISI-52100) (C-1.13, Cr-1.25, Mn-0.50, Si-0.28, S-0.04, P-0.05, Fe-balance), electrolytic iron powder (-400 mesh) and electrolytic copper powder (-400 mesh) were used in the present investigation. The degreased steel swarf was subjected to hydrogen charging to embrittle the swarf under the conditions reported earlier (1). After embrittlement treatment the swarf was thoroughly rinsed and dried in a vacuum oven at 40°C. The swarf was then refrigerated till attritor milling. The operating conditions for the attritor milling are the same as those reported in reference (1). After a milling of one and four hours, the powders were subjected to sieve analysis to get the required fractions.

A sieve fraction (-200 +400) mesh powder produced from the swarf and the electrolytic iron powder (-400 mesh) were annealed for one hour at 650°C in hydrogen atmosphere, to improve the compressibility. Annealed powders were blended in different proportions for 15 minutes so as to yield homogeneously mixed batches as shown in Table I.

The powders were consolidated into rectangular specimens of 55 x 15 mm by applying pressures in the range of 260 to 380 MPa. The sintering of the specimens were carried out in hydrogen atmosphere. The sintering schedule for compacts without copper consisted of sintering at 1300°C for 1 hr, furnace cooling, repressing at 950 MPa, resintering at 1300°C for 1 hr and furnace

TABLE I : BATCH COMPOSITIONS

Specimen Designation	Constituents (wt %)		
	AISI-52100 Powder	Fe Powder	Cu Powder
30EN	30.0	70.0	-
45EN	45.0	55.0	-
45EC	45.0	53.0	2.0
60EN	60.0	40.0	-

cooling. The copper containing specimens were subjected to the sintering sequence of sintering at 1150°C for 2 hrs, furnace cooling, repressing at 950 MPa, resintering at 1250°C for 1 hr and then furnace cooling.

Heat treatment of the samples were carried out by heating them to intercritical temperatures of 800°C or 820°C for half an hour and quenching them into water. All the quenched samples were tempered at 150°C for one hour.

The metallographic preparation sequence of the specimen consist of etching in 4% picral, rinsing with water and lapping on cloth several times until all disturbed metal was removed from the surface. After this procedure the specimens were finally etched in 4% nital and 4% picral (1 to 1 mixture) for 6 seconds. This method is found to improve the clarity of microstructure in terms of ferrite grain boundaries, martensite islands and carbide particles. The volume fraction of non-ferritic phase (NFP) in dual-phase sintered steels was determined by Leitz Tas plus image analyser. For this purpose specimens were subjected to controlled etching with picral so as to delineate only NFP.

Tensile testing of the specimens (gauge length 16.5 mm) were carried out on the Instron universal testing machine (Model 1195) using a cross-head speed of 1 mm/min. The Vickers hardness of the specimens were measured at 10 Kgf load.

Results and Discussions

Attritor Milling: According to Schuhmann (2) the process of comminution can be expressed as

$$Y = 100 \left(\frac{X}{K} \right)^m$$

where Y = cumulative weight percentage finer

X = average particle size in um

K = size modulus, and

m = distribution modulus.

The effect of the attritor milling time on the particle size distribution of the hydrogen embrittlement AISI-52100 steel swarf is shown in Figure 1 in terms of Schuhmann plots (i.e. log Y vs log X curves). Most of the data points lie on straight lines indicating the validity of the Schuhmann relation in the process of attritor milling. From the slope and the intercept on X-axis of the straight lines, values of m and K have been calculated and are also shown in Figure 1. It is seen that longer milling duration results in lower K and higher m. The size modulus (K) represents the maximum particle size in a powder whereas higher value of m implies narrower particle size distribution.

Compositions and Densities: The chemical composition, density and porosity of the sintered steels are given in Table II.

Microstructural Characterization: Optical micrographs of the dual-phase steels are shown in Figure 2 (a,b,c). All of them show the ferrite matrix with dispersed islands of martensite (dark etching). The presence of a few tiny carbide particles can also be observed. Figure 3 clearly shows the uniform distribution of martensitic phase. Side view of the polished and etched tensile specimen in fracture condition is shown in Figure 4. It is apparent that the tensile failure in dual-phase sintered steels occurs by decohesion of the martensite/ferrite interface. Examination of tensile fracture surface by SEM (Figure 5) shows the ductile mode of fracture by the microvoid coalscence. It can be seen that the dimples are formed due to the martensite/ ferrite interface decohesion.

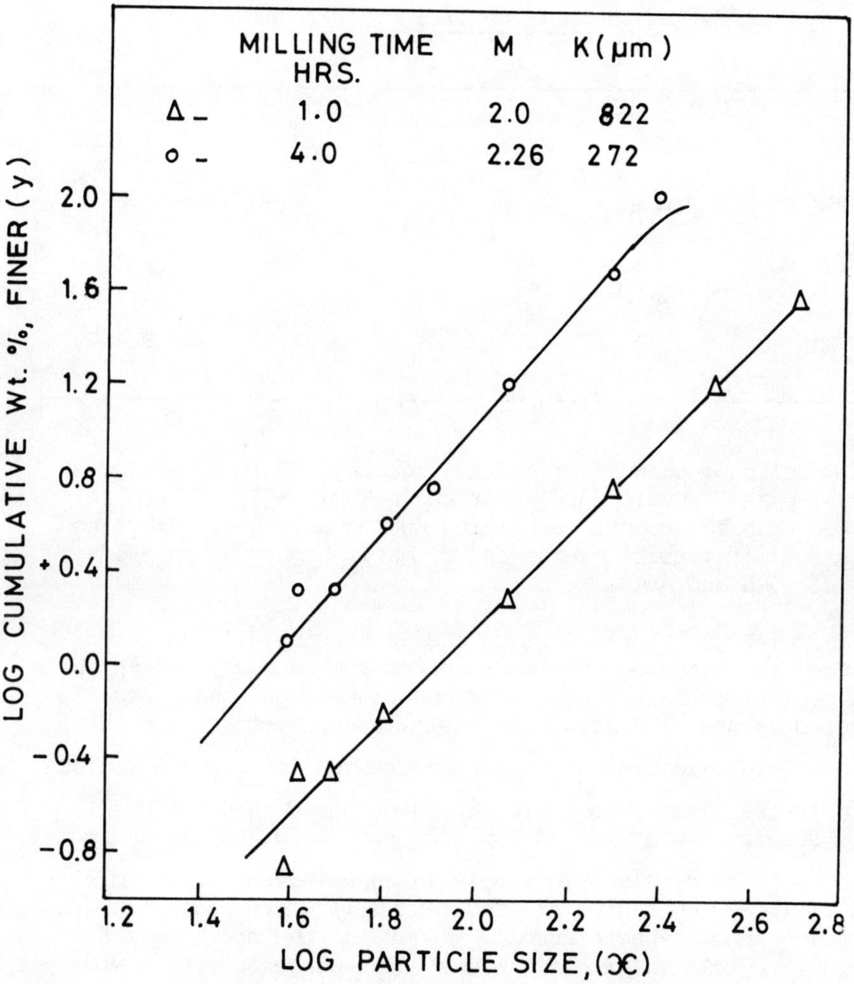

FIG. 1 _ SCHUHMAHN PLOTS FOR ATTRITOR MILLED AISI-52100 STEEL SWARF

TABLE II : COMPOSITION, DENSITY AND POROSITY OF SINTERED STEELS

Specimen	Chemical Composition (wt %)		Density g/cm³	Porosity %
	C	Cr		
30EN	0.10	0.37	7.52	2.3
45EN	0.15	0.56	7.46	3.3
45EC	0.15	0.56 Cu = 2.0	7.40	4.4
60EN	0.20	0.75	7.47	3.4

Results of quantitative metallography and image analysis of dual-phase steels are reported in Table III. It can be noticed that the copper containing dual-phase steel (45EC) contains smaller ferrite grains and slightly higher amounts of NFP (cf 45EN and 45EC).

Mechanical Properties: During tensile testing all the water quenched and tempered steels exhibited continuous yielding which is a typical characteristic of conventional dual-phase steels. Measured values of 0.2% offset yield strength ($\sigma_{0.2}$) and the ratio of yield to ultimate tensile strength ($\sigma_{0.2}/\sigma_u$) are given in Table IV. From this table it can be noted that the dual-phase steels show a reduced $\sigma_{0.2}/\sigma_u$ ratios (0.29-0.58) as compared to those of sintered steels in annealed condition (0.55-0.70). These low ratios are indicative of wider range of work hardening in dual-phase steels. Regarding the mechanism of continuous yielding of conventional dual-phase steels, a widely accepted explanation is that it is due to the accommodation strains induced in the ferrite by the austenite-to-martensite transformation. This also explains the low yield strength (3). It is reasonable to assume that the above explanation holds good for P/M dual-phase steels also.

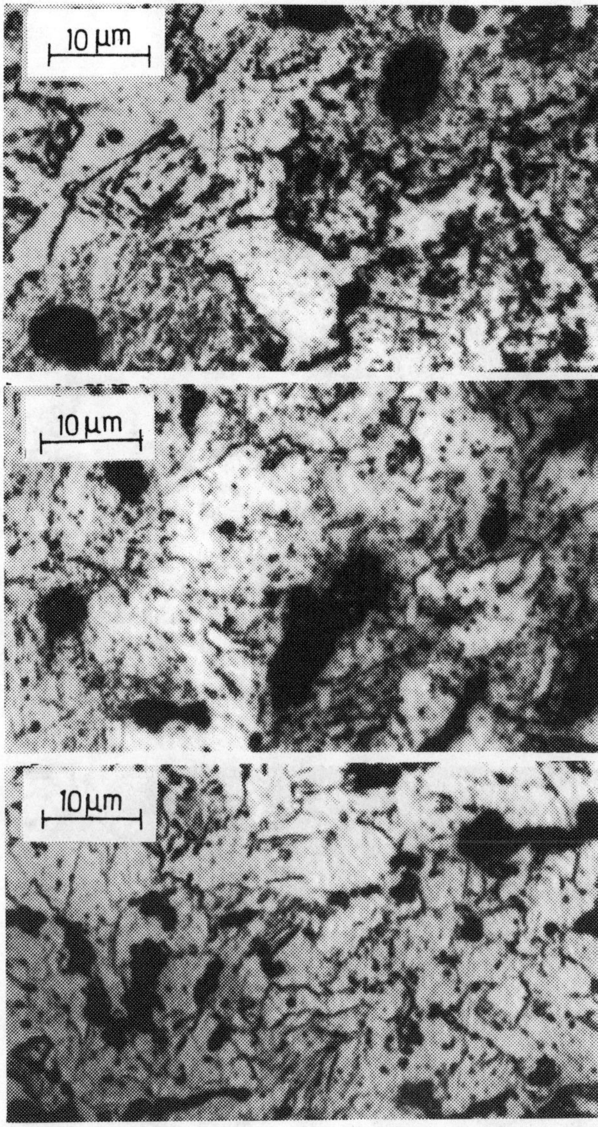

FIG.2 : Optical micrograph of dual-phase steels (etchant - 4% nital + 4% picral)

(a) 45EN (water quenched from 800°C)
(b) 45EN (water quenched from 820°C)
(c) 45EC (water quenched from 820°C)

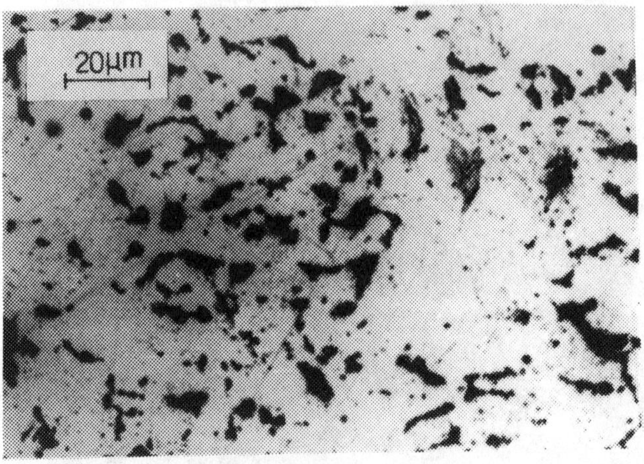

FIG. 3 : Optical micrograph of a dual-phase steel
(45EN, 4% picral etch)

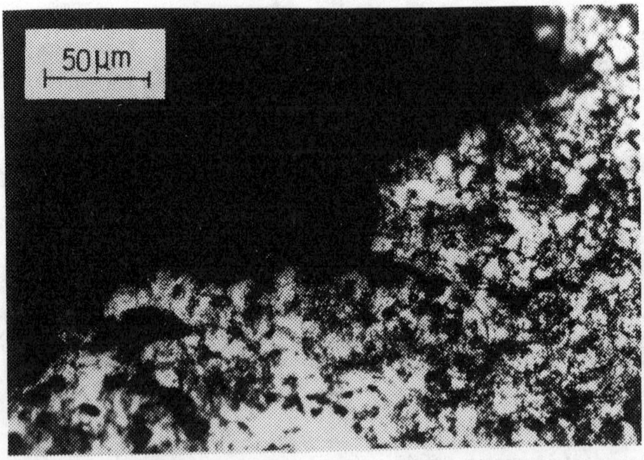

FIG.4: Side view of the fracture surface in a tensile specimen
(polished and etched in picral)

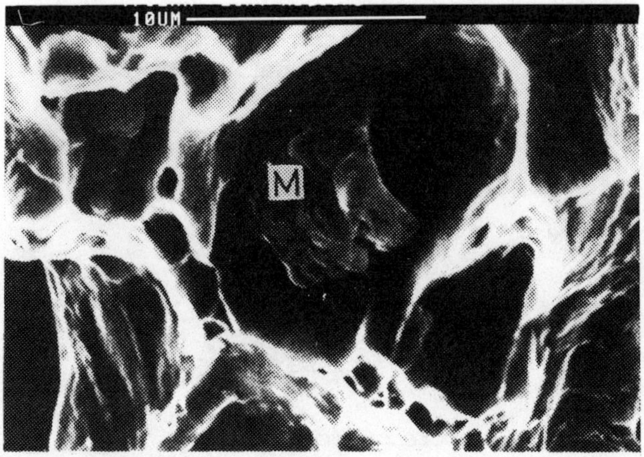

FIG.5 : SEM of a typical tensile fracture surface showing martensite particles (M) inside dimples.

TABLE III : QUANTITATIVE METALLOGRAPHIC DETERMINATIONS

Quenching temperature	Parameter	Specimen designation			
		30EN	45EN	45EC	60EN
800°C	Ferrite grain size (μm)	13	12	8	15
	Mean diameter of martensite islands (μm)	4	4	5	3
	NFP (vol %)	21	25	28	33
820°C	Ferrite grain size (μm)	13	12	11	17
	Mean diameter of martensite islands (μm)	5	5	5	5
	NFP (vol %)	24	35	39	43

The strain hardening behaviour has been studied by evaluating strain hardening exponent (n) defined by Ludwik(4) equation:

$$\sigma_t = \sigma_y + k\epsilon^n$$

where σ_t = true stress

ϵ = true strain

σ_y = yield stress (= $\sigma_{0.2}$)

k = strength coefficient

Plotting of the tensile data for the dual-phase steels (water quenched from 820°C) on a log $(\sigma_t - \sigma_y)$ versus log ϵ basis (Figure 6) did not give a straight line, but the data can be adequately represented by two straight lines. The slope of the straight line representing initial stages of plastic strain (ϵ upto 4%) is always greater than that of the line representing later stage of uniform plastic strain upto necking. This implies that the work hardening characteristics require atleast two values of strain hardening exponent n (i.e.non-single n behaviour) and initial strain hardening is quite high for dual-phase steels. Flow curves for steels quenched from 800°C were similar to those reported in Figure 6. Calculated values of n (ϵ = 0-4%) for the steels in as sintered and dual-phase conditions are reported in Table IV. From this table it can be seen that as compared to sintered steels (n = 0.55 to 0.70), dual-phase steels show higher capacity for strain hardening (n = 0.74 to 0.89). The value of n also rises as the martensite (NFP) content of steel is increased.

In order to study the effect of second phase i.e. NFP which is essentially the amount of martensite, ultimate tensile strength (σ_u) and uniform elongation (e_u) for the dual-phase steels have been plotted as a function of NFP volume percent and reported in Figure 7 and 8 respectively. In both cases data points seem to follow linear relations. By linear regression analysis,

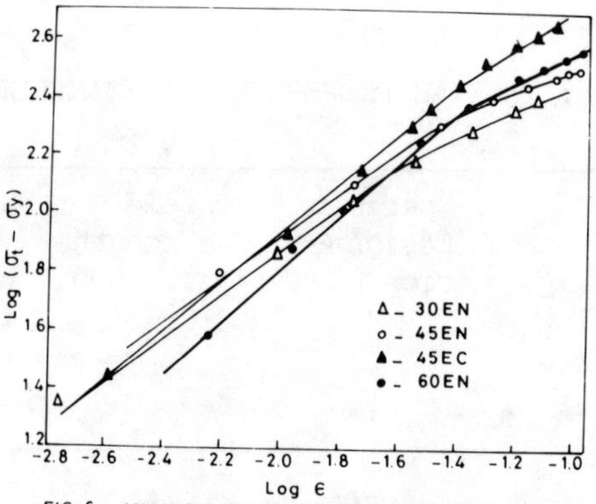

FIG. 6 _ ANALYSIS OF TRUE STRESS AND STRAIN CURVES ACCORDING TO THE LUDWIK EQUATION FOR DUAL PHASE STEELS (WQ FROM 800°C)

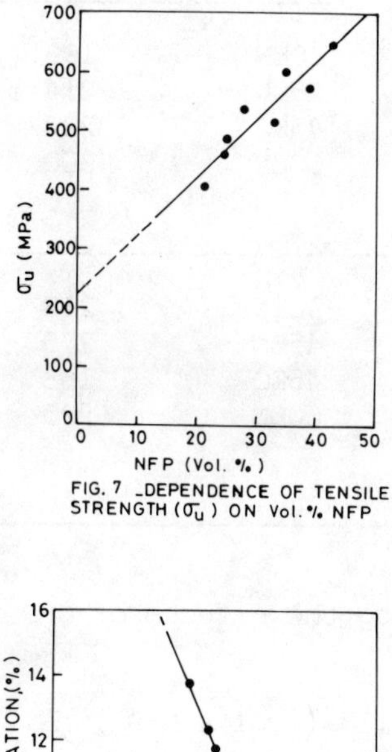

FIG. 7 _DEPENDENCE OF TENSILE STRENGTH (σ_u) ON Vol.% NFP

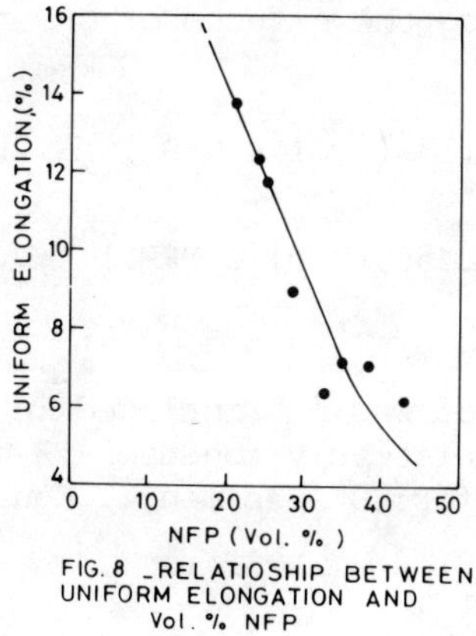

FIG. 8 _RELATIOSHIP BETWEEN UNIFORM ELONGATION AND Vol. % NFP

TABLE IV : TENSILE PROPERTIES OF SINTERED STEELS

Condition	Specimen Designation	Yield strength $\sigma_{0.2}$ MPa	Ratio of yield to ultimate tensile strength $\sigma_{0.2}/\sigma_u$	n
Sintered (Ferrite + pearlite)	30EN	180	.57	.66
	45EN	217	.65	.70
	45EC	278	.70	.55
	60EN	178	.55	.63
Water quenched from 800°C (Ferrite and martensite)	30EN	199	.49	.74
	45EN	214	.44	.77
	45EC	224	.42	.85
	60EN	300	.58	.87
Water quenched from 820°C (Ferrite and martensite)	30EN	218	.36	.80
	45EN	275	.46	.83
	45EC	168	.29	.86
	60EN	225	.45	.89

the equations expressing the effects of NFP in P/M dual-phase steels are:

$$\sigma_u = 220 + 10.13 \text{ (vol \% NFP)} \quad \ldots \quad (1)$$

$$e_u(\%) = 23.1 - 0.455 \text{ (vol \% NFP)} \quad \ldots \quad (2)$$

A grain size factor was not included in the equations (1) and (2) since it was relatively constant 12 to 17 μm for EN specimens and 8 to 11 for EC specimens. Equations (1) and (2)

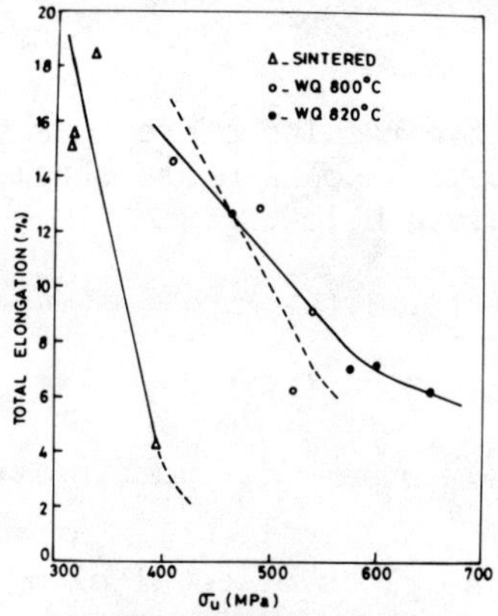

FIG. 9 _COMPARISON OF TENSILE STRENGTH VS. ELONGATION FOR AS SINTERED AND DUAL PHASE STEELS

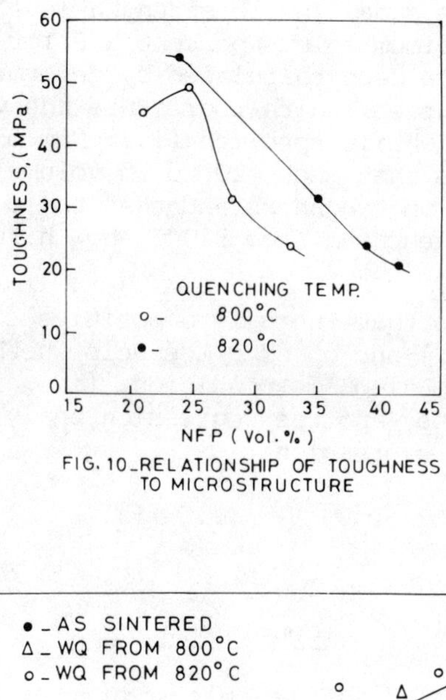

FIG. 10_RELATIONSHIP OF TOUGHNESS TO MICROSTRUCTURE

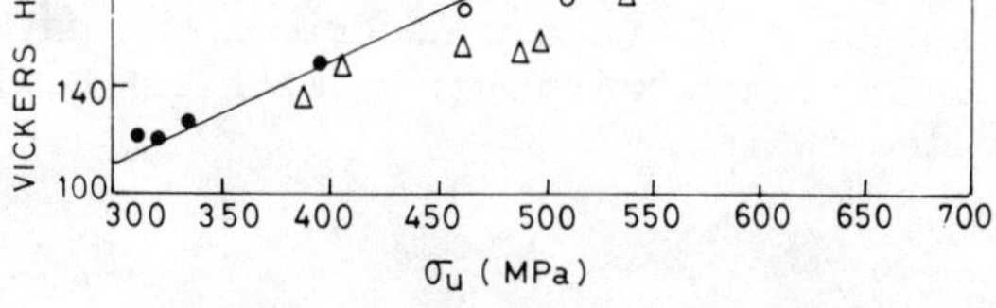

FIG. 11 _RELATIONSHIP OF VICKERS HARDNESS TO TENSILE STRENGTH

can be compared with the results of regression analysis of experimental data on conventional dual-phase steels by Hansen and Pradhan (5) and are given below:

$$UTS = 453 + 12.13 (\% M)$$

$$e (\%) = 20.21 - 0.23 (\% M)$$

where M stands for martensite.

Figure 9 shows the plots of total elongation versus σ_u for P/M steels in sintered and dual-phase conditions. This diagram clearly shows that the elongation values (i.e. ductility) of dual-phase steels are vastly superior to those of sintered steels when compared at the same tensile strength level. In order to determine the optimum volume percent NFP for maximum toughness, toughness values have been calculated by determining the area under the tensile curves, which represents the work done per unit volume of the tensile specimen (i.e. MNm/m^3 or MPa). From Figure 10 it appears that about 20 to 25 volume percent NFP results in the maximum toughness and that the dual-phase steels produced by water quenching from 820°C show higher toughness.

Figure 11 shows that there is a positive correlation between Vickers hardness (HV) and σ_u of P/M steels whether they are sintered or in the dual-phase condition. (i.e. ferrite-pearlite or ferrite-martensite) and the regression analysis of the data gives the following expression:

$$HV = 3.67 \sigma_u \quad ... \quad (3)$$

Conclusions

1. Ball bearing steel (AISI-52100) swarf can be converted to powders suitable for P/M applications by the process involving hydrogen embrittlement and attritor milling.

2. Schuhmann relation adequately describes the particle size distribution during comminution by attritor milling of the embrittled swarf.

3. A 2 wt % addition of Cu not only reduces sintering temperature but also refines ferrite grain size and slightly increases martensite content in dual-phase steels.

4. The P/M dual-phase steels exhibited a continuous yielding behaviour with a very high initial work hardening rate (strain hardening exponent n = 0.74 to 0.89), and low ratio of yield to tensile strength between 0.29 to 0.58.

5. The mechanical properties can be expressed as:

$$\sigma_u = 220 + 10.13\ (\text{vol}\ \%\ \text{NFP})$$

$$e_u = 23.1 - 0.455\ (\text{vol}\ \%\ \text{NFP})$$

$$HV = 3.67\ \sigma_u$$

ACKNOWLEDGEMENTS

The authors are grateful to National Science Foundation, U.S.A., for their support through the grant No.INT-8511412.

REFERENCES

1. R. Anand,A.N. Tiwari and P. Ramakrishnan, presented at the Int.Conf. on Powd. Met. and Related High Temp.Matls., Dec 1987, I.I.T. Bombay, India.

2. E.J. Pryor, Mineral Processing, Mining Pub.Ltd., London (1960) P.167.

3. A.M. Sarosick and W.S. Oowen, Script.Metall. 17, No.2, 227-31 (1983).

4. G.E. Dieter, Mechanical Metallurgy, (II Ed.) McGraw Hill (1981), P.341.

5. S.S. Hansen and R.R. Pradhan, in Fundamentals of Dual-Phase Steel (Ed. by R.A. Kot and B.L. Bramfitt), Met.Soc. AIME, Warrendale Pa. (1981) P.113-44.

Chapter Five

Powder Production: Methods and Properties

NICKEL COATED POWDERS VIA THE
CARBONYL NICKEL GAS PROCESS

E.L.LL. REES, F.W. HECK & G.A. DIBARI

INCO EUROPE LTD. / NOVAMET SPECIALTY PRODUCTS
CLYDACH, WALES. CORP. WYCKOFF, N.J.

Abstract

The decomposition of nickel carbonyl gas has been used for many years to produce high purity nickel. Different decomposer designs can produce either pellets or powder, and variations in powder decomposer conditions (e.g. temperature flowrate, concentration, additives etc.) can lead to a variety of powder types. The extention of this technique to decomposition in fluidised beds has created the potential for nickel coating a wide variety of substrates.

INCO has produced a number of different coated products in small scale fluidizers and larger quantities in more sophisticated, pilot scale equipment. Materials that have been coated include graphite, glass, copper, tungsten, silicon, tungsten carbides and others. Using a similar process, a free flowing spherical nickel powder has also been produced.

Any level of nickel can be deposited fairly accurately. Complete encapsulation, which may or may not be required depending on the application, is largely dependent on the substrate shape, size and size distribution.

This paper outlines the production process and discusses a variety of applications for potential products. The limitations of the current equipment and the development of new equipment to permit the coating of finer sized particulates will also be discussed.

Introduction

High purity nickel has been produced at INCO's refinery at Clydach, Wales, by the carbonyl process since the beginning of the century. More recently this process has also been adopted by INCO's refinery at Copper Cliff, Ontario. The carbonyl process is based on discoveries made by Carl Langer and Ludwig Mond, where, under certain conditions, carbon monoxide combines with nickel to form the volatile nickel carbonyl. When this is heated to temperatures above 200°C at atmospheric pressure, the reverse reaction takes place, releasing carbon monoxide and depositing pure nickel.

Outline of INCO's carbonyl process

At Clydach, nickel is volatilised to nickel carbonyl from the impure reduced matte at atmospheric pressure in a continuous process in slowly rotating kilns. These have internal cooling which limits the reaction temperature to around 50°C. Although cobalt and iron carbonyls exist, the conditions employed at Clydach ensure that only nickel is volatilized, and the gas leaving the kilns consists of carbon monoxide containing up to about 16% nickel carbonyl. This is fed to a separate plant where it can be decomposed to pure nickel without further purification other than dust removal by filtration.

The Copper Cliff process utilises a batch pressure process similar to one previously employed at Clydach but subsequently discontinued. Metallic granules obtained by processing impure nickel sulphide in top blown rotary converters are batch reacted with carbon monoxide at pressures up to 1000 p.s.i. and temperatures up to 180°C in rotating reactors. Most of the nickel is volatilized and in this case some of the iron, and pure nickel carbonyl is obtained by distillation of the condensed mixed carbonyls.

Powder Production

When a mixture of nickel carbonyl vapour and carbon monoxide is passed into an empty chamber heated to temperatures above 200°C, rapid decomposition takes place and very fine nickel powder is formed. This principle is used to produce nickel powder in electrically heated hollow cylindrical decomposers, and by varying the decomposer conditions, e.g. temperature, flowrate, concentration, trace additives to the inlet gas etc. various powder morphologies can be obtained, e.g. INCO Type 123, which consists of spiky discrete particles, or the INCO Type 200 series of filamentary chain-like powders.

Pellet Production

When nickel carbonyl passes over a heated surface it decomposes and plates nickel on to it. By this principle nickel pellets are produced in decomposers which are of a totally different design from the powder units. A bed of nickel pellets, ranging in size from a few hundred microns to around 7-8 mms, is continuously recirculated down through a shell and tube preheater into a reactor vessel, into the base of which a nickel carbonyl/carbon monoxide mixture is introduced. The gas percolates up through the moving bed of hot pellets, decomposing and plating nickel on to them. A bucket elevator recycles the pellets discharged from the base of the reactor to the top of the preheater and the cycle is repeated. A device known as the segregator and screen removes and discharges pellets that have grown to a certain size and some nuclei is added every day as required.

The economics of these production routes are dependent upon the very large volumes of nickel refined by the carbonyl process by INCO. Techniques of very advanced environmental engineering and process control have also been developed as these are essential to contain and direct these reactions.

Fluid bed decomposition

This principle of carbonyl decomposing and plating nickel on to a hot surface has been extended to fluidized beds. Nickel carbonyl, fed into a heated bed of fluidized particles decomposes, depositing nickel on to the particles. For the benefit of those not familiar with fluidization, a fluidized bed consists of particles contained in a vessel and usually resting on a porous base, which has been designed to contain the particles but to allow a fluid (either gas or liquid) to pass up through it. At a very low flowrate, the fluid would pass through the particle bed with minimum disturbance, and at a very high flowrate would cause the particles to be blown out of the vessel. In between these two extremes a situation called fluidization exists, where the passage of the fluid through the particles causes them to rapidly move about and induces excellent mixing. Increasing the fluid flow obviously increases the particle agitation and conditions can vary from a gently quiescent bed to a wildly fluidizing spouting bed. The degree of agitation achieved also depends on the fluid viscosity, density and temperature on one hand and size, morphology and density of the particles on the other.

Most powders, metallic or otherwise, that can be fluidized and withstand carbonyl decomposition temperatures can be nickel coated in this manner. Materials that have been coated include

Figure 1: Typical small scale glass fluidizer

graphite, both flake and granular, glass particles, aluminium, alumina, magnesium, sand, Brazilian quartz, silicon carbide, chromium, titanium and many others.

Most of the above mentioned powders have been prepared in sample quantities of 1 or 2 lbs or so in simple small scale 3 or 4" diameter glass fluidizers, having a carbon monoxide/ nickel carbonyl gas mixture entering through a water cooled metal venturi base cone. Figure 1. Heat is provided through the glass walls using gas ring burners. Each powder has its own individual fluidization characteristics, and the glass walls permit visual observation of the bed, which is useful in determining the suitability of a powder for nickel plating by this technique. Plating of the glass walls is avoided to a large extent by positioning the gas burners towards the top of the bed, where most or all of the carbonyl has been decomposed to carbon monoxide.

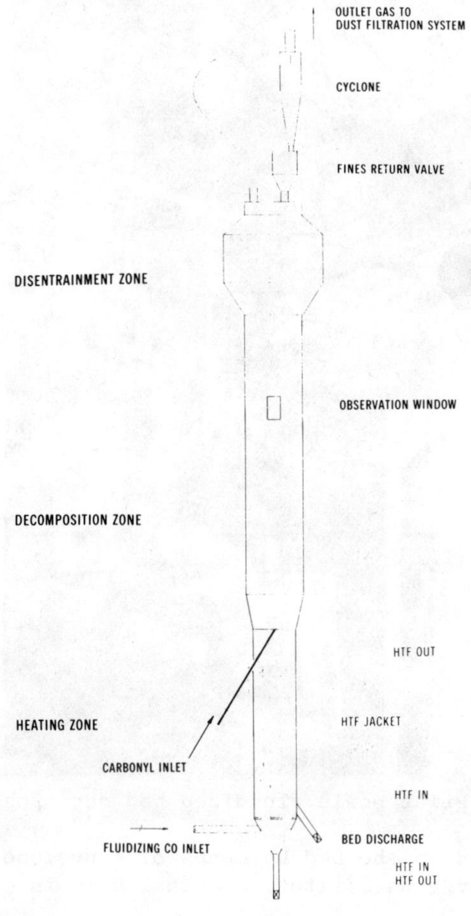

Figure 2: Diagrammatic representation of larger fluidizer decomposer

Larger quantities, ranging from around 150 kg to around 1 tonne of product (depending on particle density and nickel required) can be prepared in a more sophicticated unit in the Pilot Plant at Clydach. Figure 2. Here the bed is fluidized by carbon monoxide entering through tuyeres in the base plate and is heated by heat transfer fluid circulating inside three bayonette heaters protruding into the bed through the windbox and base plate. Additional heat is provided by jacketing the outside walls. Nickel carbonyl is introduced into the bed via a water cooled nozzle above this heating zone, the upper section of the bed being the decomposition or reaction zone. Entrained particles are removed from

Figure 3: Pilot scale fluidized bed decomposer

the off gas and refed to the bed by means of a cyclone and flap valve assembly, followed by a bag filter. Actual unit is shown in Figure 3.

Nickel coating is obviously a batch process but the apparatus can be used continuously to produce a free flowing spherical nickel powder, particle size up to 300 microns. This powder is in effect nickel coated nickel as, of course,are nickel pellets, and has up to now been used as nuclei for our nickel pellet decomposers.

Plans are in hand to build a larger scale unit based on this design.

Any level of nickel coating within reason can be deposited fairly accurately on a substrate. Complete encapsulation of the individual particle, which may or may not be required depending on application, is largely dependent on substrate shape, size and

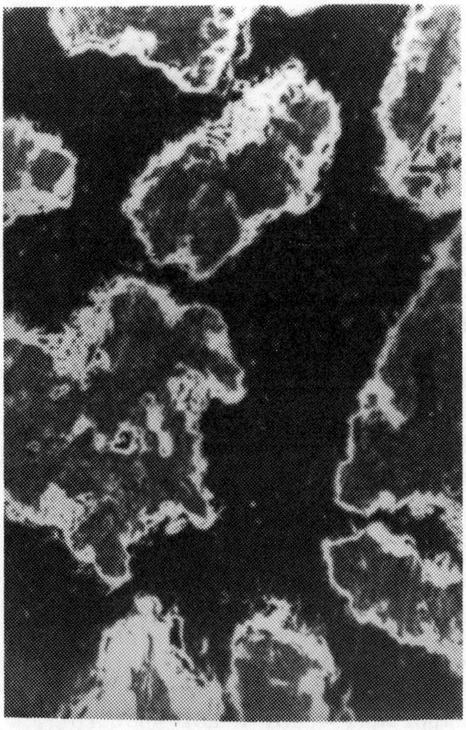

Figure 4:
Nickel coated 250 micron graphite (x250)

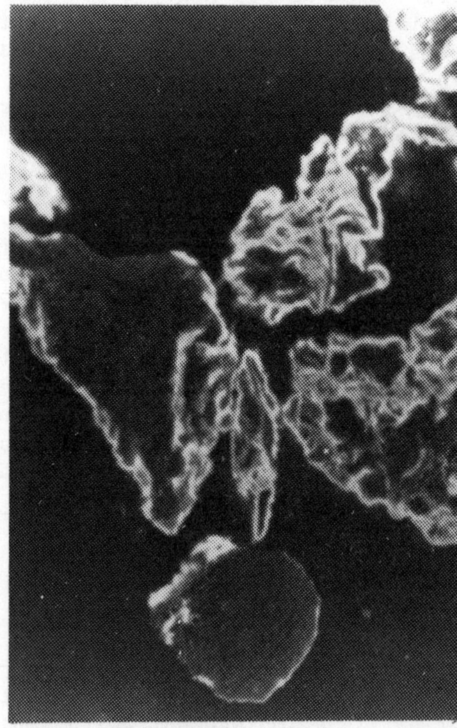

Figure 5:
Nickel coated -50 micron graphite (x2500)

size distribution. Completely encapsulated fine particles usually require more nickel than coarser particles, due to their increased surface area, and in a bed of mixed sizes, preferential nickel deposition on the finer particles occurs. This has been demonstrated by the analysis and microscopic examination of nickel coated granular graphite after being sieved into its respective size fractions. The coarser fractions, 250 microns, were well encapsulated at around 39% Ni in contrast to the finer fractions, -50 microns, having considerable gaps in the nickel coating at 67% Ni. Figures 4 & 5 respectively.

Problems Encountered

Not all powders can be successfully coated. Currently the optimum size range appears to be between 300 and about 50 microns, although powders outside this range can sometimes be coated.

Too coarse and/or heavy a powder usually requires a fairly high fluidizing gas velocity to maintain adequate particle agitation. This leads to a low carbonyl residence time in the bed resulting in considerable concentrations of undecomposed carbonyl in the off gas. On the other hand, excessive elutriation can occur with a bed of fine particles, leading to overloading of the fines filtration and return system.

Defluidization and/or the formation of "rat-holes" in the bed during coating can occur under certain circumstances, mostly with fine particles. Nickel plating on a particle begins with small nodules of nickel deposited at certain locations; these grow, join up and eventually completely cover the particle. These nodules cause increased friction between the agitated particles and sometimes when these frictional forces exceed the particle inertia forces, complete or partial defluidization takes place. Stationary or sluggishly moving particles in contact with each other tend to be cemented together by the deposited nickel, thus brisk fluidization should be maintained at all times.

Worth mentioning is an unusual problem which occasionally manifested itself when coating certain metallic particles such as Al and Mg. During the operation, a violently exothermic reaction suddenly started between the metallic particles and encapsulating nickel - the whole bed became red hot and eventually fused into one solid mass. This happened when the heating was either electrical or by flame impingement, with ample opportunity for a hot spot to develop, triggering off the reaction which spread uncontrolled throughout the bed. Although it has not been attempted, we are confident that this phenomenon would not occur in our new design of heat transfer fluid heated fluidizer, as the fluid temperature is controlled to well below the critical temperature of the reaction.

It should be pointed out that even the most unsuitable powders can usually be coated - but at a cost. It has been found that the introduction of other gases into the fluidizing gas often prevents agglomeration of the particles being coated, but on the production scale, this would lead to contamination of the works gases and would be unacceptable. Low melting point plastics have also been coated by fluidizing with hydrogen into which liquid nickel carbonyl had been vaporised. The absence of carbon monoxide permits decomposition of the carbonyl at a much lower temperature. However, as most of these techniques used are impractical or prohibitively expensive, they are usually not pursued.

Other methods of coating powders

A technique, attempted some time ago to coat very fine powders, was to use a recirculating powder decomposer. This was an adaptation of the Powder Decomposers described earlier. This unit had facilities whereby the powder to be coated could be incorporated into the carbonyl stream entering the decomposer chamber. The carbonyl would then decompose on to and nickel coat the entrained powder. The discharge was fed via a feed bunker and spiral into an ejector and recycled into the decomposer by the carbonyl feed gas. Although several powders were "coated" in this apparatus, the method was only partially successful as considerable free nickel powder was formed alongside the coated product. Further work on this apparatus was subsequently discontinued.

Experiments with a vibrating bed decomposer were also discontinued. In this, the carbonyl gas was passed through and decomposed upon a heated bed of agitated particles, but in this case the agitation was induced not by the gas flow passing up through the bed but by the vibration of the walls of the vessel. A modified vibratory ball grinding mill (with the steel balls removed) was used for these trials. Unfortunately, the excellent mixing found in conventional fluid beds did not occur in this apparatus and uneven nickel deposition took place, the particles in the vicinity of the carbonyl inlet receiving considerably more nickel than those near the gas outlet.

Applications

A wide variety of substrate materials are being nickel coated, both for developmental purposes and in production sized quantities; and at a range of nickel levels. These are summarised in Table 1 (overleaf).

The more widely known of these are the nickel coated graphites which, for years, have been thermally sprayed to create an abradable seal at the blade tip in jet engines.

Cast aluminium and zinc based composites containing particles of either graphite, carbides or oxides have been successfully produced by first nickel coating the substrates. Enhanced properties such as gall resistance, lower thermal expansion, greater wear resistance and resistance to abrasive attack have been realised. Experiments with aluminium and copper based composites produced with nickel coated silicon carbide appear to be interesting candidates for security systems because of their higher resistance to machinability.

Many types of electronic equipment are subjected to extraneous radiofrequency and electromagnetic interference (R.F.I. & E.M.I. respectively) either by natural phenomena or by emissions from other electrical equipment. Coated glass spheres have been compounded into plastics to produce conductive resins which are used to

TABLE 1

SUMMARY OF NICKEL COATED PRODUCTS AND APPLICATIONS

SEED MATERIALS	APPLIED NICKEL COATING (WT.%)	APPLICATION
Aluminium	50	Thermal spraying
Alumina,Corundum	15-40	Cutting wheels;Electrolytic grinding wheels;thermal spraying
Chromium	10	Metal injection molding
Copper	10-20	Conductive gaskets;Aerospace
Iron	4-8	P/M
Glass Frit	50	Automative;Grinding/polishing materials
Glass Spheres	20-60	EMI/RFI Shielding;plastic injection molding; Conductive fillers; Gasket fillers; Automotive
Graphite	25-85	Abradable seals, Welding rods; P/M;Automotive
Lithium Aluminate	10	Fuel cell electrodes
Silica	10-50	Cutting Wheels
Silicon Carbide	5-65	Polishing tool dispersions; Metal injection molding; Abrasive/wear resistant composites.
Titanium	45	Automotive
Tungsten	4-8	P/M;Projectiles
Tungsten Carbide	5-10	Hard-facing materials; Abrasive/wear resistant composites; Projectiles; Cutting Tools
Tungsten/Titanium Carbide	15	Hard-facing materials; Abrasive/wear resistant composites;Projectiles; Cutting Tools
Zirconia	20-30	Thermal spraying;Aerospace

injection mold computer housings to provide EMI/RFI shielding. Injection molded boxes and carrier plaques of conductive resins are used during the transportation of microchips to hold the chips securely during shipment and to prevent build up of static electricity that could damage the chips. Similarly, these coated glass spheres are used as a filler in conductive gaskets and sealants. Because nickel coated graphite also exhibits good conductivity characteristics it has been combined with nickel flake, compounded into a resin and molded into housings for electronic equipment.

There is increased interest in nickel coated hard particles for grinding, polishing and cutting materials. In electrolytic grinding, grinding occurs when an electrical charge is passed through an electrolyte that is between the grinding wheel and the part. Normally made of silver, alumina and a resin, introduction of nickel coated alumina particles into the wheel maintains the conductive characteristics of the wheel and lowers the need for silver which can be lost when these wheels are run under dry conditions.

Nickel coated graphite, iron and tungsten particles are being conducted in the areas of pressed parts, direct rolled strip and cutting tools. A nickel coated graphite particle is used in the manufacture of some welding products, such as cored wire where sufficient carbon is deposited to promote graphite precipitation and reduce the shrinkage stresses that can cause weld-cracking.

Summary

Many powders, metallic or otherwise, have been successfully coated with nickel using the carbonyl nickel gas process. Each material has shown to behave in its own unique way during coating. This has been a learning experience with many achievements, but also with some that have not been successful. Investigations continue at the Inco facilities, at Clydach and Copper Cliff, to define more clearly the coating criteria required for different materials.

Acknowledgements

The authors would like to thank the Novamet Specialty Products Corporation for promoting the distinctiveness of these nickel coated powders and the presentation of this paper; and the Inco Clydach Refinery for their efforts in producing these powders.

Oil Atomization - A Method for the Production of Rapidly Solidified Iron-Carbon Alloys

Karl Ulrich Kainer and Barry Leslie Mordike

Institut für Werkstoffkunde und Werkstofftechnik
der TU Clausthal, Agricolastraße 2,
3392 Clausthal-Zellerfeld

Abstract

Oil atomization of iron-alloys is an alternative means of producing rapidly solidified iron-carbon alloy powders. The main advantage of this method is the combination of high quenching rates - comparable with water atomization - with low oxygen content. Three alloys were chosen for fundamental investigations. The first alloy was a low carbon medium alloy steel with 0.4 % C, the second alloy was a medium carbon steel (1 % C) and the third a high carbon alloy (3 % C). The production and treatment of the powders is described and the properties discussed. A comparison with water atomized and gas atomized powders is made. The change in the microstructure is dependent on the atomizing parameters and subsequent treatment. In the case of medium and high carbon alloys no difficulties with removal of oil was experienced. Only a slight change of the microstructure occurs. The oxygen content remains unaltered at values of 300 ppm. If the melt is degassed a lower oxygen content is possible. Any change in carbon content due to atomizing can be taken into account. It is difficult to maintain low carbon contents in the low carbon alloys without accepting an increase in oxygen content and coarsening of the microstructure.

Introduction

The method of production of iron-alloy powders is determined by the properties required for the powder. Unalloyed iron powder must, for example, be cheap, easily worked ie easily pressed and with the lowest possible oxygen content /1/. The requirements for alloyed iron powder are manifold depending on the alloy composition working techniques, possible consolidation techniques and proposed application. These requirements can often only be fulfilled if the powder is produced by atomization of the melt. Water atomization has a high quenching rate and is relatively cheap /2-4/; gas atomization produces spherical powder with a low oxygen content /4/. There are disadvantages which must be set against these advantages eg in water atomization the dissociation of the water results in an extremely high oxygen content (up to 1 % as shown in Table 1). This oxygen, which is in the form of surface oxygen or oxide can only be removed by annealing in reducing atmosphere which, as a result of the high annealing temperature and long annealing times required, eliminates the advantages of the rapid quench /2/.

atomizing prozess	G	G	G	W	W
atomizing medium	Ar	N_2	N_2	H_2O	H_2O
atomizing atmosphere	Ar	N_2	N_2	N_2	Luft
quenching medium	Ar	N_2	H_2O	H_2O	H_2O
oxygen content ppm	100	100	1000	1000	10000

Table 1: Atomizing process and the attainable oxygen contents in untreated powder (W = water atomization, G = gas atomization) /4/.

In the case of elements showing a great affinity for oxygen and forming high melting point oxides (CrO_2, Al_2O_3, TiO_2 etc.) a reduction is virtually impossible due to the high reduction temperature /1/. In gas atomization the oxygen content can be kept low but the quenching effect of the liquids is absent; hence the cited quench rates of two orders of magnitude less than for the water atomized powder /2,4/. The advantages of the rapid quench eg fine microstructure, homogeneous distribution of phases and elements with depending on the composition a high degree of supersaturation and metastabile phases can thus hardly be exploited.

Oil atomization offers a solution to the problem of rapid quenching combined with low oxygen content for medium and high alloy iron alloys. Based on earlier works on the production of low oxygen soft iron powder /5-9/ where the carbon content was reduced by a decarburising anneal experiments were devised to powder high carbon content iron alloys (> 0.4 %).

Powder Production

The water atomizer Type D 5 shown in Fig. 1 was manufactured by Davy McKee. It has a capacity of 30 kg iron and was modified as shown in Fig. 2 for oil atomization.

Fig. 1: Powder atomizer - Institut für Werkstoffkunde und Werkstofftechnik.

The alloy is melted in the furnace (1) and then poured through a preheated tundish (2) into a holding crucible. The exit nozzle is in the base of the crucible. The metal flows at a steady rate into the atomizing chamber where it is atomized by 4 high pressure oil jets.

After a settling time of ∿ 3 - 4 hours the oil is drained and collected for additional sedimentation (6) for oil cleaning. Removal of excess oil can be carried out in tank (5) using compressed air. The oil powder compact produced must be treated further to attain full cleanliness (7 and 8).

The oil used for atomization must have several properties in common with water. In particular the viscosity must be low at room temperature. In addition, the flash point should not be too low. Evaporation on contact of the oil with the atomized liquid droplets is then low. Other points to be considered were availability, price, recyclability, resistance to ageing, non toxic (dioxin free). Hardening oils were thus considered most satisfactory. Shell Voluta AK R was chosen from these because of the low sulphur content and low viscosity /10/.

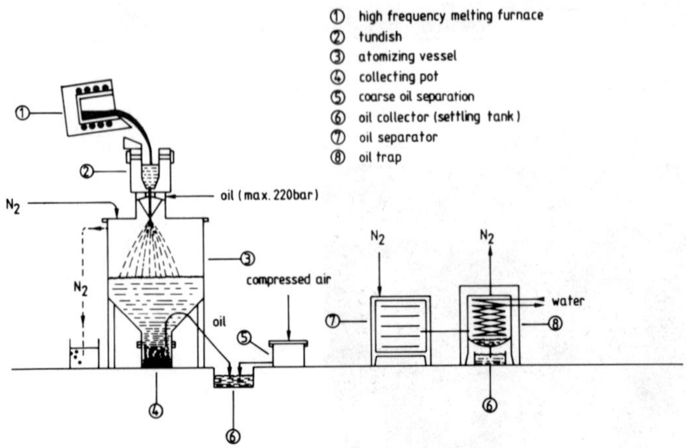

Fig. 2: Flow chart for oil atomization of iron powder.

Table 2 lists the compositions of the various melts. The sample was taken shortly before atomizing. It can be seen that it was possible to maintain the chemical composition with narrow tolerances in each case. Table 3 lists the experimental parameters for atomization.

No.	C	Si	Mn	Cr	Mo	Ni	V	S
1	0.28-0.30	0.17-0.19	0.22-0.30	1.32-1.35	<0.05	0.09-0.10	<0.01	<0.018
2	0.37-0.41	0.17-0.19	0.22-0.30	1.39-1.45	<0.05	0.10-0.12	<0.01	<0.018
3	0.96-0.98	0.17-0.18	0.22-0.30	1.38-1.46	<0.05	0.12-0.13	<0.01	<0.018
4	0.33-0.36	0.71-0.72	0.22-0.26	4.51-4.53	1.09-1.12	0.06-0.13	0.35-0.36	<0.003
5	2.83-2.93	0.45-0.54	0.48-0.64	1.57-1.67	<0.04	<0.04	<0.01	0.11-0.15

Table 2: Chemical composition weight-% of atomized alloys (composition of the melt).

alloy No. (see Table 2)	atomizing/ quenching medium	pouring temperature °C	tundish temperature °C	nozzle diameter mm	atomizing pressure MPa	max. temperature of quenching medium
1	oil	1700	1100	5.5	14.0	107
2	oil	1700	1100	5.5	14.0	105
3	oil	1660	1100	5.5	15.5	114
3	water	1660	1100	5.5	21.0	75
4	oil	1750	1100	5.5	15.5	100
4	water	1720	1150	5.0	18.0	78
5	oil	1600	1100	5.5	16.0	120
5	water	1500	1100	5.5	18.0	80
5	gas[1)]	1560	1150	4.0	8.0	293

Table 3: Atomizing parameters

1) atomized at a pilot plant at Davy McKee, Sheffield

Treatment of Powder

Iron powder still contains after the initial oil removing process about 2 - 3 weight-% oil. This must be removed before the powder can be employed. The removal process must ensure that there is no residue, that there is no or only a minimal change in microstructure and also that there is no increase in oxygen content. There are the following possibilities:

- removal of oil using solvents
- heating in air
- heating in inert gas (dry N_2)
- heating in reactive gases (eg H_2)

Table 4 shows the result of various cleaning treatments on the carbon and oxygen content and compares the values with those for gas and water atomized powder. The carbon content of squeeze cleaned powder is 0.33 - 1.89 % higher than that of the original melt or of gas and water atomized powder. Some of this carbon is in the form of an oil film on the powder and some has been absorbed into the powder particle. The oil is cracked by the liquid metal drops and some carbon diffuses into the molten drop. Absorption of carbon ceases when the droplet solidifies. Heating the powder to temperatures sufficiently high to remove the oil but not high enough for diffusion to take place shows that carburisation is dependent on both the carbon and alloy content. At high carbon contents (alloy 5) virtually no carburisation occurs. Oil can be used as atomizing medium and removed from the surface with no change in composition. This is not the case for low C-contents. If the amount of carbide forming elements in the alloy is low then the degree of carburisation is also low (alloy 1,2,3). When carbide forming elements are present in any quantity then carburisation is more marked (alloy 4).

All powders can be completely cleaned without an increase in the oxygen content. It is not necessary to use reactive atmospheres, using protective atmospheres but it can be seen that methods using solvents or heating in air are unsatisfactory. If cleaning is carried out in an inert atmosphere (nitrogen) the oil can be collected and dried and reused. If a reduction in carbon content is required then the powder must be heated in a H_2O containing atmosphere, which causes an increase in the oxygen content /6-10/. It is possible by appropriate selection of operating parameters and composition of the melt to obtain powder with the desired carbon content without increasing the oxygen content providing the desired carbon content is not less than 0.6 - 0.8 % together with a high content of strong carbide forming elements.

alloy No. (Table 2)	atomizing medium	melt	as quenched precleaned	air 200°C	washed with solvent and dryed	N_2 650°C	H_2 650°C	H_2 800°C	untreated extruded 1)
				carbon content in weight-%					
1	oil	0.30	0.63	0.39	-	0.36	-	-	0.40
2	oil	0.40	0.93	0.44	-	0.59	-	-	0.46
3	oil	0.96	2.48	1.28	1.27	1.48	1.49	1.44	1.38
3	water	0.97	0.97	-	-	-	-	-	0.96
4	oil	0.34	2.21	-	-	1.41	1.42	1.40	1.48
4	water	0.36	0.36	-	-	-	-	-	0.35
5	oil	2.92	4.29	-	3.03	3.01	-	-	2.98
5	water	2.93	2.93	-	-	-	-	-	2.92
5	N_2	2.90	2.90	-	-	-	-	-	2.89
				oxygen content in weight-%					
3	oil	0.009	0.034	0.316	0.134	0.034	0.033	0.029	0.033
3	water	0.009	0.240	-	-	-	-	-	0.190
4	oil	0.012	0.050			0.041	0.039	0.040	0.045
4	water	0.015	0.240	-	-	-	-	-	0.195
5	oil	0.002	0.069			0.073	0.072	0.068	0.070
5	water	0.002	0.400	-	-	-	-	-	0.120
5	N_2	0.002	0.110	-	-	-	-	-	0.080

Table 4: Changing of the carbon and oxygen content during treatment of the powders.
1) Untreated powder filled in cans, deoiled by slowly warming up 650°C in self-forming protective atmosphere and extruded at 950°C.

Microstructure and Properties of the Powder

The shapes of oil atomized powder are similar to those of water atomized powder; the fraction of spherical particles appears to be higher. Fig. 3 shows a comparison between oil, water and gas atomized powder of the composition 2.9 % C, 1.5 % Cr.

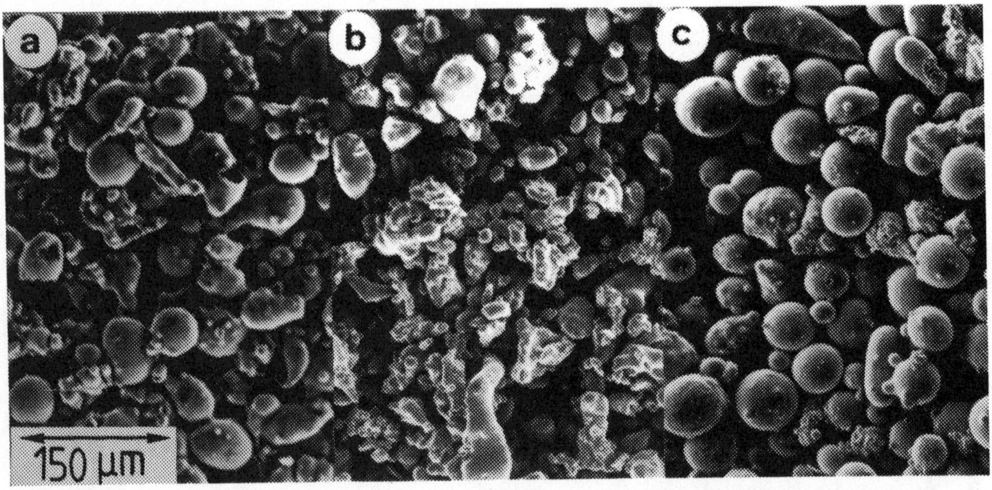

Fig. 3: Comparison of shape of a. oil, b. water and c.gas atomized powders (alloy 5, 2.9 % C; 1.5 % Cr as quenched).

The microstructures obtained depend on the chemical composition, in particular on the carbon content and content of carbide forming elements. Fig. 4 shows the microstructures of oil, water and gas atomized powders of various compositions. Cellular or dentritic solidification occurs depending in the composition. The fineness of the microstructure depends on the grain size and atomizing process. Oil and water atomized powders solidify more rapidly and hence have the finest microstructures. The quenching rate can be estimated from the secondary dendrite arm spacing /3/. Typical values are shown in Fig. 5 for the alloy 2.9 % C; 1.5 % Cr.

The quenching rate for oil atomizing and water atomizing are of the same order of magnitude. They are somewhat higher than for gas atomization in nitrogen. The phases which can arise depend on the composition and atomizing process viz austenite, ferrite, martensite and carbide. Fig. 6 shows the changes which ocurr during heating to remove oil (350°C and 650°C). The phases identified are listed in Table 6. alloy 3 with 1.39 % C in the oil atomized state consists of 85 % austenite, excluding the metal carbides. The austenite contains ∿ 1.3 % C. the remaining 15 % is martensite. Fe_3C can also be detected. After the heat

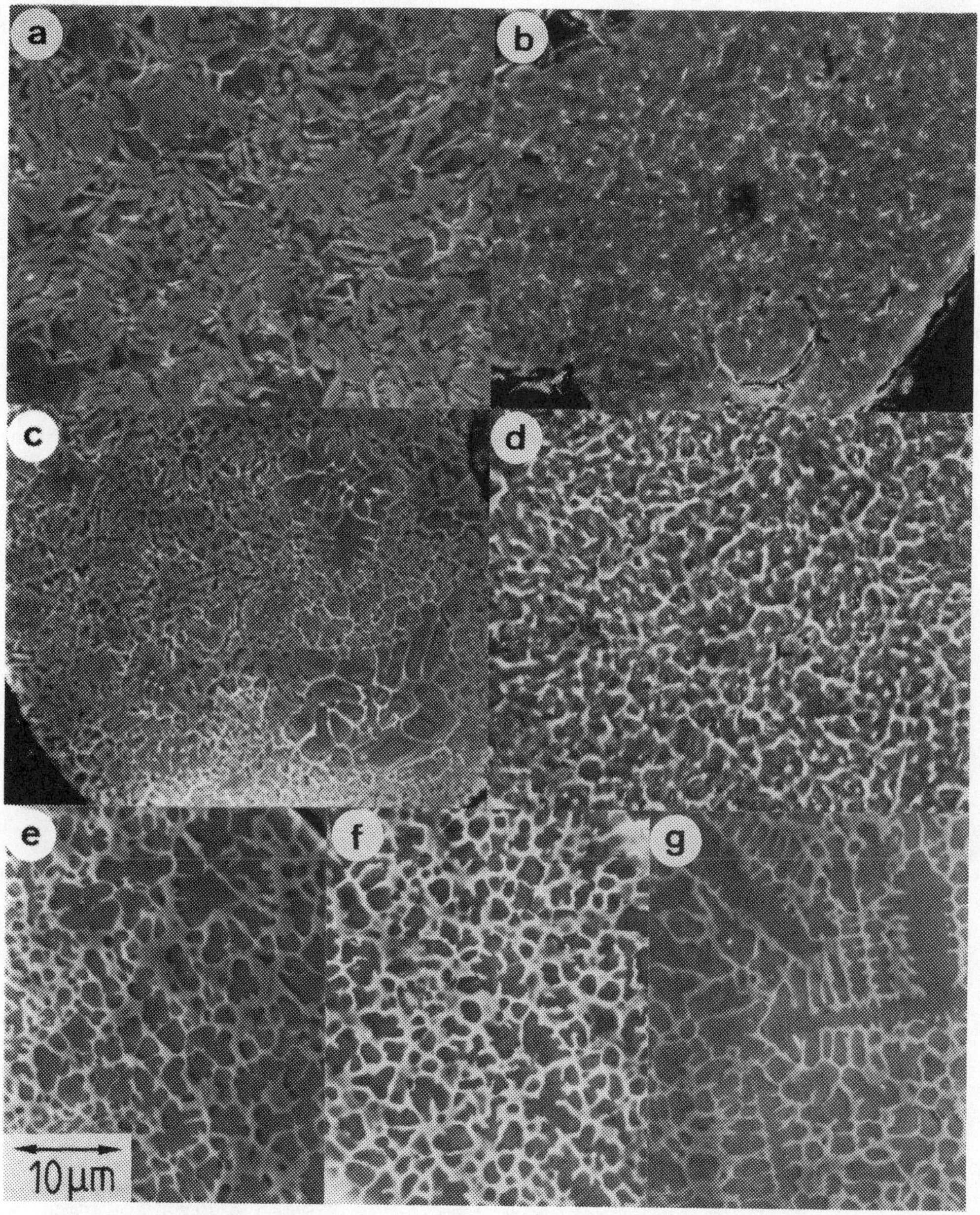

Fig. 4: SEM-observation of atomized powders (as quenched), (d ∿ 60 µm);
a. 1.33 % C-1.34 % Cr oil atomized;
b. 0.97 % C-1.33 % Cr water atomized;
c. 1.3 % C-4.5 % Cr-1.1 % Mo-0.35 % V oil atomized;
d. 0.33 % C-4.5 % Cr-1.1 % Mo-0.35 % V water atomized;
e. 2.9 % C-1.5 % Cr oil-,; f. water-,; g. gas atomized

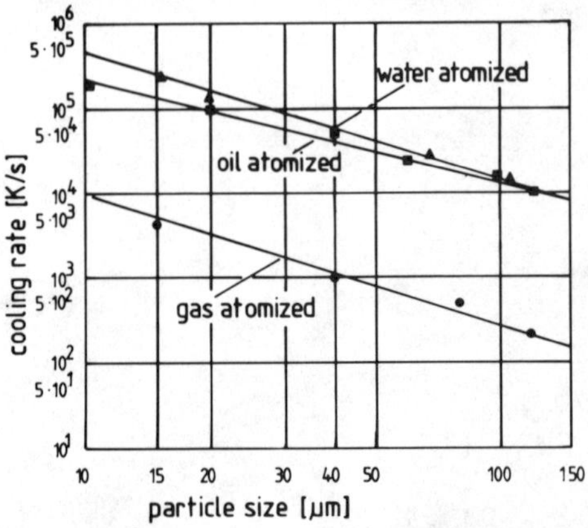

Fig. 5: Calculated quenching rate as a function of the particle size for different atomizing processes (alloy 2.90 % C; 1.5 % Cr). Used equation: $\lambda = B \cdot \varepsilon^{-n}$; λ = secondary dendrite arm separation, B and n are constants (used B = 60, n = 0.4), ε = quenching rate /3/.

treatment only ferrite and carbide are observed, the latter occurring on the grain boundaries as also within the grains (Fig. 6a). Water atomized powder (0.97 % C) in the as quenched state contains only 8 % austenite. It was not possible to determine the carbon content. The remainder of the austenite is martensite.

Carbides could not be identified by X-rays, only in the REM (Fig. 4b). After heat treatment only ferrite and carbide was observed. An oil atomized alloy 4 (1.3 % C) only supersaturated austenite with a content of ∿ 1.4 % C with a little martensite is observed. Carbides cannot be detected by x-rays (Fig. 4c). After heat treatment only ferrite and carbides including special carbides can be detected (Fig. 6b). Water atomizing alloy 4 (0.33 % C) produces martensite with a small quantity of carbides (Fig. 4d). Again heat treatment leads to the formation of ferrite and carbides. Alloy 5 (2.9 % C) leads on quenching to supersaturated austenite with carbon contents 1.6 - 1.7 % C and lamellar Fe_3C eutectic irrespective of the quenching method (Fig. 4e-g). On heat treatment ferrite is formed and carbon from the austenite reacts to form carbide on the spherodized carbides (Fig. 6e).

A mechanical property which can be measured in powders is the microhardness. This can be used to monitor the influence of microstructure and heat treatment. Fig. 7 shows the microhardness in the as atomized state and Fig. 8 the microhardness

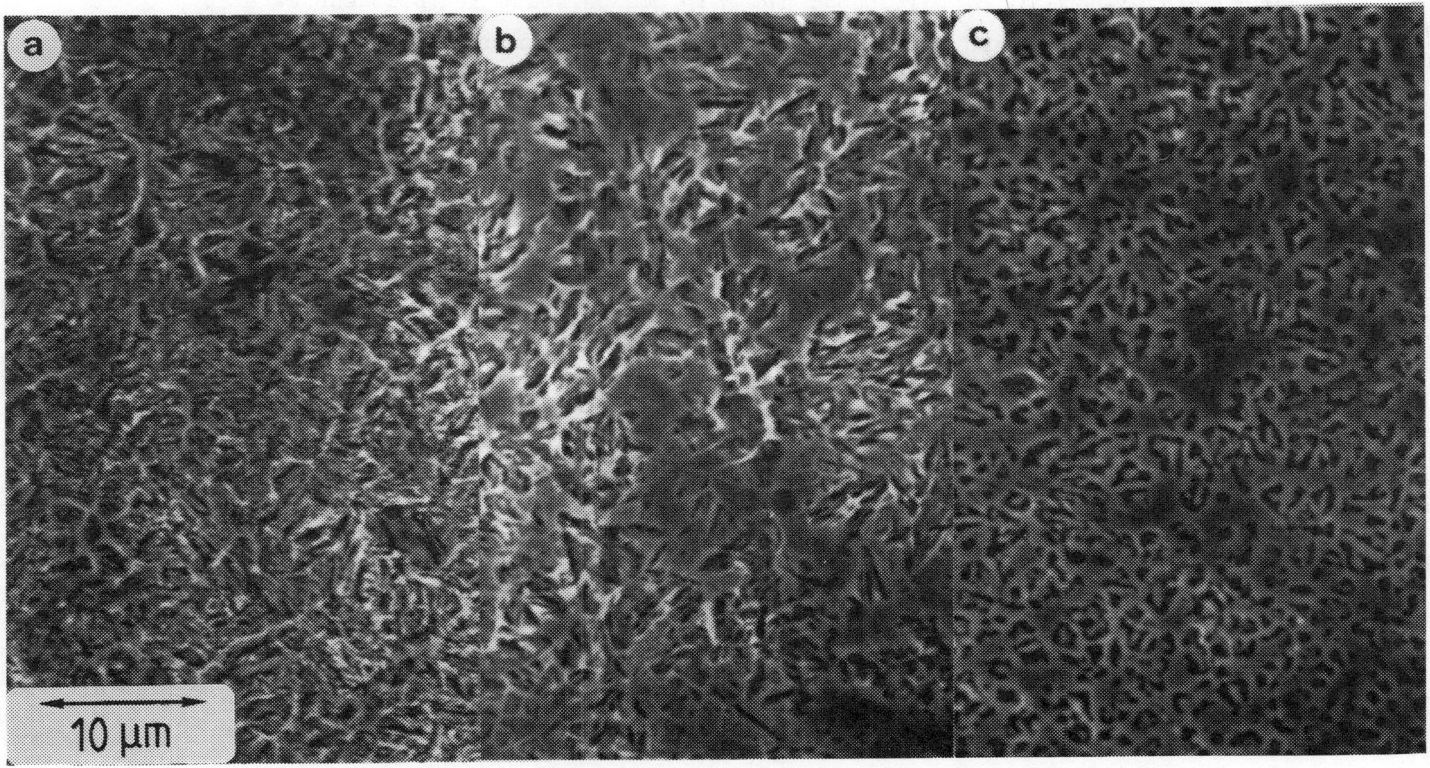

Fig. 6: Microstructure of treated oil atomized powders (treatment 650°C N_2/30 min)
a. 1.3 % C-1.34 % Cr; b. 1.33 % C-4.5 % Cr-1.1 % Mo-0.35 % V; c. 2.9 % C-1.5 % Cr.

					identified phases					
					austenite			ferrite		carbides
alloy No.	atomizing medium	weight% carbon melt	weight% carbon powder	treatment[1)]	content[2)] %	lattice parameter a_γ (Å)	carbon content % (from a_γ)	content[2)] %	lattice parameter a_α (Å)	
3	oil	0.97	1.27	a.q.	85	3.6216	1.3	15	n.a.	present
3	oil	0.97	1.27	a.	-	-	-	100	2.8670	present
3	water	0.98	0.98	a.q.	8	n.a.	n.a.	92	2.8700	traces
3	water	0.98	0.98	a.	-	-	-	100	2.8674	traces
4	oil	0.34	1.48	a.q.	~98	3.6261	1.4	traces	n.a.	traces
4	oil	0.34	1.48	a.	-	-	-	100	2.8649	present
4	water	0.36	0.36	a.q.	-	-	-	100	2.8680	traces
4	water	0.36	0.36	a.	-	-	-	100	2.8718	traces
5	oil	2.92	2.95	a.q.	100	3.6300	1.6	-	-	Fe_3C
5	oil	2.92	2.95	a.	-	-	-	100	2.8651	Fe_3C
5	water	2.93	2.93	a.q.	100	3.6339	1.7	-	-	Fe_3C
5	water	2.93	2.93	a.	-	-	-	100	2.8707	Fe_3C
5	gas	2.90	2.90	a.q.	100	3.6300	1.6	-	-	Fe_3C
5	gas	2.90	2.90	a.	-	-	-	100	2.8684	Fe_3C

Table 6: Results of the X-ray structure examination of the atomized powders (Mo-$K_{\alpha 1}$: $\lambda = 0.70926$ Å)
1) a.q. = as quenched; a. annealed 650°C; 2) in proportion to the non-carbide phases.

after heat treatment. The cross hatching shows the range of hardness as result of the difference in particle size and thus of quenching rate. Lower values are measured in large particles and high values in small. The hardness values of water atomized alloy 3 (100Cr6) are very high due to the formation of martensite. A higher proportion of carbide in the same alloy with 1.3 % C in oil atomized state compensates for the decrease in the proportion of martensite. The same applies to alloy 4 (X38CrMoV51)in which the martensite with 0.33 % C is relatively soft in the water atomized alloy. Oil quenching increases the proportion of carbide (1.48 % C) and hence high hardness. The influence of the quenching rate can be better appreciated in alloy 5 (300Cr6) because the carbon content remains virtually constant. The hardness increases from gas to oil to water atomization. After heat treatment the differences are still present but not so marked (Fig. 8). In other alloys the different iron carbide fractions and metal carbide formation is more important than the fineness of the microstructure and hence quenching rate.

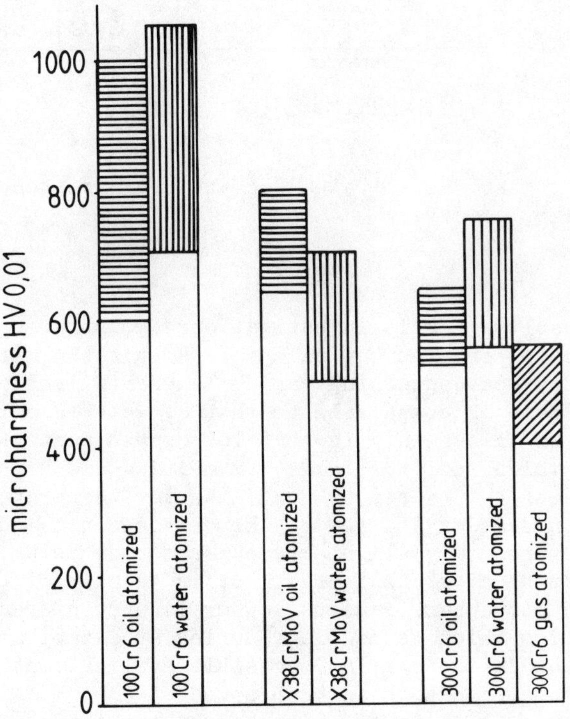

Fig. 7: Microhardness of atomized iron-carbon powder.

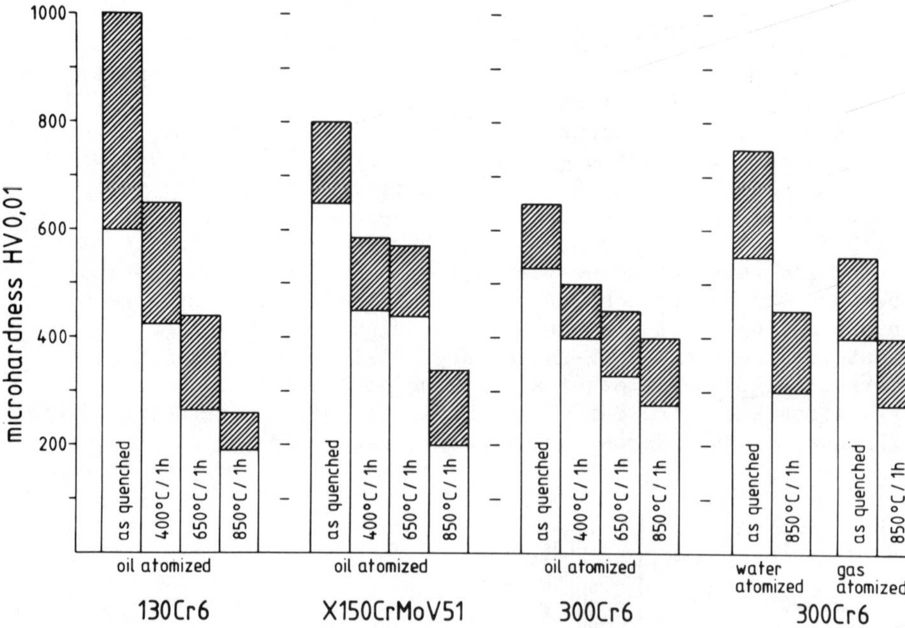

Fig. 8: Microhardness of atomized and heat treated iron-carbon alloy powder.

Conclusions

The results of this investigation demonstrates that oil atomization of iron carbon alloys is a suitable method of producing carbon containing rapidly quenched iron powder. The quenching rate is comparable to that of water atomization. The advantage of this method is that low oxygen contents can be achieved, which despite non vacuum melting and high alloy contents does not exceed 500 ppm. The low temperature heat treatment nescessary to remove the oil does not modify the microstructure, only slight grain coarsening being observed. The oxygen content after this treatment in N_2 or self produced controlled atmosphere remains low or is even decreased. The carburisation which takes place during atomization can be taken into account during melting. Possible applications involves all high carbon steels such as high speed steels, bearing steels as segregation free solidification can be obtained together with low oxygen content. Low carbon steels (< 0.6 % C) particularly if containing a high quantity of carbide forming elements present same difficulties as the necessary decarburisation in moist hydrogen results in an increase in the oxygen content.

References

/1/ W. Schatt (edt.), Pulvermetallurgie, Sinter- und Verbundwerkstoffe, VEB Deutscher Verlag für Grundstoffindustrie, Leipzig (1977).

/2/ H. Jones, Rapid Solidification of Metals and Alloys, Institution of Metallurgists, London (1980).

/3/ H. D. Burchards, PhD-Thesis, TU Clausthal (1987).

/4/ L. L'Estrade and R. Koos, "Properties of Gas Atomized Metall Powders", Proc. 1986 Int. Powder Metallurgy Conference: The Future of Powder Metallurgy, W. A. Kaysser and W. J. Huppmann (edt.), Verlag Schmidt, Freiburg (1986), 22 - 28.

/5/ M. Umino et al., Sumitomo Metal Industries Ltd., US Patent 4437891, 24.02.1981.

/6/ M. Umino et al., Sumitomo Metal Industries Ltd., GB Patent 2094834, 23.02.1982.

/7/ E. Tamura et al., Sumitomo Met. 37 (1985), 367 - 372.

/8/ E. Tamura et al., Proc. 1985 Annual US P/M Conference, Vol. 41, Sanderow et al. (edt.) (1985).

/9/ Steel Powder with Low Oxygen Content and High Compressibility, Technical Information No. 82121, Sumitomo Metal Industries (1986).

/10/ Technical Information, Shell Oil, Hamburg, Jan. 87.

CHARACTERIZATION OF CARBONYL IRON POWDER - PARTICLE SIZE DISTRIBUTION BY IMAGE ANALYSIS AND LIGHT SCATTERING TECHNIQUES

Joseph E. Japka

GAF CHEMICALS CORPORATION
1361 Alps Road
Wayne, New Jersey 07470

Abstract

Characterization of this spherical iron powder less than 10 microns in diameter was performed by SEM and light scattering techniques. The results from each method are presented as histograms of percent in range vs. particle size, and calibration of one method to another will be detailed. Data on powder manufactured in very narrow particle size ranges, ± one micron in the 0.2 to 10 micron range is presented. New microstructure data on the internal onion skin structure by electron microscopy is shown.

Introduction

Among iron powders, carbonyl iron is unique in that it is the only one made by chemical vapor deposition (from iron pentacarbonyl liquid) and the individual particles are almost perfect spheres. The internal particle structure is also unusual because it consists of a series of onion skin layers (see Figure 1). Electron diffraction analysis and TEM photos show no crystallographic explanation for the ring structure.

Carbonyl iron powder contains approximately .8% carbon, .3% oxygen, .8% nitrogen in a metastable solid solution with other elements below 10 ppm. The structure is very highly strained, with particle micro hardness ranging from 800 to 900 kg/mm ^{2}. No x-ray crystallographic evidence has been found to indicate that Martensite or intermetallic compounds form during deposition.

In the manufacture of carbonyl iron powder (CIP), iron pentacarbonyl liquid is vaporized and the vapor further heated in a large vessel until it decomposes, according to the reaction:

$$Fe(CO)_5 \text{-----}> Fe + 5CO$$

The powder falls or drifts to the bottom of the vessel where it is separated roughly according to size. The liberated carbon monoxide is recycled and reused to form fresh iron pentacarbonyl. Other finishing operations are performed on the powder which result in the production of various grades. These operations include milling, screening and annealing at elevated temperature to yield a recrystallized, higher purity product.

Carbonyl iron powder has been used primarily for its electrical properties. These are characterized by complex variables related to magnetic losses. For years the particle size of the iron powder was measured by indirect means involving the pressure drop across a packed bed of powder. This technique, called Fischer Subsieve Analysis, yielded a single number although it was known that a range of sizes existed.

The most common grades of carbonyl iron powder are E, TH and SF. Figures 2, 3 and 4 are SEM photomicrographs of these powders. Note that they are progressively finer in size. The E grade is prepared by milling the raw coarsest powder (Grade KX) from the decomposer, while the TH and SF are recovered as produced. The major purpose of this milling is to break up the agglomerates that form during the decomposition.

Particle Size Analysis by Standard Light Scattering Techniques

The technology for particle size analysis by laser light scattering is well known. It has been used at GAF as a standard quality control tool on the carbonyl iron powders. The particle size analyzer (Leeds and Northrup Microtrac) utilizes the phenomenon of Fraunhofer forward scattering of light from a laser beam projected through a stream of particles suspended in water. The angular distribution and intensity of the scattered light is a complex function of particle size. These measurements are analyzed by a microcomputer which reports the particle size distribution. The algorithm used to analyze the scattering data and derive the particle size distribution assumes the particles to be spherical. In the case of most of the carbonyl iron, this assumption is reasonable.

The results reported by Microtrac analysis include a table of cumulative volume percent finer as a function of particle size, the volume average particle size and the 10th, 50th and 90th percentiles of the cumulative curve. A histogram of the volume percent in range between sizes as a function of size is also given. The spacing between reported data points increases, with each successive size being $\sqrt{2}$ times the next lowest one. Unfortunately, in processing the cumulative percent finer data to obtain a distribution, the algorithm does not properly account for the increasing spacing between data points. This results in an erroneous skewing of the distribution towards larger sizes. Normally the volume average particle diameter is reported and used for quality release purposes.

Figure 5 shows several alternate ways of presenting the size distribution data generated by the Microtrac Analyzer. Curve A is the conventional curve which is simply the difference in cumulative weight percent finer values for two adjacent sizes plotted at the log average of the two sizes. Here the data is plotted on a linear scale instead of the logarithmic scale normally presented by the Microtrac Analyzer. The sum of these values equals 100. Curve B is for the case where the differences in cumulative percent fines values have been divided by the channel width and adjusted so that the sum of all such values equals 100. The latter adjustment was made to show the skewing toward coarser sizes in Curve A more clearly. Curve C represents an approximate distribution function for the powder in question. It was derived by dividing the differences in the cumulative percent finer values for adjacent sizes by the channel width and plotting that result at the log average for the two sizes involved. It is "approximate" because it was not derived by differentiation of a smooth function but by determining the slopes of straight lines connecting the data points. It does, however, account for the increase in channel width with increasing size.

Particle Size Distribution by Image Analysis

A more direct measure of particle size distribution can be obtained by taking a series of SEM photomicrographs of the powder and counting the particles as a function of size. Care must be taken to obtain a representative sample. Specimens were prepared for scanning electron microscope observation by mixing a sample of powder in heptane and ultrasonically agitating it to obtain a uniform dispersion. A sample of a sonified liquid slurry was taken with a pipette, placed on a microscope stub and allowed to dry.

A gold film was sputtered onto the sample to avoid charging effects in the SEM. A uniform layer of particles was deposited easily with this technique. A photomicrograph was taken of each sample to be analyzed, usually at 2000x magnification.

Figures 2, 3 and 4 show typical specimens of E, TH and SF powder. The packing of particles is reasonably dense, making for efficiency in terms of the number of photos required for a statistically significant sample of, say, 1000 particles. At the same time, overlapping of particles is low enough to enable agglomerated particles to be distinguished.

In this study a Donsanto Micro-Plan II Image Analysis System, coupled with an IBM Model XT Personal Computer, was used. As shown in Figure 6, the system has a digitizing tablet that allows individual particle diameters to be measured with a pen cursor. Normal practice was to take three photomicrographs of each of two samples with data being taken from the total of six photos. Each photograph was magnified by 1.5x on a copying machine set to increase contrast and enhance the definition of the particle edges. Although automatic imaging devices are available for use with scanning electron microscopes, these require complete separation of the particles by a clear field and, hence, a large number of fields must be examined to obtain enough particles for a statistically significant sample. Having a "human-in-the-loop", as was the case with our system, allows the use of a denser sample and fewer photographs. Variability from operator to operator was found to be small with differences in average particle sizes varying by less than 5 percent even among relatively inexperienced operators.

It is necessary to calibrate every element of the image analysis system in order to assure accurate and reproducible measurements. Two different standards of known size were photographed in the SEM and measured in the image analysis system. The data collected on glass spheres of 20 microns nominal average diameter, traceable to the National Bureau of Standards, is shown in Figure 7. The average value of 22.2 microns measured by image analysis was closer to the nominal 20 micron value than that obtained by light scattering (see Table I). Since the spheres did have a distribution with a finite width, additional calibrations were performed using scribed grids. These were photographed in the SEM and one hundred values of the grid spacing were measured for each grid. The measured values in this case (see Table I) are within 6% of the nominal spacings of the grids. With the establishment of calibration procedures, precise analysis of the various grades of powder was undertaken.

The results of image analysis on grades E, TH and SF are compared with those from laser scattering analysis in Figures 9, 10 and 11 respectively. The light scattering results, plotted as a line graph, are those derived by the approximate differentiation method described earlier (Figure 5, Curve C) rather than the data reported by the Microtrac Analyzer. For comparison, Figure 8 is a linear plot as presented by the Microtrac Analyzer for E powder, without any corrections. The original data is reported by the Microtrac Analyzer in the form of percent in range (vertical axis) versus particle size diameter, microns, on a logarithmic scale.

Image analysis results are plotted as a bar graph. Agreement between the image analysis and laser scattering results are reasonably good for the grade E except that more fine particles are reported by laser scattering than by image analysis. For the TH powder, however, the results reported by light scattering are significantly broader than those obtained with image analysis. The laser scattering curve for SF shows the poorest agreement of the three, with far more fines reported by laser scattering than by image analysis.

Table II lists the volume average particle size determined by both methods along with the standard deviations. Note that the averages for the E and TH determined by image analysis are significantly smaller than those determined by laser scattering and that the standard deviations are smaller as well. Because of the greater proportion of particles between 0 and 1.5 microns shown by the laser scattering for SF, the average value reported by laser scattering is much smaller than that reported by image analysis.

In order to elucidate the reasons for the differences between the two methods of analysis they were applied to a series of narrow distributions prepared by a proprietary size separation process. The results are presented in Figures 12 and 13. In both cases the differentiated data curves from the laser scattering are broader than for those generated by image analysis. The effect is more pronounced for the finer, 0.5 to 2 micron, powder. Note that while the true size distribution of the narrow cuts is not known, the laser scattering consistently reports a wider curve. It is hard to imagine some error in the sampling procedure used for preparing the image analysis specimens selectively removing both fine and coarse particles from two distributions with such widely varying sizes. It is reasonable, therefore, to conclude that the laser scattering tends to report a broader distribution of particles than that actually present.

Comparison of Analytical Methods

The image analysis of spherical particles of carbonyl iron powder by measurements of the particle diameters converted to a plot of percent per micron in range (by volume) versus particle size is a direct method of measurement. Other methods, such as laser light scattering and sedimentation analysis, are indirect methods. These latter methods are indirect because they employ physical laws of light scattering or the Stoke's Law phenomenon of particles settling in a viscous fluid to correlate with particle size.

Image analysis of photomicrographs obtained by calibrated SEM techniques appears to be a more accurate and precise method of particle size analysis on spherical powders than laser scattering. However, the necessity of preparing and examining SEM photographs for the image analysis makes this particular process time consuming and costly. This makes it more suitable as an R&D tool than as a production control method.

Careful examination of the data reported in Table II with the curves in Figures 9, 10 and 11 shows that such simple numbers as average particle diameter and standard deviation can be misleading as characterizations of particle distributions. These values implicitly assume that particle size distributions have a Gaussian form about a mean value, which is not always the case. This is especially true for SF powder (see Figure 11). It is necessary to examine the full curve in Figure 11 to obtain a true idea of this powder rather than relying on the simple values in Table II.

The data presented here were somewhat limited in resolution since they have been collected into .5 micron wide "bands". With a very large sample size of particles counted, a greater degree of resolution could be obtained.

Conclusions

1. Image analysis is a powerful tool for determining the particle size distribution of carbonyl iron powder, and it appears to be more accurate than laser forward scattering techniques. However the laser scattering method is more convenient as a quality assurance technique and approximate differentiation of the curve produces a more representative plot than the raw data.

2. Carbonyl Iron Powder has been shown by various methods of particle size measurement to have a size distribution lying between 0.2 and 20 microns. In typical commercial grades most of the particles lie between 1 and 6 microns.

3. Detailed particle size determinations by image analysis can be performed with a resolution of about 0.2 microns. Accuracy of the technique is traceable to a Bureau of Standards sample with a tolerance of about $\pm$.2 microns.

TABLE I

CALIBRATION OF PARTICLE SIZE MEASUREMENT TECHNIQUES

Sample	Light Scattering		Image Analysis	
	Volume Average (microns)	Std. Deviation (microns)	Volume Average (microns)	Std. Deviation (microns)
20 micron calibrated glass spheres	25.07	7.96	22.2	4.2
NBS 2.366 micron standard grid	-	-	2.5	.06
NBS 3.494 micron standard grid	-	-	3.30	.04

TABLE II

COMPARISON OF SIZE DATA ON THREE GRADES OF

CARBONYL IRON POWDER DETERMINED BY LIGHT SCATTERING AND IMAGE ANALYSIS

Powder Grade	Light Scattering		Image Analysis	
	Volume Average (microns)	Std. Deviation (microns)	Volume Average (microns)	Std. Deviation (microns)
E	5.26	2.62	3.78	1.55
TH	5.22	3.16	3.81	1.23
SF	2.37	1.45	2.87	1.01

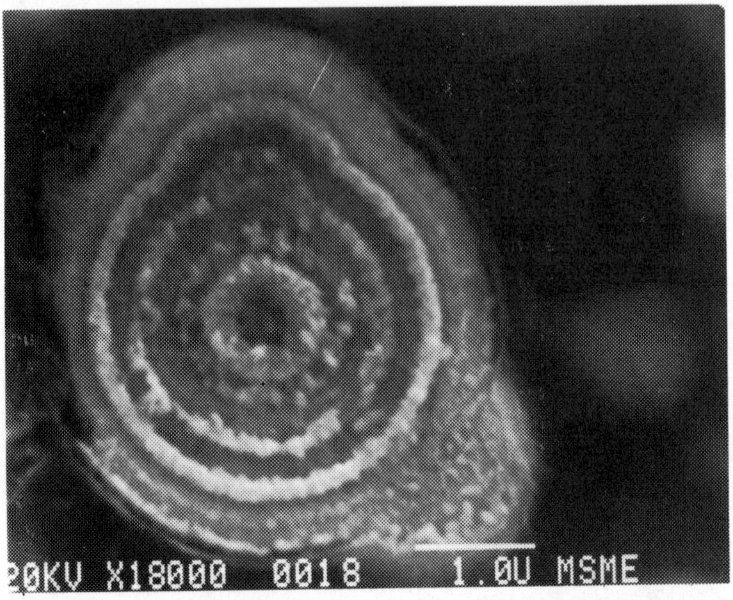

FIG 1 Etched cross section of Carbonyl Iron Particle - Note "onion skin structure"

FIG 2 Grade E Carbonyl Iron Powder produced by milling a coarse decomposer product

FIG 3 Grade TH Carbonyl Iron Powder

FIG 4 Grade SF Carbonyl Iron Powder

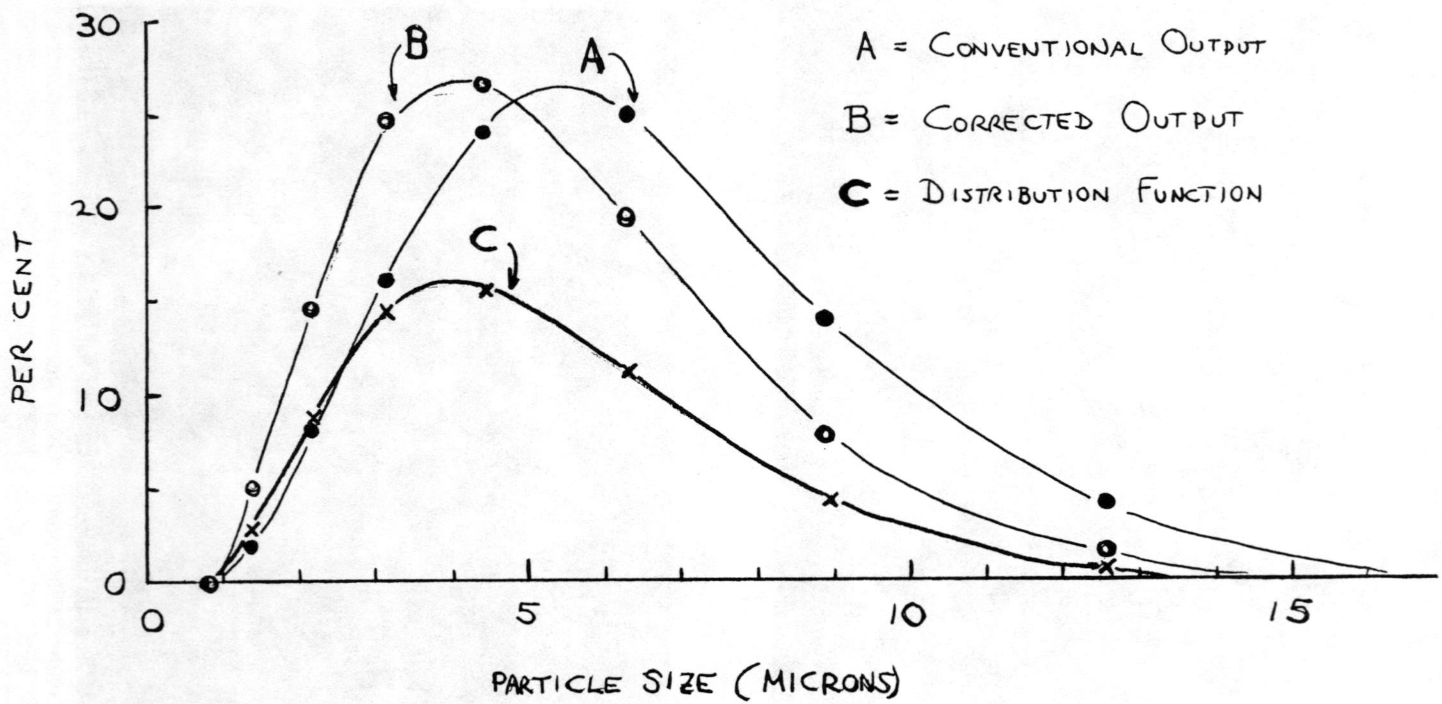

A = CONVENTIONAL OUTPUT
B = CORRECTED OUTPUT
C = DISTRIBUTION FUNCTION
A
B
C
PER CENT
30
20
10
0
0
5
10
15
PARTICLE SIZE (MICRONS)

FIGURE 5

FIG 6   Digital planimeter in use, measuring carbonyl iron microphotograph, logging data into the computer.

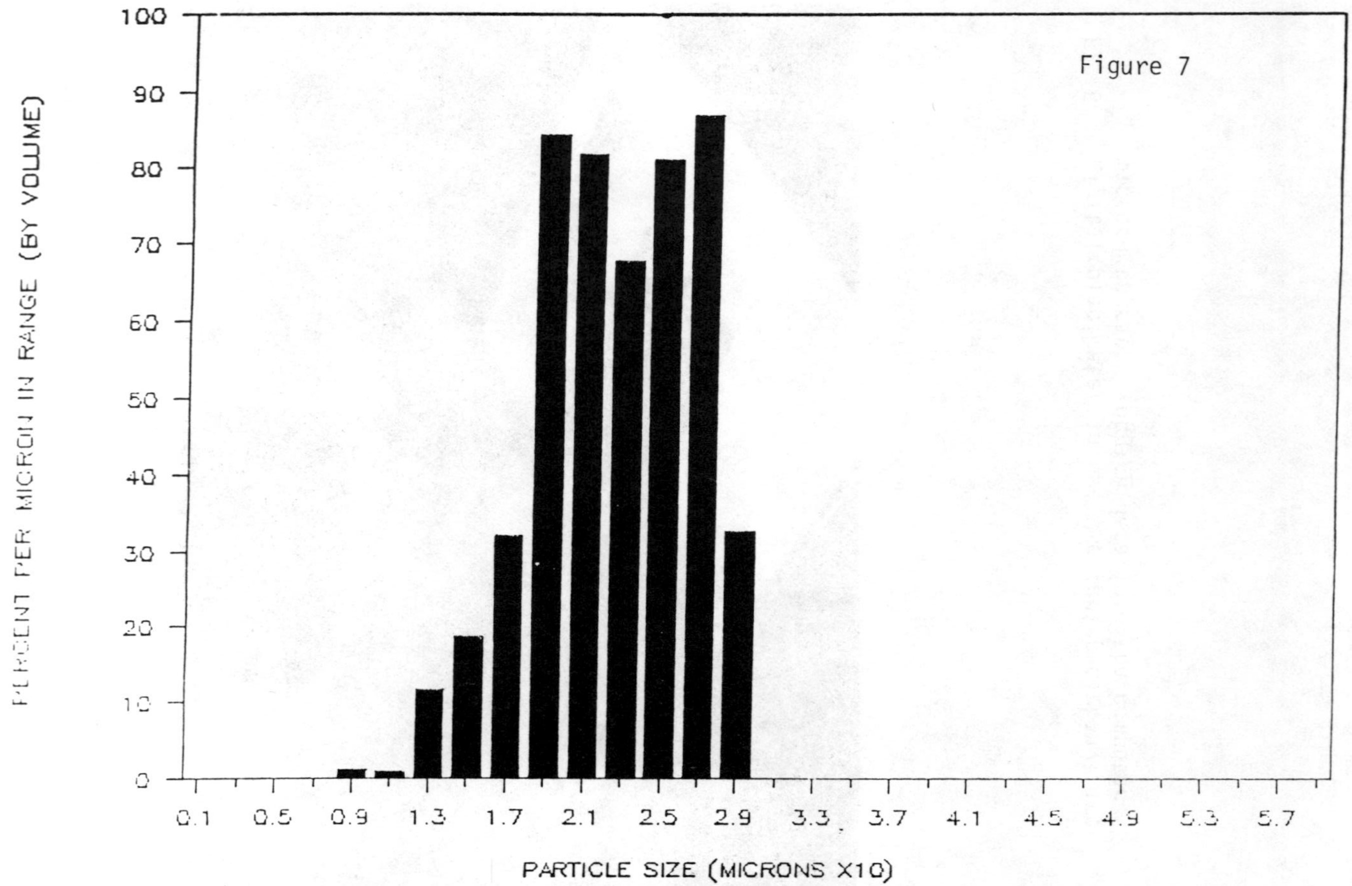

Figure 7 Particle Size Distribution of 20 micron Standard Glass Spheres Determined by Image Analysis

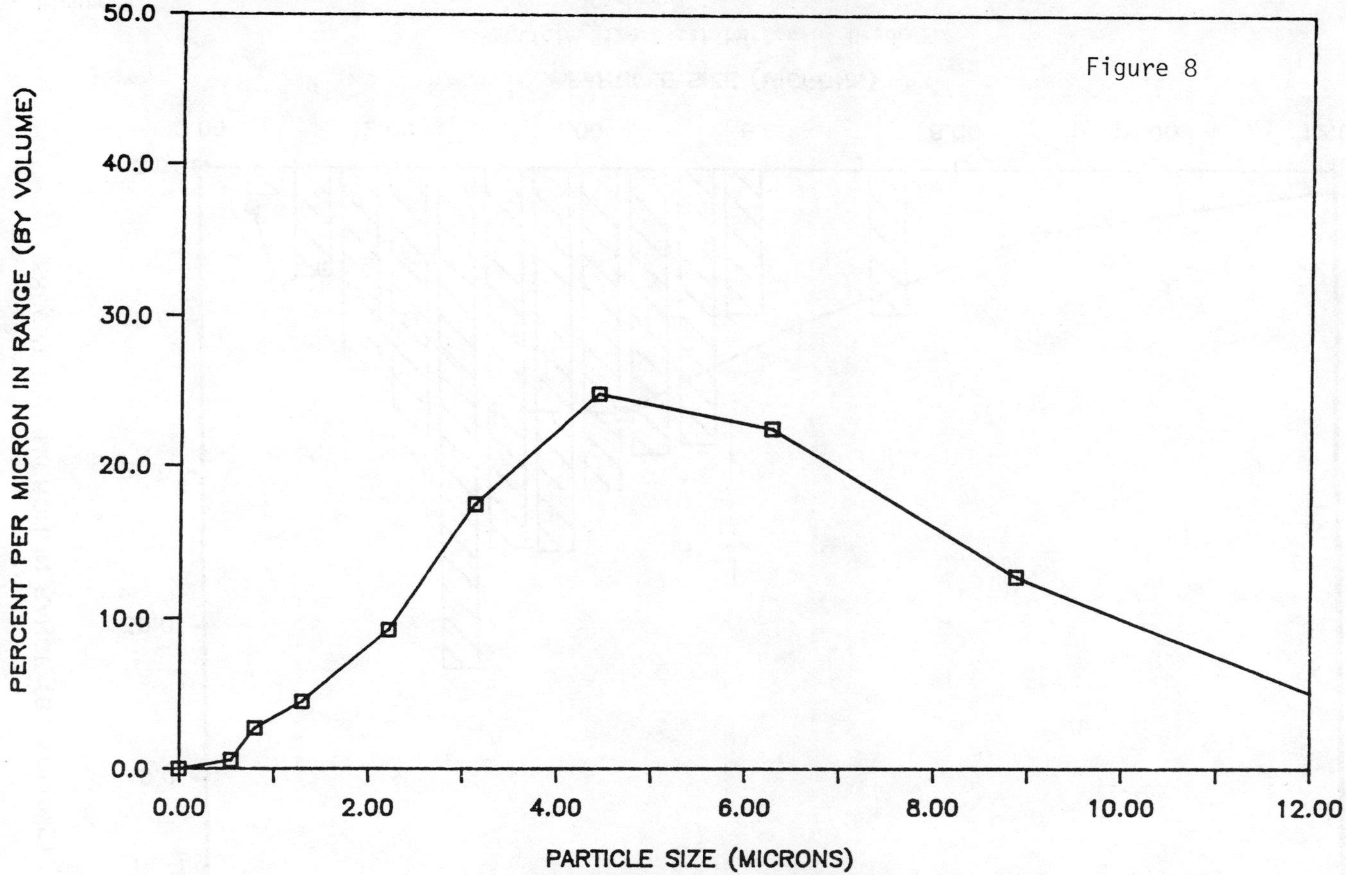

Laser Light Scattered Data for E Powder in Linear Form From Original Data

Figure 8

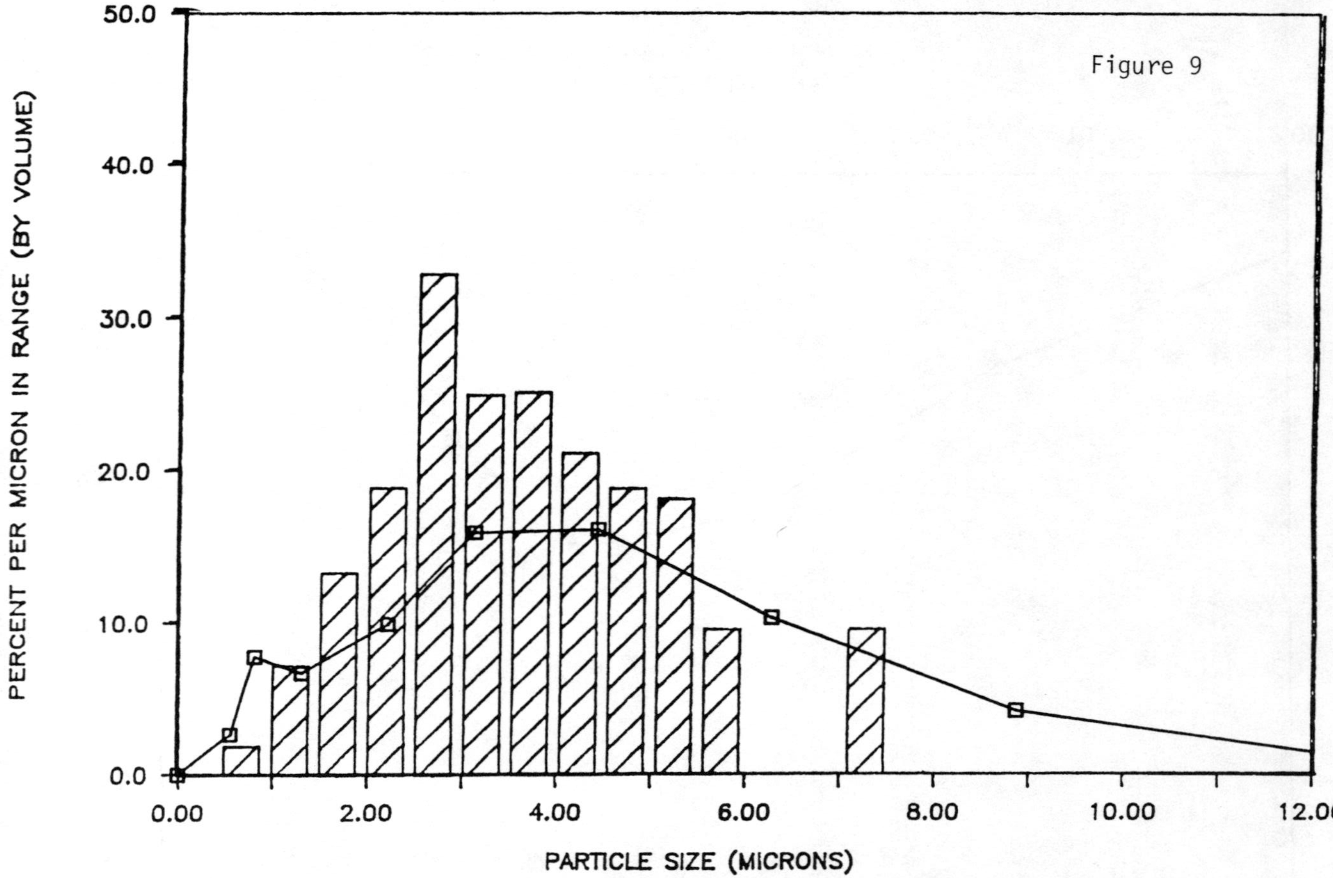

Figure 9

Particle Size Distribution of Grade E Carbonyl Iron Powder

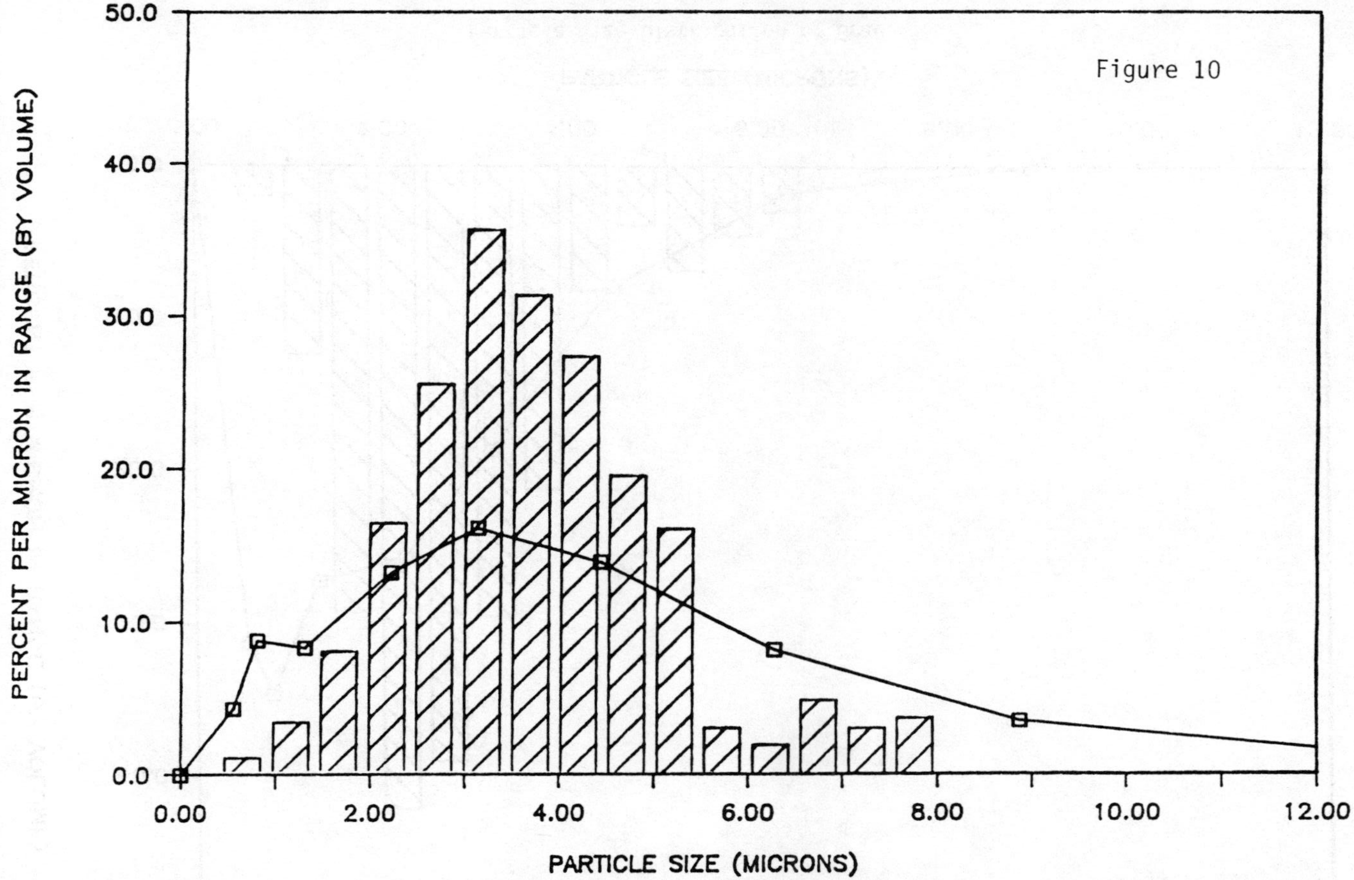

Particle Size Distribution of Grade TH
Carbonyl Iron Powder

Figure 10

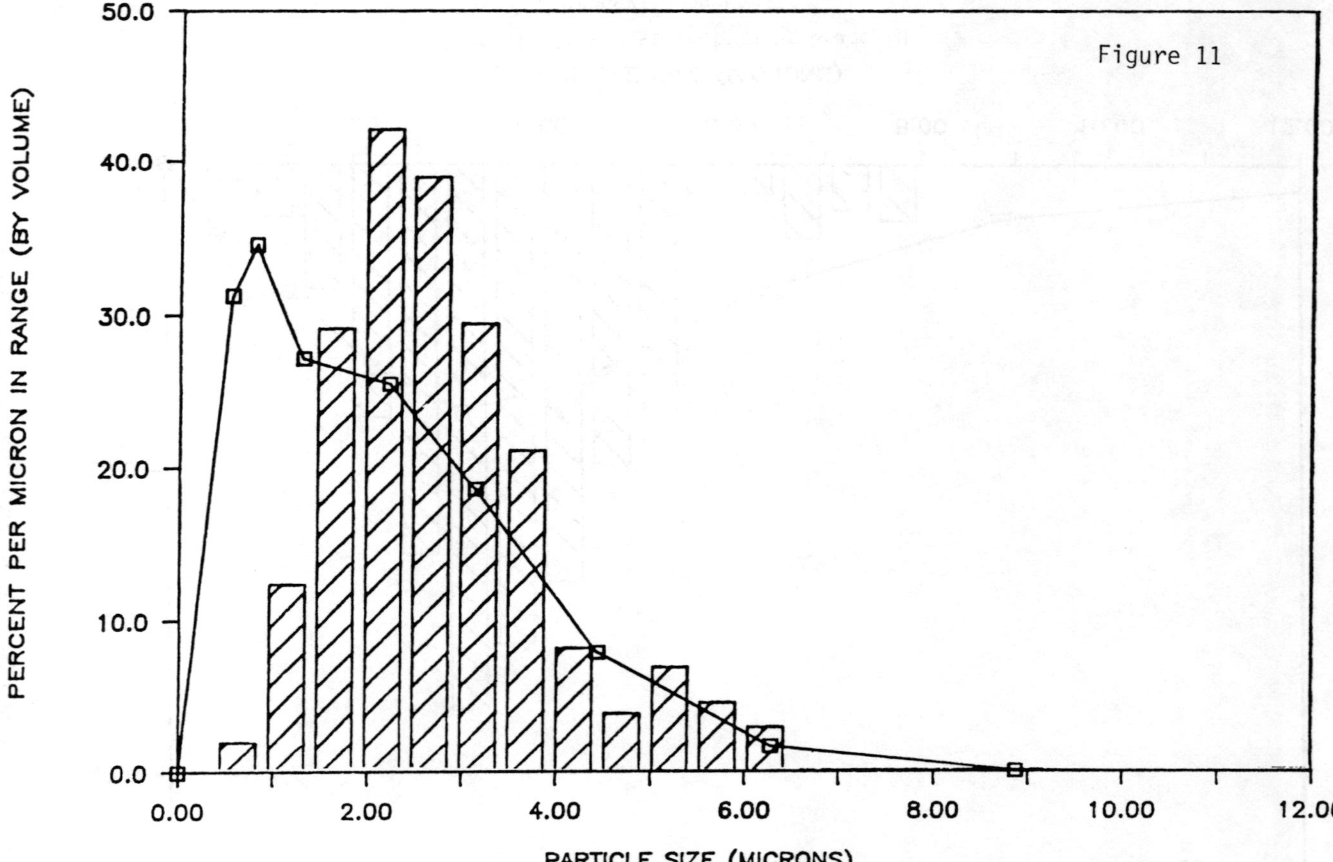

Particle Size Distribution of Grade SF
Carbonyl Iron Powder

Figure 11

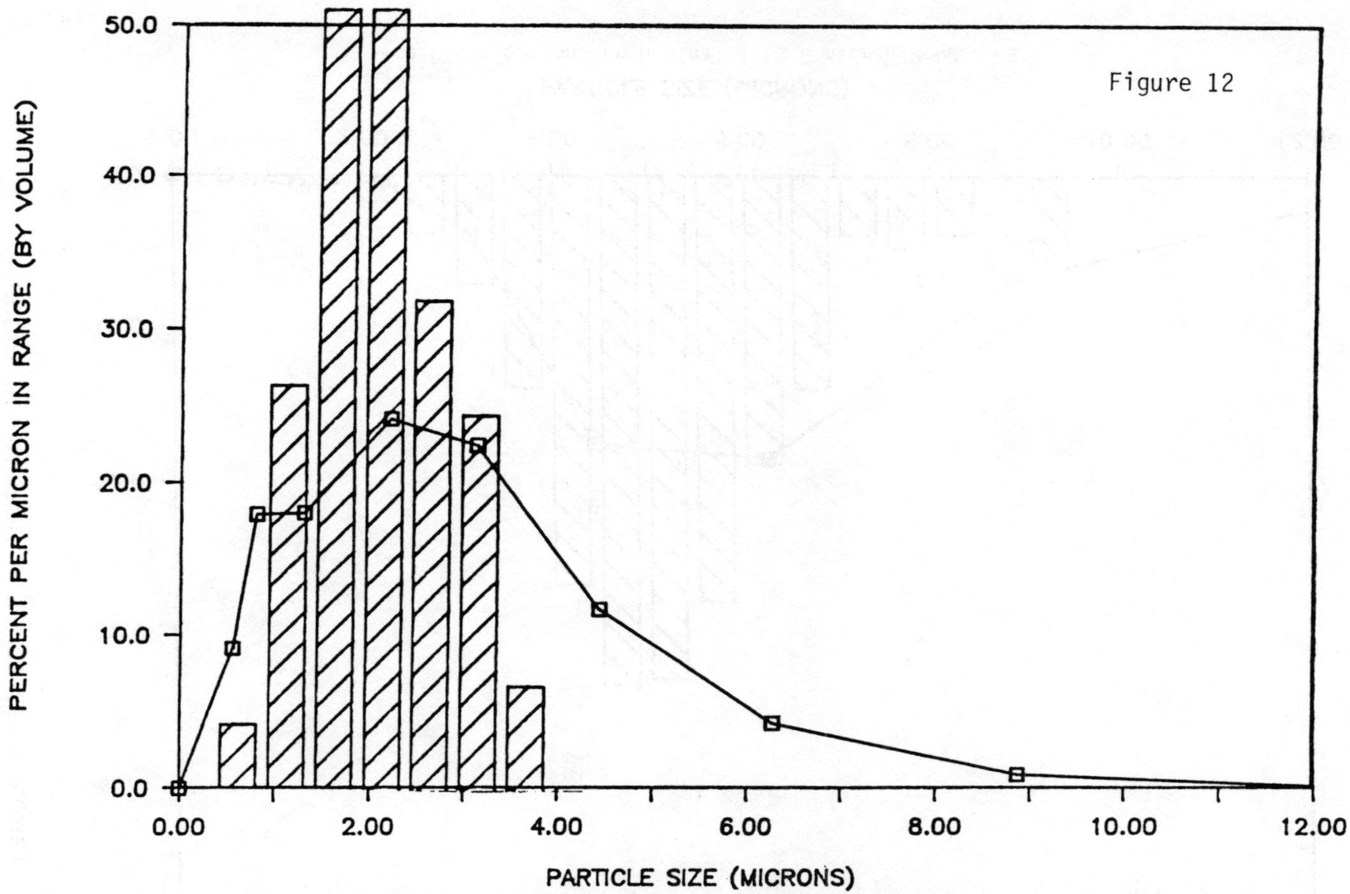

Size Distribution for Nominal 0.5 to 2 micron Powder

Figure 12

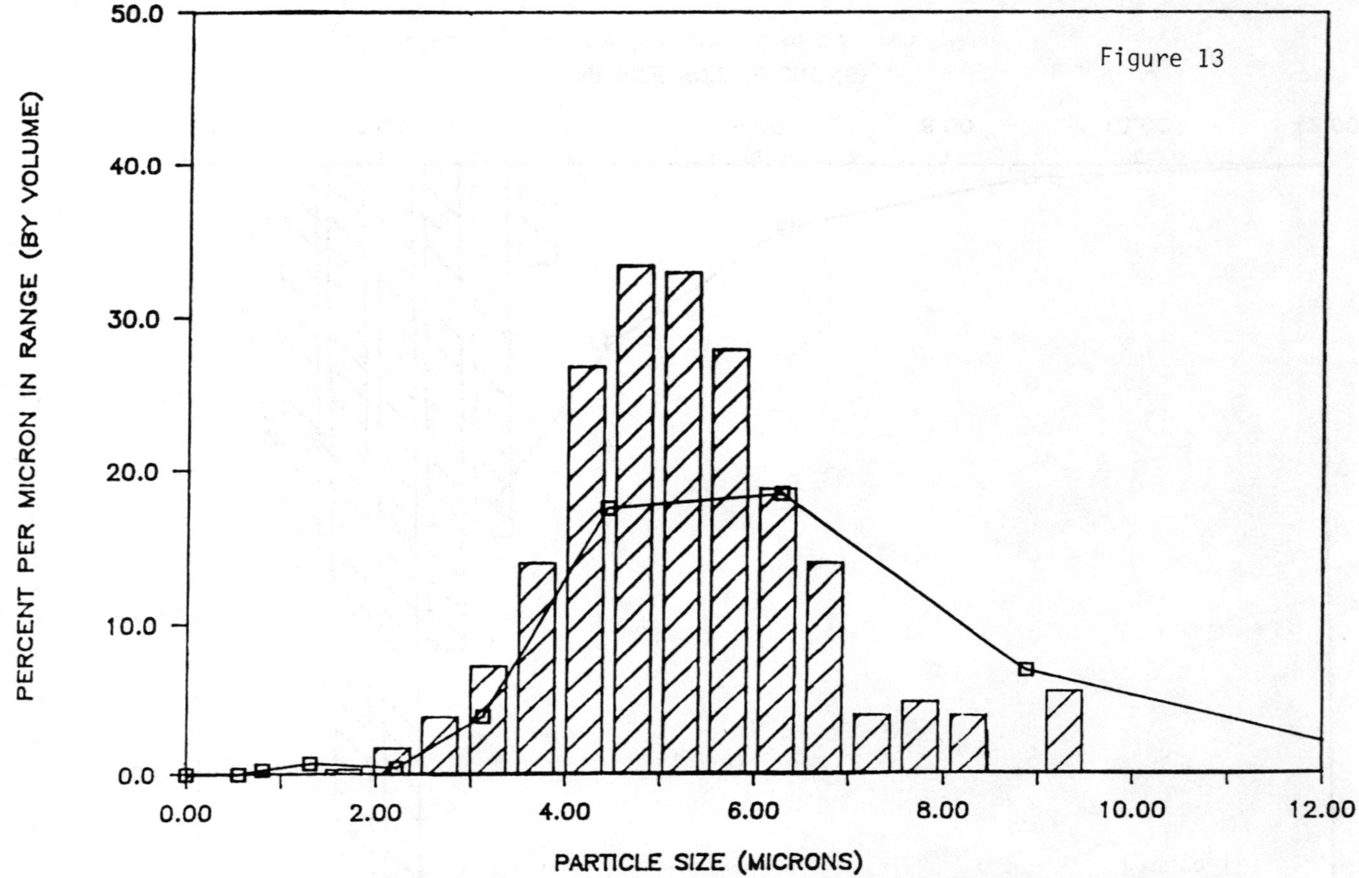

Size Distribution for Nominal 4 to 6 micron Powder

Figure 13

Chapter Six

P/M Composite Materials

THE DEVELOPMENT OF A NEW DISPERSION STRENGTHENED VITALLIUM® ALLOY FOR MEDICAL IMPLANTS

K. K. Wang, L. J. Gustavson and J. H. Dumbleton

Howmedica
Rutherford, New Jersey 07070

Abstract

A new dispersion strengthened Vitallium® alloy made by gas atomization was developed. Alloy powder with a small amount of dispersed oxides/nitrides was produced by a carefully controlled gas atomization process. The atomized powder was consolidated by extrusion. The extruded bar was further hot rolled, swaged and forged into implants. After sintering (~1300°C), a required process for producing a porous coating to enhance implant fixation, this gas atomized dispersion strengthened (GADS) Vitallium® alloy exhibits excellent room temperature fatigue strength (621 MPa at 10^7 cycles) about twice of that shown by the conventional cast and forged Vitallium® alloy with a similar high temperature exposure. In addition, GADS alloy has excellent hot formability and room temperature ductility. The corrosion resistance of GADS alloy is equivalent to that of cast or forged Vitallium® alloy. This paper will present the development and characterization of this new dispersion strengthened Vitallium® alloy.

Introduction

Vitallium® alloy is a cobalt-based alloy with chromium and molybdenum as major alloying elements. It has been used as an implant material in cast as well as wrought forms because of its good fatigue strength, resistance to wear, corrosion resistance to body fluid and biocompatibility. However, with the advent of porous coating to

Vitallium - Trademark of Howmedica

enhance implant fixation, the room temperature fatigue strength of conventional Vitallium® alloy (i.e. cast, forged and P/M) was reduced due to high temperature exposure (~1300°C) required for sintering powder particles used to form a porous coating. Therefore, there was a need to develop an alloy that meets the Vitallium® specification, ASTM F75 (Table 1) and can retain its room temperature strength after high temperature exposure.

It has been known for sometime that the strength of metals at high temperature could be increased by the addition of fine dispersed insoluble refractory oxides. One method that can produce such a dispersion is mechanical alloying. Studies(1) have shown that although a mechanically alloyed Vitallium® alloy had excellent fatigue strength (656 MPa at 10^7 cycles) after sintering, there was insufficient ductility for conventional hot working.

As an alternative powder metallurgy (P/M) and rapid solidification (RS) using gas atomization was viewed as a potential method for producing an alloy with a small amount of uniformly dispersed oxides/nitrides by controlling the composition of atomized powder and applying suitable thermomechanical processing.

This study was conducted to demonstrate the feasibility of producing a dispersion strengthened alloy by gas atomization.

TABLE 1 - CHEMICAL REQUIREMENTS (ASTM F75)

Element		Co	Cr	Mo	Mn	Si	Ni	Fe	C
Composition(%)	min	Bal.	27.0	5.0	-	-	-	-	-
	max	Bal.	30.00	7.00	1.00	1.00	1.00	0.75	0.35

Experimental Procedures

Processing of Experimental Alloy

Five 30-45kg heats of Vitallium® alloy with various aluminum and lanthanum contents were melted in a vacuum induction furnace and then gas atomized. The furnace was evacuated to a pressure of 250μm and then flooded with nitrogen at 0.5 atmosphere. Figure 1 shows a schematic cross-section drawing of a gas atomizer used for this study. The raw materials, except for the aluminum and lanthanum, were first melted in the furnace. Atomization was conducted immediately after aluminum and lanthanum (the final additions) were melted into the molten bath. The first heat was atomized with argon gas. All the other heats were atomized with nitrogen gas.

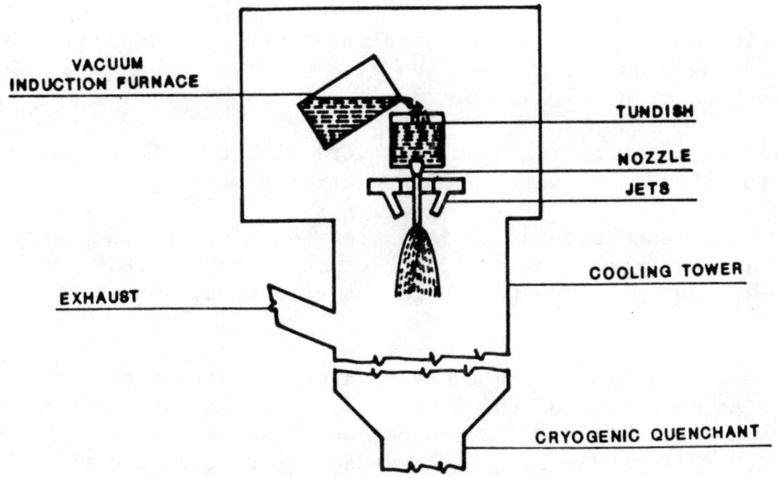

Figure 1. Schematic cross-section drawing of a gas atomizer

The chemical and screen analyses were conducted on the atomized powder after screening to remove the coarse +60 mesh (U.S. Standard) powder. The -60 mesh powders of each heat were packed into mild steel extrusion cans which were then evacuated at 204°C (400°F) to a leak rate of 10 micron per minute and sealed. Two sealed cans were produced from each atomized heat and were extruded at 1121°C (2050°F) and 1177°C (2150°F) respectively with an extrusion ratio of 9:1 to 38.1 mm (1.5 in) diameter bar. In each case, the extrusion cans were heated four hours at the extrusion temperature prior to extrusion.

The extruded bars were pickled to remove the can material to prepare for hot working. Each of decanned bars was cut into two equal pieces and they were swaged at 1066°C (1950°F) and 1121°C (2050°F) to 21.0 mm (0.825 in) diameter respectively for the first heat of GADS Vitallium® alloy, and at 1121°C (2050°F) and 1177°C (2150°F) to 16.9mm (0.655 in) diameter respectively for the remaining heats of GADS alloy. After swaging, only the first heat of GADS alloy was further forged into hip stems using the same sequence of operation currently employed in producing forged hip prothesis from forged high strength (FHS®) Vitallium® alloy at the same swaging temperature. The swaged bars of heats 2-5 were forged to 12.7mm x 15.9mm (0.5 in x 0.625 in) bar product using the same swaging temperature.

Testing of Experimental Alloy

Microstructual examinations were conducted on as-forged and forged plus sintered samples. The metallographic samples were etched in 95% hydrochloric acid and 5% hydrogen peroxide (30%).

Tensile tests were performed using round specimen with a 4.06mm (0.16 in) gage diameter and a 16.51 mm (0.65 in) gage length.

Rotating and cantilever bend fatigue tests were conducted on a Krouse rotating beam fatigue machine at $A = \infty$, 167 Hz and on a Sonntag fatigue machine at $A = 1$, 30 Hz.

All mechanical testing specimens were either as-forged or sintered to 1300°C and finish machined after sintering.

Corrosion (anodic polarization) tests on sintered samples were conducted in deaerated 0.9% sodium chloride at 37°C. FHS® and cast Vitallium® alloy were also tested in the same condition as a reference.

In order to characterize the nature of the dispersion strengthening mechanism of the GADS alloys, Transmission Electron Microscopy and STEM EDS were performed on the as-forged and sintered GADS samples (alloys 2 & 5). Both extraction replica and thin foil techniques were used for this study. For comparison, the FHS® and P/M Vitallium® alloy were examined. The FHS® alloy was produced by the conventional ingot metallurgy while the P/M alloy was made by gas atomization but without the addition of aluminum and lanthanum.

Results and Discussion

1. Composition

The chemical composition of the atomized powder of each heat is given in Table 2. Major alloying elements are within the range of ASTM F75-76. The differences in composition between the atomized Vitallium® powder and conventional FHS® alloy are the presence of aluminum and/or lanthanum, the oxide forming elements which were intentionally added to the molten alloy prior to atomization. The nitrogen content of alloy No. 1 was somewhat lower than the desired optimum (about 0.22%) and this was due to the unexpected nitrogen loss through argon atomization. The nitrogen content of atomized powders of alloys 2 to 5 was maintained within the range of 0.21 to 0.26 using nitrogen atomization.

2. Screen Analysis

The screen analysis of gas atomized powder is shown in Table 3. It appears that the powder of alloy No. 1 is coarser than that of other alloys. Photomicrographs of the powder revealed spherical particles with dendritic structure Figure 2.

TABLE 2. CHEMICAL COMPOSITION ATOMIZED VITALLIUM® ALLOY

Heat/ Alloy No.	1	2	3	4	5	ASTM F75
Co	Bal	Bal	Bal	Bal	Bal	Bal
Cr	26.47	27.24	27.27	27.00	26.81	27.0-30.0
Mo	5.50	5.97	5.97	6.01	5.97	5.0-7.0
Mn	.78	.74	.73	.75	.74	<1.00
Si	.71	.71	.70	.70	.70	<1.00
Ni	.002	.053	.048	.187	.094	<1.00
Fe	.006	.215	.199	.359	.520	< .75
Al	.40	.45	.55	.81	.47	-
La	0.04	0.02	-	-	0.15	-
C	.09	.09	.09	.11	.09	< .35
O	.0160	.0038	.0035	.0038	.0142	-
N	.10	.21	.26	.21	.22	-

TABLE 3. SCREEN ANALYSIS OF THE GADS VITALLIUM POWDER

Mesh Size (U.S. Standard, %)

Heat/ Alloy No.	-60/100	-100/200	-200/325	-325
1	50.0	26.67	13.33	10.00

Heat/ Alloy No.	-60/80	-80/100	-100/140	-140/230	-230/325	-325
2	22.54	14.71	15.11	15.00	15.11	14.77
3	17.92	14.15	17.17	19.23	14.10	14.26
4	22.46	15.71	17.00	15.20	12.80	12.67
5	21.43	13.65	14.40	16.32	15.36	17.34

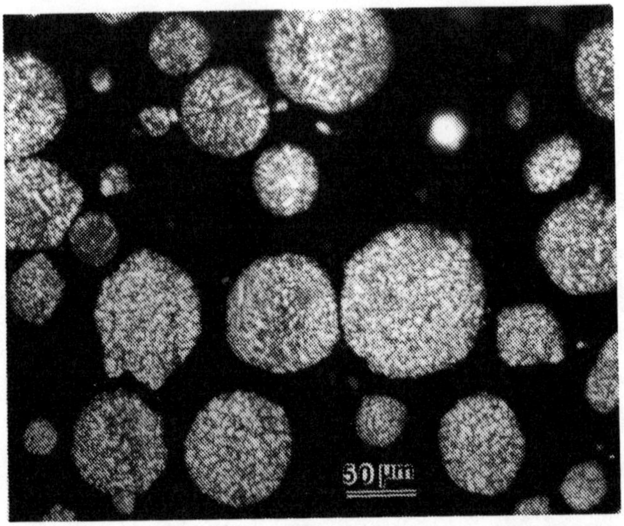

Figure 2 Microstructure of as-atomized GADS Vitallium® powder

3. Consolidation

The GADS Vitallium® alloy was successfully consolidated by extrusion and thermomechanically processed by swaging and forging into final hip stems or test bars. This indicated that this alloy has good hot workability.

4. Microstructures

Metallographic examination showed that the as-forged GADS alloy had a very fine grain size, ASTM #10. It was found that the grain size of as-forged GADS alloy was little affected by the processing temperature or the various lanthanum and aluminum contents.

The microstructures of as-forged GADS Vitallium® alloy were compared with those of cast and FHS® Vitallium® alloy (Figure 3). Both of the as-forged GADS and FHS® Vitallium alloy exhibited a very fine equiaxed grain size ASTM #10 and 9 respectively containing deformation twins indicative of the recrystallization that alloy undergoes during hot working. In contrast, the cast alloy has a very coarse dendritic structure characteristic of investment casting.

After sintering at 1300°C, the difference between the three alloys becomes most significant (Figure 4). Both the FHS® and cast Vitallium® alloy undergo dramatic microstructural changes following exposure to the sintering temperature while GADS alloy remains virtually unchanged.

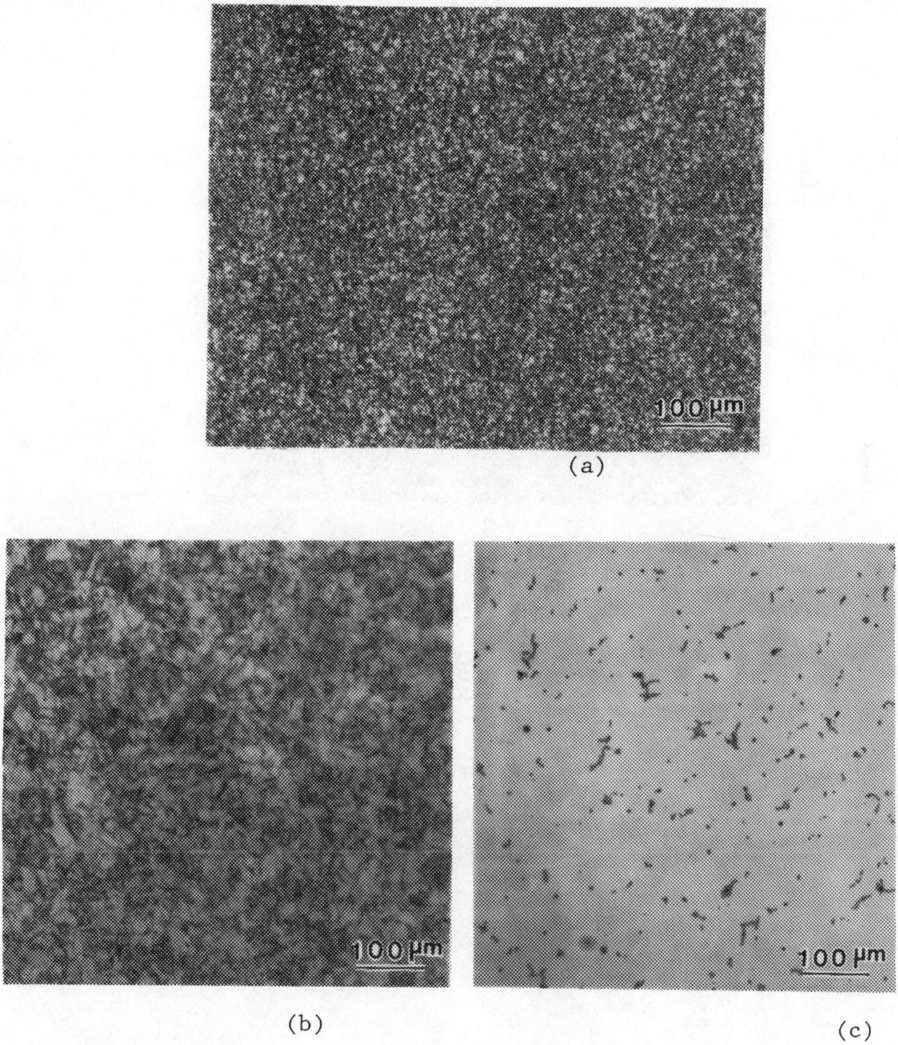

Figure 3. - Microstructures of (a) as-forged GADS, (b) as-forged FHS® (c) solution treated cast Vitallium® alloy

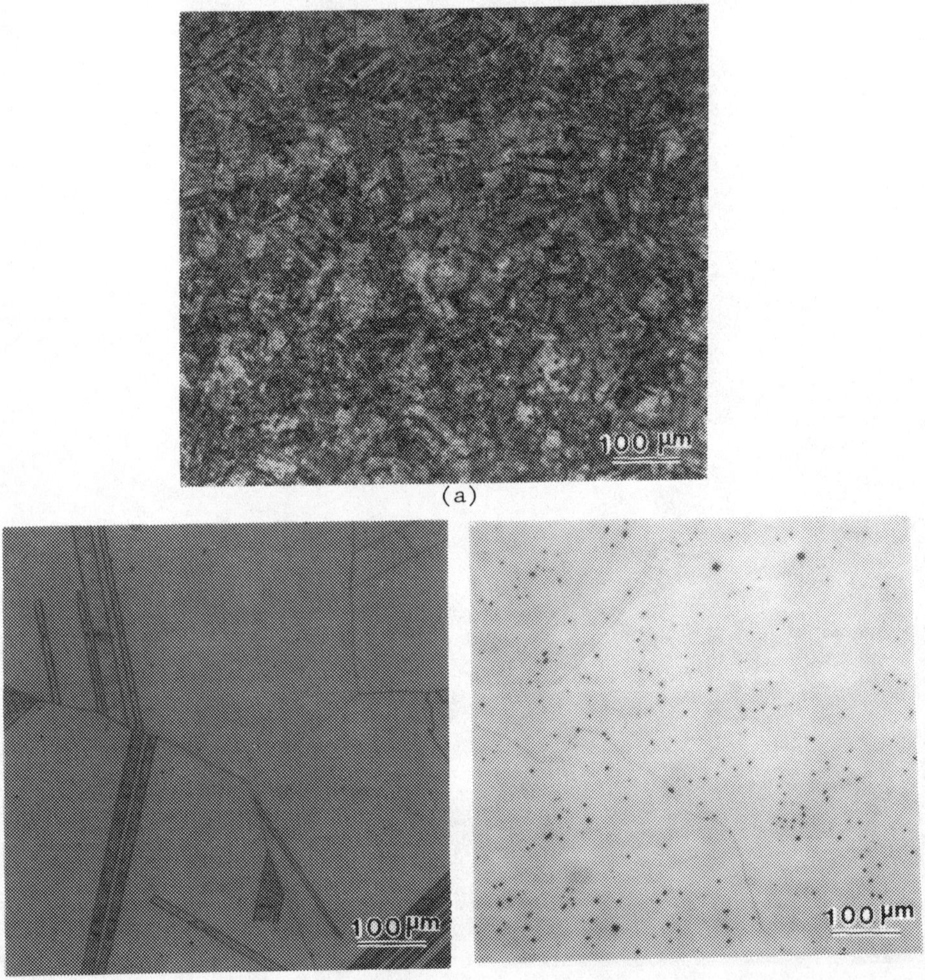

(a)

(b) (c)

Figure 4. - Microstructures of sintered (~1300°C) GADS, FHS® and Cast Vitallium® alloy

The GADS alloy (samples 1-4) still exhibited a fine, equiaxed grain structure (ASTM #8). The grain size of alloy No. 5 having a higher lanthanum content is somewhat even finer (ASTM #8-9). However, significant grain growth occurred in the FHS® Vitallium® alloy (ASTM #1). As to the cast Vitallium® alloy, the dendritic carbides which strengthened the cast material were solutioned into the matrix. Study also showed that the GADS alloy could be annealed at 1343°C (2450°F) with no significant grain growth.

It was found that Al rich dispersoids tended to cluster in the grain boundaries of alloys No. 3 & 4 having high aluminum content. (Figure 5)

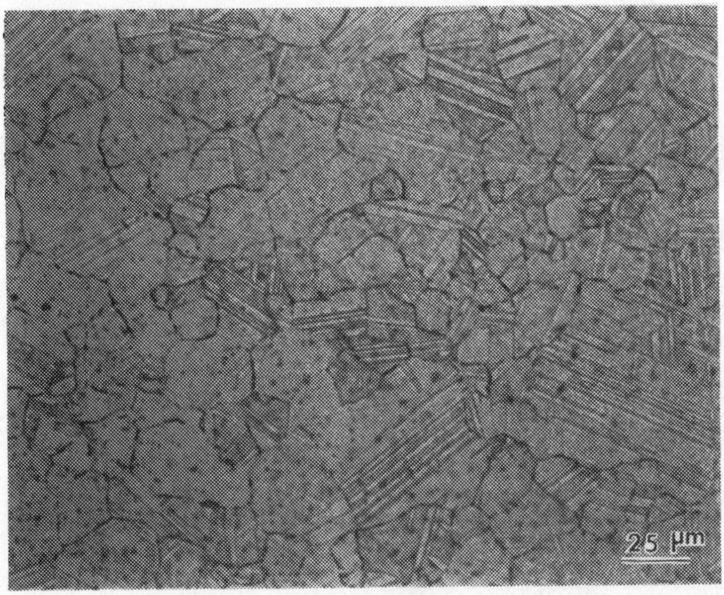

Figure 5 - Al-rich particles seen in the grain boundaries of GADS alloy having high Al content (.81wt%).

The grain size of post-sinter GADS alloys was not significantly affected by the processing temperature.

5. Tensile Properties

The results of tensile tests on the GADS alloys are presented in Tables 4 and 5. The 1121°C forged alloy both the as-forged and the sintered, exhibited higher strength than the 1177°C forged ones Alloys 2 and 5 are stronger than alloys 3 and 4. This was most likely contributed by the lanthanum oxide. The significant improvement in the strength of alloy 2 as compared to the alloy 1 was mainly due to its nitrides level. As shown in Table 4, the tensile ductility of as-forged alloys 2 to 5 is much lower than that of the alloy 1. From careful examination on the surfaces of fractured tensile samples, it was found that the samples machined from the alloys 2-5 had deep circumferential machining marks which would account for the low elongation and reduction in area values. As shown in Table 5, the alloys 2 to 5 have excellent post-sinter ductility (46.4 - 58.7%).

TABLE 4. TENSILE PROPERTIES OF GADS VITALLIUM®ALLOY

Heat/ Alloy No.	Condition	U.T.S. (ksi)	0.2% Y.S. (ksi)	EL(%)	RA(%)
1(a)(c)	As-forged	189.5	117.9	32.0	27.5
2(a)	As-forged	190.5	152.6	9.6(d)	11.6(d)
2(b)	As-forged	191.0	143.1	12.3(d)	14.0(d)
3(a)	As-forged	179.8	142.4	7.9(d)	11.0(d)
3(b)	As-forged	180.3	142.7	7.9(d)	10.9(d)
4(a)	As-forged	183.4	137.0	7.4(d)	10.1(d)
4(b)	As-forged	176.0	124.2	8.8(d)	10.9(d)
5(a)	As-forged	193.2	147.4	11.8(d)	14.3(d)
5(b)	As-forged	176.4	127.3	7.1(d)	10.1(d)

(a) Swaged and forged at 2050°F.
(b) Swaged and forged at 2150°F.
(c) Samples were machined from forged hip stems.
(d) The low elongation and reduction area were mainly due to the deep circumferential machining marks.

TABLE 5. TENSILE PROPERTIES OF GADS VITALLIUM®ALLOY

Heat/ Alloy No.	Condition	U.T.S. (ksi)	0.2%Y.S. (ksi)	EL(%)	RA(%)
1(a)(c)	Sintered	164.8	77.0	37.2	30.2
2(a)	Sintered	184.9	95.8	58.7	36.2
2(b)	Sintered	180.2	84.2	54.6	36.5
3(a)	Sintered	179.0	90.0	55.3	33.9
3(b)	Sintered	174.1	83.1	49.2	32.9
4(a)	Sintered	178.3	84.1	50.5	33.4
4(b)	Sintered	177.1	78.8	46.4	33.4
5(a)	Sintered	186.2	95.9	51.0	33.2
5(b)	Sintered	182.0	89.1	51.0	34.1

(a) Swaged and forged at 2050°F.
(b) Swaged and forged at 2150°F.
(c) Samples were machined from forged hip stems.

6. Fatigue Properties

Fatigue data indicate that the different swaging and forging temperatures 1121°C and 1177°C (2050°F and 2150°F) have little effect on the cantilever fatigue strength of the alloy. Cantilever fatigue tests (Table 6) show that the alloys 2 and 5 have higher fatigue strength than the alloys, 1, 3 and 4. The fatigue strength of alloy 5 is slightly stronger than that of alloy 2. The only significant difference between alloys 2 and 5 is that alloy 5 have higher lanthanum content than that of alloy 2. The alloy No. 4 has lower strength than that of alloy 3. X-ray EDS analysis on the alloys 3 and 4 showed that aluminum rich particles tends to cluster in the grain boundaries. The continuous aluminum rich particles in the grain boundaries could be prone to initiate a fatigue crack, therefore, further reducing the strength of alloy 4.

The smooth rotating beam results (Table 6) indicate that alloys 2-5 are stronger than alloy 1 and no significant difference in strength was found in alloys 2-5. This result demonstrates that nitrogen (nitrides) can increase the post-sinter fatigue strength of the GADS alloy.

TABLE 6. POST-SINTER FATIGUE STRENGTH OF GADS VITALLIUM® ALLOYS AT 10^7 CYCLES

Heat/Alloy No.	Cantilever(1)		Rotating Beam(2)	
	MPa	ksi	MPa	ksi
1	931	(135)	483	70
2	965-1000	(140-145)	621	90
3	931-965	(135-140)	621	90
4	896-931	(130-135)	621	90
5	1000-1034	(145-150)	621	90

Notes: (1) Data obtained from both the 1121°C (2050°F) and 1177°C (2150°F) forged samples.

(2) Data obtained from the 1121°C (2050°F) forged samples.

Rotating beam fatigue data of the alloys 2 and 5 compared with those of cast and FHS® Vitallium® alloy are listed in Table 7. The post-sinter fatigue strength of alloys 2 and 5 is much higher than those of cast and FHS® Vitallium® alloy. This corresponds to approximately 64% strength improvement over cast Vitallium® alloy which was hot isostatically pressed following the sintering thermal cycle.

TABLE 7. ROTATING BEAM FATIGUE PROPERTIES OF GADS, CAST AND FHS® VITALLIUM® ALLOYS

	Fatigue Strength at 10^7 cycles	
Cast Vitallium® Alloy	MPa	(ksi)
Solution Heat Treated	455	(65)
Sinter Cycle Treated + HIP	380	(55)
FHS® Vitallium® Alloy		
As-forged (1950°F)	758	(110)
Sinter-Cycle Treated	324	(47)
GADS Alloy (Alloys 2 & 5)		
As-forged	758	(110)
Sinter Cycle Treated	621	(90)

7. Transmission Electron Microscopy (TEM)

TEM studies revealed that the as-forged alloys 2 and 5 contained uniformly dispersed particles which were not found in the as forged FHS® or P/M material (Figure 6). Most of these particles were from 0.1μm to 0.5μm. Two types of dispersoids were identified by the selected area electron diffraction. Diffraction patterns showed that the type 1 and 2 particle was a cubic crystal structure with a = 4.74 Å and a hexagonal crystal structure with a = 2.63 Å , C = 7.84 Å respectively. STEM EDS analysis indicated that type 1 particle has 15% Al and 85% La metallic composition and type 2 particle has 88% Al and 12% La metallic composition. Therefore, the dispersed particles in GADS alloy are most likely to be aluminum lanthanum oxides/nitrides rich in aluminum or lanthanum.

Deformation twins, stacking faults and dislocations were found in all these materials. Randomly tangled dislocations were normally found in these samples. Based on samples examined, it appeared to have no obvious differences in the density of stacking faults, twins and dislocations between these as-forged samples.

As to all sintered samples, the distinct difference of the GADS material from the FHS® and P/M material was also the presence of dispersed particles. The particle size seems to be slightly bigger than that of as-forged material (Figure 7). The dislocation and small faults were normally found to be interlocked with the particles as shown in Figure 8. These extracted particles appeared to be aluminum lanthanum oxides/nitrides same as those found in the as-forged samples.

Stacking faults, twins and only few dislocations were found in all these materials. The stacking faults were found on all {111} shear planes for GADS and P/M materials. The faults tended to be formed on a preferred {111} shear plane for the FHS® material.

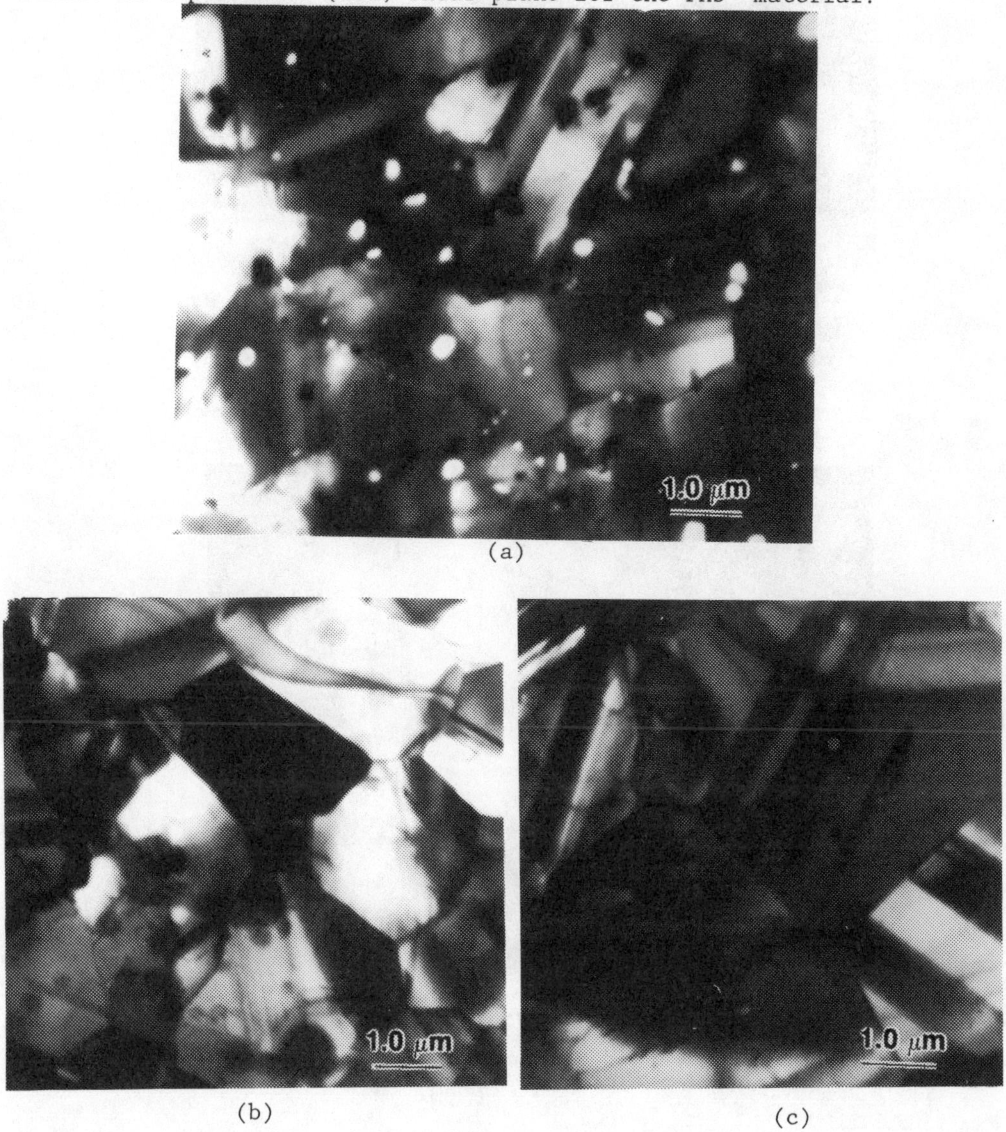

Figure 6 - TEM micrographs of (a) as forged GADS, (b) FHS® (c) P/M Vitallium® alloy

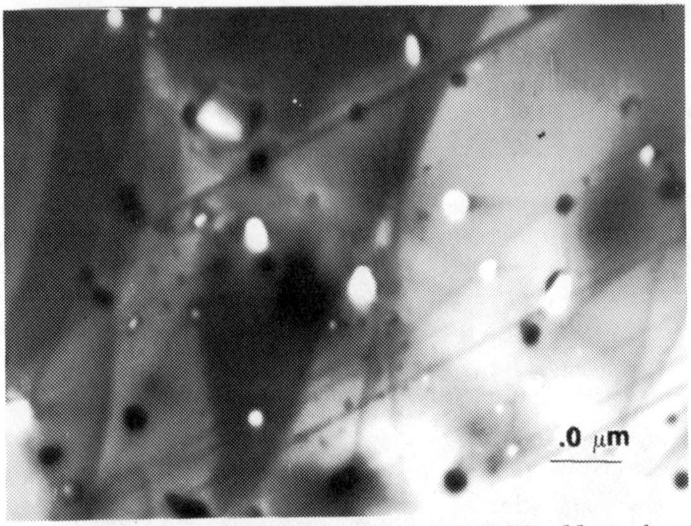

Figure 7 - TEM micrographs of sintered GADS alloy showing uniformly dispersed particles

Figure 8 - TEM micrograph of sintered GADS alloy showing dislocation and stacking faults interlocked by the dispersed particles.

8. Corrosion

The potential vs normalized current data for GADS alloy are presented in Figure 9. The average scans of GADS alloy were almost identical with those of cast and FHS® Vitallium. A small hysteresis was seen in each of these three alloys during reverse scan which indicates effective film repair.

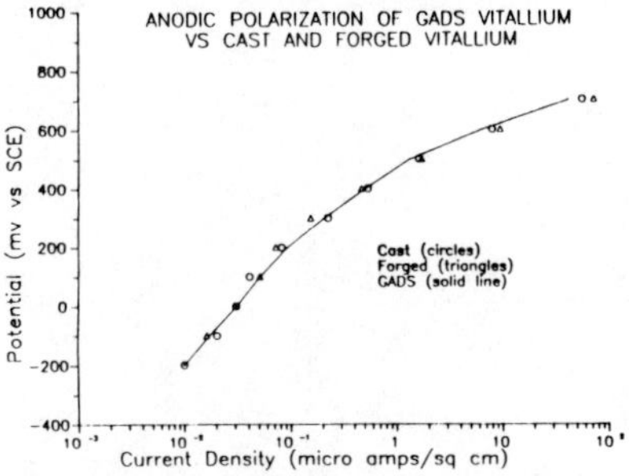

Figure 9

General Discussion

Table 7 shows that fatigue strength of the GADS alloy is much stronger than that of FHS® alloy after sintering. This corresponds to approximately 91% improvement over the FHS® alloy. Based on the above microstructural findings, the strengthening mechanism for sintered GADS alloy can be explained as follows: (1) dispersed particles (oxides and nitrides) act as obstacles to inhibit grain growth during sintering, (2) dispersed particles interfer with the progress of stacking faults and twinning during sintering, thus, the fault and twinning growth take place on all the low energy {111} shear planes instead of one preferred {111} plane as seen in the FHS® material, (3) the presence of dispersed particles interlocked with the stacking faults and dislocations.

One interesting finding concerning the sintered GADS alloy was that the grain size had little effect on its fatigue strength. This would indicate that the faults and twinning may act like grain boundaries, and the presence of dispersed phases interlocked with dislocations and faults play important roles in strengthening the sintered GADS alloy.

Based on the above results, the GADS alloy having .03-.15%La, .3-.5%Al and .21-.26%N was found to be optimal.

Conclusions

1. The feasibility of producing a dispersion strengthened Vitallium® alloy was demonstrated by gas atomization.
2. The GADS alloy is much stronger than cast or FHS® alloy after high temperature exposure.
3. The hot workability of GADS alloy is comparable to that of FHS® Vitallium® alloy.
4. The corrosion resistance of GADS alloy is comparable to that of cast and FHS® Vitallium® alloy.

Reference

1. H.L. Andrews and G.E. Gardiner, "Cobalt-Chromium Superalloy" U.S. Patent Number 4, 631, 082.

MECHANICAL AND PHYSICAL PROPERTIES OF IRON-BASED CERMETS

N. Petersen
Institute of Manufacturing Engineering
Technical University of Denmark, Bldg. 425
DK-2800 Lyngby
Denmark

ABSTRACT

A literature search was made in the area of cermets for wear resistance purposes. This resulted in the preparation and testing of iron-containing cermets.
Cermets containing iron matrix and Cr_3C_2 or TiB_2 as hard particles were made by powder technology. The content of hard particles varied from 10-30 percent by volume. The sintering conditions were varied by means of temperature and time. The temperatures were 1120 and 1150 degrees centigrade and the time was 15 and 30 minutes.
The materials were tested for density, tensile strength and hardness.
The obtained densities were from 76-81% of theoretical density for the Cr_3C_2 cermets and from 73-93% for the TiB_2 cermets. The tensile strength varied from 37-113 MPa for Cr_3C_2 and from 70-133 MPa for TiB_2.
Iron with 10% TiB_2 was also tested for wear resistance by simulating adhesive and abrasive wear on a specially designed tribometer.
Addition of TiB_2 increased the coefficient of friction but decreased wear loss. At a load of 200 N and a speed of 2.5 m/s with water as lubricant wear was decreased from 0.45 to $9.8 \cdot 10^{-5}$ mm^3/m.

INTRODUCTION

According to ASTM a cermet is a material which has "a hetorogeneous combination of metal(s) or alloy(s) with one or more ceramic phases in which the latter constitutes approximately 15 to 85% by volume and in which there is a relatively little solubility be-

tween metallic and ceramic phases at the preparation temperature" (1).
The idea of the cermet is to combine the good properties from both the ceramic and the metal, ending up with a material which has certain properties exceeding the ones of the materials alone.
The first cermets were made a century ago. Edison made a cermet when he mixed Pt-powder and zirconia. Other examples to be mentioned are metal-bonded diamonds for dental tools, magnets made of iron and iron-oxide and molybdenum-zirconia for heat-resistant purposes (2).
Many of these early cermets are now in daily use, but many of the attempts failed because of the obvious problems in bonding two totally different materials together. Aluminium has been studied intensively in order to find suitable additives. This is because of its favourable low density as opposed to its low strength and wear resistance.
The additives are e.g. mica (3) alumina (4,5), alumina-siliconcarbide (6,7,8), zirconia (9) and high-alloy steel powder (10). Iron based cermets have also been made for wear-resistant purposes. Roesen (11) added alumina, iron carbide and silicon carbide to iron powder in order to make a material for harrow tines. Champagne and Angers (12) added crushed powder of cemented carbides to iron in their production of blades in snow ploughing equipment.
Several patents have been issued for materials with iron as matrix: An American patent describes how to obtain high-density materials by means of elevated sintering temperature. The mentioned carbides are those of W, Mn, Cr, Ti, V, Nb, Ta and B_4C (13,14).
Another patent having Chromium and Carbon as main constituents is to be especially excellent for wear purposes (15,16).
Generally speaking, most of the iron-based cermets which have been made until now have either chromium carbide (Cr_3C_2) or titanium diboride (TiB_2) as the ceramic phase.
In the ceramic rich end of the cermets we find the titanium-diboride based cermets (17,18,19). The TiB_2 cermets contain about 80% boride and serve as substitutes for hard metals (cemented carbides). The carbide containing cermets have been made for wear purposes. Apart from these materials, cermets such as Fe-TiC (20,21,22) and complex boride compounds (Mo_2FeB_2) have been studied (23).

Creation of a new material necessiates that the designer specifies the requirements to the material. In the case of cermets an obvious requirement is that the bonding between the phases is good.

Ceramics are characterized by high hardness, by resistance against corrosion and wear and by excellent heat resistance. Some also have unique electrical properties.
By adding the ceramic to the metal some of these properties will hopefully improve the metal or vice versa.
In this work the special property of interest is wear resistance.

Wear can be defined as the removal of material from surfaces in contact under relative motion by mechanical and/or chemical processes.
However, the mechanisms of wear are various with the most common types being adhesive, abrasive, corrosive and fatigue. The two latter types will not be treated in this work.
Adhesive wear is caused by cold welding of the surfaces. It is favoured by structural similarities between the sliding couple.
Abrasive wear occurs when the counterpart consists of hard particles which can actually plough or scuff the softer material. Consequently abrasive wear can be avoided by choosing a material which is harder than the abrasive.

An obvious question to ask is: 1) Are ceramics as wear resistant as they are thought to be ? 2) Will the ceramics also act wear resistant in connection with a ductile metal matrix ?
There seems to be clear evidence that ceramics are wear resistant, but the wear resistance is not connected to the hardness (24).
Samsonov claims that the wear resistance of carbides falls with increasing brittleness (25). Kruschov's rule (26) predicts that the wear resistance of a multiphase solid is the sum of the wear resistances of the components (taken by the volume percentages).
This rule, however, only holds for wear which is proportional to the load. As this is not the case for ceramics, application of this rule to metal-ceramic composites yields too high values of wear resistance.
The wear reducing mechanism is probably much more complex in relation to composites. A correlation between the ductile phase and the hard phase seems probable.
Osburn (27) claims that the matrix acts as a crack-stopper in cemented carbides, and the more ductile the matrix the more efficient its crack-stopping properties are. In relation to this he also finds that larger grains (more than 10 μm), provide more ductility because of larger distances between the grains for the same volume percentage of hard particles. On the other hand, fine grained materials will tend to be much more wear resistant to some forms of wear like frictional wear or erosion. Concerning abrasive wear, the size of the hard particles must be larger than the abrasives in order to prevent wear of the ductile phase.

Obviously a content of hard particles in metals can decrease the amount of abrasive wear. This concept is utilized in high speed steels which contain large amounts of VC, WC, Cr_3C_2 and Mo_2C (28).
A factor which is especially important to sintered objects is porosity. The bending strength and wear resistance are very much dependent on porosity. As little as 5% porosity can mean up to 270% higher wear loss (18).
Concerning adhesive wear, ceramics must be advantageous with metals as counterparts since the dissimilarity of the materials will prevent cold welding. An important factor to remember in relation to wear resistance is that wear can only be defined for a system.

The properties needed to avoid wear of the cermet will therefore depend on the counterpart.

A way to predict the bonding strength of a two-phase system is by calculating the adhesion strength, W_a.
From the following equation, it can be seen that the adhesion strength is proportional to surface tension at the interface of the liquid and gaseous phase and dependent on the contact angle, θ.

$$W_a = \sigma_{LV}(1+\cos\theta)$$

θ : contact angle
σ_{LV}: surface tension at the liquid/vapour interface
σ_{LS}: surface tension at the liquid/solid interface
σ_{SV}: surface tension at the solid/vapour interface

It is quite obvious in the case of liquid phase sintering since a contact angle exceeding 90 degrees will cause sweating, that is, the liquid will flow out.
The contact angle can be determined experimentally or from the following equiation:

$$\cos\theta = \sigma_{SV} - \sigma_{SL}/\sigma_{LV} \Rightarrow W_a = \sigma_{SV} + \sigma_{LV} - \sigma_{SL}$$

Naidich (29) has collected wetting data for several metalceramic systems. Table 1 shows some examples.

Table 1. Selected values of contact angles for different systems. Naidich (29).

Compound	Metal	t °C	Atmosphere	Contact angle, degrees
Cr_3C_2	Ni	1500	Argon	0
Mo_2C	Ni	1380	Vacuum	0
	Fe	1490	-	0
WC	Cu	1100	-	30
		1200	-	20
		1100-1200	-	20-7
	Ni	1380	-	0

It should be noted, that very different values can be obtained for the same system. Several authors have treated this problem. Naidich (29) points out the importance of equilibrium conditions. Another important factor is the roughness of the measured surface (30). Still the values give a hint of whether there is good or poor wetting.

Generally ceramics are wetted better in the order; nitride-boride-carbide. An oxide will be wetted by metals with large affinity for oxygen.

When improving the wetting, one should recall the equation for the work of adhesion, W_a, which should also be high in order to achieve good bonding. In order to satisfy both equations, σ_{LV} should be minimized. This can be done by addition of alloying elements in the melt.
The alloying will cause a decrease in σ_{LV} (31,32). The alloying element in the Fe-TiB_2 systems have been Mo (17), Ni/Cr (19).
Ni often has small contact angles to ceramics.
Other possibilities are to use metal as coating on the ceramic particles (32) or addition of elements which will cause chemical reactions at the metal-ceramic interface.

Prediction of the bonding strength of a system is only partly done by determination of the contact angle. Other factors such as formation of brittle compounds during sintering or cracking due to large differences in thermal expansion can still spoil the result. These factors must be estimated separately for each system.

Hard particles such as ceramics can be used in order to increase wear resistance (especially abrasive) of ductile metals. Possible ceramics could be: carbides, nitrides or borides of transition metals from the subgroups 4-6.
Some oxides and B_4C, BN and Si_3N_4 are also very hard.
The success of the material largely depends on the bonding between the phases. This can be improved e.g. by adding alloying elements to the metal.
In order to improve wear resistance it is very important that the materials have almost full density. Sinterforging is a possible method of obtaining this.

EXPERIMENTAL METHODS

The following materials were used:

Iron powder, ASC 100.29 from Höganäs
Titanium diboride, 50-100 μm
Chromium carbide, 10-50 μm
Graphite powder
Acrawax, lubricant
Ball Bearing Steel, 100Cr6, AISI E 52100
High Speed Steel, UHB ASP 30.

The iron was mixed with 1% graphite and 0.6% lubricant for 10 minutes in a twin shell. The hard particles were added and mixed for another 15 minutes. A hydraulic press was used for pressing the specimens and the compression ratio was about 2.2 for cermets with TiB_2 and 2 for cermets with Cr_3C_2. The force needed was approximately 7.6 t/cm^2 (50t) for TiB_2 and 5.4 t/cm^2 (35t) for Cr_3C_2. The geometry tested was that of the standard tensile strength testing bar of powder metallurgy materials.
The green bodies were sintered in pure hydrogen in a batch oven for 15 or 30 minutes at 1120 or 1150 degrees centigrade.

Lubricant was burned off at 400 degrees centigrade.
Density, tensile strength and Vickers hardness of the materials were measured according to standards.
For the wear testing, cubic bodies were prepared in the same way but only with addition of 10% TiB_2. The test specimens were sintered at 1120 degrees for 15 minutes. Obtained density was 5.6 g/cm^3.

Measurement of Wear Loss

No standard procedures were available for measurement of wear loss of composite materials. Very few standards for wear measurement are available at all. The ASTM B611 deals with abrasive wear of cemented carbides.
A special tribometer which is well suited for simulating different kinds of wear (adhesive and abrasive) during dry or lubricated conditions was used for the testing.
The tribometer is described in a report by Frees (33). The test procedure uses a block-on-ring type apparatus and the test specimens are showed in Figure 1.

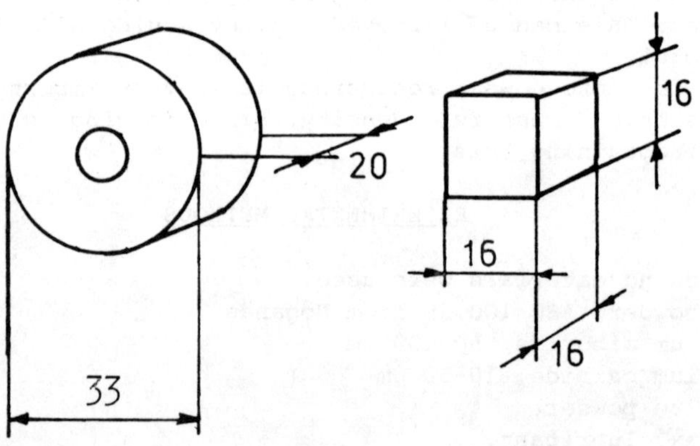

Figure 1. Test specimens for wear testing. All dimensions are in mm.

In order to get an overall view of the wear and friction properties and a trend of the working limits of the material, different lubrication conditions were tested.
The lubricants were water, oil-water emulsion (3% oil), and oilwater emulsion with silica (maximum size 25 or 250 μm) in the proportion of 0.026 g grit/ml.

The block was made of iron with 10% TiB_2 as described above. Blocks made of pure P/M steel (ASC 100.29 added 1% C) and ASP30 were also tested for reference. The ring was made of ball bearing steel.
The test was run for 20 minutes at loads of 100, 200 and 400 N and at a speed of the ring of 1 or 2.5 m/s.
The coefficient of friction was calculated during the experiment and plotted as a function of time. The wear of the blocks was calculated by microscopically measuring the width of the scar and thereby estimating the wear volume. The wear rate was given in mm^3/m.

RESULTS AND DISCUSSION

Mechanical properties

The tensile strength as a function of volume content of hard particles can be seen in Figure 2. Each point is a mean of four measurements.

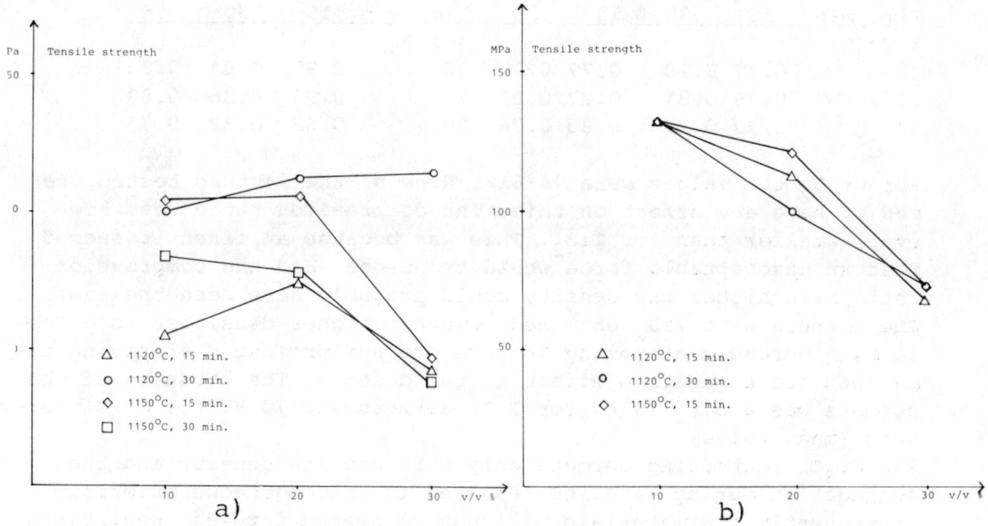

Figure 2. Tensile strength (MPa) as a function of volume content of hard particles. a) Cr_3C_2, b) TiB_2.

For Cr_3C_2 (Figure 2.a) no clear connection between the factors can be seen. Though it seems that the best results can be obtained either by prolonging the sintering time or by increasing the temperature. It must be noticed that the variance of the results is remarkable. The standard deviation is between 5 and 47% and is probably due to segregation during handling.
For the TiB_2 cermets (Figure 2.b) it is quite clear that tensile strength decreases with increased amounts of ceramic particles,

and long sintering time also seems to have a negative effect. Still, the standard deviation is high, between 4 and 36%. The production conditions are obviously not completely reproducible. Generally the tensile strength obtained are not satisfactory compared to the theoretically obtainable values. Höganäs declares that the sintered specimens of ASC 100.29 will have a tensile strength between 200 and 350 MPa depending on porosity and carbon content.

Evaluation of the microstructure revealed that most of the carbon had disappeared during sintering and the hardness testing showed that the core was approximately 30% harder than the surface. The decarbonization could be due to water vapour in the sintering atmosphere. Also the porosity was in some cases high. These facts account for the low values obtained.
The densities compared to the theoretical density can be seen in Table 2.

Table 2. Density compared to theoretical density. Each value is a mean of four measurements. a) Cr_3C_2, b) TiB_2.

Temp. ^{o}C	1120		1150	
Time min.	15	30	15	30
% Cr_3C_2				
10	0.77	0.78	0.77	0.80
20	0.79	0.81	0.77	0.81
30	0.77	0.77	0.80	0.76

a)

Temp. ^{o}C	1120		1150
Time min.	15	30	15
% TiB_2			
10	0.90	0.84	0.84
20	0.93	0.88	0.82
30	0.83	0.82	0.73

b)

For Cr_3C_2 the values were 76-81%. None of the factors tested seemed to have any effect on this. The compression ratio was, however, smaller than for TiB_2. This was because at first it seemed that an unacceptable force would be needed. Had the compression ratio been higher the density could probably have been the same. The cermets with TiB_2 obtained somewhat higher densities (see Table 2). Increased sintering temperature and prolonged sintering time both had a negative effect on the density. The hardness of the cermets was about 130 HV for TiB_2-dermets and 70 HV for Cr_3C_2 cermets (mean value).
The Cr_3C_2 containing cermets only obtained low density and the segregation during handling resulted in inhomogeneous materials. Consequently the materials will not be tested for wear resistance.

If segregation can be avoided e.g. by choosing another particle size, and better sintering conditions will result in high density, then these materials will be further investigated later. The TiB_2 containing cermets showed high density and very reproducible results for the tensile strength and the microstructure showed that the hard particles were homogeneously embedded in the iron matrix.

However, cermets containing more than 10% of TiB_2 had a tendency to grow bubbles on the sides during sintering. This was probably because of a reaction between iron and boron. The melting point of the eutectic between iron and boron at 1161 degrees centigrade was, however, not reached during sintering. Because of these facts only the cermet with 10% TiB_2 was chosen for the wear resistance experiments.

The temperature and sintering time did not seem to have an effect on tensile strength, but the highest density was obtained with 15 minutes and 1120 degrees centigrade during sintering. This combination was therefore chosen for further experiments.

Testing of wear and friction behaviour

The cermet was tested towards ball bearing steel in water. This only resulted in little wear on as well the block as the ring. Since water can be considered as a bad lubricant and the working limits of the material were sought, there was not any need to repeat the experiments in oil. However in order to minimize the adhesive wear when abrasives were added, the experiments with abrasives were run in oil-water emulsion with pure oil-water emulsion as a reference.

Friction properties

When stating the friction properties one must bear in mind, that the friction is not only dependent on the system (i.e. block-ring-lubricant-additives) but also on speed, load and time. Figure 3 shows typical examples of the coefficient of friction as a function of time in different lubrication regions.

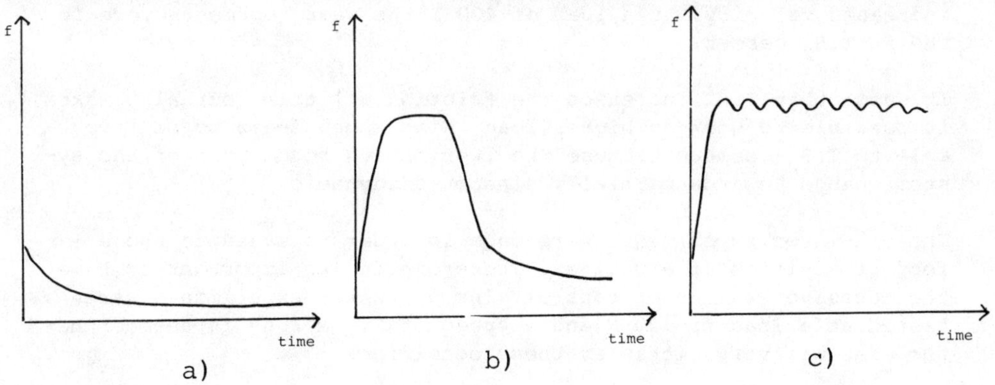

Figure 3. Coefficient of friction versus time in different lubrication regions.

a) Partial elastohydrodynamic lubrication.
b) Boundary lubrication.
c) Virtually unlubricated contact.

Figure 3.b is the typical curve of the coefficient of friction where boundary lubrication exists. The friction is high in the beginning, but running-in wear results in a low level of friction (0.1-0.3) after a few minutes. When elastohydrodynamic lubrication exists, the running-in time is so short, that it can not always be detected. In the case of virtually unlubricated conditions the coefficient of friction will stay at a high level (0.3-1) throughout the experiment. For more details about wear and friction, see de Gee, Begelinger and Salomon (34).

Here the coefficient of friction will be defined as the stable level that is reached after running-in.
Table 3 shows the coefficient of friction for iron and iron-TiB_2 cermet towards ball bearing steel in water.

Table 3. Coefficient of friction for a) iron-TiB_2 cermet and b) P/M iron in water at different loads and speeds. Iron was not tested at a higher load than 200 N since the system began to scuf after only 1 minute at P=200 N and v=2.5 m/s.

a)

P\v	1 m/s	2.5 m/s
100 N	0.34	0.06
200 N	0.47	0.06
400 N	0.56	0.47

b)

P\v	1 m/s	2.5 m/s
100 N	0.3	0.03
200 N	0.4	(0.3)

The friction increases with increasing load, but decreases with increased velocity. At a load of 400 N the wear becomes severe for the Fe-TiB_2 cermet.

It seems that TiB_2 increases the friction a little, but also makes it possible to work at higher load. High speed seems to be favourable to TiB_2-cermets because the lubrication conditions of the system change towards partially elastohydrodynamic.

The oil-water experiments were made in order to evaluate the effect of addition of abrasives. Therefore it was important to have the adhesive wear under control. The oil-water experiments were tested at a load of 120 N and a speed of 2.5 m/s as it seemed that the wear was very little at these conditions.

In oil-water the lubrication conditions were nearly elastohydrodynamic, but the coefficient of friction was somewhat higher than in water. Table 4 shows the coefficients of friction for the tested systems.

Table 4. Coefficient of friction for P/M iron, iron-TiB_2 cermet and ASP30 steel towards ball bearing steel at a load of 120 N and a speed of 2.5 m/s in different lubricants.

	water	oil-water	oil-water silica 25 μ	oil-water silica 250 μ
Fe	0.03	0.05	0.06	0.11
Fe-TiB_2	0.06	0.095	0.12	0.1
ASP 30	0.02	0.03	0.09	-

Table 4 shows that the coefficient of friction is lowest in water, and increases in oil-water emulsion and with addition of abrasives. The size of the abrasive did not seem to have much influence on the results.
Again ASP30 achieved the lowest value and the TiB_2-cermet the highest. When silica was added the coefficient of friction tended to fluctuate. The fluctuation was especially severe for the iron-test, so the result was not reliable.

Wear

Table 5 shows the wear rates for tests in water.
Table 5. Wear rates in $mm^3/m \cdot 10^4$. The wear for the Fe-TiB_2 cermet at 1 m/s and 200 N was not measured because of misalignment of the test specimen.

	100 N		200 N		400 N	
	1 m/s	2.5 m/s	1 m/s	2.5 m/s	1 m/s	2.5 m/s
Fe	4.9	0.41	39	$4.5\ 10^3$	-	-
Fe-TiB_2	1.5	0.98	-	0.98	168	28
ASP 30	-	0.001	-	-	-	-

At high loads and high speed the advantages of adding TiB_2 to iron are remarkable. Still, the special steel ASP30 which was tried as a reference, is far better.
In oil-water the results are as follows (see Table 6).

Table 6. Wear rates in $mm^3 \cdot 10^4$ for TiB_2 cermet and ASP30 in oil-water with addition of silica. 100 N.2,5 m/s.

	water	oil-water	oil-water silica 25 μ	oil-water silica 250 μ
Fe	0.41	0.042	11	11
Fe-TiB_2	0.98	0.28	3.3	0.41
ASP 30	0.001	$1.5\ 10^{-5}$	0.12	-

Compared to water, the oil-water emulsion resulted in much less wear. In both lubricants ASP30 was superior, and iron somewhat better than iron-TiB_2. When silica was added wear was increased. The increase was less for Fe-TiB_2 than for pure iron.
However, it was rather difficult to estimate wear because the wear scars were not always quite symmetrical. This was due to the fact that it was very difficult to place the specimens right. Also the surface was reason to some problems because it had to be planar and it was very hard to grind the cermets.

Further work

Actually it seemed that the silica did not act as a abrasive but only as erosive. Probably another test method must be chosen for the evaluation of the abrasive properties. This could be either the ASTM standard B611 or a pin-on disc method where the disc is grinding paper. It is also necessary to test other references.

It could be interesting to investigate the wear properties of the cermet at more different conditions e.g. at low speed, high load, in oil, and with severe abrasives. The wear resistance and mechanical properties of the cermet could be improved by obtaining higher density by sinterforging. Experiments with comparable composite materials made at the Technological Institute of Denmark have shown that porosity can be reduced to approximately 3% by means of sinter forging (35).
Another way of improving especially the tensile strength could be by alloying elements to the iron.

Acknowledgement

Thanks to the Danish Technical Research Council, which supported the work financially.

References

1. Tinklepaugh, J.R.; Crandall, W.B. Cermets, Rienhold, 1960.
2. Kieffer, R.; Eipeltauer, E.; Gugel, E. Neue Entwicklungen auf dem Gebiet der Cermets, Ber. Dtsch. Keram. Ges. 46, H.9 (1969), p. 486-492.
3. Nath, D.; Rohatgi, P.K.; Biswas, S.K. Wear characteristics and bearing performance of Aluminium-mica particulate composite material, Wear, 60 (1980), p. 61-73.
4. Surappa, M.K.; Prasad, S.V.; Rohatgi, P.K., Wear and abrasion of cast Al-alumina particulate composites, Wear, 77, (1982), p. 295-302.
5. Anand, K.; Kishore On the wear of aluminium-corundum composites, Wear, 85, (1985), p. 163-169.
6. Surappa, M.K.; Rohatgi, P.K. Preparation and properties of cast aluminium-ceramic particle composites Jour.Mat.Sci. 16 (1981), p. 983-993.
7. Bhansali, K.J.; Mehrabian, R. Abrasive wear of Aluminium-Matrix Composites, Jour. Met. Sept. (1982), p. 30-34.
8. Hosking, F.M.; Portillo, F.F.; Wunderlin, R., Mehrabian, R. Composites of aluminium alloys: Fabrication and wear behaviour, Jour.Mat.Sci., 17, (1982), p. 477-498.
9. Banerji, A.; Prasad, S.V.; Surappa, M.K.; Rohatgi, P.K. Abrasive wear of cast aluminium alloy-zircon particle composites, Wear 82 (1982), p. 141-151.
10. Jangg, G.; Oppenheim, H.; Reichmayr, J. Verbesserung des Verschleissverhaltens von Al-werkstoffen, Z. Werkstofftech.18, (1987), p. 87-99.
11. Roesen, L. Udvikling af slidfaste P/M-materialer til jordbearbejdning, Technological Institute, Denmark, 1986.
12. Champagne, B.; Angers, R. Laboratory and Field Testing of Wear-resistant Composites, Wear of Materials, Vancouver Canada, 14-18 Apr., 1985, ASME, p. 142-148.
13. Method for Producing Iron-base Sintered alloys with high Density, U.S. Patent No.: 3,859,085 1975.
14. Krawitz, A.; Drake, E.F.; Peck, J. Metallography of Tungsten Carbide Iron-Nickel Base Cemented Carbides. Int.Jour. Powd. Metal. Powd.Tech., 15,1 (1979), p. 67-71.
15. Fe-based sintered material excellent in wear resistance. Japan Patent No.: 59-16951 (A), 1984.
16. Fe-based sintered material excellent in wear resistance. Japan Patent No.: 59-16952 (A), 1984.
17. Kitsai, A.A.; Tsyganova, T.V.; Ordan'yan, S.S. Infiltration of Titanium carbide and titanium diboride base cermets with Metallic melts, Sov.Powd.Metal.Met.Ceram., Vol. 24, No.7, (271) 1985, p. 58-62.
18. Champagne, B.; Dallaire, R.H. Wear resistance of TiB_2-Fe Cermets, Wear of Materials, Vancouver, Canada, 14-18 Apr. 1985,

ASME. p. 33-38.
19. Yuriditskii, B.Yu.; Pesin, V.A.; Ordan'yan, S.S. Change in the fine structure of the titanium diboride occuring during the sintering of a TiB_2-Fe (Mo) cermet, Sov.Powd.Metal.Met.Ceram, 21,(1982), p. 280-282.
20. Sleptsov, V.M.; Kirilenko, S.N.; Naumenko, V.Ya.; Verkhoturov, A.D.; Neschetnyi, V.A. Some properties of titanium carbide-steel cermets with molybdenum additions, Sov.Powd. Metal.Met.Ceram, No. 8 (140), Aug. 1974, p. 45-49.
21. Kiparisov, S.S.; Khocholava, Z.D.; Narva, V.K. The Influence of Titanium Carbide and Bonding Phase Composition on the Properties of Titanium Carbide-Steel Cermets, in Sintering Theory and Practice, Elsevier, 1982, p. 511-518.
22. Kitsai, A.A.; Kozlovskii, L.V.; Kuchuk, N.V.; Ordan'yan, S.S. Production of TiC-base Cermets of Different Binder Concentrations in Alternating Layers, Sov.Powd.Met.Metal Ceram., No. 7 (235), July 1982, p. 39-42.
23. Takagi, K.; Ohira, S.; Ide, T.; Watanabe, T.; Kondo, Y. New P/M Iron Containing Multiple Boride Base Hard Alloy, Modern Develop.Powd.Metal. 16,17-22 June, (1984), p. 153-166.
24. Zum Gahr, K.H. Abrasivverschleiss grob-zweiphasiger metallischer Werkstoffe, Zeitschr. Metallkunde, Bd.75 (1984) H.11, p. 881-889.
25. Artamonov, A.Ya.; Bovkun, G.A. Some Aspects of the Abrasive Wear of Transition Metal Carbides, in Refractory Carbides ed. by Samsonov, G.V. Consultants Bureau, 1974.
26. Garrison, W.M. Kruschovs Rule and the Abrasive Wear of Multiphase Solids, Wear, 82 (1982), p. 213-220.
27. Osburn, H.J. Wear of Rock-cutting Tools, Powd. Metal. vol. 12 1969, p. 471-502.
28. Brewin, P.; Coleman, D.S. The wear behaviour of sintered high speed tool steels, Proc. of the 1986 Int. Powd. Metal Conf. and Exh. Düsseldorf, 7-11 July 1986, p. 229-232.
29. Naidich, Ju.V. The Wettability of Solids by Liquid Metals, Prog.Surf.Memb.Sci. vol.14 (1980), p. 353-484.
30. Passerone, A. Interfacial Phenomena in Metal-Ceramic Systems, Mat.Chem.Phys., 15, (1986), p. 263-279.
31. Delannay, F.; Froyen, L.; Deryttere, A. Review - The wetting of solids by molten metals and its relation to the preparation of metal-matrix composites, Jour.Mat.Sci.22 (1987), p. 1-16.
32. Banerji, A.; Rohatgi. P.K.; Reif, W. Role of wettability in the preparation of Metal-Matrix Composites (a Review), Metall, (1984), p. 656-661.
33. Frees, N. Wear reisstant surface layers - the solving of industrial wear problems. Lab. of Proces and Production Engineering, Tech. Univ. of Denmark. Danish Ph.D. Thesis. 1986.
34. de Gee, A.; Begelinger, A.; Salomon, G. Failure Machanisms in Sliding Lubricated Concentrated Contacts, Proceedings of the 11th Leeds-Lyon Symposium on Mixed Lubrication and Lubricated Wear, 1984, Butterworths, London.

Duplex 4600 Powder Forged Structures
Processing, Properties and Mechanics Analysis[1]

M.N.H. Ahmed*, M.J. Koczak* and M.I. Negm**

*Drexel University, Department of Material Engineering
Philadelphia, Pennsylvania, 19104
**Al-Azhar University, Mech.Eng.Dept., Cairo, Egypt

Abstract

Forming of duplex structures from laminated structure offers the promise for producing components where each constituent makes synergistic contribution, i.e.where the combined action transcends the sum of the individual properties. The layered structure concept can provide for improved performance unattainable by the conventional powder forging techniques. By the selective utilization of high performance powders, i.e.wear resistance surfaces , economical components can be manufactured. Several forming processes, e.g. extrusion, forging, rolling, can be applied to manufacture duplex or laminated billets and/or cylindrical shapes.

Two structures are presented in the present work; (i) a coaxial cylindrical duplex structure subjected to axial loading represents a symmetrical duplex structure condition and (ii) a laminated beam duplex structure subjected to flexure loading represents an asymmetrical duplex structure condition. In each case the subject will be considered from three point of views, e.g. processing, evaluation of the apparent mechanical properties and a structural mechanics analysis.

The duplex structures are made using Hoeganaes Ancorsteel powder 46OOV. The graphite addition is provide for a 4620 and 4660 alloy chemistries. The volume fraction of the 4620 and 4660 alloys will also be a variable, as well as, the effect of the alloy constituent placement. In the coaxial cylindrical duplex structure, the case and core of both the 4620 and 4660 will be considered. For the duplex layered beam structure, during flexure, the tensile and compressive response of the 4620 and 4660 surfaces will be ascertained.

Introduction

The performance and cost effective demands of metallic components imposed by competitive technologies have become diverse, that it is necessary to combined several metals or alloys into duplex structure where each constituent makes an individual synergistic contribution. Highly developed powder forged (P/F) technologies provides for an

[1]This paper is based on a thesis to be submitted by M.Nour.H. Ahmed in a partial fulfillment of requirements for the degree of Ph.D. at Al-Azhar University, Cairo, Egypt.

approach for the manufacture of such duplex components.

Researchers have published results under different terminologies, e.g. duplex, laminated, two layers, dual property, bi-metallic and macrocomposites have been used to identify such components. Investigators have considered the manufacture techniques of duplex components [1,2,3], joining interface structure [4,5], strength [5,6,7] and forming of post compacted duplex preforms [8,9]. A large variety of manufacturing techniques, e.g. forging [8], extrusion [9], HIP [1,2], high speed petro-forging [8,9], cold sintering [5], infiltration [7,10] etc. have been used.

The subsequent duplex structure provides for economical or performance benefits. The economical considerations mainly concern material saving and the selective use of expensive high performance alloys. The performance considerations can be structural [1], physical [11] or selected mechanical properties [2] concepts.

A classification for studying the duplex structure can be considered by constituent arrangement, i.e. symmetrical or asymmetrical structures and the elastic compatibility, i.e. the equality or inequality of the modulus of elasticity of the two constituents. Additional considerations involve coefficient of thermal expansion, phase transformation volume changes, yield strengths and work hardening coefficients. Of natural importance is the type of loading coupled with the probable applications of the specified duplex structure. Most of the suggested applications of the laminated beam duplex structures (e.g. turning tools, bi-metallic elements, etc.) are subjected to flexure or lateral loading, i.e. bending, in the plane of symmetry. For duplex laminated sheets the biaxial state of stress is to be considered. In contrast, the suggested applications of the coaxial cylindrical structures are subjected to either axial loading (e.g. rods, bolts, etc.) or torsional loading (e.g. reamers, end-milling tools, etc.). In the case of hollow coaxial duplex cylinder (e.g. tubes, high pressure pipes, etc.) with high internal pressure a triaxial state of stress must be considered. For flat coaxial duplex disc structure (e.g. gears, turbine wheels, horizontal milling tools, disc-saw, etc.), hub and radial residual stresses may be created. Extension of the present study to asymmetrical triplex and sandwich structures involves the case of asymmetrical lamella, e.g. surface heat treated duplex structure, and cantilevers duplex structures involving gear teeth.

A key ingredient for duplex structures is the cross section area ratios of the individual component and their placement. The area ratio "f" governs the material cost and is defined as the cross sectional area of a one constituent "A_i" to the total cross sectional area of the duplex structure "A_t" and a second area ratio "$\bar{f}$" is similarly defined

$$f = \frac{A_i}{A_t} \quad \text{and} \quad \bar{f} = \frac{A_t - A_i}{A_t} = 1 - f \qquad (1)$$

According to the above definitions, the area ratios "f" and "$\bar{f}$" vary from $0 \rightarrow 1$

In the case of coaxial cylindrical duplex structure, as shown in Figure 1a, in defining the area ratio "f", the numerator is selected to represent the area of the core, and the area ratios defined by Equations (1) are to be reduced to,

$$f = \frac{d^2}{D^2} \quad \text{and} \quad \bar{f} = \frac{D^2 - d^2}{D^2} \qquad (2)$$

where, d = the diameter of the core constituent, and
D = the outer diameter of the duplex structure.

In the case of layered beam duplex structure, as shown in Figure 1b, the area ratios "f and $\bar{f}$" are reduced to,

$$f = \frac{t}{b} \quad \text{and} \quad \bar{f} = \frac{b - t}{b} \tag{3}$$

where, t = the thickness of constituent "A", and
b = is the total thickness of the duplex structure.

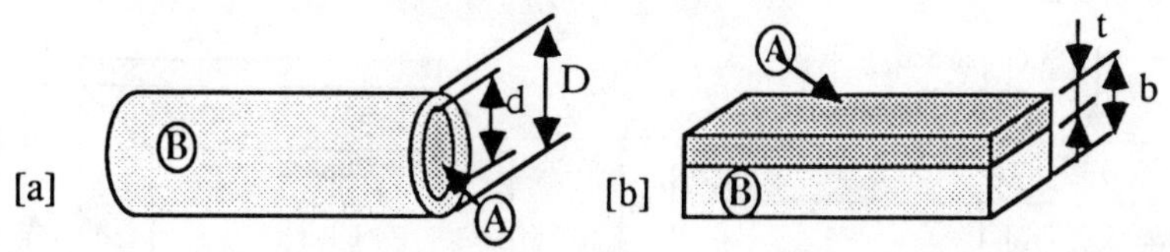

Figure 1. Symmetrical [a] and asymmetrical [b] duplex structures

For convenience in defining "f" the numerator "t" will be selected such that, constituent "A" has highest amount of expansion e_a.

The coaxial cylindrical duplex structure will be processed by the hot extrusion of a machined and assembled duplex preforms. The duplex beam-like structure will be obtained by the hot repressing of two layer presintered preforms.

Duplex Processing and Heat Treatment

Processing coaxial preforms by extrusion:

The powder preforms were compacted at 420 MPa (60 ksi)with Hoeganaes prealloyed Ancorsteel 4600 powders with graphite additions for the 4620 and 4660 blends. Presintering was performed for 30 minutes at 1120°C (2050°F) in dissociated ammonia atmosphere. The preforms of 50mm x 65mm x 125mm (2" x 2.5" x 5") had a density of 6.5 gm/cm^3. The duplex preforms were machined and assembled from two presintered isotropic billets as shown in Figure 2. The cylindrical preforms were canned and outgassed at 400°C (750°F) and sealed. Preheating of the canned preforms at 1200°C for extrusion, results in can swelling, possibly due to the trapped air. In a second trial, cans were outgassed at 500°C (930°F), back filled with argon and sealed. Prior to extrusion the sealed cans were opened to atmosphere and placed in a nitrogen preheat atmosphere. Preheating at 1200°C (2190°F) ranged from 60 to 90 minutes before extrusion. A liner of 51.8mm (2.040") and a die of 12.5mm (0.500"), at 425°C (800°F) were used for all extrusions. The extrusion speed was 0.0275 m/sec (65 ipm) with Polygraph as a lubricant. There were no incompatibility problems in the processing and extrusion.

Processing laminated preforms by compaction and hot repress forging:

The die was first half filled by the 4660 powder, tapped slightly and then completely filled using 4620 powder. The duplex briquettes were compacted at 420 MPa (60 ksi). Presintering was for 30 minutes at 1120°C (2050°F) in dissociated ammonia atmosphere. Preforms with dimension of 49.8mm x 63.5mm x 121.9mm (1.96" x 2.5"x 4.8") with a

6.5 gm/cm^3 density were obtained. The billets were subsequently hot repressed at 980°C (1800°F). The final forged billet dimensions are 52.1mm x 51.3mm x 127.25mm (2.05" x 2.02" x 5.01") all of area ratio f ≈ 0.5. For complete densification some lateral flow must allowed for improved ductility and toughness [11,12,13].

Repressing was selected as an initial investigation for the forging process , since a minimum of flow will provide for a minimum of forming incompatibility. The compaction, sintering, forging and chemical analysis were provided by Hoeganaes Corp. Pilot Forging Plant, Riverton, N.J.

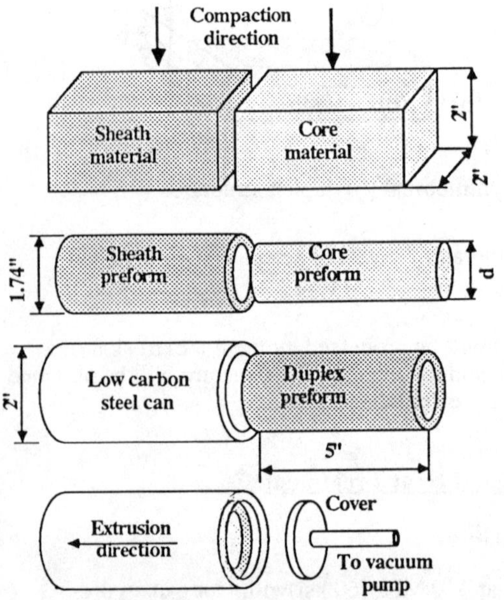

Figure 2 Assembling and canning a duplex preform for extrusion.

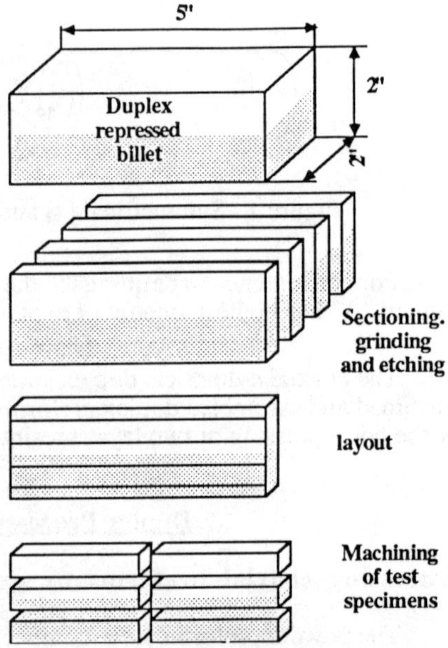

Figure 3 Preparation of isotropic and duplex specimens.

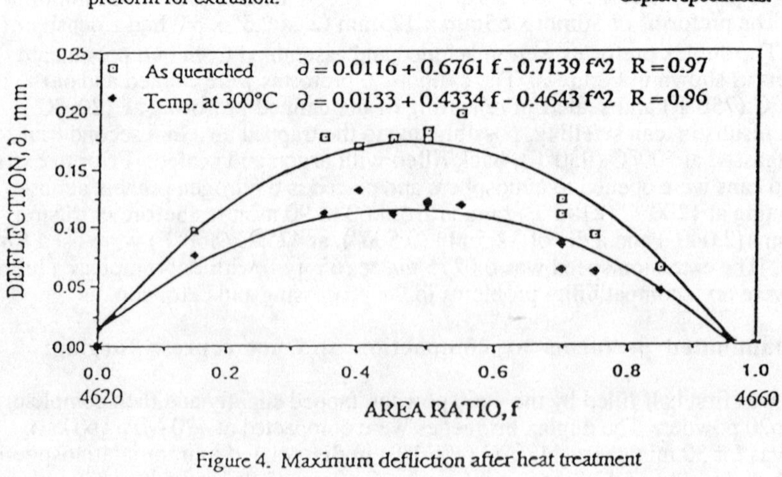

Figure 4. Maximum deflection after heat treatment

Heat treatment and specimen preparation coaxial duplex structures:

The properties of the coaxial cylindrical duplex structure were evaluated by a uniaxial compression tests. Symmetric structures were consider with a 4620 core and 4660 case as well as reverse geometry.

Extrusion samples of 150mm (6") length were austenitized at 850°C (1560°F) for 30 minutes, oil quenched and tempered at 200°C (390°F). After heat treatment, the compression test specimens had been prepared by removal of the can material by cylindrical grinding to the predetermined outer diameters. Two groups, one with a 4620 case with a 4660 core and the reverse situation were evaluated with different area ratios. The austenitizing and tempering temperatures for optimum combination of properties in the P/F 4600 system had been determined separately [14]. These two groups were tempered at 200°C. Aspect ratio, i.e. height to diameter ratios (H/D), equal to one were selected for all of the compression specimens. Three specimens of H/D = 10 mm/10 mm and another three of H/D = 6.5mm/6.5mm. had been prepared to cover the range of the area ratio "f" from 0 to 1.0. Additionally, two isotropic, 4620 and 4660 specimens, i.e. where f = 0 and 1.0 respectively were also evaluated.

Heat treatment and specimen preparation laminated duplex structure :

The mechanical properties of the layered beam duplex structure were evaluated by a three point bend test. The specimens were 10mm x 10mm x 55mm (0.4" x 0.4" x 2.2") standard unnotched Charpy specimens. The forged billets were sectioned as shown in Figure 3. and finished by surface grinding. An etchant, e.g. 2% Nital, has been used to distinguish the 4620 and the 4660 layers. The machined specimens were marked to show its exact position in the forged billet.

All specimens were austenitized for 30 minutes at 850°C (1560°F) in a tube furnace using an argon atmosphere and oil quenched and through hardened successfully. The effect of the residual stresses, i.e. curvature of the duplex specimens, due to the difference in volume change, i.e. different $\alpha \rightarrow \gamma$ phase transformations temperatures, was measured using traveling indicator of 0.0025mm (0.0001") sensitivity. The deflection measurement is plotted vs. the corresponding area ratio in Figure 4. The specimens were tempered in muffle furnace for one hour in stainless steel packs at 300°C. The deflection was measured again and shown in Figure 4. All the specimens were ground from all sides to the final dimensions to insure a completely parallel surfaces and to remove the effect of possible decarburization during either the austenitizing or the tempering processes. Also, edge rounding was necessary to reduce any stress concentrations.

Experimental Procedures and Results

Microstructure and hardness profile:

Two groups of extruded specimens had been prepared. The first group had been annealed at 850°C (1560°F) for one hour and furnace cooled. The second group had been austenitized at 850°C (1560°F) for 30 minutes and quenched in oil. As shown in Figure 5 limited amount of carbon interdiffusion has occurred,from the 4660, to the lower carbon 4620 level. The level of the carbon redistribution as determined by radial microhardness profiles is shown in Figure 6 for the annealed and as quenched conditions. These microhardness results determined the exact area ratios of the 4620 and 4660 constituents for the extruded rods. Similar procedures were performed with the laminated repressed specimens.

[a] 4660 sheath, X500

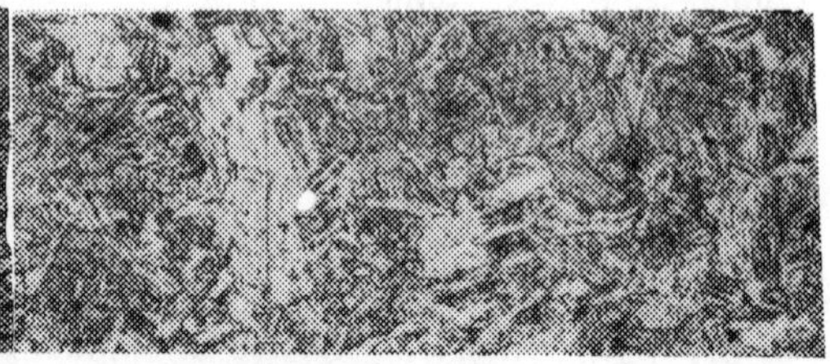

[b] 4620 core, X500

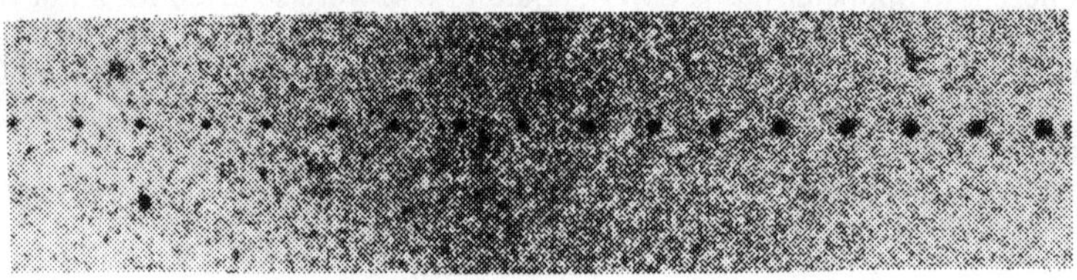

[c] Transition zone, X32

Figure 5. Microstructure of 4660 [a], 4620 [b] and transition [c] in the as quenched condition

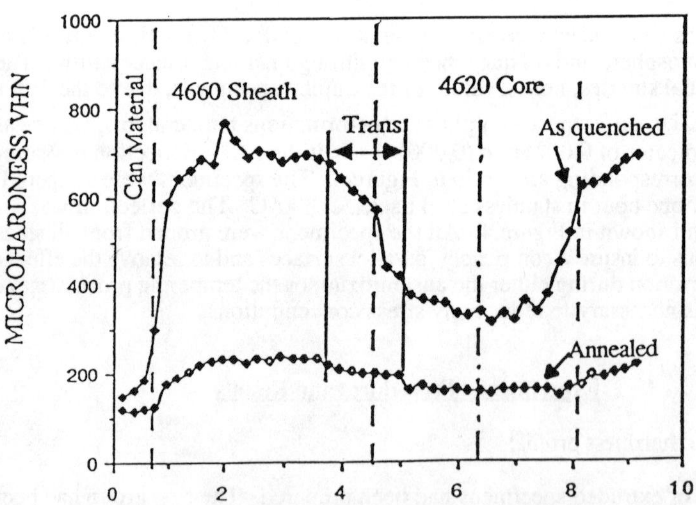

Figure 6. Microhardness profile of extruded duplex rod, cross section taken with the can material.

Compression response of the coaxial duplex specimens:

In order to determine the compressive stress strain behavior of the coaxial duplex structure, a uniaxial compression test was used. The compression test, between two smooth flat dies, has been carried out on an Instron testing machine with a cross head speed of 1 mm/min. The diameteral and axial strains were measured after each incremental loading. True stress true strain curves of different area ratio are shown in Figure 7. The 0.2 % yield strength "σ_y" is plotted vs. the area ratio "f" in Figure 8. For the case of 4660 core and 4620 sheath and the case of 4620 core and 4660 sheath, the latter case is considered with negative area ratio "f". In this type of representation The 4620 chemistry occupies the (f = 0.0) position, while the 4660 chemistry occupies the (f = 1 and f = -1) positions.

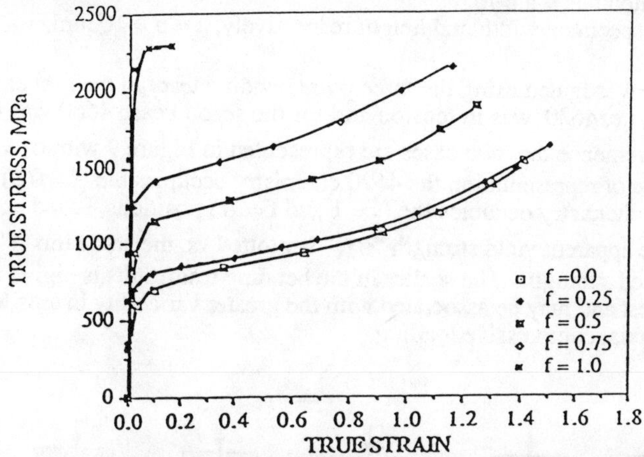

Figure 7. True stress true strain curves for extruded 4620 sheath covers 4660 core, H/D = 6.5/6.5. loaded to a max. load of 22.5 tons.

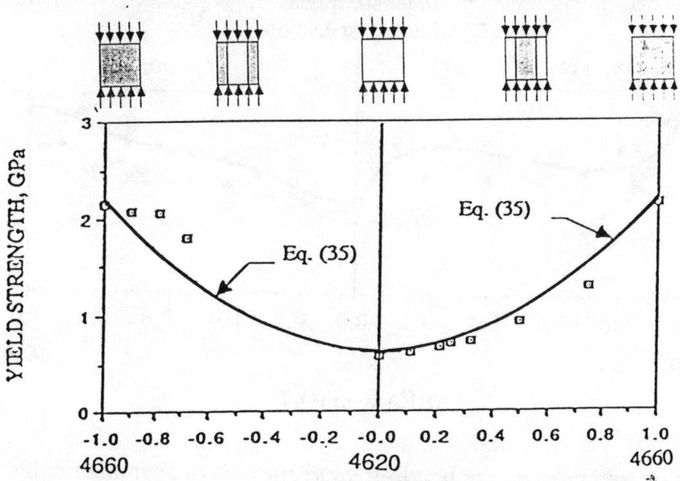

Figure 8. The effect of the area ratio on the apparant yeild strength of coaxial cylindrical duplex structure supjected to compressive loading.

Flexure evaluation of laminated structure:

The three point bend tests have been carried out on an Instron testing machine with a span of 40 mm and cross head speed of 1 mm/min. The loading was in the same direction of the forging load.

The flexure yield stress in bending "σ_{yb}" was calculated by,

$$\sigma_{yb} = M / Z \tag{4}$$

where,

M : the bending moment = P_{yb} L/4
P_{yb} : bending yield load.
L : specimen span = 40 mm
Z : section modulus = a $b^2/6$
a and b : the specimen width and height respectively, a = b = 10 mm.

Two cases will be investigated using the three point bending test. In the first case the more ductile constituent, i.e. 4620, was in tension and for the second case 4660 was under tensile loading. For convenience the two cases are represented in Figure 9 with σ_y vs. area ratio "f". In this type of representation the 4620 chemistry occupies the (f = 0.0) position, while the 4660 chemistry occupies the (f = 1 and f = -1) positions, based upon specimen geometry. The apparent yield strength "σ^*_y" is plotted vs. the area ratio "f" in Figure 9 with the estimated strength. The scatter in the bending test results is higher than that in the compression test and may be associated with the greater variability in tensile failure in flexure vs. uniaxial compressive loading.

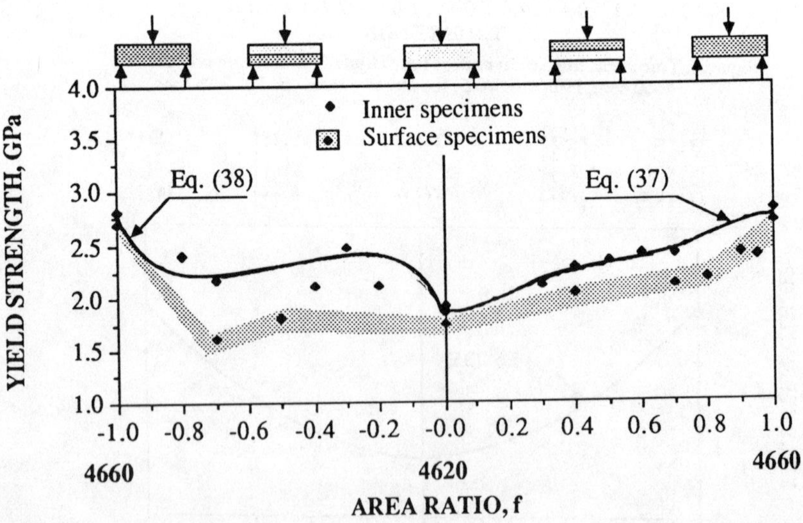

Figure 9. The effect of area ratio on the bending yield strength of asymmetrically laminated duplex beam.

Theoretical analysis and discussion

The present analysis considers the case of elastically compatible duplex structure independent of the heat treatment regime, which is an initial simplified condition. In this analysis, the modulus is considered constant since it is only slightly affected by alloying additions, heat treatment or cold work [15]. A general case, where the modulus of elasticity is considered as a variable, will be reported in future study [16].

The mechanics of the duplex structures are aimed at estimating the apparent mechanical properties of the duplex structure from the known mechanical properties of its two constituents. The presence of residual stresses can influence the reaction of a material to external applied stresses. In the tension test, the region of a specimen containing high tensile residual stresses will yield plastically at a lower value of applied stress than a specimen without residual stresses. Conversely, compressive residual stresses will increases the yield stress.

Since P/F steel structural parts are generally used in the hardened and tempered conditions, the two constituents of the duplex structure will possess different volume changes and give rise to residual stresses. In duplex structures, there are considerable residual stresses after any heat treatment processes, e.g. austenitizing, quenching, tempering, etc. These residual stresses are due to the difference of the coefficient of thermal expansion and $\alpha \rightarrow \gamma$ transformation volume change of the two constituents. These residual stresses can be responsible on delaying or promoting the failure of the duplex structure. For example, the apparent yield strength of the duplex structure depends on the algebraic sum of the residual stresses and the applied stresses. The rule of mixtures may be applied to estimate the apparent strength.

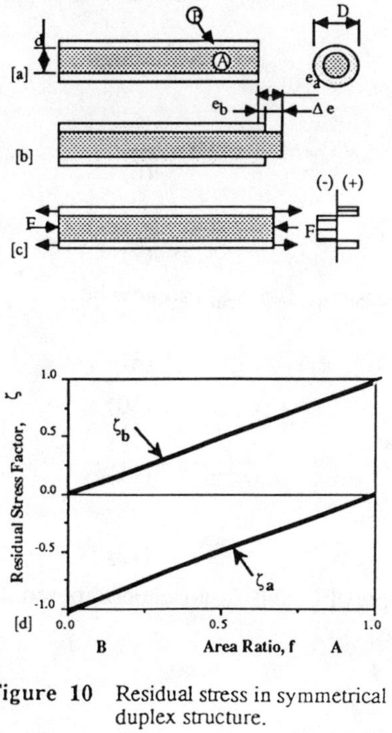

Figure 10 Residual stress in symmetrical duplex structure.

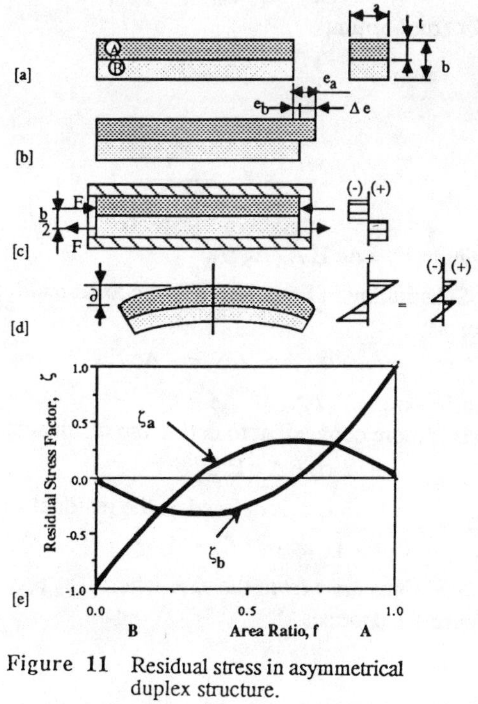

Figure 11 Residual stress in asymmetrical duplex structure.

An elementary analysis [17] can be made by assuming first that the two constituent (A and B) of the duplex structure are not bounded. When subjected to the volume change, they will elongated by different amount, as in Figure 10.b and Figure 11.b, producing the strain difference "$\Delta e = e_a - e_b$ ", i.e. "Δe" is the longitudinal difference per unit length due to phase transformation (e.g. austenite → martensite), thermal contraction (cooling down from the austenitization temperature).and/or densification (if quenching takes place directly after the sintering process). In tempering (martinsite→ tempered martensite + bainite), a considerable relaxation will expected.

It will now be assumed that this difference is to be eliminated by applying an imaginary axial force "F" to each constituent. For equilibrium ($\Sigma F=0$), the two forces must be equal and have opposite signs since there are actually no external applied forces. In the case of symmetrical duplex structure these forces will cause an axial strain, while is the asymmetrical structure there will be also a bending strain.

Symmetrical duplex structures:

In this case, the axial stresses exerted on the inner and outer cylinders, i.e. σ_{ra} and σ_{rb} respectively are related to the strain through Hook's law

$$e_a = \sigma_{ra} / E_a = F_a / E_a A_a \qquad (5.a)$$

$$e_b = \sigma_{rb} / E_b = F_b / E_b A_b \qquad (5.b)$$

But as mentioned before only the case of equal modulus of elasticity will be considered i.e., $E_a = E_b = E$

but, $\Delta e = e_a - e_b$

and for equilibrium,

$$F_a = - F_b = F$$

Hence,

$$\Delta e = \frac{F}{E} \left(\frac{A_t}{A_a A_b} \right) \qquad (6)$$

and yields, $F_a = \Delta e \, E \, A_a A_b / A_t$ (7)

Substituting in Equations (5), the axial residual stress (σ_{ra} and σ_{rb}) can now be written as,

$$\sigma_{ra} = F_a / A_a = - \Delta e \, E \, \bar{f} \qquad (8)$$

Similarly $\sigma_{rb} = F_b / A_b = \Delta e \, E \, f$ (10)

It is more convenient to define the residual stress as,

$$\sigma_r = \Delta e \, E \, \zeta \qquad (11)$$

where, ζ : is defined as the residual stress factor

then, $\zeta_a = - \bar{f}$ and $\zeta_b = f$ (12)

The signs arise from the case where, ($e_a > e_b$). Plots of ζ_a and ζ_b are superimposed in Figure 10. d verses "f"

Asymmetrical duplex structure:

Due to asymmetry, the axial forces will created ,in addition, a moment "M = F b/2 ", if the imaginary constrains in Figure 11 c were to be removed, the curvature produced by this moment could be found as detailed below,

$$\frac{1}{R} = \frac{M}{EI} = \frac{F\,b}{2\,EI}$$

$$= \Delta eE \; \frac{A_a A_b}{A_t} \times \frac{1}{E} \times \frac{b}{2} \times \frac{12}{a\,b^3}$$

hence,

$$\frac{1}{R} = \frac{\Delta e}{b}\;(6 f \bar{f}) = \frac{\Delta e}{b}\;\Omega \qquad (13)$$

where, $\Omega = 6 f \bar{f} = -6 f^2 + 6 f$ (14)

Equation (14) defines the curvature coefficient Ω in the studied case of $E_a = E_b$

Equation (13) reveals several interesting features, For example, the curvature is a linear function of the difference in total change in length (Δe) after the complete heat treatment processes. The actual values of E_a and E_b does not affect the curvature. Actually, only the ratio "$n = E_b/E_a$" is involved in "Ω" if "$E_a \neq E_b$". The curvature is inversely proportion to "b", showing that the thickness has a strong effect on the curvature.

Differentiating Equation (14) and equating to zero, it is clear that the maximum curvature takes place at,

$$f_{\text{at max}\,\Omega} = 0.5 \qquad (15)$$

Substituting in Equation (13) the maximum curvature factor could be found, $\Omega_{max} = 1.5$

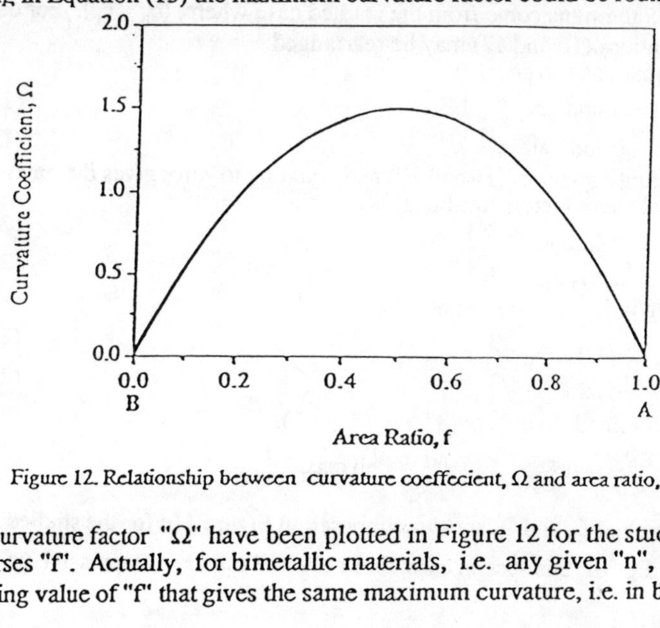

Figure 12. Relationship between curvature coeffecient, Ω and area ratio, f.

The curvature factor "Ω" have been plotted in Figure 12 for the studied case where $E_a = E_b$ verses "f". Actually, for bimetallic materials, i.e. any given "n", there will be a corresponding value of "f" that gives the same maximum curvature, i.e. in bi-metallic strips thermostats [17]. The situation thus produced is shown in Figure. 11d. The requirement of $\Sigma F=0$ are satisfied, but $\Sigma M\neq 0$. The moment $M=Fb/2$ must be eliminated. A fictitious moment "M", equal in value and opposite in sign to $M=Fb/2$, must be applied by the material of the two constituent. The residual stresses produced at the surface of the constituent "A", i.e. σ_{ra}, due to both the compressive axial stress " $-F_a/A_a$" and the bending strain "$\varepsilon = \varepsilon_{bending} = y/R$" , where "$y = b/2$" is the distance from the centroid, can be derived as,

$$\sigma_{ra} = \varepsilon \; E - \frac{F}{A_a}$$

$$= \frac{1}{R}\frac{b}{2} \; E - \Delta e \, E \frac{A_b}{A_t}$$

$$= \frac{\Delta e}{b} \; \Omega \; \frac{b}{2} \; E - \Delta e \, E \, \bar{f}$$

as before,

$$\sigma_{ra} = \Delta e \, E \, \zeta_a$$

Where in this case,

$$\zeta_a = \bar{f} \, (3 f - 1) \tag{16}$$

similarly, $\sigma_{rb} = \Delta e \, E \, \zeta_b$

Where in this case,

$$\zeta_b = f \, (1 - 3 \bar{f}) \tag{17}$$

Notice that, the signs come from the studied case where, ($e_a > e_b$). For other cases, the terms of Equations.(16 and 17) may be rearranged.

Solving Equations (16 and 17).

$$\zeta_a = 0 \text{ at } f = 1 \text{ and at } f = 1/3 \tag{18}$$

and $\zeta_b = 0$ at $f = 0$ and at $f = 2/3$ (19)

Differentiating Equations (16 and 17) and equating to zero, gives the maximum and minimum residual stress factors locations,

$$f_{at \; \zeta_{a \; max.}} = 2/3 \tag{20}$$

and, $f_{at \; \zeta_{b \; min.}} = 1/3$ (21)

Substituting in Equations (16 and 17),

$$\zeta_{a \; max.} = 1/3 \tag{22}$$

and $\zeta_{b \; min.} = - 1/3$ (23)

Within the defined domain of "f", i.e. ($0 \rightarrow 1$),

$$\zeta_{a \; min.} = - 1 \quad \text{and} \quad \zeta_{b \; max.} = 1 \tag{24}$$

A plot of each of ζ_a and ζ_b are superimposed in Figure 11e for the studied special case when ($E_a = E_b$). In this special case, both σ_{ra} and σ_{rb} are compressive up to ($f = 1/3$)

and tensile for all ($f > 2/3$), while in the range of ($1/3 < f < 2/3$) they have opposite signs. For values of "n" other than one there is a similar trend, but with a wider negative variation [16].

Equations(11, 12, 16 and 17) reveals several interesting points. For example, the higher the modulus of elasticity "E" the higher the residual stresses. The residual stresses are a linear function of the difference in the total change in length of the two constituent "Δe". The total thickness of the duplex structure "b" has no effect on the residual stresses, but, of course, the area ratio "f" which involves in ζ_a and ζ_b does.

Determination of the residual stresses:

It has been proven, e.g. Equation (11),that the residual stresses are a linear function of, geometrical factor "ζ", material factor "E" and chemistry and thermal treatment factor "Δe". It is possible to determine the residual stress after the heat treatment of certain structure by measuring Δe. Actually, Δe is a small quantity and can be subject to error during measurement.

From the analysis of the asymmetrical duplex structure, the radius of curvature could be expressed as in Equation (13) or in terms of the deflection ∂ and the length of the curved beam L as,

$$\frac{1}{R} = \frac{\Delta e}{b} \quad \Omega = \frac{8\,\partial}{L^2}$$

Hence,

$$\partial = \Delta e \frac{L^2}{8b} \quad \Omega \tag{25}$$

As has been shown from Equation (15), for an elastically compatible duplex specimen, i.e. $E_a = E_b$, the maximum deflection occurs when $f = 0.5$., i.e., $\Omega = 1.5$. Substituting in Equation (25), for Ω and,

if, $L \approx 50$ mm, $b \approx 10$ mm, hence, $\partial \approx 50\,\Delta e$ mm

and If, $L \approx 100$ mm, $b \approx 2$ mm, hence, $\partial \approx 1000\,\Delta e$ mm

From that analysis, it is possible to determine "Δe" in higher accuracy by measuring the maximum deflection of the specimen after an stimulating heat treatment regime.

Substituting For Ω from Equation (14) in Equation (25),

$$\partial = K f - K f^2 \tag{26}$$

Where, $K = 3\,\Delta e\, L^2/4b$

Hence, a polynomial of the second degree should fit the experimental results of Figure 5.

Using either of Equation (25) or Equation (26) it is possible to calculate Δe as,
in the as quenched condition, $\Delta e = 0.0045$
after tempering at 200°C, $\Delta e = 0.0034$
after tempering at 300°C, $\Delta e = 0.0029$

To show how this small extension affects the residual stresses, consider the case of coaxial cylindrical duplex structure where $f = 0.5$. From Equation (12), $\zeta = f = 0.5$. Substituting in Equation (11) and taking E = 200GPa, hence the residual stresses in each constituent are in the magnitude of 450MPa in the as quenched condition and drops to only 290MPa if tempered at 300°C (570°F). The residual stresses will naturally be compressive in one constituent and tensile in the other.

Estimation of the yield strength:

The effects of residual stresses on the compression and bending yield strength is discussed in the present work while the residual stress effects on other mechanical properties, e.g. UTS, TRS, ductility and toughness will be discussed in the future. If a residual stress rises in a duplex structure, as shown in Figure 13a and Figure 14a, the apparent yield strengths "σ^*_y" of the duplex structure may be determined if the yield strength of the two constituents "σ_{ya} and σ_{yb}" are known. The relation between the material cost represented by the area ratio "f" and performance represented by the apparent yield strength of the duplex structure "σ^*_y" could be found from the rule of mixture;

$$\sigma^*_y = \sigma_{yb}\, f_b + \sigma_{ya}\, f_a \tag{27}$$

where $f_a = A_a/A_t = f$, $\quad f_b = A_b/A_t = \bar{f} = 1 - f$,

Any increase in "σ^*_y" than that defined by Equation (27) is a gain while any decrease is a loss. In the case of uniaxial loading;

$$\sigma^*_y = \sigma^*_{yb}\, f_b + \sigma^*_{ya}\, f_a \tag{28}$$

Where $\sigma^*_{ya} = \sigma_{ya} \pm \sigma_{ra}$ and $\sigma^*_{yb} = \sigma_{yb} \pm \sigma_{rb}$ (29)

In the case of flexure loading,

$$\sigma^*_y = \sigma^*_{yb}\, i_b + \sigma^*_{ya}\, i_a \tag{30}$$

where $i_a = I_a/I_t = 4f^3 - 6f^2 + 3f$ (31)

and $i_b = I_b/I_t = -4f^3 + 6f^2 - 3f + 1 = 1 - i_a$ (32)

If there is no residual stresses, hence;

In the case of uniaxial loading,

$$\sigma^*_y = \sigma_{yb} + (\sigma_{yb} - \sigma_{ya})\, f \tag{33}$$

which is identical to Equation (27). In the case of flexure loading,

$$\sigma^*_y = 4\,(\sigma_{ya} - \sigma_{yb})\, f^3 - 6\,(\sigma_{ya} - \sigma_{yb})\, f^2 + \sigma_{yb} + 3\,(\sigma_{ya} - \sigma_{yb})\, f \tag{34}$$

Since the residual stress σ_r is a known function of "f", hence the relation between "σ^*_y" and "f" is also known by applying the rule of mixtures, Equation (28 or 30). If $\sigma_{ya} > \sigma_{yb}$ and $e_a > e_b$;

In the case of uniaxial compression,

$$\sigma^*_y = (\sigma_{yb} + \sigma_{rb})\, f_b + (\sigma_{ya} - \sigma_{ra})\, f_a$$

Substituting for σ_r from Equation (11), and for ζ from Equations (12)

hence, $\sigma^*_y = (2\,\Delta e\, E)\, f^2 + \{(\sigma_{ya} - \sigma_{yb}) - 2\,\Delta e\, E\}\, f + \sigma_{yb}$ (35)

Similarly, in the case of uniaxial tension,

$$\sigma^*_y = (\sigma_{yb} - \sigma_{rb})\, f_b + (\sigma_{ya} + \sigma_{ra})\, f_a$$

hence, $\sigma^*_y = -(2\,\Delta e\, E)\, f^2 + \{(\sigma_{ya} - \sigma_{yb}) + 2\,\Delta e\, E\}\, f + \sigma_{yb}$ (36)

In the case of bending, if constituent "B" on the tensile side,

$$\sigma^*_y = (\sigma_{yb} + \sigma_{rb})\, i_a + (\sigma_{ya} + \sigma_{ra})\, i_b$$

Substituting for σ_r, ζ from Equations (16 and 17), i_a and i_b from Equations (31 and 32), hence,

$$\sigma^*_y = -(24\,\Delta e\,E)\,f^5 + (60\,\Delta e\,E)\,f^4\,\sigma_{yb} + \{4\,(\sigma_{ya} - \sigma_{yb}) - 58\,\Delta e\,E\}\,f^3 - \{6\,(\sigma_{ya} - \sigma_{yb}) - 27\,\Delta e\,E\}\,f^2 + \{3\,(\sigma_{ya} - \sigma_{yb}) - 5\,\Delta e\,E\}\,f \quad (37)$$

Similarly, if constituent "A" on the tensile side,

$$\sigma^*_y = (\sigma_{yb} - \sigma_{rb})\,i_a + (\sigma_{ya} - \sigma_{ra})\,i_b$$

hence,

$$\sigma^*_y = (24\,\Delta e\,E)\,f^5 - (60\,\Delta e\,E)\,f^4\,\sigma_{yb} + \{4\,(\sigma_{ya} - \sigma_{yb}) + 58\,\Delta e\,E\}\,f^3 - \{6\,(\sigma_{ya} - \sigma_{yb}) + 27\,\Delta e\,E\}\,f^2 + \{3\,(\sigma_{ya} - \sigma_{yb}) + 5\,\Delta e\,E\}\,f \quad (38)$$

An estimated plot for the symmetrical and asymmetrical cases shown in Figures 13a and 14a are presented on Figure 13b and 14b respectively, the reverse cases are also shown in the same figure.

Substituting for σ_{y4660}, σ_{y4620} and Δe from the preceding section and E = 200 GPa the estimated apparent yield strength could be plotted verses "f". The estimated properties are superimposed on the experimental results in Figures 8 and 9. The estimated curves

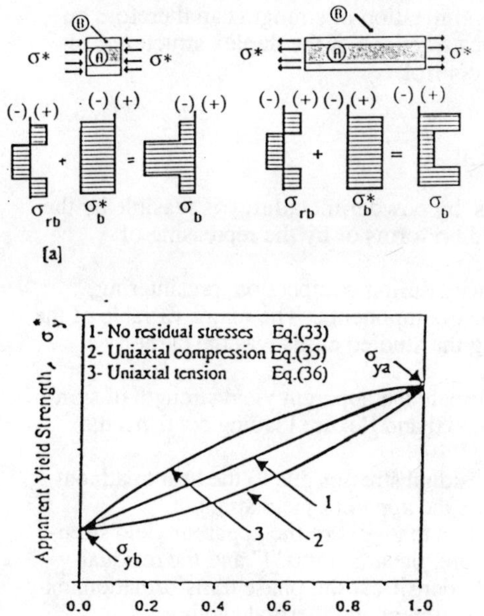

Figure 13 The effect of residual stress on the apparent yield strength of symmetrical duplex structure subjected to uniaxial loading.

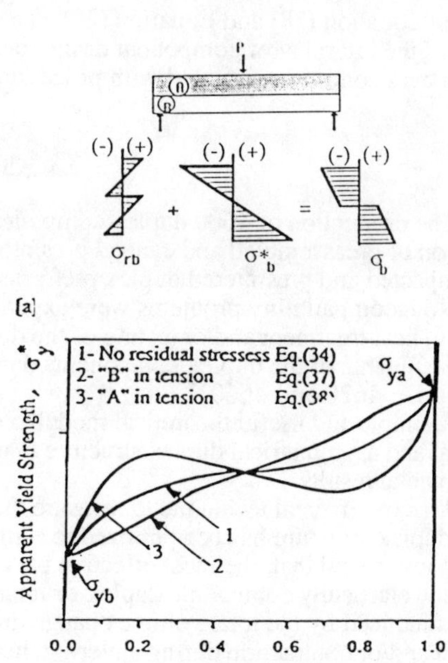

Figure 14 The effect of residual stress on the apparent yield strength of asymmetrical duplex structure subjected to flextural loading.

fairly well fits the experimental data. Neglecting the tangential and radial residual stresses in the case of uniaxial compression, an error may result in estimating the apparent strength.

Several interesting points are revealed from the previous analysis. For example, in the case of uniaxial loading of symmetrical coaxial structure the residual stresses cause a gain or loss in strength, i.e. the difference between Equations (35 or 36) and Equation (27 or 33). At "f = 0.5", a decrease of 340 MPa in the compressive yield stress. An increase by similar amount is expected in the tensile strength. The case will be reversed if $e_a < e_b$ while the condition $\sigma_a > \sigma_b$ still remain, i.e. negative Δe. The selective choice of "f" leads the additional strength for the duplex structure "$\Delta\sigma^*_{yd}$", i.e. the difference between Equation (34) and Equation (27), and the gain from the residual stresses "$\Delta\sigma^*_{yr}$", the difference between Equations (38) and Equation (34). By differentiation and substitution;

$$\Delta\sigma^*_{yd\ max} = 0.19\ (\sigma_{ya} - \sigma_{yb}) \quad \text{at} \quad f = 0.21$$

and reflects an increase of 170 MPa.

Also,at $f = 0.16$

$$\Delta\sigma^*_{yr\ max} = 0.31\ \Delta e\ E$$

reflects an increase of 210 MPa. Since Equation (34) and Equation (38) are not in phase the maximum total increase in strength will depend on both $(\sigma_{ya} - \sigma_{yb})$ and Δe E. The total increase in strength "$\Delta\sigma^*_y$" is defined as; $\Delta\sigma^*_y = \Delta\sigma^*_{yr} + \Delta\sigma^*_{yd}$, and it is the difference between Equation (38) and Equation (27). The optimization of strength can therefore be based on the interplay of component design or area fractions of the duplex structures, residual stress conditions coupled with processing control.

Conclusion

1 - The production of 4600 duplex components by powder metallurgy is feasible by the extrusion of preassembled and canned presintered preforms or by the repressing of precompacted and presintered duplex preforms.

2 - No incompatibility problems were experienced during compaction, presintering, forming, heat treatment and/or testing of the duplex components. The major variable of the two constituents of the duplex components during the studied cases was the carbon content, i.e. 4620 and 4660.

3 - A simple and useful theoretical model to estimate the apparent yield strength of symmetrical and asymmetrical duplex structure in uniaxial and flexure loading confirms the experimental results

4 - An experimental technique to measure the residual stresses due to the heat treatment of the duplex structure has been utilized to estimate the apparent yield strength.

5- It was found that, the most effective parameters that govern the apparent yield strength of the elastically compatible duplex structures are; the area ratio "f" and the residual stresses induced by the total volume change due to densification, phase transformation, expansion and/or contraction during sintering, heat treatment and thermal cycling.

6- Material savings and improved surface wear or corrosion resistance can be realized from the duplex structure. In addition a synergistic contribution of the two constituent of the duplex structure could be obtained by;

- the optimum design of the duplex component, i.e. constituent placement and determination of "f" .
- controlling the placement and magnitude of residual stresses, by the selective choice of the constituent chemistry, processing and heat treatment sequence.

Acknowledgment

The authors gratefully acknowledge program support from the Egyptian government and the U.S. Agency for International Development and the powder and processing assistance from Hoeganaes Corporation

References

[1] Moll, J. H. , Schwertz, J.H. and Chandhok, V. K. ,"P/M Dual-Property Wheels for Small Engines," Progress in Powder Metallurgy, vol. 37, 1981, pp. 303-318.
[2] Kasak, A. and Dulis, E. J. ,"Powder-Metallurgy Tool Steels," Powder Metallurgy, vol. 10, no. 2, 1978, pp. 114-123.
[3] Schreiner, H., "Two Layer Powder Compacting," Powder Metallurgy International, vol. 10, no. 4, 1978, pp. 198-201
[4] Negm, M. I.,"Metallographic and Fractographic Analysis of Extruded Iron/Steel Duplex Powder Preform," Current Advances in Mechanical Design and Production, Second Cairo Univ MDP Conference, 27-29 December, 1982, Cairo, Egypt, pp.605- 612.
[5] Gotman, I. ,Gutmannas, E. Y., "Joining of P/M T15 High Speed Steel with 4640 Steel by Cold Sintering," P/M International, vol. 19, no. 4, 1987, pp. 11-14
[6] Tabata, T., Masaki, S. ,"A New Way of Joining P/M Compacts,"The International Journal of Powder Metallurgy & Powder Technology, vol. 15, no. 3, 1979, pp. 239-244.
[7] Oyane, M. ,Okimoto, K. , Shima, S. , "Strength and Ductility of Bonded PM Products," Powder Metallurgy International, vol. 13, no. 3, 1981, pp. 136-139.
[8] Marx, J.B. ,"The Forging of Powder Preforms," Ph. D. Thesis, Birmingham Univ., 1971.
[9] Negm, M. I.," The Extrusion of Duplex Metal Powder Systems," Current Advances in Mechanical Design and Production, Second Cairo University MDP Conference, 27-29 December, 1982, Cairo, Egypt, pp. 869-880.
[10] Okimoto, K., Satoh, T., "Joining of Iron Powder Compacts by an Infiltration Method," The Int. Journal of Powder Metallurgy, vol. 23, no. 3, 1987, pp. 163-169.
[11] Schreiner, H., "Two Layer Compact-Sinter-Infiltration Technique for Producing Contact Materials for Power Engineering Purposes," Powder Metallurgy Int., vol. 7, no. 1, 1975, pp. 21-24
[12] Antes, H. W., Stockl, P. L, "The Effect of Deformation on Tensile and Impact Properties of Hot P/M Formed Ni-Mo Steels", P/M, vol. 17, no. 33, 1974, pp.178-192.
[13] Moyer,K. H., "The Effect of Residual Porosity and Flow on The Mechanical Properties of Hot Upset 1.9 Ni-0.5 Mo Steel Powder Preforms", Progress in Powder Metallurgy, 1975, pp. 111-118
[14] Ahmed, M. N. H. and Koczak, M. J., "Optimization of the Properties by Processing and Heat Treatment Powder Forged 4600 Alloys," this proceeding.
[15] Mack, D.J., Trans. AIME, 1946, vol. 58, pp. 68-85.
[16] Ahmed, M. N. H. , Vedula, M. and Koczak, M. J., "Composite Ferrous P/M structures: Mechanical Properties and Stress Analysis," to be presented at TMS Fall Meeting 25 Sep. 1988
[17] Timoshenko, S. , "Analysis of Bi-Metal Thermostats," Journal of the Optical Society of America, vol. 11, 1925, pp. 233-255
[18] Savolainen, U.U. , Sears, R.M. , "Thermostat Metals," , Composite Engineering Laminates, The MIT press, 1969 pp. 212-226.

WEAR RESISTANCE OF Al_2O_3-REINFORCED HIGH SPEED TOOL STEEL

R.A. Queeney and R.E. Masters
The Pennsylvania State University
University Park, PA 16802

R.J. Beltz and J.D. Dankoff
Kalumetals, Inc.
Latrobe, PA

Abstract

The present study examines the feasibility of enhancing the wear resistance of a selected high speed tool steel (M3) by mechanically loading it with a stable high hardness reinforcement: at the same time, the possibility of maintaining acceptable strength levels is examined. In particular, alumina, Al_2O_3, with particle diameters less than 10 microns, was blended into a suitable high speed steel powder with composition adjusted to provide an M3 grade after all heat treating. Cold pressed and sintered composites, with some samples HIPped for full densification, were compared with conventional wrought forms of several tool steels for their abrasive wear resistance. Four point bend tests were also conducted to ascertain conventional strength values for the composites. Reinforcement levels were introduced up to ten percent by weight; all reinforced composites proved to be superior in wear resistance to the conventional wrought alloys. Bend strength levels could be maintained at acceptable levels when the composite was fabricated to full density. Finally, a model is proposed to explain observed composite wear behavior.

Introduction

The wear resistance of ordinary martensitic carbon tool steels cannot be maintained when these alloys are subjected to service temperatures in the 300°C to 650°C range, as their hardness deteriorates rapidly. "High Speed Steels" (HSS) were developed to circumvent thermal instability in tooling; service speeds high enough to generate softening temperatures in traditional tool steels could be tolerated by HSS. The latter tool steels derive their thermally resistant hardness from relatively stable metal carbides. Carbide forming elements such as tungsten, molybdenum, chromium, and vanadium are alloyed in typical HSS formulations in substantial amounts to generate the relevant metal carbide constituents. Melt/cast/forging technology, leading to conventional wrought HSS alloy forms, is limited, mainly, by hot workability constraints, to carbide contents found in commercial alloys. Improved carbide particle distributions can be obtained via P/M technology and induced strength and toughness improvements over wrought alloy forms have been noted (1).

Further hardening of HSS can only be accomplished by treating the HSS as a composite matrix material, and reinforcing the matrix with ceramic constituents. Composite fabrication can be carried out by blending HSS and ceramic powders, with the final consolidated product achieved via P/M methods. Composite performance expectations can be formed on a preliminary basis by examining the hardnesses (Vickers) of undissolved carbides found in quenched and tempered HSS(2), viz: (Mo, W, Fe)$_6$ C (15 GPa) and VC (24.5 GPa). The hardness of quenched and tempered HSS is approximately 8 GPa. Some types of steel matrices have been reinforced by carbide addition: however, carbide availability, or wetting behavior, or other considerations not related to hardness alone, have determined the choice of reinforcement. Accordingly, both TiC (3) and WC (4) composites have been fabricated and reported upon. Finally, non-carbide ceramics might also be employed as reinforcements, wherein reduced solubility in the matrix might render them more stable during elevated temperature fabrication. Aluminum oxide (Al_2O_3), with a Vickers hardness of 24 GPa, is a candidate reinforcement; its potential was examined earlier (5) in a HSS matrix in which it was incorporated. The present work is a further study of the promising HSS/Al_2O_3 system quoted (5).

Material Preparation

The intent of the composite evaluation program was to produce Al_2O_3 reinforced HSS alloys corresponding to a conventional M3 grade with added Al_2O_3. The preparation of the starting alloy powder, with composition adjusted to provide the desired final content after heat treating, has been previously described (5). The blended powders were pressed in a double acting ram die at a compressive stress of 700 MPa. Wear testing specimens were cylinders 1 cm thick by 3.5 cm in diameter; bend bars were 0.8 cm by 1.3 cm in cross section and 5.0 cm long. The die sets were lubricated with zinc stearate suspended in an acetone vehicle. All samples were sintered at 1220°C, for one hour, in a vacuum of $5x10^{-4}$ Torr. The complete heating and cooling histories have been previously described in detail (5). Previous experience had indicated that as reinforcement content increased beyond approximately 5 wt.%, sintering to full density without encountering the liquid phase was not practical; the reinforcements act to restrain grain boundary motion and prevent full densification, but also prevent grain coarsening. Sintered specimens, with closed porosity, but not quite fully dense, were containerless HIPped at 1160°C for 3 hours in Argon at 100 MPa. The samples were then annealed at 890°C and slowly cooled to 620°C, then air cooled, prior to machining of near net shape test samples. After machining, all samples were preheated to 890°C, then placed for 5 minutes in a salt bath at 1225°C and quenched into salt held at 610°C; specimens were finally air cooled, then triple tempered at 2 plus 2 plus 2 hours at 570°C. Final sample preparation was by wet grinding. Compositions and Rockwell C-Scale hardnesses are given in Table 1. Commercial samples of conventional non-P/M D2, M2, M4, and T15 alloys were also prepared, with standard heat treating utilized.

Material ID Code	C	W	Mo	Cr	V	Co	Al_2O_3(Wt.%)	R_c
	Composition (Wt.%)							
A	0.90	5.9	6.1	4.0	2.0	-	1.25	65.4
B	0.98	5.8	4.8	3.9	1.9	-	2.0	65.9
C	0.94	5.8	4.7	3.9	1.9	5.0	2.0	66.5
D	0.96	6.2	6.3	3.8	2.0	-	2.5	65.4
E	0.95	5.7	6.0	3.9	1.8	-	5.0	66.3
F	0.80	5.7	5.8	3.5	1.8	-	10.0	45.0
D2	AS RECEIVED						None	53
M2	AS RECEIVED						None	66
M4	AS RECEIVED						None	66
T15	AS RECEIVED						None	67

Table 1. Compositions and Hardnesses of Composite and Conventional HSS Investigated

Experimental Studies

Cylindrical specimens suitable for wear testing were ground from the heat treated stock to 3.0 cm diameter and 0.7 cm thick. Abrasive wear resistance was determined in a pin-on-disk apparatus similar to one described elsewhere (6), using a WC-Co pin with a tip radius of 0.05 mm and conical point with a total angle of 90 degrees. The disk was driven at one revolution per second, corresponding to a relative disk speed of 6.3 cm/second under the pin. The pin was weighted at 3N for all tests. The pin/disk system was run in for a thousand revolutions before wear rate data was actually recorded.

Bend tests were conducted with a four point bending apparatus as depicted in Figure 1. The crosshead speed was 1.27 mm/minute and all testing was carried out in air at room temperature. The Modulus of Rupture was calculated from the maximum load.

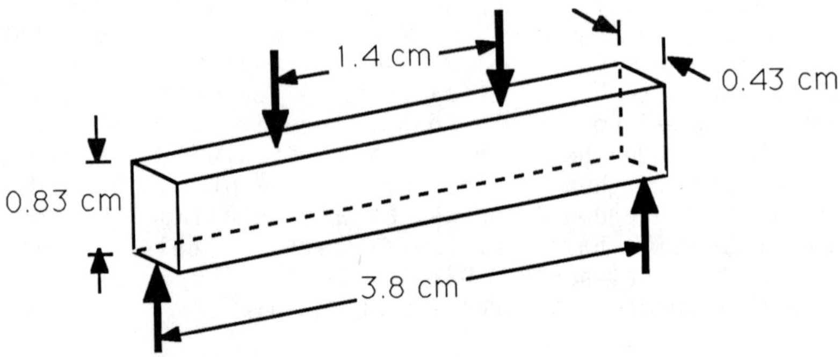

Figure 1. Bend Test Configuration

Experimental Results

A Typical wear rate curve is indicated below, with indicated rates occurring after the run-in period.

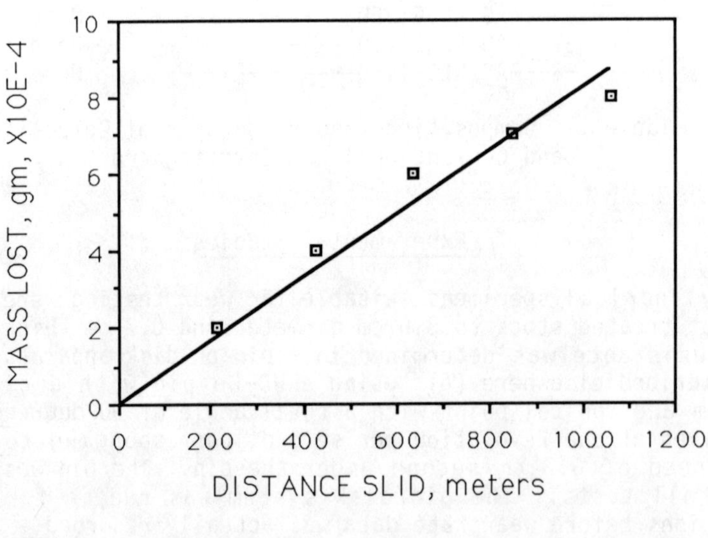

Figure 2. Wear Response for T15 HSS

The wear rates for all composites, as well as the rates for the conventional alloys, are depicted in Figure 3.

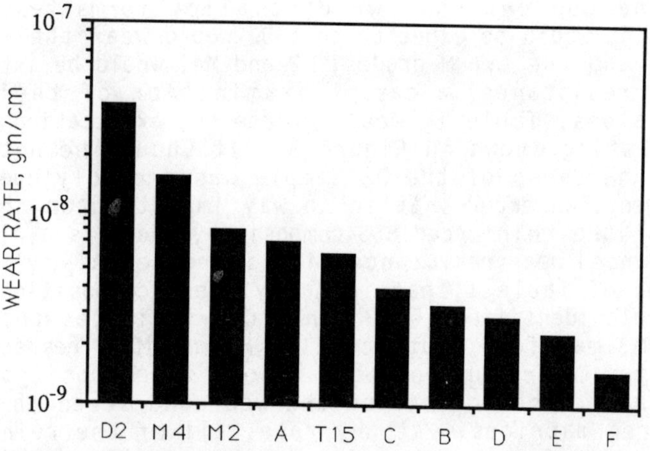

Figure 3. Composite and Conventional HSS Abrasive Wear Rates

Finally, bend test results are given in figure 4. The Modulus of Rupture values were averages taken for three or four specimens.

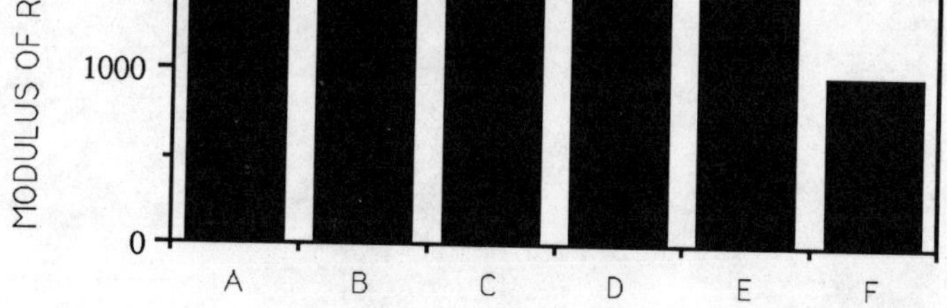

Figure 4. Bend Test Results for Al_2O_3 Reinforced HSS Composites

Analysis of Results

For the conventional wrought alloy forms tested for wear resistance, it would be expected that D2 would wear the fastest, T15 the least, and the two M grades, M2 and M4, would be intermediate in their wear resistance; a casual examination of their Rockwell hardness values, Table 1, would create the expectation borne out by the wear testing shown in Figure 3. It should be noted that the rather low hardness of the D2 sample was probably due to improper heat treatment, an error that in no way impacts upon the other results. For the reinforced HSS composites, there is a clear association of enhanced wear resistance with increased Al_2O_3 content. An examination of Table 1, particularly those compositions associated with materials identified as "B" and "C", indicates that these are not truly M3 matrices, but actually M2 and M35, respectively. The writers wished to try out some deviations in composition for other purposes, and chose to include the wear and strength results for these altered matrices. At any rate, all of the reinforced composites, except for the minimally reinforced composite "A" with 1.5% Al_2O_3, exceeded the best conventional wrought alloy, T15, in wear resistance; the minimally reinforced material provided superior wear resistance to the other three wrought alloys, D2, M2, and M4.

Figure 5 below is a scanning electron micrograph of a wear track in the 10% Al_2O_3 composite after 2.2 hrs of wear testing, or the test termination time. The wear track evidences wearing of all components - steel matrix, metal carbides, and alumina, the latter white in the image.

Figure 5. Scanning electron micrograph of a wear track in M3 steel reinforced with 10% Al_2O_3, after 2.2 hrs test time (x320).

Also evident in Figure 5 is porosity at the bottom of the wear track. The porosity may be due to incomplete densification (see discussion of bend strength below), or may be the result of carbide and alumina reinforcement being pulled from the matrix after the latter is worn away, or both; the present study does not address the eventual loss of either carbide or alumina as the matrix is diminished around them.

In an effort to examine more precisely the roles played by both carbides and alumina in resisting the cutting action inherent in abrasive wear, a single-pass wear track was produced on an as-ground pristine surface; Figure 6 is a scanning electron micrograph of that track. Upon examining

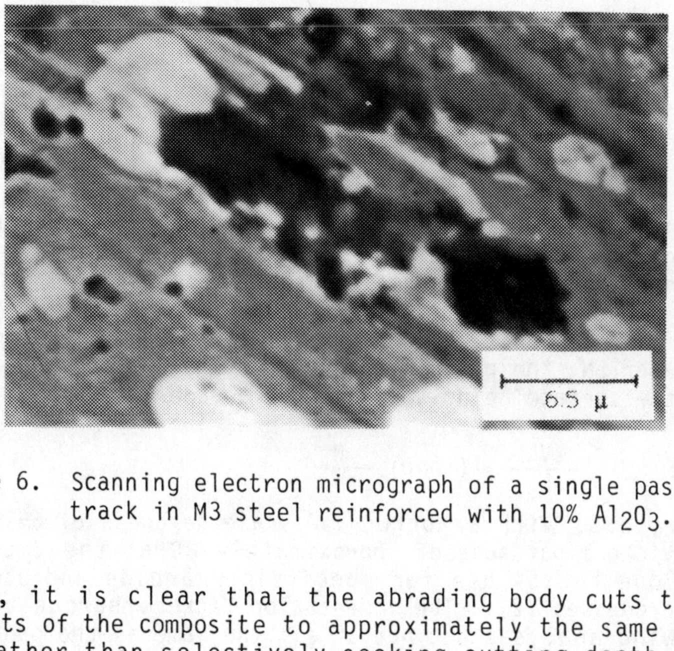

Figure 6. Scanning electron micrograph of a single pass wear track in M3 steel reinforced with 10% Al_2O_3.

Figure 6, it is clear that the abrading body cuts through all constituents of the composite to approximately the same extent and depth, rather than selectively seeking cutting depth based upon constituent hardness. Figure 6 suggests that the aggregate does behave as a true composite, at least insofar as damage imparted by the present abrader is concerned. The volume fraction of the Al_2O_3 constituent is equivalent to the areal fraction on a random plane; Figure 7 is the experimental wear rate as a function of the volume, or areal, fraction of alumina for the M3 matrix only, composites A, D, E, and F.

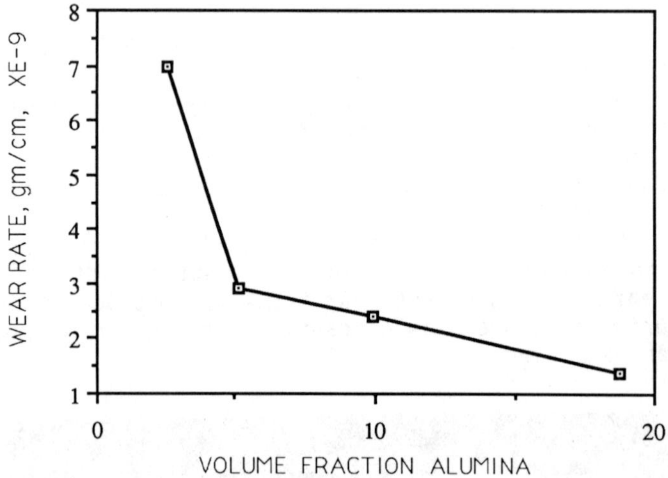

Figure 7. Wear rate and alumina content in an M3 matrix.

Clearly from Figure 7, the frequency of appearance of the alumina constituent on a wearing surface is not the sole factor in determining composite wear resistance. A simplistic model for abrasive wear rates is extant (7) and relates the volume of material worn, per unit distance the abrader slides, $\Delta V/\Delta L$, to the normal load on the abrader, N, the punch hardness H of the abradee, and the angle θ between the surface being worn and the abrader, here 45°: viz.,

$$\frac{\Delta V}{\Delta L} = (\tan\theta) \frac{N}{\pi H} \quad (1)$$

An ordinary HSS, with a Rockwell C-scale hardness of 65 or 66 would possess a Vickers hardness of approximately 8GPa; the Vickers scale is chosen due to its use for specifying carbide and other ceramic compound hardnesses (2). The indentation Vickers hardnesses used in the following analysis are not at all the same as the punch hardness H in equation (1), but there should exist a simple linear correlation between the two. Since the micrograph observations suggest composite behavior, it is reasonable to assume that the composite hardness might obey the rule of mixtures, or:

$$H_c = H_m V_m + H_r V_r \quad (2)$$

where V refers to volume fraction and the subscripts "m" and "r" refer to matrix and reinforcement, respectively. Assuming that the matrix has an indentation hardness of 8GPa, and the Al_2O_3 a hardness of 24 GPa, then equation (2) can be employed to calculate composite hardness values H_c, as summarized in Table 2. Examining the inverse of H_c,

Vol.% Al_2O_3	H_c(GPa)	H_c^{-1}(GPa)$^{-1}$
0	8.0	0.125
2.6	8.42	0.119
4.1	8.66	0.115
9.9	9.58	0.104
18.8	11.71	0.085

Table 2. Calculated Composite Hardnesses H_c

and assuming it is directly proportional to the punch hardness of equation (1), one can readily ascertain that there is only a predicted decrease of inverse hardness of around 32 percent, comparing the case for Al_2O_3 content of 18.8% (10 wt.%) to the unreinforced M2 or M4. There is an even smaller decrease in inverse hardness when the 1.25 wt.% Al_2O_3 reinforced material is compared to the 10 wt.% reinforced case. While the inverse composite hardness is varying by 15% in the latter case, the wear rate, possibly proportional, is decreasing five-fold.

An accurate predictive model for the composite wear rate must adopt a more comprehensive view of the hardness of the reinforcement, an "effective" hardness. The values of hardness used in the above speculative calculations were typical bulk values. When the reinforcement is bound into a metallic matrix, at least two major alterations to its ability to resist penetration must be considered. First, a state of plane residual stress will exist, of a compressive nature, acting upon the reinforcement. These residual stress values are functions of the thermoelastic response parameters of matrix and reinforcement, as well as the thermal history of the composite and the spatial distribution of reinforcement (8,9). The residual stresses will alter the indentation resistance of the reinforcement - enhance the resistance - but in a manner that has yet to be quantitatively analyzed. Secondly, the underlying elastic/plastic matrix will alter the response of the indentation, as has been observed in a similar context (10).

Finally, as can be seen in Figure 7, the large rate of gain in abrasive wear resistance upon reinforcement up to 2.5% Al_2O_3 is not maintained with reinforcement levels of 5% and 10% Al_2O_3. Although some difficulty was encountered in densifying the 10% reinforced composite fully, the same problem was not encountered with the 5% material. Again, the role of the matrix - its ability to generate residual stress and its elastic/plastic constraint on the reinforcement - must determine the extent of wear behavior improvement.

Strength levels of most of the composites, as indicated in Figure 4, remained above the 2000 MPa level - roughly speaking, the 300 ksi level - for the M2, M3, and M35 matrix compositions, a notable improvement over Modulus of Rupture (MOR) (in bending) strengths reported earlier (5). Exceptions were the composites "E" and "F", containing 5% and 10% Al_2O_3, respectively. Given the high strength/low ductility response character of all quenched and tempered HSS, including the reinforced materials of the present study, the MOR strengths evidenced are determined by two material parameters. The actual cohesive strength of the composite, including the mechanical strength of all interfaces, is an obvious determinant of strength. In addition, given the flaw sensitivity of solids of high-strength/low ductility, the reinforcement population and its interface with the matrix, and residual porosity also play an important role in fixing MOR values. In particular, the composite "F", or that with 10% Al_2O_3, was not sintered to the point of pore closure: therefore, subsequent containerless HIPping caused little densification. In fact, liquid-displacement densitometry did not prove successful due to liquid ingestion by open pores. Composite "E", that with 5% Al_2O_3, possessed a final density of 7.70 gm/cm^3, close to the value of 7.85 gm/cm^3 for composite "D", yet its MOR value was lessened by 25%; its density difference is almost entirely due to the lower density of the added Al_2O_3. The fracture resistance of the composites, and the dependence of the resistance on the material microstate, remain to be probed and analyzed.

Summary and Conclusions

It is clearly possible to reinforce a high speed steel such as M3 (or, probably M2 as well) with fine particulate Al_2O_3 and effect significant gains in abrasive wear resistance. A ten-fold wear rate resistance gain is possible for only a 2% reinforcement, a degree of reinforcement that does not demand a significant sacrifice in nominal strength. Further reinforcement leads to additional gains in wear resistance, but deleterious losses in strength were noted. A comprehensive composite wear resistance model is needed but has not been proposed, although the outlines of its content are suggested. Reasons for eventual strength loss, and methods to circumvent the loss, remain to be analyzed.

References

1. P. Beiss, R. Wahling, and D. Duda, Modern Developments in Powder Metallurgy, 17, 331-357 (1985).

2. G.A. Roberts, Hamaker, and Johnson, Tool Steels, ASM (3rd Ed.), P 197 (1962).

3. J.L. Ellis, P/M International, 16, 53-55 (1984).

4. Z.-H. Wan, Modern Developments in Powder Metallurgy, 17, 193-219 (1985).

5. R.J. Beltz, J.D. Dankoff, and R.A. Queeney, Progress in Powder Metallurgy, 41, 235-250 (1985).

6. R.M. Hosking, F.F. Portillo, R. Wunderlin, and R. Mehrabian, J. Mat. Sci., 17, 477-498 (1982).

7. F.A. McClintock and A.S. Argon, Mechanical Behavior of Materials, 670, Addison Wesley, Reading, Mass. (1966).

8. P.J. Fritz and R.A. Queeney, Composites, 14, 393-394 (1984).

9. P.J. Fritz and R.A. Queeney, J. Composite Materials, 17, 224-230 (1983).

10. J. Larsen-Basse, Wear of Materials, 39-44, ASME (1985).

MECHANICAL BEHAVIOUR OF COLD SINTERED HIGH
SPEED STEEL-CARBIDES COMPOSITES

E.Y. Gutmanas and D. Zak

Department of Materials Engineering
Technion, Haifa 32000 Israel

Abstract

Cold sintering or high pressure consolidation of powders was employed for preparation of full dense high speed steel- carbides composites. Attrition milling was used to obtain homogeneous distribution of added carbide particles. Treatment in reducing atmosphere improved the bonding integrity between metal particles and between metal and carbide particles. A number of compositions showed good combination of hardness, strength and abrasion wear resistance. Possible ways of design of high speed steel and wear resistant materials with higher concentration of hard particles are briefly discussed.

Introduction

One of the main advantages of powder metallurgy is the possibility of the fabrication of materials that cannot be produced using ingot metallurgy. This applies to most of composite materials and particularly to wear resistant materials [1]. High speed steels with homogeneous distribution of fine carbides are produced by hot processing of water and gas atomized powders [1-3]. The most wear resistant tool steels produced by conventional and P/M methods contain 10-15 v/o of hard carbides in a martensite matrix [2,4-6]. Addition of hard particles (carbides, nitrides) may increase wear resistance of tool steels and narrow the gap between the most wear resistant tool steels and cemented carbides. Such materials may be advantageous if they are tougher than cemented carbides and their price is lower.

Different methods used in fabrication of metal-ceramic composites require stringent conditions and relatively high processing temperatures. Addition of hard particles to high speed steel powders and processing (consolidation, heat treatments) at high temperatures may result in partial dissolution of these particles in the matrix. Such change in phase composition will result in deterioration of the properties of the steel matrix. At lower processing temperatures neither consolidation of high speed steel powders to full density nor good bonding betweeen metal matrix and ceramic particles are achieved. High pressure consolidation of powders at ambient temperature, or cold sintering, offers wide possibilities in the design of composite materials. Densities close to theoretical were achieved for cold sintered metal powders and metal-ceramic composites [7-10]. Post-consolidation heat treatments of cold sintered materials at relatively low temperatures improve their mechanical properties. Cold sintering of high speed steels was reported elsewhere [11]. It was shown that removal of oxide layers from metal powder particles results in improvement of strength of cold sintered high speed steel [12]. A new approach to producing metal-ceramic composites was reported for stainless steel-chromium oxide [13] and for steel-vanadium carbide composites [14]. Compacts with interconnected system of pores were treated in hydrogen flow, which resulted in partial reduction - surface metallization of ceramic particles. Cold sintering of such compacts to full density resulted in improved bonding integrity between metal and ceramic particles. In the present work a similar approach was applied to high speed tool steel-carbide composites.

Experimental

Water atomized T15 high speed powder purchased from SCM Metal Products were used. The particle size was -325 mesh (< 44μm). Fine tungsten carbide, WC, and vanadium carbide, VC, from H.S.T. Starck (2-4μm) were used for preparation of composites. These contained 10 and 20 w/o (~ 5.5 and 11.5 v/o) of WC or 10 w/o (~ 13 v/o) of VC. In some cases carbon was added to T15 powder as fine graphite. Attrition milling was used for mixing to achieve homogeneous distribution.

Compacts were prepared at pressure of about 0.7 GPa to provide density of approximately 80%, and had an interconnected system of pores. Compacts were treated in hydrogen flow to provide removal of oxide layers from the surface of metal particles and partial reduction (surface metallization) of carbide particles. Subsequently, the compacts were cold sintered in rigid dies at a pressure of 3 GPa. Cold sintered samples were heat treated to obtain high hardness and strength.

Hardness, R_c, and transverse rupture strength, σ_{TRS}, of cold sintered and heat treated specimens were measured. An Instron testing machine was used for σ_{TRS} measurements, the cross head speed being 0.1 mm/min. The distance between the outer points in three point bending was 16mm for a sample size of 4.5 x 4.5 x 18 mm^3. Volume loss of cold sintered and heat treated specimens was measured in abrasive wear test. Alumina grinding wheel (~ 200mm dia.) was used in these tests with the following test conditions: 2000 RPM, pressure on the specimen 0.1 and 0.2 MPa.

Microstruscture of polished and thermally etched specimens and of fracture surfaces was characterized by means of scanning electron microscopy (SEM).

Experimental Results and Discussion

T15 is a highly alloyed tool steel (≈12W,5V,5Co,5Cr,1.6C). High concentration of carbide forming elements W,V and Cr and of carbon results in high volume concentration of carbides in quenched and tempered T15 high speed steel and high wear resistance [2,5-7]. The effect of austenitizing temperature (1 hr exposure) and of tempering procedure (2 hrs at 550°C x 2) on hardness and σ_{TRS} of cold sintered T15 and T15 with additions of 0.3 w/o C was investigated. Prior to cold sintering compacts were reduced in hydrogen flow at 900°C for 30 min. Such treatment partially removes oxide layers from the surface of tool steel particles and results in pronounced increase of σ_{TRS} of cold sintered specimens [12,15]; the treatment does not cause decarburization of the material. Dependences of hardness, R_c, and σ_{TRS} on austenitizing temperature for T15 and T15+0.3C are shown in Fig.1 and Fig.2 respectively for oil quenched (curve 1) and for quenched and tempered (curve 2) specimens. For both compositions σ_{TRS} increases with austenitizing temperature and is significantly higher for tempered specimens (up to ≈ 2000 MPa). Lower hardness values after tempering for specimens quenched from lower temperatures is the result of partial decomposition of martensite. The drop of R_o for as quenched specimens which were austenitized at higher temperatures is the result of retained austenite. The tempering procedure used is typical for high speed tool steels and results in precipitation of secondary carbides and transformation of retained austenite to martensite. The highest hardness and σ_{TRS} were observed for specimens quenched from 1200°C and tempered. Even higher temperatures, 1240-1250°C, are used for austenitizing T15. Such temperatures are very close to the solidus line and even a slight increase of carbon content, or higher nitrogen content, which ia typical for high speed steel powders, may result in appearance of liquid phase and undesirable austenite grain growth [3]. No grain growth was observed after 1h austenitization at 1200°C. Grain growth and coarsening of carbides was observed at longer exposures at 1200°C (see Fig.3), carbide size being still smaller than for inert gas atomized powder and HIPed material. Addition of 0.3C results in higher concentration of carbides in hard martensite matrix: image analysis in SEM shows an increase from 12.8 v/o of carbides for T15 to 16.3 v/o of carbides for T15+0.3C.

An increase in volume concentration of hard carbides was achieved also by addition of fine particles of WC (10 and 20 w/o or ~ 5.5 and 11.5 v/o) and of VC (10 w/o or ~ 13 v/o)>. To improve the bonding integrity of added carbides and particles of T15 powder compacts of T15-VC were treated in hydrogen flow at 950°C for 30 min and compacts of T15-WC at 1100°C for 30 min prior to cold sintering. Such treatments result in partial reduction-surface metallization of VC and WC, correspondingly, [14,15] and in noticeable increase of σ_{TRS} for cold sintered specimens. Treatment of T15 in hydrogen flow at 1100°C results in partial decarburization of about 0.1 w/o [12] which may lead) to deterioration of the properties of T15 matrix. When studying the reduction of free carbon (graphite) in hydrogen flow no reduction was observed at temperatures up to 1200°C. To neutralize the effect of decarburization during reduction of T15-WC composites at 1100°C, 0.1C (graphite) was added to the composition prior to mixing and compaction. For composites T15 carbides prepared by cold sintering from compacts reduced in hydrogen flow fracture is transgranular and goes through the matrix (Fig.4, for example).

Mechanical properties σ_{TRS}, R_c and volume loss in abrasive wear test for T15, T15+0.3C, and T15 - carbides composites are presented in Table 1. For comparison, the data obtained for HIPed from inert gas atomized powder T15 (Crucible) high speed steel is also given. All compositions were austenitized for 1 hour at 1200°C, oil quenched and double tempered for 2 hours at 550°C.

Table 1. Mechanical Properties of T15 and T15-Carbides Composites.

Composition	Treatment	Hardness Rc	Transverse rupture strength σ_{TRS}, MPA	Abrasive wear volume loss,[1] $10^{-6}cm^3/Slcm^3$ at pressure P, MPa	
				0.1	0.2
T15	C.A. P-3GPa 1h 1200°C oil +2x2hx550°	64	2000	0.32	0.68
T15+0.3C w/o	"	64	2000	0.24	0.52
T15+10WC w/o	"	65	1800	0.078	0.40
T15+20WC w/o	"	64	1600	0.062	0.36
T15+10VC w/o	"	64	1500	0.058	0.35
T15 Crucible	1h, 1200°C oil q + 2x2x550°C	67	3100	0.35	0.92

1 - volume loss per unit area (S), per unit length (l) of abrasion on the alumina wheel.

It can be seen that the resistance to abrasive wear is higher for T15 with addition of carbon, which has higher concentration of fine carbides. Addition of WC and VC results in increase of wear resistance, significant low stress, P=0.1 MPa, and less pronounced at P=0.2 MPa. This may be a result of nonuniform distribution of added particles of carbides which are concentrated mainly on the interparticle boundaries of T15 powder and are pulled out at higher stresses. Example of wear tested surface for T15-10WC specimen is shown in Fig.5. Higher wear of hot processed T15 (Crucible) at P=0.2 MPa as compared to cold sintered water atomized T15 is probably a result of coarser carbides.

Higher mechanical performance of high speed tool steel- carbides composites fabricated employing cold sintering of compacts treated in reducing atmosphere may be expected if more homogeneous distribution and higher volume concentration of fine carbides are obtained. One of the possibilities to obtain such compositions is by using very fine elemental iron powder with additions of fine carbides [15] or of graphite and fine metal powders. High energy attrition milling should result in homogeneous distribution of alloying elements. Relatively low processing temperatures used in cold sintering may provide conditions for retention of high volume concentration of hard particles in tool steel based matrix.

References

1. F.V. Lenel, Powder Metallurgy, Principles and Applications, MPIF, Princeton, N.J. (1980).
2. A. Kasak and E.Y. Dulis, Powder Met. 221 (1978) 114.
3. M.T. Podob, L.K. Woods, P. Beiss and W. Huppmann, in: Modern Dev. in P/M, MPIF, Princeton, N.J. 13 (1981) 71.
4. C.G. Goetzel, Cermets, in Metal Handbook, 9th Ed., Vol.7, ASM, Metals Park, OH (1984) p.799.
5. G.A. Roberts and R.A. Cary, Tool Steels, 4th Ed., ASM, Metals Park, OH (1980).
6. F. Kayser and M. Cohen, Metal Progress, 61, No.6 (1952) 79.
7. M. Cohen, Trans ASM, 41 (1949) 35.
8. E.Y. Gutmanas, A. Rabinkin and M. Roitberg, Scripta Met. 13 (1979) 11.
9. E.Y. Gutmanas, Powder Met. Int. 15 (19983) 129.
10. E.Y. Gutmanas, in Modern Dev. in P/M, MPIF, Princeton, N.J. 15 (1985) 175.
11. E.Y. Gutmanas and A. Lawley, in Progress in P/M, MPIF, Princeton, N.J. 29 (1984) 653.
12. D.B. Goldman, J.B. Clark, S. Hart and E.Y. Gutmanas, in: Modern Dev. in P/M, MPIF, Princeton, N.J. V.17 (1985) 427.
13. E.Y. Gutmanas, D.B. Goldman, J.B. Clark and S. Hart, in: Progress in P/M, MPIF, Princeton, N.J., V.41 (1986) 401.
14. E.Y. Gutmanas, D.B. Goldman, S. Hart and D. Zak, Powder Met. Int. 18 (1986) 401.
15. E.Y. Gutmanas and A. Lawley, in: Modern Dev. in P/M, (1988) submitted.

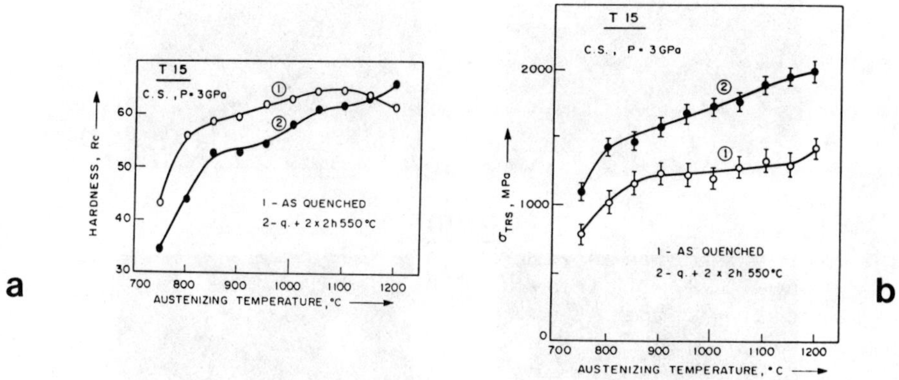

Fig.1. Dependence of a) hardness, Rc, and b) transverse rupture strength, σ_{TRS}, on austenizing temperature (1 h exposure) for T15, cold sintered at P=3GPa, curve 1 for as oil quenched specimens, curve 2 for quenched and tempered 2x2hx550°C specimens.

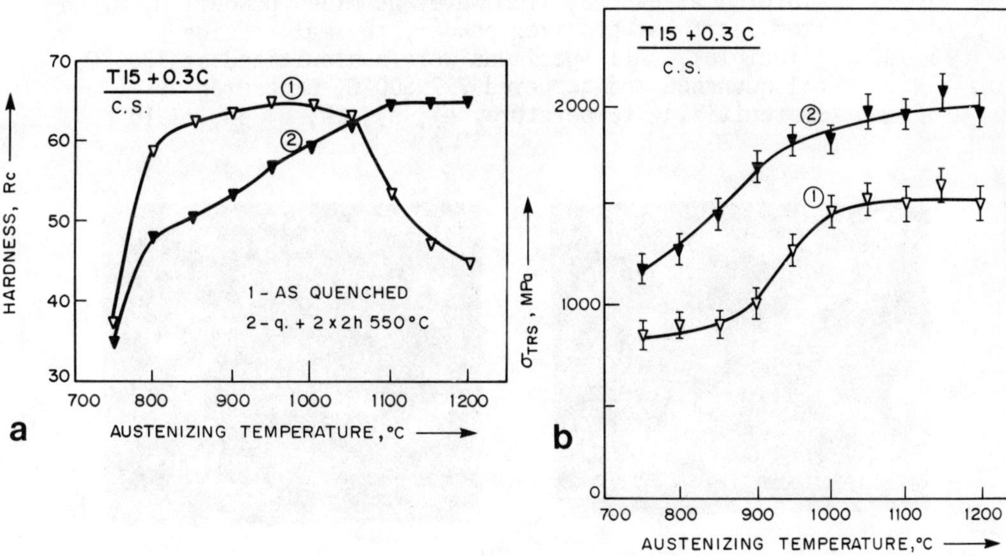

Fig.2. Dependence of a) hardness, Rc, and b) transverse rupture strength, σ_{TRS}, on austenizing temperature 91 h exposure) for T15+0.3C, cold sintered at P=3GPa. Curve 1 for as oil quenched specimens, curve 1 for quenched and tempered 2x2x550°C specimens.

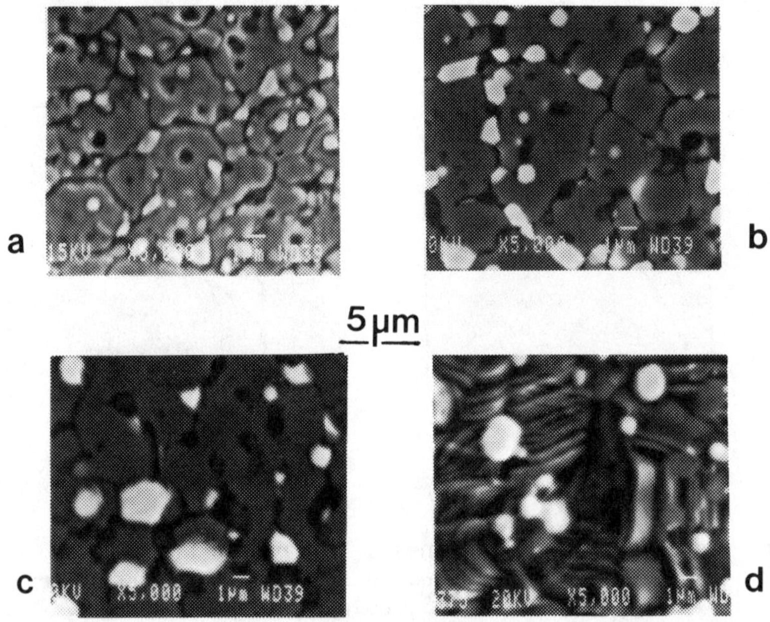

Fig.3. SEM micrographs of Ti5P/M high speed steel: a)-c) cold sintered at P=3GPa, from water atomized powder; d) HIPed from inert gas atomized powder, thermal etching (Crucible). All specimens were austenitized at 1200°C, oil quenched and tempered 2x2x500°C, exposures at austenitizing temperature: a), d), 1h, b) 3h, c) 10 h.

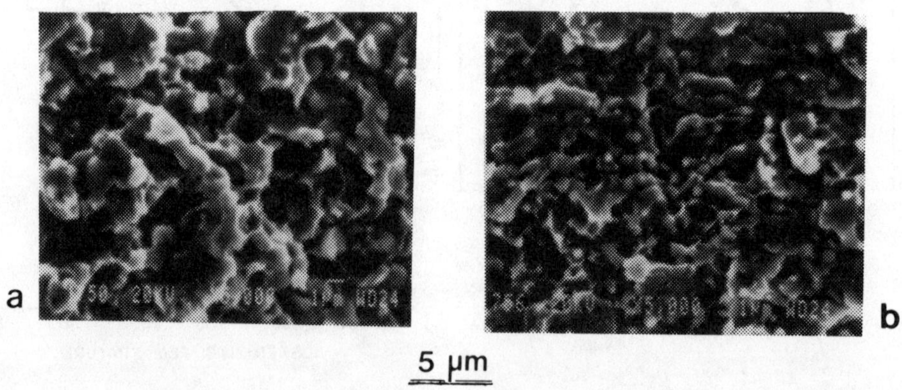

Fig.4. SEM fractographs for cold sintered at P=3GPa, austenitized for 1h at 1200°C, oil quenched and tempered 2x2hx550°C, a) T15, b) T15+10WC.

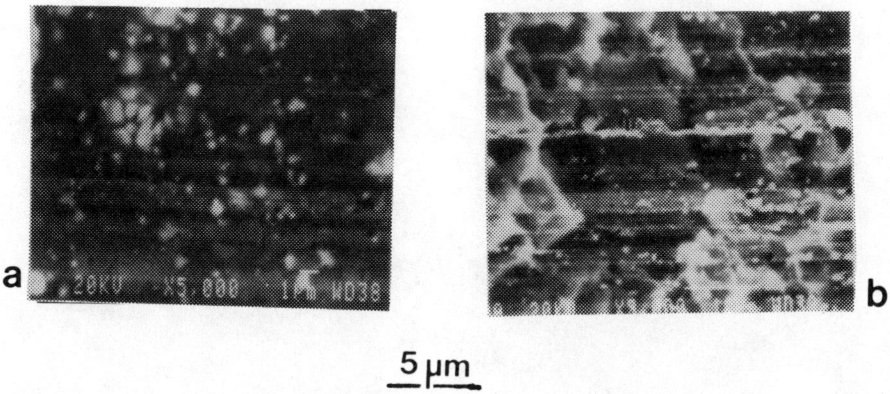

Fig.5. SEM micrograph of wear test surface for cold sintered and heat treated (1200°C, 1h, oil quenched, 2x2hx550°C). a) T15+10WC, b) T15+10V

SINTERED STEEL WITH HIGH CARBIDE CONTENT

F.Thümmler, R.Oberacker, R.Klausmann *)
Nuclear Research Center and University of Karlsruhe
D 7500 Karlsruhe 1, Postfach 3640

Abstract

Sintered steels with 10-30 vol-% of special carbides have been developed by mechanical alloying, pressing and liquid phase sintering. The best results have been achieved with systems, where the carbide phase is added as fine carbide powder, which does not take part in liquid phase formation, but is provided by addition of P or P+C. Especially in Fe-NbC-P and Fe-NbC-P-C systems very high densification and equal and fine distribution of the carbide phase in the heat treatable matrix have been reached. These alloys provide some advantages compared with HSS T15 as well as Astaloy E with 2.5% C.

Introduction

The most important wear resistant metallic alloys are the high speed steels, which contain 20-30% of various carbide forming alloying elements. Part of them are manufactured by PM methods, providing advantages with respect to fine and equal carbide distribution (Figure 1) and to manufacture near-net-shaped parts (1,2) without milling procedure. On the other hand, a very exact and costly sintering regime (±2°C, vacuum) has to be provided. A different PM-development are multicomponent alloys used in Japanese integrated camshafts (3). These alloys containing Cr, Cu, Si, Mo and C can be sintered in common belt type furnaces. They are offered as a prealloyed powder (ASTALOY E) by a European powder manufacturer (4). The microstructure of this material consists of coarse and irregular carbide phases in a martensitic matrix (Figure 2).

---

*) Now with W.C.Heraeus, Hanau

In this work, a systematic study for developing of wear resistant PM steels by using transition metal carbides is described (5). It follows the lines of:

- high densification by sintering under normal protective gas atmospheres, and at temperatures up to max. 1300°C
- fine and even distribution of carbides in a hardenable matrix
- mechanical and wear properties, which meet the requirements for working of "elastohydrodynamic contacts", where adhesive wear, abrasive wear and surface fatique take place (5).

Manufacturing and properties of the newly developed alloys shall be compared with the commercial materials mentioned above in order to look after advantages of the new systems.

Starting materials and selection of systems

The basic iron powder was water atomized ASC 100.29 from Höganäs, the hard phase was added in form of MC(or M_2C) by powder mixing technique (M = transition metal). The alternative way was to add M (in form of ferroalloy powder) and C separately, with in-situ-formation of MC (or other types of carbide) during sintering. Green density after the pressing (500 MPa) was only about 80% of theoretical, therefore a liquid phase sintering process was necessary to achieve high densification.

In the Fe-M-C systems, permanent liquid phase formation at moderate temperature occurs, e.g. in the following: Fe-Cr-C, Fe-V-C, Fe-W-C, in most cases due to eutectics, which show melting points between 1080°C (Fe-20% W-1.5% C) and 1290°C (Fe-8.2% VC) (6). From these systems, Fe-Mo-C (resp. Fe-Mo_2C) has been choosen for detailed investigation.

Fe-M-C systems, which do not form liquid phases below ∿1300°C, are e.g. Fe-Nb-C and Fe-Ti-C (6). For sintering to high densities in an "acceptable" temperature range, they need additives like B, P or P+C. In this case, transient liquid phases are formed which enhance sintering drastically.

Sintering studies

In order to investigate the principle sintering behaviour, dilatometer experiments with powder compacts have been performed in H_2-atmosphere (dew point <-50°C) up-to 1300°C.Figure 3 shows α-γ-transformation and thermal expansion in Fe-Mo_2C-compacts below 1290°C. At this temperature, the occurring liquid phase leads to additional swelling, but no shrinkage occurs during isothermal sintering at 1300°C. In Fe-Ferromolybdenum-C compacts with the same nominal composition one observes no α-γ-transformation due to solid solution formation during heating-up period. Above 1050°C carbide formation (M_3C and $M_{23}C_6$) compensates thermal expansion. At 1260°C liquid phase formation leads to some volume increase and during isothermal sintering to strong shrinkage, i.e. high densification. Figure 4 shows, that thermal expansion and isothermal sintering behaviour is strongly influenced by the carbon content (0.7-2.0%). The densification curves are pretty complicated, mainly in the range of heating-up, and the isothermal sintering behaviour is very sensitive on composition, particle characteristics and temperature.

In this system, the permanent liquid phase, occurring above a certain temperature, is formed by the main alloying constituents. The formation of mixed carbides as well the carbide solubility in the matrix results in a not defined content of carbide phase in the microstructure. The system is easy to oversinter, showing dimensional instability or, as the opposite, incomplete densification. Furthermore, decarburisation takes place during H_2-sintering, which can be overcome only by addition of hydrocarbons (CH_4) in the protective atmosphere.

In Figure 5 dilatometer curves of Fe-Nb-C and Fe-NbC-compacts are presented. Only thermal expansion and γ-α-transformation is visible during heating-up in Fe-NbC,in Fe-Ferroniobium-C additionally a volume effect from NbC formation takes place. In fact no strong reaction between matrix and NbC or Nb, resp., occurs. Only little shrinkage can be seen during isothermal sintering. Compacts of Fe-Ti-C and Fe-TiC behave similar, but even more indifferent (5). No liquid phase occurs in both systems as the eutectic melting points are 1420 and 1380°C, resp. (6).

The addition of B or P can change the densification-behaviour completely which was elaborated mainly in the system with NbC. With B, a permanent liquid phase is present above 1060°C with P a transient one >1050°C. The addition of P as a sintering aid was more successful than B with respect to the resulting microstructure. Figure 6 show the dilatometer curves of the basic powder compacts*) ASC 100.29, PASC 60 (=ASC with 0.6% P, added as Fe_2P) and mixtures with NbC. PASC exhibits a larger thermal expansion than ASC 100.29, very little α-γ-transformation due to partial Fe-P-solid solution, but strong shrinkage above 1200°C and during isothermal sintering at 1300°C. The same is valid for the PASC-NbC-compact. The relatively good wetting behaviour of NbC by the P-containing liquid phase (5) results in a sintered density of 93.5%, compared with 94% without NbC.

---

*) These compacts were pressed at 600 MPa to green densities of ∿90%.

In order to get a hardenable matrix, the addition of C is advisable, which is also very favourable for sintering. In Fe-P-C-system a permanent liquid phase occurs, the amount depending on C-content. In Figure 7, the thermal expansion and sintering behaviour of Fe-NbC-P-C can be seen. In these experiments the isothermal sintering was adjusted to the C-content, ranging from 1320 to 1280°C. The onset of strong shrinkage is within the heating-up period in all compositions and lies as low as 1160°C in the compacts with 0.9% C. Final densities between 98 and 98.5% of theoretical have been achieved.

It has to be emphasized that there is very little solid solubility of NbC in the Fe-based liquid phase end, consequently, NbC does not essentially take part in the liquid phase formation. The dilatometer curves are easy to survey and to explain. Variations of sintering temperature of ±10°C have significant influence on final density as well as on microstructure, unlike in Fe-Mo_2C-system and in HSS. The amount and type of carbide phase in the sintered product is well defined.

Microstructural optimisation by mechanical alloying

Figures 8 and 9 show the microstructures achieved with samples in the Fe-Mo-C and Fe-NbC-P-C systems in the as-sintered state. They contain grain boundary carbides, uneven carbide distribution and large residual pores in the Fe-Mo-C system. Although it was possible to improve the microstructure of Fe-Mo-C-specimen by heat treatment to some extent, it was decided to focus the following experiment on Fe-NbC-P-C due to its principle advantage in prosessing as mentioned above. Mechanical alloying was choosen for microstructural improvement.

For mechanical alloying a 1 h-treatment in a laboratory attritor in n-heptene has been applied. Tank, rotor and balls were made from unalloyed steel. In order to get an optimized distribution of NbC in Fe, the application of Carbonyl Iron instead of ASC was necessary, with mechanical addition of ferro phosphorous. A post-annealing of the powder at 600°C, followed by treatment in a tumbling device was advisable for improving the cold pressing behaviour, because the compressibility and tap density was decreased by attritoring as a consequence of deformation of the powder particles to thin plates and due to particle-particle welding effects.

After cold pressing (600 MPa) and sintering 1 h at 1280°C densities of >98% have been reached, independent on NbC-content between 10 and 30 vol-%. As shown in Figure 10, microstructure has been improved dramatically and meet the requirements much better than without attritor treatment. Similar results have been obtained with TiC as the hard component, but the densification is somewhat more difficult due to the worse wetting behaviour compared with NbC.

Properties of Fe-NbC-P-C-alloys

Figure 11 shows some mechanical properties of mechanically alloyed Fe-10% NbC-0.6% P-C systems (C = 0.45; 0.6; 0.9%) in dependence on C-content and annealing temperature. There is a strong influence of carbon content on strength properties and the higher carbon containing materials exhibit very good properties after 450°C treatment. In the unannealed state a maximum in hardness occurs at 0.6% C, due to the higher amount of residual austenite with higher C-content. In Figure 12 two examples of long term strength data are shown as Wöhler diagrams. Endurance limits of 335 and 385 MPa have been measured, which are pretty high values for sintered steels. Mechanical alloying is indispensable for high endurance limit, as seen in this figure. Preliminary abrasive wear and Vickers hardness data are collected in Figure 13 which shows an adequate low level of wear for the new alloys, generally less than Astaloy E.

Conclusions

Fe-MC-systems can be sintered to high densities, when a liquid phase takes place in the process. In case when the MC phase does not take part in liquid phase formation, P or P+C addition can take over this role, being very effective for densification. Fe-10-30% NbC-P-C alloys have been manufactured by powder route including mechanical alloying and investigated with respect to a perspective use as wear resistant material. Systems like this are not sensitive for change in MC-content and small variations in sintering temperature. Microstructure and mechanical properties of this system are promising for further investigations.

Literature

(1) R.Wähling, P.Beiss, W.J.Huppmann: Sintering Behaviour and performance Data of HSS-Components, PM-Group-Meeting, Harrogate, Oct. 1984, 151

(2) P.Beiss: PM-Methods for the Production of High Speed Steels, Metal Powder Report 38 (1983) 185

(3) T.Suganuma: Development of Sintered Integral Camshaft, Metal Powder Report 38 (1983) 433

(4) Höganäs-Information, Astaloy E, Feb. 1985

(5) for more detailed description see:
R.Klausmann: Entwicklung von carbidhaltigen Sinterstählen mit hoher Verschleißfestigkeit (Dr.-Ing.-Dissertation) KfK 4137, Oct. 1986, Kernforschungszentrum Karlsruhe

(6) H.Holleck: Binäre und ternäre Carbide und Nitride der Übergangsmetalle und ihre Phasenbeziehungen, KfK 3087 B, January 1981, Kernforschungszentrum Karlsruhe, Gebr. Borntraeger Verlag Berlin-Stuttgart, 1984

Figures

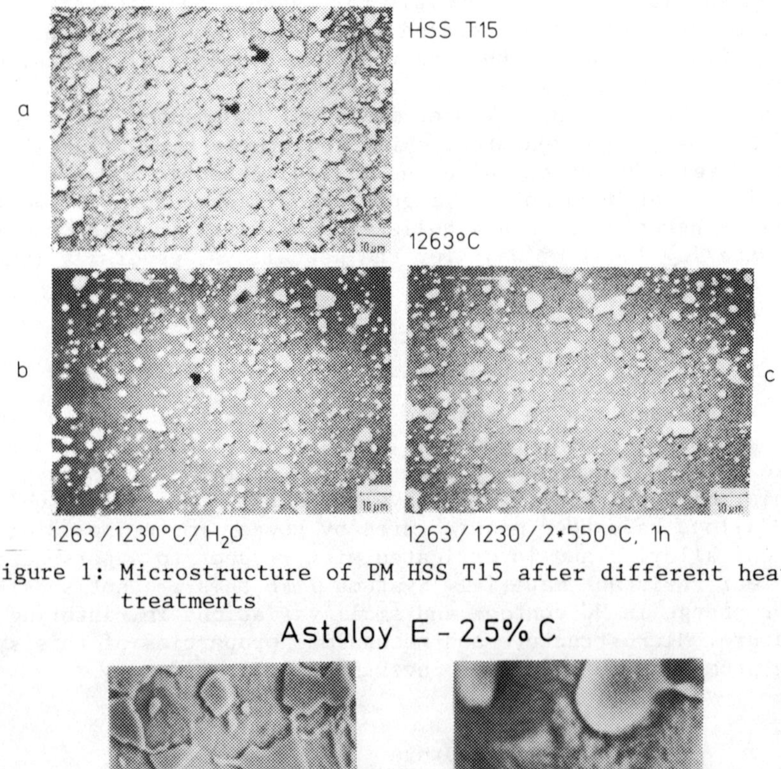

Figure 1: Microstructure of PM HSS T15 after different heat treatments

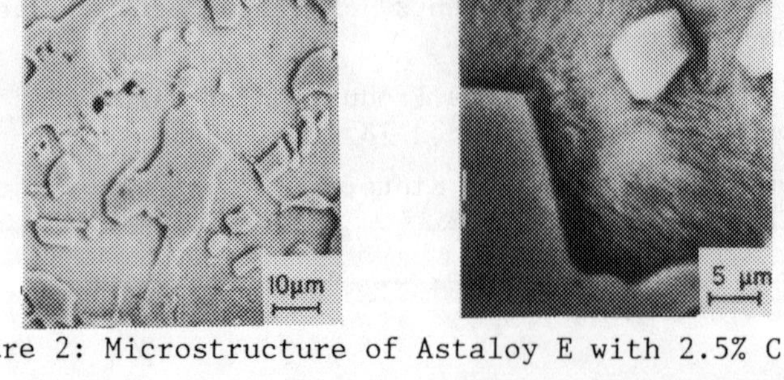

Figure 2: Microstructure of Astaloy E with 2.5% C

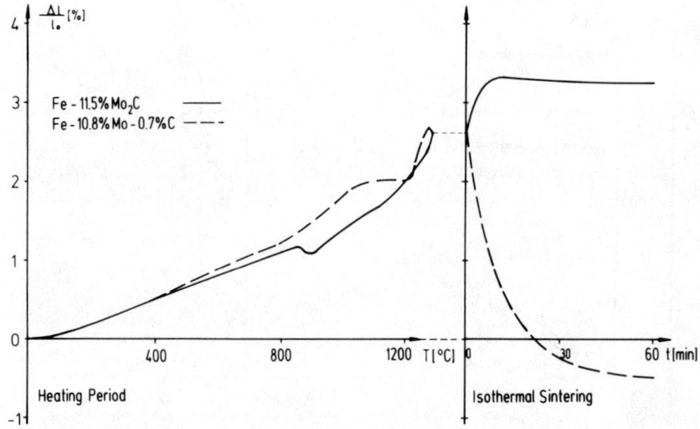

Figure 3: Dilatometer experiments on $Fe-Mo_2C$ and Fe-Mo-C mixtures

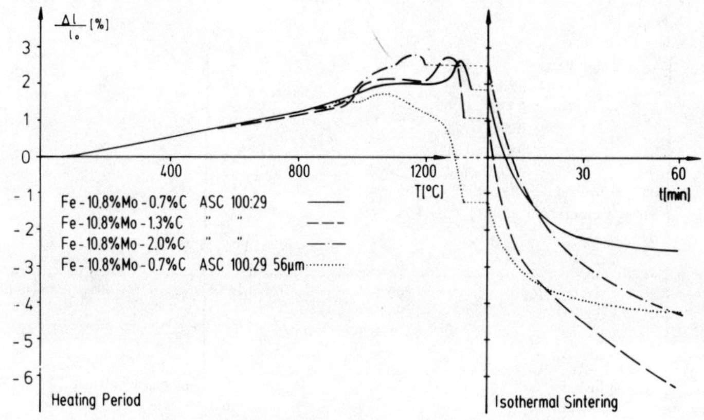

Figure 4: Dilatometer experiments on Fe-Mo-C mixtures with different C-content

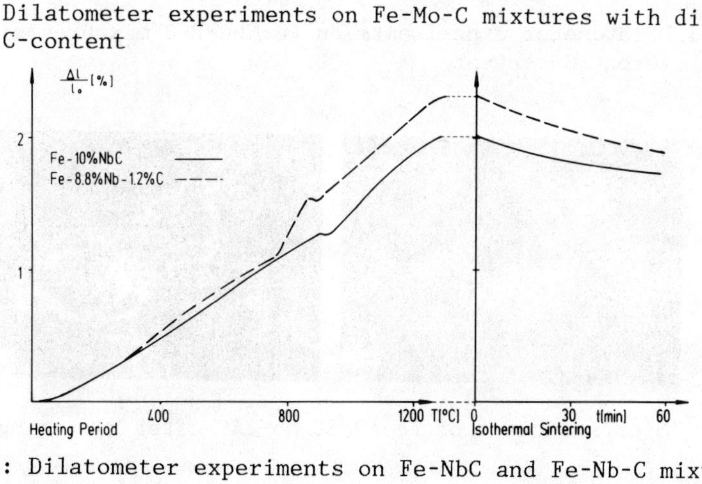

Figure 5: Dilatometer experiments on Fe-NbC and Fe-Nb-C mixtures

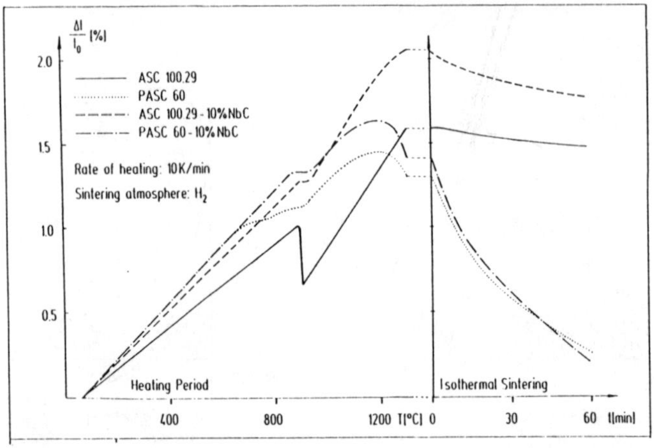

Figure 6: Dilatometer experiments on Fe, Fe-P and Fe-NbC-P

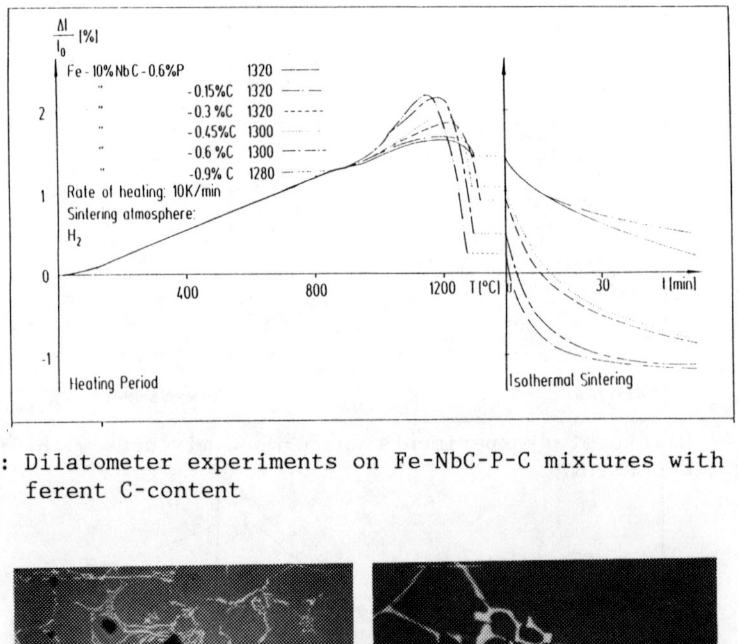

Figure 7: Dilatometer experiments on Fe-NbC-P-C mixtures with different C-content

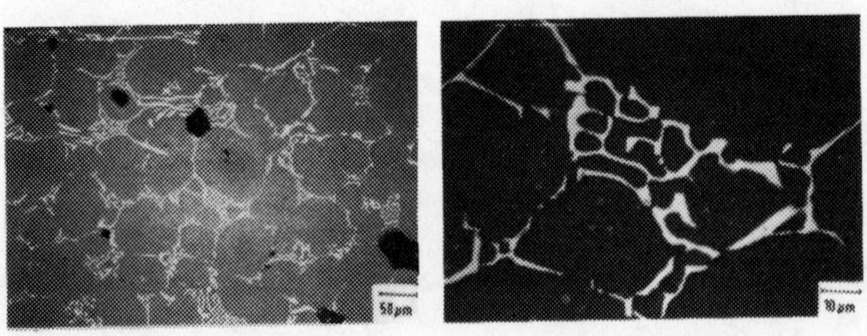

Figure 8: Microstructure of Fe-10.8% Mo-2%C after sintering at 1170°C

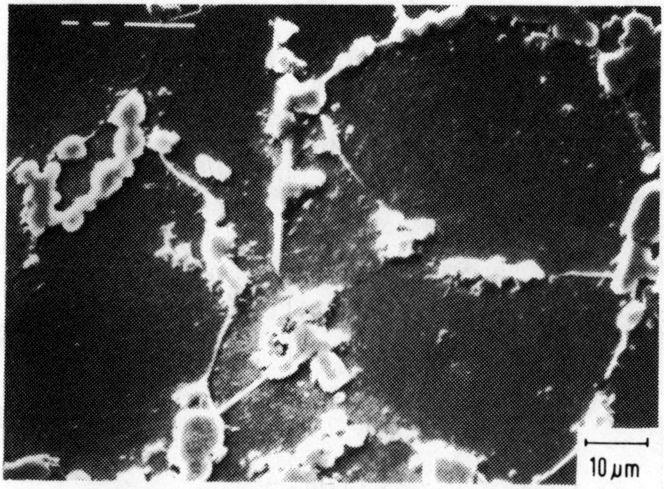

Figure 9: Microstructure of Fe-NbC-P (PASC 60 + 10% NbC) after "normal" mixing, pressing and sintering (1 h/1320°C) procedure

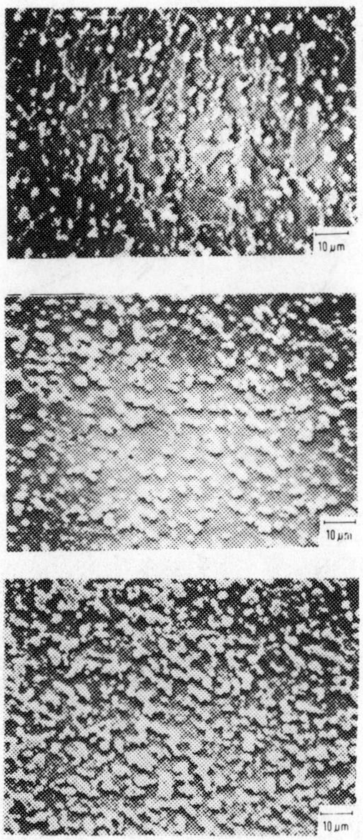

10% NbC

20% NbC

30% NbC

Figure 10: Microstructure of mechanically alloyed Fe-NbC-0.6%-P-materials

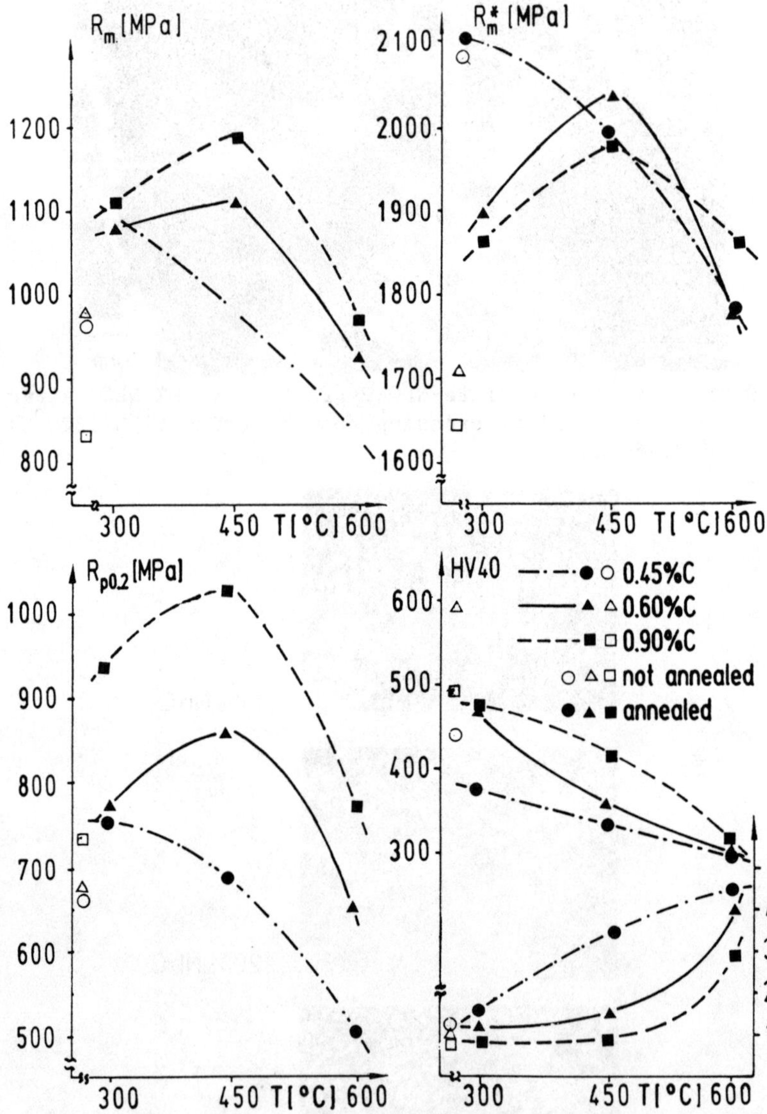

Figure 11: Tensile strength, bend strength, yield strength, hardness and elogation at fracture of mechanically alloyed Fe-10% NbC-0.6% P-C-systems with different C-contents without and after heat treatment

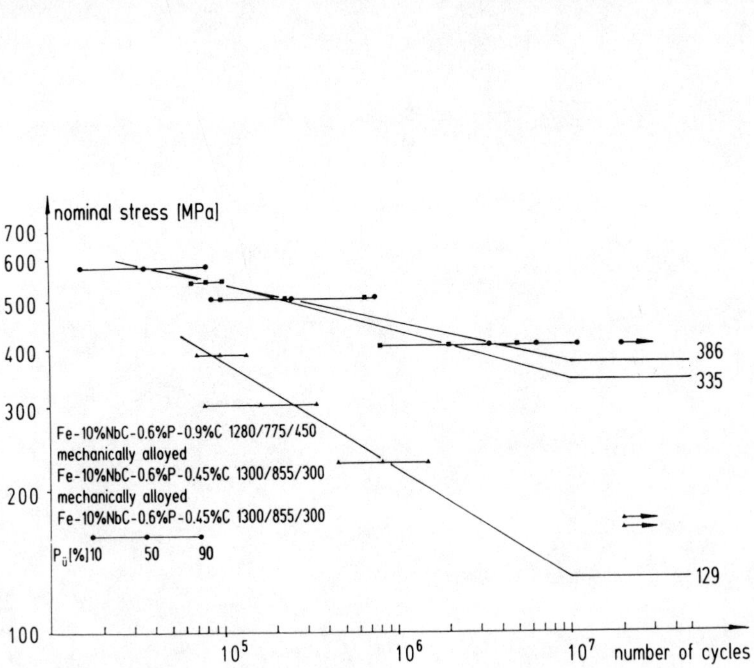

Figure 12: Wöhler diagrams of Fe-NbC-C-materials with and without mechanical alloying

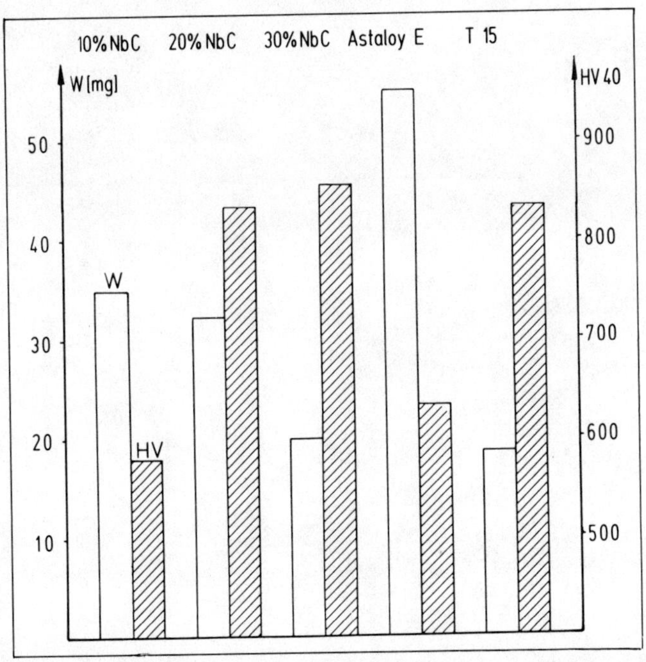

Figure 13: Abrasive wear and Vickers hardness of Fe-NbC-0.6% P-0.9% C-alloys, compared with Astaloy E and PM HSS T 15

INTERFACE BONDING
IN A DIAMOND/METAL MATRIX COMPOSITE

Zengdong Lin* and R. A. Queeney**

*Beijing Research Institute of P/M
Beijing, China
**The Penn.State University, USA

Abstract

The present study examines the interface established between diamonds and Cu-Sn metal matrix in which they were imbedded. Significant changes were induced in the interface by adding a carbide-former, titanium, into the Cu-Sn matrix. Interface reactions involved the formation of adherent carbides, bonding the diamonds to the matrix with a greatly strengthened intermediate layer. The chemical composition of the composite was examined using electron energy dispersive microanalysis. Ti was found to have been concentrated at the diamond/matrix interface; titanium exhibited a fivefold enhancement there relative to the general matrix value. Conversely, copper was depleted in regions near the diamonds relative to the matrix occurrence to that element. Tin, alloyed with the Cu for matrix strength, was also concentrated near the diamond interface. However, transmission electron microscopic (TEM) studies indicated that TiC was the sole carbide present.

Introduction

Diamond reinforced, or impregnated tools are widely used in industry, being found in drilling bits, cutting saws, grinding wheels, and conventional wheel dressers. These tools are diamond reinforced composites in actuality. The wear resistant diamon particles are held in a metallic matrix; the agglomerate is formed by blending the diamonds with a powder form of the matrix alloy and consolidated by cold pressing, sintering, and sometimes hot pressing for full densification. Understanding the particulars of the interface formed between the diamonds and the matrix is central to composite design and diamond tooling development progress.

A strong bond between the reinforcement (diamond) and the metallic Cu alloy matrix depends upon the ability of the latter to "wet" the former. For powder fabrication wholly within the solid state, the matrix is never actually molten: however, at elevated sintering temperatures, successful bonding implies interdiffusion of contacting species with matrix species mobility approaching that associated with the liquid phase. Appreciable bonging, or wetting, implies a low value of interfacial free energy between the diamond and metallic matrix in question. Copper, and many other potential matrix metals and alloys, exhibit high values of interfacial energy when in contact with diamond. Interfacial energies cab be markedly altered by slight compositional changes. Naigein and Kigikov (1,2) have studied alloy systems appropriate to wetting both diamond and graphite. Their studies indicated that alloying Cu with carbide formers such as Ti, Zr, Cr,V, etc., improved the wettability of copper alloys markedly.

Figure 1 illustrates the wetting of a diamond surface by a Cu base alloy, and the subsequent effect of alloying with carbide formers. A thin layer of carbide is formed on the diamond surface and is compatible with the latter due to carbon transfer. In addition, the carbide film is compatible with the contacting Cu-Ti alloy, due to interdiffusion of the Ti between the film and the matrix alloy. An intermediate layer, the carbide, provides the overall bonding due to its relatively high compatibility with chemical species in both the reinforcement and the matrix.

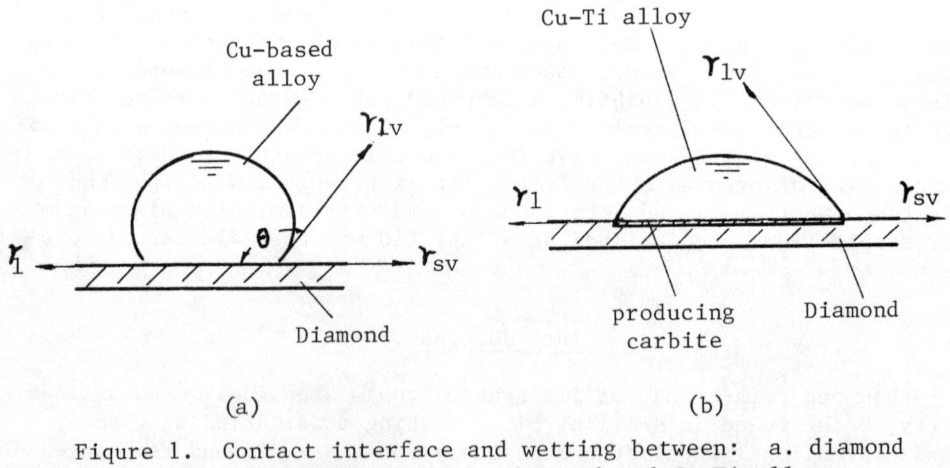

Fiqure 1. Contact interface and wetting between: a. diamond and Cu alloy, and b. diamond and Cu-Ti alloy.

Little information is extant on the microstructure of bonding interfaces between diamond and contacting metals and alloy. Kigikov (2), Mortimer (3), and Scott (4) have provided some insight into interface microstructure by performing actual molten sessile drop experiments. In addition to examining the effect of concentration of carbide former on interfacial free energy (2,3), electron microprobe analysis established figh concentrations of the carbide former within the interface (3,4). Optical light metallography established the existence of an interface larger, but its precise composition was not determined due to methodologies employed (3,4). Mortimer(3) did establish that chromium contents were close to those of Cr_3C_2 in the interface (84.3 wt % found, 86 wt % expected).

The system diamond/Cu-Ti has been examined and discussed in more detail (5,6,7). Improved mechanical response has been reported (8) wherein more than one type of diamond coating has been systematically characterized. Diamond/Cu-Ti composite reponse, while superior to diamond/Cu strength, nonetheless was surpassed by composites in which the diamond was first coated with the carbide former Ti. The present work examines interface chemistry and structure of these systems in further detail.

Composite fabrication

Four types of diamond reinforced, Cu-Sn matrix composites were fabricated for the experimental studies: as shall be seen below, not all were suitable candidates but, nonetheless, illstrate the gains to be realized by process control. Figures 2a through 2d, below, illustrate four different approaches to composite fabrication, those approaches conventionally taken in some forms of commercial practice. Those composites shown in Figures 2a and 2c are conventional in that the matrix was a 75 wt % Cu- 25 wt % Sn alloy, except that 2c contained several percent Fe by weight. The diamonds were 40/50 mesh size; compacts were sintered at 900 °C for one hour in a vacuum of 10^{-5} torr. As expected, adhesion of the matrix, even with Fe present, to the diamonds is poor; interfatial voids are clearly evidenced. In Figure 2b, the overall composite appearance has been improved because, instead of merely sintering the compact, it was hot pressed at 900 °C for one hour at 10^{-5} torr and a mild uniaxial pressure (approximately 7 MPa) in a graphite die. However, despite the

a) Composite in sintering

b) Composite in hot press

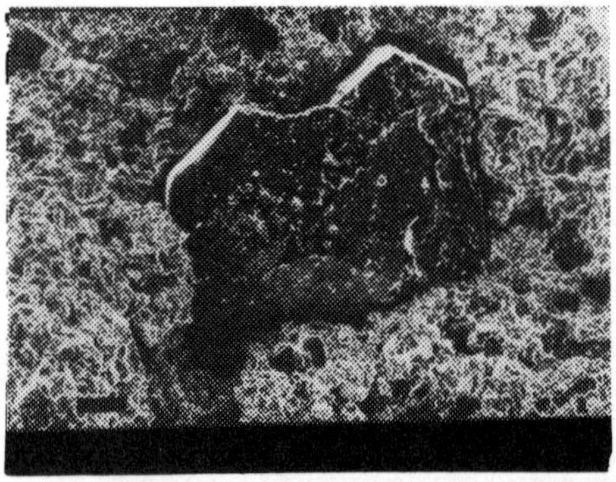

c) Composite in infiltrating With Cu-Sn alloy

d) Composite in infiltrating with Cu-Sn-Ti alloy

Figure 2. Bonding states between diamond and matrix

competent appearance of the fully dense composite,the diamonds particulates could easily be plucked out of the matrix; the 75 Cu - 25 Sn matrix alloy without carbide formers acts as a passive binder only, transferring loading to the reinforcement only because it grips the latter physically, with a possible frictional grip enhancement.

Figure 2d represents an example of a thorourhly bonded diamond composite. The matrix alloy was Cu - 10 Sn - 15 Ti and the diamonds were again 40/50 mesh size. The composite was sintered at 900 C for one hour at 10^{-5} torr as was done for the other composites. The usual methods of cutting and ion-thinning sections of composite before transmition electron microscopy (9) could not be employed for the diamond composites: not onlydid the diamond constituent resist mechanical cutting admirably but it also ion-milled much more slowly than the matrix. Thin section suitable for viewing could not be achieved. Instead, a single diamond (1/2-carat) "composite" was prepared by placing Cu - 10 Sn - 15 Ti powder on a ground and polished facet, sintering this assemblage for the usual one hour at 900 °C and 10^{-5} torr. Figure 3 shows the interface thus obtained, with Ti distribution maps superimposed.

Interface composition

The distribution of Ti in the interfacial region is shown clearly in figure 3a. As reported by others (4), Ti content is localized in the interface environs. figure 3b represents Ti concentration measured along an outwardly directed normal from the diamond surface. Quantitative assessment of Ti content, along the outwardly directed normal, is given in figure 4. Between zero and three μ-meters from the diamond surface , Ti content was constant at 89 wt %. Thereafter Ti concentration fell off rafidly, and approashed the average matrix content of 15 wt % when 40 um from the diamond surface.

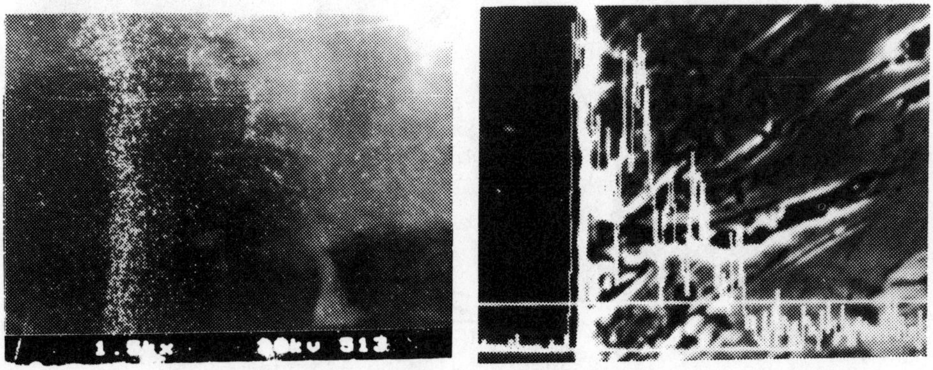

a) Ti element area distribution b) Ti element line distribution

Figure 3. Ti concentration near diamond/Cu-Sn-Ti interface

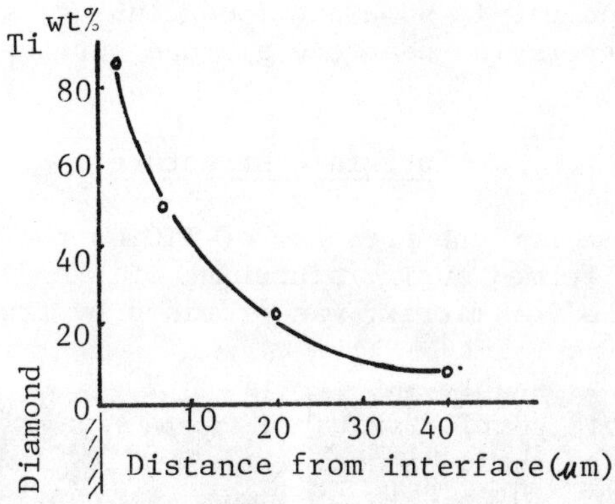

Figure 4. Titanium content along outwardly directed normal from diamond surface (15 wt % nominal matrix concentration)

Similar spectral microanalysis results were obtained in the actual composite specimen of Figure 2d, except that nearest neighbor diamond particles made their presence felt. Figure 5 shows the spatial distrabution of Cu, Sn, and Ti between nearest neighbor diamond separated by 100 μm. As expected from the above results, titanium is distinstly concentrated at the two diamond surfaces, although the aapparent magnitude of species concentration is slightly less than for the idealized single diamond case. Exactly the inverse, relative to titanium, is found for the distribution of Cu: copper is most prevalent near the center of the diamond-to-diamond separation,

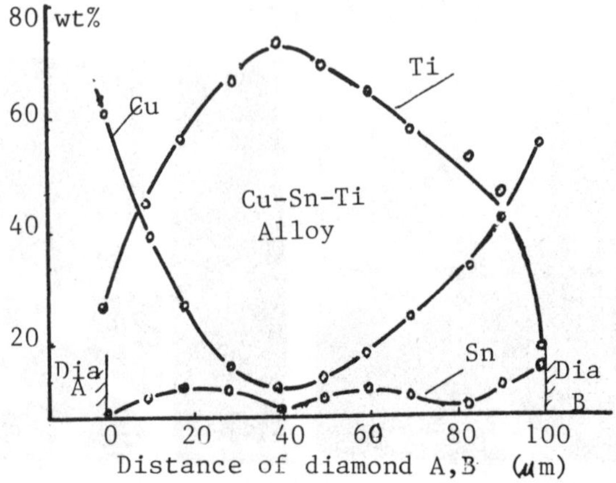

Figure 5. Cu, Sn, and Ti content between two nearest neighbor diamonds

least in evidence near the diamond surface and displaces Cu there. The Sn distribution map is somewhat inconclusive, as it decreases in magnitude of concentration near one diamond surface but increases the other.

Interface structure

Direct evidence for the presence of TiC on the diamond surface, areaction product formed during sintering of the diamond in Cu - 10 Sn - 15 Ti powder alloy matrix, was provided by transmission

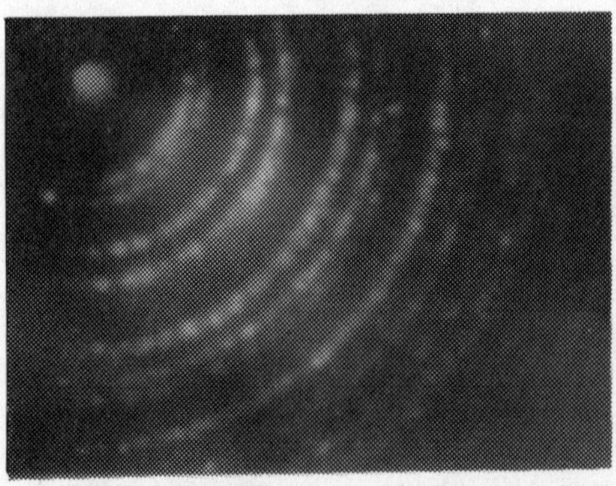

Figure 6. Selected area diffraction rings from interface

Figure 7. Profile of alloy film on diamond surface

electron microscopy (TEM). The usual TEM mode of transmitting through a thin film could not be employed, as noted earlier. Instead the electron beam was reflected at a very low angle of incidence layer was positioned nearly parallel to the electron beam. Figure 7. is a shadowgraph of the alloy film on the diaond surface. with the microscope in the diffraction mode, aselected area diffraction image was obtained for the interface region,resulting in figure 6. The interface region contribution to diffraction is seen to be polycrystalline, due to the rings, with extremely small crystallite sizes apparently present. Table 1, below, represents the results of calculated versus measured ring spacing, on the assumption that the interface is TiC. Table 1 establishes the interface region to be TiC.

Radius of Diffraction Ring (mm)	Interplanar Separations (Å) Calculated	Expected	percent Difference
12.5	2.517	2.510	0.2
14.0	2.240	2.179	3.1
20.5	1.535	1.535	0
24.0	1.311	1.311	0
25.0	1.258	1.255	0.2
28.0	1.123	1.036	3.4
29.5	1.066	0.997	6.9
31.5	0.998	0.997	2.1
35.5	0.884	0.884	0
36.5	0.839	0.833	0.7

Table 1. Interplanar TiC distances calculated from diffraction ring measurements compared to standard values

Summary and conclusion

Incorporating Ti in a typical Cu-Sn bronze matrix used to contain and hold diamond will provide an interfacial TiC layer on the diamond particulates when the aggregate is subjected to typical powder metallurgical processing schenes. The layer, approximately three to five microns thick, allows sufficient interdiffusion with the diamond (carbon) on one side, and the Ti in the matrix alloy on the other side to provide for full wetting and bonding.

References

1. U. B. Naigein and G. A. Karesulenko, Powder Metallurgy (Russian) No. 1 (1963): No. 3 (1964).
2. E. D. Kigikov, Metal Science and Heat Treatment (Russian) 71, 61-65 (1975)
3. D. A. Mortimer, and M. Nicholas, J. Mat'ls. Science, 8, 640-648 (1973).
4. P. M. Scott, M. Nicholas, and B. Dewar, J. Mat'ls. Science, 10, 1833-1840 (1973).
5 Zengdong Lin and Z. Xu, Synthetic Diamond (Chinese), 1-15, (1978).
6 Zengdong Lin and Z. Xu, Synthetic Diamond (chinese), No3, (1980).
7 Qiaojun Gao, Acta Met. Sinica, 19, 6 (1983).
8 Zengdong Lin and R. A. Queeney, Pow. Met. Int'l., 18, 76-79 (1986).
9 M. Vedula and R. A. Queeney, Interfaces in Metal Matrix Composite Dhingra and Fishman, Eds., TMS/AIME, Warrendale, PA (1896).

TITANIUM DIBORIDE AND SILICON CARBIDE PARTICULATE REINFORCED ALUMINIUM ALLOY METAL MATRIX COMPOSITES

by

B. Roebuck and A.E.J. Forno
Division of Materials Applications
National Physical Laboratory, Teddington, UK

ABSTRACT

A comparative study was made of 2014 aluminium alloy metal matrix composites (MMCs) reinforced with titanium diboride (TiB_2) or silicon carbide (SiC). Composites were manufactured by blending powders, followed by cold isostatic pressing and hot extrusion. The effects of additions of two different grades of TiB_2 were examined with nominal particle sizes of 2 and 8 µm respectively. The silicon carbide particle size was 5-10 µm. The microstructures were characterised using colour interference metallography for phase identification and particulate volume fraction measurements by image analysis. Mechanical properties were assessed by measurement of hardness and tensile properties including fractography. Careful attention was paid to obtaining accurate values of Young's modulus.

INTRODUCTION

Additions of particulate SiC have been shown to be an effective method of increasing the specific strength and modulus of aluminium alloys (1). However, the precise metallurgical reasons why the use of SiC is so advantageous are not yet clear. The objective of the current work was to compare SiC reinforcement with that conferred by the use of particulate TiB_2 in order that the relative importance of various microstructural characteristics should be made more clear. TiB_2 has a high Young's modulus, 570 kN/mm^2, compared with 450 kN/mm^2 for SiC. Also like SiC, its use in aluminium metal matrix composites (MMCs) is not likely to result in interfacial reaction products (2). Examination of a number of SiC/Al MMCs (3,4) has shown that fracture of the composite results in considerable amounts of particulate fracture. Thus the intrinsic strength of the particulate reinforcement could be an important parameter in determing composite properties. However, in other SiC/Al MMCs, particularly with small sizes of reinforcement particulates few broken particles were observed (5), indicating that fracture mechanisms could be dependent on reinforcement size. Early work (6) on particulate reinforcement of copper alloys showed that the ductility was independent of particle size but decreased with increasing volume fraction of particulate, whereas the amount of strengthening conferred by the added particles was dependent on the strength of the particle/matrix interface. Also it was found that a good bond was probably needed for an increase in strength to be observed. The strength of the interface is also likely to play a part in fracture initiation since particle/matrix decohesion should enhance void nucleation in the matrix whereas a strong bond would be more likely to lead to particle fracture (7). For SiC/Al MMCs it has been established (1,5) that Young's modulus is independent of particle size whereas (5) strength increases as the SiC particle size decreases (at constant volume fraction).

The metal matrix composites were manufactured by a powder metallurgy route. 2014 grade aluminium alloy powders were blended with approximately 17 wt% of the particulate then cold isostatically pressed and hot extruded to rod, with an extrusion ratio of about 23 to 1. The powders were not degassed before extrusion. Further experiments are being performed to examine the effects of degassing. Two extrusion temperatures were used, 365 ^{o}C and 500 ^{o}C. Two grades of TiB_2 were chosen for the experiments in order to examine the effect of TiB_2 particulate size. One grade had a nominal particle size of 8 μm, similar to that of the 5-10 μm of the SiC; in the other grade the particle size was much smaller, around 2 μm.

Ageing experiments at 150 ^{o}C and 180 ^{o}C showed that both the different types of particulate reinforcement increased the hardness substantially compared with the base 2014 alloy. The composites containing TiB_2 were slightly harder than those containing SiC. Tensile tests confirmed the increases in strength although ductility was reduced, as has been shown in other SiC particulate reinforced MMCs (1). The magnitude of the increase in strength was dependent on the extrusion temperature. Slightly stronger materials were obtained after extrusion at 365 ^{o}C. Since the particulate was added as equal weight fractions, and TiB_2 is more dense than SiC, the mechanical tests showed that for an equivalent volume fraction the TiB_2 appeared to be more effective than SiC as a strengthening addition. The Young's moduli of the composites were carefully measured and comparisons made for equivalent volume fractions of particulate. It was found that for similar volume fractions the TiB_2 conferred a larger increase in modulus. TiB_2 has a higher modulus than SiC so the relative magnitudes of increased increments in modulus indicated that load was transferred as effectively to the TiB_2 as it was to the SiC.

The volume fraction measurements were obtained by quantitative image analysis. In order to improve the accuracy of the measurements metallographic samples were interference film coated (8) in order to improve contrast between the particulate and matrix and to aid identification of phases other than the SiC or TiB_2 reinforcement.

MATERIALS AND MANUFACTURE

The 2014 aluminium alloy powder had a nominal particle size of 20 µm and had been produced by water atomisation. The SiC powder was 600 grit (5-10 µm) Carborundum grinding grade powder. The TiB_2 powders were obtained from H C Stark Ltd, grade A and grade D. The particle size characteristics quoted by the manufacturer are shown in Table 1. The value of the BET specific surface area, S, can be used to calculate an equivalent spherical particle size, d, from

$$d = \frac{6}{S\rho} \tag{1}$$

where ρ is the density. The density of TiB_2 is 4.51 kg/m^3. Therefore equivalent spherical particle sizes calculated using equation (1) results in a value of about 1.33 µm for grade A and 2-6 µm for grade D. These values agree reasonably well with the quoted Fisher subsieve particle sizes (Table 1).

Three batches of aluminium alloy powder were blended with 16.7 wt% of each of the particulates SiC, TiB_2-A and TiB_2-D. The blending was performed at the UK National Engineering Laboratory by mechanical tumbling in a proprietary blending machine for 1 hour. The blended mixtures were compacted by cold isostatic pressing to about 400 N/mm^2 in polyurethane moulds. The resulting "green" billets were reasonably strong and suitable for machining and preheating for extrusion. Extrusion billets of 36 mm diameter and 40-50 mm long were prepared and preheated for about 75 min; one set at 365 °C and another set at 500 °C. The billets were extruded at NPL on a 700 tonne vertical extrusion press to 8 mm diameter rod using a die with a 120° conical entry angle (extrusion ratio of 22.6 to 1).

RESULTS AND DISCUSSION

For convenience in discussing the results and diagrams the SiC, TiB_2-A and TiB_2-D MMCs extruded at 365 °C are coded S3, T3A and T3D respectively. The MMCs extruded at 500 °C were given codes S5, T5A and T5D.

Microstructures

Metallographic samples were prepared by a method, using diamond abrasives, specially developed (8) to produce flat surfaces in polished sections of MMCs. The latter are difficult to prepare metallographically due to the disparate hardness values of the matrix and reinforcement. Flat surfaces are needed for accurate measurement of particulate size, volume fraction and distribution. Conventional black and white micrographs of the composites are shown in Fig.1. It can be seen in the lower magnification micrographs that both the SiC and TiB_2 are uniformly distributed with no undue agglomeration or clustering of the particulate reinforcement. No apparent difference can be seen as a result of the different extrusion temperatures and the structures were porosity free. At higher magnification the differences between the three batches can be more easily seen. The particulate in the TiB_2-D MMC was similar in size to the SiC but rounder and less angular. In contrast the particle size of TiB_2-A was smaller than either the SiC or TiB_2-D and was more angular than the latter and contained fines.

The micrographs in Fig.1 show that the particulate and matrix were in intimate contact. There was no obvious debonding or interfacial reaction products. The TiB_2 appeared to be similar to the SiC in this respect.

The conventional micrographs show reasonable contrast between matrix and reinforcement and look to be adequate for image analysis. However, in order to examine their suitability for this purpose an additional metallographic technique was used which can further improve contrast. This technique is called interference film metallography (IFM) and has been shown to be particularly useful for MMCs (8). A thin (submicrometre) metal oxide film is sputtered onto the surface of a metallographically prepared surface. The latter has to be clean and flat to avoid artefacts. When viewed with an optical microscope the various phases in the sample appear different colours and the colours can alter with changing thickness of film. The optical properties of film and underlying substrate determine the precise colour (8). In the case of MMCs the colours result in improved contrast between the matrix, dispersoids within the matrix and particulate. Also additional phases can be identified which cannot easily be seen by conventional metallography. This is particularly important in SiC/Al MMCs where impurities are often present in the SiC grit and which can be more easily identified using IFM (8). These impurities frequently have the same morphology and size as the SiC reinforcement. Large particles of phases other than TiB_2 were not seen in the TiB_2/Al MMCs confirming that the additional phases present in the SiC/Al MMC were contaminants from the SiC grit and not intermetallics or impurities from the 2014 powder. The IFM technique was used on both the SiC and TiB_2 reinforced MMCs in order to measure the volume fraction. The results of measurements on IFM as-extruded bar are shown in Table 2 together with results obtained on conventional black and white (BW) micrographs. Fifty areas were measured for each structure at a magnification of X500. There were about 100 reinforcement particles in each picture frame. There were significant differences between the results obtained by the two methods. Also measurements on tensile test-pieces (Table 2) did not give the same volume fraction for the TiB_2 reinforcement as the measurements on the as-extruded bar. Further work is needed to determine the source of the discrepancy. However, for the SiC reinforcement the average value of the volume fraction measurements was 0.152 which was similar to the calculated value of 0.149, assuming a density of 3.2 kg/m^3 for the SiC. For the TiB_2 reinforcement, assuming a density of 4.51 kg/m^3, the calculated volume fraction was 0.111 which was consistent with the measurements obtained on the extruded bar.

Ageing characteristics

Ageing experiments on MMCs were performed by Vickers hardness measurements (HV5) on samples that were solution treated (505 ^{o}C, 1 h), cold water quenched and aged at 150 ^{o}C or 180 ^{o}C. The results are shown in Fig.2. The experiments were repeated to confirm the results and it was noticed that after repeated solution treatment blistering and porosity occurred, particularly on the samples extruded at the lower temperature of 365 ^{o}C. The porosity was probably due to the omission of a degassing step in the manufacturing route prior to extrusion. The samples extruded at 365 ^{o}C had porosity levels of 3-8%. After extrusion at 500 ^{o}C the SiC/Al MMC contained about 3% porosity whereas in the TiB_2/Al MMCs no porosity was observed. The porosity levels appeared to increase with each solution treatment and ageing experiment. As a result the hardness values obtained after the first set of experiments on ageing at 150 ^{o}C were measured on samples with only a small amount of porosity. Because of the presence of porosity in some of the samples the results in Fig.2 must be interpreted carefully. Thus, the absolute differences cannot be assumed to give a reliable indication of absolute values of hardness. However the hardening increments show that the TiB_2 was as effective as the SiC in increasing hardness and the finer particle size TiB_2-A was more effective than TiB_2-D. This trend of increasing hardness with decreasing size of particulate reinforcement has also been observed in SiC/Al MMCs (5).

Modulus measurements

Tensile tests were performed on rectangular dog-bone test pieces. The width of the samples in the gauge length region was 3 mm. Two batches of tensile specimens were prepared for testing, one with a thickness of 1.5 mm and the other of 3 mm. About half the samples were heat treated before manufacture into test-pieces by a solution treatment at 505 ^{o}C for 1 h, followed by a cold water quench and ageing at 180 ^{o}C for 10 h (nominally peak aged). Strain was measured by strain gauges mounted on each side of the specimens in order to obtain accurate values of Young's modulus and proportional limit.

The results were acquired using a computer-based system and the modulus values were calculated using a localised least squares analysis of strain and stress data between stress values of 20 and 100 N/mm^2. The results for representative samples are shown plotted in Fig.3 against direct current electrical resistivity measurements. Measurements of the latter can be used as an indication of particulate volume fraction. The results could be divided into two groups. As a whole, the aged samples had higher values of resistivity than the extruded materials. This was due to the presence of porosity in the former. The samples as-fabricated and extruded at 500 °C had consistently lower resistivities than those extruded at 365 °C, probably reflecting differences in matrix condition. The resistivity values of the TiB_2 MMCs were independent of particle size. Because of the porosity in the aged samples only the modulus values of the as-fabricated MMCs can be taken as an indication of the effects of different reinforcements. There was some scatter in the results due probably to variations in reinforcement content of individual tensile test-pieces. However the averages for MMCs containing SiC, TiB_2-A and TiB_2-D respectively were 88, 87 and 85 kN/mm^2. McDanels (1) performed an extensive series of tests on SiC/Al MMCs and his results on measurements of Young's modulus for particulate volume fractions, V_f, less than about 0.25 can be expressed as

$$E_c = E_o + \frac{E_p V_f}{4} \qquad (2)$$

where E_c E_o and E_p the moduli of the composite, matrix and particulate respectively and E_o and E_p have values of 70 and 450 kN/mm^2 for the matrix and particulate SiC respectively. Then, for a V_f of 0.15, E_c has the value of 87 kN/mm^2 which agrees reasonably well with the measurements on samples S3 and S5. A similar expression can be applied to the TiB_2 MMCs, assuming that the volume fractions of particulate in the latter were 0.11. Thus, using equation (2), E_c is calculated to be 86 kN/mm^2 if E_p is 570 kN/mm^2. The value of 86 kN/mm^2 agrees well with the values of 85 and 87 kN/mm^2 obtained in the measurements of the Young's modulus of the TiB_2 MMCs.

Tensile stress versus strain

Typical stress versus strain curves are shown in Figs.4a and 4b for the as-extruded and heat-treated conditions respectively. The results are given in Table 3.

In the as-extruded condition the strengths of the samples extruded at 365 °C were higher than those of the samples extruded at 500 °C. No obvious trends in ductility could be seen with measured elongations of 2.5-4%. In the TiB_2/Al MMCs smaller reinforcement particles resulted in larger strength increments and increased work hardening rates in agreement with tests on SiC/Al MMCs (5). The SiC and TiB_2 MMCs extruded at 500 °C had similar strengths and since the volume fraction of TiB_2 was about 2/3 that of the SiC, it indicates that the TiB_2 reinforcement was more effective than SiC in increasing the strength of the composite. However, in contrast, in the samples extruded at 365 °C the SiC MMC was stronger than either of the TiB_2 MMCs, as might be expected from their respective particulate volume fractions. These differences are probably due to differences in the condition of the Al alloy matrix, which almost certainly varies from sample to sample in the as-fabricated condition due to differences in temperature history and reinforcement content. In practice, differences in deformation behaviour should be easier to compare in tests on peak-aged material. However in this work the results on aged samples shown in Fig.4b are not easy to interpret unambiguously because of the variability in porosity levels that arose after solution treatment and ageing. For the two TiB_2 MMCs T5A and T5D which had little or no porosity Fig.4b shows that the effect of particle size seen in the tests on as-fabricated material was confirmed. Sample T5A, with the smaller TiB_2 reinforcement was considerably stronger than T5D although the ductility of the latter was slightly higher than that of T5A. TiB_2 MMCs were on the whole stronger than the SiC MMCs. Since these trends were similar to those observed in the hardness tests it is likely that they are correct in type, if not in magnitude, even allowing for differences in porosity levels.

Fractography

Scanning electron microscopy was used to examine the fracture surfaces of the broken tensile test-pieces. In a few of the samples, very low elongation values of about 1% were measured. Fractography showed that in each case there was an obvious fracture origin which was a disc shaped cluster of particulate which had not been penetrated by the matrix in the extrusion process. More of these defects were seen in the samples extruded at the lower temperature.

Use was made of matching fracture faces, Figs.5a-c, to examine the fracture topography in each MMC. The use of this technique enables firmer conclusions to be drawn regarding the precise nature of the fracture path (3,4) than can be obtained from examination of a single face. Figs.5a-c show matching fracture surfaces in the SiC, TiB_2-D and TiB_2-A samples respectively. Fig.5a is characteristic of fracture surfaces in SiC/Al MMCs with particle sizes in the 5-10 µm range showing extensive particulate fracture and matrix failure by void nucleation and growth. However in contrast the TiB_2/Al MMCs have a different fracture appearance. In the MMCs reinforced with the coarser TiB_2 particles, grade D, most particles have debonded, Figs.5b, 6b, although some particle fracture has occurred. In contrast, in the TiB_2 MMC reinforced with the finer particles, grade A, the failure is matrix dominated (Fig.6a) and it is difficult to see any particulate debonding or fracture; although a small amount of the latter was observed, Fig.5c. A similar observation of matrix dominated failure has been seen in SiC/Al MMCs (5) with a small size of particulate size reinforcement. However although obvious differences in fracture topography were observed these differences were not reflected in differences in ductility in the tensile test. The latter is notably difficult to correlate directly with microstructural variations. Further tests on samples that have been suitably degassed before processing should be more likely to give results that are easier to interpret.

SUMMARY

It has been shown that a powder processing route can be used to produce homogeneous TiB_2/Al MMCs. There were no obvious interfacial reaction products at the matrix/reinforcement boundary. Due to its higher Young's modulus TiB_2 was more effective in increasing the modulus of a 2014 aluminium alloy than SiC. For example a volume fraction of 11% of TiB_2 resulted in the same composite modulus as about 16% of SiC. Equation (2) can be used together with values of the composite density, ρ_c, to calculate the specific stiffness for each MMC. These are shown plotted in Fig.7 against volume fraction of particulate. It can be seen that the TiB_2 and SiC MMCs have similar specific stiffnesses, particularly at volume fractions less than 0.15. For equivalent volume fractions the TiB_2/Al MMCs appeared to be harder and stronger than the SiC/Al MMCs although the results need further confirmation because some of the samples contained porosity after ageing treatments. Of the two grades of TiB_2 the batch with the smallest particle size resulted in a bigger increase in strength. The tensile ductilities were less than would be measured in tests on matrix alloys but did not appear to be sensitive to the size or type of addition. However, as the ductilities, and by inference toughness, were insensitive to the type of particulate addition then since different fracture topographies were observed it would imply that a detailed examination of the importance and the nature of the matrix/interface bond requires mechanical property test measurements which are more sensitive to microstructural variation than ductility in the tensile test.

ACKNOWLEDGEMENTS

The aluminium alloy powder was obtained from UKAEA, Harwell. We are very grateful for the practical help and valuable advice given by Murdo McLean at the UK National Engineering Laboratory where the blending and cold isostatic pressing of the powders was performed. Thanks are also due to R Jones for the microstructural characterisation of the MMCs.

REFERENCES

1. D.L. McDANELS, Met. Trans., 1985, 16A, 1105-1115.
2. R.H. BIDDULPH, in Proc. "1st European Symposium on Engineering Ceramics" Oyez STS Ltd, London, 1985, 45-61.
3. C.P. YOU, A.W. THOMPSON and I.M. BERNSTEIN, Scripta Metall., 1987, 21, 181-185.
4. B. ROEBUCK, J. Mater. Sci. Letters, 1987, 6, 1138-1140.
5. Y. FLOM and R.J. ARSENAULT, in Proc. ICCM VI, Vol.2, Elsevier Applied Science, 1987, 189-198.
6. B.I. EDELSON and W.M. BALDWIN, Jr., Trans ASM, 1962, 55, 230-250.
7. W.H. HUNT, Jr., O. RICHMOND and R.D. YOUNG, in Proc. ICCM VI, Vol.2, Elsevier Applied Science, 1987, 209-223.
8. E.G. BENNETT, J.D. LORD and B. ROEBUCK, to be submitted to Composites, 1988.

TABLE 1

TiB_2 particle size characteristics

	BET specific surface area, S m^2/g	Fisher subsieve particle size µm	Size distribution limit (from Sedigraph 5000 D) µm
Grade A	> 1	1.5 - 2.5	95% < 4-5
Grade D	0.2 - 0.6	4 - 10	95% < 10

TABLE 2

Volume fraction of reinforcement (image analysis)

Sample *		Particulate volume fraction	Sample **		Particulate volume fraction
TiB_2-A	(BW)	0.118	TiB_2-A	(BW)	-
	(IFM)	0.107		(IFM)	0.121
TiB_2-D	(BW)	-	TiB_2-D	(BW)	-
	(IFM)	-		(IFM)	0.123
SiC	(BW)	0.150	SiC	(BW)	
	(IFM)	0.160		(IFM)	0.152

* - extruded bar ** - tensile test piece

(BW) - Conventional black and white micrographs
(IFM) - Interference film micrographs

TABLE 3

Tensile Test Results

Sample *	Thick -ness mm	Strength kN/mm^2 0.1% proof	Strength kN/mm^2 UTS	Elong-gation %	Thick -ness mm	Strength kN/mm^2 0.1% proof	Strength kN/mm^2 UTS	Elong-gation %
E/T5A	1.5	0.20	0.31	3.0	3.0	0.17	0.29	2.5 †
A/T5A	1.5	0.41	0.45	2.0	3.0	0.42	0.46	0.7 †
E/T5D	1.5	0.17	0.27	3.2	3.0	0.18	0.29	4.1
A/T5D	1.5	0.34	0.43	1.9	3.0	0.35	0.46	3.2
E/S5	1.5	0.19	0.26	1.9	3.0	0.19	0.28	3.6
A/S5	1.5	0.25	0.33	1.7	3.0	0.29	0.38	1.3
E/T3A	1.5	0.20	0.31	3.9	3.0	0.20	0.30	2.7
A/T3A	1.5	0.35	0.36	0.6 †	3.0	0.33	0.40	1.4
E/T3D	1.5	0.19	0.27	2.4	3.0	0.19	0.29	3.4
A/T3D	1.5	0.28	0.37	1.4	3.0	0.27	0.29	0.6 †
E/S3	1.5	0.22	0.30	2.4	3.0	0.22	0.32	3.4
A/S3	1.5	0.28	0.36	1.8	3.0	0.31	0.31	0.5 †

* E/- as extruded: A/- aged.
T5A, T3A; Al/TiB_2-A extruded at 500 °C and 365 °C respectively.
T5D, T3D; Al/TiB_2-D extruded at 500 °C and 365 °C respectively.
S5, S3; Al/SiC extruded at 500 °C and 365 °C respectively.
† Large-defect-initiated failure.

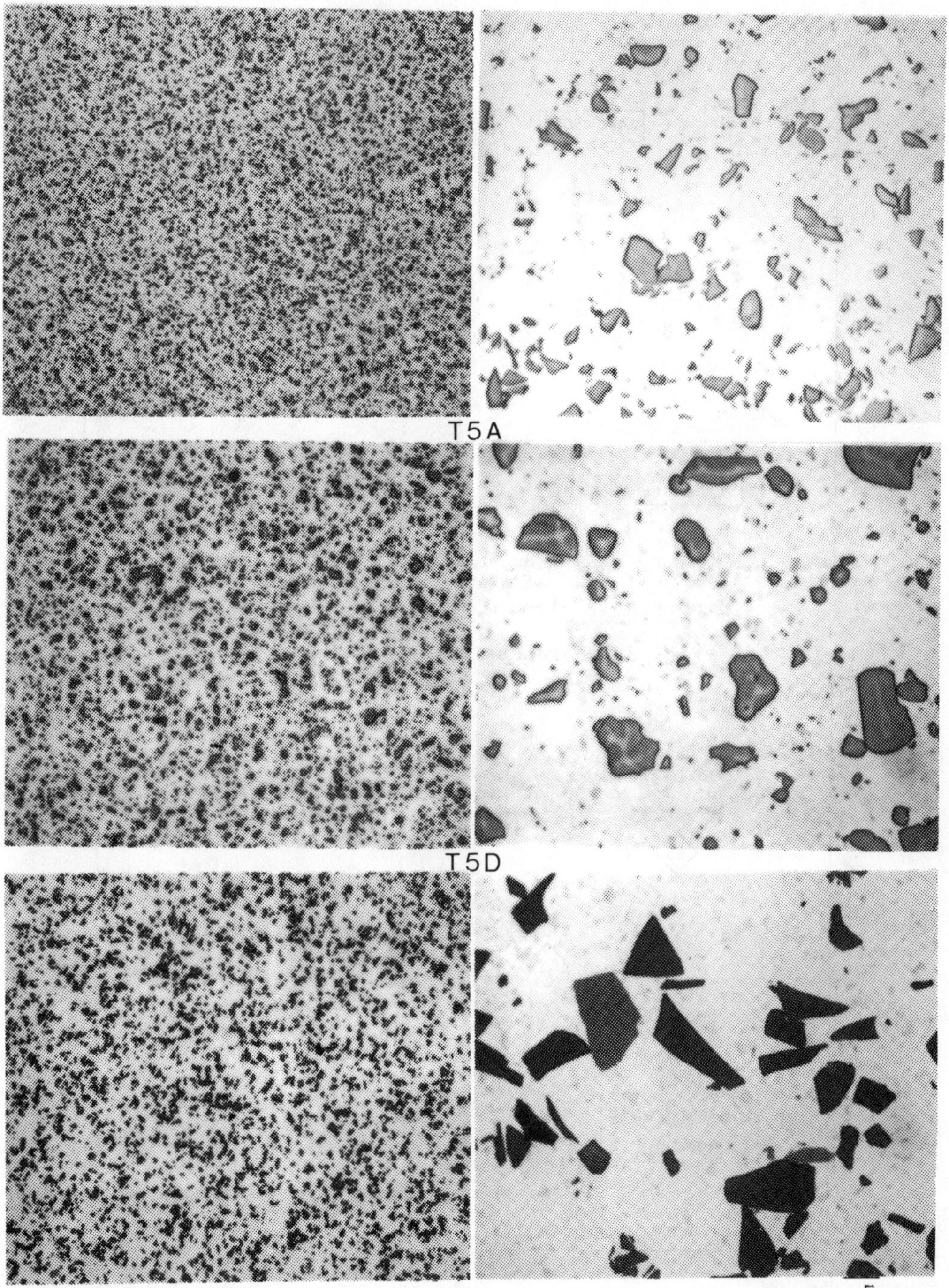

Fig. 1. Optical micrographs of SiC and TiB_2 MMCs

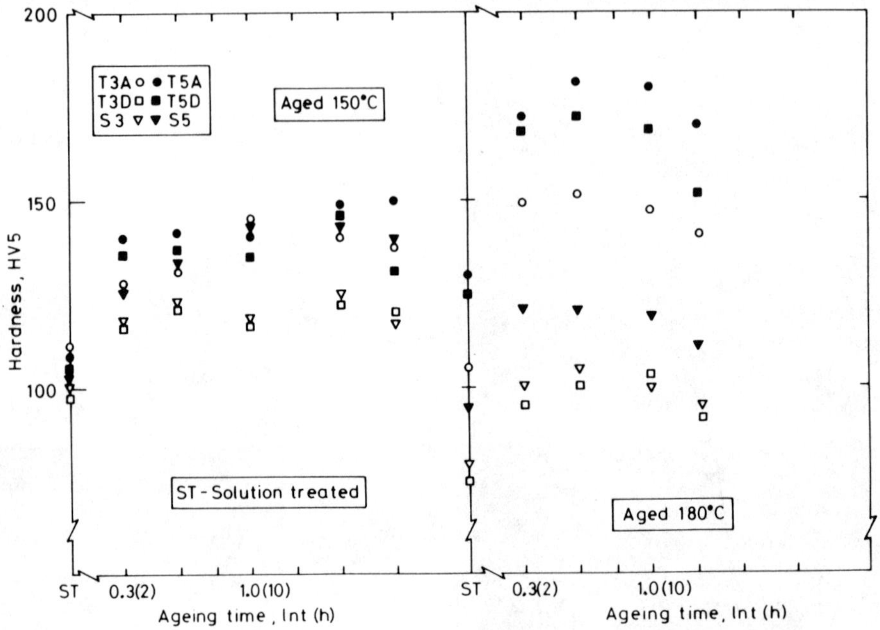

Fig. 2. Hardness versus ageing time

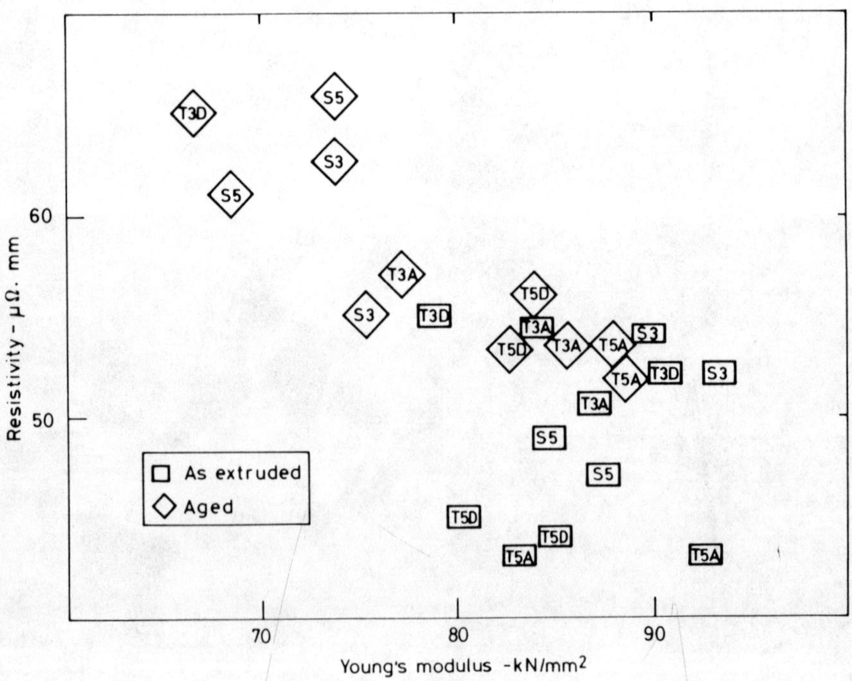

Fig. 3. Electrical resistivity versus Young's modulus

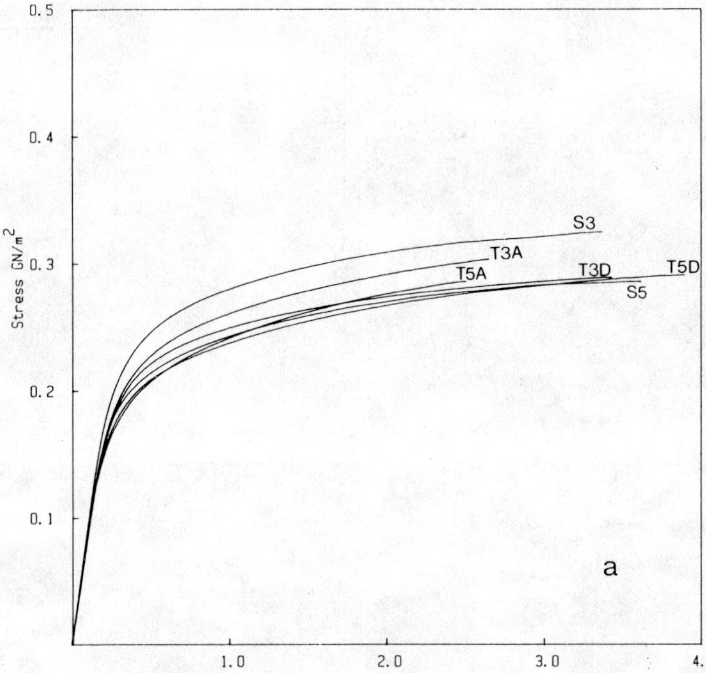

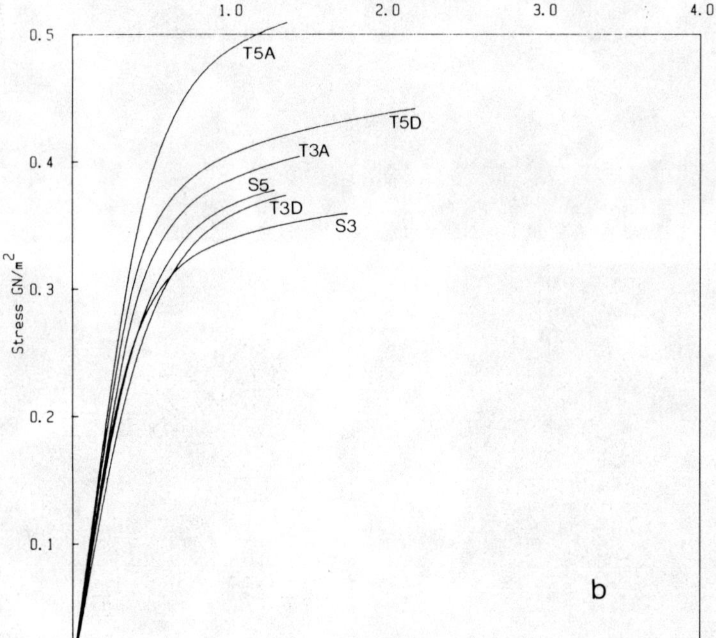

Figs. 4a and b. Stress versus strain curves for tensile tests on samples that were a) as-extruded and b) aged

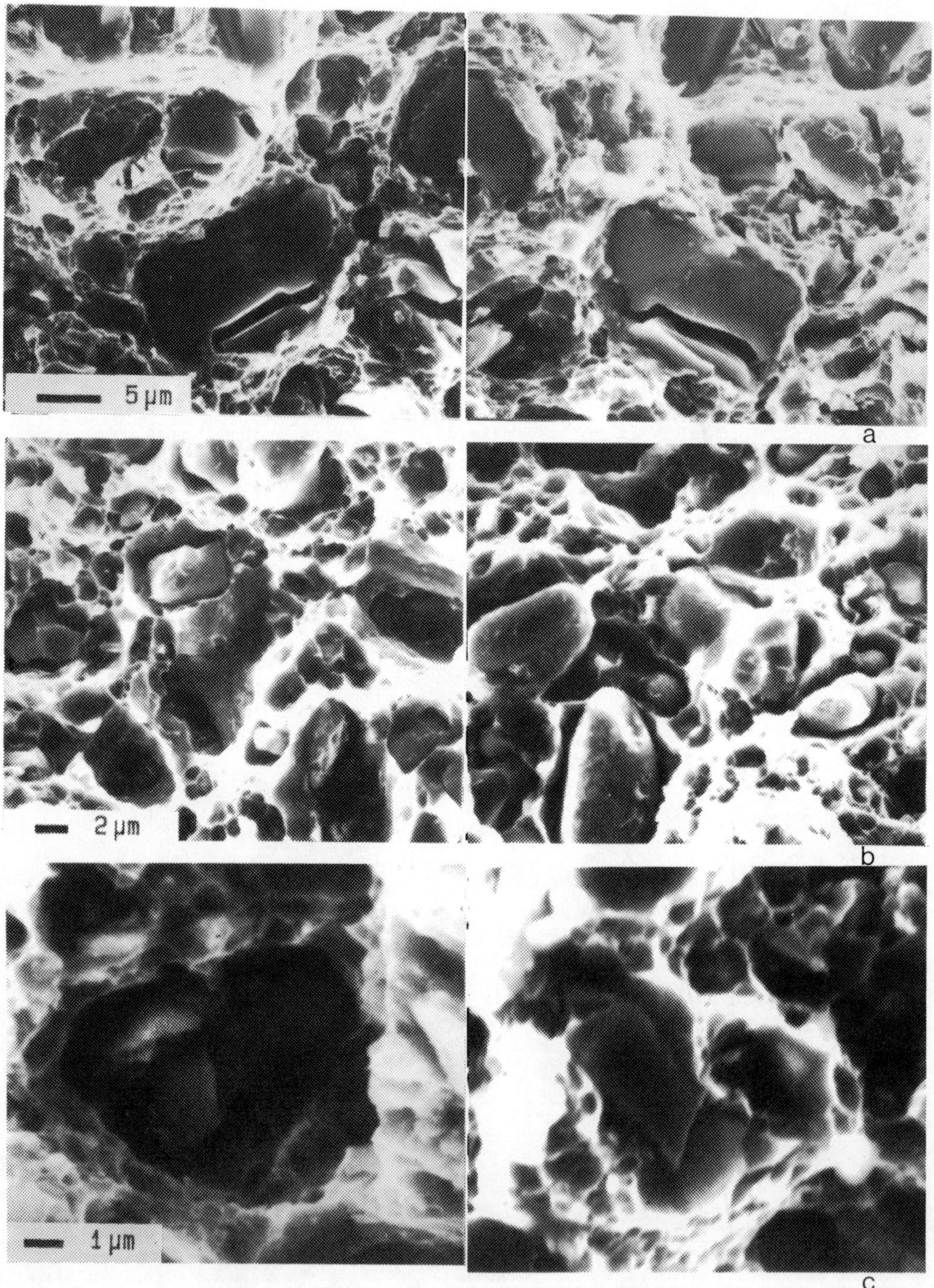

Figs. 5a-c. Scanning electron fractographs (matching halves) from broken tensile test-pieces (a, S3; b, T5D; c, T3A)

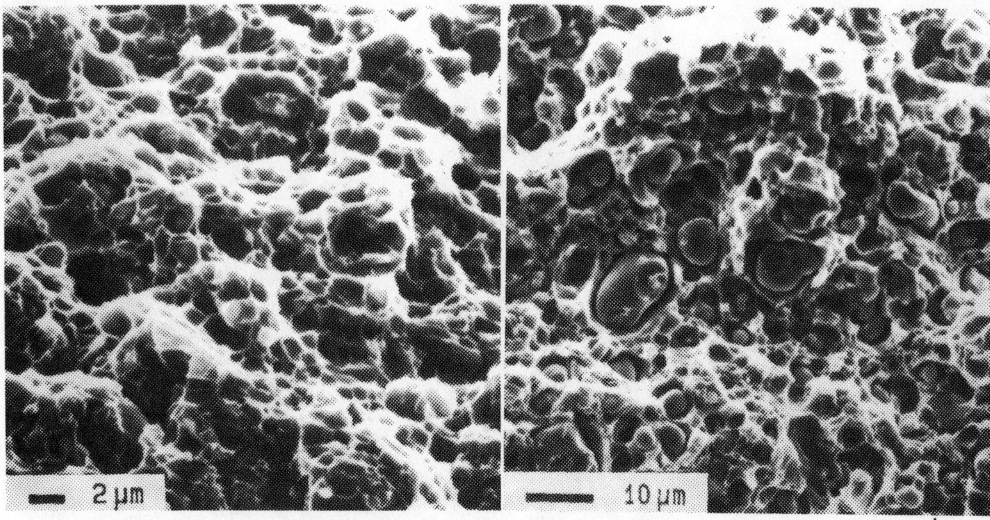

a b

Figs. 6a and b. Scanning electron fractographs from fractured test-pieces (a, T5A and b, T5D)

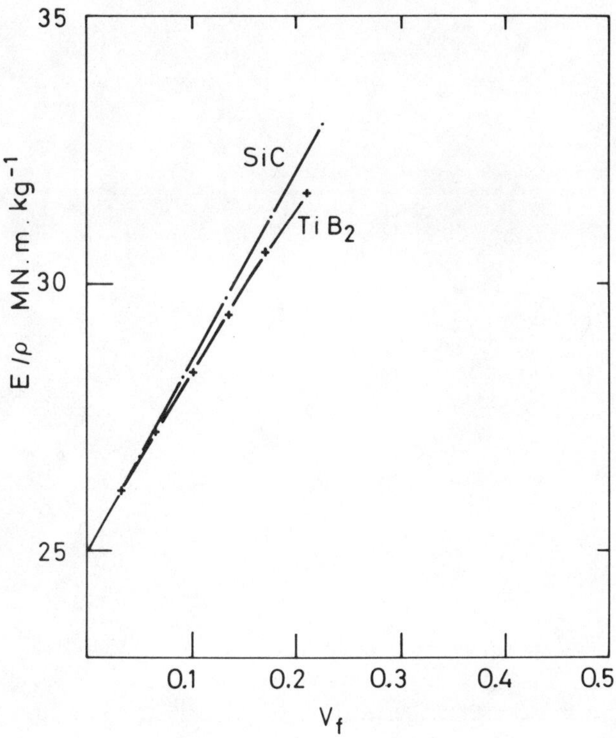

Fig. 7. Specific Young's modulus versus volume fraction of particulate

DESIGN AND P/M PROCESSING OF NEW HIERARCHICAL MATERIALS STRUCTURES

E. Y. Gutmanas* and A. Lawley**

*** Department of Materials Engineering**
Technion, Haifa 32000 Israel
**** Department of Materials Engineering**
Drexel University, Philadelphia, PA 19104, USA

Abstract

It is now possible to design metallic alloys and composite systems that possess hierarchical structures utilizing singly or in combination new P/M processing technologies. The unique microstructures intrinsic to these materials provide enhanced properties and in-service performance compared to conventional ingot metallurgy practice. Two generic examples are cited, namely the production of metal-ceramic composites, and hierarchical alloy and composite structures. In the former, a metal-ceramic compact with interconnected porosity is prepared from fine powders, treated in hydrogen to remove oxide from the metal powder particles and to reduce or metallize the surface layers of the ceramic particles, and consolidated to full density by cold sintering. The application of this approach to the processing of Fe-VC and Co-WC composites is described. In the second example, materials with hierarchical structures are processed by the attrition milling of fine elemental metal powder, with or without carbides, followed by hydrogen treatment and cold sintering. This approach is illustrated for Co-W and Fe-W-C alloys, and Co-W-WC and Fe-WC-VC composites. Attendant microstructures and mechanical properties confirm the viability of these methods of materials synthesis and illustrate the design flexibility intrinsic to P/M processing.

Introduction

The mechanical behaviour of materials with complex microstructures is controlled by the superposition of various strengthening mechanisms, namely grain and subgrain boundary hardening, solid solution hardening and dispersion/precipitation hardening. Refinement of grain size and of second phase dispersoids or precipitates, homogenization of microstructure, and bonding integrity between second phase particles and the matrix provide a combination of high strength and ductility. Such fine scale microstructures may be achieved using advanced P/M processing technologies. For example, rapid solidification of powders results in highly supersaturated solid solutions, fine scale precipitates, a fine grain size, and a high degree of homogeneity of the microstructure (1-3). Alternatively, mechanical alloying or high energy attrition milling of metal powders and ceramic particles gives rise to a uniform distribution of fine scale dispersoids in the metal matrix (4-6).

The refined microstructures obtained by rapid solidification or by mechanical alloying may coarsen, and metastable constituents may dissolve in the matrix during hot consolidation of the powders (7). Actual consolidation temperatures for hot forging and hot extrusion are frequently higher than the service temperatures. Thus, in order to retain fine scale microstructures and metastable constituents, it is necessary to shorten the time at temperature during hot consolidation, or to develop consolidation methods which are capable of achieving full density at temperatures which are lower that the normal service temperatures.

An alternative ambient temperature consolidation method for rapidly solidified powders and for metal-ceramic composites is cold sintering or high pressure consolidation (8-10). In cold sintering, plastic deformation of the powder particles results in full density. The intimate physical contact of particles free from oxides and other contaminants results in strong bonding. Retention of fine scale microstructures following cold sintering has been confirmed in rapidly solidified aluminium alloys (11-13), and high speed steels (14).

A new approach to the P/M processing of fully dense metal-ceramic composites has been developed. Briefly, a metal-ceramic compact containing interconnected pores is treated in hydrogen in order to remove oxide layers from the metal powder, and to bring about a partial reduction or metallization of the surface of the ceramic particles. Subsequently, the compacts are cold sintered to full density. This processing sequence improves bonding integrity between the metal matrix and the ceramic particles, with attendant improvement in composite strength. To-date, composites of stainless steel-chronium oxide (15) and steel - vanadium carbide (16) have been evaluated. Relatively coarse ceramic particles (40-90μm) were used and this resulted in low ductilit In the present study, we present results for Fe-VC and Co-WC composites processed as described above using fine metal and ceramic powders.

The high energy attrition milling of mixtures of fine elemental metal powders, with or without fine carbide powders, followed by hydrogen treatment and cold sintering offers an attractive processing alternative for the design of new high hierarchical structures. Diffusion distances are shorter, and compressibility is higher than for prealloyed powders. Examples of Co and Fe-base alloys and composites processed in this way are considered here.

Experimental Procedures

Materials and Sample Preparation

Elemental fine carbonyl iron powder (1-2μm), fine cobalt powder (1.5-3μm) prepared by a combination of pyrolysis and the reduction of cobalt oxalate, and fine oxide-reduced tungsten powder (0.5μm) were used in this study. Fine VC and WC powders (2-4μm) were purchased from H.C. Starck, Berlin, W. Germany.

Mixtures of elemental powders, with or without carbide powder additions, were prepared by means of high energy attrition milling. Green compacts of about 80% theoretical density were prepared from the iron-base and cobalt-base materials, with the addition of tungsten and/or carbides, at pressures in the range 0.5 to 1.0 GPa, Figure 1(a).

The open interconnected pore structure provides access for hydrogen in order to remove oxide layers from the surface of the metal particles and for partial reduction or metallization of the surface layers of the ceramic particles, at elevated temperature, Figure 1(b). Hydrogen purity was >99.99% with a water content <1ppm, oxygen <1ppm and nitrogen <10ppm. Reduction diagrams were used to establish the temperature limits over which surface oxides on the metal powders and on the vanadium carbide and tungsten carbide powders can be reduced. Reaction with the hydrogen was monitored by means of a thermal conductivity detector (TCD); details are given elsewhere (15, 16). The signal from the TCD is proportional to the concentration of the gases evolved.

After treatment in hydrogen, the compacts were cold sintered in rigid dies made from high speed steel at a pressure of 3 GPa, Figure 1(c). Densities close to the theoretical level were obtained. Similarly, milled mixtures of elemental metal powders, with and without carbide powder additions, were reduced in hydrogen and cold sintered to full density. Mechanical properties were determined with and without aging of the consolidated material.

Mechanical Testing

Transverse rupture strength was measured in three-point bending using a sample size 4.5mm x 18mm x 4mm, and a distance between the outer points of 16mm. Yield stress was determined at temperature on compression specimens 4.5mm x 4mm x 8mm., at a cross head speed of 0.1mm/min. in a standard Instron machine.

Microstructural Characterization

The microstructure of the hydrogen reduced and cold sintered material was examined by means of scanning electron microscopy (SEM) with energy dispersive analysis (EDS), and by x-ray diffraction. Polished specimens were thermally etched under a flow of hydrogen.

Results and Discussion

1. Metal-Carbide Composites

The effect of the reduction temperature on subsequent room temperature transverse rupture strength (σ_{TRS}) of Fe and Fe-30w/o VC is shown in Figure 2. Treatment in hydrogen was for 1 hour at each temperature. In the cold sintered Fe, the σ_{TRS} begins to increase at about 400°C; this correlates with the peak corresponding to the removal of oxygen, Figure 3. There is a two-step increase in σ_{TRS} for the Fe-30w/o VC composite. The first step at about 400°C is associated with reduction of the oxide from the iron powder particles. The second-step increase above about 900°C correlates with the reduction peak observed for VC, Figure 3.

After treatment in hydrogen at the highest temperature examined (1100°C), a decrease in the intensity of the main x-ray diffraction peak of VC is observed. This is believed to reflect the partial reduction (surface metallization) of the VC particles, with attendant improvement in bonding at carbide particle - metal particle interfaces.

For the Fe and Fe-VC composites prepared from fine powders and reduced in hydrogen at temperatures above 900°C prior to cold sintering, crack propagation occurs exclusively in the matrix, Figure 4(a). The microstructure of a Fe-30w/o VC composte is shown in Figure 4(b) after thermal etching. A similar fracture morphology is observed in Fe-VC composites prepared from coarse powders (VC=100μm; Fe~ 44μm), with the additional characteristic of cleavage in the coarse VC particles (16).

Response of the composites prepared from fine Co and WC particles is similar to that of the Fe-VC. The ambient temperature response of transverse rupture strength following reduction in hydrogen is shown in Figure 5 for the cold sintered Co and the Co - 40w/o WC composites. In the cold sintered Co, the σ_{TRS}, begins to increase above about 300°C, which correlates with the peak corresponding to the removal of oxygen, Figure 6. In the Co-40w/o WC composite there is a two step increase in σ_{TRS}, Figure 5. The first step above 300°C is associated with the reduction of the oxide from the cobalt powder particles. The second step above 1000°C correlates with the reduction peak observed for WC, Figure 6.

Representative micrographs (SEM) of the fracture surfaces of composite transverse rupture specimens are shown in Figure 7 for Co - 40w/o WC and Co - 80w/o WC. For both carbide levels, crack propagation is confined to the Co matrix. This confirms the integrity of the matrix-carbide bond resulting from cold sintering following hydrogen reduction at 1100°C.

The mechanical properties of the Fe-VC and Co-WC composites processed in accordance with the scheme shown in Figure 1 can be further improved by subsequent heat treatment following cold sintering. For example a σ_{TRS} of 1200 MPa is achieved in the Fe-30w/o VC composite following a 1 hour anneal at 1000°C. Similarly, a 1 hour anneal at 1100°C gives a σ_{TRS} of 1600 MPa. in Co-40w/o WC.

2. Hierarchical Structures

Several Co and Fe-base alloys and composites processed by attrition milling, hydrogen treatment and cold sintering of powder were included in this study. W and WC additions were made to the Co powder. Additions to the Fe powder were W, C(graphite), WC and VC powders.

The temperature dependence of the compressive yield stress of the Co-base material is shown in figure 8. Curve number 1 is for pure Co; the green compact (~80% theoretical density) was treated in hydrogen at 1000°C for 1 hour, cold sintered at 3 GPa, annealed for 1 hour at 1000°C, and finally repressed at at 3GPa. During annealing, grain coarsening was observed. The final grain size was in the range 8-10μm. A finer grain size (1.5-3μm) can be retained by using a lower temperature for the hydrogen treatment (600°C) and for annealing (800°C). In this condition, strength versus temperature is given by curve number 2 in Figure 8.

The effect of the addition of W powder to the Co is illustrated by reference to curve number 3. Strength data pertain to a Co-20w/o W alloy subjected to the same processing history as the pure Co, i.e. curve number 1 in Figure 8. It is seen that the addition of fine W powder results in a significant component of solid solution hardening. Further, the tungsten retards grain growth in the Co matrix. For example, for pure Co, grain growth is observed after a 1 hour anneal at 1000°C, wheareas fine grains are retained is the Co-20w/o W alloy after the same heat treatment. This retardation in grain coarsening is attributed to the diffusion of W into the surface of the Co powder particles. W concentrations as low as 4w/o were found to be effective in controlling or restricting grain growth.

Aging of the Co-W alloy results in a further strength increment. As an example, curve number 4 is for the Co-20w/o W composition given the same processing history as the material corresponding to curve number 3, but with the addition of a two-hour aging step at 700°C. Aging results in the precipitation of Co_3W, with attendant hardening of tl Co matrix.

Incorporation of WC particles in the Co-W matrix gives rise to further hardening. Thus, in Figure 8, curve number 5 shows the strength of a composite of Co-20w/o W reinforced with a fine dispersion of 40w/o WC. The effect of the WC is seen by comparing curves number 5 and number 3, since the latter reflects the same processing history. Curve number 6 shows the strength response for the Co-20w/o W-40w/o WC composite after aging for 2 hours at 700°C. To assess the effect of the WC in the aged condition, curve number 6 should be compared with curve number 4. It is seen that the dispersion of WC provides additional hardening.

The ability to develop a fine grain microstructure in the Co-20w/o W - 20w/o WC composite is illustrated in Figure 9. Corresponding strength data are given on curve number 5 in Figure 8.

Development of a fine-scale microstructure in Fe-12w/o W - 1w/oC processed from a mix of fine powders of Fe, W, and C (graphite) is shown in Figure 10(a). Similarly the fine-scale microstructure in a composite of composition Fe-10w/oWC - 5w/oVC, processed from a mix of fine powders of each constituent, in illustrated in Figure 10(b). Processing steps were treatment in hydrogen at 900°C for the Fe-12w/o W - 1w/o C alloy, and at 1100°C for the Fe-10w/o WC - 5w/o VC composite, followed by cold sintering and a 1 hour anneal at 1200°C. Fine precipitates of W-rich carbides are presen in the matrix of the Fe-12w/o W - 1w/o C alloy, Figure 10(a). In the Fe-10w/o

WC-5w/oVC composite, there is a homogeneous distribution of the light etching WC particles, and the dark etching VC particles, Figure 10(b). A composite of the same composition was processed under identical conditons, except that the starting iron powder was a coarse water atomized product (< 44μm). This resulted in a coarser microstructure with carbide particles concentrated at interparticle boundaries, Figure 11. A comparison of Figures 10(b) and Figure 11 clearly reveals the differing scales of the microstructure.

From a more limited study, we have found a similar response when fine W and Fine WC powders are added to Ni powder. The fine-scale microstructures exhibit strength increases similar to those achieved in the Co-base and Fe-base materials.

The ceramic particulate must be chemically compatible with, and stable in, the metal matrix during post consolidation heat treatment. If there is any dissolution of the ceramic particles and/or reaction with the metal matrix, it becomes necessary to coat the ceramic particles with a barrier layer to minimize/prevent diffusion.

As an example, Si_3N_4 forms silicides by reaction with Co, Ni and Fe (17,18). The silicides exhibit low moduli and are incapable of promoting dispersion strengthening. Coating of the Si_3N_4 particles with a thin layer of V or Cr results in the formation of a protective nitride at particle-matrix interfaces (19). Thus, a combination of mechanical alloying by the attrition milling of micron and submicron metal and ceramic powders, the surface treatment of the powders, and subsequent consolidation to full density by cold sintering provides virtually unlimited flexibility in the design and P/M processing of new hierarchical materials structures.

Conclusions

1. Metal-ceramic composites of Fe-VC and Co-W with fine scale microstructures and attractive mechanical properties can be processed from fine powders by a sequence involving compaction, hydrogen treatment, and cold sintering.

2. Hierarchical alloy and composite structures can be processed by the attrition milling of fine elemental powder, with or without carbides, followed by hydrogen treatment and cold sintering. This approach has been demonstrated for Co-W and Fe-W-C alloys, and Co-W-WC and Fe-WC-VC composites.

3. The design flexibility inherent in these P/M processing regimes allows for optimization of material performance in terms of the superposition of operative strengthening mechanisms. These fine-scale materials are expected to offer improved wear resistance and performance at elevated temperatures.

References

1. M. Cohen, B.H. Kear and R. Mehrabian, in "Rapid Solidification Processing: Principles and Technologies II", Eds. R. Mehrabian, B.H. Kear and M. Cohen, Claitor's Publ. Div., Baton Rouge, LA, 1980, p.1.

2. S.J. Savage and F.H. Froes, J. of Metals, 36, 4 (1984) 20.

3. A. Kasak and E.J. Dulis, Powder met., 21, (1978) 14.

4. J.S. Benjamin; , Met. Trans., 1A (1970) 2943.

5. J.S. Benjamin and M.J. Bomford, Met. Trans., 8A (1977) 1307.

6. P.S. Gilman and W.D. Nix, Met. Trans., 12A (1981) 813.

7. E.Y. Gutmanas, M. Premkumar and A. Lawley, in Progress in Powder Metallurgy, Eds: H.S. Nayar, S.M. Kaufman and K.E. Meiners, MPIF, Princeton, NJ 39, (1983) 669.

8. E.Y. Gutmanas, A. Rabinkin and M. Roitberg, Scripta Met. 13, (1979) 11.

9. E.Y. Gutmanas, Powder Met. Int. 15, (1983) 129.

10. E.Y. Gutmanas, Y. Schefer and D. Zak, P/M-82 in Europe. Int. Powder Met. Conference, Associazine Italiana di Metallurgia, Florence, Italy, (1982) p. 643.

11. O. Botstein, E.Y. Gutmanas and A. Lawley in Modern Developments in Powder Metallurgy, Eds: E.N. Aqua and C.I. Whitman MPIF, Princeton, N.J. 15, (1985) 761.

12. O. Botstein, E.Y. Gutmanas and A. Lawley in Progress in Powder Metallurgy Eds: H.L. Sanderow, W.L. Giebelhausen and K.M. Kulkarni, MPIF Princton, N.J. 41, (1986) 123.

13. O. Botstein, E.Y. Gutmanas and A. Lawley, Submitted to Met. Trans.

14. E.Y. Gutmanas in Modern Developments in Powder Metallurgy Eds: E.N. Aqua and C.I. Whitman, MPIF, Princeton, N.J. 15, (1985) 175.

15. E.Y Gutmanas, D.B. Goldman, Y. B. Clark and S. Hart, in Progress in Powder Metallurgy Eds: H.L. Sanderow, W.L. Giebelhausen and K.M. Kulkarni, MPIF, Princeton, N.J. 41, (1986) 631.

16. E.Y. Gutmanas, D.B. Goldman, S. Hart and D. Zak, Powder Met. Int. 18 (1986) 401.

17. F. Edelman, E.Y. Gutmanas, A. Katz and R. Brener, J. Appl. Phys. Let. In press.

18. F. Edelman, E.Y. Gutmanas, A. Katz and R. Brener, to be published.

19. I. Gotman and E.Y. Gutmanas, to be published.

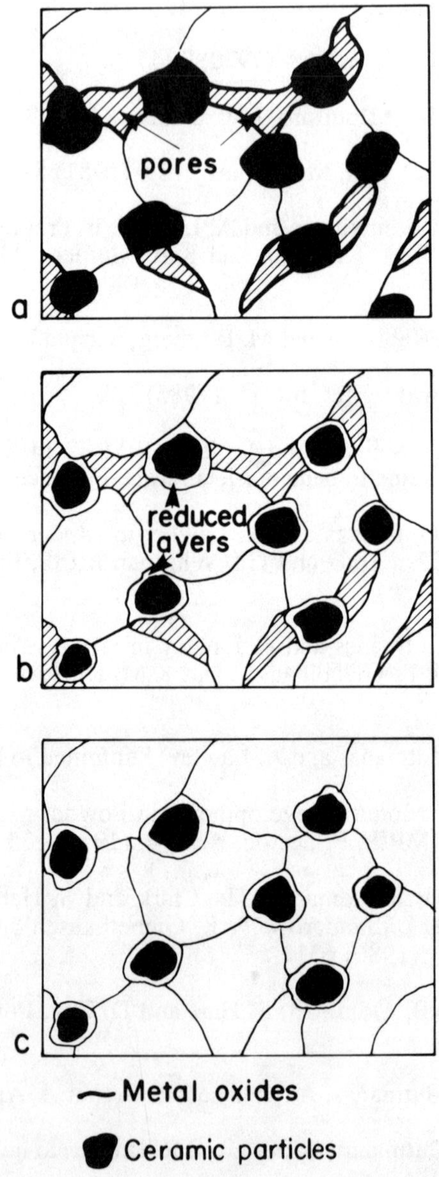

Figure 1: Schematic of processing sequence for metal-ceramic composites (a) compact to 80% theoretical density, (b) treatment in hydrogen, (c) cold sinter to full density.

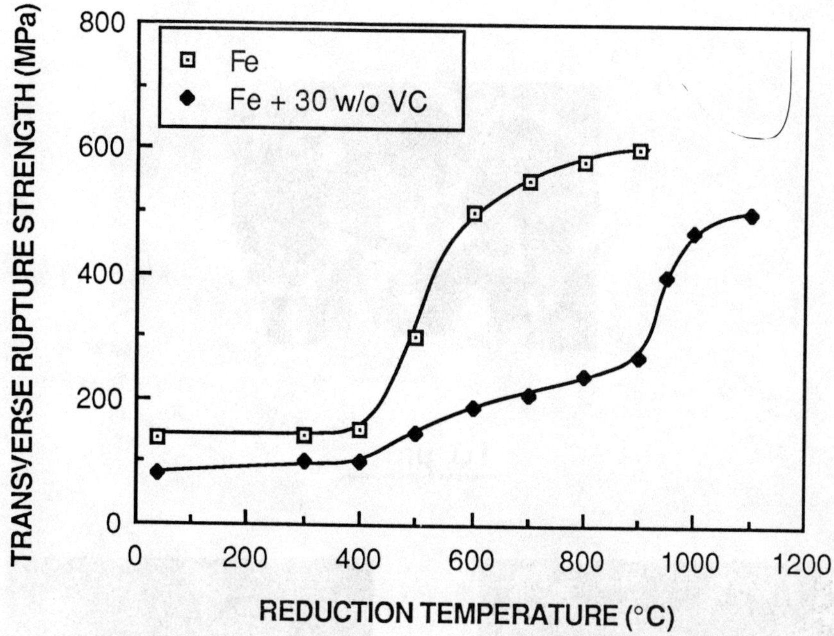

Figure 2: Room temperature transverse rupture strength of cold sintered Fe-base materials-as a funciton of green compact reduction temperature.

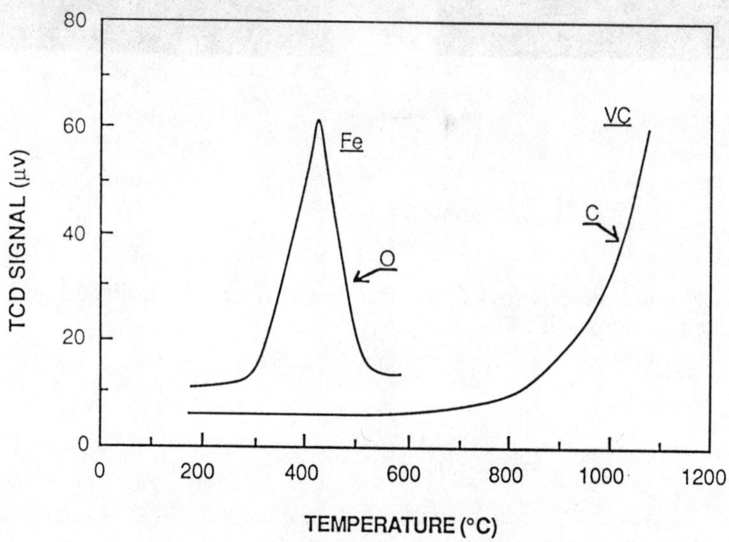

Figure 3: Reduction diagram for Fe and VC powders.

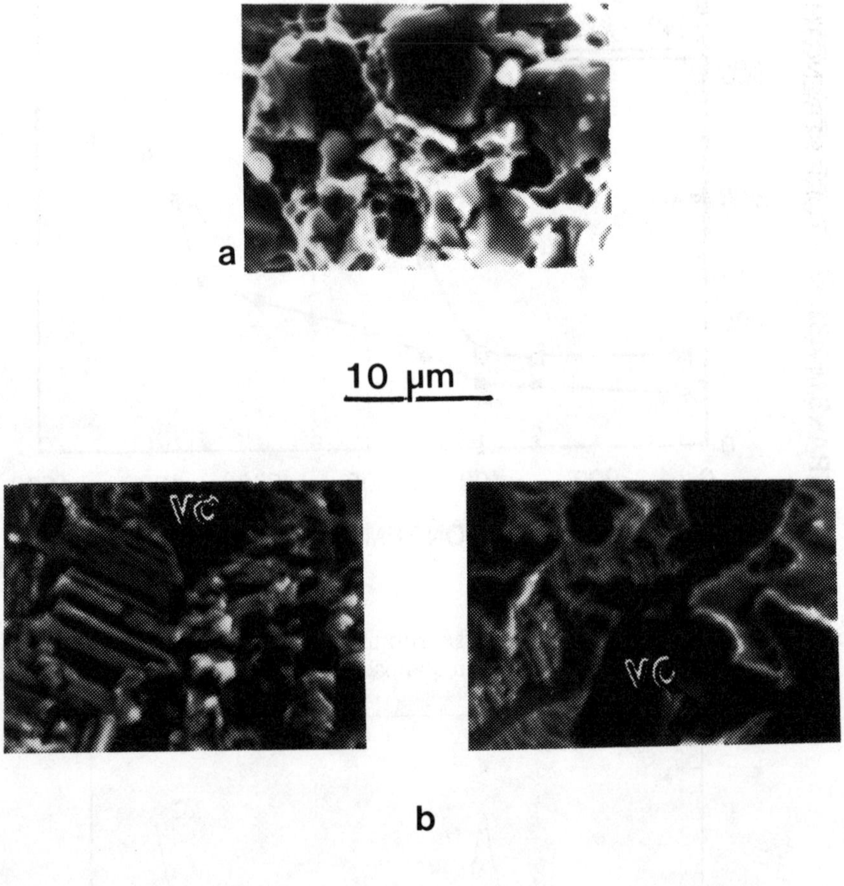

Figure 4: Fully dense Fe-30w/o VC (a) fracture surface, (b) polished and thermally etched sections - SEM.

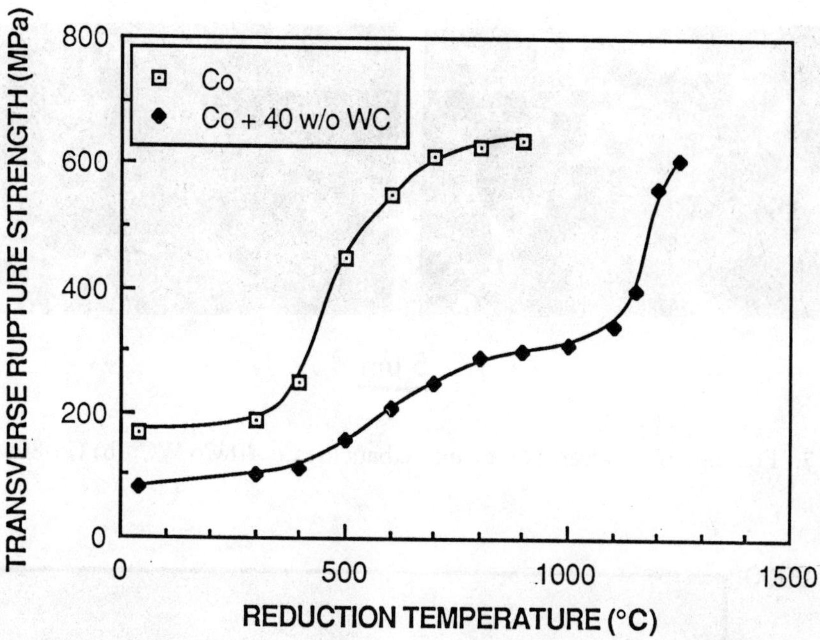

Figure 5: Room temperature transverse rupture strength of cold sintered Co-base materials - as a function of green compact reduction temperature.

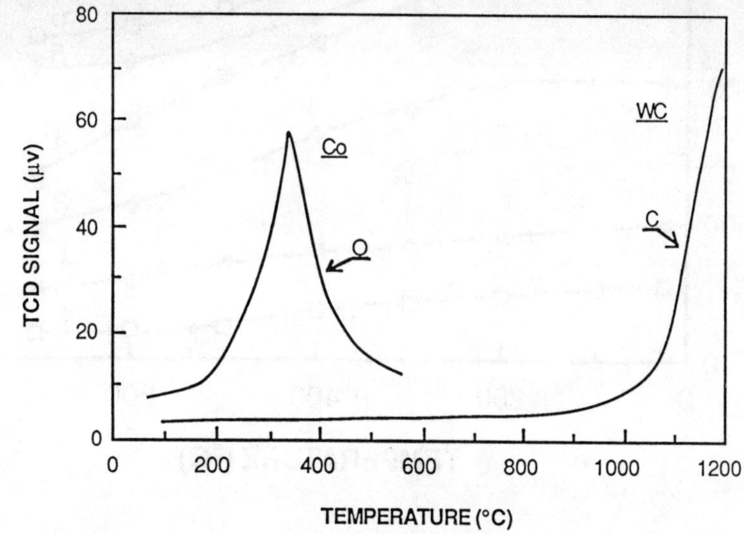

Figure 6: Reduction diagram for Co and WC powders.

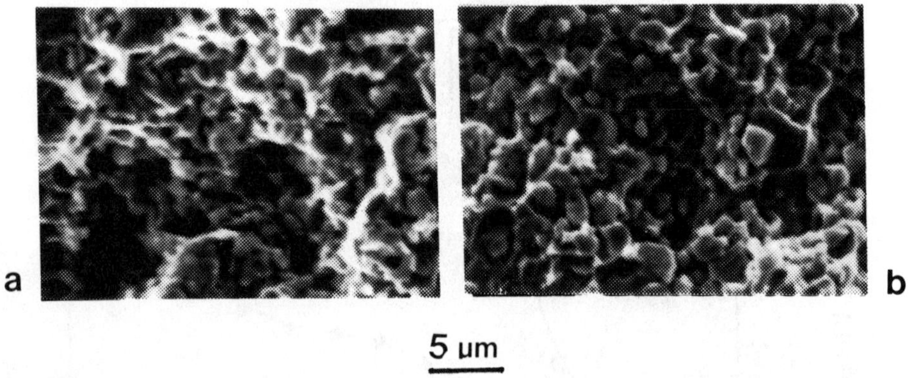

Figure 7: Fractographs of transverse rupture bars (a) Co-40w/o WC, (b) Co-80w/o WC.

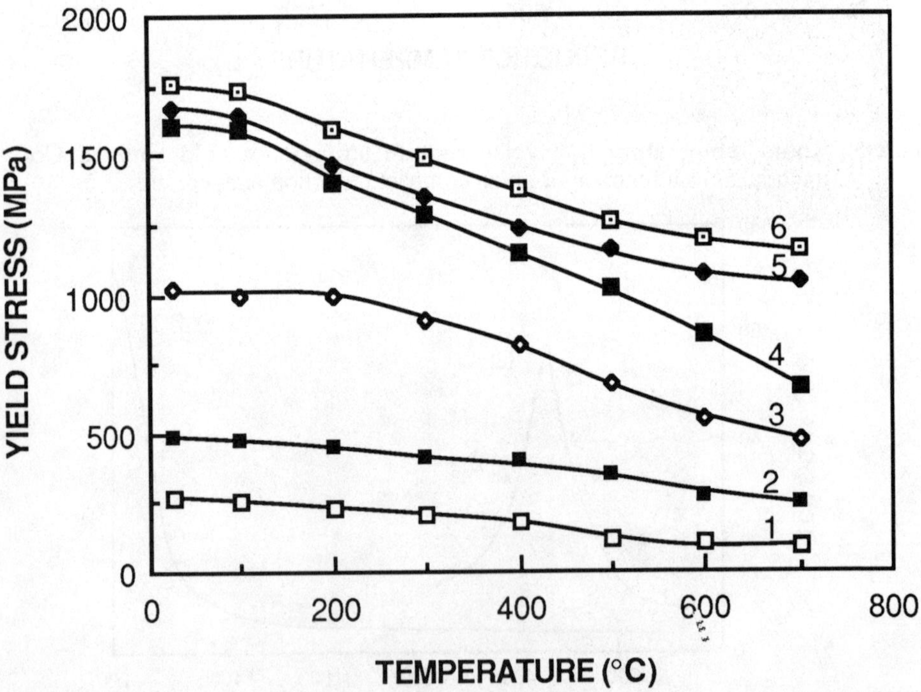

Figure 8: Compressive yield stress of Co-base materials as a function of temperature, and processing history (see text).

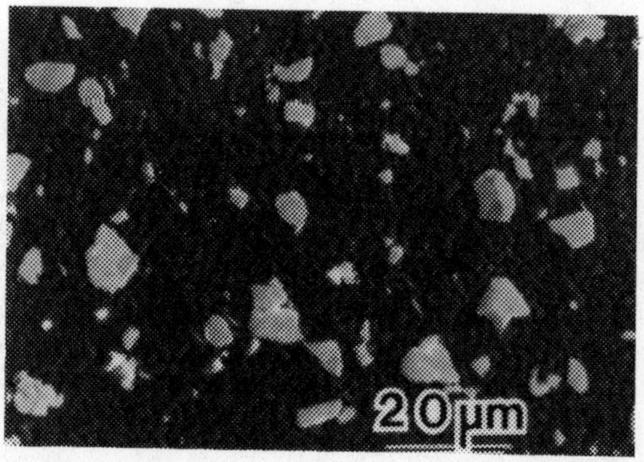

Figure 9: Micrograph of polished and thermally etched Co-20w/o W-20w/o WC; SEM

Figure 10: Micrographs of polished and thermally etched Fe-base materials (a) Fe-12w/oW -1w/oC (b) Fe-10w/oWC - 5w/oVC; SEM.

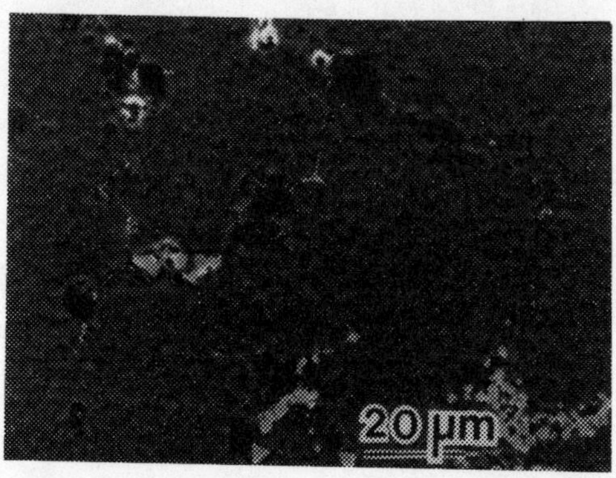

Figure 11: Micrograph of polished and thermally etched Fe-10w/oWC - 5w/oVC; SEM

METAL-POLYMER COMPOSITE MATERIAL

M. S. Stoichev

Institute for Metal Science and Technology,
Bulgarian Academy of Sciences, Bulgaria

Y. T. Simeonov, Zh. H. Zhelazkov

Central Physical-Chemical Mechanics Laboratory,
Bulgarian Academy of Sciences, Bulgaria

S. B. Milkov

High Institute of Architecture and Construction,
Bulgaria

Abstract

Metal-matrix composites are currently experiencing active development in Europe, US and Japan. Envisioned uses range from high performance, high value-added military and aerospace applications to high volume, low-cost commercial applications. While significant progress is being made concerning high-cost and better mechanical properties, new production methods are still desirable.

This paper outlines a method for the production of composite materials by infiltration of a sintered metal matrix (iron base) with polymers. Results of examining the effect of the polymer type, powder particle shapes and the nature of porosity on the properties of the composite material are shown.

Introduction

Metal-polymer composite materials are a combination of metals and polymers and they belong to the multicomponent systems. Their substantial feature is the presence of boundary surface between the phases.

Terminology of multicomponent materials exactly determined and adopted by everyone is not known yet today. That is why we accept that the term complex material is wider and includes the term composite material in itself.

Metal-polymer materials and coatings, metal-polymer constructions and metal-polymer compositions may fall under the term complex metal-polymer materials (systems). The metal-polymer compositions comprise separate components as a result of compounding of metals and polymers (e.g. by mixing). One composition may be transformed into a composite material, e.g. by crystallization.

For the first time the term "metal-polymers" has been proposed by E.M.Natanson /1/. These materials cover two-phase stable systems produced by forming of highly dispersible metal particles in a polymer medium. These two-phase systems are homogeneous and physically stable. V.A.Belliy /2/ interprets the term "metal-polymer" material as consisting of polymer and metal bonded rigidly in-between by atom-molecular forces. The metal-polymer coatings are multi-layer coatings comprising a metal and a polymer.

The present paper considers the properties of metal-polymer materials produced by infiltration of a powder-metallurgy matrix with monomers and oligomers, followed by thermocatalytic polymerization.

Composite material production

The schedule for the production of a metal-polymer composite material is different from similar schedules used for infiltration of non-metallic materials with polymers, mainly because of the absence of the drying stage for the material.

For this initial study the metal matrix materials used were iron base produced by cold pressing and sintering. ASC 100.29, NC 100.24 and ПЖ-К iron powders were used. The last one was obtained by reduction of iron super-concentrate with solid carbon (coke). The properties of the initial powders are shown in Table 1, and the particle shape and morphology in Figure 1.

Monomers and oligomers were used as infiltrating compositions. The viscosity of the compositions varied from 5 to 35 seconds, according to B3-4 /3/, depending on the oligomer content.

The aim of the research has been to study the influence of:

- the type of the polymer (the composition for infiltration);
- the type of the initial metal powder (shape and structure of the particles, the grain size);
- porosity of the sintered matrix

on:

- the infiltration process;
- the composite material properties.

Cylindrical specimens with dimesions of 20 mm in diameter and 20 mm in height were made with the three types of iron powder with similar and different general porosity. In order to obtain similar porosity we used equal amounts of powder and pressed them to similar volumes. The general porosity was determined in accordance with Bulgarian State Standard 8440-86. The pore size was measured by mercury infiltration.

TABLE 1: PROPERTIES OF THE POWDERS

Properties	ASC 100.29	NC 100.24	ПЖ-K
1. Chemical composition (%)			
C	0,01	0,01	0,05
Si	0,03	-	-
Mn	0,05	0,08	0,12
P	0,01	0,01	0,01
S	0,01	0,01	0,01
O_2(H_2loss)	0,20	0,29	0,31
2. Sieve analysis (%)			
- 200 + 160	5,0	traces	0,56
- 160 + 100	26,8	29,61	22,28
- 100 + 0,80	28,4	17,23	25,96
- 0,60 + 0,63	18,2	18,29	30,35
- 0,63	21,6	34,93	20,85
3. Apparent density (g/cm^3)	2,92	2,44	2,40
4. Flow (sec/50 g)	24,2	30	40
5. Compressibility (g/cm^3)			
Comp.pressure 4,2 MPa	6,80	6,48	5,90
Comp.pressure 6,00 MPa	7,00	6,70	6,50

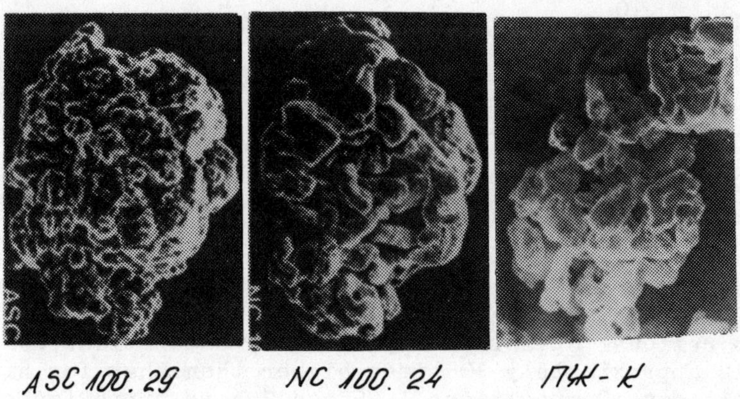

ASC 100. 29 NC 100. 24 ПЖ- K

Figure 1: Macrostructure and shape of iron powder particles ASC 100.29, NC 100.24 and ПЖ-K.

The shape and sizes of the sintered matrices for the tension and elasticity tests were in accordance with Bulgarian State Standard 8402-82. The infiltration was conducted under identical conditions with different infiltrating compositions.

Results

Experiments with seven infiltrating compositions were carried out for studying the influence of the type of infiltrating composition on the process of infiltration. The infiltrating composition 1 which is a monomer features a low viscosity and a higher volatility and compositions 2 through 7 are mixtures of another monomer and oligomers. The matrices with porosity of approximately 27% were prepared with ПЖ-K powder. The infiltration was performed in vacuum.

TABLE 2: INFILTRATING COMPOSITION INFLUENCE ON THE POLYMER CONTENT

Infiltrating composition No	Viscosity, sec.	Polymer content, %	Degree of filling with polymer, %	Residual open porosity, %
-	-	-	-	27,08
1	5	21,6	75,6	4,82
2	7	20,47	80	1,30
3	8	22,46	83	0,60
4	10	23,73	88	0,42
5	32	23,28	86	0,34
6	9	20,39	75	0,22
7	15	20,12	74	0,16

From the results presented in Table 2 it can be seen that along with the increase of the infiltration composition viscosity (1 to 5) the degree of filling the pores increases too. It is obvious that with low-viscosity compositions 1 and 2 during polymerization part of the infiltrating composition evaporates from the matrix surface. This is confirmed also in the determination of the residual open porosity by water saturation. For composition 1 it is approximately 10 times higher than that for high viscosity compositions. Compositions 1, 6 and 7 have approximately the same degree of pore filling, but the residual porosity expressed by water saturation differs considerably. A less viscous infiltrating composition fills the pores of the matrix better and is also more readily evaporated during polymerization from the matrix surface which determines the increased surface residual porosity. On the contrary, the infiltrating compositions 6 and 7 are of higher viscosity, fill the volume of the matrix pores with greater difficulty and evaporate less readily during polymerization. That is why the surface layer of the composite material is more saturated with polymer, i.e. the surface residual porosity is lower.

The difference in the shape and macrostructure of particles (Figure 1) and in the grain size (Table 1) of the powders result in obtaining matrices of different porosity character. Those made of ASC 100.29 (with dense round particles of close diameters) show pores with higher uniformity of dimensions, (Figure 2). Those made of NC 100.24 and ПЖ-K (dendritic structure) have pores with greater deviation of sizes (10^2 to 10^4 nm). The utilization of copper, nickel and carbon alloyed powder mixtures reflects also on the porosity character. The pores obtained are more uniform. The increase of the pore size along with the degree of increase in density is common for the three types of sintered matrices.

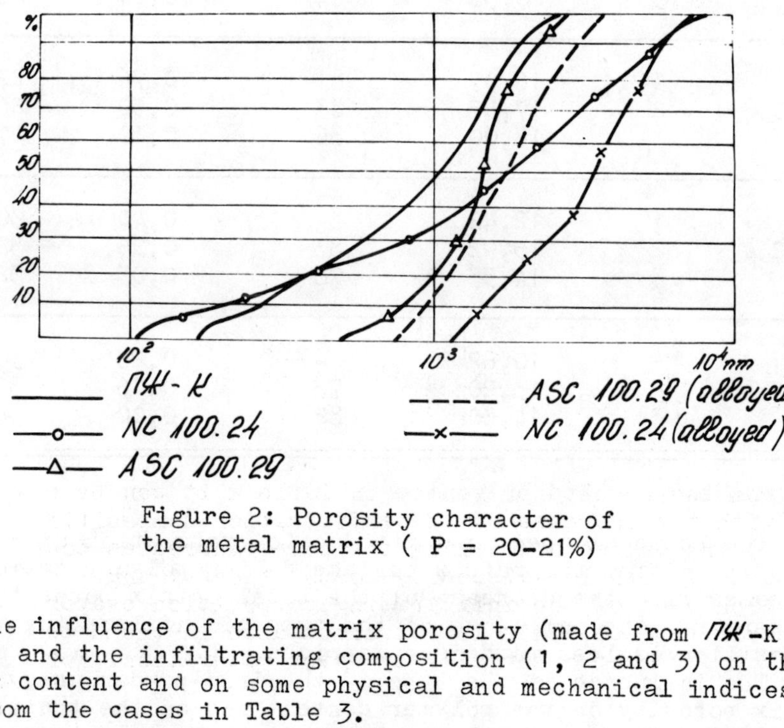

Figure 2: Porosity character of the metal matrix (P = 20-21%)

The influence of the matrix porosity (made from ПЖ-K powder) and the infiltrating composition (1, 2 and 3) on the polymer content and on some physical and mechanical indices is seen from the cases in Table 3.

The conclusions mentioned earlier with regard to the data in Table 2 are valid here as well. With the decrease of matrix porosity the content of polymer decreases but the degree of filling the pores is of the same order.

TABLE 3: INFLUENCE OF THE MATRIX POROSITY AND THE TYPE OF INFILTRATING COMPOSITION ON THE PHYSICAL AND MECHANICAL INDICES (ПЖ-К)

Porosity, %	Infiltrating composition, No	Polymer content, %	Degree of filling with polymer, %	Residual open porosity, %	Compression proportionality limit, MPa
28,4	1	23,51	84	2,08	185
	2	25,65	90	1,16	240
	3	25,17	89	0,15	238
24,5	1	20,05	82	1,17	232
	2	21,42	87	0,81	248
	3	20,36	84	0,21	267
21,0	1	16,97	81	0,76	225
	2	17,47	83	0,50	267
	3	18,00	86	0,10	278
19,2	1	12,42	-	0,40	300
	2	15,92	83	0,26	320
	3	14,53	78	0,05	323
17,5	1	10,62	-	0,23	300
	2	13,20	75	0,16	320
	3	14,30	82	0,00	345

The residual open porosity due to evaporation of filling composition from the matrix surface during polymerization decreases with the decrease of the total matrix porosity. Furthermore, the influence of the type of infiltrating composition is less in denser matrices.

The influence of the type of the iron powder used and of matrix porosity on the polymer content and on the tensile strength of the metal-polymer composite materials is seen in Table 4. The strength increases at an average of 10 to 20% as a result of the infiltration and the following polymerization. When using powder ПЖ-K the effect of strengthening is more strongly expressed than with the use of the other powders (ASC 100.29, NC 100.24), increase of 1,5 to 3,5 times in dependence on the matrix porosity.

TABLE 4: THE INFLUENCE OF THE TYPE OF POWDER AND MATRIX POROSITY ON THE POLYMER CONTENT AND THE TENSILE STRENGTH

Powder	Matrix No	Polymer content, %	Degree of filling, %	Residual open porosity, %	Tensile strength MPa		Effect of strengthening, times
					PM matrix	Composite material	
ПЖ-К	1	38,8	93,2	0,40	15,9	54,7	3,45
	2	35,9	88,4	0,63	17,1	52,9	3,09
	3	29,7	96,3	1,19	44,6	81,8	1,83
	4	25,2	95,2	0,60	55,8	87,7	1,60
	5	22,9	88,1	0,37	65,9	100,0	1,52
ПЖ-К alloyed	1	25,61	87,9	0,86	74,0	87,3	1,18
	2	25,11	90,6	0,60	74,3	78,0	1,05
	3	18,41	80,5	0,77	99,2	119,0	1,20
NC 100.24	1	27,21	83,0	0,91	91,8	97,6	1,06
	2	24,91	88,1	0,74	104,2	111,0	1,09
	3	23,91	86,0	0,62	106,3	120,0	1,13
	4	17,2	93,0	0,43	165,8	159,0	-
NC 100.24 alloyed	1	18,4	79,6	2,93	188,2	205,0	1,09
	2	14,6	82,2	1,65	262,5	266,0	1,01
	3	8,3	67,6	1,24	326,8	356,0	1,09
ASC 100.29	1	21,6	94,3	0,83	97,2	132,0	1,36
	2	14,52	85,5	0,44	132,0	152,0	1,15
	3	12,6	85,7	0,26	143,5	161,0	1,12
	4	11,6	88,9	0,11	182,8	186,0	1,02

Conclusions

1. Infiltrating compositions on monomer basis, modified with oligomers are suitable for producing metal-polymer composite materials. The effect of infiltration and polymerization on the physical and mechanical properties is increased with the increase of the sintered metal matrix porosity.

2. Because of its effect on the matrix porosity the type of powder exerts an influence on the degree of filling the pores with polymer, on the water saturation and on the tensile strength.

3. The infiltration of P/M matrix with polymers improves a number of physical and mechanical indices:
   - the tensile strength is increased by 10 to 20% up to three times for more porous materials;
   - the compression proportionality limit is increased by 20 - 30%;
   - the residual porosity decreases 30 - 60 times.

4. The infiltrating effect on metal-polymer materials produced of sintered matrices from ПЖ-K powder is more strongly expressed than with the use of the other powders (ASC 100.29, NC 100.24).

REFERENCES

1. E.M. Natanson, Coloidal metals and metal-polymers, KIEV, Naukova dumka 1971, 348.
2. Metal-polymer materials and parts, MOSCOW, CHEMISTRY, 1979, 317.
3. Method for determination of the viscosity by means of B3-4 device.

Chapter Seven

Intermetallic Compounds

OXIDE DISPERSION STRENGTHENED ALUMINIDES

K. Vedula*, G.M. Michal* and A.M. Figueredo**

*Associate Professors and **former Graduate Student, respectively
Department of Materials Science and Engineering
Case Western Reserve University
Cleveland, Ohio

**currently Graduate Student, Massachussetts Institute of Technology.

Abstract

Aluminide intermetallics are attractive as potential materials for high temperature structural applications. Oxide dispersion strengthening (ODS) is a Powder Metallurgy (P/M) process which has been very effective in refining microstructures and improving properties of conventional alloys in the past. This paper will review attempts to modify the microstructures and properties of several aluminides by oxide dispersion strengthening. The review will include some of the recent work conducted at Case Western Reserve University on ODS FeAl and NiAl alloys.

Introduction

The aluminides of titanium, nickel and iron are under consideration as potential structural materials for high temperature applications. These intermetallic compounds are attractive because of their low densities, excellent oxidation resistance, reasonably high melting points as well as their ordered crystal structures. Reviews of the potential for aluminides as high temperature structural materials are available in several papers in recent symposia proceedings [e.g. Refs. 1,2]. Table I provides a comparison of the physical properties of some of these aluminides of interest.

Low temperature brittleness as well insufficient creep strength at temperatures of interest are severe drawbacks limiting the potential use of alloys based on these aluminides. This is particularly the case for alloys based on B2 NiAl, which has a very large density advantage over conventional superalloys. There have been attempts to improve the creep properties of these aluminides by solid solution strengthening as well as by precipitation strengthening. The strength improvements obtained in this manner, however, appear to be somewhat limited due to coarsening of the precipitates. For example, in the case of NiAl, the strengthening precipitates which include Ni_2AlNb, Ni_2AlTa, NiAlNb and NiAlTa, coarsen fairly rapidly at 1300 K under an applied stress [3,4,5].

Dispersions of stable oxides such as Y_2O_3 hold better promise for high temperature strengthening since they show good stability against coarsening and have been very effective in strengthening superalloys in the form of Oxide Dispersion Strengthened (ODS) alloys [6,7,8]. In order for the dispersion of oxide particles to be effective for high temperature strengthening, the particle sizes should be in the range of 10 to 50 nm, interparticle spacings in the range of 50 to 300 nm and volume fractions in the range of 0.004 to 0.05. The choice of the oxide dispersoid is dictated by its stability and, typically, Y_2O_3 is the most successful dispersion for applications around 1300 K.

The oxide particles are commonly dispersed in the matrix by the mechanical alloying (MA) process, which is a dry, high energy ball milling technique. Figure 1 illustrates how the different constituents of the powder blend react to the ball-powder-ball collisions. The ductile elemental powders get flattened, whereas the brittle phases disintegrate. Figure 2 illustrates how the ductile phases and brittle phases become alloyed by a series of cold welding and fracturing processes. Although initially, the composite powders form layered structures, eventually the alloying is on an atomic scale and each powder particle represents a homogeneous alloy composition with a uniform distribution of fine dispersoids. The particle size distribution of the blended powders goes through significant changes in the initial stages , and typically, a balance between the processes of welding and fracturing results in a steady state particle size distribution when mechanical alloying is complete. Of particular significance, however, to the mechanical alloying of intermetallic compounds is that these compounds are brittle and, hence, are always subject to further degradation.

ODS of Ti_3Al and Ni_3Al

Konitzer and Fraser [9] used rapid solidification techniques to produce a refined dispersion of oxide in Ti_3Al. Buttons were arc melted with oxygen as the impurity and laser surface melting was used to produce a disordered solid solution of Ti and Al with a supersaturation of Er and O. A refined dispersion of Er_2O_3 was formed on heat treatment at 973 K. This oxide is found to be stable at elevated temperatures up to 1173 K. No attempts were made to evaluate the strengthening due to these oxide dispersoids.

Rowe and coworkers [10,11] have carried out studies of dispersoids in Ti_3Al-Nb alloys prepared with additions of Ce, O and S. The alloys were rapidly solidified by melt spinning. After annealing at 1023 K, 10 to 50 nm dispersoids of Er_2O_3, Ce_2S_3 or $Ce_4O_4S_3$ were identified. The ribbons were consolidated by hot isostatic pressing at 1123 to 1223 K and hot extruding at 1198 K. Consolidated alloys showed preferential dispersoid depletion and coarsening at grain boundaries. Bend tests on ribbons of the as-melt spun and aged alloys indicated that bend ductilities of alloys with disperoids were lower than those of the same alloys without dispersoids. No elevated temperature property comparisons were reported.

Lauf and Walls [12] obtained Al_2O_3 dispersoids in Ni_3Al, by mechanically alloying the Ni_3Al-base powder with 1 wt% Al_2O_3 powder using a Spex mill for 10 hours. The Ni_3Al base particles showed layers of Al_2O_3 powder folded into them. After hot pressing at 1573 K for 15 minutes in vacuum at 4000 psi (27.6 MPa), the resulting compact was fully dense, had a fine grain size and a uniform dispersion of Al_2O_3 particles. The room temperature hardness for the dispersion strengthened compact was 488 ± 27 DPH compared to 225 ± 10 DPH for the Ni_3Al without the dispersion. Hot microhardness tests showed that the hardness of the ODS Ni_3Al decreased much more rapidly with increasing temperature than that of Ni_3Al, so that the ODS alloy was actually softer at 1273 K. This has been attributed to the smaller grain size of the ODS alloy.

Donnelly and Koch [13] have studied mechanical alloying of Ni_3Al-base alloy powders containing 0.5 at% Hf and 500 wppm B. Oxide dispersoids of Al_2O_3, Y_2O_3 and ThO_2 were studied. A process control agent was added to some of the powder blends in order to prevent excessive cold welding. Mechanical alloying was carried out until a steady state was reached in terms of hardness and particle morphology. The powder was canned in mild steel and hot isostatically pressed at 1573 K for 3 hours. Samples without the process control agent could be swaged. In the samples to which the process control agent was added, hard, brittle, decomposition products were observed and swaging was not possible. The mechanically alloyed samples showed a large increase in room temperature hardness compared with samples of the base alloys.

ODS of B2 FeAl

Seybolt [14] investigated an Fe-40at% Al alloy containing a large volume fraction (12 vol%) of Al_2O_3. The alloy was prepared by ball milling. The dispersoids had sizes ranging from of about 0.1 to 0.4 μm. The ODS alloy showed a four-fold increase in stress-rupture strength over that of the binary cast and wrought alloy, even though the particle sizes were not as fine as those typically recommended for strengthening by this means.

Strothers and Vedula [15,16] have dispersion strengthened Fe-40at%Al-0.1at%Zr-0.2%B powders with 1 vol% Y_2O_3, by attritor milling prealloyed powders of Fe-40at%Al-0.1at%Zr with B and Y_2O_3 powders. They obtained powders which had layers of Y_2O_3 in the FeAl powder particles. Hot extrusion at temperatures of 1250 K and 1450 K resulted in very fine grained microstructures compared with those of non-ODS alloys. Extruded grain sizes were in the range of 3 to 5 μm and the Y_2O_3 was found to be finely distributed throughout the matrix. The results of their tensile tests at room temperature and at 1100 K for the as-extruded alloys are reproduced in Figure 3. Room temperature tests revealed that the ODS fine grained alloys had tensile elongations at room temperatures of upto 10% which is the highest ductility observed in this class of alloys. In addition, the ODS alloys were much stronger at room temperature as well as at 1100 K. At elevated temperatures, the ductility of the ODS alloys is found to deteriorate due to the fine grain size. However, it was found that the fine grained microstructures can be coarsened to large elongated grains by suitable heat treatments and these columnar grains may prove beneficial for high temperature creep resistance. Further investigation of these alloys is in progress.

ODS of B2 NiAl

Seybolt [14] also investigated the oxide dispersion strengthening of NiAl. Alloys were prepared by ball milling mixtures of Ni_2Al_3, Ni and the oxide. The oxide was in the form μm size powders of Al_2O_3, ThO_2 or Y_2O_3. Significant quantities of Al_2O_3 were picked up even in the base alloy due to the processing. The powder mixtures were hot extruded and the volume fraction of dispersoids ranged from 3 to 6% and the particle sizes ranged from 50 nm to 0.5 μm. The bend transition temperature was raised by about 150 C due to the presence of Y_2O_3 in the alloy containing 5 vol% Y_2O_3. The embrittling effect of dispersoid oxides was confirmed by high temperature tensile tests as well. Although significant amounts of strengthening were reported at 1123 K, compared with the non-ODS alloys, the strength increments decreased with increasing temperature.

The remaining portion of this paper presents the results of some attempts at Case Western Reserve University to produce NiAl alloy powders dispersed with Y_2O_3 by mechanical alloying.

The powders utilized for this work consisted of prealloyed intermetallic powders with nominal compositions of Ni-24 at% Al and Ni-45 at% Al and particle size less than 44 μm. Elemental powders were used in the form of atomized aluminum powder with particle size less than 44 μm and carbonyl nickel powder with particle size less that 10 μm. The Y_2O_3 was in the form of pseudomorphs with crystallite size less that 10 μm . Table II summarizes the compositions of the blends prepared from these powders for mechanical alloying. Two of these blends contain prealloyed intermetallic powder as the major component, a third blend contains less of the intermetallic powder and the fourth blend contains no prealloyed intermetallic powder.

Szegvari attritors were used to process the powder blends. All blends were processed in the 01 type attritor at CWRU, except one (#3 blend), which was processed in the 1S model at Union Process Inc. (Akron, Ohio). Both attritors contained stainless steel tanks with a sealed cover and grinding was carried out under a protective (argon) atmosphere. The volumes of the tanks were approximately 0.2 and 1.5 gallons for the 01 and 1S models, respectively. The grinding media was 6.3 mm stainless steel balls and the ball-to-powder ratio (wt. basis) was approximately 15:1 in the 01 model and 28:1 in the 1S model.

Initially, blends consisting of Ni-24at% Al and Ni-45at% Al prealloyed powders, with Al and Y_2O_3 additions to make the overall composition 50at% Al, were processed in the 01 attritor. These are Blends # 1 and # 2 in Table I. Due to the relatively small volumetric content of Al in these blends and due to the brittleness of the prealloyed powders, no composite particles were obtained. These blends were unsuccessful in producing significant amounts of cold welding and the reason for this appears to be that there was not enough ductile powder in the blend.

In order to increase the volume fraction of ductile powders, the formulation of the powder blends was changed to include carbonyl Ni, atomized Al powder and Y_2O_3 powders in addition to the prealloyed Ni-45 at% Al powder. The amount of the prealloyed aluminide powder was initially close to 30 wt% and was further reduced to 20% in Blend # 3. The overall composition was maintained at 50 at% Al.

The increase in the amount of ductile powders improved the processing of these blends (# 3 and # 4), i.e. the composite particle formation was promoted and the average size of the composite particles was much larger than the average size of the starting powders. For these blends (# 3 and # 4), despite the improved composite particle formation, it was noticed that after a certain time of processing, usually a few hours, degradation of the composite particles began to take place. Figure 4 shows SEM micrographs of the starting materials and the powders obtained in the 1S attritor after processing Blend # 3 for 0.5 h and 2 h. After 0.5 h, the formation of composite particles is obvious from Figure 4(b). However, as seen in Figure 4(c) a significant degradation of the former composite particles occurs and a steady state particle size distribution is not reached. The larger composite particles degrade progressively before the remaining powders in the blend can be properly mechanically alloyed.

At the other extreme, blends of the pure metals, i.e. carbonyl Ni and atomized Al powders with Y_2O_3 additions and the overall composition of 50 at% Al and 2 vol% Y_2O_3 were processed in the 01 attritor (Blend # 4). Additions of 1 to 2 wt% methanol were made to the powder blend just before starting the powder milling. Methanol acts as a process control agent. It prevents the excessive cold welding of the aluminum particles during the initial period of composite particle formation, helping to achieve a balance between fracture and cold welding. A sample taken after 3.5 h processing revealed a narrow particle size distribution, shiny composite particles and no evidence of degradation. However, extending the processing time beyond 4h led again to a marked degradation of the composite particles. The starting powder blend and the powders obtained after 3.5 and 4.5 h processing are shown in Figures 5(a), 5(b) and 5(c) respectively.

The composite particle degradation could be attributed to the formation of a brittle phases. In fact, x-ray diffraction revealed significant presence of NiAl peaks in the powder samples obtained by processing Ni and Al powders (Blend # 4) taken after 3.5 and 4.5 h. It is interesting that other intermediate phases were not observed and the total oxygen content, determined for the Blends # 3 and # 4 was relatively low, namely 1.55 and 1.80 wt% respectively. It was noticed that the intensity of the peaks corresponding to the pure metals decreased and those of the NiAl peaks increased with processing time.

Hence, initially, composite particle formation occurs easily but with time, degradation sets in. It seems that the optimization of the content of the ductile phases in the starting blends provides some improvements in the mechanical alloying process. However, since the end product, NiAl, is a brittle phase, as soon as considerable amounts of NiAl are formed, the degradation of the composite particles begins. This degradation limits the fineness of the dispersion of Y_2O_3 which can be incorporated in the intermetallic matrix.

Although the mechanical alloying of brittle NiAl presents processing problems, it appears possible to obtain reasonable amounts of composite particle formation before severe degradation occurs. Hence, by careful monitoring of the process it should be possible to stop the mechanical alloying process at the proper time. The distribution of the oxide, however, may not be as uniform as desired. An approach to getting a better distribution of the oxide while minimizing the degradation problem would be to prepare Y_2O_3 dispersed aluminum and nickel separately and then blend the resultant powder by attrition. This may be possible since it relatively easier to mechanically alloy Y_2O_3 with nickel or aluminum by themselves.

Summary

Mechanical alloying of aluminides is difficult and hence very little success has been reported in literature. The main problems are the high aluminum contents and the brittleness of the resultant compounds. Although milling of prealloyed powders has been shown to be reasonably effective at least in some aluminides, it may not provide a uniform distribution of the oxide particles.

It appears that significant amounts of the ductile elemental powders are needed in order to obtain sufficient cold welding and fracture and thereby a uniform fine distribution of the oxides. This requires the use of a process control agent to prevent excessive welding.

Since the product of the mechanical alloying is a brittle intermetallic phase, the composite powders begin to disintegrate as soon as the compound phase has been formed. Hence, although composite particles are formed at some intermediate time during the alloying process, the powder begins to degrade very rapidly, particularly in the case of NiAl.

In spite of these problems, the authors feel that it should be possible to obtain ODS aluminides with very careful process control.

Acknowledgements

The financial support from General Electric Co. (Evendale) for the research on NiAl, is gratefully acknowledged. Mr. Figueredo further acknowledges the support from Fundacao de Amparo a Pesquisa and IPEN-CNEN/SP, Brazil. The authors are grateful to Mr. John Becker of Union Process Inc. for his advice and for use of some of their facilities.

References

1. MRS Symposia Proceedings, High Temperature Ordered Intermetallic Alloys, Vol. 39, 1985.
2. MRS Symposia Proceedings, High Temperature Ordered Intermetallic Alloys II, Vol. 81, 1987.
3. V.M. Pathare, Ph.D. Thesis, Case Western Reserve University, 1987.
4. M. Sherman and K. Vedula, J. Materials Science, Vol. 21, p. 1974, 1986.
5. K. Vedula, V.M. Pathare, I Aslanidis and R.H. Titran, MRS Symposia Proceedings, High-Temperature Ordered Intermetallic Alloys, Vol. 39, p. 411, 1985.
6. J.S. Benjamin, Mechanical Alloying for the Preparation of Superalloys in Adv. High Temp. Alloys - Processing and Properties - N.J. Grant Symposium - S.M.Allen, R.M. Pillouz and R. Widmer, eds. ASM 1986.
7. R.C. Benn, J.S. Benjamin and C.M. Austin, Design of Oxide Dispersion Strengthened Superalloys. Proc. of Conference sponsored by the Met. Soc. of AIME and the Metals and Ceramics Div. of ORNL., held at Bethesda, Md, Apr. 9-11., 1984.
8. J.S. Benjamin and T.E. Volin, " The Mechanism of Mechanical Alloying", Met. Trans. 5, 1974, p. 1129.
9. D.G. Konitzer and H.L. Fraser, MRS Symposia Proceedings, Vol.39, 1985, p.437.
10. R.G. Rowe, J.A. Sutliff and E.F. Koch, in Rapidly Solidified Alloys and their Mechanical and Magnetic Properties, ed. B.C. Giessen, D.R. Polk and A.I. Taub, MRS Symposia Proceedings, Vol. 58.
11. J.A. Sutliff and R.G. Rowe, ibid.
12. R.J. Lauf and C.A. Walls, ECUT Quarterly Report, July 1-Sept.30, 1985, ORNL, p. 56.
13. S.G. Donnelly and C.C. Koch, ECUT Quarterly Reports, Sept.1, 1983 to Sept. 1, 1986, North Carolina State University.
14. A.U. Seybolt, Trans. ASM, Vol.59, 1966, p.860.
15. S.D. Strothers and K. Vedula, presentation at the Fall TMS-AIME Meeting, Cincinnati, September, 1987.
16. S.D. Strothers and K. Vedula, Progress in Powder Metallurgy, Vol.43, 1987, p.597.
17. J.S. Benjamin, Ni-Mechanical Alloying in Encyclopedia of Materials Science and Engineering, V.4, p. 3194, M.B. Bever, ed. MIT Press, Cambridge MA 1986.

18. J.S. Benjamin, "Dispersion Strengthened Superalloys by Mechanical Alloying", Met. Trans. 1, (1970), p. 2943.

Table I. Comparison of Physical Properties of Some Aluminides

Compound	Crystal Structure	Melting Temperature	Density g/cc
Ti_3Al	DO_{19}	1873 K	4.2
TiAl	$L1_o$	1733 K	3.9
Fe_3Al	DO_3	1813 K	6.7
FeAl	B2	1603 K	5.6
Ni_3Al	$L1_2$	1663 K	7.5
NiAl	B2	1913 K	5.9

Table II. Composition of the powder blends processed in the attritors.
The overall matrix composition is 50 at% Al for all blends.

Blend	Prealloyed Powder		Elemental Powder		Yttria	Attritor
#	Ni-24a%Al -44 μm	Ni-45at%Al -44 μm	Al -44 μm	Ni -10 μm	-10 μm	
	wt%	wt%	wt%	wt%	vol%	
1	78.46	-	21.54	-	1.0	01
2	-	94.28	5.72	-	1.0	01
3	-	20.00	26.02	53.98	2.0	1S
4	-	-	31.49	68.51	2.0	01

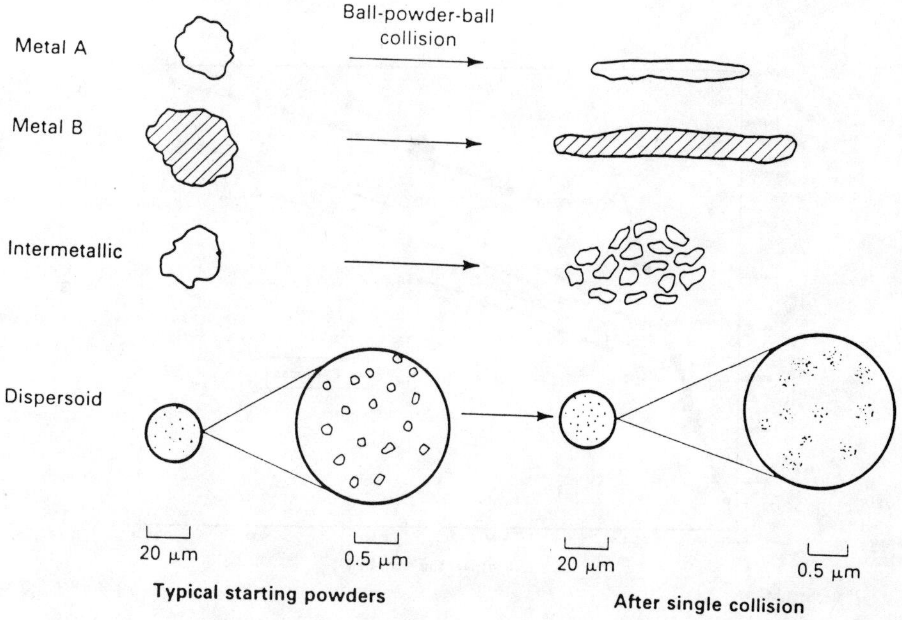

Figure 1. Schematic illustration of how individual powders change during ball-powder-ball collisions in mechanical alloying.

(a) Cold welding. (b) Powder fracture. Courtesy of International Nickel Co., Inc.

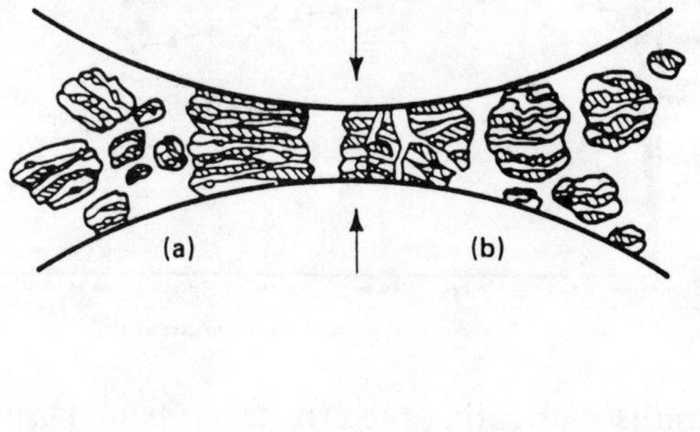

Figure 2. Schematic illustration of cold welding and fracture during mechanical alloying.

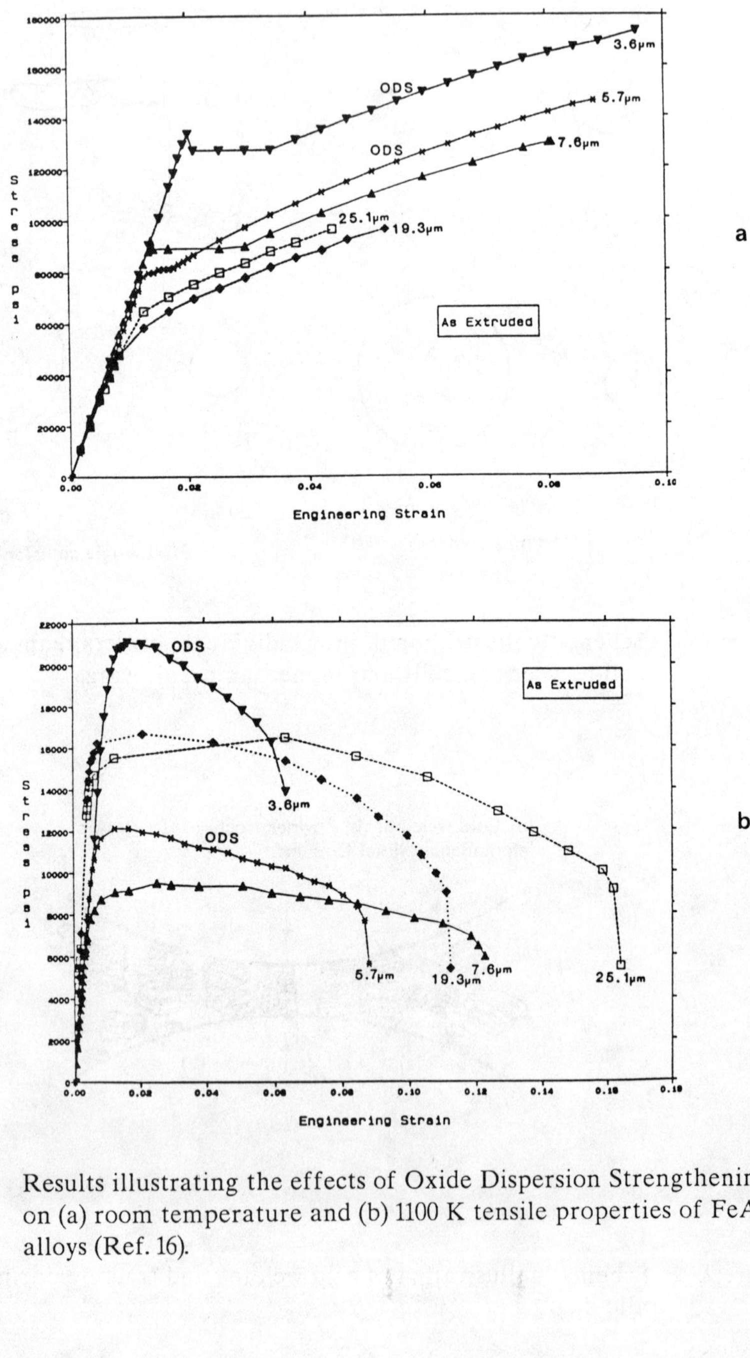

Figure 3. Results illustrating the effects of Oxide Dispersion Strengthening on (a) room temperature and (b) 1100 K tensile properties of FeAl alloys (Ref. 16).

(a)
Starting Powder Blend

(b)
After 0.5 h in attritor

(c)
After 2 h in attritor

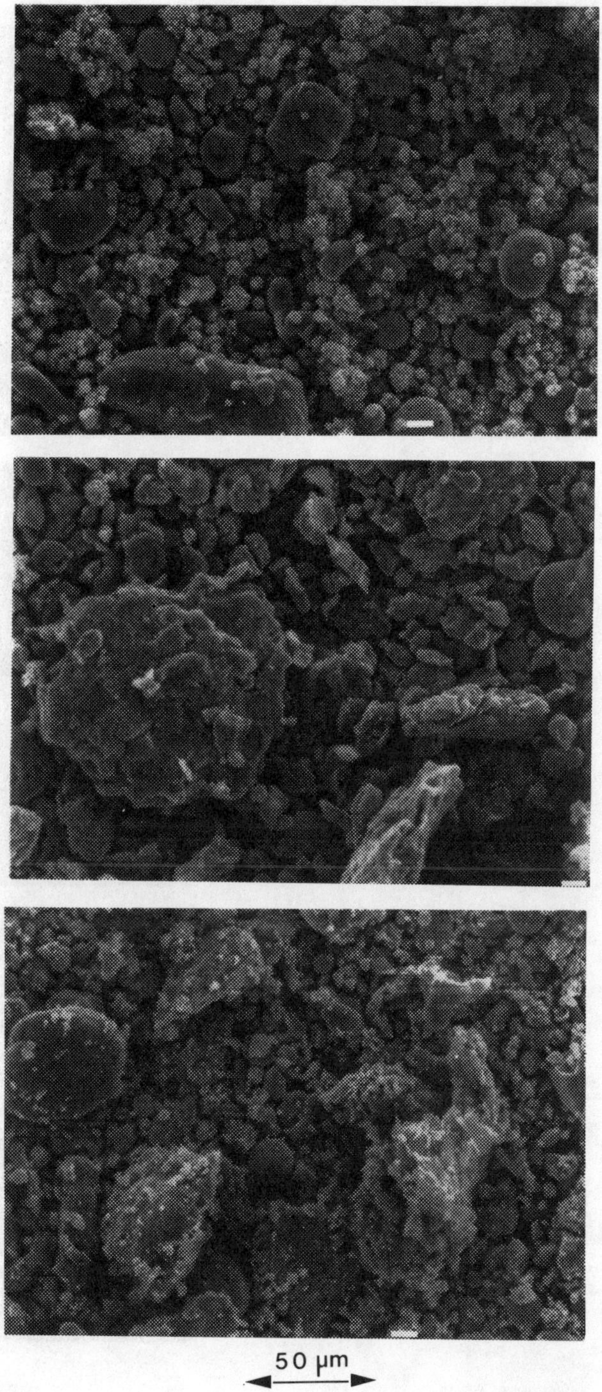

Figure 4. Morphologies of powder blend # 3 containing NiAl + Ni + Al + Y_2O_3

(a)
Starting Powder Blend

(b)
After 3.5 h in attritor

(c)
After 4.5 h in attritor

Figure 5. Powder morphologies for Blend # 4, Ni + Al + Y_2O_3

PRODUCTION OF A PRECIPITATION STRENGTHENING Cu-0.6wt%Zr ALLOY BY HOT EXTRUSION OF A RAPIDLY SOLIDIFIED POWDER

A.M. Hallén, L. Arnberg, N. Bäckström and U. Backmark*)
Swedish Institute for Metals Research
Drottning Kristinas väg 48, S-114 28 Stockholm, Sweden
*) Present address: Flygt Pumpar AB, Box 1309
S-171 25 Solna, Sweden

Abstract

A precipitation strengthening copper alloy has been produced by hot extrusion of rapdily solidified Cu-0.6 wt% Zr powder. The influence of powder size and extrusion conditions on the mechanical properties have been investigated as well as the effect of thermomechanical treatment of the extruded alloy. It is shown that powder particle size, in the range studied, has little effect on the tensile properties while an increased extrusion temperature results in a lower strength. Cold drawing the alloys to 90% area reduction and subsequent heat treatment results in a room temperature strength above 600 MPa and a ductility of ~10%. Elevated temperature tensile testing and creep testing of the as-extruded alloy shows little softening up to 400°C. The electrical conductivity is of the order of 85-90% IACS for the as-extruded material.

Introduction

Previous work (1-3) has shown that rapid solidification processing has strong potential for development of high strength-high conductivity copper alloys. A Cu-0.5wt%Zr alloy prepared by hot extrusion of ultrasonically helium atomized powder has shown interesting tensile properties in combination with high electrical conductivity (3). These properties are attributed to finely dispersed intermetallic particles precipitated on a fine scale in the rapidly solidified powder, or precipitated from extended solid solution during consolidation. The mechanical properties of the alloy are expected to be highly dependent on the extrusion parameters. No previous systematic investigation has been made on the influence of these parameters on the mechanical properties of this Cu-Zr alloy and the present investigation was therefore undertaken with the following aims:

- to investigate the influence of cooling rate on the tensile properties by extruding different powder size fractions
- to investigate the influence of extrusion temperature on the tensile properties by extruding a given powder size fraction at different temperatures
- to investigate the influence of thermomechanical treatment (cold work and subsequent annealing) on the tensile properties of the extruded alloy.

Tensile test data at temperatures up to 600°C, and 100 hours stress rupture data have been obtained and will be presented and compared with commercially available high strength copper alloys.

Experimental

A master alloy with the composition Cu-0.6wt%Zr was produced by induction melting high purity elements under argon atmosphere. The alloy was atomized by helium at a dynamic pressure of 6.9 MPa from a superheat of 250K in an ultrasonic gas atomization unit (USGA) described in reference 4. The powder was characterized by optical microscopy and X-ray diffractometry. Powder sieved to different size fractions was filled into 45 mm diameter copper extrusion cans. Some of the powder batches were completely processed under inert atmosphere conditions, while some were exposed to the laboratory atmosphere during part of the processing. The filled cans were evacuated, sealed and precompacted by cold isostatic pressing at 500 MPa to ~80% density. The billets were preheated in a furnace at the extrusion temperature for at least one hour and then extruded at an area reduction of 20:1 into round bars. The experimental conditions for the different extrusion experiments are summarized in Table I.

TABLE I. The designation and contents of the extrusion cans, the packing density before CIP, and some extrusion parameters.

Designation	Powder size (μm)	Packing density (%)	Extrusion temp. (°C)	Pre-heating time (min)	P_{max} (MPa)	P_{stat} (MPa)
1	-50	69	560	105	869	852
3	-25	61	560	125	673	639
4	-50	67	540	130	1184	639
5	-50	66	560	95	984	929
10	-50	64	640	80	980	937
11	90-180	59	560	100	835	852

Parts of the extruded bars were cold drawn to 90% area reduction. Some of the cold drawn materials were then heat treated for 1h at 500°C. Room temperature tensile properties were measured on as-extruded, cold drawn and cold drawn + heat treated materials while elevated temperature tensile and creep properties were measured only on the as-extruded alloys. The electrical conductivity was measured on the cold drawn material.

Results and discussions

The solidification cell size of powders in the -40μm size range, varied between 2-3μm. This corresponds to an approximate cooling rate of 10^5 K/s, calculated from data given in Reference 1. This cooling rate corresponds well with cooling rates from other powders atomized under similar conditions (4).

The zirconium concentration in solid solution, obtained from transmission electron microscopy is about 0.2% in the as-extruded material. In the as-atomized powder zirconium is present as precipitates in the grain boundaries, and it is seen that further precipitation occurs in the material during the consolidation process.

Tensile properties at room tempeature and elevated temperatures of the as-extruded alloy are presented in Figures 1 and 2, and compared with MZC-copper (HIP:ed and extruded). MZC-copper has been developed for gas-turbine applications and contains 0.05%Mg, 0.17%Zr and 0.35%Cr (5).

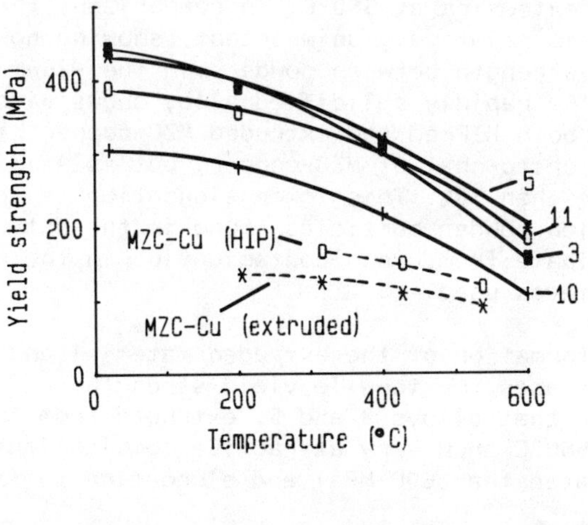

Figure 1. The yield strength as a function of temperature for the as-extruded Cu-0.6wt%Zr alloy. No 3, 5, 10 and 11. Data for MZC-copper are also presented for comparison (5).

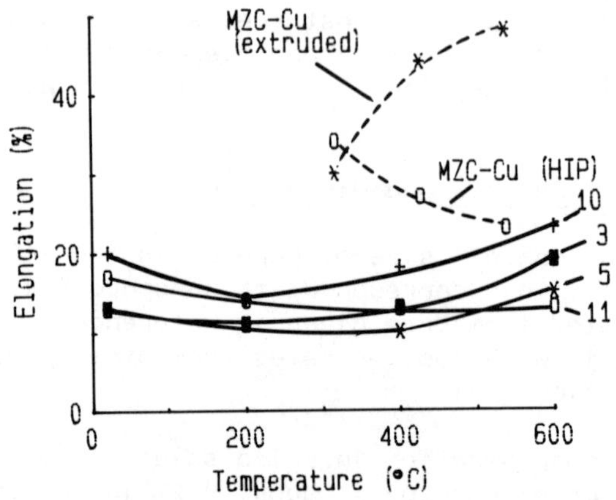

Figure 2. The elongation as a function of temperature for the as-extruded Cu-0.6wt%Zr alloy, No 3, 5, 10 and 11. Data for MZC-copper are also presented for comparison (5).

The results from elevated temperature tensile testing showed the rapidly solidified alloy has god retention of strength up to 400°C, at which temperature the alloy starts to soften more rapidly. It is clear that the extrusion temperature has a strong influence on the tensile strength; extrusion at 640°C gives a significantly lower strength than extrusion at 560°C. In comparison, the effect of powder particle size is relatively unimportant, showing no significant difference in strength between powders in the -25μm and 90-180 μm size ranges. The rapidly solidified alloy shows significantly higher strength than both HIP:ed and extruded MZC-copper. The elongation is however inferior to that of MZC-copper, but is still acceptable and always greater than 10%. This lower elongation is thought to be due in part to alien powder particles found in the extruded alloy, which probably originate from cross-contamination in the atomisation unit or from the sieves used.

Cold deformation of the extruded material and subsequent heat treatment increases the tensile yield strength, as shown in Figure 3. It can be seen that alloys 3 and 5, extruded from the fine powder fractions at 560°C show very attractive combinations of tensile strength (greater than 600 MPa) and elongation to fracture of about 10%.

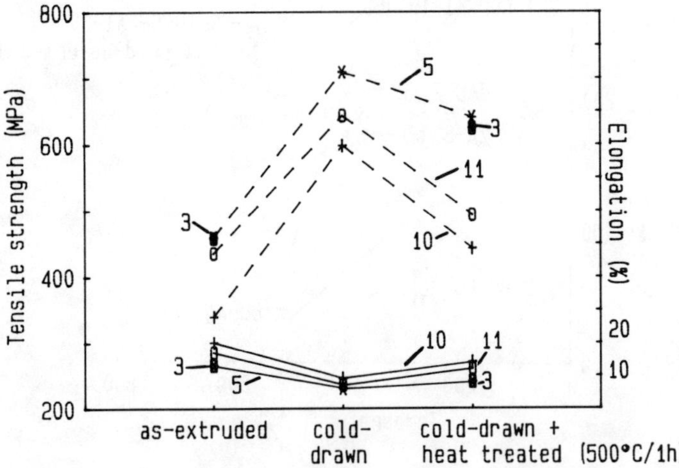

Figure 3. The tensile strength (dotted lines) and the elongation (continuous lines) for the Cu-0.6wt%Zr alloy, No 3, 5 10 and 11, in the as-extruded, cold-drawn plus heat treated conditions.

One hundred hour stress rupture data of the present alloy (nr 1 and 4, in the as-extruded conditions) are shown in Figure 4 and compared with other commercial copper alloys. It can be seen that the Cu-0.6wt%Zr alloy has attractive creep properties, but is still inferior to dispersion strengthened Cu-Al_2O_3 alloys (GLIDCOP Al-15 and Al-60) (6). The rapid decrease in strength for the present alloy at higher temperatures is assumed to be caused by coarsening of the intermetallic Cu-Zr particles whereas the oxide dispersons resist coarsening at higher temperatures better.

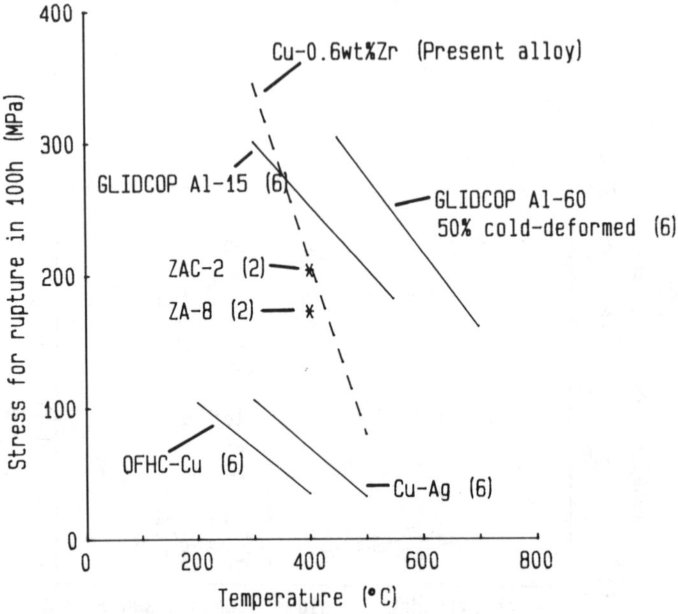

Figure 4. One hundred hour stress rupture data of the as-extruded Cu-0.6wt%Zr alloy, and corresponding data for other copper alloys.

The measured electrical conductivities of cold drawn Cu-0.6wt%Zr alloys are shown in Table II. Corresponding conductivities of the as-extruded alloys have not been measured, but are estimated to be a few percent higher. The highest value of conductivity is obtained for the material which was extruded at the highest temperature. The explanation for this is that a smaller amount of zirconium will remain in solid solution in the matrix after the higher temperature extrusion than for the other materials.

Table II. The electrical conductivities of cold-drawn Cu-0.6wt%Zr alloy.

No	% IACS
3	82
5	82
10	87
11	84

Conclusions

This investigation has shown:

* The cooling rate of ~10^5 K/s experienced by these powders is not sufficient for complete metastable solid solution of 0.6wt%Zr in copper. After atomization some precipitates of intermetallic Cu-Zr-particles occur on a fine scale in the powder particles.

* Extrusion temperature has a strong influence on mechanical properties and therefore the alloy should be extruded at as low temperature as possible.

* Initial powder particle size has little effect on the mechanical properties of the as-extruded alloy. Material extruded from the finer powder factions (-50μm) seem, however, to respond more favourable to thermomechanical treatment.

* The as-extruded alloy has good yield and tensile strength up to 400°C and retains good stress rupture strength.

* The room temperature tensile strength can be increased by cold work. Subsequent annealing results in an attractive combination of high strength (>600 MPa) and acceptable ductility (~10%).

* The electrical conductivity of the cold drawn alloy varies between 82 and 87% IACS depending on processing conditions. Corresponding values for as-extruded material are ~85-90% IACS.

Acknowledgements

The present work has been financed by the Swedish Board for Technical Development.

References

1. K. V. SARIN and N. J. GRANT, Metall. Trans. 3 (1978) 875.
2. K. V. SARIN and N. J. GRANT, Powder Metall. Int. 11 (1979) 153.
3. L. ARNBERG, U. BACKMARK, N. BÄCKSTRÖM and J. LANGE, Mat. Sci. and Eng. 83 (1986) 115.
4. U. BACKMARK, N. BÄCKSTRÖM and L. ARNBERG, Powder Metall. Int. 18 (1986), 338, ibid. 422.
5. L. G. PETERSON, D. W. SEAVER, and A. RIFFEN, Final Report INCRA Project 337, (1983).
6. Data sheet from Glidden Metals.

PROCESSING EFFECTS ON DENSIFICATION IN REACTIVE SINTERING OF NICKEL-ALUMINUM POWDER MIXTURES

B. H. Rabin*, A. Bose** and R. M. German**
*EG&G Idaho, Inc., P.O. 1625
Idaho Falls, ID 83415
**Department of Materials Engineering
Rensselaer Polytechnic Institute
Troy, NY 12180-3590

ABSTRACT

The reactive sintering behavior of nickel-aluminum powder mixtures containing 12 to 30 wt.% aluminum has been studied with emphasis on the formation of stoichiometric nickel-aluminide Ni_3Al. During heating, solid-state reactions at interparticle contacts lead to localized melting and the initiation of an exothermic reaction. Under appropriate conditions, very rapid densification and compound formation occur during this exothermic heating. The effects of several process variables on sintered density have been examined including stoichiometry, heating rate, atmosphere, particle size, and green density. Densification results are explained by the influence of these variables on the quantity and distribution of the transient liquid phase, as well as the role of entrapped gases. Optimum processing conditions allow near full density Ni_3Al to be produced directly from elemental powder in very short times.

INTRODUCTION

Reactive sintering involves simultaneous exothermic reaction and densification within a powder compact through the formation of a transient liquid phase [1,2]. The exothermic reaction between mixed powders gives the spontaneous formation of a stable intermediate compound. Variations on this process have been referred to in the literature as gasless combustion sysnthesis (GCS) and self-propagating high temperature synthesis (SHS), depending upon whether the sample was heated uniformly to initiate the exothermic reaction or whether the reaction was initiated locally and allowed to propagate through the sample, respectively. In either case, the process has produced a variety of compounds including carbides, borides, nitrides, silicides, and aluminides [3,4]. In most cases the reaction product was porous and could be made into powder for subsequent consolidation. Recent interest has focused on using reactive sintering to produce high density reaction products as a potential low- cost fabrication technique for high temperature structural materials [5].

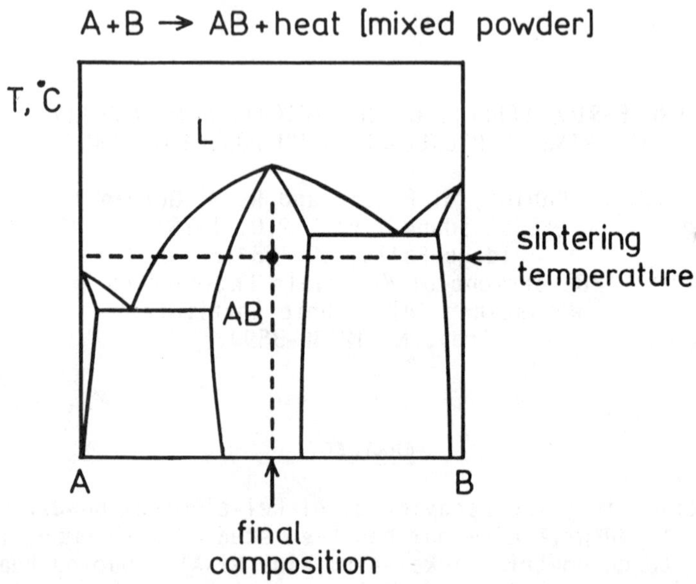

Figure 1. Schematic binary phase diagram for a potential reactive sintering system. A stoichiometric mixture of A and B powders reacts to give the compound AB by forming a transient liquid phase.

Figure 1 shows a schematic binary phase diagram for a potential reactive sintering system wherein a stoichiometric mixture of A and B powders reacts to form the compound AB. Upon heating, a rapid self-propagating reaction is initiated with the formation of the first liquid, in this case a eutectic. The compact temperature increases rapidly as a result of heat liberated by compound formation. The highest possible compact temperature can be calculated, assuming adiabatic conditions, from the heat capacity and enthalpy of formation of the product phase. This temperature may be above or below the melting points of the reactants and products; in general, the reaction and densification kinetics, as well as the resultant microstructure, will depend upon the nature of the reaction zone.

The nickel-aluminum phase diagram is shown in Figure 2 [6,7]. There are five intermetallic compounds and a low melting eutectic occurs about 640°C. Several authors have analyzed the mathematical basis for the combustion process [8-11], and there are several prior reports on the exothermic characteristics of nickel-aluminum powder mixtures [5,12-19]. Compound formation occurs in the solid-state during heating, beginning with the aluminum-rich phase $NiAl_3$. The

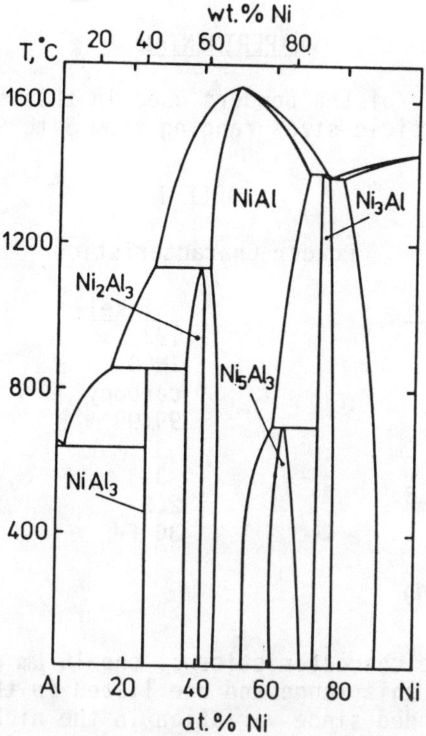

Figure 2. The nickel-aluminum phase diagram which displays the characteristics favorable for reactive sintering.

onset of the exothermic reaction, the magnitude of the heat release, the reaction velocity, and the product phases depend on a number of factors including stoichiometry, particle size, green density, and heating rate. Alternatively, there have been few reports of the effects of these variables on densification. An exception is the case of the stoichiometric NiAl, where the adiabatic temperature exceeds the melting point of the product phase and a cast structure results. For the Ni_3Al stoichiometry, Philpot et al. [18] reported a decrease in porosity from 28 to 16% as the heating rate increased from 1 to 5 K/min. The addition of 0.5 at.% boron further decreased the porosity to 7%; however, no explanation for this behavior was provided.

In comparison with conventional powder metallurgy processing routes for intermetallics, such as hot isostatic pressing or hot extrusion of prealloyed powders, reactive sintering has potential as a low-cost fabrication technique. It uses elemental powders, relies on self-heating, involves very short processing times, and does not require expensive equipment [5]. An understanding of the densification mechanism which occurs during exothermic compound synthesis is required if these advantages are to be realized. This study focuses on isolating the process variable effects on densification for the Ni_3Al composition.

EXPERIMENTAL

The characteristics of the powders used in this study are shown in Table I. Aluminum particle sizes ranging from 3 to 95 μm

TABLE I

Powder Characteristics

property	nickel	aluminum
designation	123	H-15
vendor	INCO	Valimet
powder type	carbonyl	gas atomized
purity, %	99.99	99.7
Fisher Subsieve size, μm	3	15
mean size*, μm	3.3	22.6
apparent density, g/cm^3	2.2	1.3
major impurity, ppm	30 Fe	1200 Fe

*laser light scattering

were investigated; the characteristics of the 15 μm aluminum powder are representative of this range and are listed in the table. Only one nickel powder is included since variation in the nickel size has less influence on the reaction unless the particle size is very large. The elemental powders were mixed in a turbula mixer for 30 minutes. The aluminum contents ranged from 10 to 30 wt.% with emphasis on compositions close to the Ni_3Al stoichiometry, 13.3 wt.% Al. Nominally cylindrical specimens 12 mm in diameter were die compacted at pressures ranging from 100 to 300 MPa giving green densities between 50 and 75% of theoretical. Zinc stearate was used as a die wall lubricant. Additional samples approximately 10 mm in diameter and 100 mm in length were prepared by cold isostatic pressing at 210 MPa.

Densification experiments were conducted in a horizontal tube furnace in either a vacuum of 7×10^{-3} Pa, flowing dry hydrogen or dry argon. Heating rates ranged from 3 to 30 K/min. Several self-propagating experiments were performed outside the furnace by locally heating one end of the cylindrical specimen with a propane torch to initiate an exothermic reaction which subsequently spread through the compact as a combustion wave front. Qualitative information about the exothermic reaction and densification were obtained from differential thermal analysis (DTA) and dilatometry. In

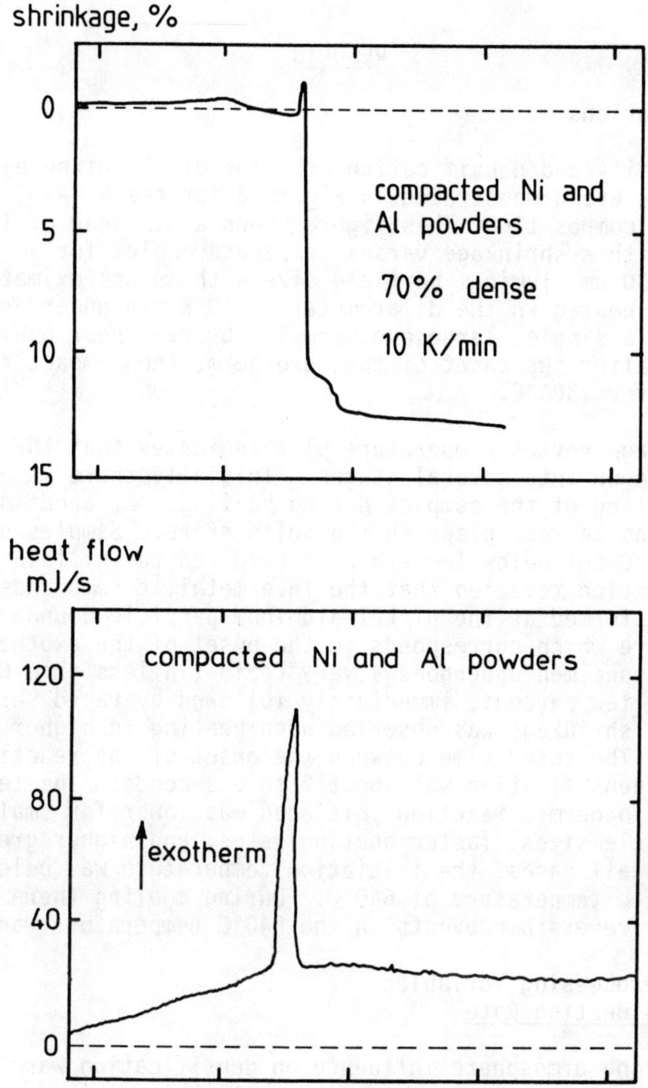

Figure 3. A typical differential thermal analysis scan and a plot of shrinkage versus temperature for a sample heated in a dilatometer. The large exothermic reaction near 600°C is accompanied by rapid shrinkage.

many cases the residual pores were isolated and the sintered densities could be determined by weighing the samples in air and in water. Microstructural characterization was carried out using optical metallography, scanning electron microscopy, transmission electron microscopy, x-ray diffraction, and electron microprobe analysis.

RESULTS

General Observations

The reactivity and densification behavior of the mixed elemental powder compacts are demonstrated in Figure 3 for the Ni_3Al stoichiometric composition. This figure shows a DTA scan of loose powder along with a shrinkage versus temperature plot for a sample with a 30 μm aluminum particle size with an approximate green density of 70% heated in the dilatometer at 10 K/min under argon. During heating a single, large exotherm is observed near 600°C. Within a few seconds after the onset of the exotherm, the compact temperature increased to over 1300°C.

The shrinkage versus temperature plot indicates that the process can be broken down into several stages. Initially there is a small amount of swelling of the compact during heating. At about 450°C shrinkage begins to take place in the solid state. Samples quenched from above 450°C but below temperatures required to initiate the exothermic reaction revealed that the intermetallic compounds $NiAl_3$ and Ni_2Al_3 had formed at the nickel-aluminum particle boundaries. At a temperature which corresponds to the onset of the exothermic reaction, the specimen undergoes a very rapid (in less than 0.5 s) expansion of a few percent, immediately followed by rapid shrinkage. No significant shrinkage was observed upon heating to higher temperatures. The total time between the onset of the reaction and the completion of densification was about 2 to 3 seconds. The temperature at which the exothermic reaction initiated was lower for smaller aluminum particle sizes, faster heating rates, and higher green densities. In all cases, the initiation temperature was below the lowest eutectic temperature of 640°C. During cooling there were no indications of reversible events in the 640°C temperature range.

Influence of Processing Variables

Atmosphere and Heating Rate

The sintering atmosphere influence on densification was investigated at two heating rates using a stoichiometric mixture of Ni_3Al prepared from 15 μm aluminum powder. The compacts all had a green density equal to 70% of theoretical. The results are shown in Table II in terms of the final porosity. For the lower heating rate of 3 K/min, the samples sintered in hydrogen and argon exhibited swelling. At a heating rate of 30 K/min densification occurred in all atmospheres. The best densification was obtained using vacuum, giving a final density over 97% of theoretical. Argon, which is an insoluble gas, resulted in the lowest sintered density.

Dilatometer experiments established that at slow heating rates more solid-state shrinkage occurred prior to the reaction. There was a corresponding increase in the initiation temperature at the slower heating rates. Clearly, rapid heating gives more liquid and more

TABLE II

Atmosphere and Heating Rate Effects on Final Porosity, %

atmosphere	heating rate 3 K/min	30 K/min
argon	47.0	6.9
vacuum	2.5	2.7
hydrogen	46.0	3.6

densification because of less solid-state interdiffusion prior to the reaction.

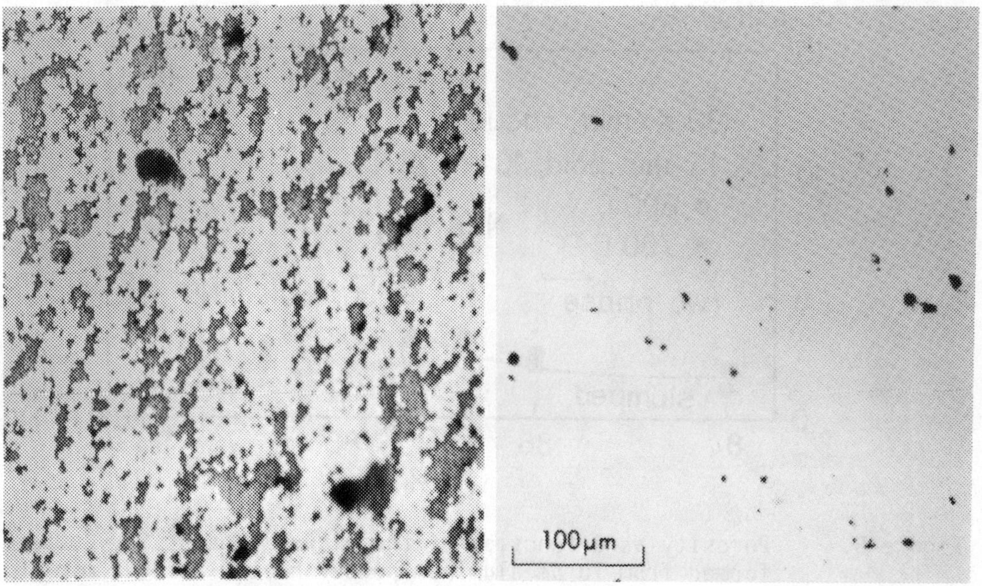

Figure 4. Optical micrographs taken from Ni-13.3 wt.% Al samples produced from 10 μm aluminum powder; a) green compact, b) reacted and annealed unetched compact.

The microstructure of the sample sintered in vacuum with a heating rate of 30 K/min is shown in Figure 4. The first micrograph shows the green compact while the second shows the reacted structure following a high temperature anneal (unetched). The grain structure is equiaxed with a mean grain intercept size of 30 μm. Chemical analysis after sintering gave the bulk composition as 87.6% Ni, 12.2% Al, 0.02% Fe, 0.01% Si, 482 ppm N, and 420 ppm C. According to electron microprobe analysis, the matrix phase is Ni_3Al with an aluminum content of of 24.3 at.%. After the reaction a second phase with an aluminum content of 34.8 at.%, corresponding to the stoichiometry of Ni_5Al_3 was present. This phase is unstable over 740°C and annealing of the reactive sintered material for one hour at 1350°C resulted in a homogeneous single phase, ordered Ni_3Al product, as confirmed by electron diffraction.

Stoichiometry

Stoichiometry effects were investigated using 10 μm aluminum powder compacts with a green density of 70% of theoretical. These compacts were heated in vacuum at 30 K/min to temperatures greater than 600°C and held for 15 minutes. The effect of the maximum temperature (600 versus 700°C) is small. Figure 5 shows the porosity after reaction as a function of nickel content. For nickel-rich samples the

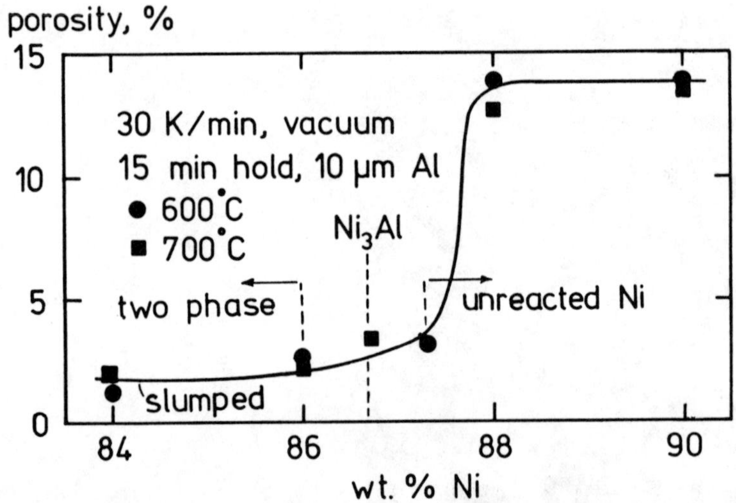

Figure 5. Porosity as a function of the nickel content for compacts formed from 10 μm aluminum powder heated at 30 K/min to either 600°C or 700°C for 15 minutes.

sintered density was low. Near the Ni_3Al stoichiometry a dramatic change in densification behavior is observed. Increasing the aluminum content above the stoichiometric composition results in near full density two phase alloys of NiAl and Ni_3Al. Micrographs of three nickel contents are shown in Figure 6 for 10 wt.% Al, 16 wt.% Al, and 18 wt.% Al. With 10 wt.% aluminum, the compact is very porous, and shows a two phase solid structure. The magnitude of the exothermic effect was greater in the high aluminum content alloys, but compact slumping was observed for the highest aluminum contents, indicating excessive liquid. Above 20 wt.% aluminum the compacts underwent complete melting and a cast structure was obtained.

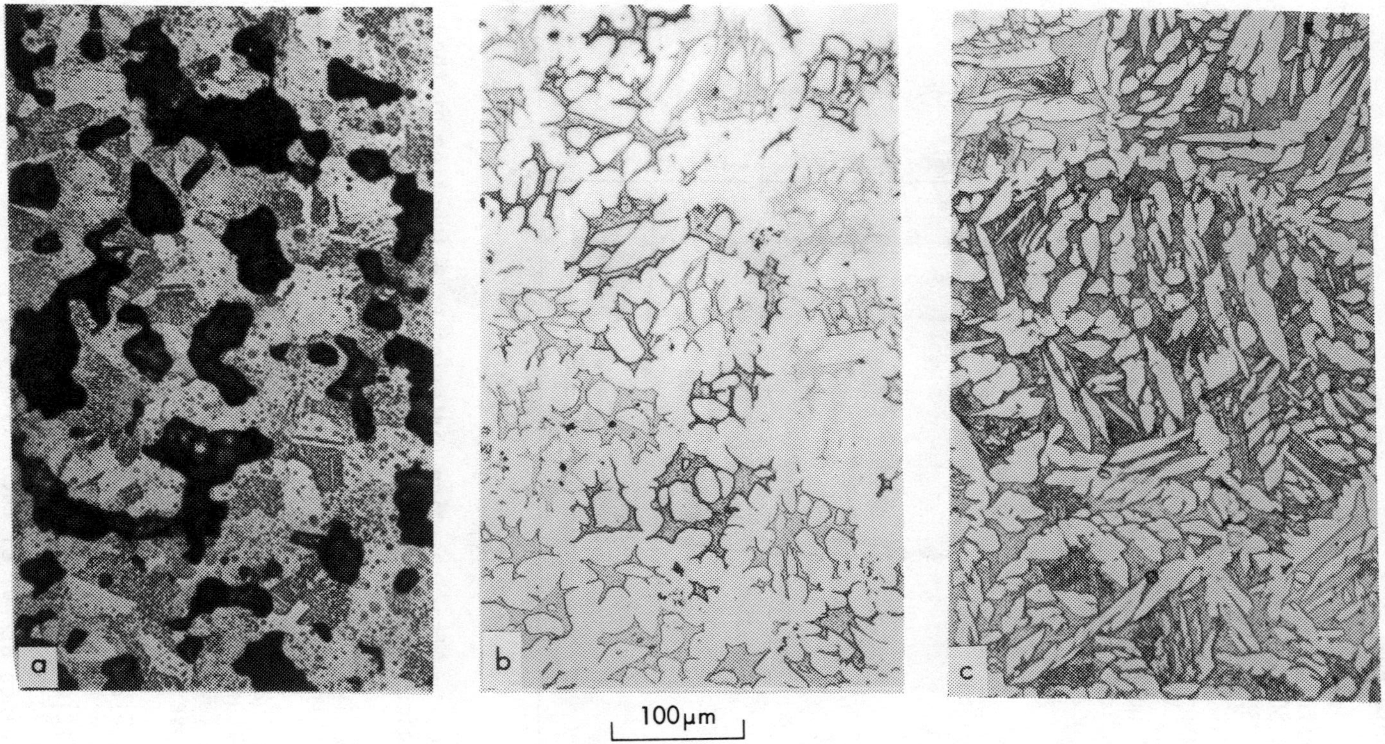

Figure 6. Optical micrographs of the reacted microstructures for samples heated at 30 K/min in vacuum; a) 10 wt.% Al, b) 16 wt.% Al, and c) 18 wt.% Al.

Sintering Temperature and Time

In contrast to classical sintering behavior, maximum temperature and time at temperature have little influence on densification. Temperatures below 500°C gave poor densification, while for temperature between 550°C and 750°C the final porosity was between 2.7 and 3.9% with a nonsystematic variation. This is a result of the very rapid, self-sustaining nature of the exothermic reaction once initiated. The dilatometer results indicated that the initiation temperature was strongly dependent upon processing variables. For the range of experimental conditions studied, a minimum temperature of at least 550°C must be reached. Figure 7 plots porosity after reactive

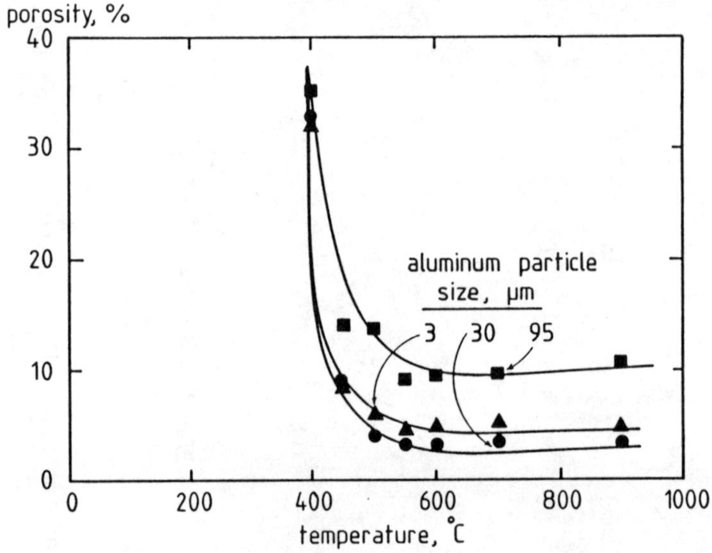

Figure 7. Porosity after reactive sintering as a function of the furnace temperature for three aluminum particle sizes.

sintering for the Ni_3Al stoichiometry versus the furnace temperature. This set of experiments was performed using three aluminum particle sizes, with a 30 K/min heating rate in vacuum. For all three aluminum particle sizes the porosity is essentially unchanged by furnace temperatures over 550°C. Because the reaction is completed on the order of a few seconds after initiation, isothermal holding is not required and overall processing times are kept to a minimum.

Aluminum Particle Size

Aluminum particle size effects were studied with a stoichiometric alloy sintered in vacuum with a heating rate of 30 K/min. The green density was maintained at 70% of theoretical. The results are shown in Figure 8. Large aluminum particle sizes resulted in the formation of

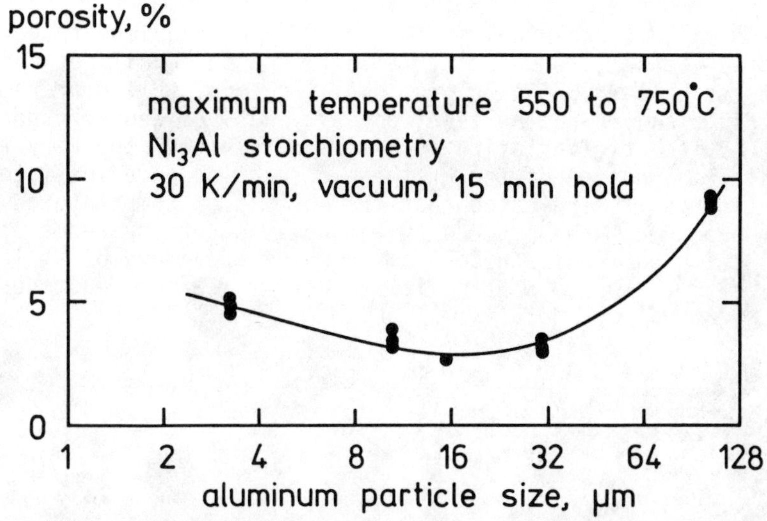

Figure 8. The role of aluminum particle size (Fisher subsieve size) on the porosity after reactive sintering for the Ni_3Al composition.

large pores at the prior aluminum particle sites due to the capillary spreading of the liquid phase. These large pores remain in the microstructure after sintering. Figure 9 demonstrates the pore generation effect with a compact formed from the 95 μm aluminum particles. The green microstructure shown in Figure 9a, and the reacted microstructure shown in Figure 9b, give a good correlation between the size and spacing of the aluminum particles and resulting pores.

With a smaller aluminum particle size the reaction initiated at a lower temperature (for example 510°C for 3 μm versus 605°C for 30 μm), but densification was optimal for an intermediate aluminum (15 μm) particle size. Quenching experiments confirmed that pores formed at the sites of large aluminum particles.

Self-propagating Reactions

Self-propagating exothermic reactions provided information regarding reaction velocities and densification mechanisms. After ignition of one end of a cylindrical compact, the reaction front propagated along the length of the specimen. In general, the reactions were self-sustaining after ignition for a distance of about 70 mm from the heated end of the specimen. Upon reaching the unheated region of the sample the reaction slowed, and often the specimen fractured ahead of the reaction front. This fracture event terminated the reaction propagation. Fracture was believed to result from thermal stresses and could be eliminated by preheating the entire sample prior to ignition

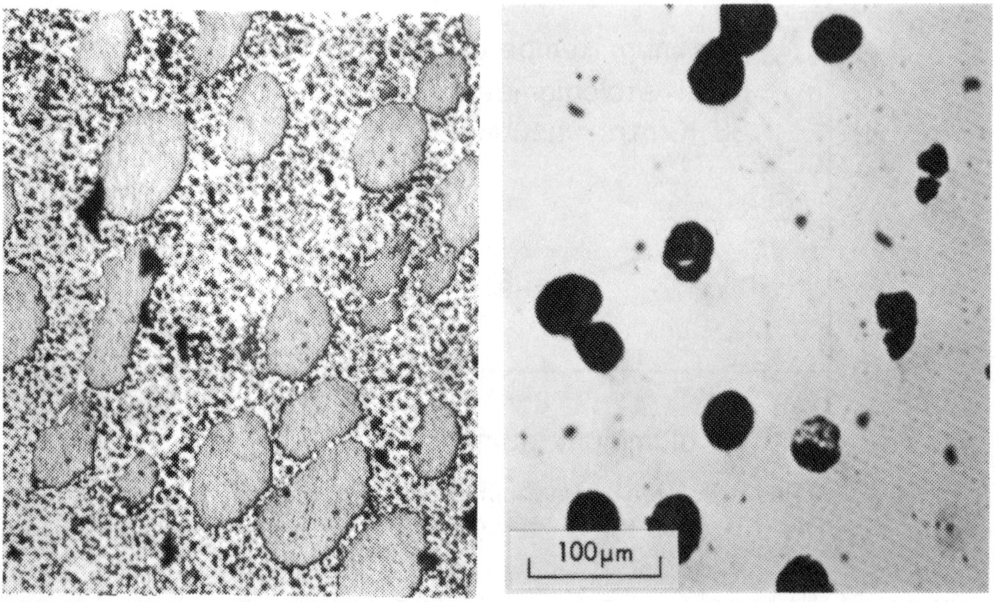

Figure 9. Optical micrographs showing the formation of large pores in Ni_3Al due to use of large (95 μm) aluminum particles; a) green, and b) reacted.

or by using a lower green density. Both steady state and pulsating reaction conditions were observed and appeared to correlate with initial powder packing inhomogeneities. For the Ni_3Al stoichiometry the steady state reaction velocity was about 15 mm/s. The velocity increased with Al content up to about 75 mm/s at 30 wt.% Al, the NiAl stoichiometry.

DISCUSSION

The mechanism for reactive sintering is shown schematically in Figure 10 for the Ni_3Al stoichiometry. Upon heating, solid-state diffusion across interparticle contacts results in the formation of the aluminum-rich compounds $NiAl_3$ and Ni_2Al_3. The heat released from the formation of these compounds increases the temperature locally above the lowest eutectic temperature, 640°C. Initially, the aluminum-rich liquid does not wet the nickel particles and swelling begins. However, immediate dissolution of the nickel allows the liquid to wet the nickel particles. Subsequent capillary spreading and the associated capillary force lead to densification. The liquid phase becomes saturated with nickel, allowing precipitation of the Ni_3Al reaction product behind the advancing liquid front. Because diffusion within the liquid phase is quite rapid, compound formation occurs quickly and heat is generated which further accelerates the reaction.

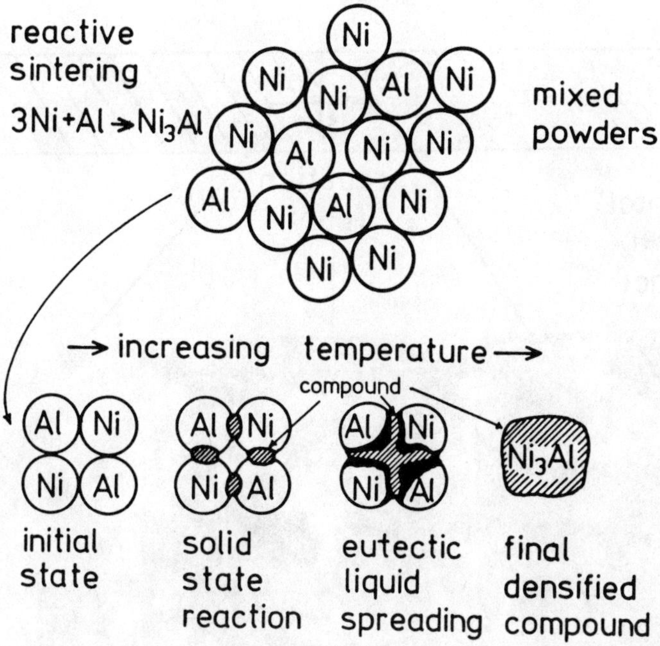

Figure 10. Schematic diagram showing the sequence of events that take place when a compact containing a stoichiometric mixture of nickel and aluminum powders is heated.

Evidence for this type of reaction mechanism is shown in Figure 11 which is a series of micrographs taken from the terminated reaction zone in a self-propagating reaction specimen. The direction of propagation is indicated in the figure along with the microstructural features. There are three distinct regions within the reaction zone. Ahead of the reaction front there is a zone of heating in which solid-state diffusion and compound formation take place. In the center of the reaction zone is a region characterized by an aluminum-rich liquid. Capillary spreading of the liquid phase and dissolution of nickel particles are evident; however, it appears that the melting temperature of nickel was not exceeded. Behind the advancing liquid front is a region where crystallization of the product phase from the liquid occurs and where additional structural changes in the solid-state take place. Compositional inhomogeneities remain after the passage of the reaction front; in particular, nickel-rich regions resulted from incomplete dissolution of the nickel particles. This specimen was reacted in air, consequently complete densification was not obtained and porosity is observed at the prior aluminum particle sites after the passage of the reaction front.

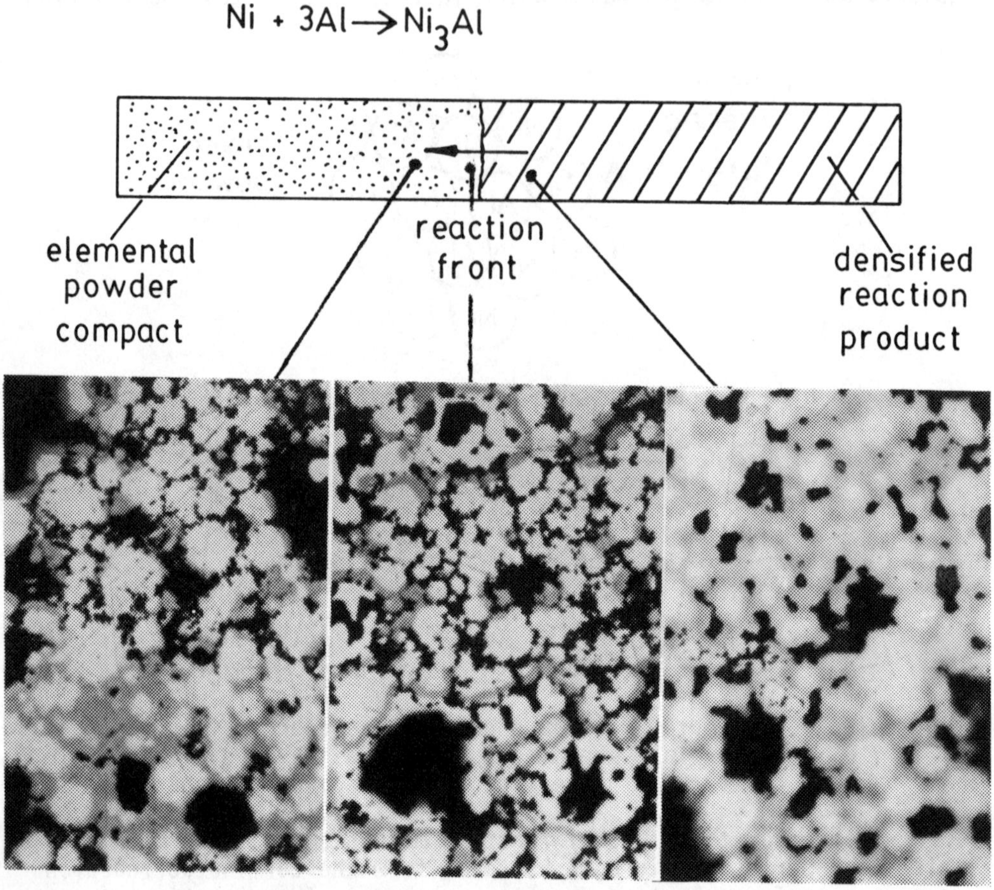

Figure 11. Optical micrographs showing the structural changes across a self-quenched propagating reaction zone for a specimen processed in air.

The adiabatic reaction temperature can be estimated from thermodynamic data. For stoichiometric Ni_3Al, the standard heat of formation at 298 K is estimated at -41 kJ/mol [20]. If it is assumed that the heat capacity is constant over the temperature range of interest, then a value of 27 J/mol-K seems reasonable. If all reactants are consumed in an adiabatic reaction, the temperature increase is calculated to be approximately 1400 K and the maximum compact temperature would exceed the melting point of Ni_3Al, 1668 K, but remain below the melting point of pure nickel, 1726 K. These calculations are consistent with the observations made in Figure 11 that the nickel particles are not melted within the maximum heating zone and that crystallization of the product phase only occurs behind this zone where the temperature is lower.

Caution should be used when interpreting these calculations because of the great number of assumptions which are implicit. Additionally, it appears from Figure 11 that not all of the reactants are consumed within the reaction zone. Nevertheless, the observations are consistent with previous studies and theoretical predictions based upon the propagation of an exothermic reaction front in gasless systems [9-11]. Steady state temperature and reactions profiles have been determined from numerical solutions to the diffusion equation for the case involving an Arrhenius temperature dependence of the reaction rate. Based upon these calculations and Figure 11, the temperature profile within the propagating reaction zone can be inferred. This is shown schematically in Figure 12.

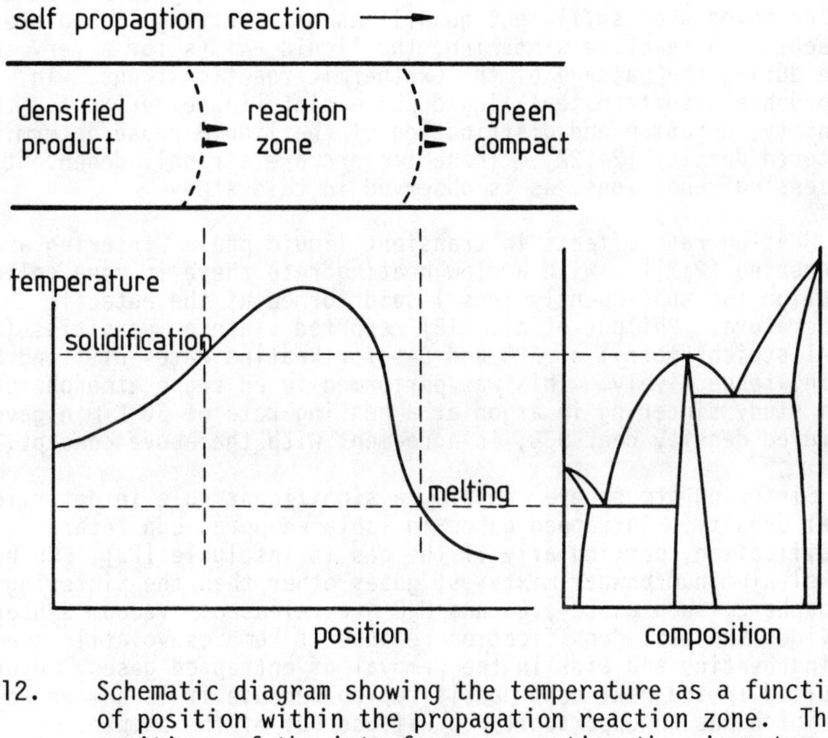

Figure 12. Schematic diagram showing the temperature as a function of position within the propagation reaction zone. The positions of the interfaces separating the microstructural regions in Figure 10 are determined from the temperature profile and phase diagram.

Specific temperatures determine the positions of the boundaries separating the different microstructural regions shown in Figure 11. The width of the reaction zone will depend on the green density, particle sizes, and initial compact temperature.

Because the exothermic reaction is initiated in the solid-state, the dependence of the initiation temperature on processing variables can be explained by their influence on the extent of interdiffusion during heating. Small aluminum particle sizes, high green densities, and high heating rates lead to lower reaction initiation temperatures. Dilatometer results confirmed these predictions. Philpot et al. [18] reported that the initiation temperature increased with increasing nickel particle size; the explanation of this result is not clear at this time. Although the initiation temperature was not the focus of this study, we have seen little effect on final porosity from variations in nickel particle sizes over the range from 3 to 37 μm. However, when the ratio of aluminum to nickel particle sizes becomes small the final porosity is large.

The densification results obtained in this study can be interpreted in terms of liquid phase sintering mechanisms. High sintered densities are achieved when sufficient quantities of a wetting liquid are present. In reactive sintering, the liquid exists for a very short time during the passage of the exothermic reaction front. In accordance with transient liquid phase sintering experiments, the quantity, duration and distribution of the liquid phase determine the sintered density [21,22]. These factors are strongly dependent upon processing conditions, as is observed in this study.

Heating rate effects in transient liquid phase sintering are well documented [2,21]. With a slow heating rate there is more solid state reaction and subsequently less liquid formed at the eutectic temperature. Philpot et al. [18] reported sintered densities (for the Ni_3Al stoichiometry) of 72% and 84% for heating rates of 1 and 5 K/min, respectively. This was performed in an argon atmosphere. In this study sintering in argon at a heating rate of 30 K/min gave a sintered density near 93%, in agreement with the above concept.

Sintering atmosphere can play a significant role in determining final density. Entrapped gases in isolated pores can retard densification, particularly if the gas is insoluble [23]. On heating nickel-aluminum powder mixtures, gases other than the sintering atmosphere, such as CO_2, CO and H_2O are released. Vacuum sintering provides the best densification because it removes volatile species during heating and aids in the removal of entrapped gases. Hydrogen resulted in the next best densification because it is a more soluble gas which has an opportunity to diffuse out of the sample.

An important consideration which influences the amount of densification is the distribution of liquid within the microstructure of the reaction zone. This is particularly important in reactive sintering because the very short duration of the liquid does not allow significant densification by classic solution-reprecipitation processes. Instead, capillary induced densification by particle rearrangement dominates. In this regard, persistent liquid phase sintering experiments have demonstrated that a uniform dispersion of the liquid throughout the microstructure aids densification. The current experimental results involving aluminum particle size and stoichiometry effects on densification support this hypothesis. Figure 13 sketches our basic concept showing the importance of green microstructures. Large aluminum particle sizes result in the formation

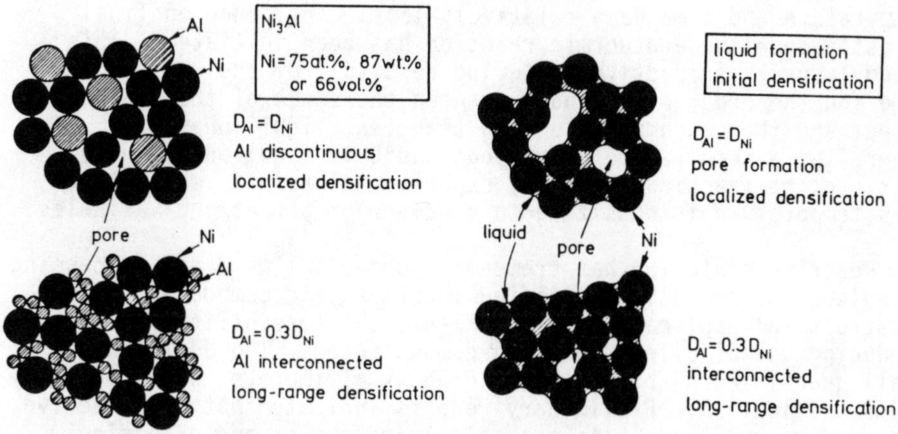

Figure 13. A schematic diagram showing the importance of green microstructure and the connectivity of the aluminum particles in attaining densification by reactive sintering. Small aluminum particles form a surrounding liquid that promotes densification while large aluminum particles lead to pore generation as the liquid wicks into the neighboring microstructure.

of isolated liquid pools and large pores at the prior aluminum particle sites. At the Ni_3Al stoichiometric composition, aluminum constitutes 34 volume percent of the solid structure. This is insufficient to form an interconnected network of aluminum, which is required for optimal densification, unless the aluminum particle size is less than the nickel particle size. In the green compact, the nickel agglomerates have a mean intercept length of approximately 30 μm, corresponding to a diameter of 45 μm. According to Kusy [24], a particle size ratio of at least 2.4:1 (major phase diameter to minor phase diameter) is required at 34 vol. % to form a connected network of the minor phase. Using this criteria an aluminum particle size of 19 μm or smaller is required. Experimentally, optimal densification occurred with the 15 μm aluminum particle size, in agreement with this value. The smaller aluminum particle sizes evidence considerable agglomeration in the green microstructure. Thus, the dominant parameters appear to be the agglomerate size and initial homogeneity, not the particle size.

A similar argument may be used to explain the stoichiometry effects on densification. Low aluminum contents decrease the amount of liquid present and the liquid connectivity. On reaction the disconnected aluminum particles result in isolated liquid regions within the compact which are ineffective in promoting densification by rearrangement.

With an excess of aluminum, there is too much liquid, consequently, slumping and shape loss are observed.

In contrast to traditional sintering treatments, sintering temperature and time have relatively little influence on final porosity once the exothermic reaction has been initiated. This unique aspect of reactive sintering results from the very short time (on the order of seconds) between the onset of the exothermic effect and the consumption of the transient liquid phase. Generally, a temperature over about 550°C is required to fully initiate the reaction; however, experimental results show that this temperature is sensitive to a number of processing variables.

Reactive sintering has tremendous potential as a new processing technique for producing densified intermetallic compounds for use in structural applications. For Ni_3Al, the feasibility of producing useful alloys has been demonstrated. The addition of small quantities of boron (about 0.05 wt.%) improves both ductility and densification. Preliminary results indicate that the reactive sintered material exhibits mechanical properties and oxidation resistance comparable to or exceeding similar compositions produced by conventional methods.

CONCLUSIONS

Heating nickel-aluminum powder mixtures which contain greater than 12 wt.% aluminum results in a strong exothermic reaction. Under appropriate conditions the compact temperature increases rapidly and densification occurs simultaneously with intermetallic compound formation in the presence of a transient liquid phase. Densification is controlled primarily by the quantity of liquid formed, its distribution throughout the microstructure, and the sintering atmosphere. High heating rates, intermediate aluminum particle sizes (near 15 to 20μm), and vacuum sintering are desirable. Under controlled conditions a near full density, single phase Ni_3Al product can be obtained directly through the initiation of the exothermic reaction in a stoichiometric powder mixture.

ACKNOWLEDGEMENTS

This research was performed under sponsorship of DARPA/ONR in the RPI composite materials research program. The authors are most appreciative of David Sims for his initial identification of this research area.

REFERENCES

1. R. L. Coble, Sintering-Theory and Practice, D. Kolar, S. Pejovnik and M. M. Ristic (eds.), Elsevier Scientific, Amsterdam, Netherlands, 1982, p. 145.

2. R. M. German, Liquid Phase Sintering, Plenum Press, New York, NY, 1985, p. 164.
3. W. L. Frankhouser, K. W. Brendley, M. C. Kieszek and S. T. Sullivan, Gasless Combustion Synthesis of Refractory Compounds, Noyes Publications, Park Ridge, NJ, 1985.
4. A. G. Merzhanov and I. P. Borovinskkaya, Comb. Sci. Tech., vol. 10, 1975, pp. 195-201.
5. D. M. Sims, A. Bose and R. M. German, Prog. Powder Met., vol. 43, 1987, pp. 575-579.
6. M. Hansen and K. Anderko, Constitution of Binary Alloys, second edition, McGraw Hill, New York, New York, 1958.
7. I. M. Robertson and C. M. Wayman, Metallog.. vol. 17, 1984, p. 43.
8. F. Booth, Trans. Faraday Soc., vol. 49, 1953, p. 272.
9. B. I. Kaikin and A. G. Merzhanov, Comb. Explos. Shock Wave, vol. 2, 1966, p. 22.
10. A. G. Merzhanov, Comb. Flame, vol. 13, 1969, p. 143.
11. A. P. Hardt and P. V. Phung, Comb. Flame, vol. 21, 1973, p. 77.
12. Y. S. Naiborodenko, V. I. Itan and K. V. Savitskii, Sov. Phys. J., vol. 11(10), 1968, p. 19.
13. Y. S. Naiborodenko, V. I. Itan and K. V. Savitskii, Sov. Phys. J., vol. 11(10), 1968, p. 89.
14. Y. S. Naiborodenko, V. I. Itan, B. P. Belozerov and V. P. Ushakov, Soviet Phys. J., vol. 16(11), 1973, p. 1507.
15. Y. Y. Naiborodenko and V. I. Itan, Comb. Explos. Shock Wave, vol. 11, 1975, p. 293.
16. Y. S. Naiborodenko and V. I. Itan, Comb. Explos. Shock Wave, vol. 11, 1975, p. 626.
17. V. M. Maslov, I. P. Borovinskkaya and A. G. Merzhanov, Comb. Explos. Shock Wave, vol. 12, 1976, p. 631.
18. K. A. Philpot, Z. A. Munir and J. B. Holt, J. Mater. Sci., vol. 22, 1987, p. 159.
19. O. A. Kashin, E. F. Dudarev, V. I. Itim, E. G. Fedorana, A. R. Gafarov, I. E. Zayats, L. A. Kornienko and A. N. Tabachenko, Soviet Powder Met. Metal Ceram., 1987, vol. 26, p. 68.
20. R. Hultgren, P. D. Desai, D. T. Hawkins, M. Gleiser, and K. K. Kelley, Selected Values of the Thermodynamic Properties of Binary Alloys, American Society for Metals, Metals Park, OH 1973, p. 191.
21. R. M. German and J. W. Dunlap, Metall. Trans. A, 1986, vol. 17A, p. 205.
22. V. N. Eremenko. Y. V. Natanzon and V. Y. Petrishchev, Soviet Powder Met. Metal Ceram., 1987, vol. 43, p. 575.
23. R. M. German and K. S. Churn, Metall. Trans. A, 1984, vol. 15A, p. 747.
24. R. P. Kusy, J. Appl. Phys., 1978, vol. 487, p. 5301.

DEVELOPMENT OF INTERMETALLIC MATRIX COMPOSITES FOR ELEVATED TEMPERATURE SERVICE

G.E. Fuchs and W.H. Kao

The Aerospace Corporation
Materials Sciences Laboratory
Los Angeles, CA 90009-2957

Abstract

The mechanical properties and the chemical compatibility of several powder processed particulate reinforced intermetallic matrix composites (IMCs) have been examined. The reinforcement chosen for this investigation were TiC and Al_2O_3 due to their low density, high strength and good thermal stability. Pre-alloyed powders of the advanced nickel-aluminide IC-221 were mixed with approximately 25 volume percent of either TiC or Al_2O_3 powders, and processed using several different techniques, including: vacuum hot pressing and hot extrusion. The particulate/matrix interfacial reaction in the range 25° - 1000°C was analyzed for reaction products and kinetics using SEM and IMMA. Although the Al_2O_3-IMC had slightly slower reaction kinetics than the TiC-IMC, both IMCs displayed good chemical compatibility with very little interfacial reaction between the particulate and the matrix. In addition, the room temperature mechanical property evaluation indicated that both of the IMCs had 20-30% higher moduli than the IC-221 matrix, as well as conventional nickel-base superalloys. The TiC-IMC had slightly higher tensile properties than the Al_2O_3-IMC at room temperatures.

Introduction

An increase in the performance and the efficiency of future aerospace systems can be achieved by increasing the operating temperature and stresses, and decreasing the density of the structural components. In order to meet these design goals, a new class of materials is needed that is lighter, stronger and more oxidation/corrosion resistant than current base-line elevated temperature materials: nickel-base superalloys. These new materials are required for further developments in propulsion, power generation, airframe and skin material for the NASP, space shuttle derivatives and advanced launch systems. Potential candidate materials for these applications are long range ordered intermetallic compounds; however, their inherent brittleness in bulk form has limited their use to strengthening phases in conventional alloys.

Recent advances in micro-alloying (B-doping) and rapid solidification rate processing has resulted in the production of ductile intermetallic alloys. For example, the B-doping of hypostoichiometric Ni_3Al has resulted in an increase in tensile elongation to 50% from essentially 0%[1,2]. The ductilization of Ni_3Al has stimulated the rapid growth of the data base for the mechanical and physical properties of nickel-, as well as titanium-, aluminides. The current generation of nickel-aluminides have alloying additions of Cr (high temperature oxidation resistance/ductility) and Zr (or Hf, solid solution strengthening)[3]. The resultant alloy is a high strength, high ductility advanced nickel-aluminide. However, a clear cut advantage of intermetallics over superalloys has not been realized, and the goals for stronger, lighter, more heat resistant materials remain.

In order to become more competitive with state-of-the-art nickel-base superalloys, the strength of Ni_3Al+B must be increased still further. The advanced Ni_3Al+B alloys have many attractive properties that make it an excellent candidate for the matrix material for a high temperature composite. For example, the nickel aluminides have an anomalous yield strength, high modulus, good oxidation resistance and a density 7% lower than superalloys. Further, the selection of appropriate reinforcement could reduce the density of the composite while increasing the mechanical properties.

Previous investigations have indicated that nickel-aluminides react extensively with SiC and B_4C[4,5]. In the present study TiC and Al_2O_3 particulate were selected as the reinforcements for the intermetallic matrix composites (IMCs) via powder metallurgy techniques including: vacuum hot pressing or VHP'ing, and hot extrusion. The chemical compatibility of the composites was examined after the IMCs were heat treated at 1100°C for 1, 10, 100, and 1000 hours. The mechanical properties of the IMCs were evaluated in air at room temperature and the fracture modes were characterized.

Experimental Procedure

The matrix alloys with composition Ni-16Al-8Cr-1Zr-0.05B (Oak Ridge National Laboratory designation IC-221) was argon gas atomized by Homogeneous Metals. The HMI powder was sieved to -200 mesh for use in this study. The TiC and Al_2O_3 reinforcements were obtained from

ICD/UMC Group and had an average particle diameter of 2.5um. In all cases, the IMCs contained a volume fraction of 25% reinforcement. The theoretical densities of the composites and the matrix are compared to that of typical Ni-base superalloys below:

ALLOY	DENSITY(g/cc)
IC-221	7.65
TiC-IMC	6.97
Al_2O_3-IMC	6.73
Ni-base alloy	8.2

Note that the composites represent a 15% (TiC-IMC) and a 18% (Al_2O_3-IMC) reduction in density in comparison to Ni-base superalloys.

The powders were vacuum dried and mixed in a v-cone blender for 8 hours. The mixtures were then vacuum hot pressed to ingots to achieve 85% of theoretical density. The ingots were subsequently hot extruded to rod. The extrusion resulted in fully dense materials. Additionally, powders of the IC-221 matrix alloy without reinforcement were processed under the same conditions so that property comparisons could be made.

Chemical compatibility studies were conducted on samples heat treated under vacuum at 1100°C for 1, 10, 100, and 1000 hours. The samples were then metallographically prepared and examined both optically and with an SEM using a windowless EDS system.

Room temperature tensile tests were performed on cylindrical samples (composites and matrix). The samples were ground from the extruded rod and had a gauge section of 5.08 x 25.4mm. The tensile tests were performed on an Instron screw driven test machine at a strain rate of 3.3×10^{-6}/sec. Fractured samples were examined optically and by SEM.

Results

a.) Metallography

The microstructure of the IC-221 matrix alloy and the composites are shown in Figure 1. Equiaxed grains were observed in the matrix as well as both of the composites. The average grain size was approximately 5-10um. Little or no reaction between the reinforcement and the matrix was observed in either composite, suggesting good chemical compatibility. SEM observation indicated a fine (<0.5um) network-like dispersion of ZrO_2 at prior particle boundaries (PPBs) in all three alloys. Presumably this ZrO_2 network reflects the existence of residual oxygen in the die cavity during hot pressing. It should be noted that the oxygen content of the powder was 121ppm, and should not be a contributing factor to the PPB precipitation. These observations of ZrO_2 PPB precipitation are consistent with those of Povirk, et al[6] on Al_2O_3 fiber reinforced Ni_3Al. The TiC in the TiC-IMC and the Al_2O_3 in the Al_2O_3-IMC exhibited substantial agglomeration of the particulate. Further, extensive particle pull-out after tensile testing was observed in the Al_2O_3, suggesting a weak bond between the Al_2O_3 and the IC-221 matrix. No instances of particle pull-out was observed in the TiC-IMC.

b.) Chemical Compatibility

TiC-IMC

The chemical compatibility of the TiC in the IC-221 matrix was examined on samples heat treated at 1100°C for up to 1000 hours. Examining the interfacial region using an SEM with a windowless EDS detector, only limited amounts of interaction was observed (Figure 2). Zr and B were observed to diffuse to the TiC particle surface and react to form a complex (Ti,Zr)(C,B) interface. Further, the Ti from the TiC diffused out into the nickel-aluminide matrix. The extent of the reaction layer, however, was less than 1um after 1000 hours at 1100°C. The ZrO_2 precipitates decorating the PPBs appeared to be stable with no change in size, shape and quantity after the same heat treatment. However, after 100 hours at 1100°C, a second blocky-type of oxide was observed to have formed in the matrix (Figure 2). EDS analysis identified this oxide phase to be Cr_2O_3. In contrast to the stable ZrO_2, the Cr_2O_3 tended to increase in size with time, growing from 1-3um at 100 hours to 5-10um at 1000 hours. Cr_2O_3 particles were not observed prior to 100 hours of exposure and their origins are under investigation. It should be noted that long term exposure of the IC-221 matrix at 1100°C did produce ZrO_2 PPB particles, but no Cr_2O_3 formation was observed.

In addition, due to the high alloy content of the IC-221 nickel-aluminide, a partial disordering (approximately 15%) of the ordered matrix was observed (Figure 3). A two phase gamma/gamma-prime web-like structure was observed in all samples. EDS analysis of the two phases indicated a difference of chemical composition, as shown below:

ordered	Ni-19.0Al-5.8Cr-1.2Zr
disordered	Ni-12.1Al-10.9Cr-0.2Zr

Similar observations were made in IC-221 matrix samples exposed to 1100°C. These observations of the two microstructures and their composition are consistent with those of Povirk, et al[6] and Liu, et al[7].

Al_2O_3-IMC

Little or no reaction between the Al_2O_3 and the IC-221 was observed for samples heat treated at 1100°C for up to 1000 hours (Figure 4). The ZrO_2 PPB decoration was also observed and found to be stable. In addition to PPB precipitation, discrete ZrO_2 nodules were observed to form along the Al_2O_3/matrix interface. The interfacial ZrO_2 were present in the as-extruded condition and did not appear to grow in size for up to 1000 hours at 1100°C. Instead, Cr_2O_3 (3-5um) was observed to form at the interface after 100 hours at 1100°C, and increase in size to 10-20um after 1000 hours at temperature. It is unclear at this time whether the interfacial ZrO_2 and Cr_2O_3 formed as a result of interaction between the matrix and the Al_2O_3 or due to enhanced nucleation at the interface. As in the TiC-IMC, Cr_2O_3 particles (1-3um) also formed in the matrix after 100 hours at 1100°C, and their size increased in size (5-10um) at 1000 hours. The two-phase ordered/disordered matrix was also observed in the Al_2O_3-IMC, with compositions similar to the TiC-IMC and IC-221 matrix.

c.) Mechanical Testing

Room temperature tensile tests were performed on initial heats of the IMCs, which suffered from particulate agglomeration and residual porosity (Table 1). Due to the inhomogeneities of the composite microstructure, the IMCs had relatively low strengths and ductilities. However, the specific modulus of the composites were 10-20% greater than the IC-221 matrix and 25-35% greater than typical Ni-base superalloys. Tensile testing of subsequent homogeneous, fully dense IMCs is currently underway.

SEM analysis of the fracture surfaces revealed that failure of the composites predominently occurred at regions of particulate agglomeration. These areas of agglomeration are weakly bonded and failure occurs at low stress and low strain. Although areas of particle pull-out were observed in the Al_2O_3-IMC, only agglomeration separation and matrix failure was observed in the TiC-IMC.

Discussion

Intermetallic matrix composites have been successfully fabricated via powder metallurgy techniques. These IMCs, based on the advanced nickel-aluminides, utilizes TiC and Al_2O_3 for reinforcement, and are a viable alternative material for elevated temperature service. Both the TiC and the Al_2O_3 are stable reinforcements within the highly alloyed nickel-aluminide matrix at temperature of 1100°C for up to 1000 hours. No reaction is observed between the Al_2O_3 and the IC-221 matrix, and tensile tests indicate that the bond between the matrix and the reinforcement is weak. Similarly, Povirk, et al[6] and Stoloff, et al[8] reported poor bonding in Al_2O_3 fiber reinforced binary and alloyed Ni_3Al+B. The fiber reinforced composites exhibited low ductility and strength, with tensile failure occurring due to decohesion at the fiber/matrix interface[6,8]. Crispin and Nicholas[9] reported that small (<1.0wt%) additions of strong oxide formers (e.g. Cr, Y, Zr) to nickel, strengthened the bond between Al_2O_3 fibers and a nickel matrix. Further additions, however, weakened the bond at the interface[9]. The high chromium (6.90 wt%) and zirconium (1.45 wt%) alloy content of the IC-221 would not, therefore, be expected to strengthen the Al_2O_3/matrix bond. In fact, the bond would be expected to be weak due to the high alloy content of oxide formers. Although interfacial ZrO_2 and Cr_2O_3 was observed to form in the Al_2O_3-IMC, their origin is thus far unclear. The ZrO_2 and Cr_2O_3 may form due to an interfacial reaction between the matrix and the reinforcement, or they may form due to enhanced nucleation at the interface. The tensile failure mode in the IMCs was particulate pull-out, and the fiber reinforced composites are reported to fail by fiber pull-out[6,8]. Therefore, if the Cr_2O_3 and the ZrO_2 form due to an interfacial reaction, the bond formed by this reaction must be weak. In addition, Povirk, et al[6] reported that cracks were observed in the interfacial ZrO_2 in TEM samples of alloyed Ni_3Al. However, no cracks were observed at the fiber matrix interface when ZrO_2 was not present or in un-alloyed Ni_3Al+B. No evidence of Cr_2O_3 formation was reported[6], but the TEM samples were exposed to high temperature for only short periods of time (<<10 hours). In general, Al_2O_3 is very stable within the IC-221 matrix. The interfacial bond is weak and the composites exhibit low mechanical strengths and ductilities. Westfall[10] reported that long term (>100 hours) heat treatment at temperatures above 1200°C strengthens the interfacial bond in Al_2O_3 fiber reinforced nickel-base alloys, and further experiments to examine this are planned.

In contrast, a limited amount of interfacial reaction is observed between the TiC and the IC-221 matrix. An interface enriched in Ti, Zr, B, and C formed and slowly increased in size with time at 1100°C. The reaction layer grew to approximately lum in thickness in 1000 hours at temperature. However, the interfacial reaction resulted in good bonding between the particulate and the matrix for load transfer. The result of the limited interfacial reaction is improved mechanical properties of the composite, as seen in Table 1. Even though both composites suffered from agglomeration, the more tightly bound TiC apparently produced superior matrix reinforcement superior in comparison to the Al_2O_3. In fact, the modulus, as well as the specific modulus, of the TiC-IMC was superior to both the IC-221 matrix and typical nickel-base superalloys. For more representative data, tensile tests on homogeneous IMCs are in progress.

Additionally, due to the high alloy content of reactive elements, such as Cr and Zr, free oxygen during processing and the oxygen content of the powder must be controlled. The ZrO_2 PPB precipitation does not seriously effect the properties of the IMCs in this study, due to the shear deformation during extrusion. However, fabrication by techniques with little or no shear deformation of the PPBs, may result in IMCs with poor mechanical properties. In addition, the formation and rapid growth of the Cr_2O_3 after 100 hours at 1100°C may limit service life and temperature. Further investigations of the formation and the effects of Cr_2O_3 on IMC properties are underway.

Summary and Conclusions

Nickel-aluminide based intermetallic matrix composites have been successfully processed by powder metallurgy techniques. The TiC and Al_2O_3 reinforcement are stable within the IC-221 matrix, with only limited interfacial reaction observed. The TiC particulate was more tightly bound to the matrix than Al_2O_3, and, therefore, the TiC-IMC exhibited superior mechanical properties. Overall, the tensile strengths of the composites were low, due to agglomeration of the particulate. However, the specific modulus of the composites was 10-20% higher than the IC-221 matrix and 25-35% higher than nickel-base superalloys.

References

1.) K. Aoki and O. Izumi, Nippon Kinzoku Gakkaishi, 43, 1979, p. 1190.

2.) C. T. Liu, C. L. White and J. A. Horton, Acta Met, 33, 1985, p. 213.

3.) C. T. Liu and V. K. Sikka, J. Metals, 38, No. 5, 1986, p. 19.

4.) W. H. Kao, unpublished results.

5.) C. T. Liu, private communication.

6.) G. L. Povirk, J. A. Horton, C. G. McKamey, T. N. Tiegs and S. R. Nutt, to be published in Acta Met.

7.) C. T. Liu, J. J. Cathcart, J. A. Horton, E. H. Lee and J. J. Campbell, Oak Ridge National Laboratory, Internal Report, June 1986, unpublished.

8.) N. S. Stoloff and R. M. German, private communication.

9.) R. M. Crispin and M. Nicholas, J. of Mat. Sci., 11, 1976, p. 17.

10.) L. J. Westfall, Bonding and Compatibility of Aluminum-oxide in Nickel-alloys, NASA TM X-3333, 1976.

TABLE 1

Tensile Properties of IMCs

Alloy	Y.S. (MPa)	U.T.S. (MPa)	$\%e_f$	Specific Modulus
IC-221	875	1525	34	28.34
Al_2O_3-IMC	450	470	<1	31.66
TiC-IMC	640	640	<1	33.15
Ni-base alloy				25.28

1a.) IC-221

Figure 1 - Optical Micrograph of the Microstructure of the Hot Pressed and Extruded Alloys.

a.) IC-221 matrix,

b.) TiC-IMC,

c.) Al_2O_3-IMC.

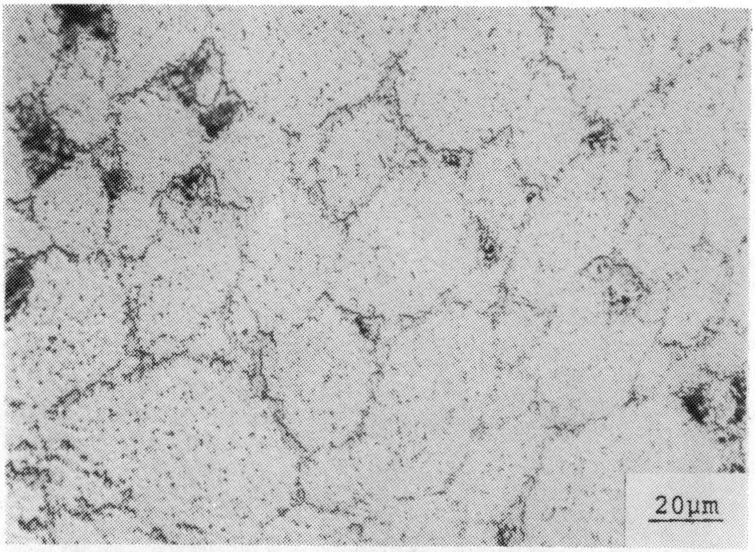

1b.) TiC-IMC

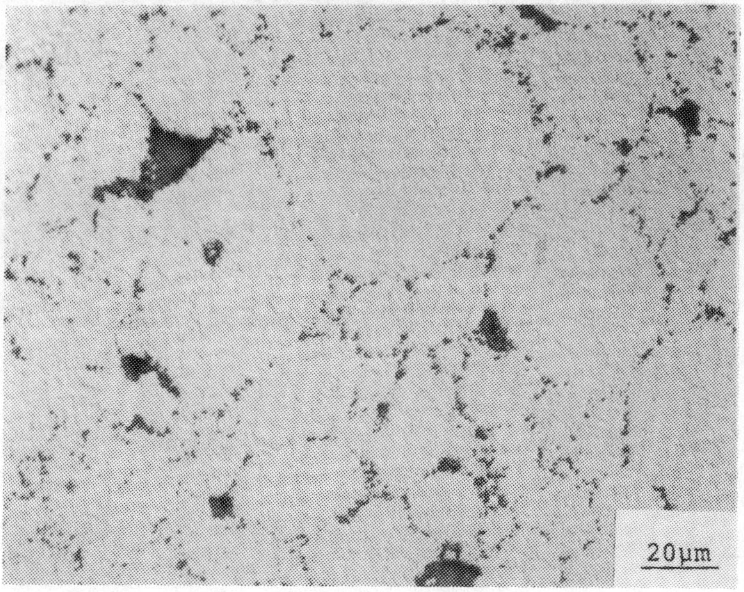

1c.) Al_2O_3-IMC

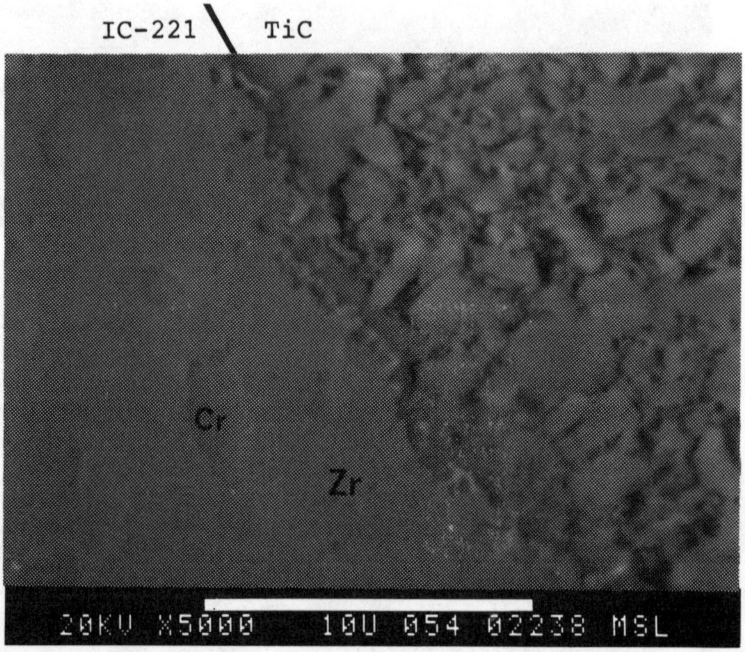

Figure 2 - SEM Micrograph of the TiC-IMC Compatibility Sample after Heat Treatment at 1100°C for 1000 hours.

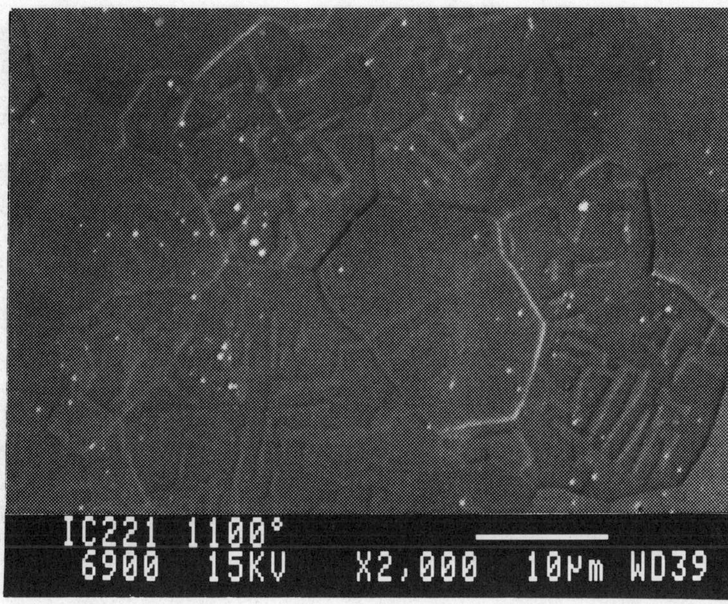

Figure 3 - Two phase microstructure of the IC-221 matrix after heat treatment at 1100°C for 100 hours. The light grey regions are disordered gamma phase, while the darker grey regions are ordered gamma-prime.

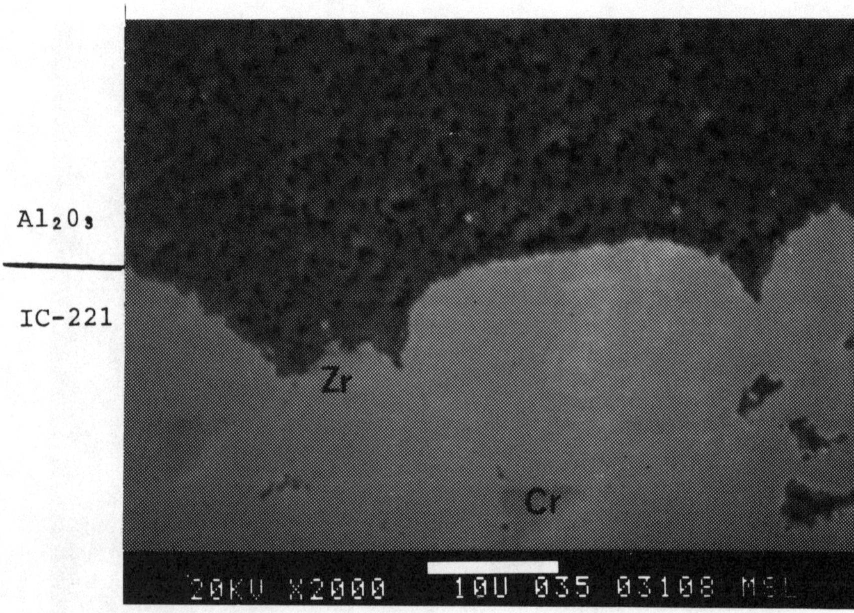

Figure 4 - SEM micrograph of Al_2O_3-IMC compatibility sample after heat treatment at 1100°C for 1000 hours.

POWDER PROCESSING OF Ni_3Al*

V. K. Sikka

Metals and Ceramics Division
Oak Ridge National Laboratory
P. O. Box 2008
Oak Ridge, Tennessee 37831

Abstract

This paper deals with the powder metallurgy processing of ductile-ordered intermetallic Ni_3Al and Ni_3Al-based alloys. Powder consolidation techniques presented include hot extrusion, hot isostatic pressing (HIP), consolidation under atmospheric pressure (CAP), and rapid omnidirectional compaction (ROC). Near-net-shape processing by superplastic forming is also discussed. Optical microstructures of powder consolidated by each of the techniques listed above are presented. Tensile properties of powder consolidated by most of the processes listed above are also presented. It is concluded that the powder processing technology is now ready for commercialization for ductile-ordered intermetallic Ni_3Al and Ni_3Al-based alloys.

Introduction

The nickel aluminide, Ni_3Al, is one of a series of intermetallic compounds having the $L1_2$-ordered crystal structure. Single crystals of Ni_3Al are highly ductile and show an increase in 0.2% yield strength with increasing temperature and peaks over 600°C. Polycrystals of Ni_3Al were extremely brittle until recent (1,2) ductilization obtained by small additions of boron (200 ppm by weight) and controlling the aluminum content between 22 and 24 at. %. Although earlier studies indicated that boron-doped nickel aluminide had good tensile ductility at elevated temperatures when tested in vacuum, recent tests of these aluminides in controlled environments at elevated temperatures showed reduced ductility when tested in air or oxygen (3-5). The reduced tensile elongation in air at 600 to 800°C is accompanied by a change in fracture mode from transgranular to intergranular. This effect was explained (3) as a dynamic embrittlement effect due to oxygen at the grain boundaries normal to the applied stress. An effective solution to this

*Research sponsored by the Office of Energy Utilization Research, Energy Conversion and Utilization Technologies (ECUT) Program, U. S. Department of Energy, under contract DE-ACO5-84OR21400 with Martin Marietta Energy Systems, Inc.

problem was devised by adding a moderate amount of chromium (approximately 8 wt %) to Ni_3Al. The chromium atoms occupy equally the nickel and aluminum sites in the ordered lattice, and, thus, its additions to Ni_3Al are made to maintain aluminum equivalence of 22 to 24 at. %. The beneficial effect of chromium is to provide a rapid formation of protective chromium or chromium-rich oxide film that obstructs the flow of gaseous oxygen along the grain boundaries. The aluminum oxide forms the protective scale at ≥800°C to avoid the oxygen-related reduction in ductility at higher temperatures. The Ni_3Al structure is an ordered phase (γ') from room temperature to the melting point. Additions of chromium to Ni_3Al introduce small amounts of disordered phase γ to the primarily ordered γ'. The amount of disordered phase increases at temperatures over 1000°C.

The nickel aluminide compositions under development and commercialization at the Oak Ridge National Laboratory, in conjunction with industry, are listed in Table 1. Both casting and powder metallurgy processes are being explored (4, 6-12) for scale-up production of these intermetallic compounds. This paper presents the status of powder processing technology possibilities for nickel aluminide alloys and their mechanical properties.

TABLE I: COMPOSITIONS OF THREE NICKEL ALUMINIDE INTERMETALLICS BEING FABRICATED AT ORNL

Element	Weight Percent		
	IC-50	IC-218	IC-221
Al	11.3	8.5	8.5
Cr	--	7.8	7.8
Zr	0.6	0.8	1.7
B	0.02	0.02	0.02
Ni	Bal.	Bal.	Bal.

Powder Production

Powders of all three nickel aluminide alloys (IC-50, IC-218, and IC-221) have been produced up to 100-kg quantities by commercial vendors. Both Homogeneous Metals, Inc. and Cytemp Specialty Steel have produced all three alloys. Carpenter Technology Corporation has produced IC-221 alloy. All powder has been produced by inert gas atomization and generally contains an oxygen content of 100 to 150 ppm by weight. In almost all cases, powder used for consolidation was of -100 mesh size.

Powder Consolidation

Several powder consolidation techniques have been tried. Each of them are briefly described here.

Hot Extrusion

In this method of consolidation, powder was canned in mild steel containers, evacuated, and extruded. The extrusion temperature was 1100°C with an area reduction ratio of ≥8:1. In most cases, round containers were extruded to round bar. Optical microstructures showing full consolidation by this process for all three alloys are shown in Figure 1. This process is scalable to large containers and production of billets for subsequent processing by processes such as superplastic forming.

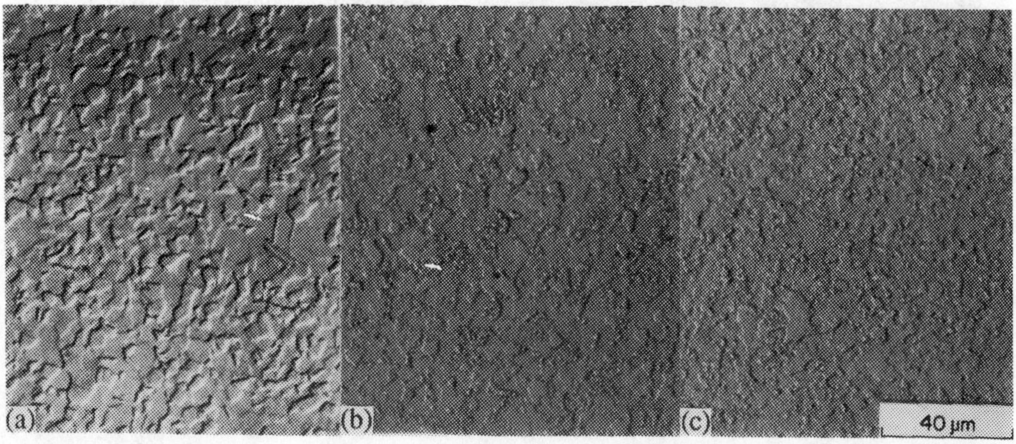

Figure 1: Microstructures of nickel aluminide powders consolidated by hot extrusion at 1100°C with a reduction ratio of 9:1; (a) IC-50, (b) IC-218, and (c) IC-221.

Hot Isostatic Pressing

In this method powder is canned, evacuated, and hot isostatic pressed at 1120°C and 103 MPa. Using this technique, a 230-mm-diam billet of IC-221 has been consolidated. Microstructure of IC-221 consolidated by HIP is shown in Figure 2. For billet fabrication out of powder, hot extrusion is probably a more economical method than HIP. However, the HIP process offers a special advantage in that it can be used for fabrication of near-net-shape parts directly from powder.

Consolidation Under Atmospheric Pressure

In this process, a glass container is filled with powder, which is evacuated and sealed. When this sealed container is heated to approximately 1150 to 1200°C, the glass softens and the powder is consolidated under atmospheric pressure (the pressure difference between the outside and inside of the glass container). This method allows the consolidation of powder into billets and a variety of complex shapes (shapes are limited to those that are possible by glass blowing). The nickel aluminide powder was consolidated by this method into billet shapes. The microstructure of IC-50 consolidated by CAP is shown in Figure 3. This figure shows that the structure produced by CAP still contains some noninterconnected porosity and, to take full advantage of properties, should be further densified by hot extrusion, HIP, or ROC, which is described in the next section.

Rapid Omnidirectional Compaction

In this process, powder is canned in a mild steel container or mold, evacuated, and sealed. The sealed can is heated to 1100 to 1150°C and subjected to a maximum stress of

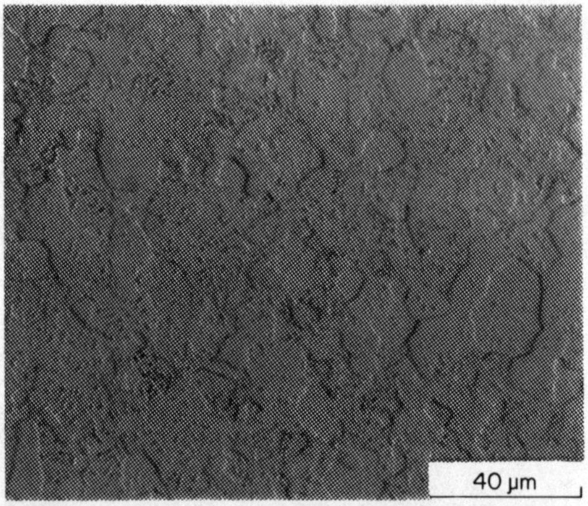

Figure 2: Microstructure of IC-221 powder consolidated into 230-mm-diam billet by the HIP process at 1120°C and 103 MPa.

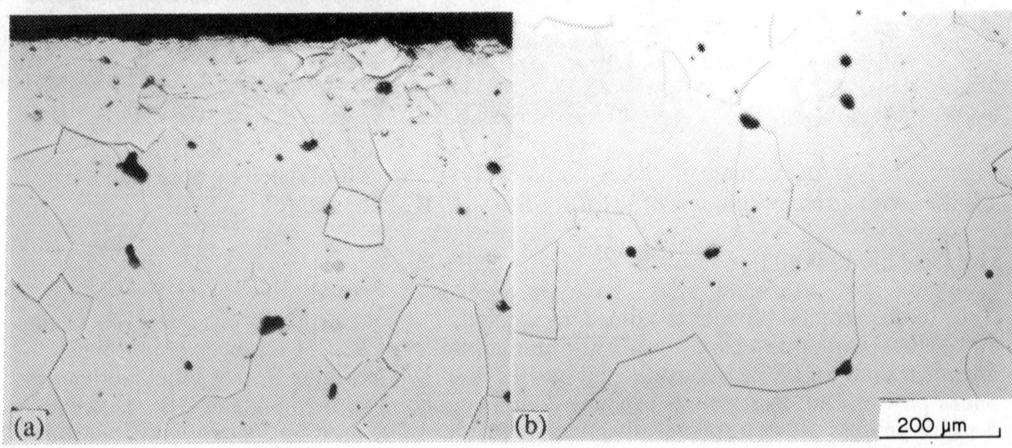

Figure 3: Microstructures of IC-50 powder consolidated into 75-mm-diam billet by the CAP process. Notice the porosity which is not interconnected; (a) edge and (b) center.

827 MPa (120,000 psi). At these temperatures, the mild steel flows like a fluid die and the uniaxial-applied stress is translated to a triaxial stress state. The applied stress is so high that time duration in the range of 3 to 5 s causes full densification of the powder. The microstructure of IC-218, produced by this method, is shown in Figure 4. This method offers excellent potential for fabrication of complex near-net-shape parts of full density, but is currently limited by the facilities to a maximum diameter of approximately 250 mm.

Other Processes

Other processes possible for consolidation of powder include hot pressing and biaxial compaction. Both of these processes have been successful for small experimental samples only. Also, instead of using an atomized stream of liquid metal for powder making, it is possible to use it for spray forming onto a substrate. Such a process is known in literature as the Osprey process or liquid dynamic compaction and has been used successfully for nickel aluminides by the General Electric Company. Such a process can be used for sheet, bar, tubing, and even near-net shapes. The spray forming produces fine-grain material but

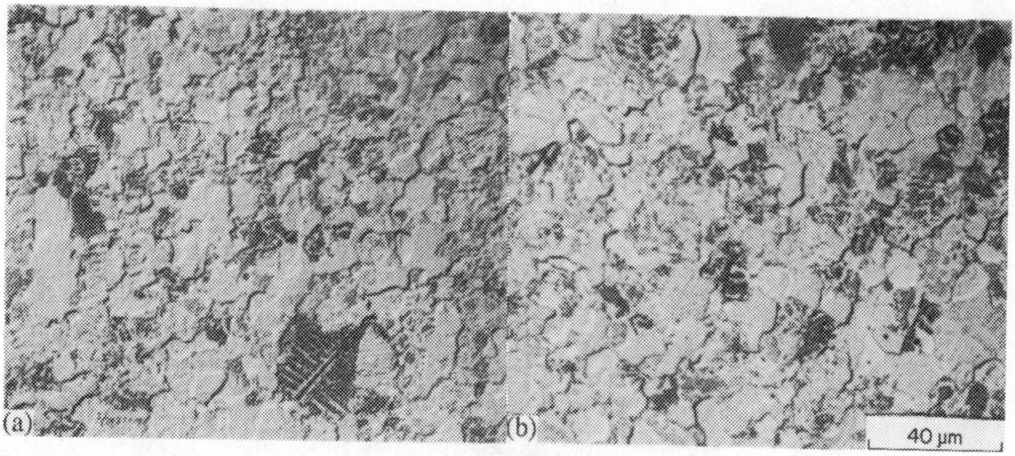

Figure 4: Microstructure of 100-mm diam × 25-mm disk of IC-218 powder consolidated by the ROC process at 1100°C and 825 MPa; (a) edge and (b) center.

contains porosity, which has to be closed by a secondary process such as HIP, ROC, or thermomechanical processing.

Superplastic Forming

Fine-grained ≤10 μm nickel aluminide alloys show superplastic behavior under proper conditions of test temperature and strain rate. The fine-grain structure can be produced either by hot working the cast ingot followed by thermomechanical processing or by powder metallurgy. An example showing the start of superplasticity in IC-218 at a strain rate of 8.3×10^{-4}/s is shown in Figure 5. The application of superplasticity to form a prototype turbine disk at 1100°C and a strain rate of 8.3×10^{-3}/s is shown in Figure 6. The superplastic forming rate used for the nickel aluminide alloy is typical of that used for many superalloys and, thus, is readily adaptable for scale-up by industry.

The transverse cross section from the center of the superplastically forged disk, shown in Figure 7, reveals excellent die-fill in a varied and complex thickness. Representative optical micrographs at various points in this cross section, shown in Figure 8, reveal a uniform fine-grain size (approximately 10 μm) with no evidence of porosity. Porosity was observed at and very near the rim surface, which would normally be machined off. Miniature tensile test specimens were also prepared from a forged disk, the results of which are presented in the next section.

Mechanical Properties

Tensile properties of IC-50 prepared by consolidating the powder by hot extrusion are listed in Table 2. Data on sheet obtained by cold working the hot-extruded bar are listed in Table 3. These tables show that consolidation of powder by hot extrusion produces high ductility at room temperature, and sheet processed from the consolidated bar is highly isotropic. Tensile data on IC-218 and IC-221 alloys produced from powder by the hot extrusion process are compared in Figure 9. This figure shows that the higher zirconium-containing alloy IC-221 is consistently stronger than IC-218 for test temperatures

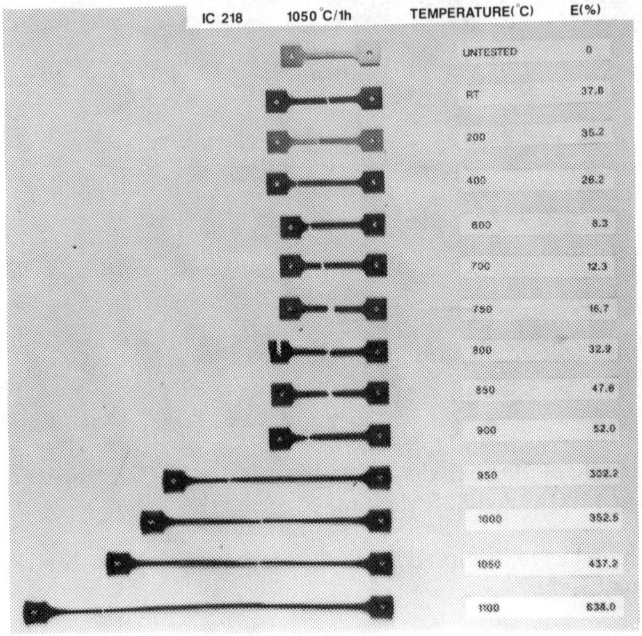

Figure 5: Effect of test temperature on tensile ductility in air of IC-218 alloy at a strain rate of 8.3×10^{-4}/s. Superplasticity occurred at temperatures ≥950°C at this strain rate.

Figure 6: Prototype turbine disk produced by superplastic deformation of powder extruded bar of IC-218 at 1100°C and 8.3×10^{-3}/s at a commercial forging source.

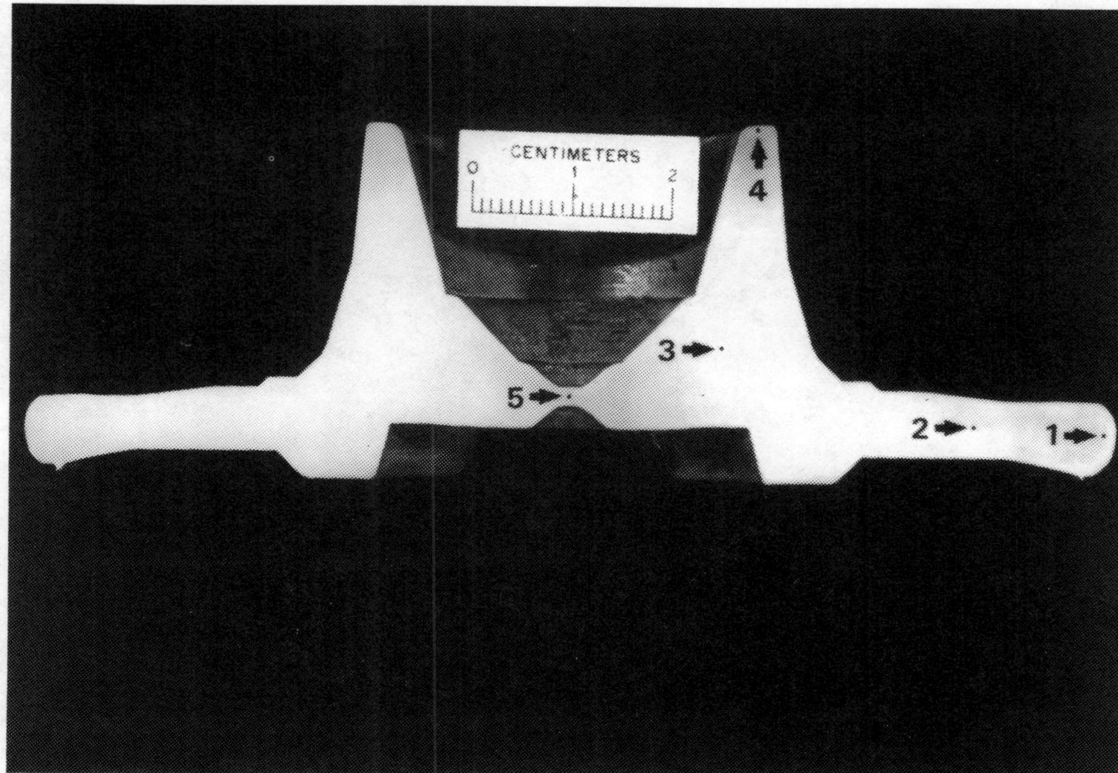

Figure 7: Cross section of superplastically-forged disk of IC-218. Numbers indicate locations where micrographs were taken for Figure 8.

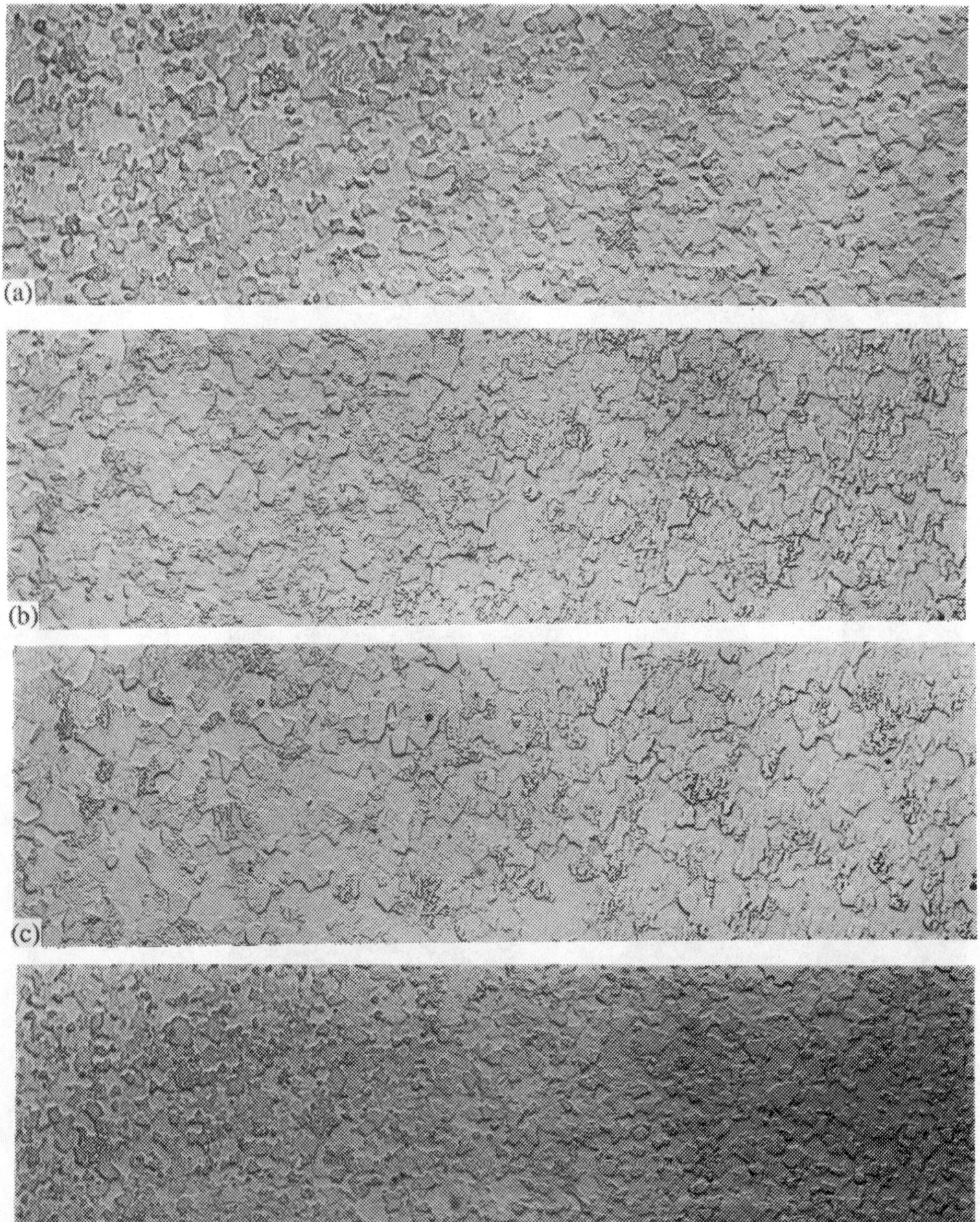

Figure 8: Optical micrographs of grain structure in various locations of the cross section of the superplastically-forged disk of IC-218 shown in Figure 7; (a) location 2, (b) location 3, (c) location 4, and (d) location 5.

TABLE II: TENSILE DATA OF IC-50 BAR PREPARED BY CONSOLIDATING THE POWDER BY HOT EXTRUSION. ALL TESTS WERE AT A STRAIN RATE OF 8.3×10^{-4}/s

Test Temperature (°C)	Strength, MPa (ksi) 0.2% Yield	Ultimate	Total Elongation (%)
Room	534 (77.4)	1455 (211.0)	47.8
1000	281 (40.8)	335 (48.6)	5.9

TABLE III: TENSILE DATA ON IC-50 SHEET PRODUCED BY COLD WORKING HOT-EXTRUDED BAR. ALL TESTS WERE AT A STRAIN RATE OF 8.3×10^{-4}/s

Specimen Orientation	Test Temperature (°C)	Strength, MPa (ksi) 0.2% Yield	Ultimate	Total Elongation (%)
RD	25	504 (73.1)	1482 (214.9)	46.0
45	25	497 (72.1)	1471 (213.3)	47.7
90	25	503 (73.0)	1491 (216.2)	48.2
RD	1000	294 (42.6)	326 (47.3)	6.8

approaching 1000°C. Total elongation of both alloys is the same (approximately 35%) at room temperature. The ductility of IC-218 remains over 10% for the entire test temperature. Total elongation of IC-221 approaches 5% at 800°C and 400% at 1100°C. The lower ductility values for IC-221 at 800°C may be caused by oxygen embrittlement along the zirconium-enriched boundaries.

Tensile properties of IC-221 prepared by the CAP process followed by hot extrusion are compared with those of powder consolidated by hot extrusion only, in Figure 10. This figure shows that comparable properties are obtained for both cases.

Tensile properties of an IC-218 disk prepared by the ROC process are compared with those for the superplastically formed disk in Figure 11. This figure shows that the strength properties of the disk produced by the two processes are nearly the same. Ductility values of the disk produced by the ROC process are somewhat lower than the superplastically-formed disk. This difference is explained by approximately twice the grain size of the ROC disk as compared to the superplastically formed disk. This point is more clearly made in Figure 12, where properties of intentionally grain-coarsened (by 15.5-h heat treatment at 1150°C) specimens of superplastically formed disk of IC-218 are compared with the fine-grained specimens. The increase in grain size from 9 to 34 μm produced a significant reduction in yield and ultimate tensile strength at test temperatures below 700°C. At a higher temperature, strength of the coarse-grained material increased with a significant decrease in ductility. Deformation is controlled by a diffusion mechanism at temperatures

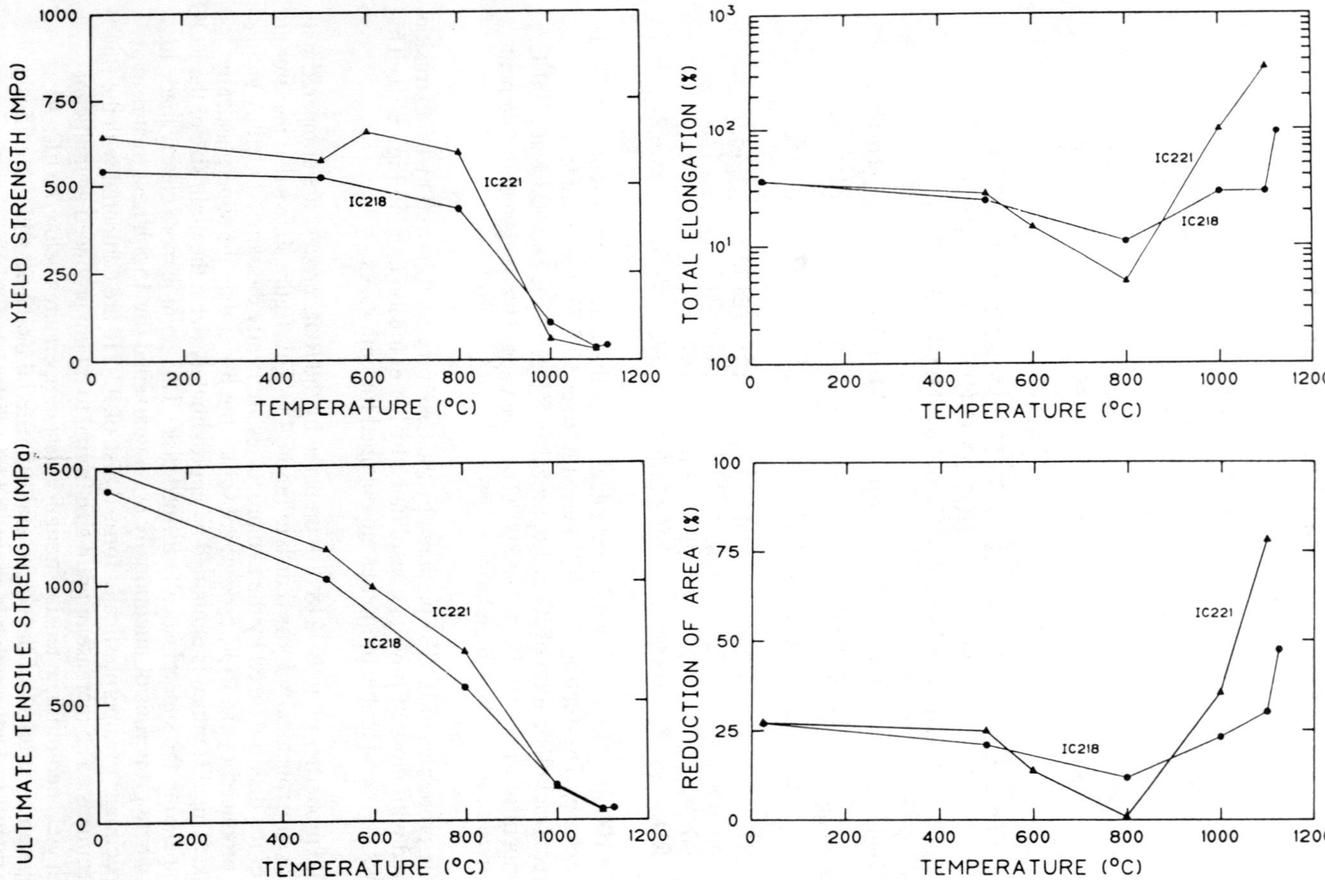

Figure 9: Tensile properties of IC-218 and IC-221 alloys prepared by consolidation of powder by hot extrusion at 1100°C at a reduction ratio of 9:1. Tensile tests were run in air at a strain rate of 8.3×10^{-4}/s.

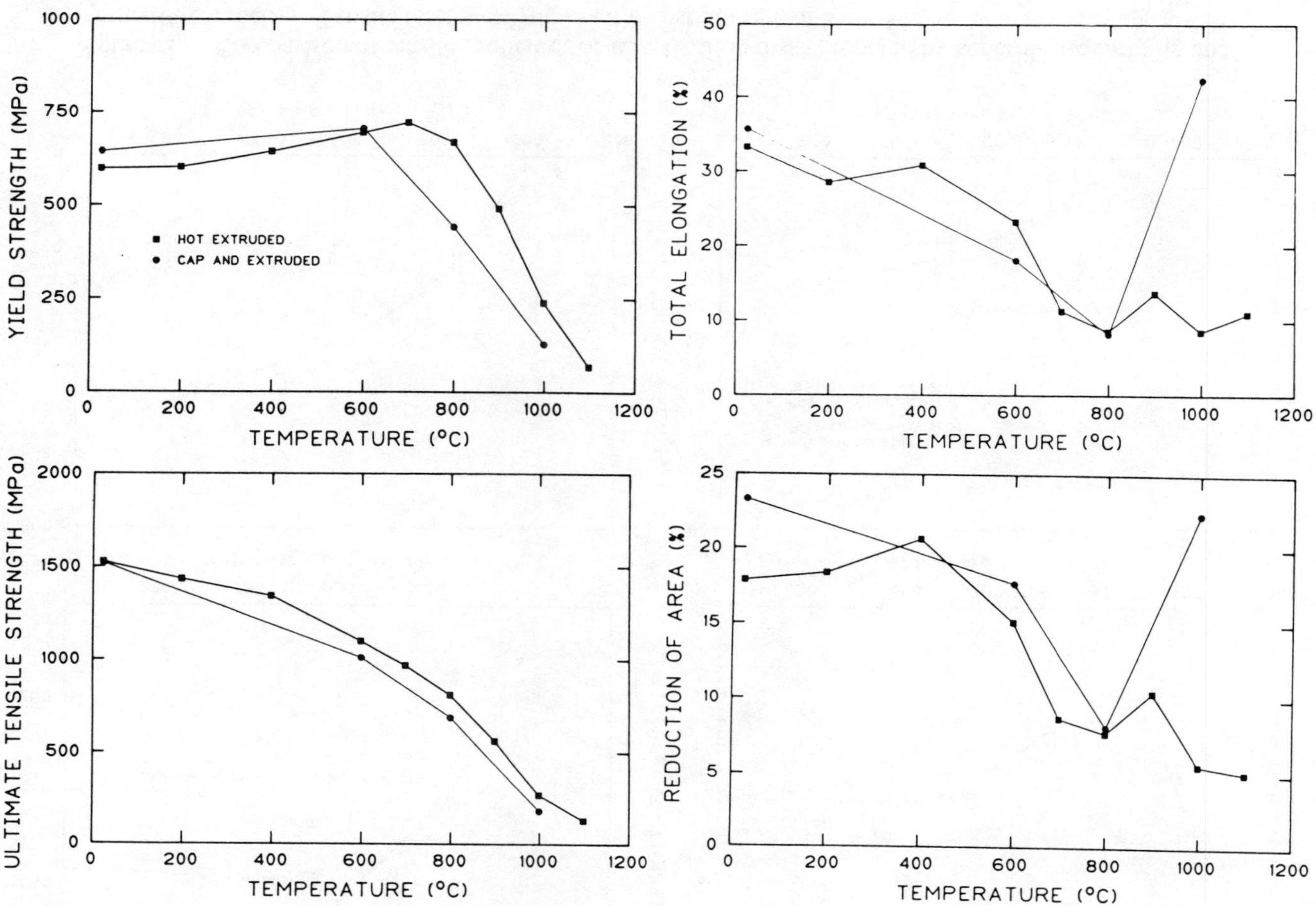

Figure 10: Comparison of tensile properties of IC-221 alloy powder consolidated by hot extrusion and CAP followed by hot extrusion. Tensile tests were run in air at a strain rate of 3.3×10^{-3}/s.

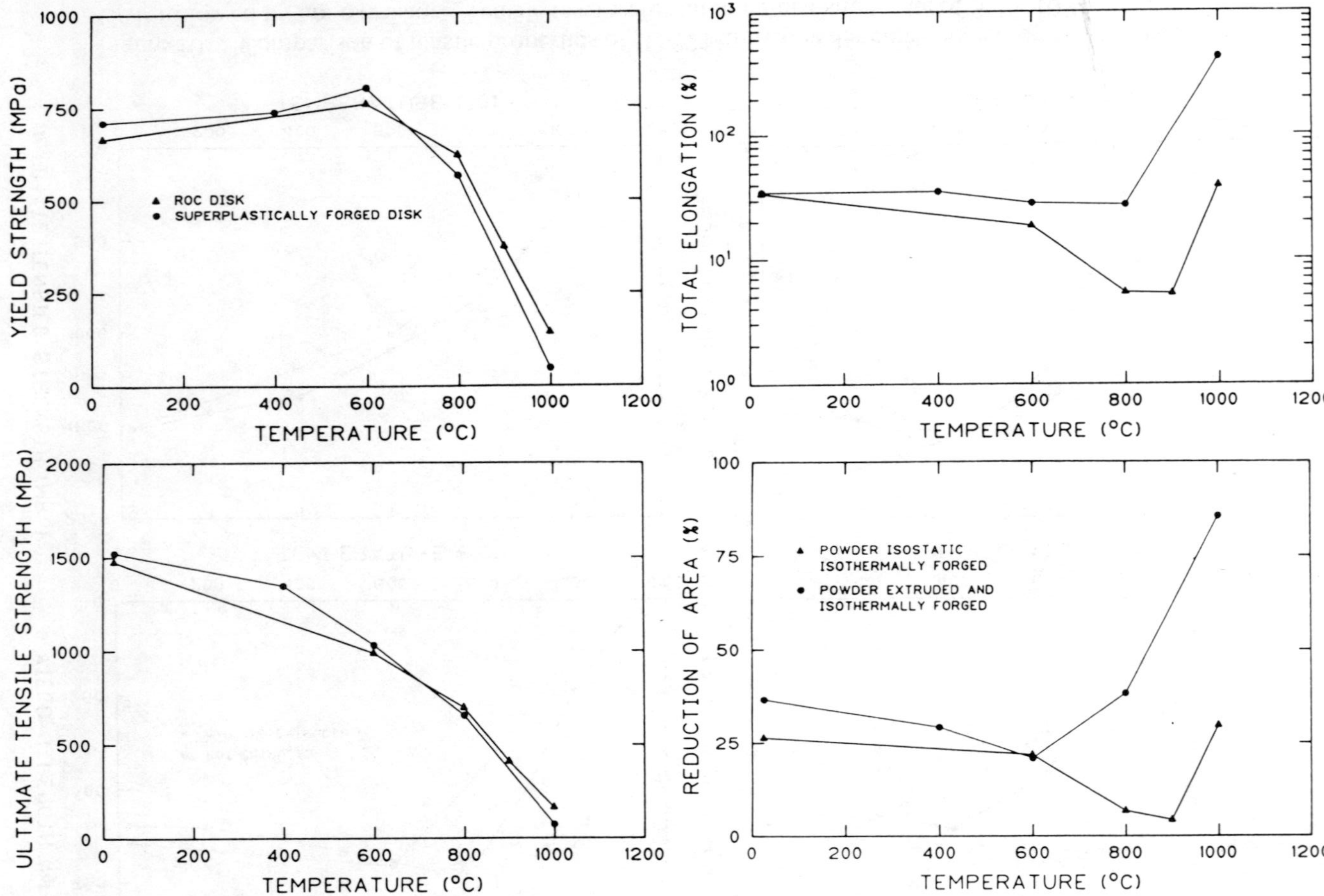

Figure 11: Comparison of tensile properties of IC-218 alloy disks prepared by superplastic forming and by the ROC process. Tensile tests were run in air at a strain rate of 8.3×10^{-4}/s.

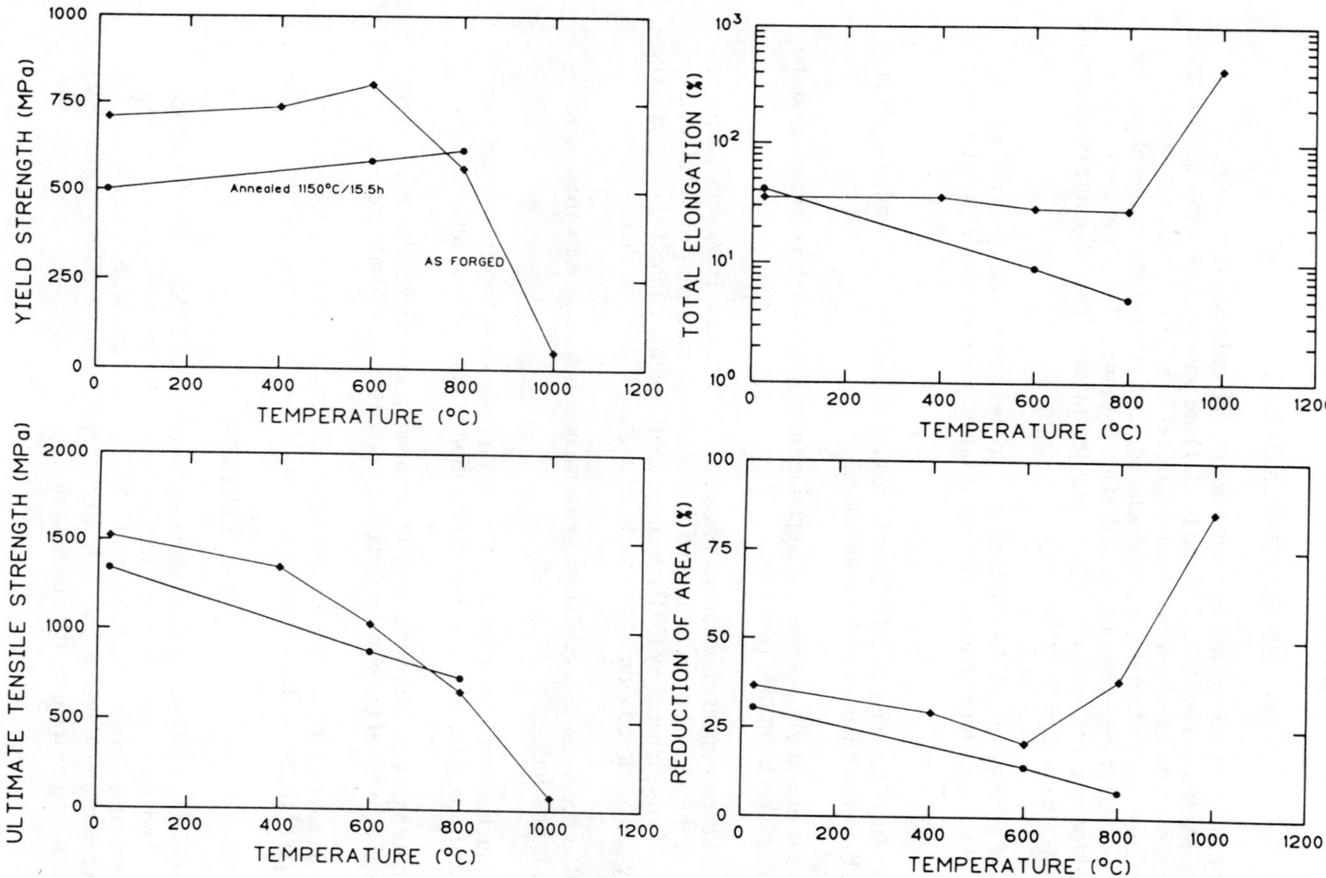

Figure 12: Comparison of tensile properties of superplastically forged disk of IC-218 in as-forged condition (grain size 9 μm) and after grain-coarsening treatment of 15.5 h at 1150°C (grain size 34 μm). Tensile tests were run in air at a strain rate of 8.3×10^{-4}/s.

above which strength and ductility are significantly affected by grain size. The deformation rate in a diffusion-controlled region is inversely proportional to the square of the grain size. Thus, even a small change in grain size produces a significant effect on properties.

Summary and Conclusions

Powder metallurgy processing techniques for ductile-ordered intermetallic Ni_3Al (IC-50) and Ni_3Al-based alloys (IC-218 and IC-221) are presented. Powder consolidation techniques described included hot extrusion, HIP, CAP, and ROC. Near-net-shape processing by superplastic forming is also discussed. Optical microstructures of powder consolidated by each of the techniques described are presented. Tensile properties were also generated for powder consolidated by various techniques. The following observations and conclusions are possible from this work:

1. Powders of all three nickel aluminide alloys have been produced up to 100-kg quantities by commercial vendors. Scaled-up quantities are possible from the same vendors.

2. Hot extrusion of powders at 1100°C and a reduction ratio of ≥8:1 is suitable for preparing fully dense billets of nickel aluminides.

3. Hot isostatic pressing is possible for preparing fully dense billets and near-net-shape parts from nickel aluminide powders.

4. The consolidation under atmospheric pressure (CAP) process is possible for the preparation of thepreparation of billets and near-net-shape parts from nickel aluminide powders. Parts prepared by CAP contained some noninterconnected porosity. This porosity can be closed either by subsequent extrusion, HIP, or ROC process.

5. The ROC process can be used to prepare dense near-net-shape parts from nickel aluminide powders.

6. All three nickel aluminide alloys of ≤10 μm grain size show superplastic behavior, which can be used for near-net-shape fabrication of parts such as turbine disks.

7. Tensile data are available over a range of temperatures for nickel aluminide alloys processed by most of the powder metallurgy processes described here.

8. It is concluded that the powder processing technology is now ready for commercialization of ductile-ordered intermetallic Ni_3Al and Ni_3Al-based alloys.

Acknowledgment

The author wishes to thank Homogeneous Metals, Inc., Cytemp Specialty Steel, Carpenter Technology Corporation and Boride Products for their cooperation in powder production and consolidation; C. E. Dunn, C. W. Holland, and K. S. Blakely for extrusion and mechanical processing; T. K. Roche and D. S. Easton for internal review; and M. L. Atchley for preparing the manuscript.

References

1. C. T. Liu, C. L. White, and J. A. Horton, "Effect of Boron on Grain Boundaries in Ni_3Al," Acta Metall., **33**(2), 213-29 (1985).

2. K. Aoki and O. Izumi, "Improvement in Room Temperature Ductility of the $L1_2$-Type Intermetallic Compound Ni_3Al by Boron Addition," Nippon Kinzoku Gakkaishi, **43**, 1190 (1979).

3. C. T. Liu, C. L. White, and E. H. Lee, "Effect of Test Environment on Ductility and Fracture Behavior of Boron-Doped Ni_3Al at 600°C," Scripta Metall., **19**, 1247-50 (October 1985).

4. C. T. Liu and V. K. Sikka, "Nickel Aluminides for Structural Use," J. Met., **38**, 19-21 (May 1986).

5. C. T. Liu, "Development of Nickel and Nickel-Iron Aluminides for Elevated-Temperature Structural Use," pp. 226-37 in Micon 86, eds., B. L. Bramfitt, R. C. Benn, C. R. Brinkman, and G. F. Vander Voort, ASTM, Baltimore, Maryland, February 1988.

6. V. K. Sikka, "Processing Technology for Nickel Aluminides," pp. 487-93 in Mater. Res. Soc. Symp. Proc., **81**, Materials Research Society, eds., N. S. Stoloff, C. C. Koch, C. T. Liu, and O. Izumi, 1987.

7. V. K. Sikka, "Advances in Processing Techniques for Nickel Aluminides," (to be published in proceedings of First ASM Europe Technical Conference on Advanced Materials and Processing Techniques for Structural Applications, Paris, France, 7-9 September 1987).

8. V. K. Sikka, "Superplasticity and Its Application to Near-Net Shaping of Intermetallic Alloys," pp. 451-67 in Interdisciplinary Issues in Materials Processing and Manufacturing, **2**, ASME, eds., S. K. Samanta, R. Komanduri, R. McMeeking, M. M. Chen, and A. Tseng, 1987.

9. V. K. Sikka and E. A. Loria, "Deformation Behavior of Powder Metallurgy Ni_3Al-Hf-B," Proceedings of International Symposium on Nickel Metallurgy, CIM, **2**, 293-308 (1986).

10. V. K. Sikka, C. T. Liu, and E. A. Loria, "Processing and Properties of Powder Metallurgy Ni_3Al-Cr-Zr-B for Use in Oxidizing Environments," Proceedings of Processing of Structural Metals by Rapid Solidification, 417-27, ASM, Metals Park, Ohio, 1987.

11. V. K. Sikka, "Engineering Processing and Properties of Nickel Aluminides," (to be published in proceedings of Second International SAMPE Metals and Metals Processing Conference, Dayton, Ohio, August 2-4, 1988).

12. V. K. Sikka, "Near-Net-Shape Casting of Sheet and Bar of Ordered Nickel Aluminide Alloys," (to be published in Casting of Near Net Shape Products, Honolulu, Hawaii, November 1988).

The Effect of Order on the Tensile Properties of as Sintered and HIP'ped Ni_3Fe Intermetallic Compound

by

B. Lograsso, A. Bose, G. Camus,
N.S. Stoloff, D.J. Duquette and R.M. German
Materials Engineering Department
Rensselaer Polytechnic Institute
Troy, NY 12180-3590

Abstract

This study describes the fabrication of Ni_3Fe by a powder metallurgy route. Solid state sintering of fine carbonyl nickel and carbonyl iron powders mixed in the desired ratio, close to the solidus, yields a 95% dense material. The interesting feature of the microstructure is the fine pores which are homogeneously distributed in the sintered microstructure. The sintering atmosphere was dry hydrogen followed by a shift to dry argon during the last ten minutes of the hold time. The shift in the atmosphere was made to bias the diffusive flow of hydrogen outward from the samples since the material is susceptible to hydrogen embrittlement. Complete disorder in the samples was achieved by heating to 800°C and holding for 120 minutes in dry argon, followed by water quenching. Ordering of the samples was done by heat treating the material at 470°C for 200 hours.

To evaluate the effects of residual porosity, the material was hot isostatically pressed to eliminate the porosity. Cold isostatically pressed rods were sintered to 95% density. The rods were then densified by containerless HIP at 1200°C for 30 minutes at a pressure of 103 MPa. Fully dense material with a large grain size of 450 μm was obtained. Heat treatments were carried out to obtain the ordered and disordered structures. Ordering of Ni_3Fe increases the yield strength, ultimate tensile strength, and hardness, with negligible effect on the ductility. Processing to full density significantly improved the yield strength for both ordered and disordered Ni_3Fe.

Introduction

Ni_3Fe is a compound which exhibits the $L1_2$ long range ordered superlattice structure below 500°C. There has been a great deal of interest in $L1_2$ intermetallic compounds to examine the effects of order on properties of interest such as strength and sensitivity to environment (for instance the hydrogen embrittlement of the deformable $L1_2$-type intermetallic compounds (1,2)).

This paper discusses the powder processing techniques used to fabricate the Ni_3Fe ordered $L1_2$ and disordered intermetallic compounds from elemental powders. In the present work, the effects of order and porosity on mechanical properties were examined at room temperature.

Ni_3Fe is known to exhibit appreciable ductility in both the ordered and disordered conditions (4,5) i.e. there is no inherent brittleness of the grain boundaries. Therefore, it represents an ideal model alloy to examine the effects of residual porosity from powder processing on the tensile properties of long-range ordered intermetallic compounds.

Experimental Procedure

Alloy Preparation

For this study an alloy of composition Ni-23.4 at.% Fe (22.5 wt.% Fe) was fabricated by solid state sintering from mixed elemental powders. This alloy composition falls in the gamma prime phase field below 500°C as shown on the phase diagram in Figure 1 (3). The characteristics of the carbonyl iron and nickel powders are given in Table I. The average particle sizes as determined by forward laser-light-scattering were 10.8 and 3.3μm, respectively. The powders were mixed for 30 minutes in a 200 g batch. The test samples were flat dogbone tensile specimens with a pressed area of 645 mm^2, a cross section width of 5 mm and a thickness of approximately 5 mm. These tensile bars were compacted at 300 MPa in a double action die without binder, using zinc stearate as a die wall lubricant. The solid state sintering cycle used is shown in Figure 2 has a maximum temperature of 1410°C. The sintering atmosphere was dry hydrogen followed by a shift to dry argon during the last ten minutes of the hold time to remove hydrogen and prevent hydrogen embrittlement of the compound. The sintering treatment resulted in Ni_3Fe compound with 5% porosity (determined by density measurement). Fully dense specimens were produced by first cold isostatically pressing rods 18 mm in diameter and 75 mm long, with 210 MPa compaction pressure. These rods were then sintered with the profile shown in Figure 2, except for 120 minutes at 1410°C instead of 60 minutes, due to their larger volume. Finally, these rods were fully densified by containerless HIP at 1200°C for 30 minutes at 103 MPa in argon. After HIP, the rods were machined into round tensile bars with a diameter of 4.0 mm and a 15.1 mm gauge length.

Order Disorder Treatments

The fully disordered Ni_3Fe structure was obtained by annealing at 800°C for 2 hours in flowing argon, followed by a water quench. The ordered intermetallic compound was obtained by encapsulating the sintered rods in a vacuum better than 10^{-5} torr, with a Zr getter, and then heat treating at 470°C for 200 hours. The degree of order, S, achieved by a similar heat treatment has been shown to be 0.95 (6).

Properties and Microstructure Evaluation

The flat and round bar tensile specimens were tested using a screw-driven Instron machine at a cross-head speed of 0.12 mm/min. Specimens were sectioned and metallographically prepared and etched for examination. Fracture surfaces were examined in a scanning electron microscope.

Results

Material Evaluation

The resulting microstructures of the as sintered Ni_3Fe and the hot isostatically pressed after the order and disordering treatments are shown in Figures 3 and 4. Figure 3 shows that the sintered and heat treated Ni_3Fe has a finer grain structure with a average size of 80 μm compared with 450 μm for the hot isostatically pressed material shown in Figure 4. Note the uniform distribution of fine porosity (approximate size is 5μm) in the sintered and heat treated material in Figure 3. From the optical micrographs the estimated porosity is 5%. The hot isostatically pressed samples showed a fully dense microstructure. The ordering or disordering treatment temperatures which followed were too low to affect the scale of the microstructure beyond the growth during the sintering or hot isostatic pressing treatments.

Mechanical Properties

The tensile properties and hardness values of the ordered and disordered compounds are shown in Tables II and III. For both sintered and sintered and hot isostatically pressed samples, the ordered condition had a significantly higher yield strength, ultimate tensile strength, elastic modulus and hardness. On the other hand, the ordered compound has only a slightly lower ductility than the disordered compound.)

For both the ordered and disordered Ni_3Fe, the hot isostatic pressing treatment to remove the residual porosity resulted in a significant increase in the yield strength. Data in Tables II and III also show very little change in the ultimate tensile strength and percent elongation for both ordered and disorderd compounds with elimination of the residual porosity by HIP. Figures 5a and 5b plot the engineering stress-strain behavior for the processed compounds and graphically show the improved strength with ordering.

The onset of necking appears to be largely increased in the case of the fully dense hot isostatically pressed material. In these fully dense specimens the strain begins to localize (necking) at about the same strain for both the ordered and disordered Ni_3Fe. However, the disordered alloy is able to avoid linkup to fracture until a greater value of total strain. For the press and sinter specimens, strain is not localized until a much greater value of total strain (0.36 -0.41) than the HIP specimens (total strain of 0.32), possibly because the uniform distribution of pores facilitates uniform intiation of and continued slip. However, once slip is localized, the porous Ni_3Fe quickly fails (Figure 5a).

Although the total overall strain is about the same for both the fully dense and the specimens with 5% porosity, the path to failure results in different ductile fracture surface characteristics as shown in Figures 6 and 7. Examination of the fracture surfaces indicates that the ordered and disordered material failed by a dimpled transgranular mode of rupture (Figure 6 and 7). The mean dimple size seems to be the same for the porous ordered and disordered conditions. However, in the fully dense material, the dimples are larger in the case of the disordered Ni_3Fe. The fully dense ordered and disordered alloys exhibited considerable necking, with the disordered alloy fracture surface showing the greatest amount of necking. The HIP disordered Ni_3Fe fracture surface shows that the voids grew to a much larger size than in the other conditions before the voids link up to failure.

The very large grains in the fully dense material shows slip bands in the deformed grains on the side of the samples near the fracture surface. The ordered alloy exhibits a large amount of planar slip (Figure 8A) while the disordered alloy shows diffuse slip (Figure 8B)

Discussion

Ni_3Fe can be fabricated from mixed elemental powders. Tensile tests have shown that the mechanical behavior is affected by ordering. Ordered alloys generally deform by the motion of superlattice dislocations, resulting in intense planar slip and restricted cross-slip, as was observed in Figure 8A for ordered Ni_3Fe. The disordered Ni_3Fe showed more diffuse slip (Figure 8B) due to easier cross slip. Data in Tables II and III show that ordering increases the yield strength, ultimate tensile strength, and hardness. These results are consistent with the results of Calvayrac and Fayard (4) who found increases in strength associated with ordering. For Ni_3Fe, ordering does not affect ductility. This alloy, even in the ordered condition, has a sufficient number of independent slip systems, as do other $L1_2$ intermetallic alloys such as Ni_3Al or Ni_3Si. However, the latter two alloys exhibit inherent brittleness. Recent studies have shown that the wide range of fractures occurring for nominally the same type of $L1_2$ intermetallic compounds can be correlated with differences in valence, electronegativity, and atomic radii between the two elements of the compounds (5). The tendency toward brittle behavior appears to increase with differences in the above factors.

Table II shows a significant decrease in yield strength due to presence of the 5 % porosity in the sintered ordered and disordered compounds. Elimination of this porosity resulted in a 37% increase in strength for ordered and 55% increase in strength for the disordered compound. Slip initiates near the individual pores as plastic zones form due to the concentrated stresses. This accounts for the lower flow stress for the sintered compounds. Also, in the range of 2 - 5 percent voids, an increase in the porosity decreases the strength but has a negligible effect on the tensile ductility (7,8). Ductility is dependent on a shear instability process involving the linking of a critical number of pores/voids and the path of high void content. Therefore, due to low probability of linking a critical number of pores/voids to failure in materials with porosity in the range of 2.5 to 5 %, the ductility of the porous ordered and disordered alloys is not affected. These materials cannot relax stress concentrations at pores as easily as other ductile metals.

The data in Tables II and III also show that as a result of the hot isostatic pressing, not only was the porosity removed but the grain size increased from 80 μm to 450 μm. Further investigation is needed to determine if a substructure is present within these large grains after hot isostatic pressing.

Conclusions

1. Ordering of Ni_3Fe increases the yield strength, ultimate tensile strength, and hardness, with negligible effect on ductility.

2. Processing to full density by eliminating residual porosity resulted in a significant increase in the yield strength for both ordered and disordered Ni_3Fe. The presence of the pores in the sintered compounds could have resulted in initiation of slip near the pores. The pores had little effect on ductility, hardness or ultimate tensile strength.

Acknowledgments

This work was sponsored by DARPA and ONR under Contract No. N00014-86 -K0700 a University Research Initiatives Program on High Temperature Advanced Structural Composites.

References

1. A.K. Kuruvilla and N.S. Stoloff, Scripta Met., vol. 19, p.83 (1985).
2. T. Takasugi and O. Izumi, Acta Met., vol. 34, p. 607 (1986).
3. *Metallography, Structures and Phase diagrams*, Vol. 8, 8th edition, American Society for Metals, Metals Park, Ohio, p. 304 (1973).
4. Y. Calvayrac and M. Fayard, Phys. Stat. Sol., vol. 17, p. 407 (1973).
5. T. Takasugi and O. Izumi, Acta Met., vol. 33, p. 1247 (1985).
6. D.G. Morris, G.T. Brown, R.C. Piller and R.E. Smallman, Acta Met.,vol. 24, p. 21 (1976).
7. R. Haynes, *The Mechanical Behavior of Sintered Alloys* , Freund Publishing House, London, UK 1981.
8. E.M. Dubensky and D.A. Koss, Metall. Trans., vol. 18A, p.

Table I. Powder Characteristics

Property	Ni	Fe
vendor	INCO	GAF
designation	123	HP
purity, pct	99.99	99.50
Fisher subsieve size, μm	2.8	3.0
mean size, * μm	3.3	10.8
BET specific surface area, m^2/g	2.19	0.88
apparent density, g/cm^3	2.15	2.20
major impurities (ppm)	Ca(10) Fe(30) Si(40)	Ca(600) Al(600) Si(300) O(300) Mn(2000)

Table II.
Mechanical Properties Ni_3Fe Pressed and Sintered to 95% Theoretical Density (Grain Size = 80 μm)

Condition	0.2% Yield Strength (MPa)	Elongation %	Tensile Strength (MPa)	Hardness (HRB)
Disordered Ni_3Fe	111	43%	490	43.4
Ordered Ni_3Fe	180	37%	627	70.4

Table III.
Mechanical Properties Ni_3Fe Pressed, Sintered and HIP'd to 99.5% Theoretical Density (Grain Size = 450 μm)

Condition	0.2% Yield Strength (MPa)	Elongation %	Tensile Strength (MPa)	Hardness (HRB)
Disordered Ni_3Fe	172	42.5%	487	35.5
Ordered Ni_3Fe	247	37%	627	70.7

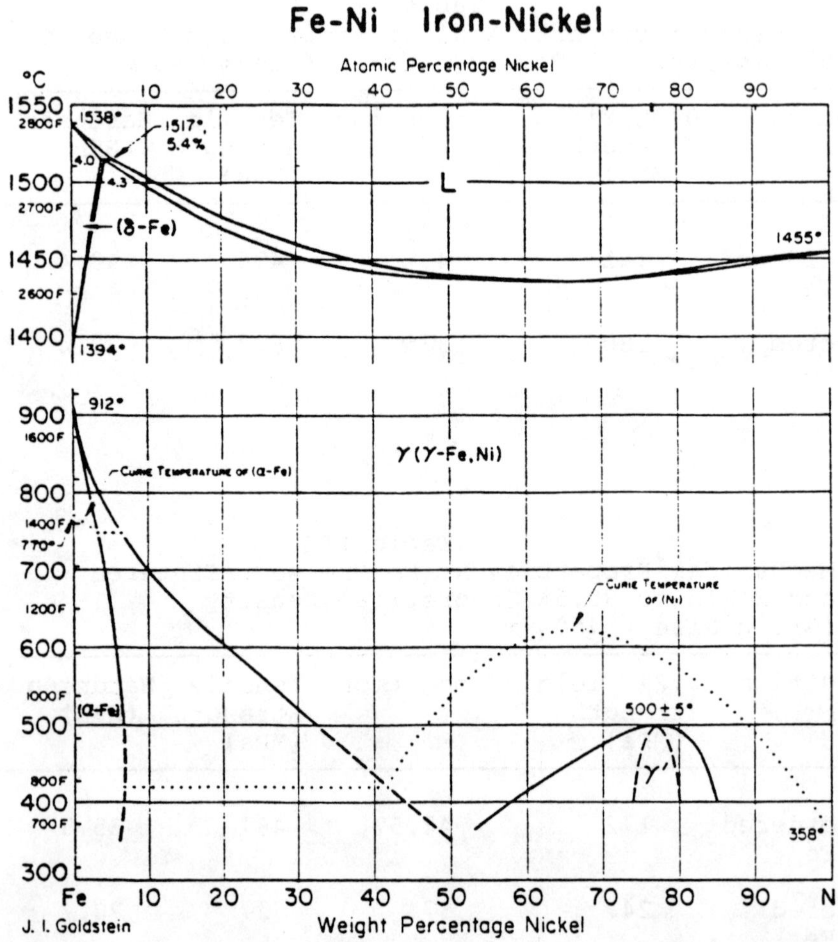

Figure 1. Ni_3Fe binary phase diagram

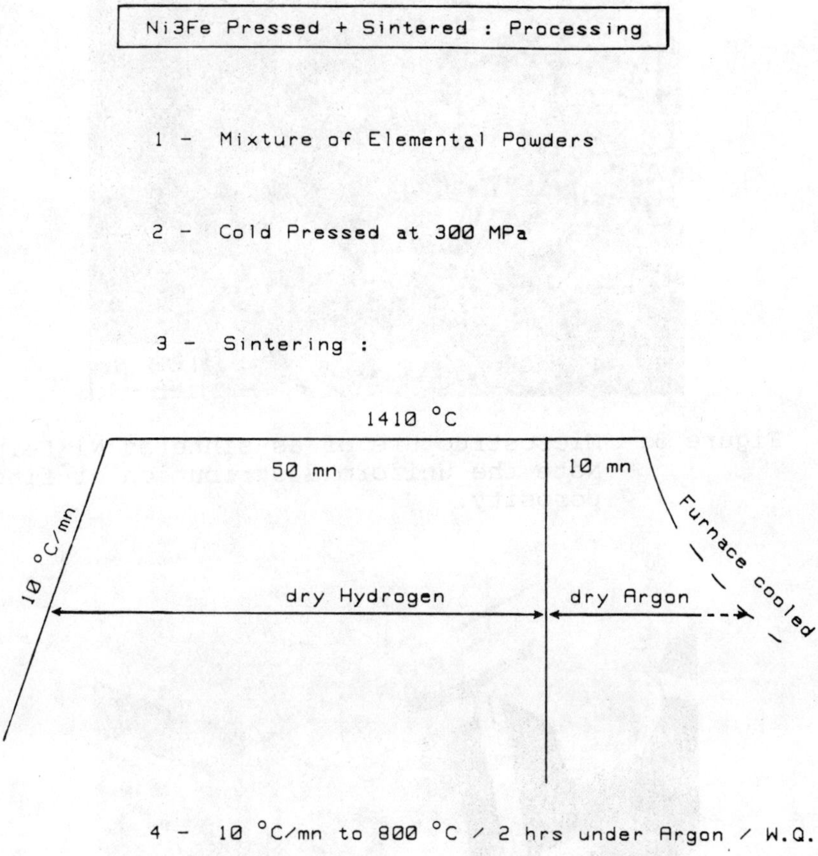

Figure 2. Solid state sintering cycle used for Ni_3Fe.

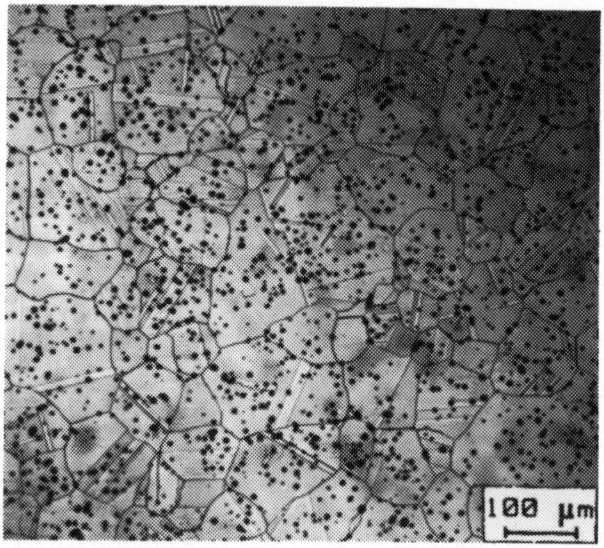

Figure 3. Microstructure of as-sintered Ni_3Fe. Note the uniform distribution of fine porosity.

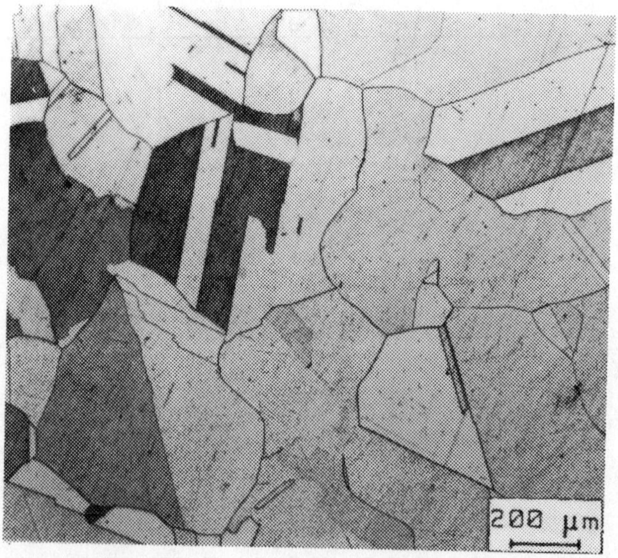

Figure 4. Microstructure of sintered and HIP'd Ni_3Fe.

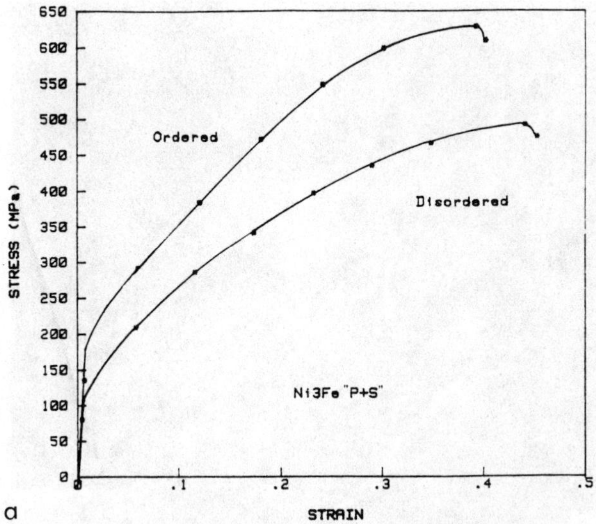

a

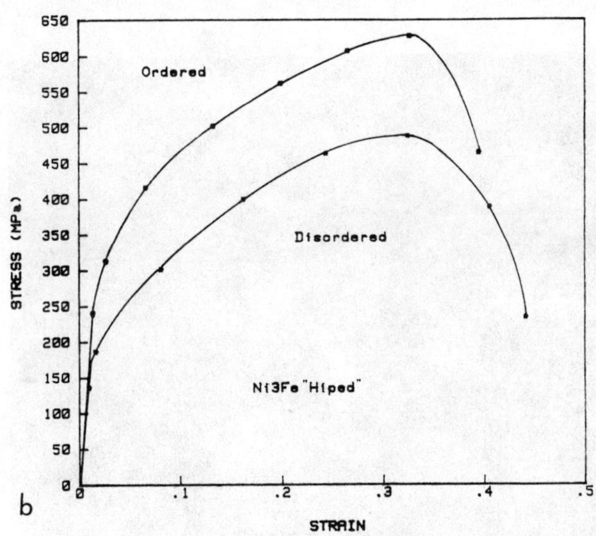

b

Figure 5. Stress-strain behavior of Ni_3Fe for both the ordered and disordered conditions for a) pressed and sintered b)pressed, sintered and HIP'd.

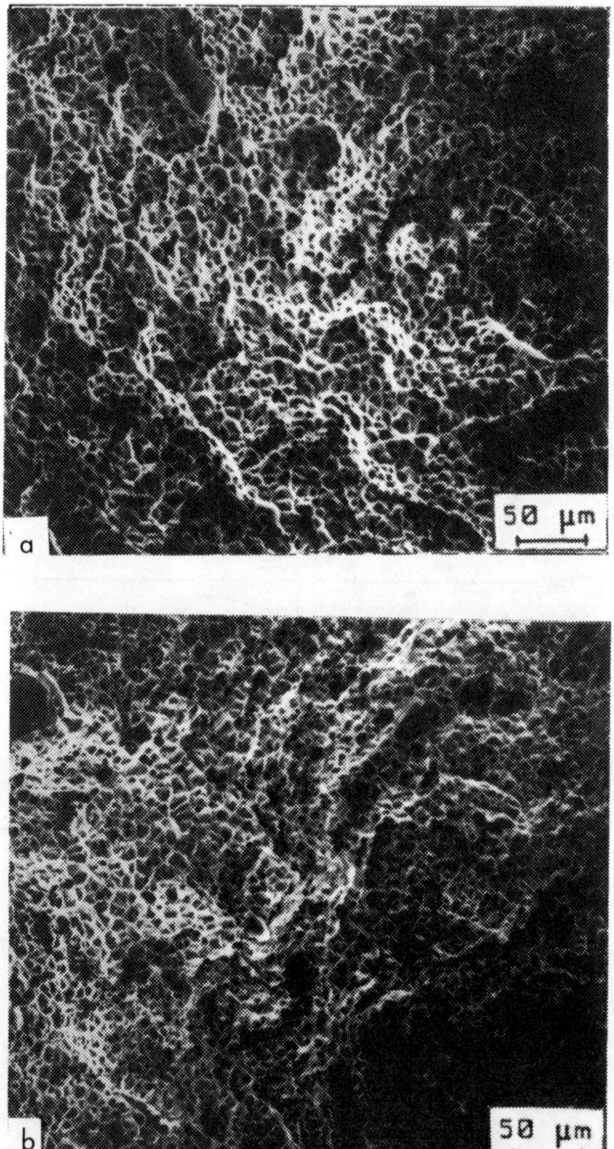

Figure 6. Scanning electron fractograph of pressed and sintered Ni_3Fe for the a) ordered and b) disordered conditions.

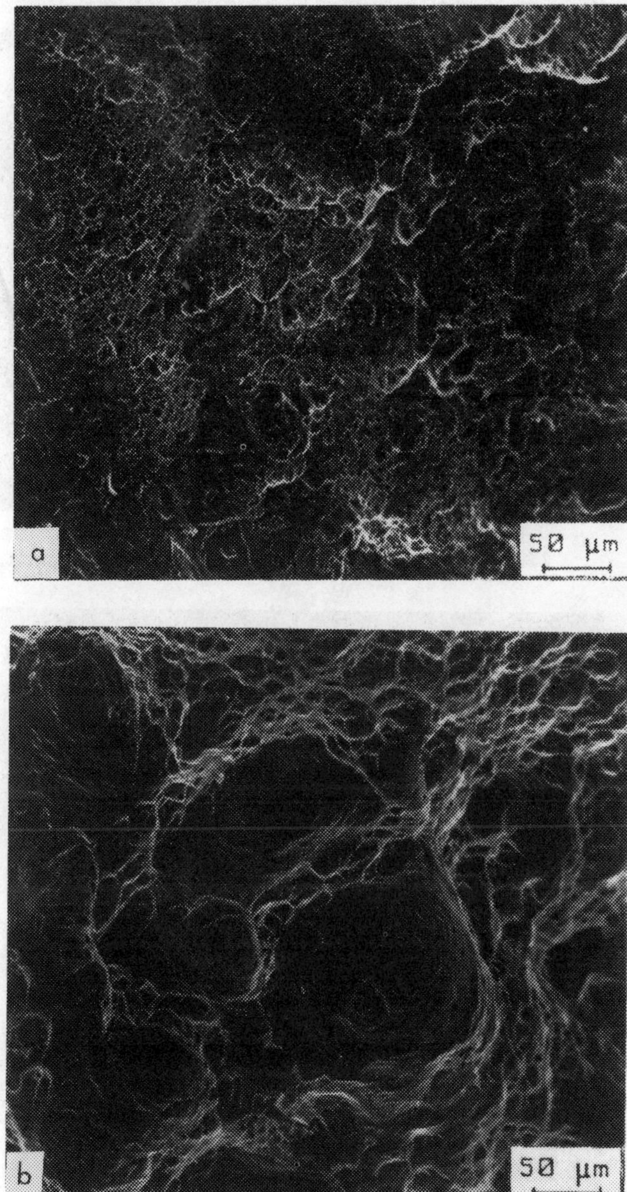

Figure 7. Scanning electron fractograph of pressed, sintered and HIP'd Ni_3Fe for the a) ordered and b) disordered conditions.

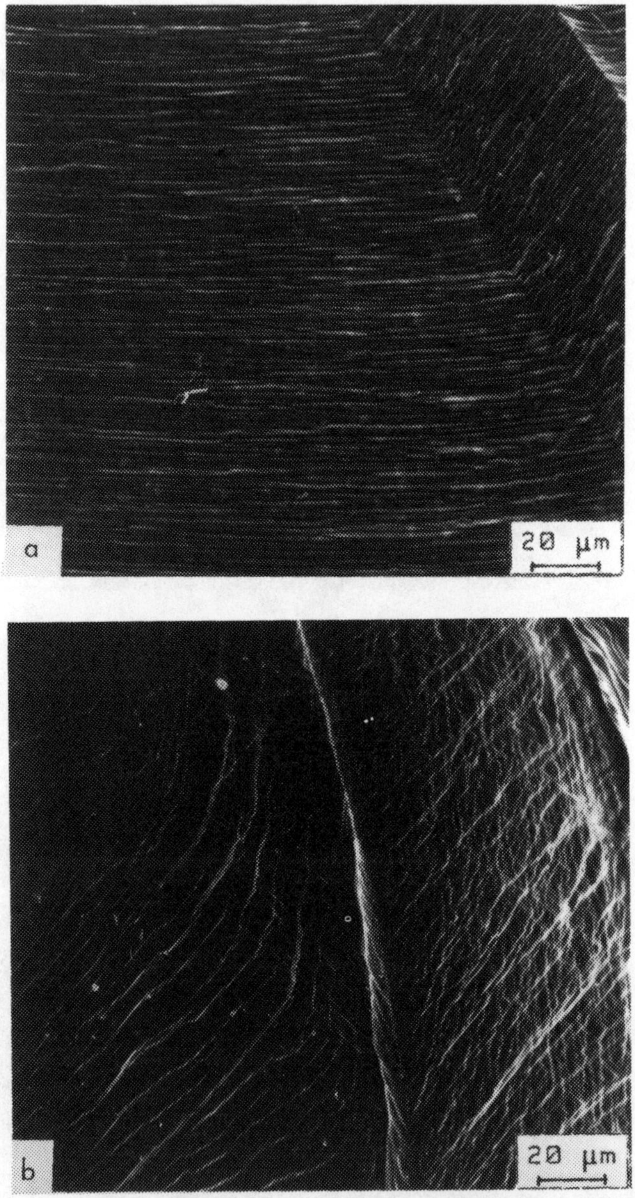

Figure 8. Scanning electron micrographs of Ni_3Fe showing a) planar slip bands in the ordered condition and b) diffuse slip bands in the disordered condition.

Chapter Eight

High-Temperature Nickel-Base Materials

PROCESS DEVELOPMENT FOR LOW-COST, HIGH-STRENGTH P/M Ni-BASE SUPERALLOY TURBINE DISK

H.Takigawa, N.Kawai, K.Iwai, S.Furuta and N.Nagata

Technical Development Group
Kobe Steel, Ltd.
Kobe, Japan

Abstract

From the viewpoint of applying a high-strength PM Ni-base superalloy of AF115 to turbine disks, the mechanical properties required of a turbine disk material were investigated. PM AF115 was found to be especially suitable for a turbine disk to be operated at comparatively high temperatures because the stress rupture strength is much superior to other alloys including conventional wrought alloys and representative PM alloys such as IN100M and René 95.

The FCPR(fatigue crack propagation rate) of the AF115 also surpassed those of other PM alloys in the stress intensity range less than 20 MPa$\sqrt{m}$, although the FCPR was primarily influenced by the grain size. The LCF is greatly influenced by manufacturing-process parameters such as powder size or the process itself.

The as-HIPed AF115 was applied to a turbine blisk for a stationary gas turbine by developing a new near-net-shape technology.

Introduction

A turbine disk material possessing high strength and good fatigue properties at intermediate temperatures is indispensable to increasing the power of a gas turbine engine and to the improvement of its thermal efficiency. Conventional manufacturing processes of the material, such as the cast and forge process are, however, limited in their application to such advanced-materials development. This is because the more alloying elements are added to increase the strength, the more densely these elements and impurities segregate to boundary regions of the microstructure during solidification of the material, resulting in degradation in mechanical properties, forgeability, etc. To solve this problem, the PM process, in which rapidly solidified fine powder with no macro segregations and fine microstructures is consolidated into a bulk material, has been applied to a material development for high-strength disks [1]. As a result, PM IN100 and René 95 disks were developed and have come into practical use for aero-gas turbine engines [2]~[4].

In this study, to apply a high-strength PM Ni-base superalloy AF115 [5] to turbine disks, first the mechanical properties of tensile and stress rupture strength, LCF life, and FCPR were investigated, correlating with some manufacturing-process parameters such as powder size, solution heat treatment temperature, and the process itself (that is, whether as-HIP process or HIP + isothermal forging process). These data were compared with similar data obtained by others for PM and wrought Ni-base superalloys such as IN100M, René 95, Waspaloy, and In718.

Then, the PM AF115 was applied to a turbine blisk (integrally bladed disk) for a small stationary gas turbine by developing a near-net-shape technology using an as-HIP process.

Mechanical Properties of PM AF115

Disk Preparation

The disk prepared for evaluating all the mechanical properties concerned in this study was an as-HIPed disk, 340 mm in diameter, 55 mm thick at the rim region, and having bosses in the center.

The disk was fabricated as follows: Ar-gas atomized powder with size less than 105 μm in diameter (-150M) was HIPed and then the disk was solution heat treated (1448K×3h/RFC) and aged (1033K ×16h/FC). The chemical composition of the HIPed disk is listed in Table 1. Its mechanical properties of tensile and stress rupture strength, LCF life, and FCPR were investigated.

In addition to the above disk, to evaluate LCF life in relation to manufacturing-process parameters, several more disks were prepared, in which smaller sizes of powders and/or isothermal forging after HIPing were applied.

Furthermore, in order to evaluate stress rupture properties and FCPR in terms of grain size, a disk for which the solution heat treatment temperature was changed, was also fabricated.

Manufacturing conditions and mechanical properties evaluated for each disk are summarized in Table 2. Besides these disks, a wrought Waspaloy disk on the market was additionally evaluated to compare its FCPR with that of PM AF115.

Test Method

Test specimens were prepared from the heat-treated disk described above. Geometry of gauge sections of each specimen and test conditions are listed in Table 3.

Tensile tests were carried out in the temperature range of

Table 1 Chemical composition of HIPed AF115

wt.%

C	Cr	Ni	Co	Mo	W	Nb	Ti	Al	B	Zr	Hf	O	N
0.04	10.6	bal.	14.9	2.8	6.0	1.7	4.0	3.8	0.02	0.05	0.77	0.0045	0.0009

Table 2 Manufacturing conditions and mechanical properties evaluated for each disk

Disk (mm)	Process	Powder Size	Heat Treatment	Grain Size (μm)	Mechanical Test*
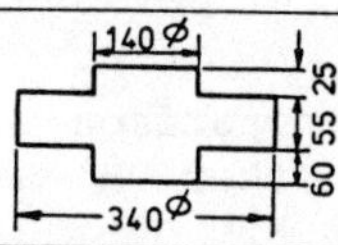	as-HIP	−150 M	1448K × 3h/RFC +1033K × 16h/AC	16.0	T, S, LCF, FCPR
$120^{\phi} \times 100^{h}$	ditto	−250 M	1423K × 3h/RFC +1033K × 16h/AC	4.7	S, LCF, FCPR
ditto	ditto	−325 M	1448K × 3h/RFC +1033K × 16h/AC	6.7	LCF
ditto	ditto	−325 M	1463K × 3h/RFC +1033K × 16h/AC	22.0	S
$165^{\phi} \times 16^{h}$	HIP + IF	−150 M	1448K × 3h/RFC +1033K × 16h/AC	5.6	LCF
ditto	ditto	−325 M	ditto	4.7	LCF

*T: tensile, S: stress rupture

Table 3 Test conditions and dimensions of test specimens

Test	Conditions	Gauge Section of Test Specimen
Tensile	R.T.∼923K, $\dot{\varepsilon}$ = 0.5%/min to 0.2% proof stress	$6^{\phi} \times 20^{\ell}$ (mm)
Stress Rupture	1033∼1173K × 245∼700 MPa	$6^{\phi} \times 20^{\ell}$ (mm)
Low Cycle Fatigue (LCF)	stress control, R_{σ} = 0.0, f = 7S/cycle (dwell time : 2S) T = 673,908K σ_{max} = 1100∼1500 MPa for smooth specimen	$7^{\phi} \times 16^{\ell}$ (mm)
Fatigue Crack Propagation Rate (FCPR)	T = 908K, constant load, f = 2 Hz (triangular waveform), R_{σ} = 0.0	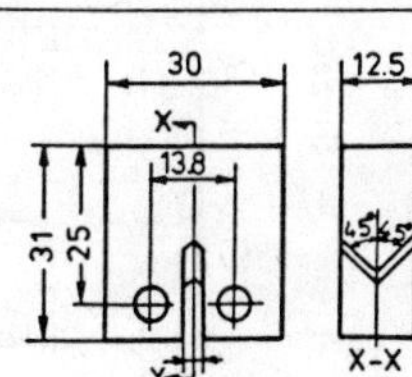

R.T.through 923K. Stress rupture tests were performed by changing the combination of temperature and stress, 1033 to 1173K and 245 to 700 MPa, respectively. Smooth test specimens with gauge sections 20 mm long and 6 mm in diameter, were used for both tensile and stress rupture tests.

LCF tests were conducted on smooth specimens at 673K and 908K under the load-control condition, using a 1/6-Hz trapezoidal waveform with a dwell time of 2 seconds and a stress ratio R_σ of 0.0. FCPR tests were carried out using CT specimens ($31\times30\times12.5^t$ mm) at 908K under constant load, 2Hz, and zero R_σ using a triangular wave, where the crack length was measured by an unloading compliance method. Both LCF and FCPR tests were performed on closed-loop servohydraulic testing machines.

Test Results

Mechanical properties of PM AF115 are shown in comparison with published data for the conventional wrought and other PM alloys.

Tensile properties

Tensile properties obtained for an as-HIPed AF115 disk 340mm in diameter are presented in Fig.1. PM AF115 shows a much higher strength than conventional wrought alloys such as Waspaloy and In718, even though the elongation is almost the same. This is considered to result both from a much higher addition of alloying elements and from a homogeneous fine microstructure inherited from rapidly solidified fine powder.

On the other hand, comparing PM AF115 with other PM materials, it is a little higher than PM IN100M but a little lower than PM René 95 both in 0.2% proof stress and in tensile strength [2] [6]. But, the elongation shows an inverse relation to the strength.

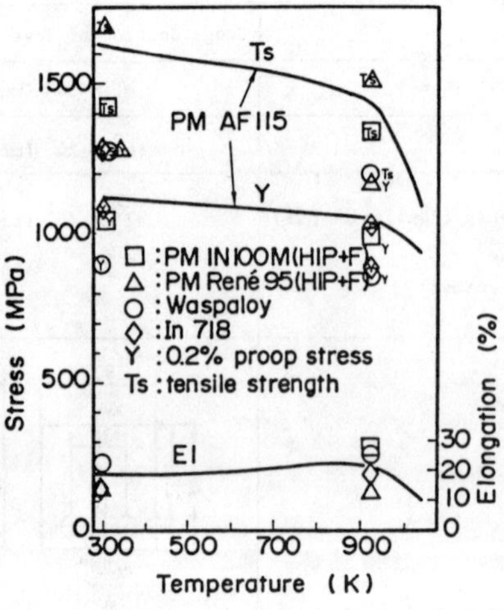

Fig. 1 Comparison of tensile properties between PM AF115 and other Ni-base superalloys

Stress rupture strength

The data of stress rupture strength are plotted in Fig.2. It can be observed that grain size has a great influence on the stress rupture strength. It also can be seen that PM AF115 is far superior to other alloys such as conventional wrought Waspaloy or PM René 95 and IN100M even if the grain size is equivalent or smaller [2][6]. For example, comparing PM AF115 with a grain size of 16μm with wrought Waspaloy, the stress rupture life is about 1.5 times longer in terms of P-value in the stress range of 400 to 800 MPa. This means that the life of PM AF115 at 1000K is about 30 times as long as that of Waspaloy.

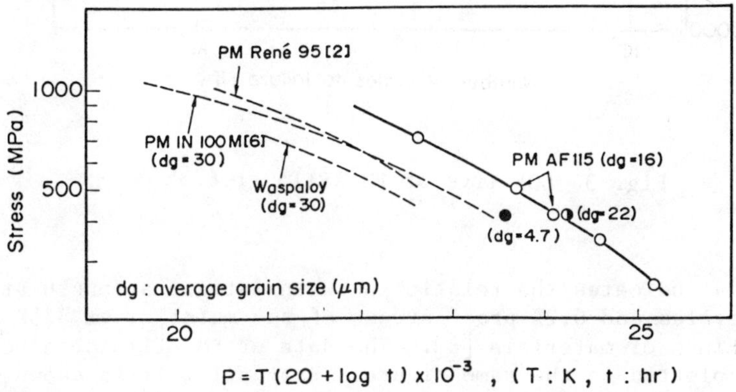

Fig. 2 Comparison of stress rupture strength between PM AF115 and other Ni-base superalloys

Low cycle fatigue life

Results obtained in the LCF test at 673K with smooth test specimens are plotted in Fig.3. At low stress range, the material consolidated from smaller powder shows a longer fatigue life. In addition, the forged disks show longer lives than the as-HIPed disks when the powder size is the same. This is because the maximum size of origins of fatigue cracks such as porosities and inclusions can be controlled much smaller for the above both cases.

The LCF lives at the higher stress range, however, show no significant difference between the materials. In this loading range, the life is essentially determined by plastic slip from a surface scratch. Defects such as pores and inclusions are scarcely noticeable on the fracture surfaces. Under such conditions, it is considered that manufacturing-process parameters have little influence on the LCF life, as shown in Fig.3. Similar data were obtained for PM René 95 by R.V. Miner and J.Gayda [7].

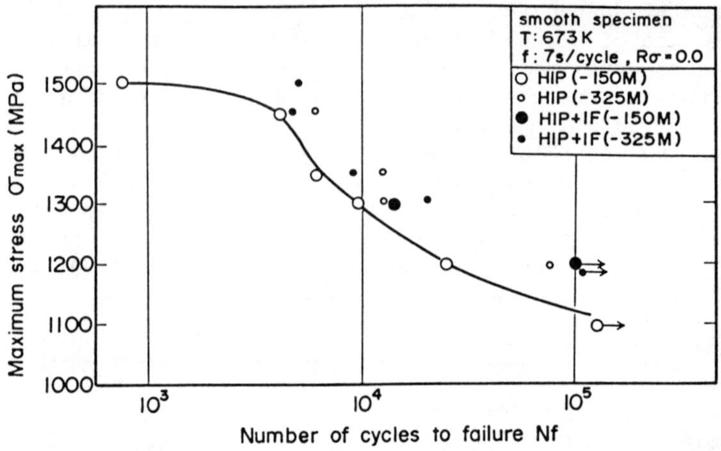

Fig. 3 LCF life of PM AF115 at 673K

Fig.4 indicates the relation between fatigue strength at 10^4 and 10^5 cycles and 0.2% proof stress of the material at 873K for various kinds of materials [8]. The data of PM AF115 obtained at 908K are plotted in the same figure. Basically, it is shown that the fatigue strength increases with the 0.2% proof stress of the material. But, as suggested in Fig.3, there is the possibility of making the LCF life of PM AF115 surpass these values to some extent by consolidating fine powder (-325M) and forging the preform.

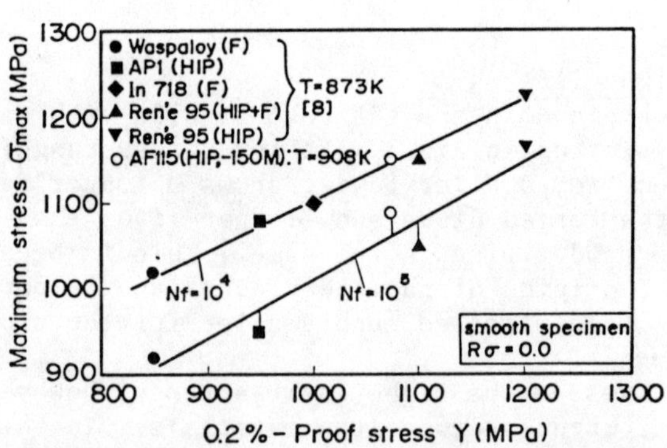

Fig. 4 Relations between LCF life at 10^4 and 10^5 cycles and 0.2% proof stress of the materials

Fatigue crack propagation rate

Fig.5 indicates the results of FCPR tests on PM AF115 and wrought Waspaloy. Similar data published elsewhere are also presented [9][10].

It is known that coarse grain has a positive effect on fatigue-crack propagation characteristics in hot air, in connection with the retardation of grain-boundary oxidation [9]. This grain-size effect can also be observed in this study, both for PM AF115 and for wrought Waspaloy.

Fig.6 shows the fracture-mode difference between the two kinds of PM AF115, differentiating by grain size, 4.7μm and 16μm. Both fracture surfaces were obtained at a stress intensity range of about 26MPa$\sqrt{m}$. The intergranular fracture mode is observed for the disk with small grain, while the disk with larger grain fractured in the transgranular mode. Due to this effect, PM alloys including PM AF115 show faster FCPRs than wrought alloys with much larger grain size, as observed in Fig.5.

Among PM materials such as IN100, René 95 and AF115, the FCPR is almost the same in the stress intensity range over 20 MPa$\sqrt{m}$. In the range just below 20 MPa$\sqrt{m}$, however, PM AF115 shows the lowest FCPR. This near-threshold fatigue crack propagation behavior is considered to be greatly influenced by microstructure and oxidation kinetics peculiar to the chemical composition of the material. Concerning this problem, more detailed research should be conducted.

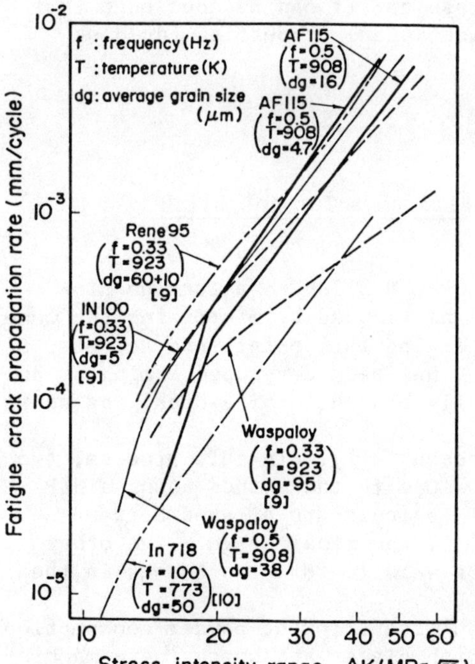

Fig. 5 Comparison of FCPR between PM AF115 and other conventional wrought alloys and PM alloys

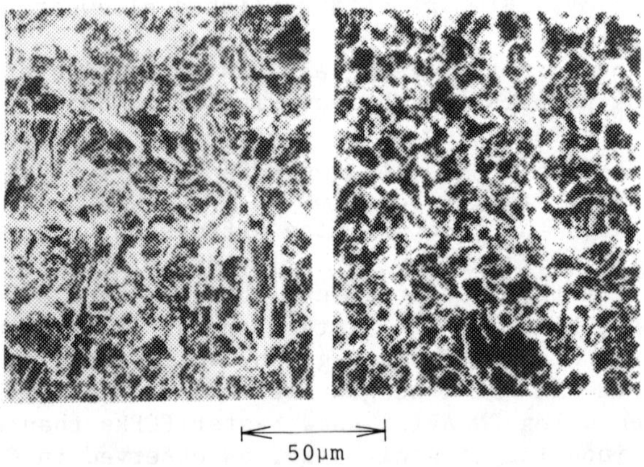

Fig. 6 Fracture mode difference in different grain sizes of PM AF115, observed at a stress intensity range of about 26 MPa$\sqrt{m}$

From the above-mentioned results, it can be concluded that PM AF115 provides appropriate mechanical properties for disk application.

Application of PM AF115 to a Turbine Blisk

Near-Net-Shape Technology

It is desirable in applying a PM alloy to a turbine disk that manufacturing costs should be reduced by minimizing the consumption of expensive powder. Taking this point into account, a unique near-net-shape technology has been developed aiming at an as-HIPed turbine blisk (integrally bladed turbine disk), as shown in Fig.7.

Fig.8 shows the process sequentially. In this process, two noteworthy techniques are used. One is the method using a HIP capsule in which the mild steel segments are arrayed between blades so as to be machined off at the final stage. The other technique is to apply a friction weld to form two bosses in the center of the disk.

These techniques enabled 60% reduction of powder consumption comparing with the method in which blades and bosses were machined from a pancake disk.

Fig. 7 PM AF115 blisk for a stationary gas turbine (360 mm in diameter)

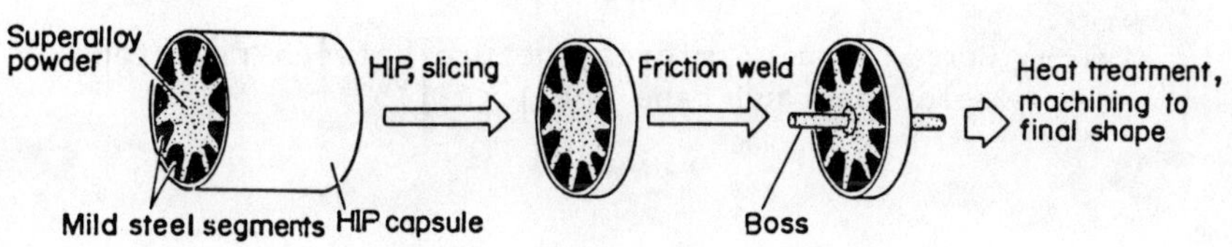

Fig. 8 Near-net-shape technology developed for manufacturing PM AF115 turbine blisk

Friction-Weld-Zone Characteristics

In order to confirm the soundness of the friction weld zone, the macro and microstructures were investigated for longitudinal sections of friction-welded round bars, each 28 mm dia. and 50 mm long. In addition, tensile, stress rupture, and LCF tests were carried out on the same joints.

Fig.9 shows an optical microstructure of the weld zone. The weld zone has larger grain size and fewer large γ' phases, which precipitate during HIP treatment, than the base metal, which is not heat affected during friction welding. This is because most γ' phases resolve in the matrix of the weld zone and then the region is rapidly cooled. It is also noteworthy that the dendritic microstructure cannot be observed distinctly.

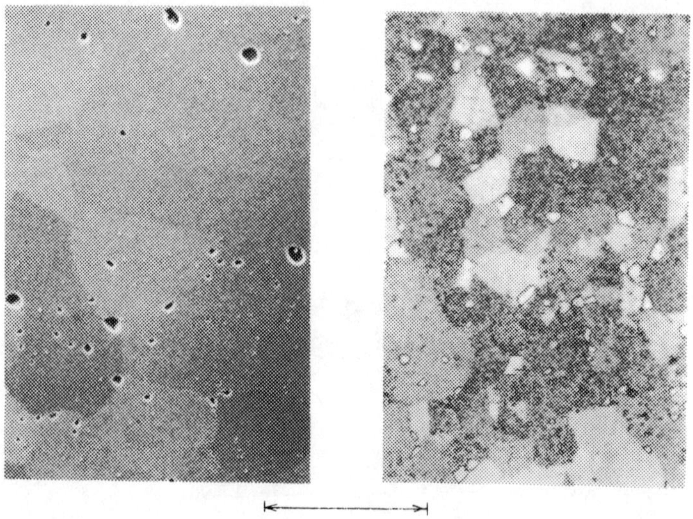

50 μm

(a)friction weld zone (b)base metal(AF115)

Fig. 9 Comparison of microstructure between friction weld zone and base metal (AF115)

For these reasons, the mechanical properties of the friction-weld joint are nearly equivalent to those of the as-HIPed base metal, as shown in Fig.10. The test specimens did not fracture in the weld zone at all.

Spin Test

In order to confirm the mechanical integrity of the turbine blisk, a spin test at 115% rated speed was executed in a vacuum chamber. The blisk was inspected by ultrasonic and fluorescent penetrant tests before and after the spin test, but no defects were found. After the spin test and the confirmation of mechanical properties, the blisk was assembled into a 750 KVA stationary gas turbine.

After 100 hours full-load operation of the engine, the blisk was checked by dye penetrant test, and no imperfections were found.

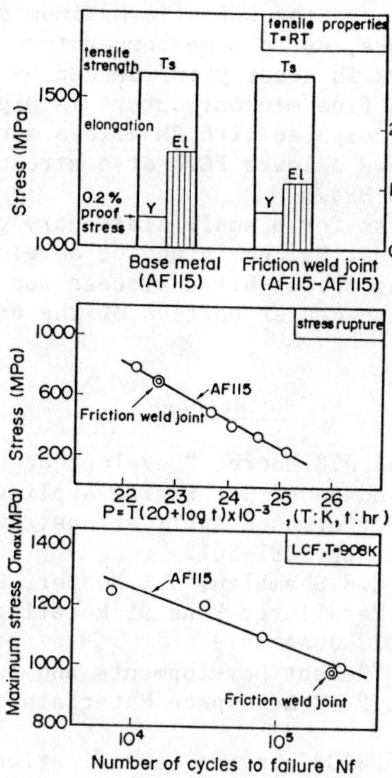

Fig. 10 Mechanical properties in friction weld zone

Conclusion

Concluding remarks both as to the investigation of mechanical properties of AF115 and the development of a turbine blisk are as follows:

1) The tensile strength from R.T. through 923K of PM AF115 is much higher than those of conventional wrought Waspaloy and In718. It is also a little higher than that of PM IN100M, but a little lower than that of PM René 95.

2) As far as the stress rupture strength is concerned, PM AF115 markedly surpasses other conventional wrought and PM alloys. This fact suggests that the AF115 may be especially suitable for disks to be operated at comparatively high temperatures.

3) PM AF115 shows almost the same LCF life as PM René 95. It is also concluded that the LCF is greatly influenced by manufacturing-process parameters. From this viewpoint, it can be

suggested that the HIP-plus-isothermal forged AF115 composed of -325M fine powder endures the LCF at a maximum stress level of around 1200 to 1300 MPa, up to some temperature higher than 673K.

4) Since the FCPR is greatly influenced by grain size, the FCPR of PM AF115 with fine microstructure is higher than that of wrought In718. But, compared with PM alloys such as IN100M or René 95, PM AF115 shows a lower FCPR at a stress-intensity range of less than about 20 MPa$\sqrt{m}$.

5) A turbine blisk for a small stationary gas turbine could be developed at low cost by combining the developed near-net-shape technology using the as-HIPed process and a friction weld to attach bosses to the center portion of the disk.

References

[1] R.L.Athey and J.B.Moore, "Development of IN-100 Powder-Metallurgy Disks for Advanced Jet Engine Application," Powder Metallurgy for High-Performance Applications, Syracuse University Press, New York, 1972, pp. 281-301.

[2] W.R.Pfouts, C.E.Shamblen, J.S.Mosier, R.E.Peebles and R.W.Gorsler, "Powder Metallurgy René 95 Rotating Turbine Engine Parts," NASA CR-159802, June 1979.

[3] R.F.Singer, "Recent Developments and Treands in High Strength PM Materials," PM Aerospace Materials, Vol.1, 1984, pp. 12-14.

[4] Anonymous, "PW2037 Engine Certification Introduces PM Superalloys to Commercial Aircraft," Metal Powder Report, Vol.39, 1984, p.101.

[5] J.L.Bartos, "Development of a Very High Strength Disk Alloy for 1400F Service," AFML-7R-74-187, National Technical Information Service, Nov. 1974.

[6] Material Research Laboratories of Kobe Steel, Ltd., "Tensile, Stress Rupture and LCF Properties of Mod. IN100," Technical data.

[7] R.V.Miner and J.Gayda, "Effects of Processing and Microstructure on the Fatigue Behaviour of the Nickel-Base Superalloy René 95," International Journal of Fatigue, No.3, July 1984, pp. 189-193.

[8] G.W.König, "Comparison between the Properties of Conventional Wrought and Powder Metallurgical Alloys for Turbine Disc Applications," AGARD Conference Proceedings No.368, June 1984, pp.6-1~6-11.

[9] J.Gayda and R.V.Miner, "Fatigue Crack Initiation and Propagation in Several Nickel-Base Superalloys at 650°C," International Journal of Fatigue, Vol.5, No.3, July 1983, pp.135-143.

[10] S.R.Holdsworth and W.Hoffelner, "Fracture Mechanics and Crack Growth in Fatigue," High-Temperature Alloys for Gas Turbines, 1982.

RAPIDLY SOLIDIFIED Ni-Mo-Fe-B ALLOY FOR HIGH TEMPERATURE WEAR RESISTANCE APPLICATIONS

Y. J. Park,

AMAX Research and Development Center
Golden, Colorado 80403

S. K. Suh, W. W. Park and C. J. Choi

Korea Institute of Machinery and Metals
Changwon, Korea 641-010

Abstract

Process variables for the production of Ni-Mo-Fe-B amorphous ribbon by melt spinning were studied. Optimum ribbon was obtained at an ejection angle of 75 degrees and an ejection pressure of 4 psi. The rapidly solidified ribbon exhibited a partially crystallized microstructure due to its low boron content. Age hardening behavior of the ribbon was investigated to understand the sequence of phase transformation. The ribbon was then pulverized and consolidated by hot extrusion. Microstructure, hardness, and wear resistance of the extruded alloy were evaluated at room and elevated temperatures and compared with those of commercially extruded RST alloys. The extruded alloy exhibited excellent hot hardness and good wear resistance, which can be attributed to the well dispersed stable Mo_2NiB_2 borides and ultra fine grain size.

Introduction

Since the rapid solidification technology (RST) was first recognized as a means of controlling microstructure in the sixties,[1] a great deal of interest has been generated on the technology with a hope to find commercial utilization of the unique properties of rapidly solidified materials.[2-5] Recently there has been a significant progress made in fabricating amorphous powder into practically useful forms. However, much of the progress still remains inside laboratories, and commercialization of RST materials seems to require a considerable amount of further research and development efforts.

Although the major effort for commercialization of RST materials has been concentrated on the electric and magnetic applications, there has also been a great interest in the development of RST materials for high temperature wear resistance applications. For example, Allied Corporation has developed Ni- Mo-B (Devitrium* 3065) and Ni-Mo-Fe-B (Devitrium 7025) alloys which exhibit high hot hardness, excellent wear resistance, and good corrosion resistance.[6] These properties make the alloys attractive candidate materials for such applications as extrusion dies, cutting tools and a variety of other high temperature applications.[6-10]

The purpose of the present study was to develop a Ni-Mo-Fe-B alloy by melt spinning and hot extrusion for high temperature wear resistance applications. In designing the alloy, boron content was reduced, as compared to Devitrium 7025, in order to study the effect of boron on the microstructure and mechanical properties of the Ni-Mo base alloy.

Experimental Procedures

Melt Spinning

A Ni-Mo-Fe-B alloy was prepared by melt spinning at Korea Institute of Machinery and Metals (KIMM). The composition of the alloy is shown in Table 1 along with the typical compositions of Devitrium alloys (the alloys were received in the as-extruded condition).

TABLE I: CHEMICAL COMPOSITIONS OF THE ALLOYS

Alloy	Composition (Wt. %)			
	Ni	Mo	Fe	B
KIMM Alloy	50.2	37.2	11.3	0.39
Devitrium 7025	53	36	9	2
Devitrium 3065	54	44	-	2

*Registered Trademark of Allied Corporation

For melt spinning, KIMM alloy was superheated to about 100°C above its melting point. The molten metal was ejected by pressurized argon gas through a 0.7-mm diameter circular section quartz nozzle onto a rotating Cu-Be substrate in a chamber filled with argon gas. The cooling rate was 10^6C/sec or higher. The gap between the nozzle and the wheel was 1 cm and the nozzle inclined to the surface of the wheel at an angle of between 60 and 90 degrees. Wheel speed of between 10 and 50 m/sec and ejection pressure of between 3 and 7 psi were employed to study the optimum process variables for melt spinning.

Age Hardening

The melt spinned ribbon samples were heat treated in vacuum or in inert gas atmosphere for 1 hour at 700-1100°C. Microstructural changes occurred during each age hardening treatment were investigated by a transmission electron microscope to study the sequence of crystallization process. Information obtained from this study was used in establishing process parameters for hot extrusion and subsequent aging treatment.

Consolidation by Hot Extrusion

The ribbon was comminuted into powder of a required particle size (approximately 150 μm) by a rapid moving hammer in a pulverizer. The pulverized powder was filled into a stainless steel can, which was then evacuated and sealed. The can measured approximately 80 mm in outside diameter by 180 mm in overall length. The can was preheated in a gas fired furnace for 2 hours at 1100°C and extruded at a reduction ratio of 5.6:1. The extrusion was performed using a 700-ton press.

The extruded bars of KIMM alloy and Devitrium alloys were solution heat treated for 1 hour at 1100°C, water quenched to room temperature, and then aged for 4 hours at 800°C to improve hardness and wear resistance. Metallographic samples were prepared from the aged materials, etched for 10 sec in 50% HNO_3, and examined with a scanning electron microscope. Microstructures were also examined by a transmission electron microscope. Vickers hardness of each material was determined with a 1 kg load at temperatures up to 1000°C. Wear resistance of the alloys was determined at room temperature and at 300°C by a wheel abrasion test. The test consists of pressing a specimen of known volume at a 2.1-kg load against a H13 steel wheel rotating at a constant speed of 0.304 m/sec and measuring volume loss of the specimen after 330 seconds.

Results and Discussion

Rapidly Solidified Ribbon

The key process variables for making amorphous ribbon are orifice diameter, ejection angle, ejection pressure, and roller- surface velocity of melt spinning.[11] In this study, optimum ribbon was obtained with a 0.7-mm diameter orifice at an ejection angle of 75 degrees (angle between roll tangent line and molten jet) and an ejection pressure of 4 psi at a roller-surface velocity of 30 m/sec. The rapidly solidified ribbon exhibited a partially crystallized structure, as confirmed by the electron diffraction pattern shown in Figure 1. This partially crystallized structure is believed to be caused by the low boron content of the alloy.

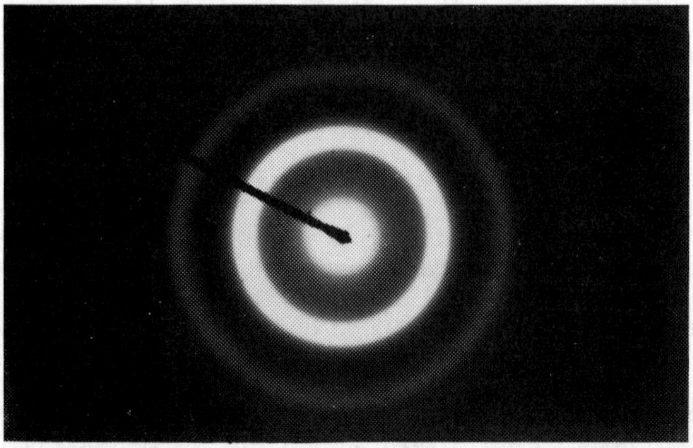

Fig. 1: Electron diffraction pattern of rapidly solidified Ni-Mo-Fe-B ribbon.

Age Hardening

Figure 2 illustrates the microstructural changes occurred in the ribbon samples after each age hardening treatment. The electron diffraction pattern of the specimen heat treated at 700°C revealed a FCC structure with d-spacing and angle equivalent to that of Ni.[12] At 800 and 900°C, metastable Ni_3Mo and beta-Ni_3Mo precipitates, which are known to contribute to the coherency hardening phenomenon,[13] as well as stable Mo_2NiB_2 particles were detected. At 1000 and 1100°C, the Ni_3Mo precipitates were completely dissolved into the Ni matrix, while the stable Mo_2NiB_2 borides were still remained in the matrix. The borides grew to submicron sizes (200 to 500 nm) after the alloy was heat treated at these high temperatures. Microstructural evaluation showed that the borides were isolated and well dispersed in the Ni matrix.

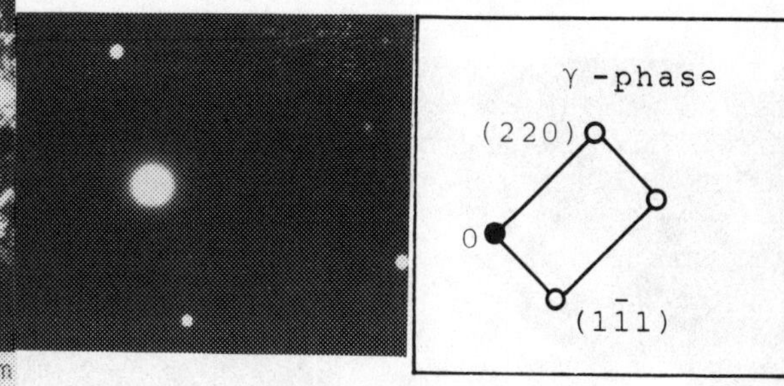

(a) (a')

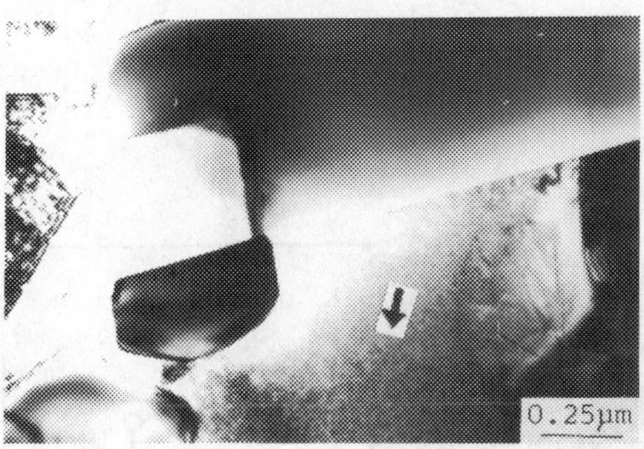

metastable phase

(b)

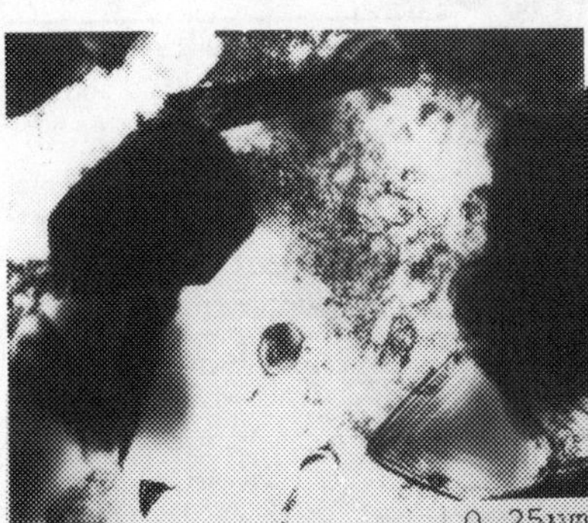

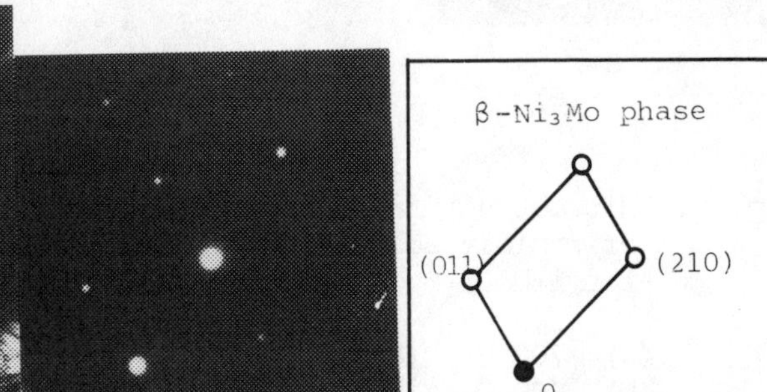

(c) (c')

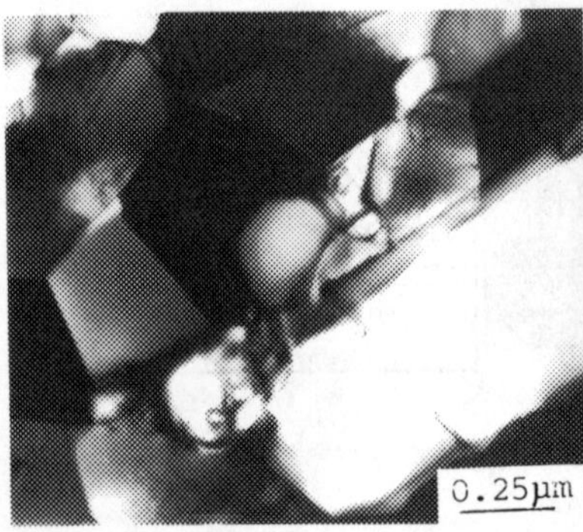

(d)

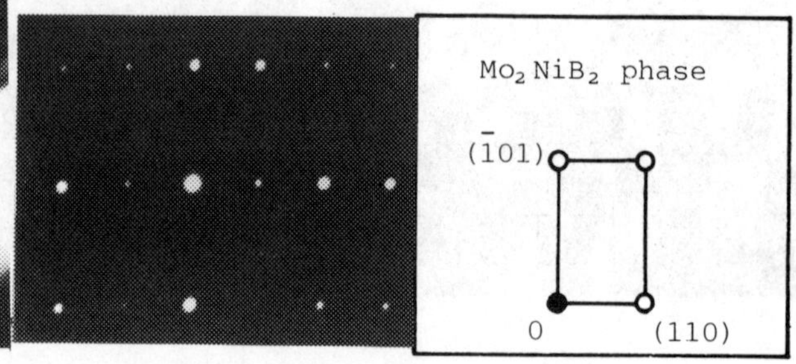

(d')

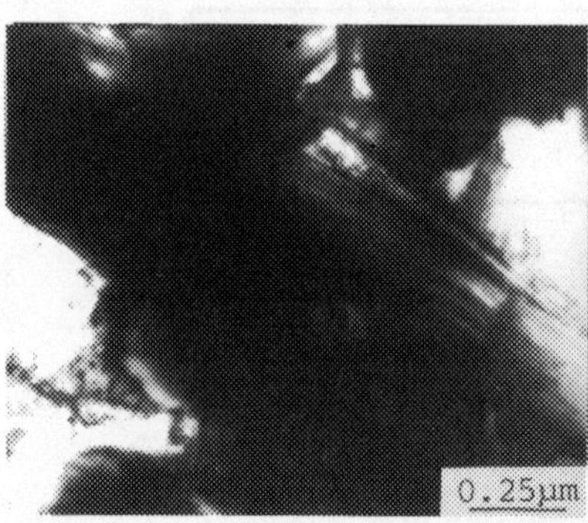

(e)

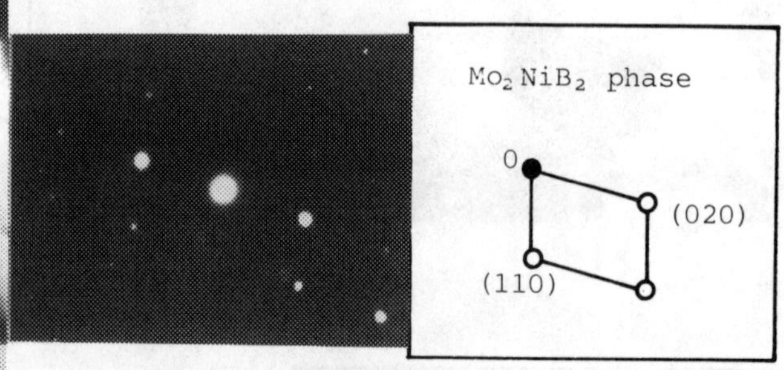

(e')

Fig. 2: Microstructural changes and electron diffraction patterns of rapidly solidified KIMM ribbon after heat treatment for 1 hour

(a) 700°C (a') ED pattern of matrix
(b) 800°C : impossible to take ED pattern of finely dispersed precipitates
(c) 900°C (c') ED pattern of stable fine particle
(d) 1000°C (d') ED pattern of stable coarse precipitate
(e) 1100°C (e') ED pattern of stable coarse precipitate

An ideal age hardening condition is to have precipitates of Ni_3Mo intermetallic compounds as well as well-dispersed and hard borides in the Ni matrix. A material with such a microstructure is expected to exhibit excellent high temperature properties. This microstructure was obtained at the intermediate aging temperatures of 800 and 900°C.

Pulverized Powder

It was noted during pulverizing that the Ni-Mo-Fe-B amorphous ribbon was extremely brittle and broke easily. Since the powder was pulverized from the ribbons which essentially experienced the same thermal history, its microstructure was uniform throughout the thickness and the lot. Figure 3 shows the shape of comminuted powder particles which have a platelet morphology with relatively smooth edges.

Fig. 3: The shape of comminuted powder

Microstructures and Mechanical Properties of Extruded Alloys

The extruded alloy exhibited a very fine grain size. Figure 4 shows the transmission electron micrographs of the three RST alloys which were consolidated by hot extrusion, solution heat treated, and aged for 4 hours at 800°C. The coarse precipitates dispersed in the matrix are Mo_2NiB_2 particles. KIMM alloy exhibits a lesser amount of Mo_2NiB_2 particles due to its low boron content than Devitrium alloys. The matrix also exhibits evidence of finely dispersed Ni_3Mo intermetallic precipitates which contribute to age hardening.

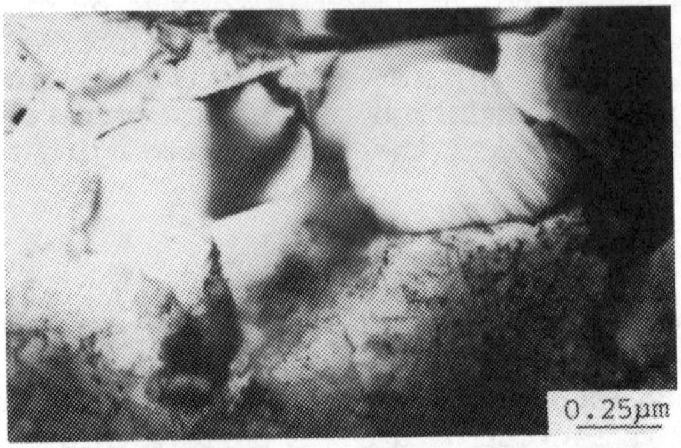

(a) KIMM Alloy

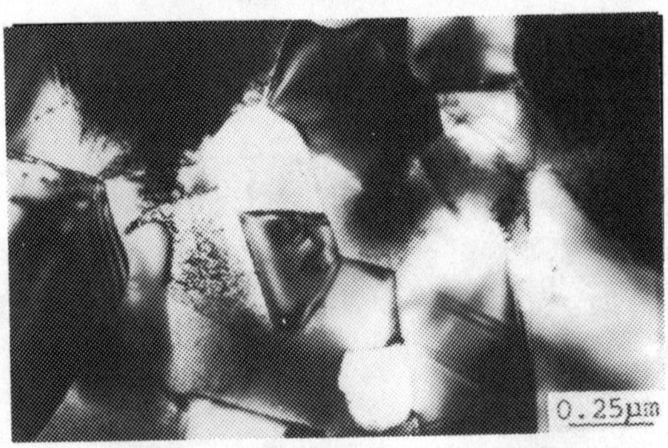

(b) Devitrium 7025

(c) Devitrium 3065

Fig. 4: Microstructures of extruded materials aged at 800°C for 4 hours.

The hot hardness values of the three extruded alloys are shown in Figure 5 as a function of temperature. Devitrium 3065 alloy shows higher hot hardness, due to a higher Mo content, than Devitrium 7025 or KIMM alloy. KIMM alloy exhibits hot hardness values very similar to those of Devitrium 7025 despite its lower boron content.

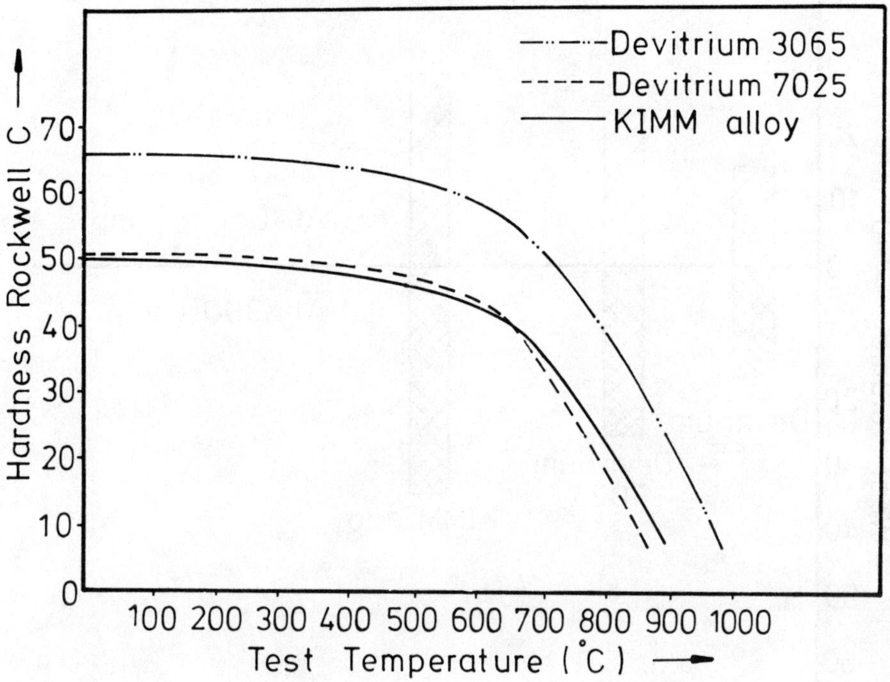

Fig. 5: Hot hardness values of extruded RST alloys as a function of temperature

The results of the wheel abrasion test are shown in Figure 6. Devitrium 3065 exhibits the best wear resistance among the three alloys, which is expected from its high hardness values. KIMM alloy shows less wear resistance than Devitrium 7025. This is believed to be caused by the fact that KIMM alloy has a lesser amount of Mo_2NiB_2 particles due to its lower boron content, which apparently does not affect hot hardness significantly but has an effect on wear resistance. It is also interesting to note here that for a given alloy, the amount of wear at 300°C is only slightly higher than that at room temperature. If this is any indication, the three RST alloys are expected to show good wear resistance up to 600°C because the alloys exhibit relatively constant hardness up to the temperature.

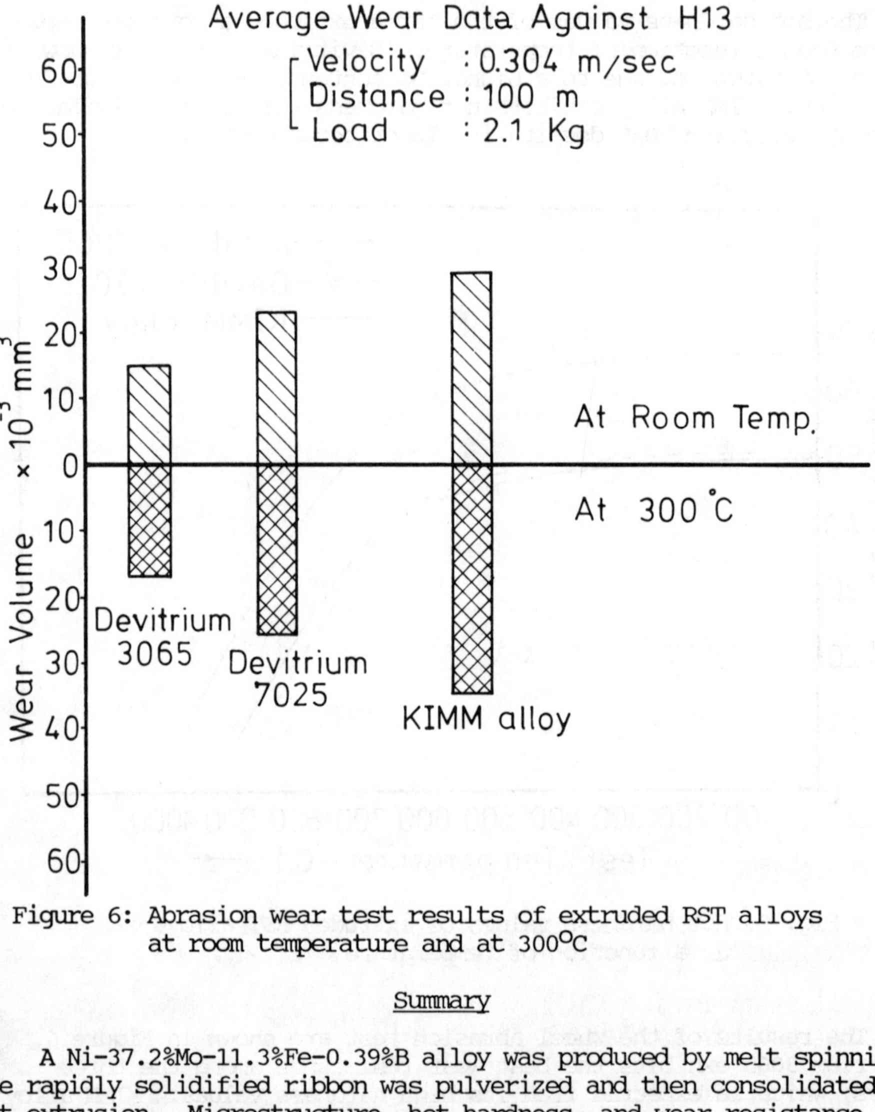

Figure 6: Abrasion wear test results of extruded RST alloys at room temperature and at 300°C

<u>Summary</u>

A Ni-37.2%Mo-11.3%Fe-0.39%B alloy was produced by melt spinning. The rapidly solidified ribbon was pulverized and then consolidated by hot extrusion. Microstructure, hot hardness, and wear resistance of the materials were evaluated, and the following observations were made:

1. The rapidly solidified ribbon exhibited a partially crystallized structure due to its low boron content.

2. When the ribbon was aged at 800 and 900°C, metastable Ni_3Mo and beta-Ni_3Mo precipitates as well as stable Mo_2NiB_2 particles were observed in the microstructure. At aging temperatures of 1000 and 1100°C, the Ni_3Mo precipitates were completely dissolved into the Ni matrix, whereas the stable Mo_2NiB_2 particles were still detected.

3. The extruded material exhibited hot hardness values similar to those of extruded Devitrium 7025 alloy despite its lower boron content, but its wear resistance was slightly less than that of Devitrium 7025. Based on these observations, it is believed that this alloy can be successfully used for high temperature wear resistance applications such as extrusion dies, cutting tools, and bearings used at high temperatures.

References

1. P. Duwez, Trans. ASM, 60 (1967) 607.

2. R.W. Jech, T.J. Moore, T.K. Glasgow and N.W. Orth, J. Metals, 36 (4) (1984) 41.

3. F.E. Luborsky, Amorphous Metallic Alloys, Butterworths Monographs in Materials, (1983) 34.

4. H.H. Liebermann, Mat. Sci. Eng., 43 (1980) 203.

5. S.J. Savage and F.H. Froes, J. Metals, 36 (4) (1984), 20.

6. Ultra Rapidly Solidified Alloys for Dies, Tools and Wear Parts, MPR 39 (5) (1984) 282.

7. A.K. Jena, J. Mat. Sci., 19 (1984), 3121.

8. C.C. Wan, Rapidly Solidified Amorphous and Crystalline Alloys, B.H. Kear ed., Elsevier Science Publishing Co., (1982) 441.

9. High Strength Devitrified Iron-Base Alloys, MPR 37 (5) (1982) 261.

10. RST Alloys, the Dramatic New Standard for Hardness at Elevated Temperatures, Climax Specialty Metals, Cleveland, OH.

11. H. Hillmann and H.R. Hilzinger, Proc. of the International Conference on Rapidly Quenched Metals, (1978) 22.

12. D.B. Snow, Rapidly Solidified Amorphous and Crystalline Alloys, B.H. Kear ed., Elsevier Science Publishing Co., (1982) 523.

13. J.M. Oblak, D.F. Pavlonis and D.S. Duvall, Met. Trans., 5 (1974) 143.

14. P.E. Price and S.P. Kohler, Metals Handbook, Ninth Edition, Vol. 7 (1984) 419.

15. R.G. Bourdeau, C. Adam and E. Van Reuth, Proc. of the 4th International Conference on Rapidly Quenched Metals, Sendai (1981) 155.

SUPERPLASTICITY OF HIPped PM SUPERALLOYS MADE FROM ATTRITED PREALLOY POWDER

N.Uenishi,Y.Takeda and K.Kuroishi

Sumitomo Electric Industries,Ltd.
1-1, Koya-Kita 1-Chome, Itami
Hyogo,664 Japan

Abstract

A powder metallurgical process which enhances the superplasticity of Ni base superalloy was investigated. The process consists of strain energizing of prealloy powder by using an attritor and HIP consolidation of the strain energized powder. The characteristics of the strain energized powder and HIPped billets were examined for IN100, Astroloy and TMP-3 alloys. Hardness of the strain energized powder was above 600mHv; that means the powder was well strain energized and as a result HIPped billets showed a good superplasticity of above 800% elongation. The analysis of flow stress and m-value showed that the processed billet could be deformed at a lower pressure than non-strain energized billets, and suggested the possibility that the processed billet could be deformed at a speed of faster or at a temperature of at least 80°C lower than the non-processed one. The strength and ductility of the heat treated billets were also examined. As a result of eliminating oxygen and iron, the specimen taken from the processed billet showed 117kg/mm² for strength and 8.8% for elongation at 760°C which are the same strength and elongation as non-strain energized billets.

Introduction

Recently there has been a strong demand for the development of high quality gas turbine engines, and to meet this demand Ni base superalloy in which the alloy elements are combined in large amounts has been developed.(1,2) However, since Ni base superalloy is not easy to machine and not easy to forge because of its high strength at high temperatures, a superplastic forging known as GATORIZING is being applied to the production of IN100 alloy turbine engine disks. In this process, a forging preform with fine grains is produced by hot extrusion of superalloy powder.(3) On the other hand, the HIP consolidation technique is also used for the production of the preform, and it has the added benefit of easily producing large preforms with uniform microstructure. However the grain size of the preform that must be extremely fine (less than 5μm) in order to perform superplastic forging tends to be larger than that of the hot extruded preform.

To make the grains finer we provided the powder with strain energy by an attritor before HIP.(4) The reason for giving powder strain energy is to make fine grains by recrystalization during HIP. The idea is based on the fact that fine grain size is obtained by recrystalization when a large amount of strain energy is given to the material.

The purpose of this work is to examine the superplastic character of the preform that is produced by HIP consolidation of strain energized superalloy powder. Since the material is used after heat treatment, the work also attempted to measure the strength and ductility of the billet after heat treatment and to evaluate the optimum condition of producing the preform.

Experimental procedure

1.Strain energizing

The chemical composition of IN100, Astroloy and TMP-3 alloy atomized powder is listed in Table 1. The TMP-3 alloy was developed in a research and development project for future fundamental industrial technology sponsored by the Agency of Industrial Science and Technology of Ministry of International Trade and Industry.

The strain energizing was carried out in a high energy ball mill, namely an attritor, under an argon gas atmosphere. The water cooled chamber of the attritor was filled with 85kg of steel balls (3/8 inch diameter) and 5kg of prealloy superalloy powder. The rotating speed of the agitator was 200rpm and processing time was from 25 min. to 3 hr. Strain energized powders were taken after attriting to investigate hardness. Observing the shape of the strain energized powder was also carried out by SEM. To compare the amount of strain energy, hardness of strain energized powders produced by other methods was also examined.

Table 1 Chemical compositions of Astroloy, IN100 and TMP-3 powder

	Ni	Cr	Co	Mo	Ti	Al	W	C	N	O
Astroloy	bal.	15.09	16.87	5.22	3.50	4.01	0.03	0.03	16ppm	125ppm
IN100	bal.	12.38	18.52	3.53	4.26	4.90	0.03	0.068	54ppm	108ppm
TMP-3	bal.	10.94	7.04	3.10	2.82	3.96	3.39	0.06	17ppm	73ppm

(wt%)

2.HIP

The strain energized powders were poured into mild steel cans, evacuated to less than 10^{-5} torr and then sealed. The sealed can were then consolidated by HIP. HIP was carried out at temperatures of 1030°C for Astroloy and IN100 alloys and 1080°C for TMP-3 alloy, at a pressure of 1500kg/cm^2 for 1 hr. After consolidation of the powder, tensile tests were carried out on the specimens machined from HIPped billets to investigate their superplastic behavior. Total elongation, flow stress and m-value (strain rate sensitivity) were measured at temperatures from 960 °C to 1080°C and strain rates from 3 X $10^{-5}s^{-1}$ to $10^{-2}s^{-1}$. Compression tests were also carried out on billets about ϕ70 X 50 to examine the forgeability.

3.Heat treatment

HIPped billets were annealed at 1220°C for 2 hr and air cooled to obtain large grains. Following this solution heat treatment, a three-step precipitation treatment of 1080°C X 4hr/AC + 843°C X 16hr/AC + 760 °C X 24hr/AC was applied. To examine the final properties of the processed billets, tensile properties were examined at 760°C.

Results and discussion

1. Strain energizing

To obtain fine grains, a lot of strain energy must be given to powder during strain energizing. If the strain energy level is low, grain size won't be small enough, therefore the improvement of material's superplasticity can't be expected. To evaluate strain energy level, the hardness of attrited powder which represents the amount of dislocation density, namely strain energy, was measured. The change in hardness of attrited powder as a function of processing time is shown in Fig.1. A sudden increase in hardness is observed in the early stages of attriting, and after that a gradual increase is observed. Hardness of the strain energized powder was observed to be above 600 mHv, which was 200mHv higher than that of original powder. So it was proved that the powder was deformed and also the powder obtained strain energy gradually. Fig. 2 shows the morphologies of attrited powders. Through collisions with balls, spherical shaped Astroloy powder is deformed and welded together to become about 180μm diam. This process takes 2hr45min. The comparison of strain energizing methods is shown in Table 2. In table 2, the method of our work is listed as dry attritor method, and other methods shown here are wet attritor method in which alcohol was agitated with steel balls and powder, and dry ball mill method in which ball mill with a rotating dram of 90 diam. was used instead of an attritor. Of these three methods, hardness showed the maximum value in the dry attritor method in spite of its short processing time. To efficiently give strain energy to powder, the dry attritor method used in this work could be considered the most appropriate method of these three types of strain energizing methods.

So far the characteristics of strain energized powders were examined, however we should also examine if the amount of strain energy is enough to obtain fine grains and to improve the forgeability of HIP consolidated billets. This will be examined in the next section.

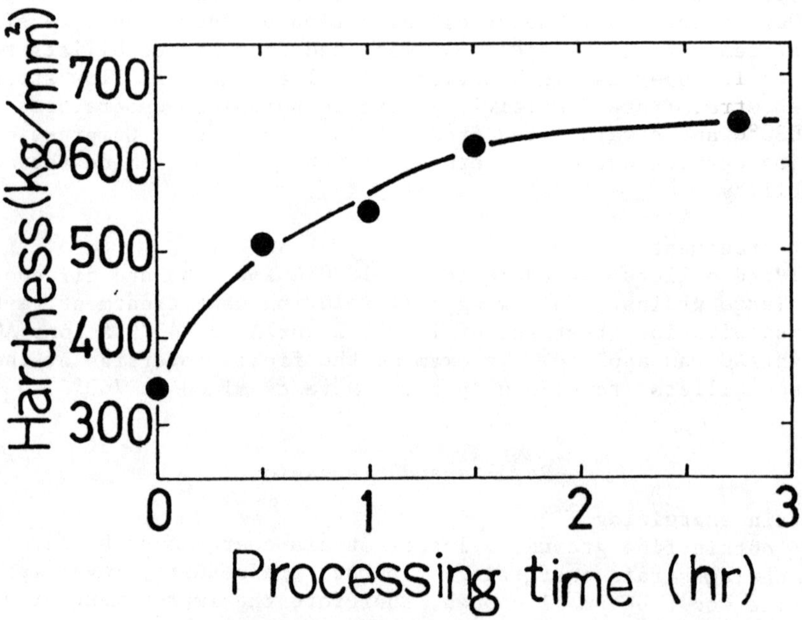

Fig.1 The change in hardness of attrited powder as a function of processing time

Table 2 Hardness of strain energized Astroloy powder

No	Strain energizing method	Processing time	Hardness (mHv)
1	before Strain Energizing	0	339
2	Dry attriter	1hr25min	602
3	Dry attriter	2hr45min	647
4	Wet attriter	10hr	462
5	Dry ballmill	50hr	601

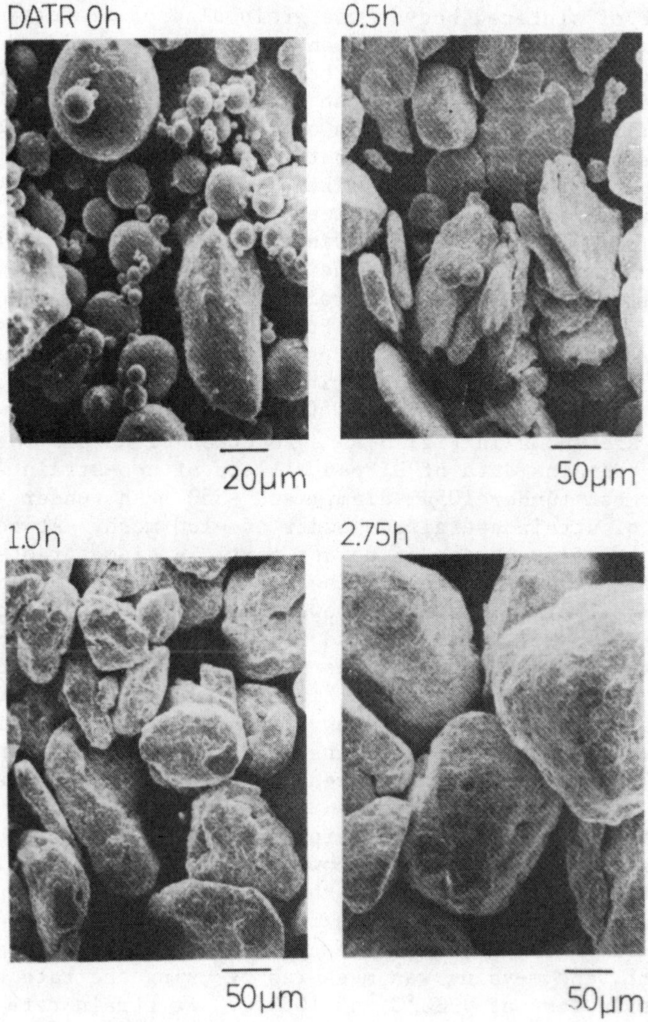

Fig.2 Morphologies of attrited powder

2. Superplasticity of HIPped billets

To make sure if the amount of strain energy is enough to obtain fine grains, model tests were carried out in which IN100 alloy strain energized powder was consolidated by CIP, annealed at various temperatures for 3 hr in a vacuum and then the grain sizes of the sintered body were measured. Fig. 3 shows the influence of annealing temperature on the grain size of sintered body. the grain size of sintered body of strain energized powder was finer than that of non-strain energized powder, and was smaller than 3μm at temperatures below 1150°C. However the size increased sharply at temperatures above 1150°C this indicates that following dissolution of the gamma prime phase, grain coarsing takes place. So HIP temperature should be below the solvus temperature. The fact that grain size of strain energized powder was finer than that of non-strain energized powder indicates the possibility of improvement of superplasticity, since superplasticity is better if grain size of material is finer. The microstructure of original atomized powder and HIPped billet of strain energized powder are shown in Fig.4. Compared with the powder, HIPped billets have a finer uniform microstructure. It doesn't retain dendritic structure because recrystalization has taken place during HIP consolidation.

The flow stress at 960°C and 1040°C of specimens machined from HIPped billet are shown in Fig. 5 as a function of strain rate for IN100 alloy. Fig.5 contains data of HIPped billets of non-strain energized powder of -100mesh (under 105μm diam.) and -350 mesh (under 44μm diam.) size and also of strain energized powder of -100 mesh. At both testing temperatures, the flow stress of powder particle diameter of -350 mesh was lower than that of -100 mesh. The reason for this phenomenon could be that HIPped billet of -350 mesh powder has finer grains compared with that of -100mesh powder. It was also found that the HIPped billet flow stress of strain energized powder was much lower than those of the other two billets. So it was proved that HIPped billet of strain energized powder was easy to deform at low flow stress and the forgeability was improved by means of strain energizing. Fig.6 shows specimens of strain energized IN100 and Astroloy after tensile tests, which showed an elongation of above 800%.

Through the improvement of superplasticity, specimens taken from HIPped billets of strain energized powder showed the possibility of high speed deformation. Fig.7 shows the change in m-value as a function of strain rate. The specimen was taken from the HIPped billet of strain energized powder of TMP-3 alloy. The processing time of strain energizing was 3 hr, and m-value was measured by using the rate changing method on tensile test at 980 °C and 1080°C. At strain rate of $5 \times 10^{-3} s^{-1}$ m-value was above 0.3 which is the minimum value needed to show good superplasticity. So the m-value shown here indicates the possibility of high speed superplastic deformation faster than $5 \times 10^{-3} s^{-1}$.

Following the above tensile tests, compression test were also carried out on HIPped billets of strain energized powders. Fig.8 shows the change in flow stress on the compression test as a function of pressing die stroke. The TMP-3 powder of mean particle diameter of 139μm and 154μm was attrited for 30 min. or 2 hr, and HIPped billet of ⌀70 X 50 mm size was forged superplastically at 1050°C and at strain rate of 2 X $10^{-4}s^{-1}$. Billets were forged successfully with no cracks, and it was also proved that the billet of strain energized powder of finer size or longer attriting time showed lower flow stresses.

Flow stress and deformation speed of the material are very important factors in the actual forging process. Flow stress should be as low as possible from the viewpoint of the capacity of the forging press and the life of die; high speed superplastic forging is profitable from the viewpoint of production cost. The above results indicate the possibility of low flow stress or high speed deformation.

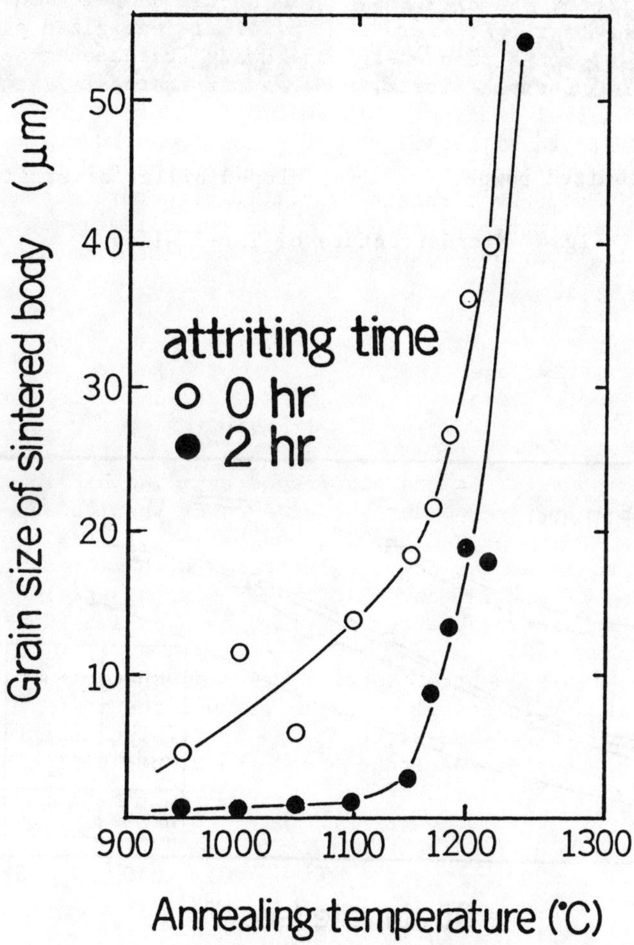

Fig.3 The influence of annealing temperature on the grain size of sintered body

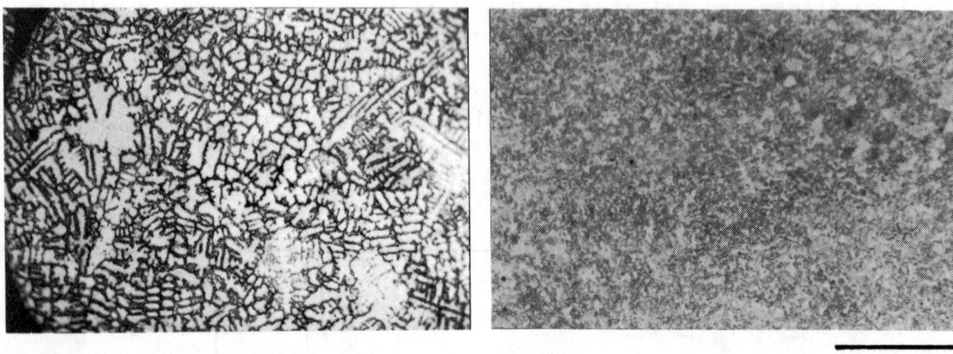

(a) atomized powder (b) HIPped billet of strain energized powder

Fig.4 Microstructure of TMP-3 alloy

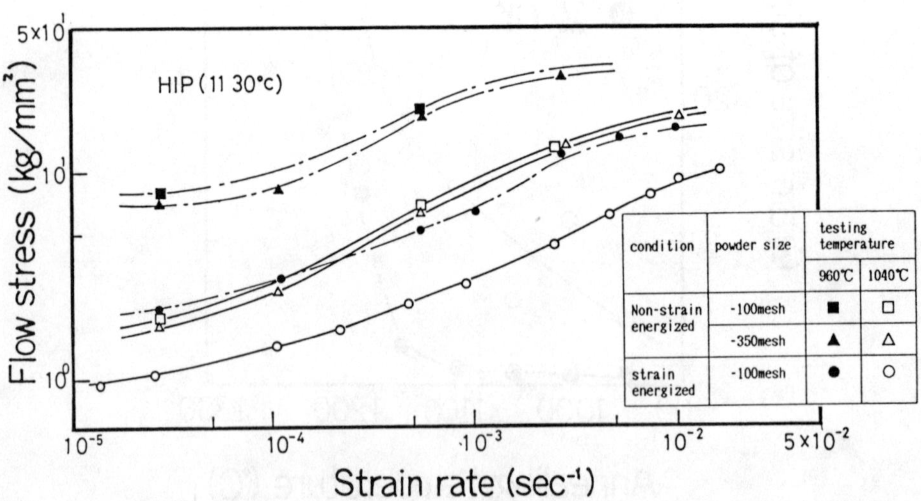

Fig.5 Dependence of strain rate on the flow stress for IN100 alloy

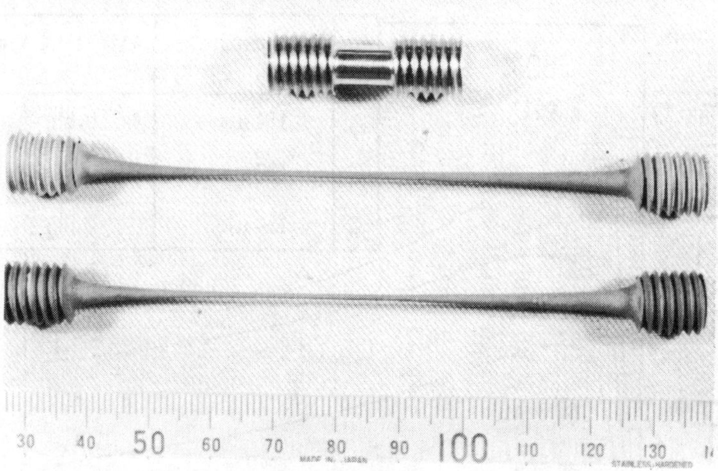

Fig.6 Specimens of IN100 and Astroloy alloy after superplastic tensile test
upper : before test
middle: IN100 after test
lower : Astroloy after test

(Attriting time: 25min for IN100 and 1hr 25min for Astroloy
HIP condition : 1030°C X 1500kg/cm² X 1hr)

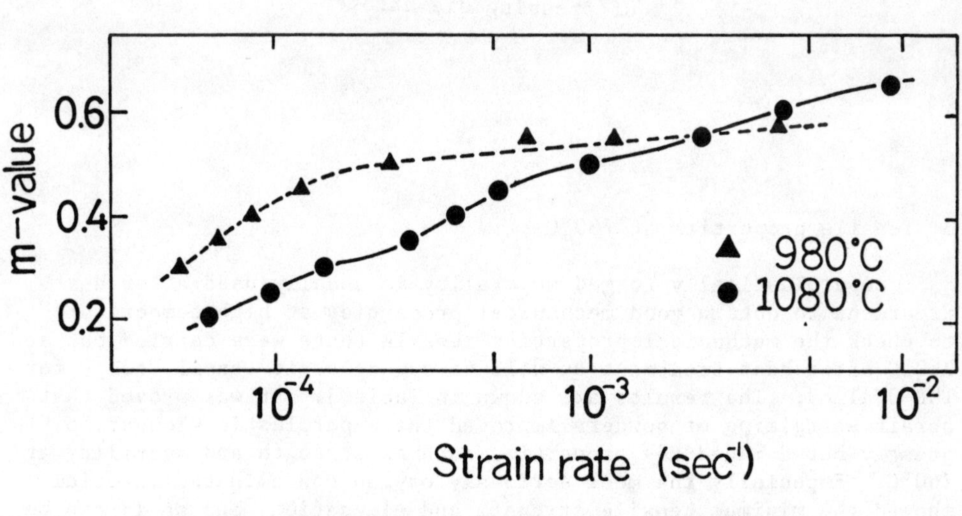

Fig.7 Change in m-value as a function of strain rate

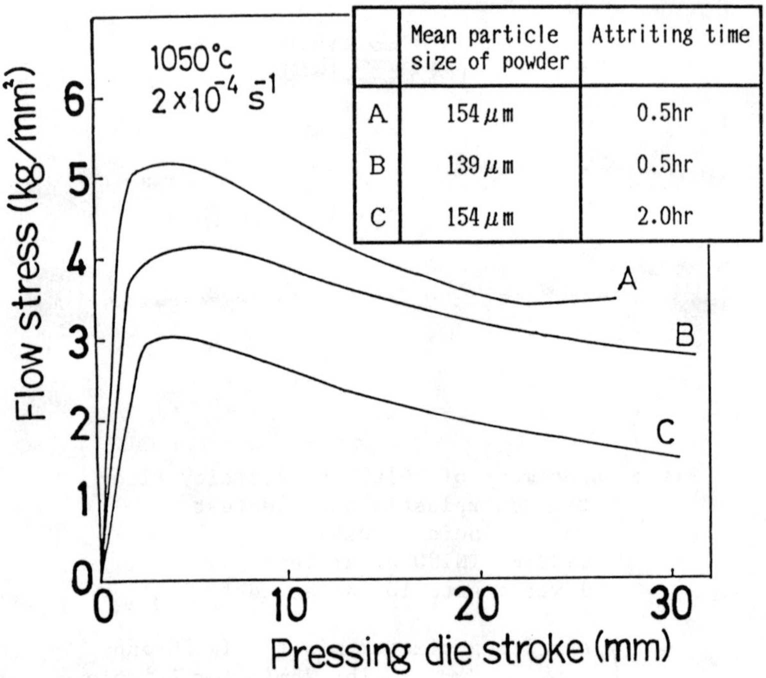

Fig.8 Change in flow stress on the compression test as a function of pressing die stroke

3. Tensile properties at 760°C

Superplastically forged superalloy is usually used after heat treatment to obtain good mechanical properties at high temperature. So to check the mechanical properties tensile tests were carried out at 760°C after heat treatment as well as superplastic tensile tests for TMP-3 alloy. The results are shown in Table 3. It was proved that strain energizing of powders improved the superplastic elongation greatly but damaged the properties such as strength and ductility at 760°C. Especially the most seriously oxygen contaminated material showed the minimum tensile strength and elongation, and so it can be thought that oxygen contamination should be minimum throughout the strain energizing to obtain good mechanical properties at 760°C. To do this, oxygen content in the chamber was reduced to 50 ppm, the oxygen contamination of powder was reduced to a level of 100 ppm. To find other factors that damage the mechanical properties, a micro auger analysis was conducted on strain energized powder. A micro auger analysis revealed an iron contamination shown in Fig. 9. This contaminant supposedly originated from steel balls and a chamber of the attritor.

To improve the mechanical properties a magnetic separation was conducted. A part of strain energized powder that was attached to the magnet and removed was seriously contaminated with iron. The mechanical properties of HIPped and heat treated materials made from three kinds of powders, magnetically cleaned strain energized powder, seriously contaminated strain energized powder and strain energized powder of no magnetic separation, were investigated. The results are shown in Table 4. Oxygen content and iron content were also measured on the three kinds of strain energized powders, of which results are shown in Table 5. As shown in Table 4, the material made from the magnetically cleaned powder showed significant improvement of high temperature strength and ductility, which is almost equal to that of non-strain energized material. On the other hand, the material made from the heavily contaminated powder showed the worst strength and ductility. To conclude, material that has prominent superplasticity and no damage in mechanical properties at 760°C was produced by means of strain energizing and magnetic separation.

Table 3 Superplastic elongations of HIPped billets and mechanical properties of HIPped and heat treated billets

Attriting time (hr)	Oxygen content of strain energized powder (ppm)	Tensile strength at 760℃ (kg/mm^2)	Elongation at 760℃ (%)	R.A. at 760℃ (%)	Superplastic elongation at 1040 ℃ (%)
0	60	117.7	6.3	9.2	< 10
1	150	116.9	1.7	4.7	135 ~ 215
2	1000	75.2	0	0	> 300
3	270	114.6	2.7	5.5	> 300

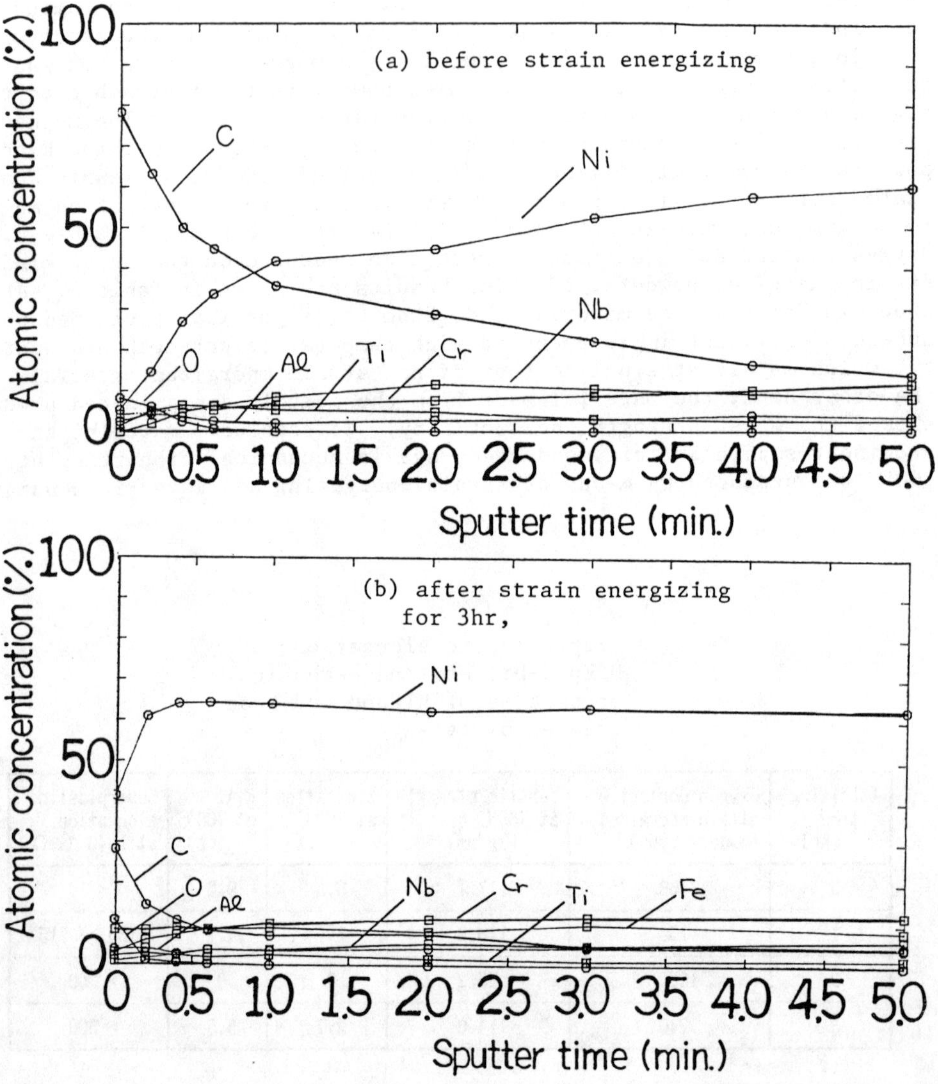

Fig.9 Concentration profile on the surface of TMP-3 powder

Table 4 Mechanical properties of HIPped and heat treated billet of TMP-3 alloy

Attriting time (min)	Magmetic separation	Magmetic attachment	Tensile strength (kg/mm)	Elongation (%)	Reduction of area (%)
20	No	—	96.0	0.7	5.3
20	Yes	attached	76.7	0.0	0.0
20	Yes	unattached	117.8	5.7	8.8
0	No	—	117.7	6.3	9.2

Heat treatment condition : 1,220℃×2hrAC/ 1,080℃×4hrAC
843℃×16hrAC/760℃×24hrAC

Table 5 Oxygen contents and iron content of three kinds of strain energized powder

condition of powder	oxygen content	iron content
powder before magnetic separation	68ppm	0.55%
magnetically cleaned powder	54ppm	0.29%
magnetically removed powder	170ppm	1.20%

Conclusions

1) The superalloy powder of IN100, Astroloy and TMP-3 were strain energized by an attritor. After strain energizing hardness of the strain energized powder was observed to be above 600 mHv, which was 200 mHv higher than that of original powder.
2) HIPped billets of strain energized powder showed good superplasticity. Specimen machined from the billet of strain energized powder showed 800% elongation and suggested the possibility that the forging pressure can be decreased or the forging speed can be increased compared with that of non-strain energizing powder.
3) Oxygen and iron contaminations of strain energized powder were found. To minimize the oxygen contamination, atmosphere in the chamber of an attritor was improved, and to minimize iron contamination magnetic separation was carried out. In the result, mechanical properties of HIPped and heat treated billets of strain energized powder were improved. Tensile strength was 117.8kg/mm^2 and elongation was 8.8% at 760°C.

Acknowledgments

This work was performed under management of Research and Development Institute of Metals and Composites for Future Industries as a part of R & D of Basic Technology for Future Industries sponsored by Agency of Industrial Science and Technology, MITI.

References

(1) Gerald I. Friedrman, George S. Ansell: The Superalloys, (1972) 427
(2) A. Lowley: Int. J. Powder Met. & Powder Tech., 13 (1977) 3,169
(3) A. Y. Kandeil, W. Wallace, J. P. A. Immarigeon : Can Metall. Q. (1980) 245
(4) J. M. Larson, F. A. Thompson : Superalloys Metallurgy and Manufacture (1976) 483

Chapter Nine

Rapidly Solidified Materials

CONSOLIDATION OF RAPIDLY SOLIDIFIED MAGNESIUM ALLOY POWDERS

S. Krishnamurthy and Young-Won Kim

Metcut-Materials Research Group
P.O. Box 33511
Wright-Patterson AFB, OH 45433-0511

Abstract

Recently there has been considerable interest in the development of improved magnesium alloys via rapid solidification/powder metallurgy (RS/PM) approach. While a number of investigations have dealt with the microstructural aspects of RS, studies addressing the consolidation of these novel alloys are limited. In the present study, the consolidation characteristics of inert gas atomized powders of three magnesium alloys have been investigated. The RS magnesium alloy powders were canned, degassed, hot pressed, and extruded using a 12:1 extrusion ratio at temperatures in the range 200-350°C (390-660°F). The extruded alloys were characterized by optical microscopy, SEM, and TEM techniques. Further, these alloys were compared by using tension and compression tests, and polarization measurements. The results of microstructural characterization, mechanical property tests, and corrosion measurements are presented, and the effects of processing parameters are discussed.

Introduction

Although magnesium is the lightest structural metal available, the use of magnesium and its alloys is limited in structural applications because of their low strength and poor corrosion resistance. Recently, efforts have been directed to improve the mechanical properties and corrosion characteristics of magnesium alloys. A significant innovation is the development of fluxless melting technique in the magnesium foundry industry to produce very high purity alloys. It has been reported that the high purity AZ91 alloy produced by this improved melting practice can resist salt water corrosion better than carbon steel or 380 aluminum [1]. Other efforts towards improving the properties of magnesium include the application of rapid solidification (RS) and powder metallurgy (PM) processing. It has been demonstrated that attractive combinations of properties can be achieved in novel Mg-Al-Zn alloys containing rare earth elements [2, 3] as well as in conventional alloys such as AZ91 [4] by using RS approach.

The application of PM methods to magnesium alloy systems is not new. Early work on PM magnesium alloys has been reviewed elsewhere [5]. Initial work examined the compaction of band saw chips and turnings. This was followed by extensive work on atomization to produce prealloyed powders and powder compaction by extrusion [5, 6]. The atomized "pellets" were powder particles which ranged between 100 μm and 400 μm in diameter. Atomization of a series of alloys including AZ31, M1, ZK60, and AMZ111 was carried out and the powders were then extruded to produce a variety of product shapes [5]. The tensile and compressive properties of these PM extrusions were generally higher than those of billet extrusions [5, 6]. Another study [7] dealt with blended elemental approach to produce Mg-Al alloys containing up to 10 wt%Al. The method consisted of blending fine powders of magnesium and aluminum followed by conventional pressing, sintering, and coining operations. This resulted in compacts having about 99% theoretical density [7]. Later work on consolidation of prealloyed ZK60A powders (-25 mesh and -35 mesh size) obtained by the rotating electrode process (REP) also showed that significant strengthening can be achieved by the PM route [8]. More recently, consolidation of RS ribbons produced by melt spinning has yielded notable property improvements in conventional as well as novel magnesium alloys [2-4, 9].

Previous studies [5, 6, 8] on PM processing of magnesium alloys employed somewhat coarse powders (-25 mesh and -35 mesh size) for consolidation work primarily to minimize problems associated with oxide present on powder particle surface. The objective of the present study was to investigate the consolidation behavior of more rapidly solidified finer powders (-170 mesh size) having more refined microstructures compared to coarse powders.

Experimental

The alloys Mg-3Nd-1Pr-2Mn, Mg-12Al-4Zn-0.5Mn, and Mg-9Li-3Al-2Si were prepared by induction melting using high purity elemental components. They were converted to powder by remelting followed by ultrasonic gas atomization (USGA) using argon. Table 1 lists the actual composition of the -170 mesh size powders used for consolidation. It is noted that the heavy metal impurities are within acceptable limits. The oxygen content of the lithium-containing alloy powder was very high.

TABLE 1. CHEMICAL ANALYSIS OF MAGNESIUM ALLOY POWDERS (-170 mesh)

Nominal Composition	Analyzed Composition, wt % (bal. Mg)												
	Nd	Pr	Al	Zn	Li	Mn	Si	Fe	Cu	Ni	Na	K	O
Mg-3Nd-1Pr-2Mn	3.14	1.09	-	-	-	1.54	0.009	0.011	0.004	<0.002	0.001	<0.001	0.14
Mg-12Al-4Zn-0.5Mn	-	-	12.25	3.93	-	0.13	0.036	0.003	0.003	<0.002	<0.001	<0.001	0.13
Mg-9Li-3Al-2Si	-	-	2.52	-	7.40	0.004	1.64	0.006	0.002	0.056	0.1	0.01	2.40

The alloy powders were filled in aluminum cans and degassed under dynamic vacuum conditions for 1 hr at 300°C (570°F). After degassing, the powders were preheated at 250°C (480°F) for 1 hr and hot pressed at that temperature. The hot pressed compacts were extruded into round bar extrusions using a 12:1 extrusion ratio at either 225°C (435°F) or 350°C (660°F). These consolidation steps are indicated in the flow chart in Figure 1.

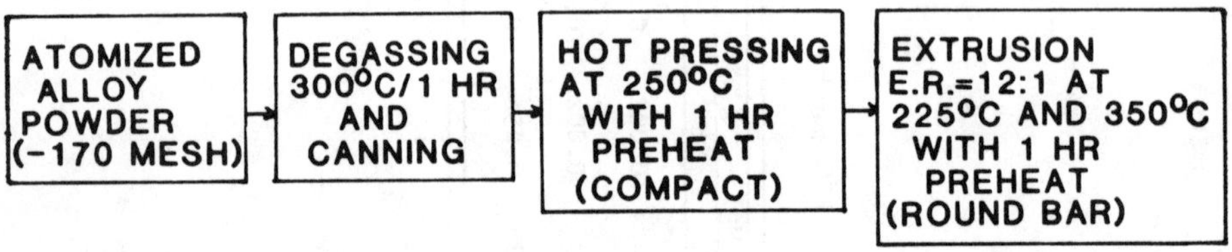

Figure 1 - Flow chart showing the various steps used in the present work for consolidation of magnesium alloy powders.

The microstructures of as-produced powders, hot pressed compacts, and extrusion bars were examined by optical microscopy, SEM, and TEM techniques. The extruded bars were machined to produce tensile test specimens having cylindrical gage sections and cylindrical compression specimens. Tensile and compression tests were conducted at room temperature using a strain rate of $6.7 \times 10^{-5}\ s^{-1}$. The corrosion behavior of the extruded alloys was studied by standard anodic polarization tests in an electrolyte of 0.01M NaCl solution. This electrolyte was chosen to limit the rate of dissolution of the alloys. The tests were conducted at room temperature by using a calomel reference electrode and sweeping over a potential range from -1.8V to 1.2V with a scan rate of 1.0 mV/s. All the specimens were equilibriated for an hour to obtain a near steady-state potential. The corrosion rates of the extruded alloys were estimated based on the Tafel slopes of the polarization curves.

Results

Figure 2 shows the characteristics of the three powders used. The size fraction selected should have particles ranging up to 98 µm diameter. However, the alloy powders containing rare earth elements are finer than the other alloy powders. Mg-12Al-4Zn-0.5Mn powders are irregularly-shaped, whereas Mg-9Li-3Al-2Si powders are nearly spherical particles with a wide size distribution. The microstructures of the powder particles displayed dendritic features, Figure 3. Degassing and hot pressing of the powders led to fully dense compacts as indicated in Figure 4. However, the hot pressed compacts fractured easily along powder particle boundaries during cutting.

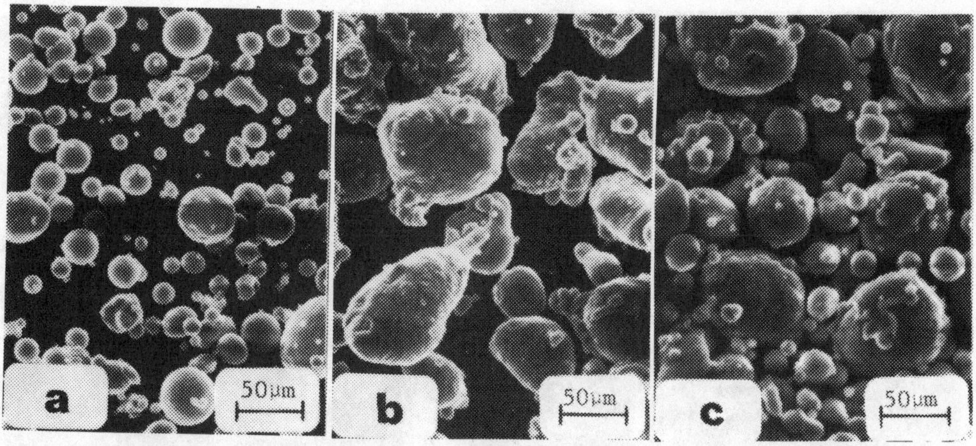

Figure 2 - SEM micrographs showing powder characteristics: (a) Mg-3Nd-1Pr-2Mn, (b) Mg-12Al-4Zn-0.5Mn, and (c) Mg-9Li-3Al-2Si.

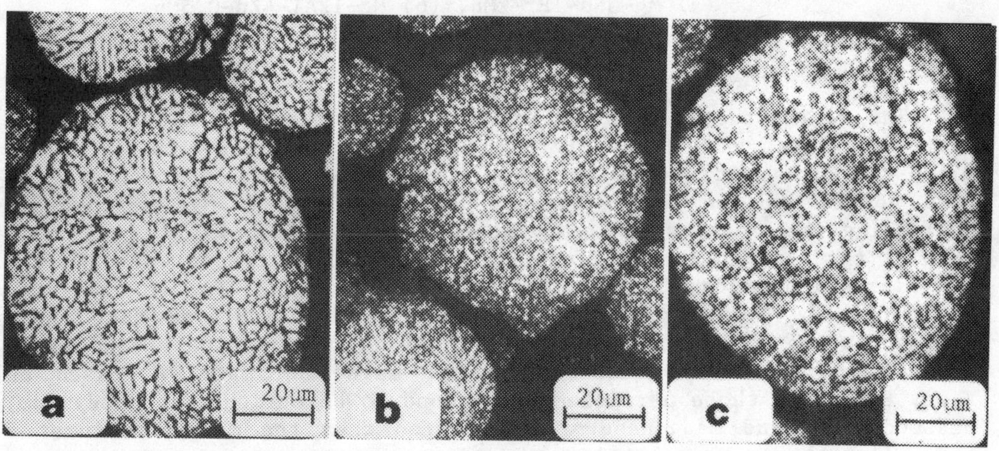

Figure 3 - Optical micrographs showing the structure of powder particles: (a) Mg-3Nd-1Pr-2Mn, (b) Mg-12Al-4Zn-0.5Mn, and (c) Mg-9Li-3Al-2Si.

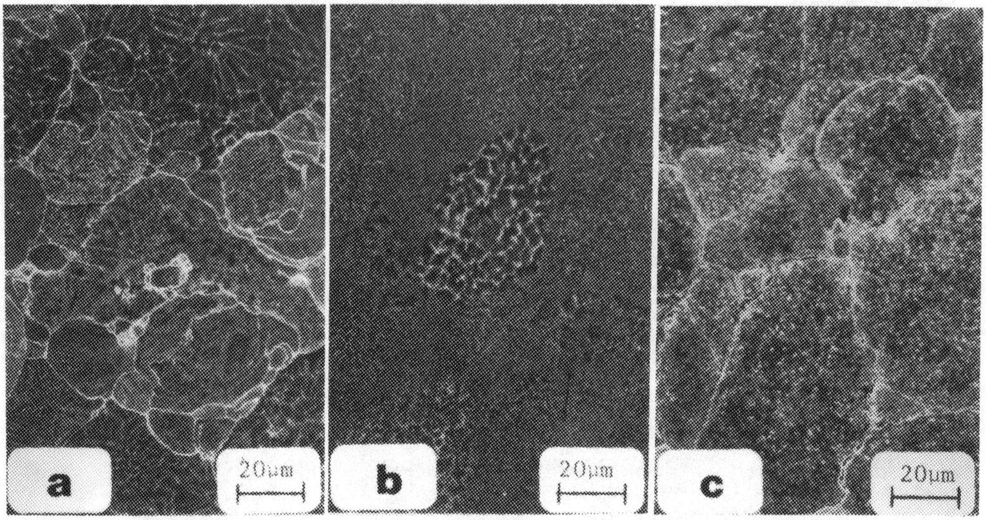

Figure 4 - SEM microstructures of hot pressed compacts: (a) Mg-3Nd-1Pr-2Mn, (b) Mg-12Al-4Zn-0.5Mn, and (c) Mg-9Li-3Al-2Si.

The microstructures of PM extrusions are shown in Figure 5. The extruded bars of Mg-3Nd-1Pr-2Mn alloy exhibited fully dense microstructures with very little evidence of voids. In contrast, the void population was noticeable in the other two alloys (Figure 5). Increasing the extrusion temperature from 225°C (435°F) to 350°C (660°F) resulted in more uniform but coarser microstructures. SEM and TEM micrographs of the alloys extruded at 225°C (435°F) are presented in Figure 6. The alloy containing rare earths displayed the finest microstructure with a large volume fraction of very fine second phase particles distributed uniformly in the magnesium matrix. Based on X-ray EDS data, these particles were found to be enriched in neodymium, praseodymium, and manganese. It is noted that the microstructure of this alloy showed very little coarsening when the extrusion temperature was increased to 350°C (660°F). The extrusions of Mg-12Al-4Zn-0.5Mn alloy consisted of a two-phase microstructure of magnesium matrix solid solution and coarse particles of a ternary phase of magnesium, aluminum, and zirconium. This microstructure coarsened further at the higher extrusion temperature. In the case of Mg-9Li-3Al-2Si alloy extrusions, the microstructure was composed of magnesium and lithium solid solutions and coarse Mg_2Si particles. This microstructure also coarsened as the extrusion temperature was raised.

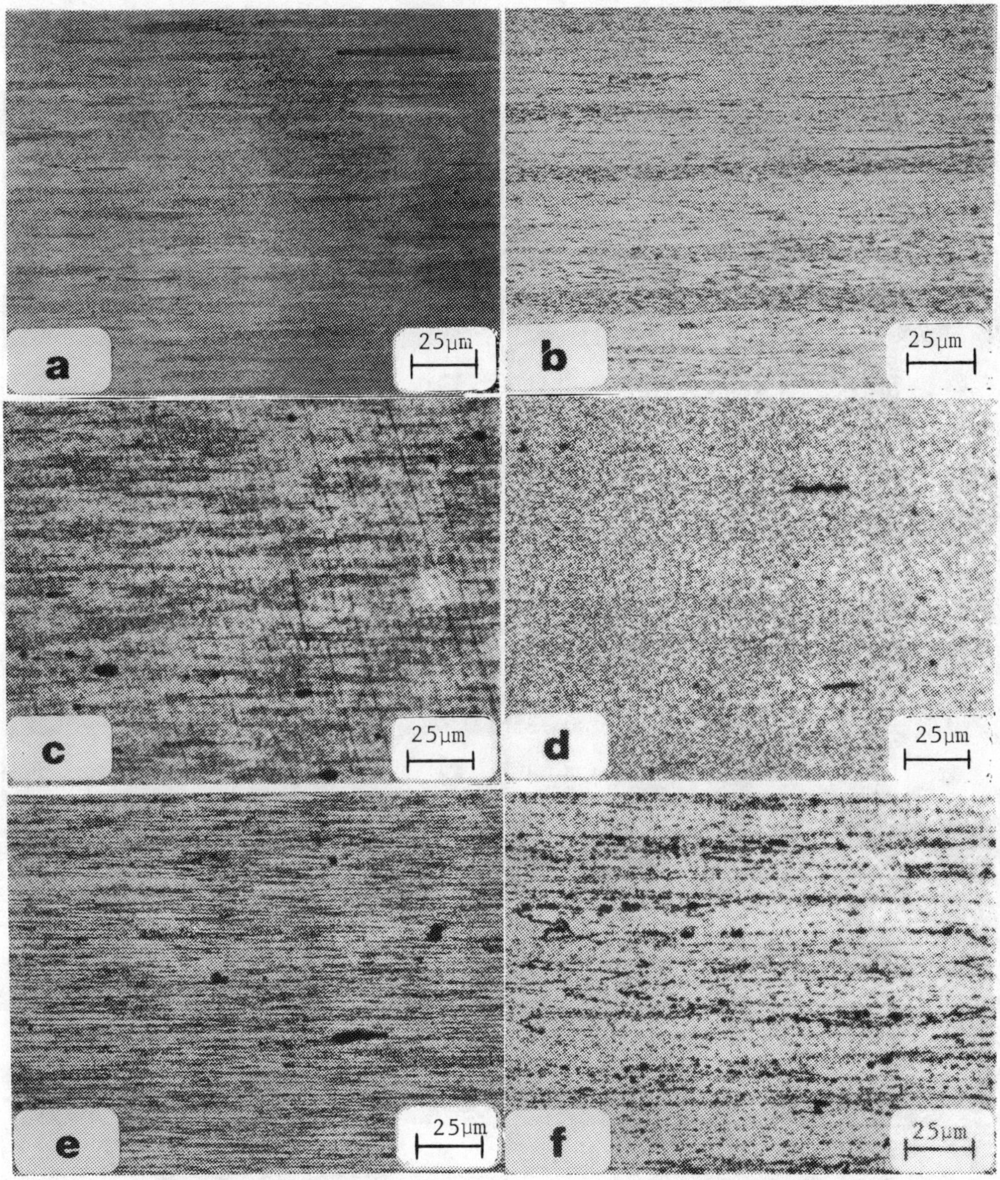

Figure 5 - Optical micrographs of Mg alloy bars extruded at 225°C (left) and 350°C (right): (a, b) Mg-3Nd-1Pr-2Mn, (c, d) Mg-12Al-4Zn-0.5Mn, and (e, f) Mg-9Li-3Al-2Si.

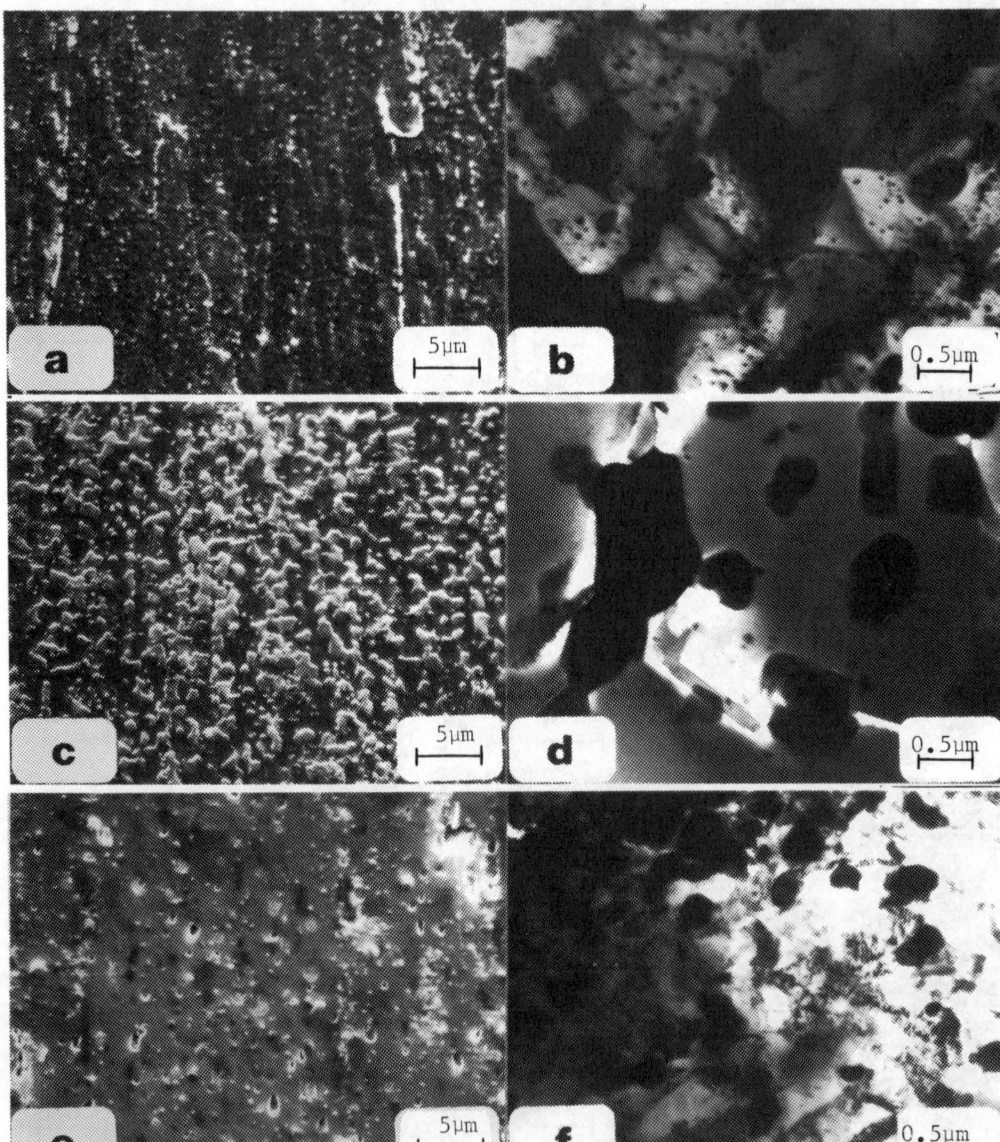

Figure 6 - SEM (left) and TEM (right) micrographs of PM magnesium alloy bars extruded at 225°C: (a, b) Mg-3Nd-1Pr-2Mn, (c, d) Mg-12Al-4Zn-0.5Mn, and (e, f) Mg-9Li-3Al-2Si.

The results of tension and compression tests at room temperature are shown in Table 2. Tensile tests of the lithium-containing alloy extrusions were not successful because of premature failure of the specimens. The Mg-3Nd-1Pr-2Mn alloy extrusions showed reasonably high yield strengths and good elongation, whereas the extrusions of Mg-12Al-4Zn-0.5Mn showed similar strength levels and considerably lower elongation values. In both of these materials, the yield strength was found to be higher in compression than in tension (Figure 7).

TABLE 2. ROOM TEMPERATURE TENSILE AND COMPRESSIVE PROPERTIES OF POWDER METALLURGY Mg ALLOY EXTRUSIONS

Alloy	Extrusion Tempera- ture, °C	0.2% Tensile Yield Strength, MPa (ksi)	Ultimate Tensile Strength, MPa (ksi)	Elongation, %	RA, %	Compressive Yield Strength, MPa (ksi)
Mg-3Nd-1Pr-2Mn	225	349 (50.6)	354 (51.4)	7.7	15.7	360 (52.3)
Mg-3Nd-1Pr-2Mn	350	310 (45.0)	325 (47.1)	13.3	15.0	342 (49.6)
Mg-12Al-4Zn-0.5Mn	225	308 (44.7)	342 (49.7)	1.7	2.7	360 (52.2)
Mg-12Al-4Zn-0.5Mn	350	291 (42.3)	338 (49.1)	2.0	3.0	322 (46.7)
Mg-9Li-3Al-2Si	225	N/A	237 (34.4)	0.5	N/A	293 (42.5)
Mg-9Li-3Al-2Si	350	N/A	206 (29.9)	0.4	N/A	254 (36.9)

[a]N/A = not available because of premature failure.

The estimated corrosion rates of the extruded alloys are listed in Table 3. Among the three compositions studied, the alloy containing rare earth additions displayed the highest corrosion resistance. In all cases, the corrosion rates were lower for the higher temperature extrusions.

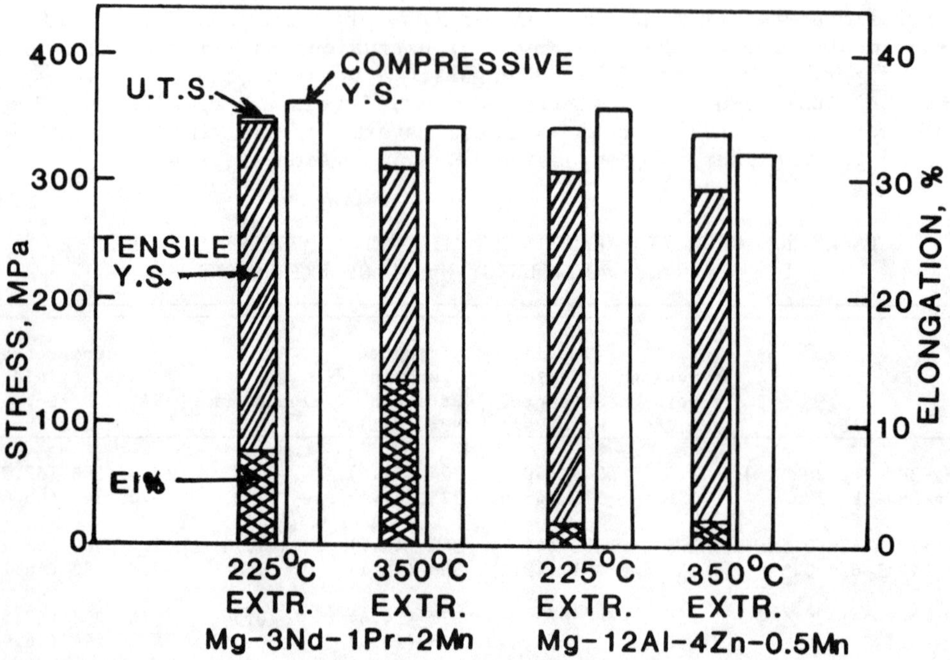

Figure 7 - Tensile and compressive properties of Mg-3Nd-1Pr-2Mn and Mg-12Al-4Zn-0.5Mn extrusions.

TABLE 3 - CORROSION RATES OF MAGNESIUM ALLOY POWDER METALLURGY EXTRUSIONS IN 0.01M NaCl SOLUTION

Alloy Composition	Corrosion Rate, mpy[a] 225°C Extrusion	350°C Extrusion
Mg-3Nd-1Pr-2Mn	12	11
Mg-12Al-4Zn-0.5Mn	69	31
Mg-9Li-3Al-2Si	127	27

[a]Based on Tafel regions of polarization curves.

Discussion

The present results have shown that consolidation processing of more rapidly solidified, fine magnesium alloy powders is essentially similar to that of coarser powders or RS ribbons. However, the role of process parameters is more critical in the case of fine powders because of increased surface area and decreased compactability. It should be pointed out that the strength levels obtained in this work are somewhat lower than those reported for the extrusions of coarser ZK60A powders (375MPa YS, 400MPa UTS, and 17% elongation) [8] and RS ribbons of various magnesium alloys [2-4] (e.g., 435MPa YS, 475MPa UTS, and 14% elongation for AZE555 alloy). Direct comparison of these data is difficult in view of the compositional differences between the alloys used. Clearly, the degassing step for processing fine powders requires further optimization, particularly when the oxygen content of the powder is high. The presence of excess oxide in the powder is deleterious as evidenced by the results for Mg-9Li-3Al-2Si extrusions (Figure 5). With regard to extrusion parameters, the role of extrusion temperature has been considered in this work. Although the microstructures coarsened somewhat at the higher extrusion temperature, the mechanical properties showed only a slight decrease (Figure 7). More importantly, the corrosion resistance was found to be higher in the alloys extruded at 350°C (660°F). Thus, this higher extrusion temperature appears to be adequate.

Mg-3Nd-1Pr-2Mn alloy extrusions exhibited a good combination of mechanical properties and corrosion resistance. Further improvements in the properties should be possible with process optimization. For example, tensile strength levels exceeding 415MPa (60 ksi) were obtained in this alloy when streamlined dies were used for extrusion [10]. The attractive corrosion resistance of this alloy is consistent with the earlier observation of pseudo-passivation and lower corrosion rates in binary magnesium-rare earth alloys [11].

Conclusions

1. Mg-3Nd-1Pr-2Mn PM extrusions displayed a desirable combination of high mechanical properties (equal to those of conventional ZK60A-T5 extruded alloy) and good corrosion resistance because of refined and homogeneous microstructures.

2. Mg-12Al-4Zn-0.5Mn PM extrusions exhibited high strength but low values of elongation and corrosion resistance. This behavior is attributed to coarse second phase regions (enriched in both aluminum and zinc) present in the magnesium matrix.

3. Mg-9Li-3Al-2Si PM extrusions contained excess oxide because of the high oxygen content of the starting powder. This led to brittleness and premature failure of tensile specimens. The microstructure consisted of magnesium and lithium solid solutions and

coarse Mg_2Si particles. The corrosion resistance of this alloy was also low.

4. In all these PM alloys, the yield strength was higher in compression than in tension.

5. Good corrosion resistance of the alloy containing rare earth additions is related to pseudo-passivation behavior observed in binary magnesium-rare earth type alloys.

6. Further improvements in properties of these alloys should be possible with process optimization, e.g., longer degassing cycle and a higher extrusion ratio.

Acknowledgments

This work was supported by the U.S. Air Force Wright Aeronautical Laboratories, Materials Laboratory, under Contract No. F33615-85-C-5072. The authors would like to thank Dr. F. H. Froes for his support during this program. Experimental assistance by W. A. Houston and manuscript preparation by Miss Karen A. Sitzman are acknowledged.

References

1. K. N. Reichek, K. J. Clark, and J. E. Ellis, "Controlling the Salt Water Corrosion Performance of Magnesium AZ91 Alloy," SAE Technical Paper No. 850417 (March 1985).

2. S. K. Das, C. F. Chang, and D. Raybould, "High Performance Magnesium Alloys by Rapid Solidification Processing," Light Metal Age (December 1986), pp. 5-8.

3. H. Jones, A. Joshi, R. G. Rowe, and F. H. Froes, "The Current Status of Rapid Solidification of Magnesium-Base and Titanium-Base Alloys," The Int. J. Powder Metallurgy, Vol. 23, No. 1 (1987), pp. 13-24.

4. G. Nussbaum, H. Gjestland, and G. Ragazzoni, "Rapid Solidification of Magnesium Alloys," in Magnesium: A Strategic Material, International Magnesium Association, McLean, VA (1988), pp. 19-24.

5. R. S. Busk and T. E. Leontis, "The Extrusion of Powdered Magnesium Alloys," Trans. AIME, Vol. 188, No. 2 (1950), pp. 297-306.

6. R. S. Busk, "The Pellet Metallurgy of Magnesium," Light Metals (July 1960), pp. 197-200.

7. S. Storchheim, "Magnesium Powder Metallurgy," The Int. J. Powder Metallurgy, Vol. 8, No. 3 (1972), pp. 115-126.

8. S. Isserow and F. J. Rizzitano, "Microquenched Magnesium ZK60A Alloy," Int. J. Powder Metallurgy and Powder Technology, Vol. 10, No. 3 (1974), pp. 217-227.

9. M. C. Flemings and A. Mortensen, "Rapid Solidification Processing of Magnesium Alloys," AMMRC Technical Report, TR-84-37 (September 1984).

10. S. Krishnamurthy, I. Weiss, and F. H. Froes, "Consolidation of Rapidly Solidified Magnesium Alloy Powder," presented at the International Conference on PM and Related High Temperature Materials (P/M-87), Bombay, India (December 1987); proceedings in press.

11. S. Krishnamurthy, E. Robertson, and F. H. Froes, "Magnesium-Neodymium Alloys Produced by Rapid Solidification," in Processing of Structural Metals by Rapid Solidification, edited by F. H. Froes and S. J. Savage, ASM International, Metals Park, OH (1987), pp. 399-408.

FABRICATION OF SHAPED CHARGE LINERS BY RAPID SOLIDIFICATION TECHNIQUES AND P/M PROCESSES

C. Gélinas*, R. Angers* and B. Voyzelle#

* Laval University, Dept of Mining & Metallurgy, Québec, Canada G1K 7P4

Defense Research Establishment Valcartier Québec, Canada G0A 1R0

Abstract

The performance of shaped charges is very dependent on the ability of the liner material to deform and elongate without fracturing during the jet formation (high ductility when subjected to dynamic stressing). Best performances are obtained with materials having fine and homogeneous microstructures and high density and/or high ductility under dynamic loading. In the present work, rapid solidification techniques and powder metallurgy processes were combined to fabricate shaped charge liners with fine and homogeneous microstructures. A rapidly solidified superplastic Zn-20.5Al-0.70Cu alloy was produced by melt-spinning in the form of metallic ribbons. These ribbons were thereafter pulverized into powders in a specially designed blade cutter mill. Powders with selected particle sizes were then isostatically cold pressed into liner preforms which were finally hot isostatically pressed. Results and details of the fabrication process are reported and microstructures are presented and discussed.

Introduction

The term "shaped charge" originates towards the end of the 18th century when mining engineers discovered that creating a cavity at the end of a high-explosive charge opposite to the point of initiation, i.e., shaping the charge, increased the penetration. This "cavity" or "hollow" effect is known as the Munroe effect in the United States and United Kingdom [1], the Neumann effect in Germany and the cumulation effect in Russia [2]. Eichelberger [3] credits R.W. Wood [4] with the recognition in 1936 of the usefulness of metallic liners in the cavity charge to enhance the penetration capability of the charge and to produce fragments of extremely high velocity. The penetration improvements of these different charges are illustrated in figure 1.

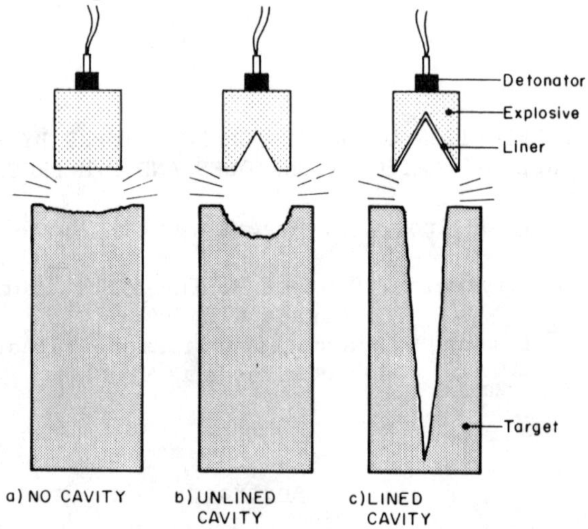

Figure 1: Penetration of various explosive charges.

Shaped charges can take various geometries. In the present work, conical shaped charges as shown in figure 2 were studied. The principle of operation of these devices is illustrated in figure 3 [5]. Upon initiation of the charge, a detonation wave propagates and impinges on the liner. The high pressures produced cause the liner to behave as an inviscid fluid and force it to collapse on the axis of symmetry. When the liner material reaches the axis of symmetry, the inner layer of material (approximately 20% of the liner mass) forms a high velocity jet, while the remainder of the mass forms a low velocity slug. As the shaped charge jet forms, it continuously lengthens due to a velocity gradient (7 to 10 km/s at the tip to 2 km/s at the tail) until it breaks up.

Figure 2: Conical shaped charge device.

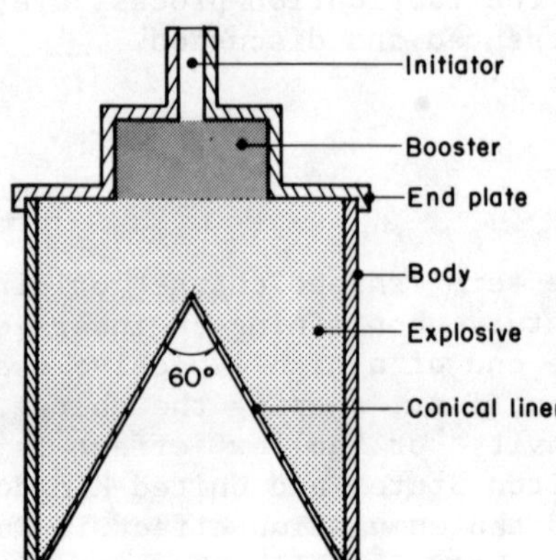

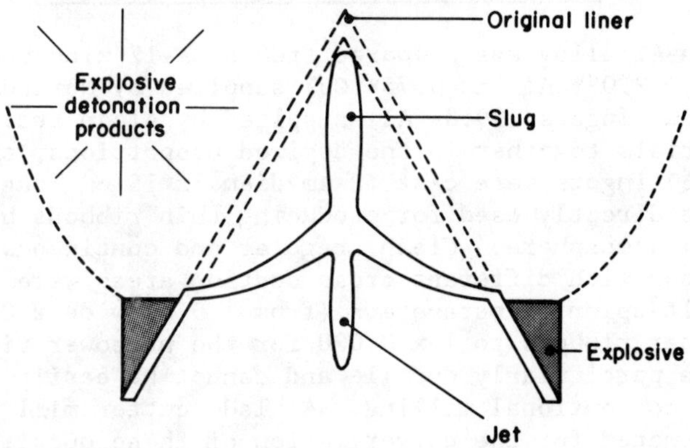

Figure 3: Shaped charge jet and slug formation.

Improvements of the shaped charges are expressed in terms of target penetration performances. Using Bernouilli's law, a simple equation describes the total penetration depth P as [1]:

$$P = (\rho_j/\rho_t)^{1/2} \ L \qquad (1)$$

where ρ_j and ρ_t are the densities of the jet and target materials and L the length of the jet prior to rupture. According to this equation, the penetration of shaped charge jets is characterized by two important parameters: density of the jet or liner material and maximum elongation prior to jet breakup. The maximum elongation is essentially dependent on the initial mechanical properties and microstructure of the liner material [6], which are themselves related to its chemical composition and to the method used for the fabrication of the liner (e.g. material preparation, heat treatments, mechanical working, etc.).

To date, the most used liner material is copper which has a relatively high density and high ductility and is also used as a reference material. An improved liner material should have a density and/or a ductility higher than copper under dynamic loading in order to delay the jet fragmentation. The latter implies finer and more homogeneous microstructures.

In the present work, rapid solidification techniques and powder metallurgy processes were combined to fabricate shaped charge liners. A superplastic Zn-Al alloy already used as a liner material has been produced in the form of rapidly solidified ribbons by the well known melt-spinning technique. The ductile ribbons were then pulverized into powders in a specially designed blade cutter mill. Powders of various sizes were thereafter selected and isostatically cold pressed into preforms which were finally consolidated by hot isostatic pressing (HIPing). Details of the fabrication process are reported and microstructures are presented.

Experimental procedures and results

A master Zn-Al alloy was prepared from a ZA-12 zinc foundry alloy (87.13% Zn, 12.08% Al and 0.75% Cu) supplied by Noranda Ltd and from aluminum ingots (99.8% Al) supplied by Alcan Ltd. By remelting these materials together in the desired proportions, small Zn-20.5Al-0.7Cu (wt %) ingots were cast (2 cm diam. x 15 cm long). These ingots were directly used for producing thin ribbons by melt-spinning in a helium atmosphere. Clean, regular and continuous Zn-20.5Al-0.7Cu ribbons with different cross section areas were obtained by varying melt-spinning parameters (from 1.5 mm wide x 0.050 mm thick for the wider ribbons to 1 x 0.020 for the narrower ribbons). These ribbons are particularly ductile and cannot be easily reduced into powders by conventional milling. A blade cutter mill apparatus was then constructed for the pulverization of these ductile ribbons. This apparatus is specially designed to reduce ductile ribbon into powders by a cutting or slashing action. It is schematically illustrated in figure 4. The blades, which are rotated at 3580 RPM by an electric motor, cut the ribbon into flaky powder particles. Deflectors are fixed on the wall of the cylindrical pulverization chamber to deviate the material towards the center of the chamber. To reduce heating of the charge, cold water is circulated through a jacket around the chamber.

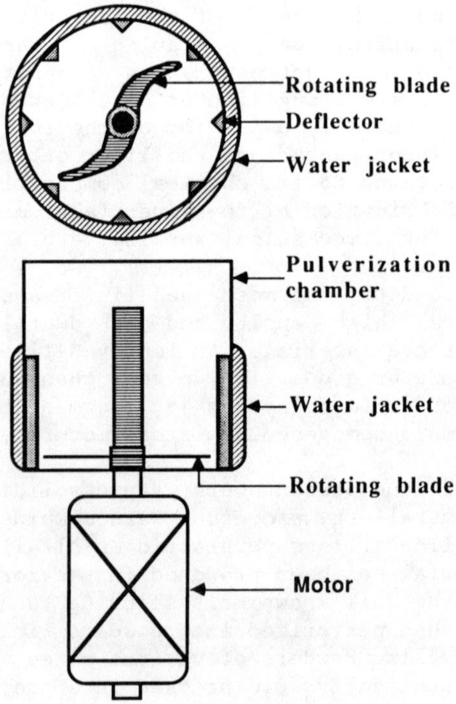

Figure 4: Blade cutter mill apparatus used to pulverize metallic ribbons into powder.

Wide and narrow Zn-20.5Al-0.70Cu ribbons were pulverized separately into powders in this blade cutter mill by processing 200 g batches at a time during two hours. The resulting particle size distributions of the powders obtained from wide and narrow ribbons are shown in figure 5. Despite very different ribbon cross section

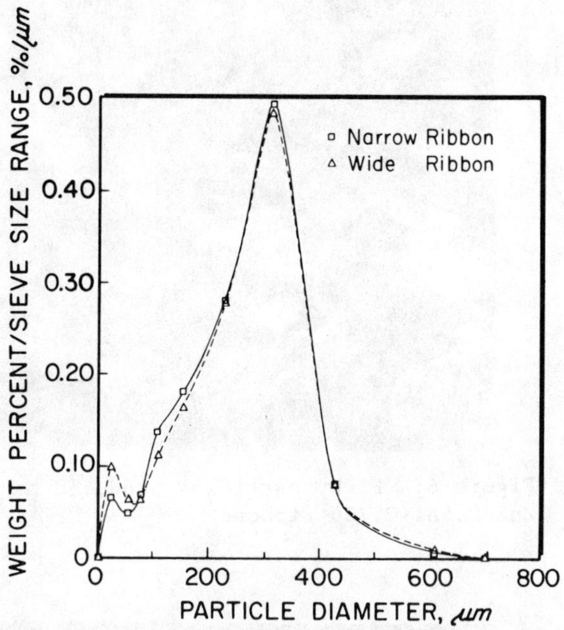

Figure 5: Particle size distributions of powders produced from wide and narrow ribbons.

areas (3.75 times larger for the wide ribbons), these two particle size distributions are practically similar. Particles between 45 and 355 μm, corresponding to 82 wt % of the total produced particles, were used for subsequent compaction. Figure 6 illustrates the shape of these flaky particles obtained by pulverization of wide or narrow ribbons. The microstructure of these particles is also shown in figure 7. This high magnification optical micrograph (500X) reveals the very fine and homogeneous microstructure obtained by rapid solidification.

The following step of liner fabrication was cold isostatic pressing of the powder into conical preforms. It was performed at 275 MPa which gave preforms with about 90 % of the theoretical density. Three different mold designs were tested for cold isostatic pressing; they are schematically illustrated in figure 8. It is the mold design #1 which was retained for the present work. It allows pressing of conical liners of uniform density and constant geometry. The conical liners were made longer than necessary to allow extraction of an annular sample for evaluation of the liners quality after final consolidation without damaging the usable part of the

Figure 6: Flaky particles produced by pulverization of Zn-20.5Al-0.7Cu ribbons.

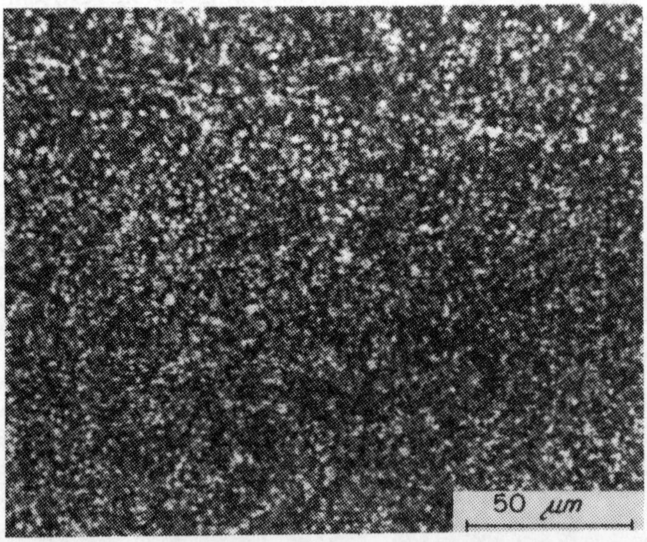

Figure 7: Optical micrograph of a Zn-20.5Al-0.7Cu particle (etchant: 50 g CrO_3, 4 g Na_2SO_4 and 1000 ml H_2O).

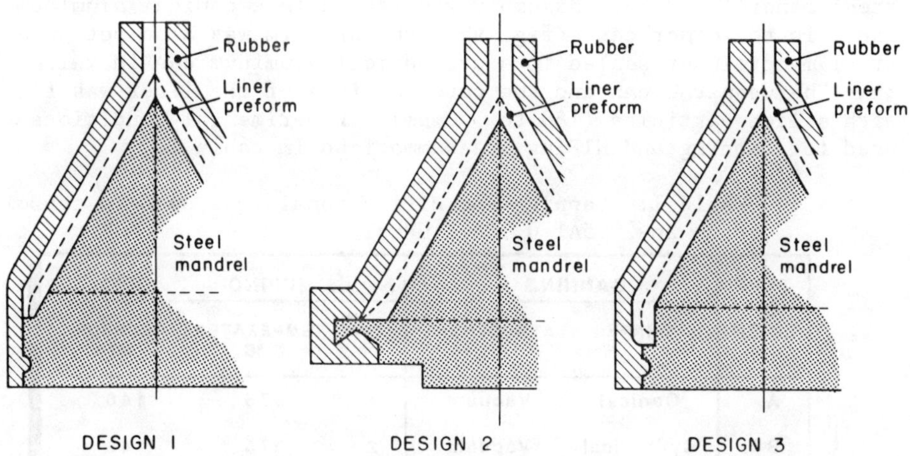

Figure 8: Three mold designs tested for cold isostatic pressing (the dashed line corresponds to the liner profile after cold compaction).

liners. As for the design #2, it was adequate for making thin wall liners but not for thick wall liners. Deformation was relatively important during pressing and the steel base of the mandrel retained the rubber mold leading to a poor densification at the liner base. Mold design #3 was rejected because it gave preforms with irregular liner base geometry. With this design, the liner base geometry is not defined by the steel mandrel because the rubber envelope inserts itself between the liner and the base of the steel mandrel.

The subsequent fabrication steps were canning of the liner preforms and HIPing. Two canning methods were tested; they are illustrated in figure 9. In one case (fig. 9a), the liner preform was put on a

Figure 9: Canning methods tested for hot isostatic pressing.

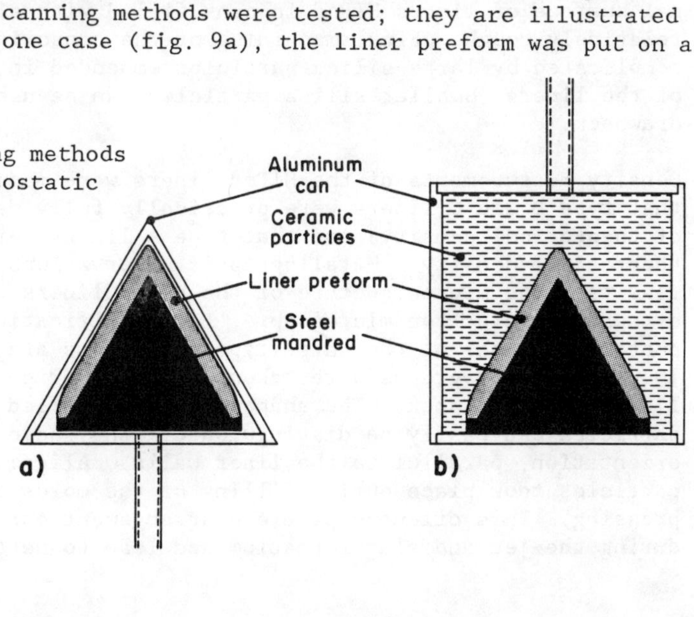

steel mandrel and this assembly was sealed in a conical aluminum can. In the other case (fig. 9b), the preform was also put on a steel mandrel but sealed in a cylindrical aluminum can in which the space between the can and the outer surface of the liner was filled with silica particles (180 to 500 μm). Experimental conditions used for canning and HIPing are summarized in table 1.

Table 1: Canning and HIPing conditions used to consolidate Zn-20.5Al-0.7Cu liners.

TYPE OF LINER	CANNING		HIPING		
	SHAPE	ATMOSPHERE	TIME h	TEMPERATURE °C	PRESSURE MPa
A	Conical	Vacuum	1	375	140
B	Cylindrical	Vacuum	2	375	140
C	Cylindrical	Argon	1	375	140

Type A and type B cans were sealed under a mechanical vacuum of about 50 Pa and type C cans were sealed with argon at 100 kPa above atmospheric pressure within the can. The samples were separately HIPed for one or two hours at 375°C and 140 MPa. Pressure was applied only when the HIPing temperature was attained to avoid breaking of the welded joints of the aluminum cans.

Liners consolidated in cylindrical cans were easily removed from the cans and the surrounding ceramic particles after HIPing. It was not so easy with the liners consolidated in the conical cans in which case machining of the aluminum cans was necessary. The external surface of liners HIPed in cans filled with silica powder was relatively rough. Final machining of the consolidated liners was complicated by large silica particles embedded in the outside surface of the liners. Smaller silica particles can be used to avoid this drawback.

Density measurements of the HIPed liners were made. They revealed that type B and C liners were practically fully dense (99.8 to 100% of theoretical density) and that type A liners were 97.5% of the theoretical density. Metallographic observations corroborated these results. The microstructure of the HIPed liners was also studied by scanning electron microscopy. Low magnification scanning electron photomicrographs of the three types of liners are shown in figure 10. These micrographs were taken in the upper part of the conical liners near the apex. The shape of the densified Zn-20.5Al-0.7Cu particles can easily be distinguished: they have a preferential orientation, parallel to the liner walls. Alignment of the flaky particles took place during filling of the molds for cold isostatic pressing. This oriented particle arrangement can be favorable during the jet and slug formatiom and lead to better ballistic

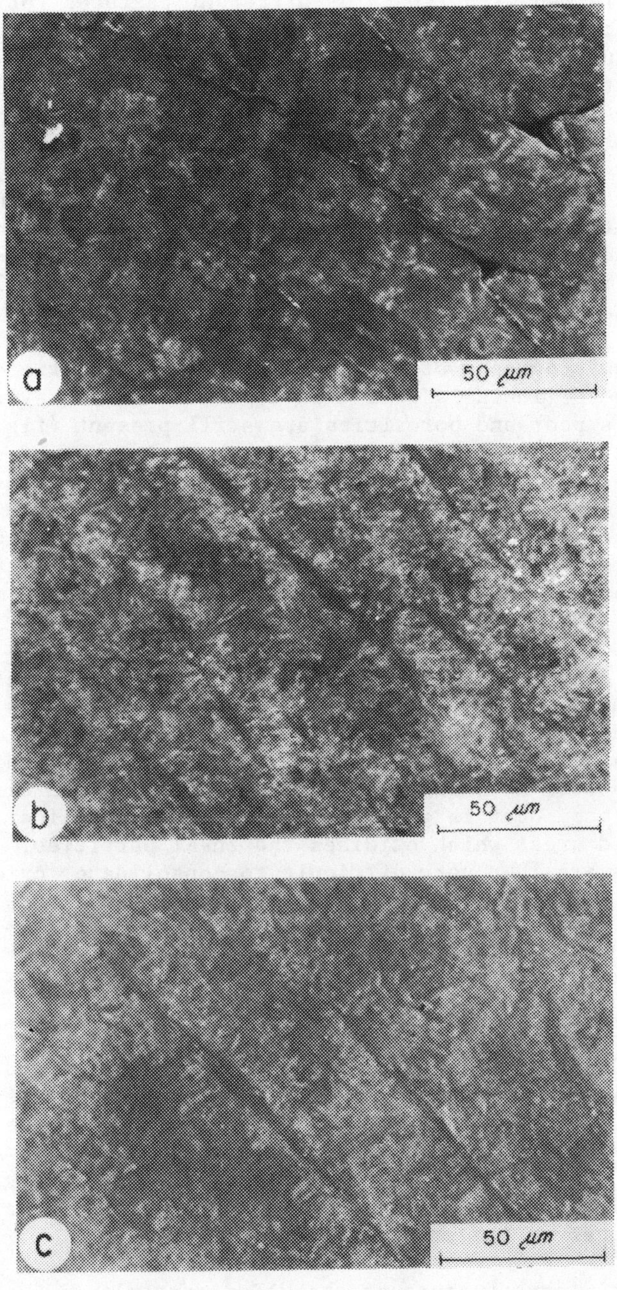

Figure 10: Low magnification SEM micrographs of liner microstructures: (a) type A liner; (b) type B liner; (c) type C liner.

performances of the shaped charges. Indeed, according to preceding tests performed with liners of different textures (oriented fiberlike grains), better results were obtained with fiberlike grains parallel to the liner axis [7]. In the present case, the particles are still better oriented since they are parallel to the liner walls.

The microstructure of the consolidated particles is very different from that of the original starting particles (see fig. 7). The heat treatment undergone by the particles during HIPing greatly affected the microstructure. The starting fine and homogeneous microstructure transformed into a lamellar eutectoid composed of an aluminum rich α phase and a zinc rich η phase. Lamellae are however still relatively fine.

The photomicrographs of figure 10 reveal that particle bonding in type B and C liners is good. In type A liners however, particle bonding is poor and porosities are still present (fig. 10a). Higher magnification photomicrographs of the particle boundaries are shown in figure 11. In type B and C liners, there is a dense boundary area (4 to 5 μm wide) in which diffusion took place producing a continuous metallurgical bond. In type A liners however, consolidation was not so good (fig. 11a): a discontinuity line is clearly visible between adjacent particles indicating a poor metallurgical bond.

The poor results obtained with type A liners can be ascribed to the canning method. Indeed, the closing of aluminum cans by welding generated a lot of heat. In the case of cylindrical cans filled with ceramic particles, heating is practically limited to the welded areas and does not affect appreciably the liner material. For conical cans, heat is easily transferred to the liner material near the welded areas which oxidises the Zn-Al particles. These oxidised particles are then very difficult to consolidate, oxidised surfaces limiting diffusion and bonding between adjacent particles.

Conclusion

The Zn-20.5Al-0.7Cu shaped charge conical liners produced by HIPing of rapidly solidified powders in the present work display interesting characteristics. The use of flaky particles (obtained by pulverization of ribbons produced by melt-spinning), led to a microstructure with the flat surfaces of the particles aligned along the liners walls. This type of microstructure should lead to better ballistic performances of the shaped charges.

The inner microstructure of the HIPed particles is constituted of a lamellar eutectoid with lamellae of an aluminum rich α phase and of a zinc rich η phase. This microstructure is not as fine as the original microstructure of the rapidly solidified particles but

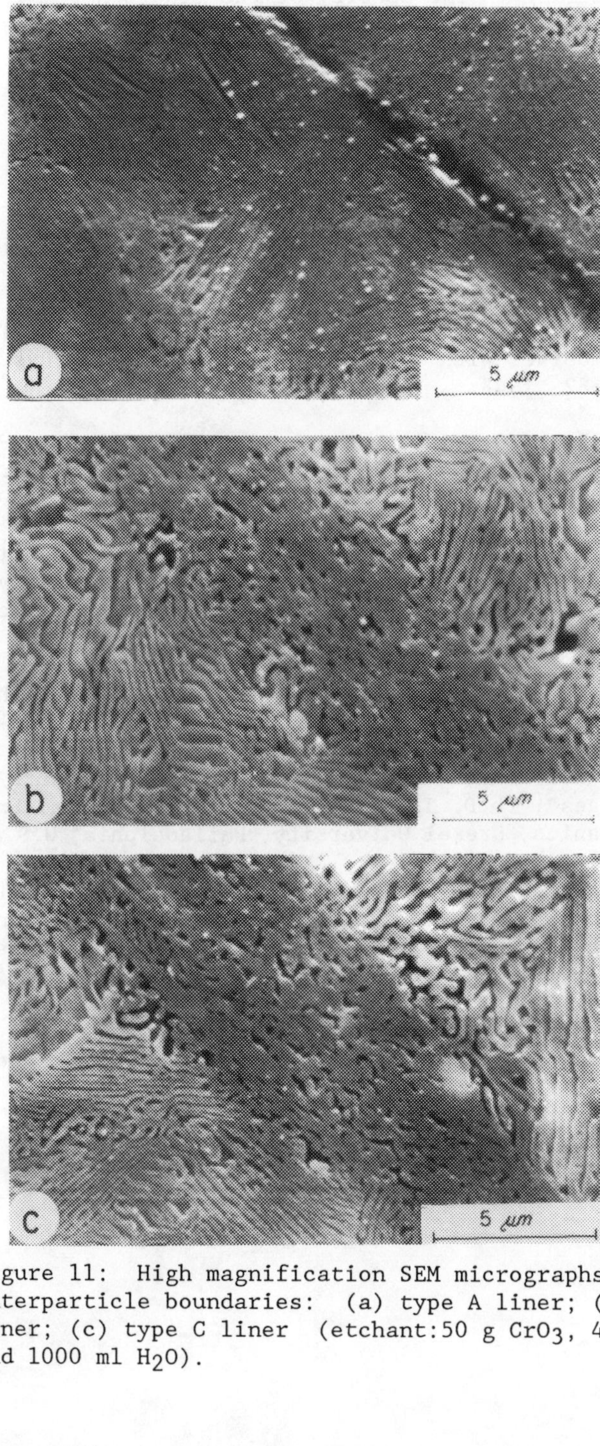

Figure 11: High magnification SEM micrographs of interparticle boundaries: (a) type A liner; (b) type B liner; (c) type C liner (etchant:50 g CrO_3, 4 g Na_2SO_4 and 1000 ml H_2O).

remains very fine compared to that of a liner fabricated by conventional metallurgy. The density of the liners HIPed in cylindrical cans was between 99.8 and 100% of the theoretical density.

References

1. G. Birkhoff, D.P. MacDougall, E.M. Pugh and G. Taylor; "Explosives with lined cavities", J. Appl. Phys., Vol. 19, pp 563-582, 1948.

2. F.A. Baum, R.P. Stanykovich and B.I. Shekter; "Physics of an explosion", (AD 400 151), p. 546, Research Information Service, New York, 1949.

3. R.J. Eichelberger; "Re-examination of the theories of jet formation and target penetration by lined cavity charges", Doctoral Dissertation, Carnegie Institute of Technology, Pittsburgh, Pa., 1954.

4. R.W. Wood, "Optical and Physical effects of high explosives", Proceedings Royal Society, London, 157A, pp 249-261, 1936.

5. C.A. Weickert; "Spinning self-forging fragments and shaped charges", Ph.D. Thesis, Faculty of Mechanical Engineering and Mechanics, Drexel University, Philadelphia, U.S.A., June 1986.

6. A. Lichtenberger, M. Scharf and A. Bohmann; "Influence of the structural and metallurgical state of a liner on the performance of a shaped charge", 6th Int. Symposium on Ballistics, Orlando, Florida, pp 237-243, 1981.

7. F. Jamet; "Etude de jets de charges creuses par diffraction X instantannée", Institut Franco-Allemand de recherches, Saint-Louis, France, (PB85-199206), 1984.

Chapter Ten

Structure and Properties of Ferrous Alloys

MEASUREMENT OF COMPRESSIVE YIELD STRENGTH AND POISSON'S RATIO OF FN-0208, HEAT TREATED, AT 6.85 G/CM^3

by

J. Charles Webster [1], Jim J. Lengyel [2],
Lu-Ling Tu [3], and Leander F. Pease III [4]

The design and development of many new components or the redesign of existing components utilize finite element analysis. The application of finite element analysis to the design of P/M components is limited due to the lack of published data of several mechanical properties. A review of MPIF Standard 35 mechanical property tables such as compressive yield strength, Poisson's ratio, Youngs Modulus and Fatigue Strength are footnoted, "Additional data in preparation will appear in subsequent editions of this standard." In addition, no data is presented for shear strength or shear modulus. The lack of sufficient mechanical property data for P/M alloys indicates a need to produce specific data. This paper will present data on the compressive yield strength and Poisson's ratio for heat treated FN-0208 at a density of 6.85 g/cm^3.

Past work done on measuring 0.1 or 0.2% compressive yield strength of P/M materials (20 to 40 years ago) involved successive compression measurement. A disc with about a 1 to 1 ratio of length to diameter was ground with flat and parallel ends. After measuring length to the nearest 0.0001", the disc was then compressed at some force, say 1000 lbs., for a .5" diameter. The stress was released and the length remeasured.

----

1 Technical Specialist - Materials, Ford Motor Company, Engine and Product Manufacturing Engineering

2 Senior Engineer, Ford Motor Company, Central Laboratory Services

3 Engineer, Ford Motor Company, Central Laboratory Services

4 President, Powder-Tech Associates

This process was continued until there was at least 0.2% permanent shortening of the disc. A plot of stress versus strain could be made, and the 0.1% yield strength was determined. The method was quite time consuming.

This paper compares determination of compressive yield strength using the successive compression versus the modern method of applying strain gages to the sides of the specimen. It also evaluates the effects of applying lubrication on the ends of specimens. Also, specimens were instrumented with strain gages for determination of Poisson's ratio.

Sample Preparation

The material selected for this investigation was FN-0208 at 6.85 g/cm^3 density. Test discs were produced using Ancorsteel 1000 Fe with 2% 123 Nickel, 0.9% SW1651 graphite and 0.75% Acrowax C. The test discs were sintered along with a production load in a nitrogen based atmosphere at a low dew point and oxygen content. The discs were turned and ground to the final dimensions. After machining the test specimens were heat treated at 1575 $\pm$ 25°F, oil quenched and tempered for one hour at 475 $\pm$ 25°F. The test specimens had a density range of 6.83 - 6.86 g/cm^3. After heat treatment, the compression yield strength test specimen had a surface hardness of HRC 25-30 and the Poisson's Ratio test specimens had a surface hardness of HRC 22-27. The unetched microstructures of both specimens, Figures 1 and 2, showed few or no original particle boundaries. This condition is considered an average or typical sinter as standardized in MPIF Standard 35. The microstructure of the 0.375 diameter compressive yield strength specimens, Figures 3 and 4, consisted of tempered martensite and was through hardened. The larger 1.0 inch diameter Poisson's Ratio test specimen only fully hardened to a depth of 0.075 inch. The microstructure of the surface of the Poisson's ratio specimen, Figure 5, consisted of tempered martensite. The case of the Poisson's ratio specimen, Figure 6, consisted of fine pearlite and tempered martensite. This partial hardening is typical of FN-0208 having a 1 inch diameter section size.

Compressive Yield Strength

Test Methods

The successive compressive yield strength specimens were 0.375 in. dia. x 1 in. long. The test method utilized a predetermined uniaxial compression load applied to the specimen with a load rate of 6,000 lbs./min. After releasing the load and cleaning off grease (if grease was applied), the length of the specimen was measured to the nearest 0.0001". The specimen was then loaded by recompressing to a higher load, then releasing the load. The length was remeasured until the permanent compressive strain exceeded approximately 0.5%.

The strain gage compressive yield strength method utilized a bonded foil resistance strain gage technique. The strain gages were affixed in opposite dimidiate pairs which were procured as a stacked ninety degree relationship, Figure 7. A full Wheatstone bridge arrangement of the strain gage pairs (2 axial & 2 transverse) provided bilateral strain information with cancellation of bending related measurement problems. Additionally, this circuit arrangement of strain gages augmented the bridge output while maintaining temperature compensation.

A compression load was applied to the specimen continuously with a load rate of 6,000 lbs./min. The load, total apparent strain and deflection values were recorded, and then, the load versus total apparent strain and the load versus deflection curves were plotted.

The results of the compressive yield strength using the successive compressive loading method are shown in Table 1 and the strain gage method results are shown in Table 2. The results show that the strain gage method resulted in compressive yield strengths 10-17% higher than the successive compression method. The use of wheel bearing grease did not affect the results and testing could be done with or without it. Figures 9 and 10 are typical stress-strain curves which show a comparison of the strain gage method and successive compression method with and without lubrication on the end faces of the test specimens. The shapes of the curves are different and the linear portions of the curves show a different slope. The difference in geometry between the two test method curves may be attributed to the fact that in the strain gage method the specimen is subjected to continuous elastic strain only, whereas the successive load method specimen is subjected to both elastic and plastic strain.

If we assume that the strain gage method is more accurate but have only data based on the successive compression method, the strain gage results can be approximated by adding 10% to successive compressive yield data. Also, the test procedure for the successive compression method is much simpler than the stain gage method. The work involved in the successive compression method is only about 40% of the work involved in the strain gage method. Moreover, the successive compression method can be done on almost any mechanical testing system, and does not require strain gages.

Poisson's Ratio

The test discs for the Poisson's Ratio specimens were processed concurrently with the compressive yield strength specimens. However, the test specimens were turned and ground to 1.000 in. dia. x 1.050 in. long. The Poisson's Ratio specimens utilized a bonded foil resistance strain gage technique. The positioning of strain gages is as follows: Eight (4 axial x 4 transverse) strain gages were attached to each specimen (see Figure 2). Two redundant horizontal

TABLE 1

Compressive Yield Strength
Successive Compression Method

With Grease

	0.1% Strain KSI	0.2% Strain KSI
	151	167
	143	165
	151	167
	153	168
	154	171
Average	150	168

Without Grease

	0.1% Strain KSI	0.2% Strain KSI
	144	162
	148	167
	143	162
	147	163
	146	166
Average	146	164

TABLE 2

Compressive Yield Strength
Strain Gage Method

With Grease

	0.1% Strain KSI	0.2% Strain KSI
	170	185
	167	187
	172	184
	170	183
Average	170	185

Without Grease

	0.1% Strain KSI	0.2% Strain KSI
	173	188
	172	182
	172	185
	168	179
Average	171	184

and vertical measurement circuits were staggered on each sample to provide a higher degree of strain resolution. Bridge compensation gages (companion elements required to complement a full Wheatstone Bridge circuit) were epoxied to a separate metal substrate to better utilize the available sample surface geography for the appropriate measurements.

Uniaxial compression tests were run on a Hydraulic Universal Testing System with a load rate of 6,000 lbs/min. Applied stress was about 25 ksi (below the proportional limit) for each specimen. Hardened steel compressing platens were used. The apparent strain values (in both loading direction and transverse direction) obtained from the strain gages were recorded in micro-strain. The Poisson's Ratio was calculated from this data. Four specimens were tested with wheel bearing grease on each end and one was tested without the grease. The results are shown in Table 3.

Table 3

	Poisson's Ratio
	0.28
	0.25
	0.23
	0.22
Average	0.25
Without Grease	0.27

Summary

The P/M Industry has a need to develop mechanical property data for use in finite element analysis of new or redesigned components. Based on this investigation, it appears that the successive loading method for the determination of compressive yield strength will produce results which are approximately 10% less than those results from strain gaged specimens. Also, it is not necessary to use any lubrication on the end faces.

The development of Poisson's Ratio will require strain gaged test specimens and is a function of specimen size.

Acknowledgments

The Ford Motor Company authors wish to acknowledge the PMPA Standards Committee which prepared the test specimens and provided the assistance of Dr. Pease in this test program.

FIGURE 1
Compressive Yield Strength Specimen
200X Unetched

FIGURE 2
Poisson's Ratio Specimen
200X Unetched

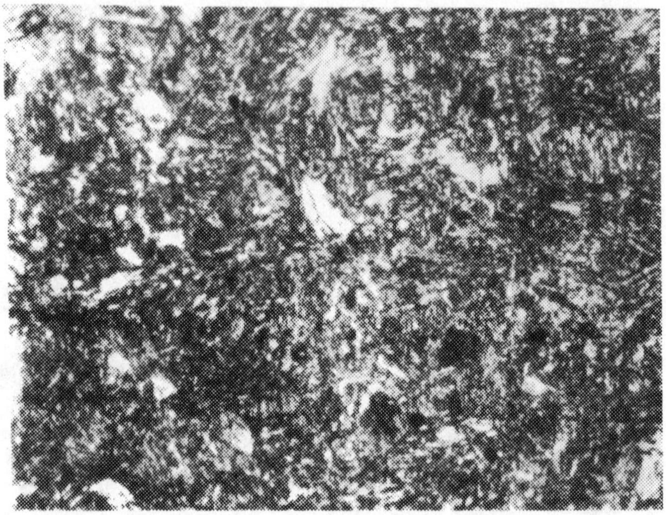

FIGURE 3
Surface of Compressive Yield Strength Specimen
1000X 3% Nital Etch

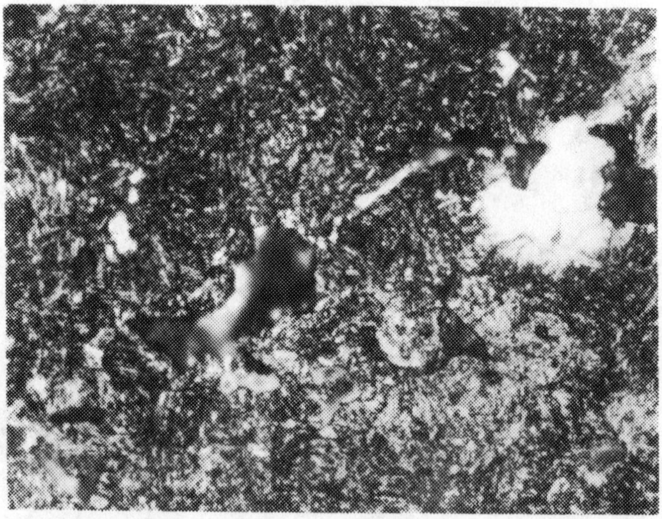

FIGURE 4
Core of Compressive Yield Strength Specimen
1000X 3% Nital Etch

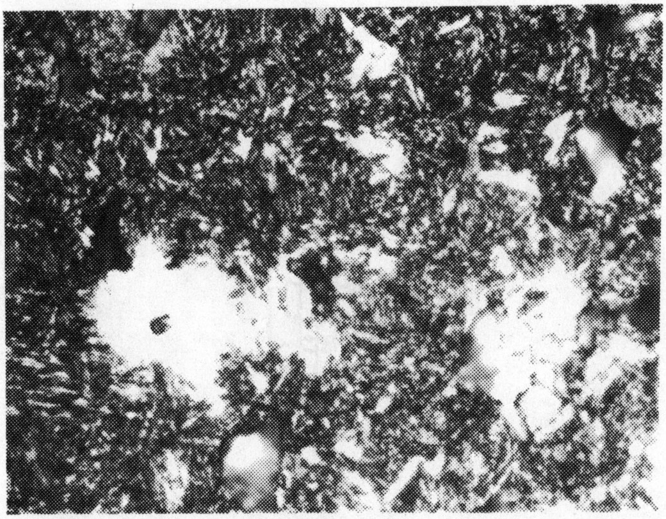

FIGURE 5
Surface of Poisson's Ratio Specimen
1000X 3% Nital Etch

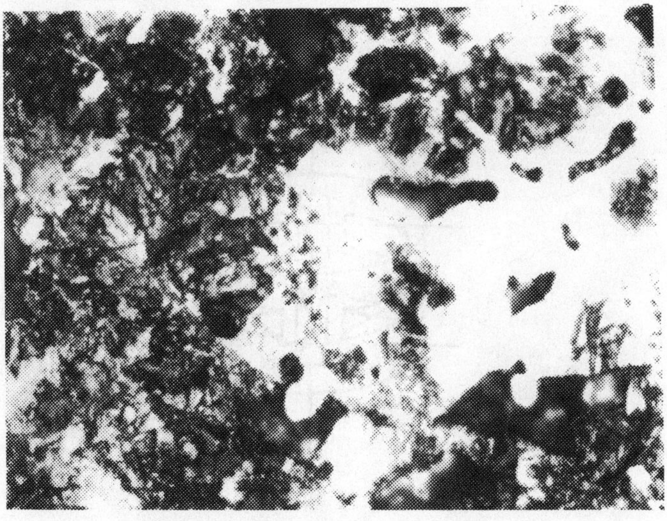

FIGURE 6
Core of Poisson's Ratio Specimen
1000X 3% Nital Etch

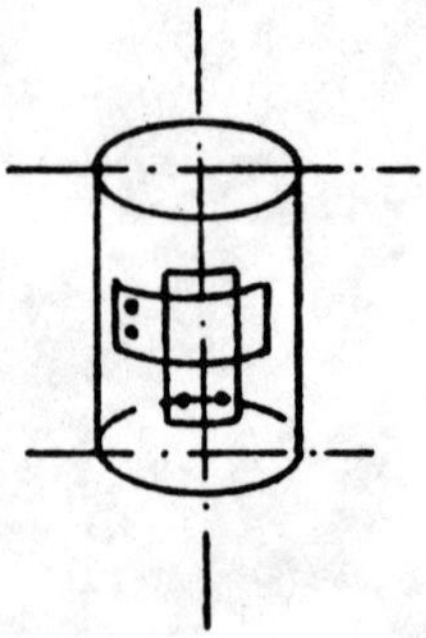

FIGURE 7
Strain Gage Placement
Compressive Yield Strength Specimens

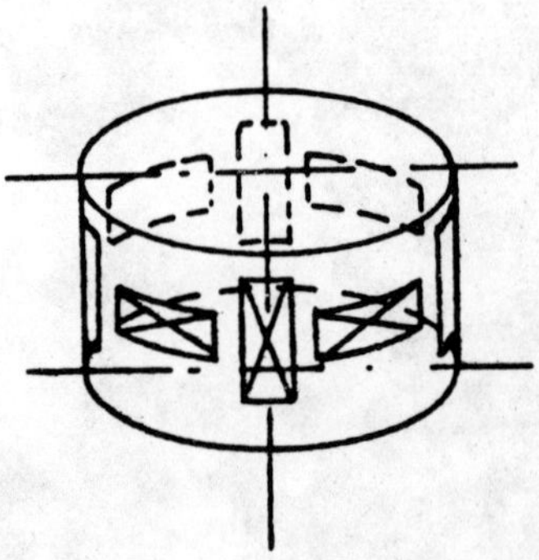

FIGURE 8
Strain Gage Placement
Poisson's Ratio Specimens

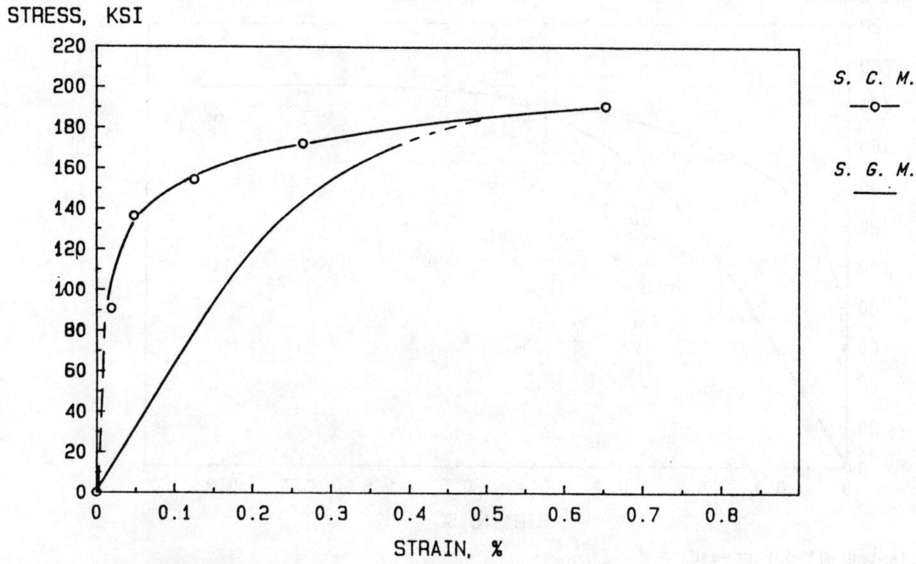

Tested with grease
S.C.M.=Successive Compression Method
S.G.M.=Strain Gage Method

FIGURE 9

Comparison of Compressive Yield Strength Curves Using Strain and Gage Method and Successive Compression Method With Grease

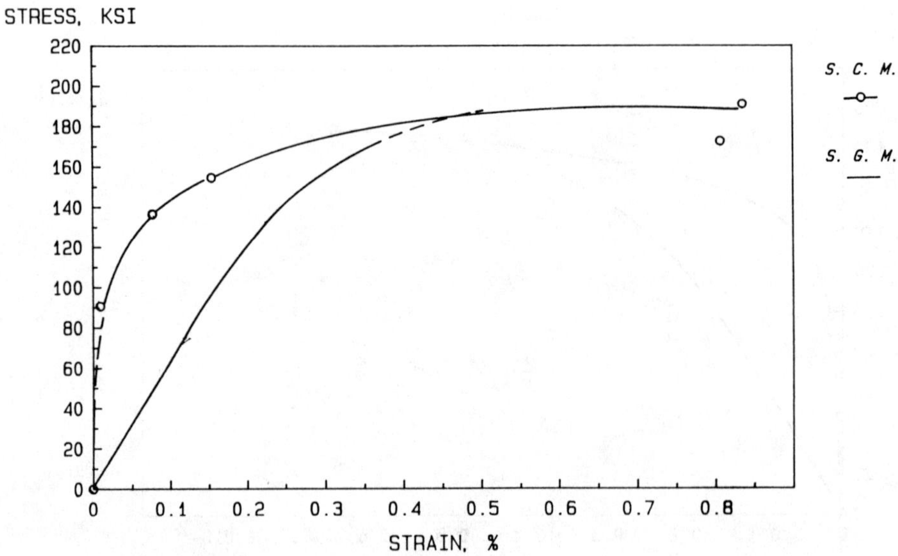

Tested without grease
S.C.M.=Successive Compression Method
S.G.M.=Strain Gage Method

<u>FIGURE 10</u>

Comparison of Compressive Yield Strength Curves Using Strain Gage Method and Successive Compression Method Without Grease

MECHANICAL PROPERTIES AND MICROSTRUCTURE OF PHOSPHORUS ALLOYED SINTERED STEEL

Per Engdahl

Höganäs AB, Höganäs, Sweden

ABSTRACT

Phosphorus alloyed sintered steels are normally characterized by an unusual combination of strength and ductility. Addition of phosphorus, up to 0.6 %, improves both strength and ductility. At higher phosphorus levels the elongation and impact energy decrease rapidly, unless carbon is added to the material.

Optical microscope studies and electron probe analysis have shown an inhomogeneous distribution of phosphorus in the structure. This distribution originates from a two-phase structure, phosphorus-rich ferrite and phosphorus-poor austenite, present during sintering. The two phosphorus levels remain after cooling, resulting in a non-uniform hardness and strength distribution which positively affects the mechanical properties, especially elongation and impact energy.

A change from two-phase (austenite and ferrite) to single-phase (ferrite sintering with increased phosphorus additions causes a transition from a ductile to a brittle fracture behavior. If the phosphorus content exceeds 0.6 %, it may be necessary to control the amount of austenite during sintering by additions of carbon.

Results will be presented concerning the influence of phosphorus and carbon content on the sintered structure and the mechanical properties of iron-phosphorus materials, i.e. tensile strength, elongation and impact energy.

1. INTRODUCTION

In conventional wrought steel, phosphorus causes temper embrittlement due to grain boundary segregation and normally the content is limited to a very low value. Phosphorus is often used in iron-based powder metallurgy and it is one of the most interesting alloying elements. Addition of phosphorus to a sintered steel, up to 0.6 %, improves the strength without any loss in ductility (1, 2, 3, 4, 5).

Many alloying techniques have been proposed (1, 3, 6, 7, 8) but the original one, described by Lenel 1939 (9) is the most commonly used. Phosphorus is added to the material by admixing fine grained ferrophosphorous, Fe_3P, with iron powder. Today, materials alloyed with 0.3–0.6 % phosphorus, often in combination with other elements, are used in industrial applications. Even higher phosphorus contents, 0.8–0.9 %, are used but then always in combination with carbon. Also, in order to reduce shrinkage of the compacts during sintering, copper can be added.

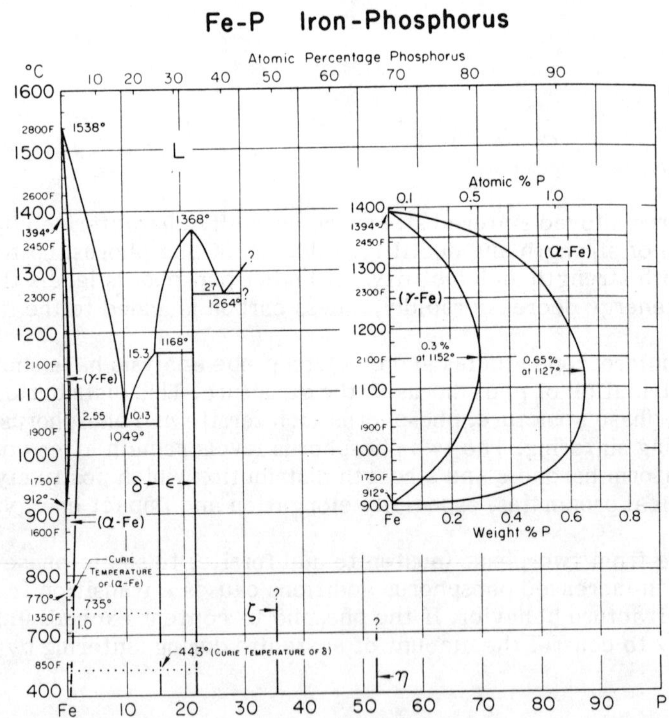

Fig. 1. The binary iron-phosphorus system (Metals Handbook, 8th ed., 1973, Vol. 8)

The beneficial effect of phosphorus on the properties of sintered steels can be explained by three different mechanisms:

1. Solid solution strengthening of the ferrite.

2. Liquid phase formation during sintering.
3. Stabilization of the ferritic phase during sintering.

Phosphorus is one of the strongest substitutional solid solution strengthener, which contributes to the high strength of sintered iron-phosphorus steels.

When the temperature reaches 1050°C during heating a liquid is formed by the peritectic reaction Fe_3P + ferrite (α)->L, see binary phase diagram in figure 1. The melt is distributed by capillary force between the particles, thus providing for a high rate of material transport.

During sintering, normally at 1120°C, Fe-P materials develop a two-phase structure of ferrite and austenite. At a phosphorus level higher than 0.65 % the structure will be completely ferritic. Since the self-diffusivity of iron is approximately two orders of magnitude greater in ferrite than in austenite the rate of sintering will be very high. The active sintering of Fe-P alloys results in rounded pores and in a reduced porosity.

2. EXPERIMENTS

The iron-powder used was the water-atomized Höganäs grade ASC100.29. Fine (<20 µm) ferrophosphorus, Fe_3P, powder with a phosphorus content of 16 % was added and 0.8 % zinc stearate was used for lubrication of the powder mix. In some mixes carbon was added (graphite ultra fine UF).

The blended powders were compacted into tensile specimens and impact test bars at 589 MPa compacting pressure. The test bars were sintered 30 minutes in a continuous belt furnace at 1120°C in dissociated ammonia, cooling rate approximately 1°C/s.

Tensile strength and elongation were determined using conventional tensile testing (geometry of the specimens according to standard ASTM E 8-69). Impact energy was determined on unnotched Charpy test specimens according to ASTM standard E 9272.

3. RESULTS AND DISCUSSION

3.1 Mechanical properties

The shrinkage increases very rapidly with increased phosphorus content due to the high self-diffusivity of iron in ferrite and the liquid phase, which is formed above 1050°C. The large shrinkage, at higher phosphorus levels, is a drawback for binary Fe-P alloys as the tolerances of sintered details may become affected. One solution to this problem is to add copper, which reduces the dimensional change (5, 8, 10). Also small additions of carbon can be used (5).

The phosphorus improves the mechanical properties indirectly by the sintered density. The compressibility of the powder is not noticeably influenced by the ferrophosphorus but the sintered density is affected by the dimensional change.

The sintered density of a material based on ASC100.29 is, after using the above described processing parameters, 7.13 g/cm^3, whereas the same material with 0.6 % phosphorus material has a density of about 7.3 g/cm^3.

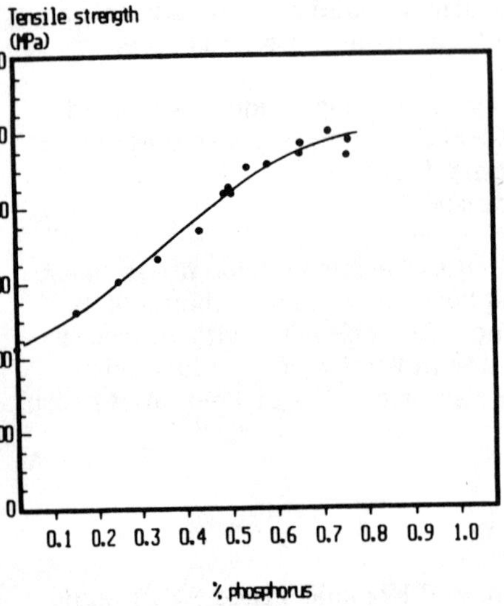

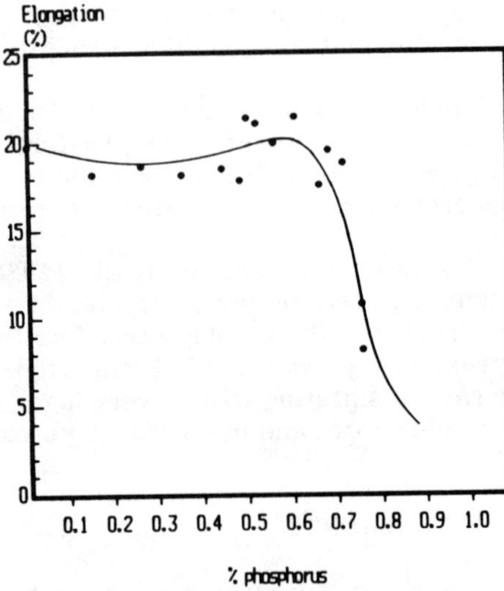

Fig. 2 and 3. Tensile strength and elongation of carbon-free material based on ASC100.29 with various contents of phosphorus

Due to the strong solid solution effect of phosphorus the tensile strength increases, nearly lineary, with the phosphorus content, see figure 2. By adding 0.6 % phosphorus the strength is doubled compared to unalloyed sintered steel.

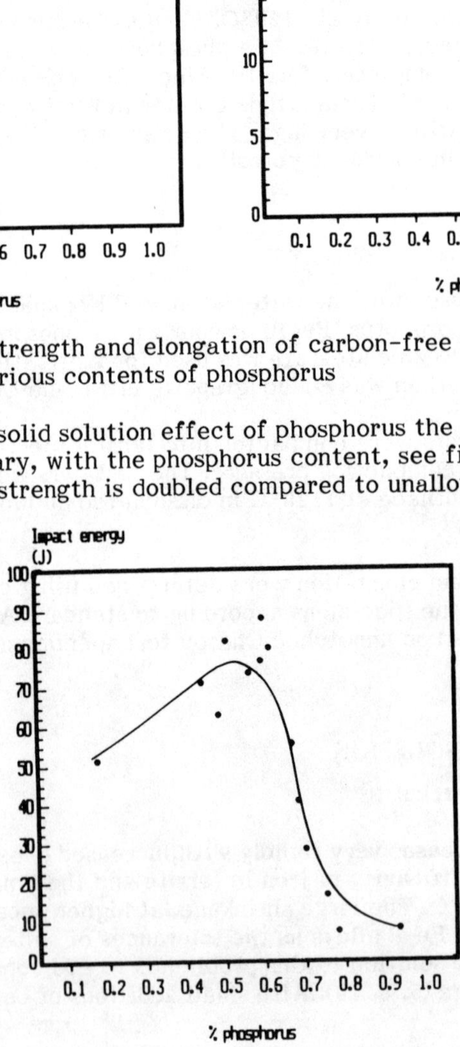

Fig. 4. Impact energy of a carbon-free material based on ASC100.29 with various contents of phosphorus

The effect of phosphorus on the elongation and impact energy is illustrated in figures 3 and 4. Addition of, up to 0.6 % P, improves the ductility slightly. This is probably a result of higher density and more rounded pores. At higher phosphorus content both elongation and impact strength decrease very rapidly. The impact energy is reduced about 10 times if the phosphorus content increases from 0.6 % to 0.8 %.

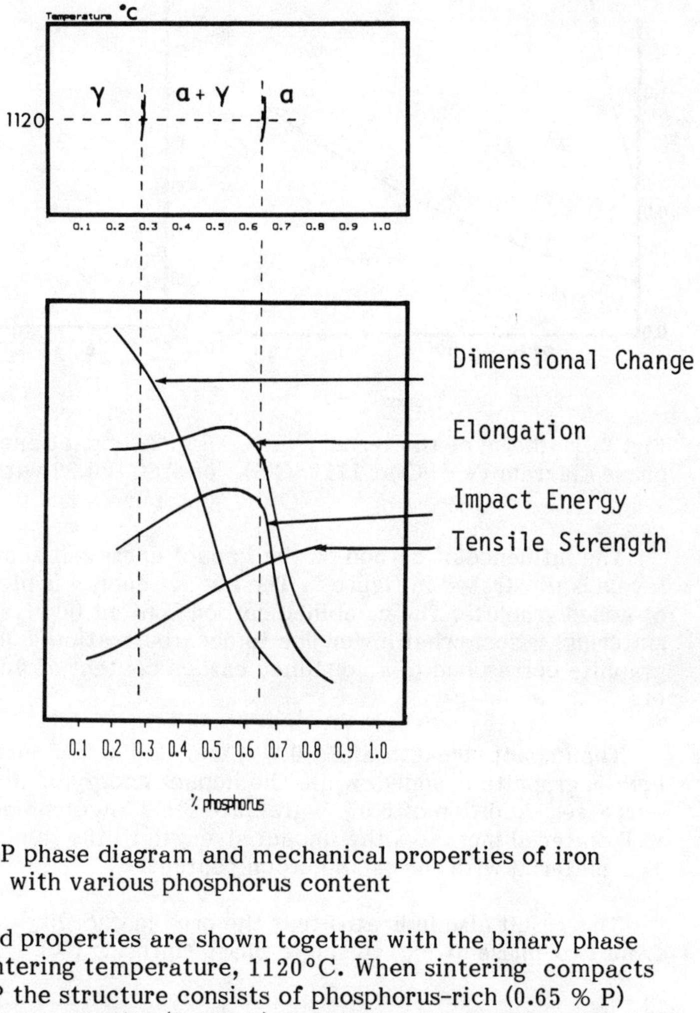

Fig. 5. Binary Fe-P phase diagram and mechanical properties of iron with various phosphorus content

In figure 5 all sintered properties are shown together with the binary phase diagram (Fe-P) at the sintering temperature, 1120°C. When sintering compacts or iron with 0.3-0.65 % P the structure consists of phosphorus-rich (0.65 % P) ferrite and phosphorus-poor austenite (0.3 % P). At higher phosphorus levels the structure will be completely ferritic. The remarkable drop in ductility occurs when the sintering structure changes from two-phase ($\alpha + \gamma$) to single-phase (α). This may indicate what causes the problem and some possible explanations will be discussed later in this paper.

By adding an austenite-stabilizing element, e.g. carbon, the fraction of phosphorus-rich ferrite can be reduced. In figure 5 a ternary Fe-P-C phase diagram at 1120°C is shown. A 0.7 % P material is completely ferritic but if 0.05 %

P is added the fraction of austenite is increased to about 25 %, which corresponds to a binary Fe-P alloy with about 0.55 % phosphorus.

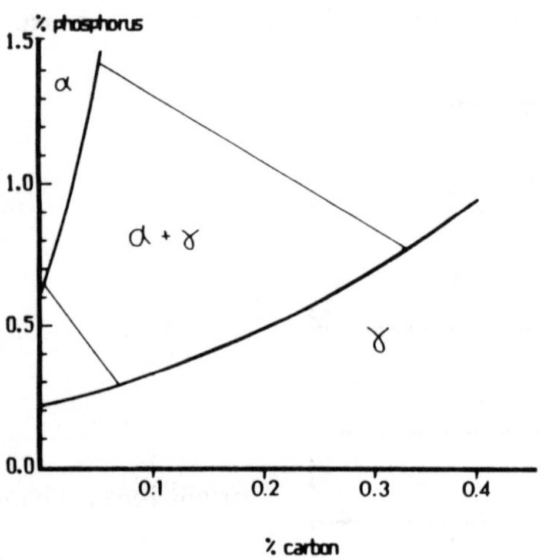

Fig. 6. Isotherm of the ternary phase diagram Fe-P-C at 1120°C(5)

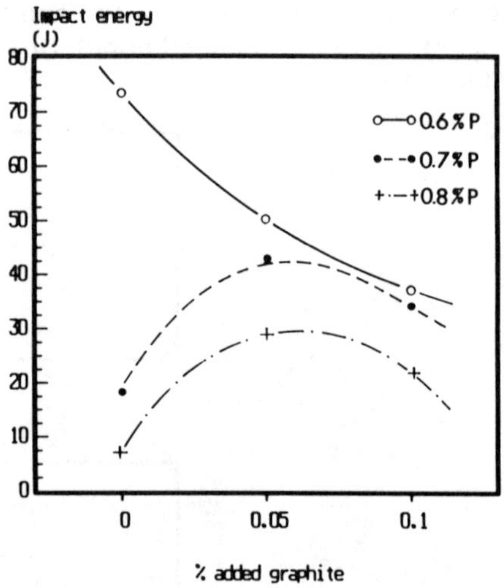

Fig. 7. Impact energy of a material based on ASC100.29 with various contents of phosphorus and carbon

The influence of carbon on the impact energy at some different phosphorus levels is illustrated in figure 7. The impact energy is plotted versus the amount of added graphite. The combined carbon content (analysed carbon content after sintering) is somewhat lower. due to decarburization. 0.05 % and 0.1 % of added graphite correspond to a combined carbon content of 0.02 % and 0.05 % respectively.

The impact strength of Fe-0.7 % and Fe-0.8 % P materials is improved when 0.05 % graphite is added, while the impact energy of the Fe-0.6 % P compact decreases. Addition of 0.05 % graphite (0.02 % combined carbon) to a Fe-0.7 % P material increases the impact strength to the same level as for a Fe-0.6 % P material with the same carbon content.

This result also indicates that the drop in ductility is caused by a change from two-phase ($\alpha+\gamma$) to single-phase sintering (α).

3.2 Metallography

Phosphorus alloys are normally sintered in the two-phase region ($\alpha+\gamma$) and the proportions are determined by temperature and content of carbon and phosphorus. During cooling the austenite transforms to ferrite but the two levels of phosphorus, corresponding to ferrite and austenite, remain. The resulting microstructure consists therefore of ferrite with high P-content (about 0.65 % and ferrite with lower P-content (about 0.3 %). The composition gradients can be visualized by using some different etchants, e.g. Oberhoffer and Klemm. The method used in this study was etching in 4 % nital and optical microscopy on an unfocused sample, see figure 8. The low-phosphorus ferrite appears as the brighter minority phase and the high-phosphorus ferrite as the darker phase. The fact that there are different phosphorus contents in the two phases has been confirmed by using a wavelength-dispersive spectrometer (WDS), see table 1.

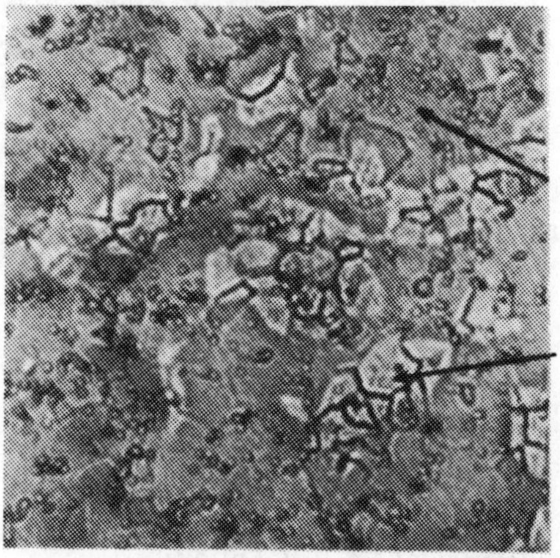

Fig. 8. Microstructure of Fe-0.6 % P material. (4 % Nital and unfocused sample).

Table 1. WDS analysis of phosphorus (Fe-0.6 % P)

Analysis No.	1	2	3	4	$\bar{x}$
Light-etching areas	0.45	0.54	0.59	0.59	0.54
Dark-etching areas	0.24	0.22	0.22	0.26	0.23

In figure 9 the microstructure of 0.6 % P and 0.7 % P materials is shown. The structures are in accordance with the phase-diagram at 1120°C. Hardly any area with low P-content is present in the Fe-0.7 % P material while the two other structures are very similar (approximately 10-20 % light etched phase). Apart from the P-poor ferrite, no difference in microstructure can be observed by optical microscopy. The grain size of the ferrite does not seem to be affected by the change from two-phase to single-phase sintering.

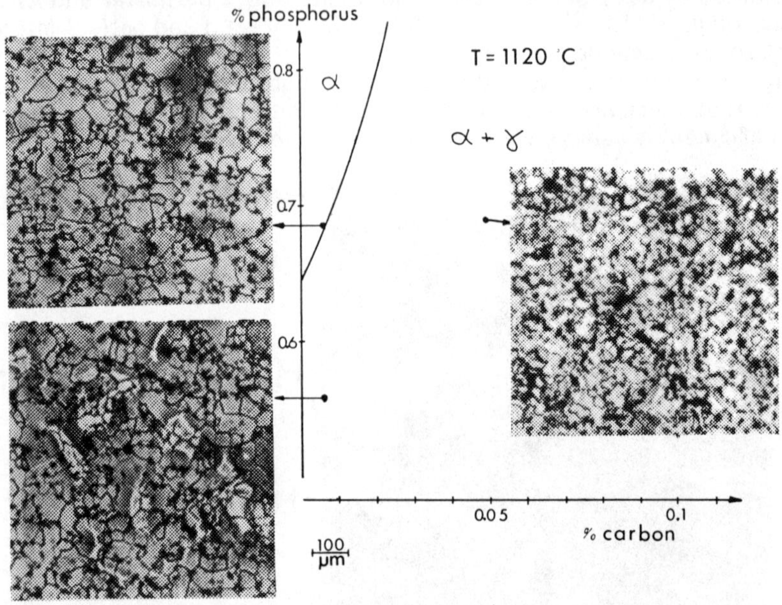

Fig. 9. Microstructures of Fe-P-(C) alloys.
(4 % Nital and unfocused sample).

Fractographs of some Fe-P-C-alloys, see table 2, are illustrated in figure 10.

Table 2. Impact energy and fracture appearance of some Fe-P-C materials. Sintering at 1120°C in dissociated ammonia for 30 min.

% P	% C	Impact energy, J	Structure (at 1120°C)	Fracture appearance
0.21	0.01	52	γ	Dimple
0.59	0.01	76	α+γ	Grain-boundary and cleavage
0.95	0.01	9	α	Grain-boundary
0.58	0.06	37	α+γ	Grain-boundary, dimple and cleavage
0.61	0.27	23	γ	Dimple

The fracture appearance is determined by the structure at the sintering temperature. Austenite sintering results in a dimple-like fracture while single phase ferrite sintering results in a brittle intergranular fracture. The fracture appearance of the two-phase sintered materials is complex and consists of all fracture types.

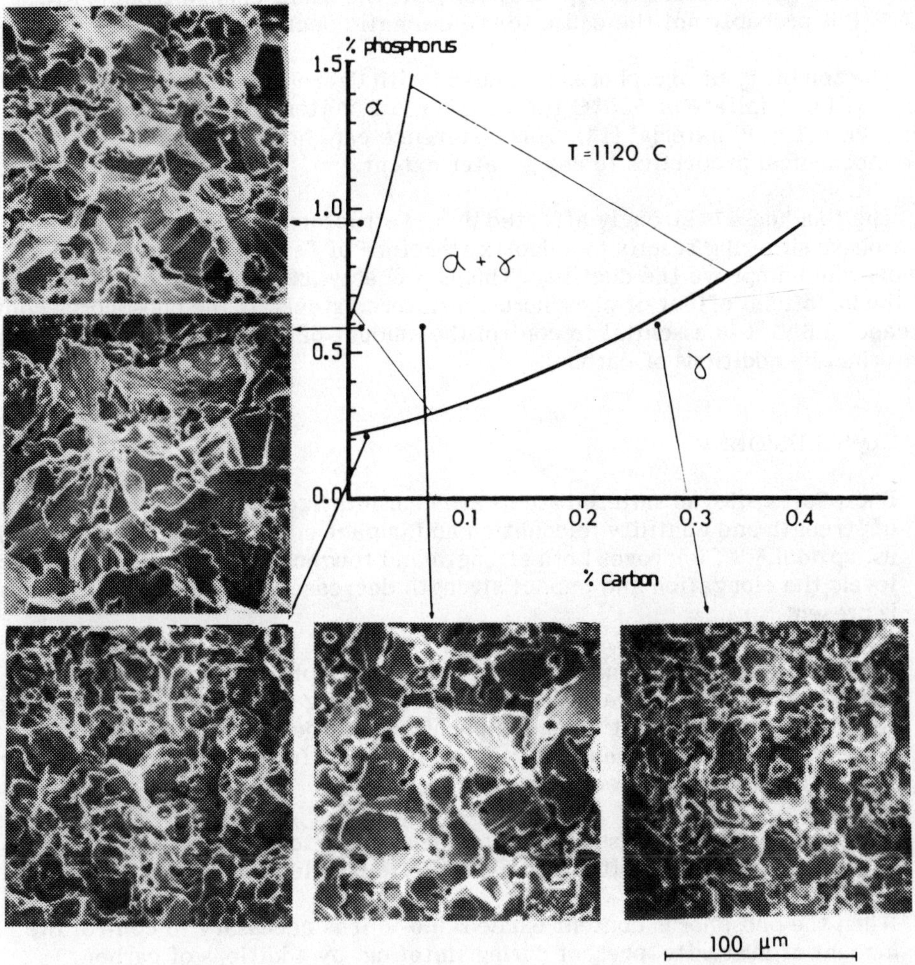

Fig. 10. SEM fractographs of Fe-P-C alloys.

The drop in elongation and impact energy of phosphorus alloyed steels can normally be explained by several mechanisms:

1. Temper brittleness (segregation of P to grain-boundaries at temperatures between 325 and 575°C)

2. Precipitation of Fe-P at grain-boundaries

3. The change from two-phase sintering (α+γ) to one-phase sintering (α)

Temper brittleness is a wellknown phenomenon in conventional low-alloyed steels. Impurity element, especially phosphorus, segregates at grain-boundaries in the temperature range 325-575°C (11, 12). Even if grain-boundary enrichment increases with increasing phosphorus content, the small change from 0.6 to 0.7 % P is probably not the cause to the dramatic decrease in ductility.

The solubility of phosphorus is reduced with decreased temperature and Fe-P will precipitate at 650 °C for a Fe-0.6 % P material and at about 700 °C for a Fe-0.7 % P material (13). This difference can, however, not influence the mechanical properties to any greater extent.

The toughness is strongly affected by the structure during sintering. The two-phase sintering results in a duplex structure of ferrite with two hardness levels which improve the ductility. This is probably an additional explanation to the beneficial effect of phosphorus in sintered steels. If the phosphorus content exceeds 0.6 % it is essential to control the amount of austenite, present during sintering, by additions of carbon.

4. CONCLUSIONS

1. Phosphorus alloyed sintered steels are characterized by an unusual combination of strength and ductility (elongation and impact energy). Addition of phosphorus, up to 0.6 %, improves both strength and toughness. At higher phosphorus levels the elongation and impact strength decrease very rapidly unless carbon is present.

2. Fe-P materials are normally sintered in the two-phase ferrite-austenite region. The austenite transforms to ferrite during cooling but the two phosphorus levels remain, resulting in a non-uniform hardness distribution. This duplex structure is probably of great importance for the mechanical properties of P-alloyed sintered steels.

3. A change from two-phase (austenite and ferrite) to single-phase (ferrite) sintering causes a transition from ductile to brittle behaviour.

4. When the phosphorus content exceeds 0.6 % it is necessary to control the amount of austenite, present during sintering, by additions of carbon.

REFERENCES

1. EISENKOLB, ,
Über phosphorlegierten Sinterstahl,
Archiv für das Eisenhüttenwesen 24 (1953) pp. 257-266.

2. BLUM, G,
Die Produktion von phosphorlegierten Sintereisen,
Fertigungstechnik 7 (1957) pp. 504-507.

3. REBSCH, H,
Versuche mit phosphorlegierten Eisenpulver,
Neue Hütte 10 (1965) pp. 355-360.

4. LINDSKOG, P and CARLSSON, A,
Sintered Alloys Based on Sponge Iron with Additions of Ferrophosphorus,
Powder Metallurgy International 4 (1972) pp. 39-43.

5. LINDSKOG, P; TENGZELIUS, J and KVIST; S-Å,
Phosphorus as an alloying element in ferrous P/M,
Modern Developments in Powder Metallurgy, Vol. 10, H.H. Hausner (ed.), Plenum Press, N.Y. 1977:97-128.

6. NAESER, G and BURMEISTER, H,
Archiv für das Eisenhüttenwesen 24 (1953) pp. 251-255.

7. DAUTZENBERG, N,
Eigenschaften von Sinterstählen aus durch Wasserverdüsung von Stahlschmelzen erzeugten unlegierten und fertiglegierten Pulvern,
Archiv für das Eisenhüttenwesen 40 (1969) pp. 351-357.

8. BOCKSTIEGEL, G,
Metallurgie III, 4 (1962) pp. 67-78.

9. LENEL, F V,
US Pat. 2,226,520 (1940).

10. EISENKOLB, F,
Stahl und Eisen, 79 (1959) pp. 1345-1352.

11. SEAH, M P,
Interface adsorption, embrittlement and fracture in metallurgy. A review,
Surface Science 53 (1975) pp. 168-212.

12. ERHART, H and GRABKE, H J,
Equilibrium segregation of phosphorus at grain boundaries of FeP, Fe-C-P, FeCrP and Fe-Cr-C-P alloys,
Metal Science 15 (1981) pp. 401-408.

13. HOFMANN, H-P; LÖHBERG, K and REIF, W,
Über die Löslichkeit des Phosphors und Eisen-Kohlenstoff-Legierungen unterhalb 1000°C,
Archiv für das Eisenhüttenwesen 41 (1970) pp. 975-982.

Chapter Eleven

Technical Briefs - Poster Presentations

Extended abstracts of works in progress, developments which may have potential for the industry and presentations where circumstances limited the submission of full technical papers for publication.

PROPERTIES OF DENSIFIED STEEL GRIT POWDERS WITH CARBIDES OR IRON ADDITIONS.

A.KOBYLANSKI, E.N.S.M., Saint-Etienne (France)
J.LEROUX, E.N.S.M., Saint-Etienne (France)
P.VU CHAT, E.N.S.M., Saint-Etienne (France)

D.HUET, WHEELABRATOR-ALLEVARD, Le Cheylas (France)

I-INTRODUCTION.

Steel shots, hypereutectoidal particles, 1 to 2.5 mm dia., used for abrasive shotting, are produced by a water atomization process.Larger shots are crushed in ball-mills for the grit production;the grinding process also fabricates small particles, dia. inferior to 200 μm, with no present commercial application.

The aim of our work was to study the use of the recuperated powders for an eventual application for P/M parts. Three laboratory studies have been conducted: 1-densification of steel grits (SG) using the current P/M techniques.

2-strengthening by carbides additions, VC, WC or TiC, to increase the wear-resistance.

3-mixing with different quantities of ductile iron powder to obtain a composite iron-steel material.

II-EXPERIMENTAL.

The raw material SG, fines produced during the ball-mill crushing, has the following chemical composition (wt.%):

C	Si	Mn	S	P	O
0.87	1.00	0.76	0.025	0.026	0.20

The average size lies between 100 and 130 μm and the angular shape results from repeated cleavage fractures.

Mixed powders (carbides,pure iron or copper additions) were prepared using a planetary mill for 24 hours, thus ensuring a full homogeneity.

Compacted samples for sintering, 10 mm high, 20 mm dia.,were obtained under an uniaxial load of 400 to 1,000 MPa with a 0.75 wt.% addition of zinc stearate.In a few cases, a hot-stamping at 1,100 °C was added for a full densification.

To obtain directly fully densified samples, the hot-swaging technique was applied.The SG powder, alone or mixed, was introduced in air-tight containers, 200 mm long, 26 mm dia..Then the material was degassed under vacuum before closing the containers and hot-swaging at 1,150 °C in 4 to 6 passes.The wear-resistance test used a Faville testing machine [1].

III-RESULTS AND DISCUSSION.

1-Steel grit compaction.

The hardness of the as-grinded powder, 1,150 HV0.3, was too high for a direct die compaction.Annealing at 800 °C for 2 hours, under a $H2+CH4$ atmosphere, lowered the hardness to 300 to 400 HV0.3, allowing a die compaction with a 0.75 % zinc stearate addition.

The green density was low, moving from 73 to 85 % when the applied pressure changed from 400 to 1,000 MPa.A slight increase, with 76 and 87 % respectively, was obtained after a 2 hours sintering at 1,200 °C.

The sintering time and temperature, the protection atmosphere (H2, H2+CH4, Ar) did not affect the density or the oxygen content noticeably.

The sintering kinetic was studied by dilatometry, examining the simple following kinetic equation:

$$\frac{\Delta l}{l_o} = (kt)^n \quad \text{and} \quad k=k_o \exp(-Q/RT)$$

where l_o : initial length
l : final length
k_o : constant
Q : activation energy
t : sintering time
T : sintering temperature.

n was found to lie between 0.43 and 0.49, characteristic of a bulk diffusion mechanism [2].Different measurements at 1,100, 1,125 and 1,150 °C led to a Q-value of 42.8 kcal/mol inferior to that of pure iron, 64.8 kcal/mol [3].This difference was attributed to the presence of stable Mn and Si oxides beside a high C content [1].

Sintered samples were then hot-stamped after heating at 1,100 °C for 15 minutes.No crack appeared during the dynamic deformation leading to a final density of 97 to 98 %.

Since the as-grinded powders could not be satisfactorily densified with the classical P/M techniques (compaction followed by sintering), we tried a direct hot-forging.A preliminar laboratory experiment gave encouraging results: a hot compression at 950 °C (45 °C/mn heating) under 40 MPa during 15 mn led to a density of 99 %.

Using the same raw powder introduced in air-tight containers, tapping increased the density from 3.05 to 4.06. The containers were then hot-swaged. Figure 1 shows the evolution of density with the reduction rate (number of passes) compared to that of pure iron. A full density is obtained at 65 % reduction in both cases. The tensile characteristics of these fully densified samples (T.S. = 910 MPa, elongation = 6 %) are close to those of a known P/M steel, FN-04XX , containing 0.8 % C and 4 % Ni [3].

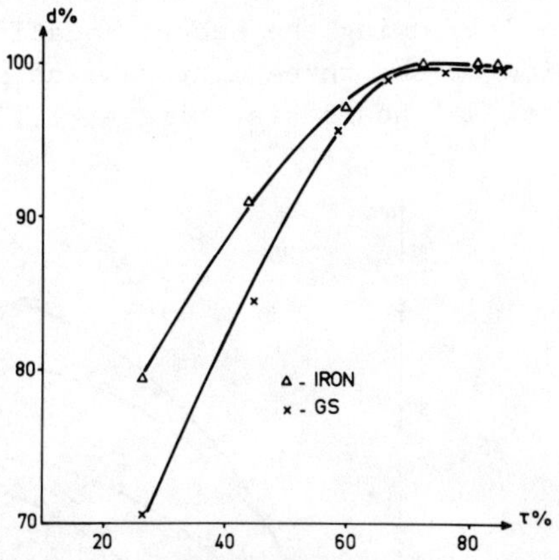

Figure 1
Evolution of density with the reduction rate

2-Carbides additions.

Previous studies [4] have shown the interest of adding carbides for the high-speed steels M2 and M4. Since our steel powder was already pre-alloyed we studied its use as a matrix for carbides additions. Six compositions, 4% or 10% of either WC, VC or TiC, were tested. 0.2% C was added before hot-swaging to compensate for the loss during operation.

The density was high in all cases and a good matrix cohesion was obtained with WC and VC. In the case of TiC, inclusions were found at the interface as it had already been observed in high-speed steels [4]. Bending tests after heat-treating (quenching at 900 °C and tempering at 300 °C) confirmed these results with, for example, the following values:

Carbide addition	Maximal strength (MPa)
4% WC	1700
4% VC	1200
4% TiC	950

The wear-resistance was measured after the same heat-treatment.The weight loss of the SG powder alone was roughly twice that of M2.A 10% WC or VC addition increased the resistance of 50% only.

3-Iron additions.

A direct industrial application of sintered or hot-swaged grits seems improbable.But steel grits mixed with a ductile commercial iron powder could give interesting characteritics for an eventual mechanical use, with a reinforcement of iron parts or the softening of steel grit parts.

For that purpose, three different mixtures were studied: 25, 50 and 75% SG (75, 50 and 25% iron respectively) with or without a 3% copper addition.

The compressibility was greatly enhanced when the iron content was increased, as can be seen on figure 3.1, using the Heckel equation [5]. Figure 3.2 compares the densification of the three mixtures and pure SG at different compaction levels, after a 2 hours sintering at 1,175 °C.

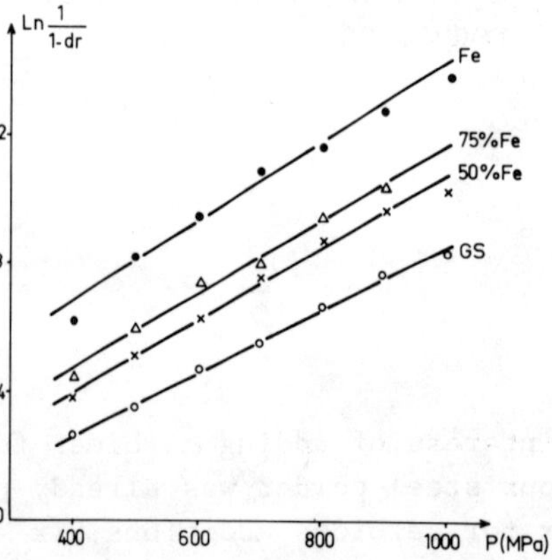

Figure 3.1
Evolution of compressibility with the applied pressure using Heckel equation

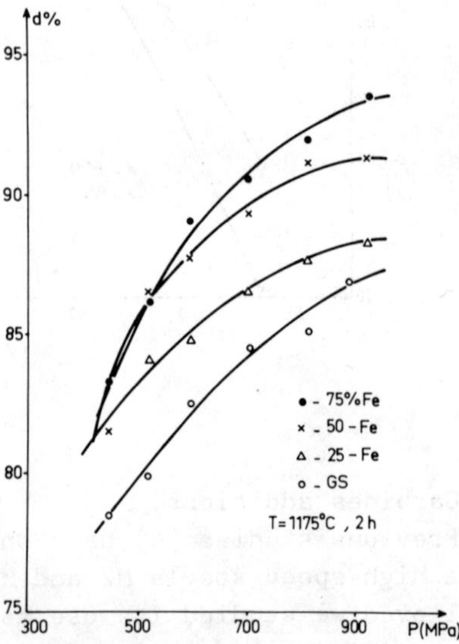

Figure 3.2
Relative density after sintering versus compaction pressure

Added to SG or the above mixtures, Copper improved the compressibility but, except for SG, left the density after sintering unchanged.

Metallographic observations showed pearlite-ferrite structures with a ratio of ferrite to pearlite equal to the iron content. Figure 3.3 shows this type of structure for a hot-stamped 50% SG-50% iron sample.

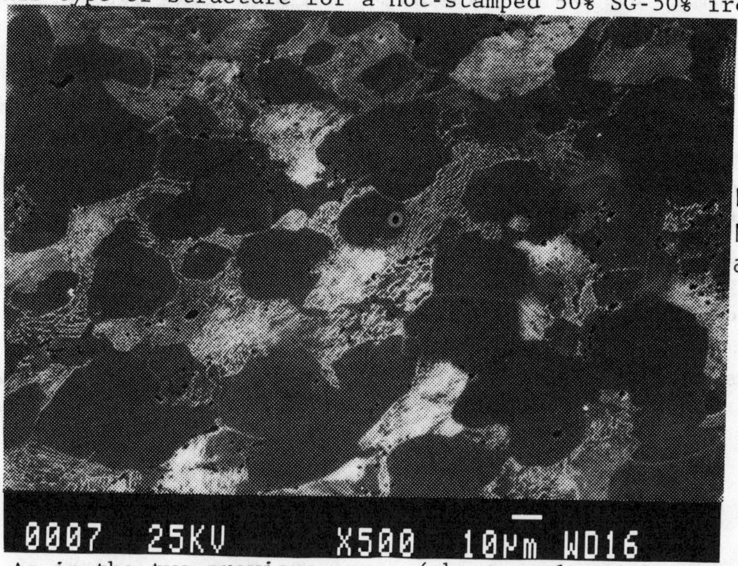

Figure 3.3. Microstructure of a 50% SG-50 % iron hot - stamped sample

As in the two previous cases (chapters 1 and 2), we obtained a full densification after hot-swaging the grit-iron mixtures. Tensile tests were applied on the three mixtures: results are reported on figure 3.4, showing the tensile strength and elongation versus composition.

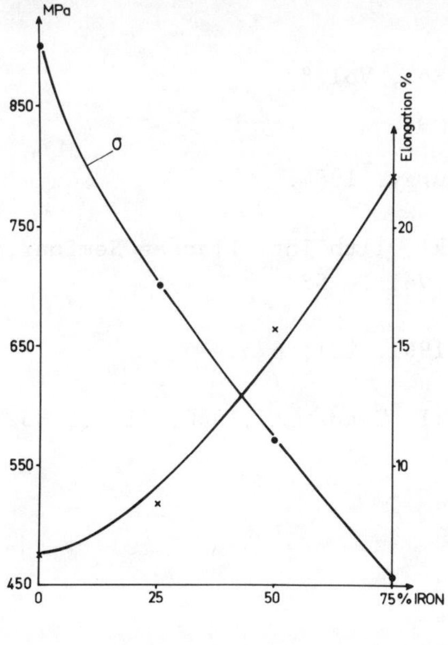

Figure 3.4
Tensile strength and elongation versus composition

Since hot-swaging is a quick operation, it was possible to leave the initial particles' structures unchanged with a short holding time at high temperatures (no great diffusion of the carbon to the iron could occur).Figure 3.3 above.

Then a heat-treatment at low temperatures, slightly above 720 °C, beginning of the α-γ transformation, followed by quenching produced structures close to the martensite-ferrite dual-phase structures.[6].

IV-CONCLUSION.

SG powders could be compacted after a short annealing.Sintered samples have shown poor densities (86%) and appeared uninteresting.Hot-stamped samples gave a better density (97%) and good mechanical values (T.S.= 910 MPa, E=6.2%).

Mixing with 3% copper increased the density of the sintered parts to 96% with the following mechanical properties : T.S.=840 MPa and E=5.3%. Carbides added samples did not show high enough wear-resistances.

The best results were obtained with iron-mixing, a 50% steel- 50% iron sample having the following tensile characteristics : T.S=570 MPa and E=16% .

Further fatigue tests should confirm the technical interest of these mixtures.

BIBLIOGRAPHY

1- P.Vu Chat, Dr Ing Thesis, 1986

2-G.Cizeron and al, Semin. Chim. Etat Sol, Vol 8, Masson SA Edit. Paris 1974,73.

3- Metals handbook, ASM, Powder Metallurgy, 1984.

4- P.Vu Chat, H.P.Aksas and A.Kobylanski, 11th Int. Plansee Seminar, Proceedings, H.Bilstein Edit., 1985, Vol 1, 587.

5- R.W.Heckel, Trans. Met. Soc. AIME, 1961, 221, 675.

6- J.Y.Koo, G.Thomas, M.J. Young, Metall. Trans. A.,1980, 11, 5, 852.

THE EFFECTS OF MICROSTRUCTURE OF GAS-ATOMIZED Zn-22%Al ALLOY POWDER ON DENSIFICATION PROCESS DURING HOT PRESSING

M.Tokizane, K.Isonishi*, M.Hashii** and K.Ameyama

Department of Mechanical Engineering, Ritsumeikan University, Kita-ku, Kyoto, Japan
* Nippon Steel Welding Products & Eng. Co.Ltd., 5-4,3-Chome, Tsukiji, Chuo-ku, Tokyo, Japan
** Graduate School, Ritsumeikan University

INTRODUCTION

In spite of extended development of hot pressing (HP) and hot isostatic pressing (HIP) technologies in industry, little is known about the kinetics of densification process during such a pressure sintering (PS). The densification process mainly progresses by elevated temperature plastic deformation of the powder particles(1). Therefor, two kinds of powders which are prepared from the same alloy and have the different microstructures, presumably show the different densification behaviors owing to the difference in the operative deformation mechanisms during PS of these powders.

In the present study, the effects of microstructure on the densification kinetics of gas-atomized Zn-22%Al alloy powder during HP were investigated.

EXPERIMENTAL PROCEDURE

Zn-22%Al alloy powder used in the present study was produced by Ar-gas atomizing process. Particles between 37μm and 105μm diameters (-145+400 mesh : JIS Z 8801) were screened from as-received powder and then prepared two kinds of samples having different microstructure. Sample-1 has as-atomized microstructure. Sample-2 was prepared by heating the Sample-1 at 623K for 3.6ks followed by furnace cooling by using a vacuum furnace. These samples were hot pressed at 523K and at various pressure.

The die assembly used for HP and its geometry are shown in Fig.1. The top and bottom plugs were loose fitted (a tolerance of 10μm) to the inner hole of the die. 8.0000g of the powder were measured and poured into the die cavity. The plugs were lubricated with molybdenum disulfide.

HP at each constant pressure was performed using a hard, mechanical screw-drive-type universal testing machine accompanied by a vacuum furnace in which the die assembly can be set up.

The changes in height of the compact were continuously measured from the movement of cross head of the testing machine and calculated the changes in relative density, ρ , and densification velocity, $\dot{\rho}$, as a function of HP time, respectively, by using a personal computer.

RESULTS AND DISCUSSION

The scanning electron micrographs of the particles of Sample-1 and Sample-2 were shown in Figs.2 (a) and (b), respectively. The particles of Sample-1 are mainly consisted of $(\alpha + \beta)$ microduplex structure while those of Sample-2 have a typical lamellar eutectoid structure.

Figures 3 (a) and (b) show the variations of the value of ρ of the compacts as a function of HP time and the relations between the values of $\dot{\rho}$ and ρ of the compacts during HP at 523K and 10MPa, respectively. From these results it is recognized that the densification of Sample-1 progresses significantly faster than that of Sample-2. The value of $\dot{\rho}$ at a given value of ρ of the compact of Sample-1 is approximately 100 times larger than that of Sample-2 as illustrated in Fig.3(b).

Such a significant difference in densification rate between Sample-1 and Sample-2 suggests that the different densification mechanisms were operative between these samples according to their individual microstructures as mentioned above. It can be presumed that, in Sample-1 used in the present study, superplastic flow which believed to progress by grain boundary sliding occurred in each particle of the powder during HP at 523K, since the particles of Sample-1 showed the $(\alpha + \beta)$ microduplex structure which closely resembles to the structure of typical superplastic Zn-22%Al alloy(2). Recently, Caligiuri, Whalen and Sherby(3) suggested that "Superplastic HP" utilizing the superplastic deformation of the particles of the powder is a possible means of accelerating deformation during HP. From such a point of view, a kinematic analysis of the densification process of Sample-1 was investigated here. According to Caligiuri(4), the densification rate, $\dot{\rho}$, during HP is represented by:

$$\dot{\rho} = C \left(\frac{1 - \rho}{\rho} P_A \right)^{1/m} \tag{1}$$

where C is constant, P_A is applied pressure and m is strain rate sensitivity exponent of the powder material during compressive deformation.

The relation between ρ and HP time and that between $\dot{\rho}$ and ρ were illustrated as shown in Figs.4 (a) and (b), respectively, from the experimental results of HP at 523K and at various constant pressure between 7 and 10MPa with respect to Sample-1. Namely, the value of $\dot{\rho}$ is relatively large during the early stage of HP at each pressure and is decreased with increasing value of ρ. Figure 5 shows the effective stress, $\sigma_E=((1-\rho)/\rho)P_A$ [4], with respect to each pressure, P_A, against $\dot{\rho}$. The results indicate that the experimental values of $\log\dot{\rho}$ and $\log\sigma_E$ can be represented approximately by the same linear relationship regardless of the value of P_A. Namely, the experimental results can be explained by Eq.(4) within experimental error. Moreover, the slope of the line which correspond to the value of m in Eq.(4) becomes approximately 0.4 as indicated in Fig.5.

From these results of the analysis, it is concluded that the densification processes of Sample-1 during HP progresses mainly by superplastic HP mechanism.

SUMMARY

The densification rate of gas atomized Zn-22%Al alloy powder consisting of ($\alpha+\beta$) microduplex structure (Sample-1) is approximately 100 times larger than that of the same alloy powder having the lamellar eutectoid structure. The experimental results of HP of Sample-1 can be explained by the derived densification rate equation from the concept of "Superplastic HP". In the present study the values of m is to be 0.4.

REFERENCES

(1) F.B.Swinkels, D.S.Wilkinson, E.Arzt and M.F.Ashby : Acta Metall., **31**(1983), 1829.

(2) P.Chaudhari : Acta Metall., **15**(1967), 1777.

(3) R.D.Caligiuri, R.T.Whalen and O.D.Sherby : Int. J. Powder Metall. & Powder Technol., **12**(1976), 9.

(4) R.D.Caligiuri : Ph.D thesis, Stanford University, (1977).

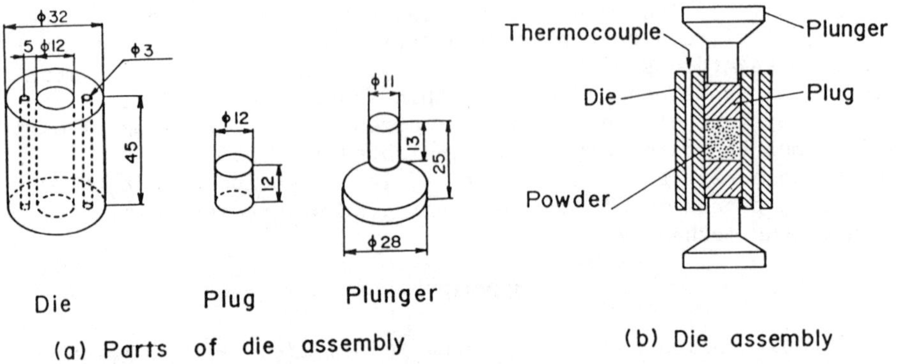

Fig.1 Dimensions of the die assembly and schematic diagram of the die assembly used for HP.

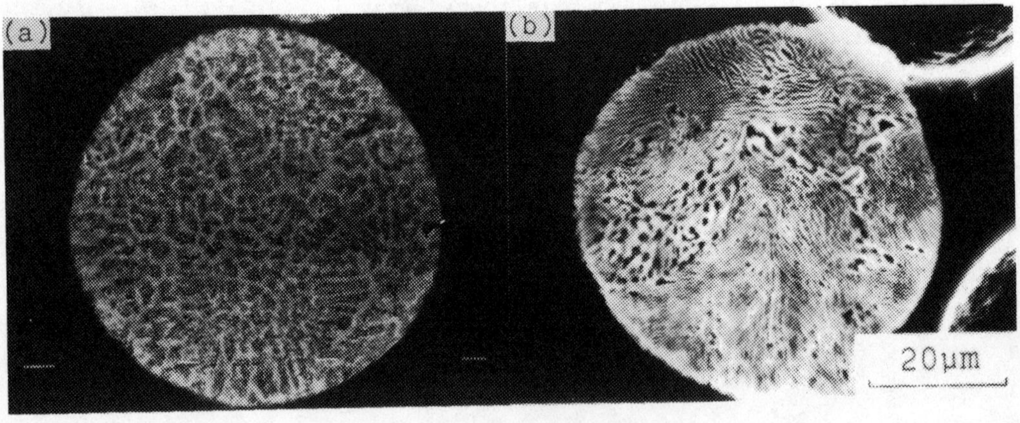

Fig.2 SEM micrographs of the cross-section of powder particles.
(a) As-atomized (Sample-1).
(b) Annealed at 623K for 3.6ks followed by furnace cooling (Sample-2).

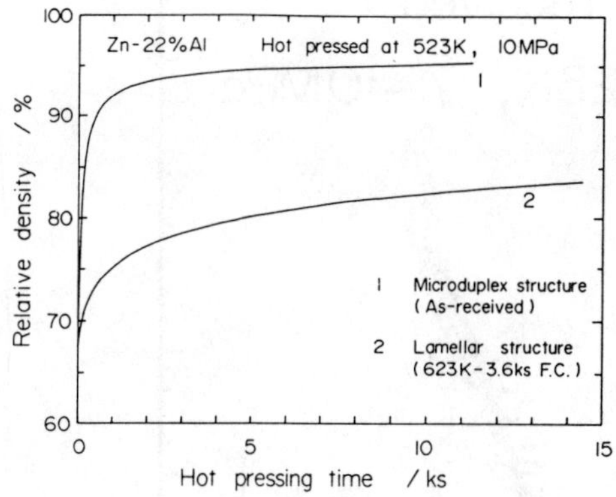

(a)

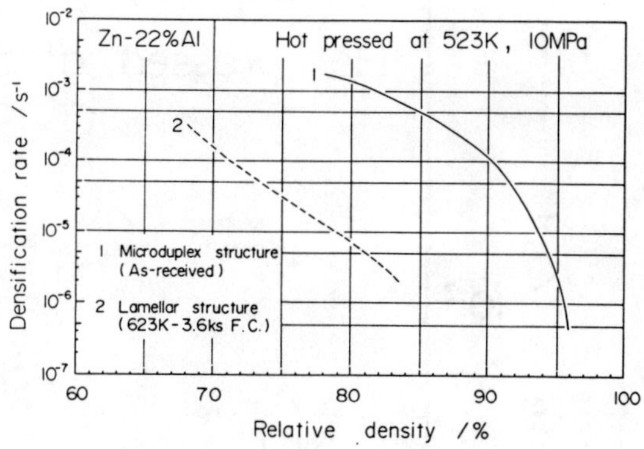

(b)

Fig.3 Comparison of densification behavior of sample-1 and -2.
(a) Relation between relative density and HP time.
(b) Relation between densification rate and relative density.

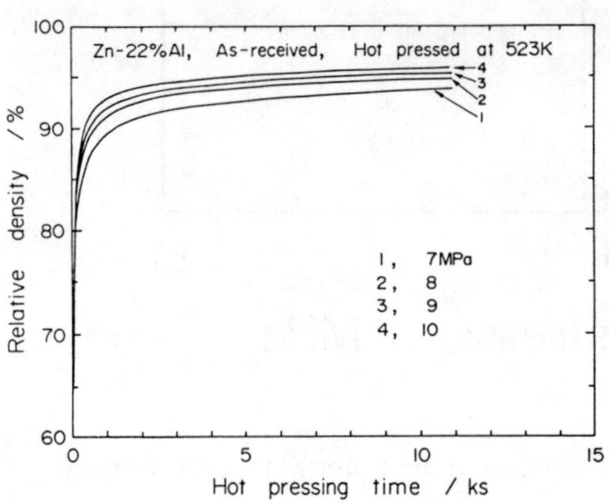

(a)

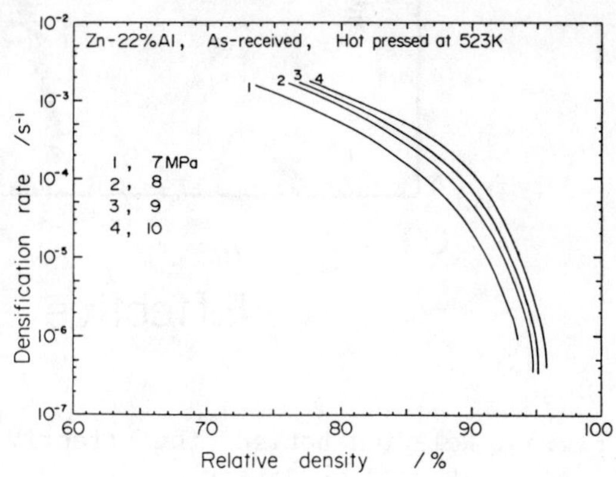

(b)

Fig.4 Effect of HP pressure on densification behavior.
(a) Relation between relative density and HP time.
(b) Relation between densification rate and relative density.

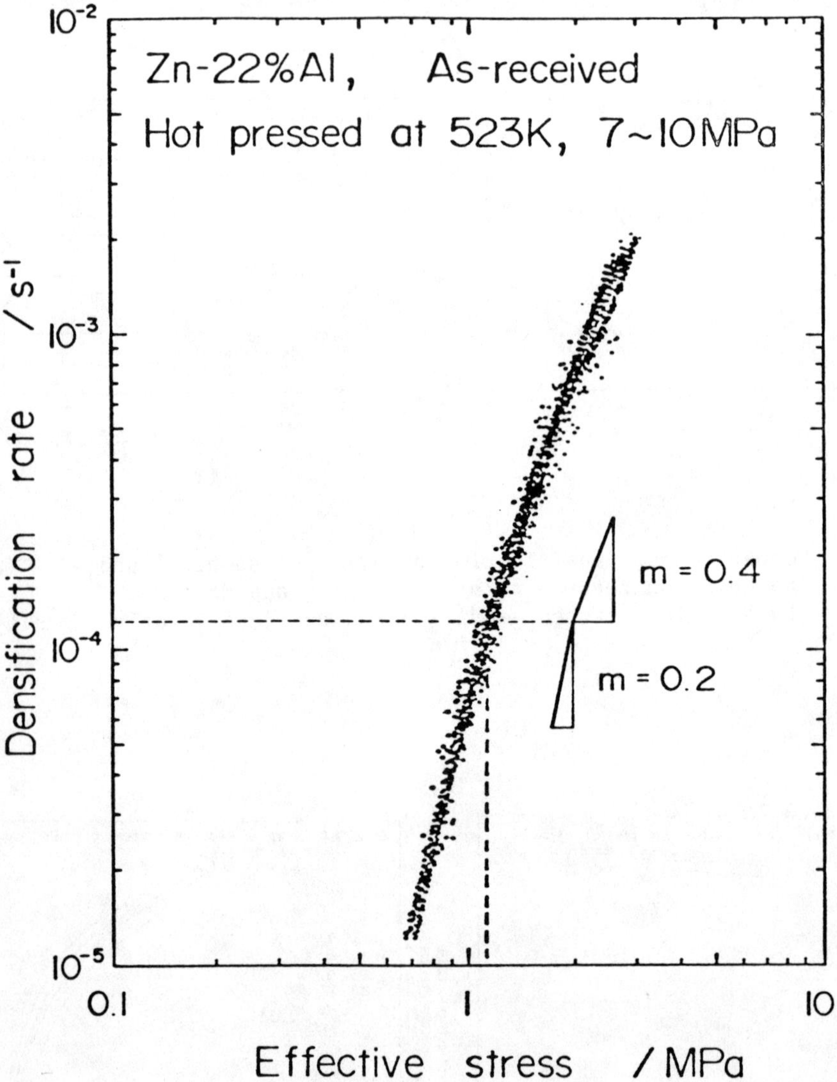

Fig.5 Relation between the effective pressure and densification rate at each HP pressure.

MANUFACTURING AND PROCESSING OF HIPPED AND/OR FORGED STEEL AND STELLITE COMPOUNDS

B. Hribernik, J. Stamberger,
H. Jäger, O. Pacher
Böhler Ges.m.b.H.

ABSTRACT

Metals depend on their analysis, manufacturing technique and heat treatmant for certain properties. These properties are interrelated for a particular material, and it is often the case that desired variations cannot be achieved as required. Values such as high wear resistance and good toughness cannot be obtained together.
With todays rapid technological progress it becomes more and more difficult to fulfill the complex demands made on materials for construction parts and tools.

One possible solution to this problem is the use of compound materials, with each steel constituting the compound being given a particular task.

Metal compounds can be produced in various ways, i.e. casting, welding, hot isostatic pressing (HIP), forging etc. Metal compounds (both parts and samples) manufactured by various methods have been investigated.

A speciality of Böhler is the production of Round Compound Materials which can also be produced by PM-techniques. The advantages of this product are shown under Example Of Applications.

Working material on the outside

Cold working die used in screw production, working material on the inside

Working material on the inside

figure 1: Compalloy-round compound steel

INTRODUCTION

The method particularly suited for the production of metallic compound materials is that of diffusion welding.

Diffusion welding carried out in a hot isostatic press entails a number of further advantages, e.g. the possibility of joining traditionally unweldable materials. Extrusion tooling, manufactured from a hot work tool steel/Co-base hard alloy combination, is just one example of the numerous practical applications of compound materials.
The compound parts are already in the solid condition prior to hot isostatic pressing.

HIP Technology also enables the production of multi-compounds, where solid and powder components are utilised. The correct choice of material allows savings to be made in both material and machining costs.

An example of this is the use of a tri-metal compound, (composed of a heat treatable steel, a Co-Base hard alloy and a free cutting steel), in the manufacture of barrels for plastic processing machines.

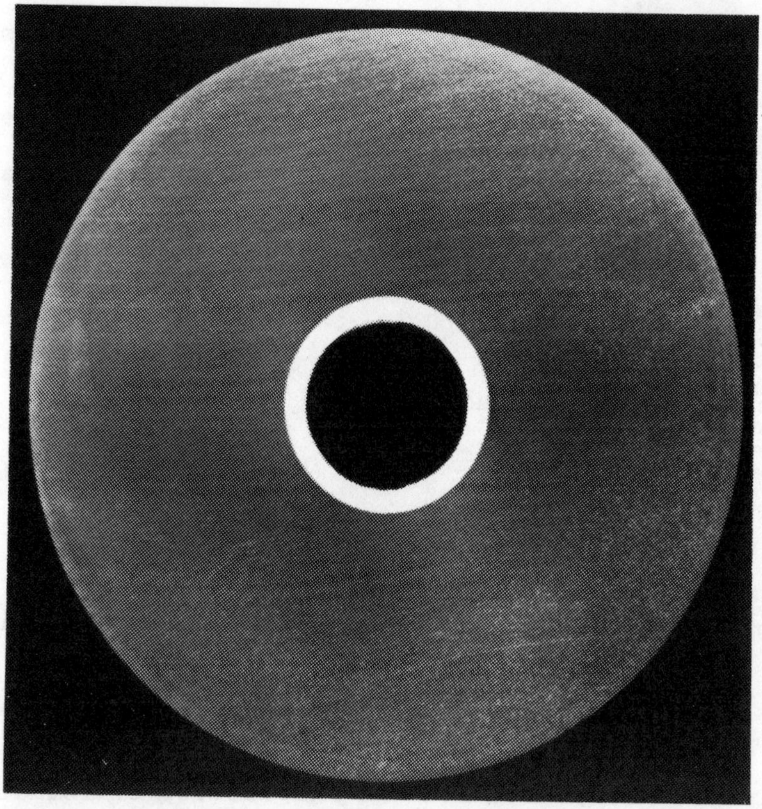

figure 2: tri-metal compound

In this case, the hot isostatic pressing of the compound material was followed by a hot forging process, which enhanced both mechanical and metallurgical properties.

In vital, highly stressed tools and components used in mechanical equipment engineering and tool manufacture the decisive properties required are cracking resistance, wear resistance and toughness.
Very frequently, a single material is not fully capable of fulfilling the complex and often contradictory requirements involved in practical application (e.g. high hardness and high toughness).

Special processes such as hardfacing, surface treatment or cladding offer improvements in certain cases.
These processes are, however costly and labour intensive.
A solution to the problem of the establishment of a metallic bond between dissimilar material types has occupied a special R&D team within Böhler for many years.

Generally, compound metals are produced in the form of flats, since these shapes do not pose any problems to the rolling and forging technologies so far employed. Hot isostatic pressing is a process involving at the same time high gas pressures and elevated temperatures. In the special steel industry, the HIP-Process is used primarily for

1. consolidation of metal powders, mainly of high speed steel and tool steel powders

2. diffusion welding

3. defect healing of special steel, Ni-base, Co-base and Ti-base castings.

DIFFUSSION WELDING

One way of joining metallic materials of similar or dissimilar chemical composition is that of difussion welding - a process in which solid workpieces are joined by the application of pressure and temperature.

Diffussion welding can be performed in air, inert atmosphere or vacuum. When diffusion welding is carried out in a hot isostatic press, it is not necessary to take into account properties such as shapability or formation of intermetallic phases. The resulting metallic bond between the contact areas is uniform and homogeneous.

Distortion is relatively low. In contrast for example to cladding by rolling, which is only possible for simple geometrical configurations, the HIP process imposes hardly any restictions as to the intricacy of the shapes to be joined.

Therefore, the HIP method offers a further step forward in the process technology of diffusion welding. In hot isostatic pressing the pressure acts uniformly from all sides, which offers several advantages over processes involving only single or two-axial pressure application.

Distortion is further reduced and surface roughness is negligible. Expensive grinding and polishing operations are rendered superfluous.

BONDING OF SOLID METALS

The simplest form of bonding different materials is the joining of a cladding plate to a backing material. Fig. 3 shows a steel plate clad with Bohler L306 (Ni-base-alloy 2.4952).

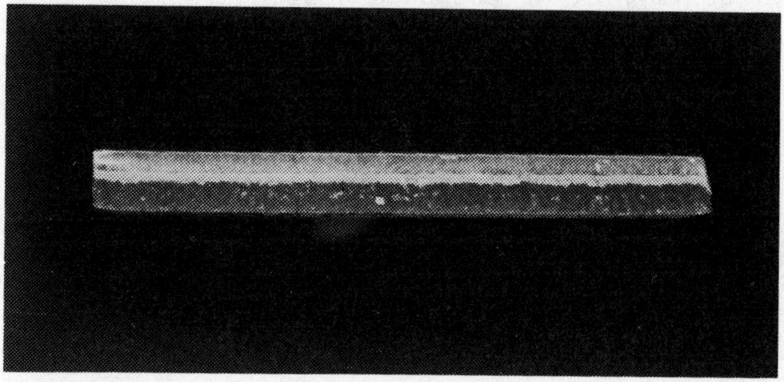

figure 3: clad steel V320 (42CrMo4) L306

The diffusion welding process imposes no restrictions as to the thickness of the cladding material other than that the two parts to be joined must fit into the HIP plant.

The encapsulation technique for this form of metal compound does not pose problems where simple configurations are required.
The most convenient case is that where the cladding material itself can be used as the capsule.

EXAMPLES OF APPLICATION

Compound steel tube production is used for various tools subjected to intensive wear stresses at elevated temperatures (e.g. fig.2).
For example, in the manufacture of dies for CU and Cu-alloy extrusions, it is the Co-base hard alloys (also called "Celsite") which are ideally suited on account of their exellent wear resistance and red hardness.

The high-temperature strength of conventional hot work tool steels is insufficient for the processing temperatures of heavy metal alloys. Co-base alloys, on the other hand, are not tough enough to withstand the high internal pressures. For this reason, the Co-base alloy tool inserts must either be hardfaced, which is rather costly, or of compound design.
The compound design allows the specific combination of the excellent high-temperature strength and wear resistance of the Co-base alloy with the high toughness of the heat treatable hot work tool steel.

This permits optimum adjustment to the relevant operating conditions. Dies of different thickness can be cut from bar by sawing, which substantially facilitates stock keeping. Bohler L290 (Stellite 21) in Compound design can be subjected to hot forming, which in turn allows an increase in the hardness values during the ageing process.

POWDER CONSOLIDATION ACCOMPANIED BY BONDING WITH A SOLID WORKPIECE

The HIP process allows powder compaction to be attained, and at the same time a metallic bond with a backing material.
L290 (Stellite 21) exhibits not only excellent high-temperature wear resistance but also extraordinary corrosion resistance (e.g. fig.4).

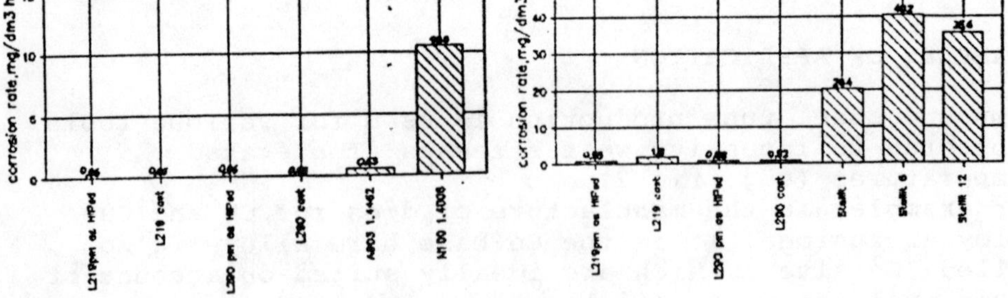

figure 4

This permits the use of L290 also for applications involving intensive corrosive and abrasive stresses of the type imposed for example on barrels for screw injection moulding machines.
As injection moulding machines may involve internal pressures of up to 3000 bars, the bursting force of the internal pressure needs to be absorbed by an outer cylinder part made from case hardening and/or heat treatable steel.

A cost reduction can be obtained in the following way:
outer part: heat treatable steel Böhler V320 (42CrMo4) in the form of bored bar

intermediate part: Böhler L219 (Stellite 6) in the form of gas-atomized powder, grain size 500 um max.

core: free-cutting steel Böhler Z952 (1.0737, 9SMnPb36)

Powder compaction and metallic bond are affected in one single HIP cycle.

In order to obtain the required overall lengths, to improve capacity utilization of the HIP plant and to enhance the mechanical and metallurgical properties, hot forming to the cross section shown in the figure 2 took place after hot isostatic pressing.

This has allowed the production of compound steel tubes measuring more than 10 m in length, which is several times the standard length of HIP plants. The boring operation is facilitated by the free-cutting steel in the bar core.

Metallographic examination of HIPped and HIPped+forged round compound steel W300 (H11) hot work steel and L219-PM (stellite 6) were carried out.

In figure 5 and 6 the etched samples are shown. The dark area near the W300-core is iron-rich. The bright zone near the stellite is Co-rich.

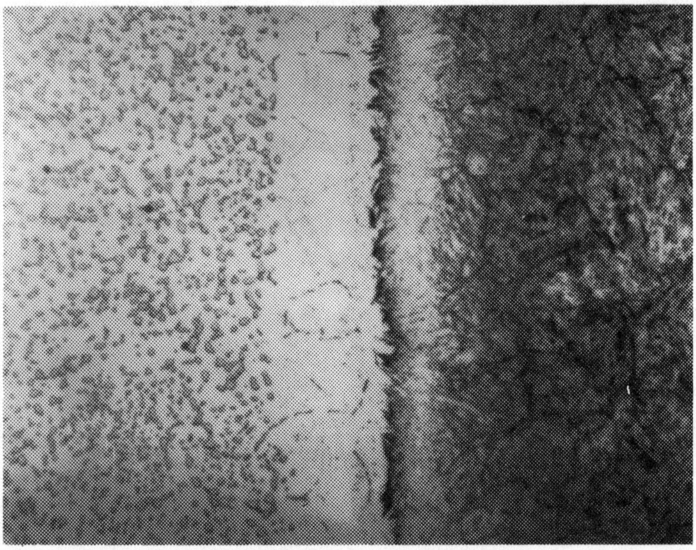

figure 5: HIP+heat treatment
40% Na OH + HNO_3

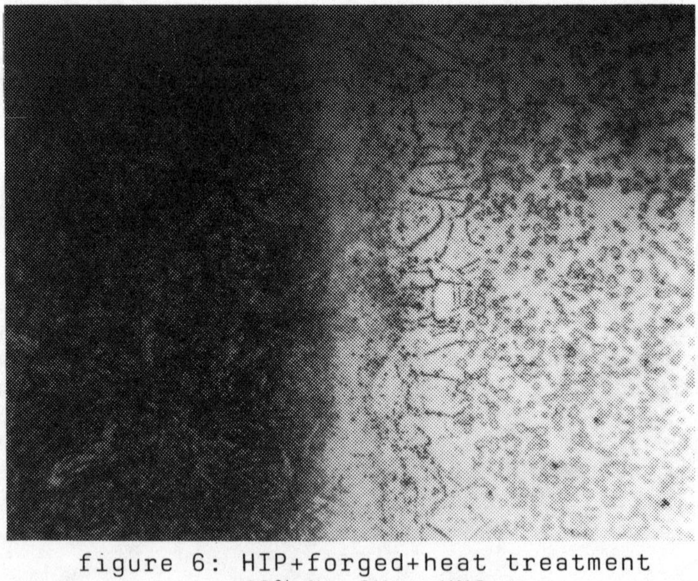

figure 6: HIP+forged+heat treatment
40% Na OH + HNO_3

The significant difference between the HIPped and the HIPped+forged compound lies in the grain size. After a forging rate if 50% the grain size of W300 drops from ASTM3 to ASTM 7-8.
The fine grained microstructure of the HIPped and forged stellite is the reason for the better mechanical properties, when compared with material in the as-HIPped condition.

DISCUSSION

The HIP process has assumed an increasingly important role in the production of compound materials. The combination of material with different properties such as

hardness - toughness

corrosion resistance - high strength

high-temperature strength - toughness

wear resistance - toughness

promises to solve a number of problems relating to material choice. Hot forming after hot isostatic pressing allows grain refinement to be brought about thus a further improvement in material properties.

Lit. (1) Landolt Börnstein, 6 Auflage, Band IV/2a
(2) British Corrosion Journ. 1986, Vol. 21, No.4, pp.257
(3) Druckschritt Chemische Beständigkeit nicht-rostender VEW-Edelstähle Al 170D, 1184
(4) Schweißtechnik 6/85 S.90-95

MECHANICAL ALLOYING OF ALUMINIUM-LITHIUM-MAGNESIUM POWDERS

Albir A Layyous*, S Nadiv** and I J Lin**

* ISCAR LTD, Box 34, Nahariya, Israel 22100
**Depts of Materials Engineering and Mineral Engineering
Technion, Israel Institute of Techology, Haifa, Israel 32 000

ABSTRACT

The production of high-purity aluminium-lithium-magnesium alloy powders, by mechanical alloying through grinding in a vibratory mill under high vacuum at room temperature, is described in detail. The source materials for the grinding mixture were aluminium-lithium alloy powder obtained by thermal vacuum-dehydrogenization of $AlLiH_4$ hydride, magnesium metal powder and chemically deoxidized aluminium metal powder. The implications of the high reactivity of the component elements are discussed, and the measures taken to overcome them are described. The procedures used for the chemical analysis and powder characterization are given.

The mechanical properties and the microstructures listed in this paper were determined with specimens of different compositions and made by hot-pressing in vacuum.

INTRODUCTION

Aluminium-lithium systems (up to 4.5 wt.% Li) alloyed with other constituents (e.g., magnesium or copper) appear to hold high promise for the avionic industry; and because of their "plasma- confinement" properties also have good prospects in nuclear engineering. They are notable for lower densities and higher moduli of elasticity compared with other, more conventional aluminium alloys. For example, the density of the Al-2.9% Li alloy is 2.49 gr/cm^3, as against 2.78 gr/cm^3 of the 2024 and 2.80 gr/cm^3 of the 7075 alloy. Its modulus of elasticity under high-strength conditions (following aging and extrusion) is 80.6 x 10^3 MPa, and per unit density (specific stiffness) is, respectively, 21% and 20% higher than those of the latter two alloys. The aluminium-lithium alloys are strengthened by a conventional aging process. Details on their microstructures and related mechanical properties can be found in relevant publications [1-7].

The present study was undertaken with a view to examining the possibilities for the production of suitable alloy powders by the mechanical alloying technology [8-13]. Mechanical alloying consists in grinding a mixture of the pure constituents in the desired proportions in a ball mill at room temperature; it is intended to some extent as a cheaper substitute for the sophisticated process based on high- temperature atomization to alloy melt droplets, immediately followed by high-rate solidification. In addition to its lower cost, mechanical alloying has specific advantages where highly reactive constituents -- such as aluminium, lithium and magnesium -- are involved under atmospheric conditions, and also when there are constituents with a high partial vapor pressure at their liquid temperatures, as is the case with lithium.

In the ball-mill, the powder particles undergo numerous repeated sequences of plastic flattening, fracturing, cold welding and strain hardening. At first their free energy is increased through an increase in the stored elastic strain energy, which in turn is converted into specific surface energy by fracturing -- both processes of activation. Next, the accumulated free energy is reduced through agglomeration and cold-welding of the particles -- a process of relaxation. Under plastic deformation, part of the energy is released in the form of heat, whereas the remainder is associated with the strain-hardening effect: as the particles' size decreases, they become progressively harder as their tendency to cold welding is reduced. A laminar microstructure thus appears, with random layers of the different constituents. Upon further grinding, the layers gradually decrease in size to below the resolution of optical microscopy and finally become amenable to conversion, by sintering, into a homogeneous solid solution with a very fine polycrystalline microstructure. These particles have irregular geometrical shapes, suitable for high packing density during consolidation; and, on the other hand, they are free of interdendritic microsegregations and pores, such as are sometimes encountered in atomized and rapidly solidified material.

The difficulties inherent in the production process described are due to the wide disparity between the three constituent metals. Pure aluminium is ductile, whereas magnesium is relatively brittle. Fine magnesium powder is pyrophoric and tends to ignite spontaneously under certain ambient conditions. Pure lithium metal can not be handled freely and must always be kept in a protecting liquid (e.g, paraffin oil); it is very soft and is characterized by a high rate of sublimation at moderately high temperatures. Moreover, the free surfaces of aluminium and magnesium particles are invariably coated with an amorphous oxide film, which interferes with the consolidation process and affects the quality of the product.

Accordingly, the technological problems to be solved are: devising a lithium-containing master alloy that can be handled freely at ambient conditions; maximum removal of the oxide film; and overcoming the differences in hardness.

EXPERIMENTAL PROCEDURE

Aluminium-lithium-magnesium powders of different compositions were obtained after grinding in a vibratory ball mill under 10^{-4} mm Hg vacuum. Full details of the process for producing the powders, especially of the use of a master alloy, are given in a previous paper [14].

The hot pressing was carried out in a special machine permitting control of the temperature, the pressure and the vacuum. Solution heat treatment was carried out in a salt bath, and aging took place in an oil bath. The conditions for the different treatments were worked out on the basis of experiments designed to achieve optimal conditions for obtaining good mechanical properties. At each stage of the production, as described in the schematic diagram, Fig. 1, the lithium content was determined by Atomic Absorption Spectroscopy (AA), so as to ensure that no sublimation of lithium had taken place.

Cylindrical specimens, 10mm long and 6.66mm in diameter, were compressed in an Instron machine with a view to determining the mechanical properties of the different alloys as well as the effects on them of changes in the Li or Mg content.

The fractures were fractographically characterized with the aid of SEM and EDAX, while the microstructure and the precipitates were examined by STEM.

Mechanical Alloying of Al-Li-Mg Powder

Direct mixing and grinding of the constituents in the proportions corresponding to the target concentration is inefficient and cannot yield a homogeneous alloy (especially when one constituent is dominant while the others contribute at most a few weight percent). The correct procedure calls for, as a first stage, a master alloy powder containing large amounts of the minor constituents; in the second stage this powder is "diluted" to its final concentration by further additions of the dominant constituent. The tendency of the soft aluminium particles to agglomerate and to adhere to the grinding medium of the ball mill is obviated by the presence of a brittle co-constituent which interposes itself between the aluminium particles and keeps them apart. Both magnesium and AlLi serve this purpose.

The alloys examined in the present experiment were expected to contain up to 4% Mg and/or Li. The master powder was produced by grinding a mixture of 4.44 g coarse magnesium power (BDH Chemicals) and 5.56 g fine AlLi powder (obtained by the thermochemical decomposition of $AlLiH_4$ for 12 hours at 10^{-4} mm Hg). The aluminium powder used as dilutent (at the rate of 10.1 g per g of the master powder) was thoroughly deoxidized in advance by immersion in a medium consisting of 20 g solid chromic acid crystals dissolved at boiling point in 35 ml phosphoric acid (85%) for 10 minutes. It was then washed with distilled water, strained, dried at 100°C, and finally cooled to room temperature in a desiccator in technical vacuum. The average weight loss due to the deoxidation treatment was 0.46% (Fig. 2).

Longer grinding is advantageous since it ensures mechanical alloying, but this advantage is offset by the powder's tendency to agglomerate. Even after agglomeration, some unalloyed AlLi is still present, as indicated by a residual reactivity towards water. Cold-welding of the particles within an agglomerate is incomplete, and some voids remain.

Mechanical Properties and Microstructures

The effectiveness of mechanically grinding the alloyed powders having different Li and Mg concentrations was reflected in the properties of the alloys after hot pressing in vacuum and solution heat treatment and aging to maximum hardness. The mechanical properties were determined by the construction of stress-strain curves from the results of the compression tests on the Instron machine.

Hot pressing in vacuum resulted in aluminium specimens having a single-phase microstructure, the two-phase structure obtained after mechanical alloying having been eliminated by hot pressing (Fig. 2).

Oxidation at the end of the manufacturing process, whether of the Li or the Mg, was minimal, and not even a low level of oxides could be detected by either XRD or Transmission Electronic Microscopy (TEM).

The higher the Li content rises above 1 wt% and up to 3 wt%, the magnesium level remaining constant (Al, 2.2 wt% Mg), the better the elastic modulus, the yield stress and the fracture stress. At the same time, however, there is a drop in ductility and in the energy to be applied until fracture (Fig. 3).

In the same series of alloys with varying Li concentrations, the material density was reduced from 2.65 gr/ccm in the alloy (Al, 1 wt% Li, 2.2 wt% Mg), to 2.58 gr/ccm in the alloy (Al, 3wt% Li, 2.2 wt% Mg). Bearing in mind the increase in the elastic modulus and the drop in material density in the range of concentrations quoted, an improvement of 30-40% is seen in the specific elastic modulus.

With constant Li concentration (Al, 2wt% Li) and Mg contents varying between 2.2 wt% and 4.4 wt%, a certain improvement in the mechanical properties is seen as the Mg content increases (Fig. 4), although it is not as pronounced as that brought about by the increase in Li concentration.

Increases in the Mg concentration have no effect on the ductility of the material, but increase its strength and the yield stress. They cause a relatively small change (reduction) in the elastic modulus, so small in fact that it may well be within the range of error. In other words, the Mg concentration has no effect on the elastic modulus in the range covered. The considerable improvement in the elastic modulus of Al-Li alloys is connected with their principal hardening mechanism, viz. the precipitation of the δ' (Al_3Li) phase, coherent with the matrix, as a result of aging. At a later stage in the aging process, the following transformation takes place [15, 16]:

$$\delta' (Al_3Li) \dashrightarrow \delta \ (AlLi)$$

With Mg present in the alloy, the transformation of the precipitates after aging is this: $\delta' (Al_3Li) \dashrightarrow s \ (Al_2MgLi)$

The s (Al_2MgLi) precipitates are not coherent with the matrix and do not contribute to the increase in hardness of the material.

The influence of alloying Al with Li was conspicuous in several respects, especially by the increase in hardness following solution heat treatment and aging. For example, the maximum micro-hardness of the alloy (Al, 3 wt% Li, 2.2 wt% Mg) is 180 kg/mm^2.

The influence of the microstructure on the mechanical properties is made manifest in the characterization of the precipitates after aging to maximum hardness. In the various diffraction patterns the appearance of streaks testifies to the formation of disc-shaped precipitates, the thin place of which lies on a plane perpendicular to the direction of the streaks. In other words, preferred precipitation has taken place on preferred planes as dictated by the composition of the material and the heat treatment conditions. In the alloy (Al, 2 wt% Li, 3.3 wt% Mg) the streaks are in the [110] direction, while the small dimension of the δ' (Al_3Li) precipitates lies on the (110) planes (Fig. 5).

Another factor that affects the mechanical properties and that can be directly linked to mechanical alloying is the fine-grained structure of the material (Fig. 6). In TEM a mixed microstructure is discerned which consists of coarse grains having a diameter of a few microns and of fine grains of ~0.5 micron diameter. The presence of these fine grains contributes to the increase in yield stress at room temperature. During grinding ("mechanical alloying"), high plastic energy is invested, and this is reflected in the reduction of the material grain size.

CONCLUSIONS

1 The addition of a brittle metallic phase, such as AlLi (which is obtained by the elimination of hydrogen from the appropriate hydride) to a powder of ductile aluminium while using a master alloy led to the prevention of agglomerations and eventually to a mechanically alloyed powder of a composition that rendered it suitable for sintering.

2 An increase in the lithium content produced an increase in the specific elastic modulus and in the material strength, as well as to a drop in the ductility to fracture.

3 The improvement in the mechanical properties is due to the precipitation of the δ' (Al_3Li) phase and to the formation of a fine-grained microstructure.

4 The proposed technology is a novel one, the proposed raw material for the process is unique, the production conditions are specific -- the three elements that make up the novelty of the present work.

REFERENCES

1 B Noble, G E Thompson, **Met Sci J,** 5 (1971), 114-120.
2 T H Sanders, E S Balmuth, **Metal Progress,** 113 (1978), 32-37.
3 H Suzuki, M Kanno, N Hayashi, **Aluminium,** 58 (1982), 120-122.
4 E A Stark, T H Sanders, I G Palmer, **J of Metals,** (August 1982), 24-32.
5 B Noble, S J Harris, K Dinsdale, **Metal Sci,** 16 (1982), 425-430.

6 B Noble, S J Harris, K Dinsdale, **J Mater Sci,** 17 (1982), 461-468.
7 T H Sanders, Jr and E A Starke, Jr, editors, **Aluminium-Lithium Alloys II**, Monterey, California (1983).
8 J S Benjamin, **Met Trans,** 1 (1970), 2943-2951.
9 J S Benjamin, T E Volin, **Met Trans,** 5 (1974), 1919-1934.
10 J S Benjamin, R D Schelleng, **Met Trans,** 12A (1981), 1827-1832.
11 R D Schelleng, S J Donachie, **MPR,** (October 1983), 575-576.
12 P S Gilman, J S Benjamin, **Ann Rev Mater Sci,** 13 (1983) 279-300.
13 (a) S J Donachie, P S Gilman; (b) P S Gilman in **Proc 2nd Int Al-Li Conf,** Monterey, April 12-14, 1983.
14 A Layyous, S N Adiv, I J Lin, **Powder Met Int,** 1 (1987), 11-13.
15 C L Angerman, **J Nucl Mat,** 1, 11 (1964), 41-52.
16 D B Williams and J W Edington, **Acta Met,** 24 (1976), 323-332.

FIGURE CAPTIONS

Fig. 1: Schematic diagram of the Al-Li-Mg mechanical alloying and sintering process.

Fig. 2: X-ray diffraction spectra following mechanical alloying for the production of Al, 2% Li, 4.4% Mg powder.
(A) Before mechanical alloying.
(B) After mechanical alloying (24 hrs).
(C) After sintering.

Fig. 3: The mechanical properties of (Al, 2.2% Mg) with different Li concentrations.

Fig. 4: The mechanical properties of (Al, 2% Li) with different Mg concentrations.

Fig. 5: TEM micrographs from Al 2% Li, 3.3% Mg after solid solution and aging.
(a,b) Electron diffraction with streaks from δ' (Al_3Li).
(c) Dark field from (200) with the precipitates.
(d) Bright field.

Fig. 6: TEM micrograph of Al 2% Li 2.2% Mg after solid solution and aging.

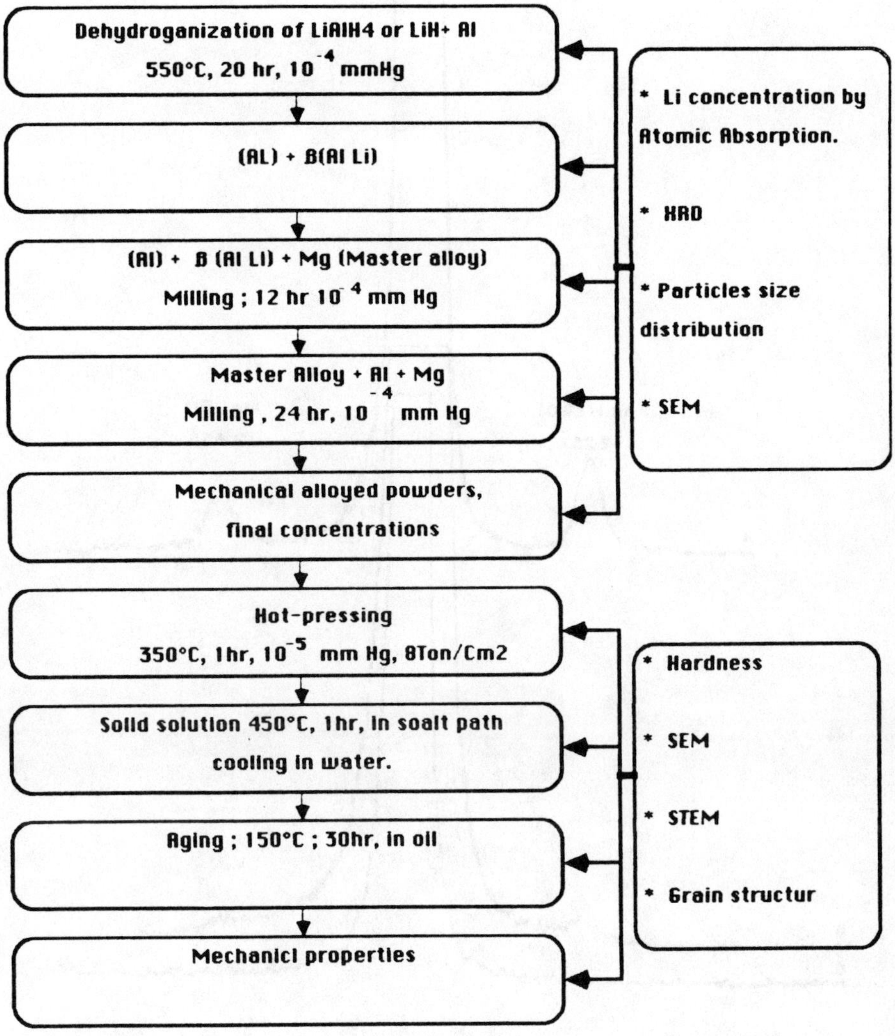

Fig. 1: Schematic diagram of the Al-Li-Mg mechanical alloying and sintering process.

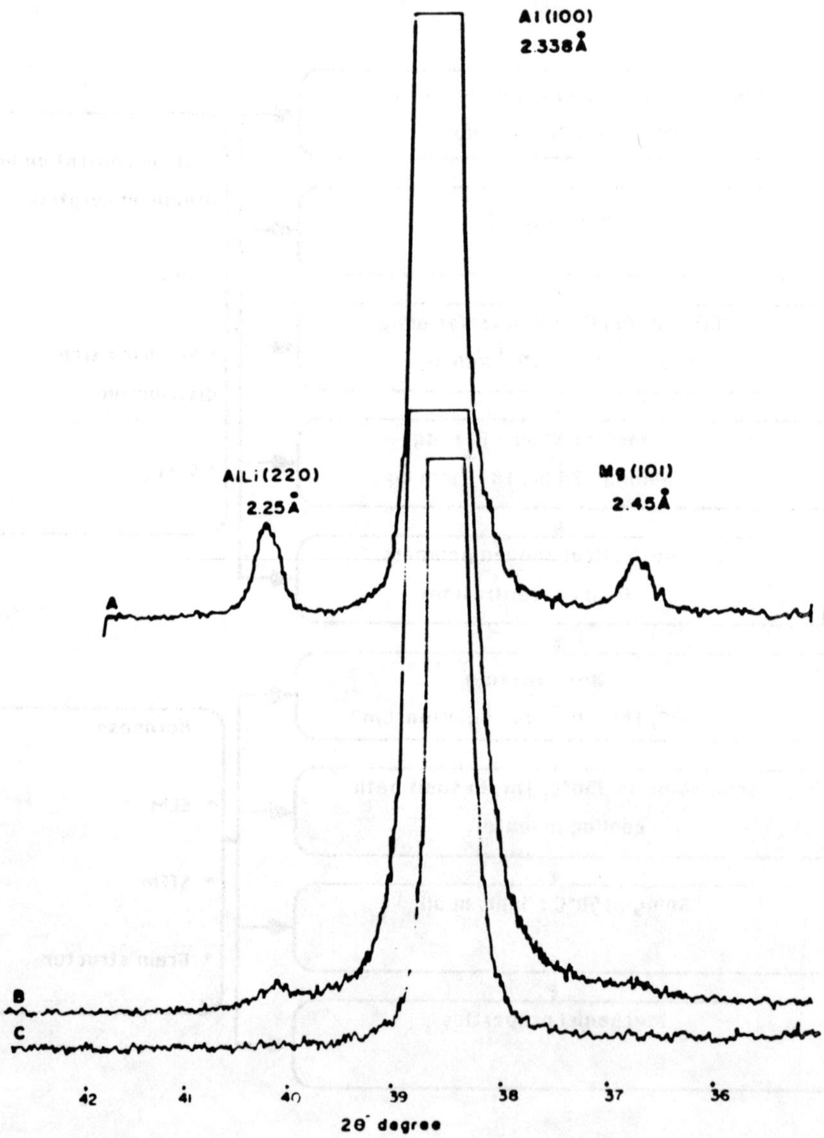

Fig. 2: X-ray diffraction spectra following mechanical alloying for the production of Al, 2% Li, 4.4% Mg powder.
(A) Before mechanical alloying.
(B) After mechanical alloying (24 hrs).
(C) After sintering.

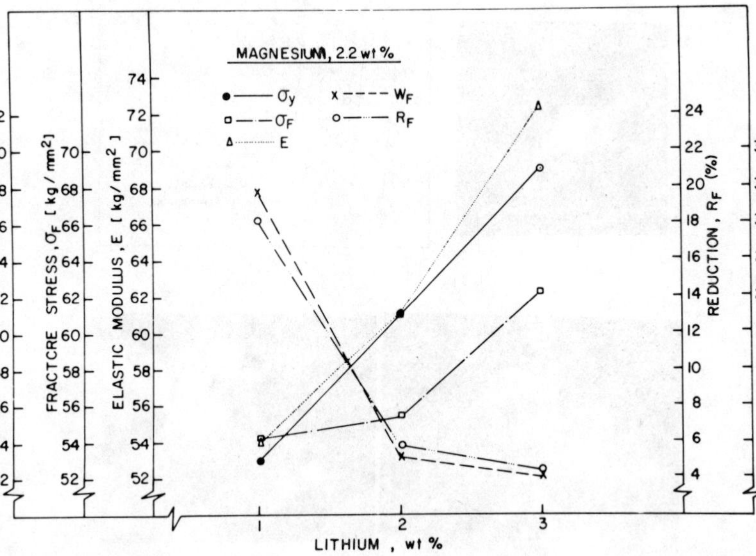

Fig. 3: The mechanical properties of (Al, 2.2% Mg) with different Li concentrations.

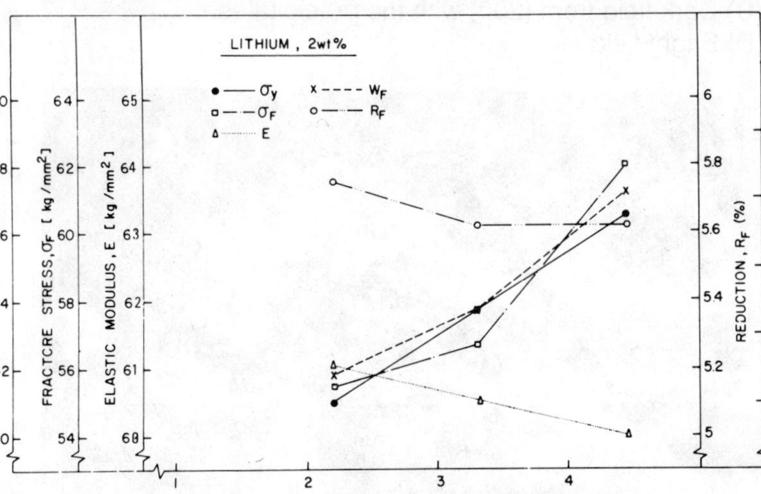

Fig. 4: The mechanical properties of (Al, 2% Li) with different Mg concentrations.

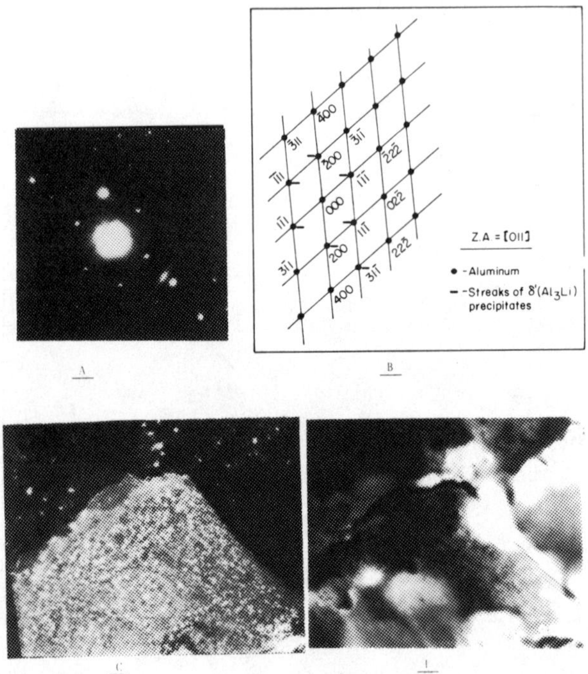

Fig. 5: TEM micrographs from Al, 2% Li, 3.3% Mg after solid solution and aging.

(A, B) Electron diffraction with streaks from δ' (Al_3Li).

(C) Dark field from (200) with the precipitates.

(D) Bright field.

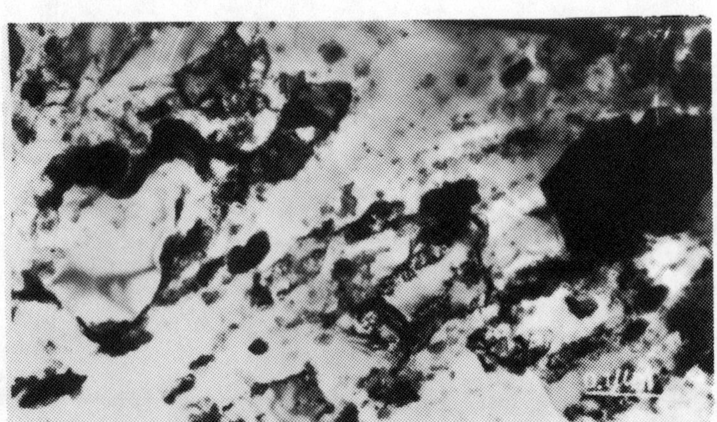

Fig. 6: TEM micrograph of Al 2% Li 2.2% Mg after solid solution and aging.

NEW APPLICATIONS OF SINTER TECHNOLOGY FOR PRODUCTION OF MACHINE ELEMENTS

Zygmunt MARCELA, Ryszard MOSZUMAŃSKI
Zbigniew POLAŃSKI, Andrzej SUŁKOWSKI
Władysław TABOR

Technical University
Cracow, POLAND

Abstract

The purpose of the work was to master production of several assortments of products of special properties with the use of unconventional methods of sinter technology. Results of performed laboratory tests confirmed usefullness of pulsatory shaping for produktion of machine elements of complex shape and better properties.

Processes of production of a large group of products of constructional materials based on Fe and non-ferrous metals were worked out. It is possible to mention as examples a lever for electronic equipment,querns and bearing sleeves for cutting devices. With the use of strong magnetic field, anisotropic ring magnets for loudspeakers and cores for transformers were produced. Dies, trimming dies and screw-taps for bolt industry were shaped at increased temperatures.

By method of isostatic pressing a table, electronic and sanitary ceramics ware shaped.

Elements of CAD/CAM were applied for design work and control of processes.

Positive results were obtained for tests performed on a sample series of products.

Analysis of production efficiency showed essential advantages after application of innovative methods of sinter technology.

Investigatin of technological processes

Investigation was carried out to observe and measure parameters of phenomena during pressing under classical and pulsating conditions. Standard samples ⌀ 25,0 and ⌀ 12,7, as well as functional products such as strainers, sleeves, shoes, rings segments and tool fittings under laboratory and semi-technical conditions. Comparisons were made for ferrous, copper, tungsten and porcelain powders. Comparative examinations of powder pressing under classical and pulsating conditions concerned |Fig.1a| feeding, compaction integration and classical ejection, then |Fig.1b| feeding, settling, compaction, integration and pulsating ejection. The input quantities are: height h and diameter D of feeding chamber, pressing pressure p_p and (for processes realized by pulsatory technique) frequency of function f and generating pressure p. The output quantities are: feeding density ϱ_z, pressing density ϱ_p and ejection density ϱ_w, ejection pressure p_w and (for processes realized by pulsating technique) size of settling Δh. The constant values for a given series of tests are: material parameters M, tool parameters N and parameters of a technological machine MT. To examine processes on a laboratory prototype of the PXP 1.00 press the measuring path was used, which consisted of assemblies of force transducers F, pressure transducers p, displacement l and frequency f. Signals were transmitted to amplifiers W_1 and W_2, meter M and voltmeter V (scale in frequency units). Recording was accomplished by means of a 4-channel measuring tape-recorder RM, set of oscilloscopes Os_1 and Os_2 , and recorder xyt.

The experiments carried out fully confirmed that it is advisable to apply pulsating force in all phases of powder pressing process. A vibrating lower stamp caused settling in 7-26 % of feeding chamber height. It made possible to introduce an upper stamp into a die without blowing powder out. During realization of pressing process a distinct decrease of pressing force value to achieve the same pressing ratio was stated. The differences amounted to 30 %. Similarily during compact ejection the lowering of maximum ejection force to 25 % was observed, as well as a clear modification of character of phenomena taking place. Self-excited vibrations (especially clear while tungsten carbide pressing, ceramic material and ferrite) vanish.

The decrease of forces in pressing and ejection processes cause the lowering of energy-consumption of processes (in spite of extra energy demand for generator driving) and unloading of tool.

The estimation of quality of a one-side presses sinter and a two-side press sinter was made, which revealed the improvement of structure uniformity in both technological instances. It proves that friction forces on a die surface and cores diminish, which enables pressing forces to influence on layers distant from the stamp surface.

Main phases of pressing of bearing sleeve are presanted in Fig.2a, final product in Fig.2b.
Exemplary courses of pressing and ejection forces, and quality parameters of product are presented in Fig.2c.

Main phases of pressing of magnetic ring are presented in Fig.3a, final product in Fig.3b.
Exemplary courses of displacement of tool (with vibratory amplitude), course of magnetic field and quality parameters of product are presented in Fig.3c.

Main phases of pressing of trimming die are presented in Fig.4a, final product in Fig.4b.
Exemplary courses of displacement of tool (with vibratory amplitude) and course of temperature are presented in Fig.4c.

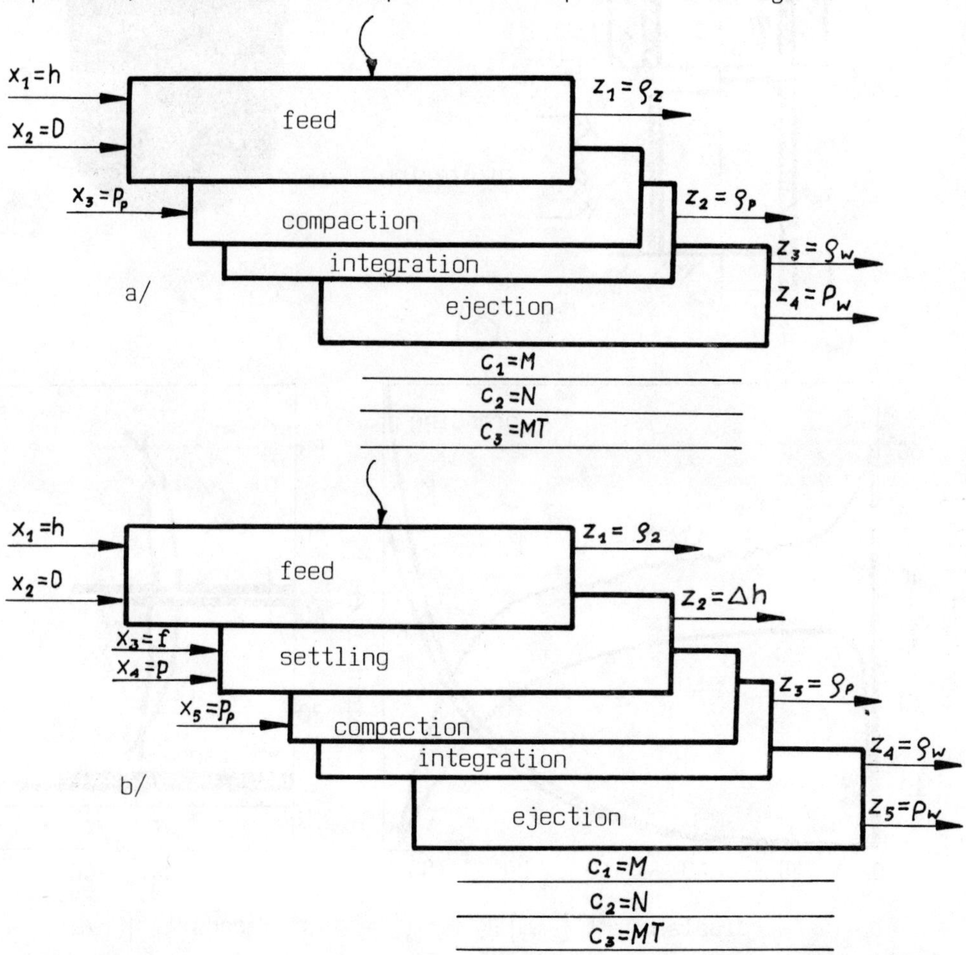

Fig.1.The characteristics of powder pressing processes under classical (a) and pulsating (b) conditions

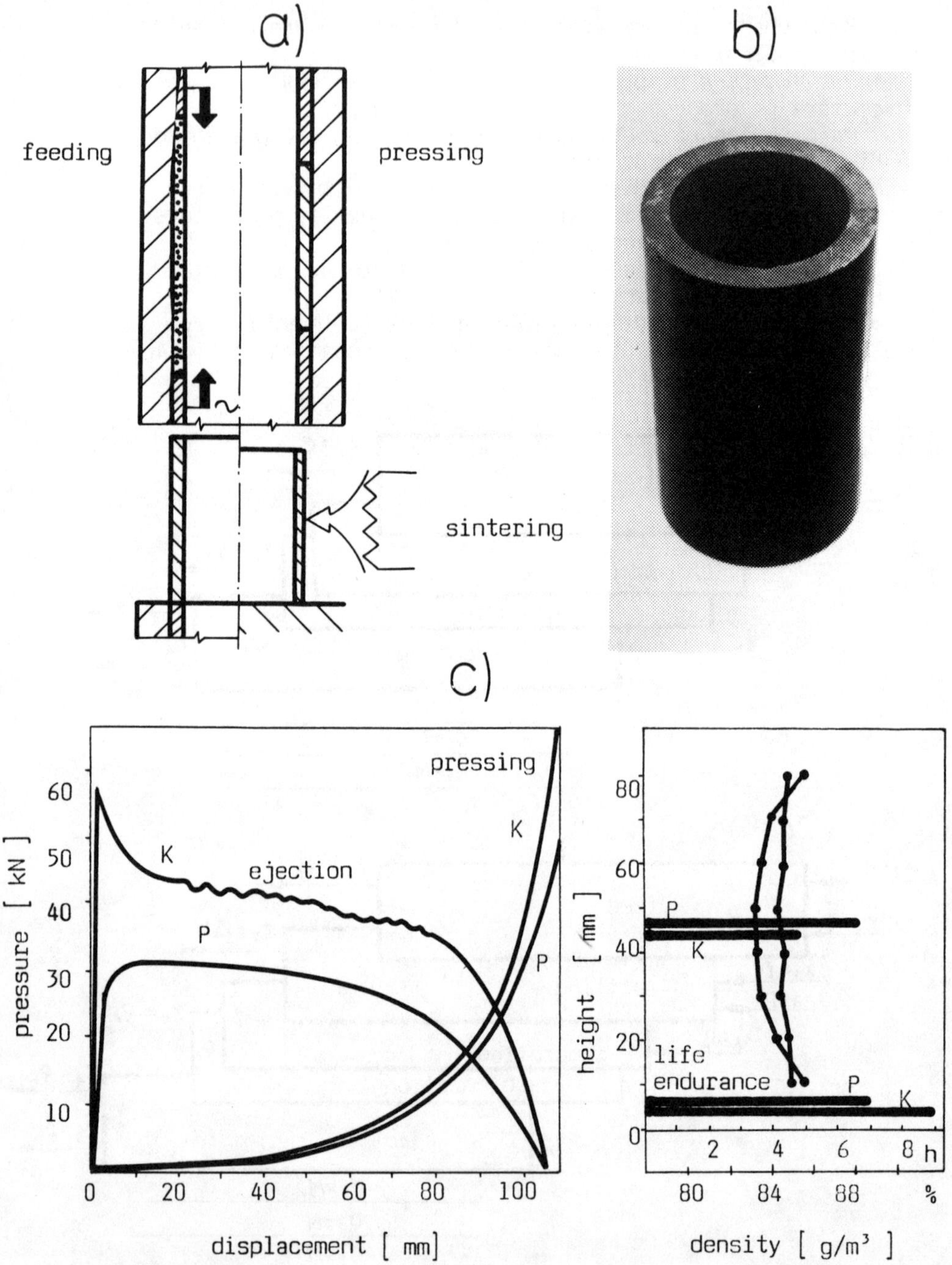

Fig.2.Main phases of pressing of sleeve bearing (a),final product (b), parameters of pressing and quality of product (c).

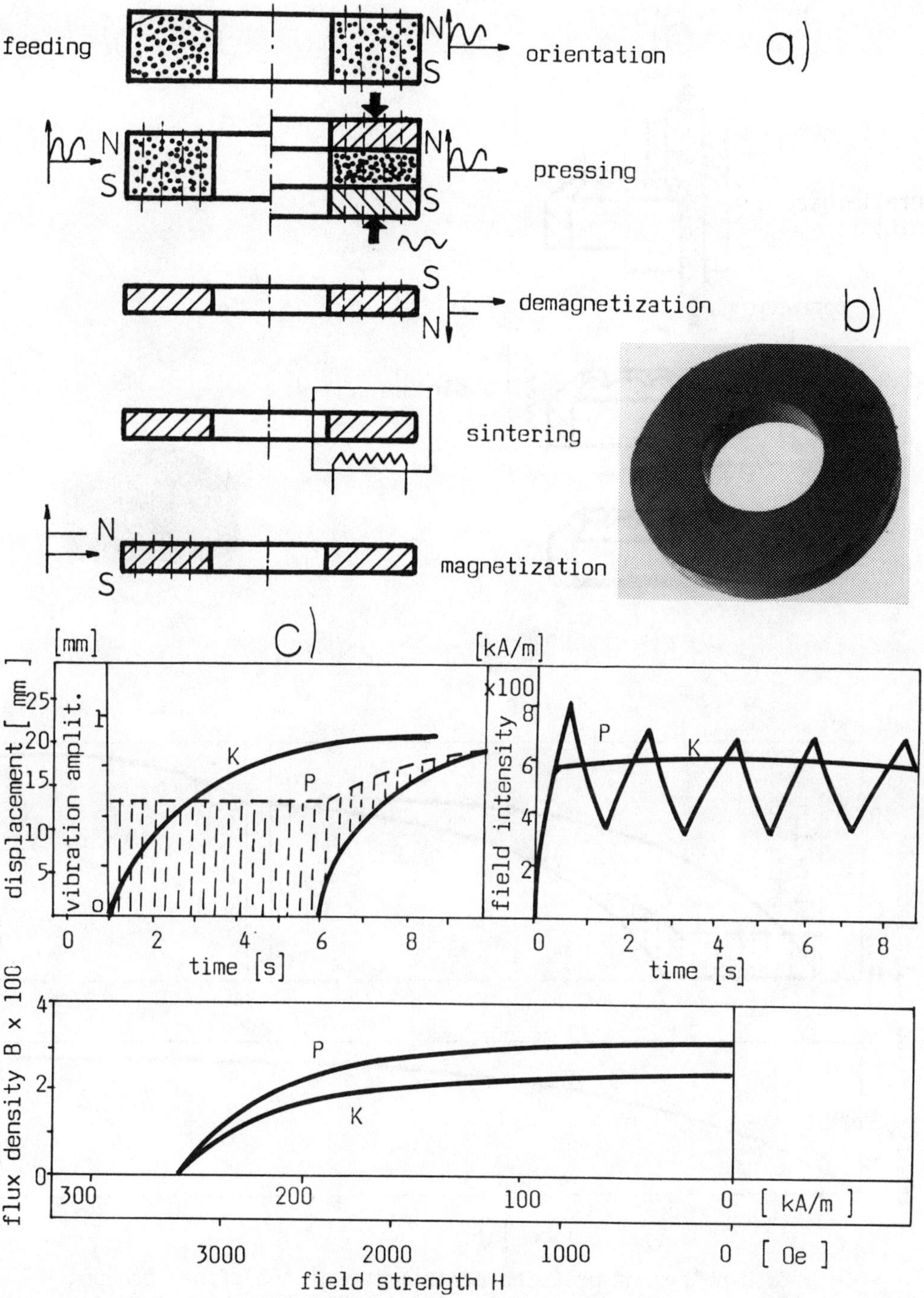

Fig.3.Main phases of pressing of magnetic ring (a),final product (b), parameters of pulsatory pressing in magnetic field(c)

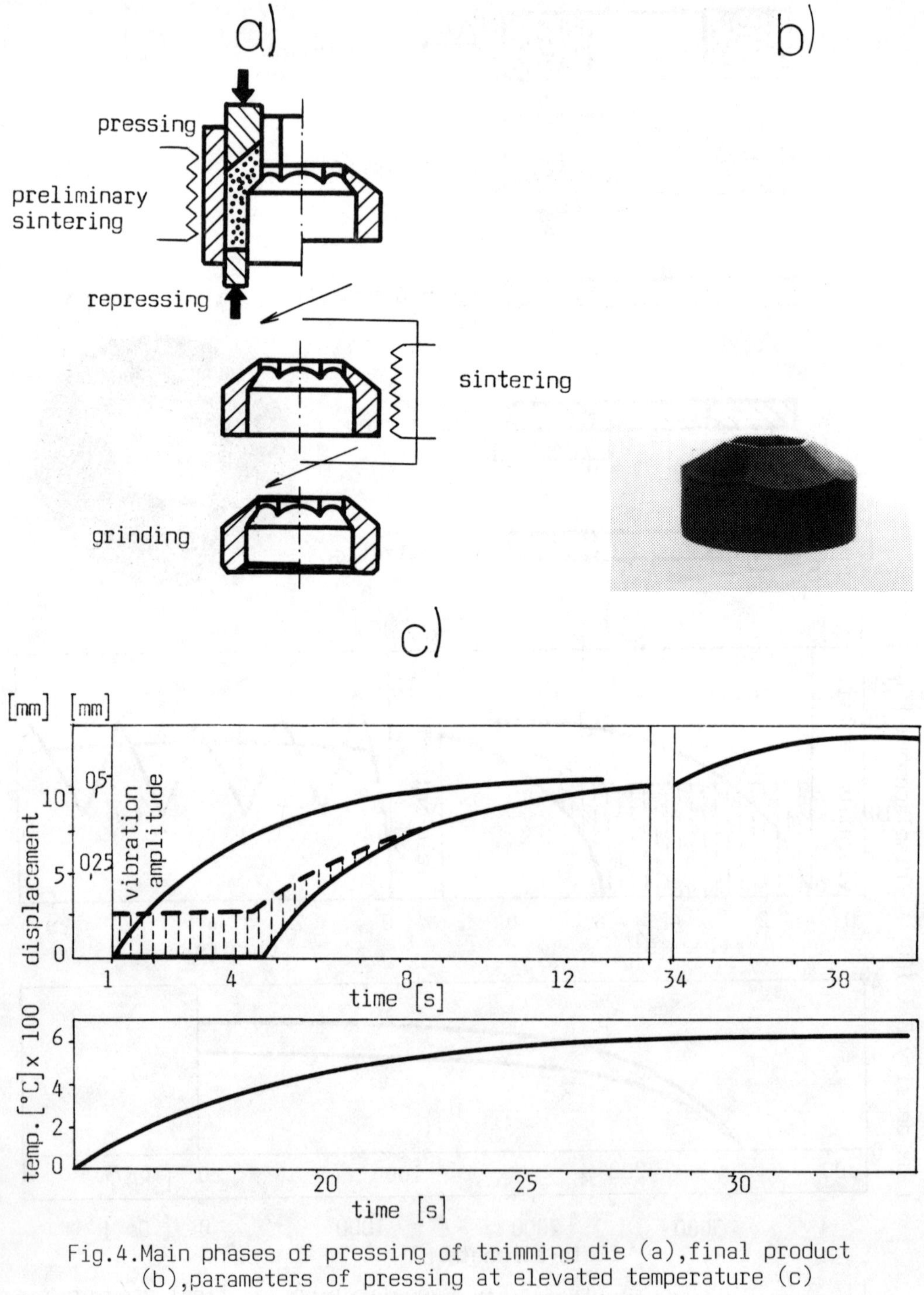

Fig.4.Main phases of pressing of trimming die (a),final product (b),parameters of pressing at elevated temperature (c)

PREPARATION OF AMORPHOUS FeTa-POWDERS BY MECHANICAL ALLOYING (MA)

G. Veltl, B. Scholz, H.-D. Kunze

Fraunhofer-Institut für angewandte Materialforschung
Lesumer Heerstraße 36, D-2820 Bremen 77, FRG

Abstract

Amorphous iron-tantalum powders were produced by mechanical alloying. The preparation of the mechanically alloyed Fe_xTa_{1-x} powders was carried out in the composition range $0.2 \leqq x \leqq 0.8$. The degree of amorphization was monitored by x-ray diffraction measurements.

The alloys provide materials with very high crystallization temperatures up to 1034 °C at the composition $Fe_{30}Ta_{70}$.

Introduction

Mechanical alloying is besides rapid quenching from the melt a suitable way to produce amorphous alloys. Koch et al. [1] succeeded in 1983 in producing amorphous $Ni_{60}Nb_{40}$ by this process. Until now many binary alloy systems were amorphized by MA by several scientists e.g. [2,3]. The powders prepared in this way were mostly of the type TM1 + TM2 (TM1: early transition metal, TM2: late transition metal). One of these combinations of elements is represented by the system FeTa, in which amorphous layers were formed by sputtering [4].

Aim of this work is to examine the possibility of amorphizing FeTa by MA.

Experimental Methods

Elemental crystalline iron (FSSS 5 μm) and tantalum (FSSS 5 μm) powders were mixed in the desired compositions, sealed under argon together with steel balls in a steel vessel and milled in a high energy ball mill for 5 h up to 25 h. The milling ball to powder weight ratio was 15:1.

The x-ray analysis was carried out in a diffractometer in reflection geometry using CuK_α-radiation.

The crystallization temperatures were determined with a Netzsch DTA 404 at a heating rate of 20 K/min.

The morphology of the powders was examined with a SEM (Cam Scan 4).

Results and Discussion

The process of amorphization by MA requires a large negative heat of mixing and a sufficiently fast diffusion of at least one element at low temperatures.

The heat of mixing can be estimated by calculations based on Miedema's model [5]. Fig. 1 shows that there exists a driving force for a solid state reaction in the middle of the composition range. The first investigations were carried out near the maximum of the driving force with the mixture of 50 at% Fe and 50 at% Ta. After 25 h milling time the powder was amorphous, as shown by the broad diffraction maximum, Fig. 2, developing with increasing milling time from the peak patterns of the crystalline metals. Fe_xTa_{1-x} ($0.2 \leq x \leq 0.8$) mixtures were milled for 25 h and the amorphization of the resulting powders was controlled by x-ray analysis, Fig. 3. The x-ray patterns depict that there are crystalline rests in powders of the boundary concentrations beside the broad diffraction maximum of the amorphous phase. This can be explained by the lower driving force for amorphization indicated by the smaller difference between the ΔG of mixed elements and the ΔG of the amorphous phase leading to imcomplete amorphization during the milling of 25 h. These results are confirmed by the metallographic micrographs, Fig. 4, showing the good resistance against etching of the $Fe_{50}Ta_{50}$ alloy due to its amorphous state. The micrographs of $Fe_{80}Ta_{20}$ and $Fe_{20}Ta_{80}$ exhibit areas of fine lamellar structures typical for the mechanical alloying of crystalline materials.

In Fig. 5 the typical morphology of FeTa powders after 25 h milling time is to be seen. For all compositions the particle size is in the range of 20 μm - 80 μm. The higher magnification emphasizes the typical plate like structure of the powder particles.

Besides x-ray analysis and morphological examinations of the powders the thermal stability of the amorphous phase was determined by DTA measurements.

In Fig. 6 the crystallization temperatures versus Ta-concentration are drawn. The crystallization temperature increases with tantalum content from 747 °C at $Fe_{80}Ta_{20}$ up to 1034 °C at $Fe_{30}Ta_{70}$ which is one of the highest ones for amorphous metals at all. At $Fe_{20}Ta_{80}$ the crystallization temperature drops again to 760 °C. The compositions of the amorphous phases of the boundary concentrations deviate from the nominal compositions, because they contain crystalline rests of the majority component.

Summary

The present investigation proves that amorphization by mechanical alloying in the system FeTa is possible. The amorphous phases formed in this system offer very high crystallization temperatures. This will provide a wide range of possible applications.

References

[1] C.C. Koch, O.B. Cavin, C.G. McKamey, J.O. Scarbrough
Appl. Phys. Lett. 43 (11), 1983, 1017

[2] E. Hellstern, L. Schultz
Appl. Phys. Lett. 48 (2), 1986, 124

[3] F. Petzoldt, B. Scholz, H.-D. Kunze
Mat. Sci. Eng. 89 (1988); 97 (1988), 25

[4] C.L. Chien, S.H. Liou, B.K. Ha, K.M. Unruh
J. Appl. Phys. 57 (1), 1985, 3539

[5] A.R. Miedema
Philips Tech. Rev. 36 (1976), No. 8, 217

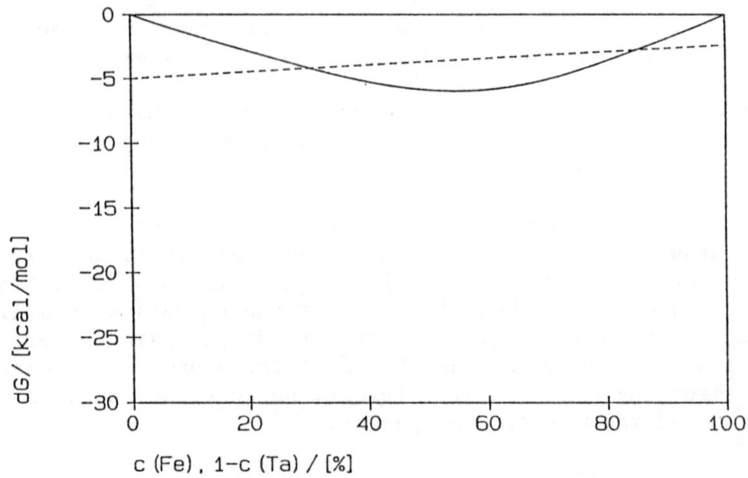

Fig. 1: ΔG curves of the system FeTa calculated after Miedema's model for 500 K.
ΔG of the mixed elements (dashed line)

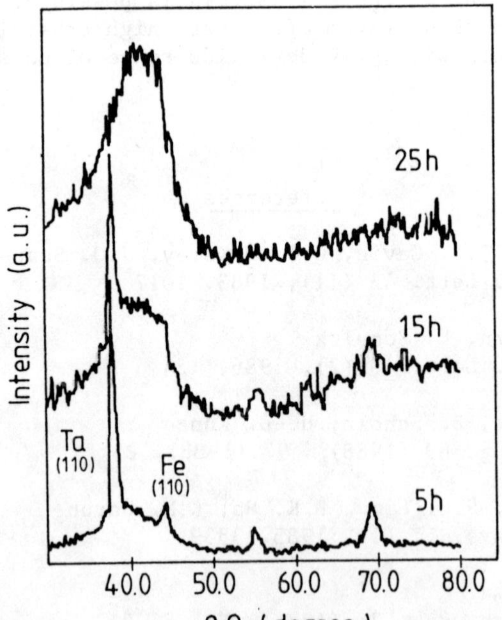

Fig. 2: X-ray diffraction patterns of $Fe_{50}Ta_{50}$ after different milling times.

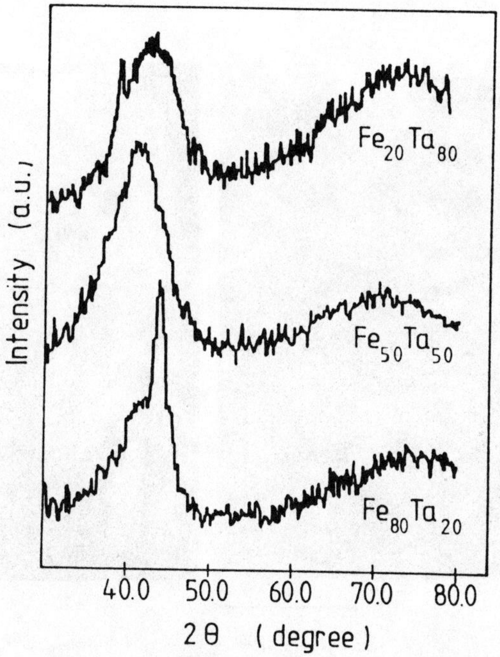

Fig. 3: X-ray diffraction patterns of FeTa alloys of different compositions after 25 h milling time

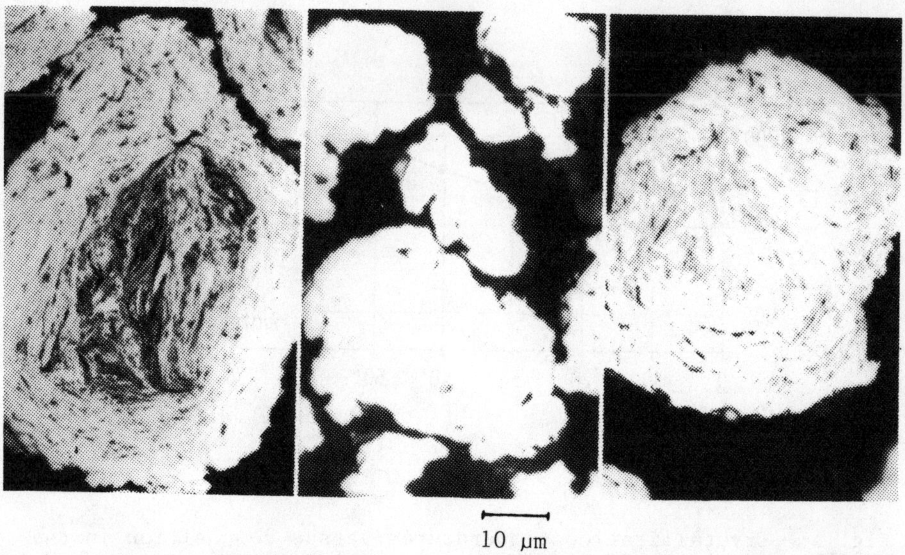

Fig. 4: $Fe_{80}Ta_{20}$ $Fe_{50}Ta_{50}$ $Fe_{20}Ta_{80}$
Optical micrographs of the MA powders after 25 h milling time

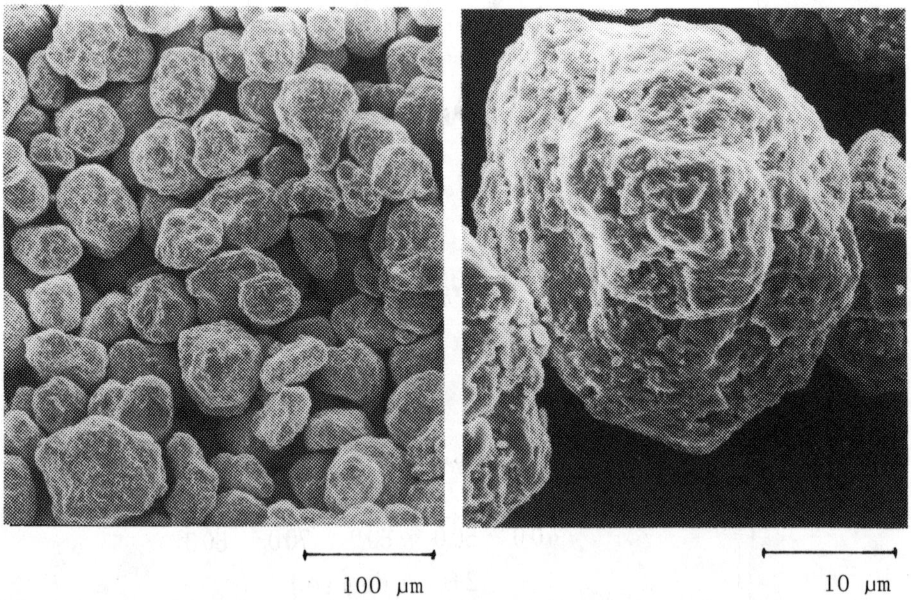

<u>Fig. 5</u>: Scanning electron micrographs of $Fe_{50}Ta_{50}$ after 25 h milling time

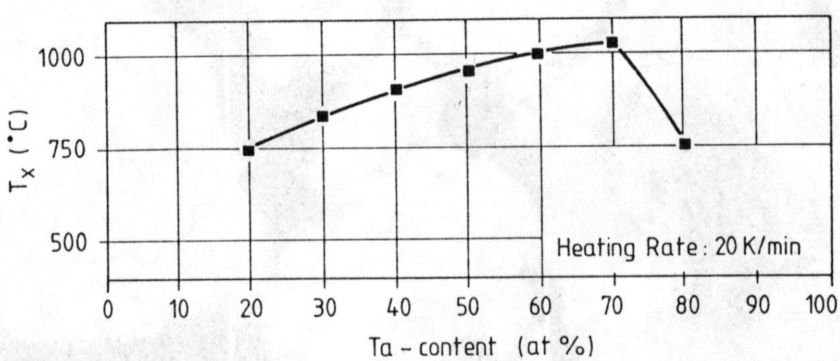

<u>Fig. 6</u>: Crystallization temperatures versus composition in the system FeTa.

AERODYNAMIC STUDY OF GAS FLOW FROM A MOLTEN METAL ATOMIZING NOZZLE

Marin S. Stoichev, Institute for Metal Science and Technology, Bulgarian Academy of Sciences, Bulgaria
Ivan S. Antonov, Institute for Mechanical and Electrical Engineering, Technical University, Bulgaria

The influence of the initial parameters on the swirled gas flow formed by annular nozzle for atomization of metal melts has been studied.

Nowadays the atomization of metal melts is the most widespread method for the production of metal powders. In our country the two-jets (two-fluids) gas and water atomization has found practical application. A considerable advantage it provides is the possibility for controlling the powder properties during the atomization process. From this point of view the operation of the atomizing nozzle is of particular interest.

The principle of the two-jets (two-fluids) atomization is described in /1/. Another paper /2/ examines the characteristics of the flow area which is generated during the outflowing of high speed jet from an atomizing nozzle, considering the influence of jet swirl. A model examination has been used and the tests have been carried out with nozzle designed by the authors. As a result three zones in the flow area have been determined and additionally studied by suitable experimental technique. The investigation parameters are in conformity with the operating conditions in practice.

In the present paper the subject of research is the aerodynamics of a swirled jet flow used in two-jets atomization of metal melts. The atomizing nozzle is of the following construction (Figure 1). The swirl of the flow is realized by the two feeding tubes tangentially located along the chamber periphery. The hole through which the metal jet flows is centrally positioned and equiaxially around it lies the ring through which the gas flows. The ring profile is designed like Laval's nozzle, so that outflow of supersonic speed gas flow (jet) is provided. The resultant flow presents tapered annular swirled jet. Inside this jet an area of low pressure is observed, its minimum values being situated along the jet axis.

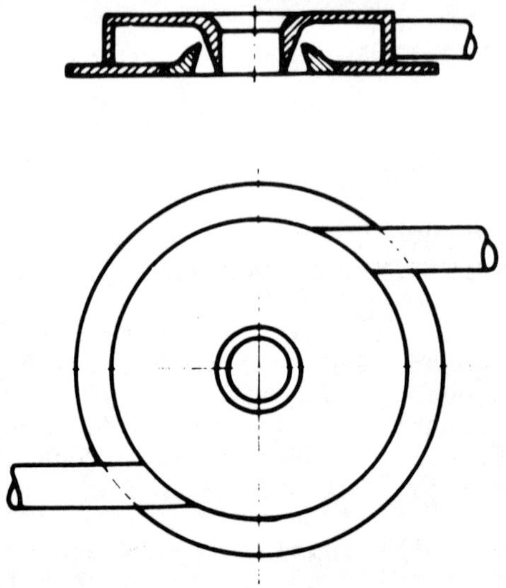

Figure 1. Diagram of the atomizing nozzle.

The investigation of the flow is carried out when the distribution of the axial velocity component U and the pressure distribution in particular cross-sections on the jet flow length being measured. The velocity component U is calculated according to the relationship

$$u = \sqrt{\frac{2}{\rho} \cdot \frac{P_f - P_{st}}{1 + \varepsilon}} \tag{1}$$

where ρ is fluid density, kg/m^3
P_f is total pressure , Pa
P_{st} is static pressure, Pa
ε is correction coefficient (giving the fluid shrinkability, Ma).

On the basis of velocity distribution an analysis of the basic geometric and dynamic parameters of the complex jet flow has been made. At first place these are the geometric parameters of the jet flow: its inside boundary r_1, the direct and return current of the circulation zone: the line giving the geometric place of the maximum velocities r_m and the jet outside boundary r_2. In Fig.2 these parameters are shown in dependence of the annular jet inclination angle to the axis of metal melt outflow (angle of attack α). The values r_1, r_m, r_2 are made dimensionless with the semi-width of the hole through which the jet flows out. In the same manner the coordinate $\bar{x} = \frac{x}{0.5\delta}$ is made dimensionless. The graphical dependence is shown in Figure 2, from where the following characteristic conformities may be specified: in the area of circulation existence zone the increase of angle α leads to

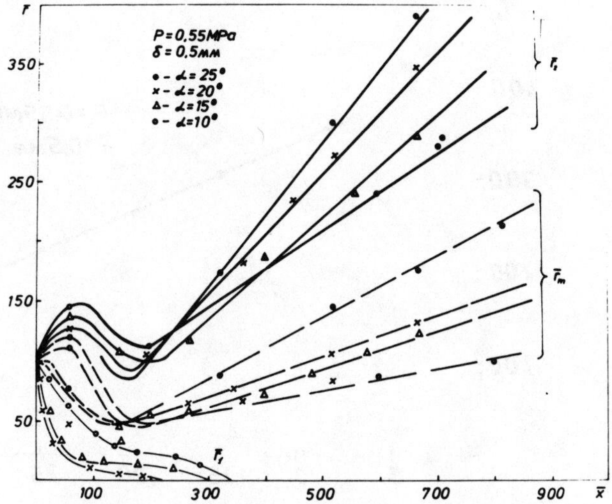

Figure 2. Effect of the attack angle on the geometric parameters of the flow

$\bar{r}_1$ - jet inside boundary

$\bar{r}_2$ - jet outside boundary

$\bar{r}_m$ - maximum velocity

constriction of the jet and its approach to the axis of symmetry. This constriction has its minimum in the range of $110 \leq \bar{X} \leq 210$, thereafter the flow "dissolves", its "empty cone" configuration disappears. In this section the influence of angle α is opposite. Its increase leads to a faster broadening of the flow. This may be explained by the fact that most rapidly the cone character disappears at the high values of α, an "impact" of directed under great angle fluid particles is obtained and as a consequence of this the jet intensively expands.

The effects of the attack angle (α) and the width of the hole (δ) on the length of the circulation zone (l_c) is shown in Fig.3. It is seen that the width of the hole (i.e. the flow rate) exerts more important influence, considerably increasing the length of the circulation zone.

The attenuation of the maximum value of the axis velocity component of the flow ($\bar{U}_m = \frac{U_m}{U_o}$) is an important flow characteristics. The effect of the attack angle (α) on the maximum value of the velocity component is shown in Fig.4. It is obvious that the increase of the attack angle (α) leads to faster U_m attenuation which is explained by the greater expansion of the jet.

Better demonstrativeness for the character of the examined flow is obtained by the constant velocity lines. In Fig.5 constant velocity lines of three regimes of flowing out are shown.

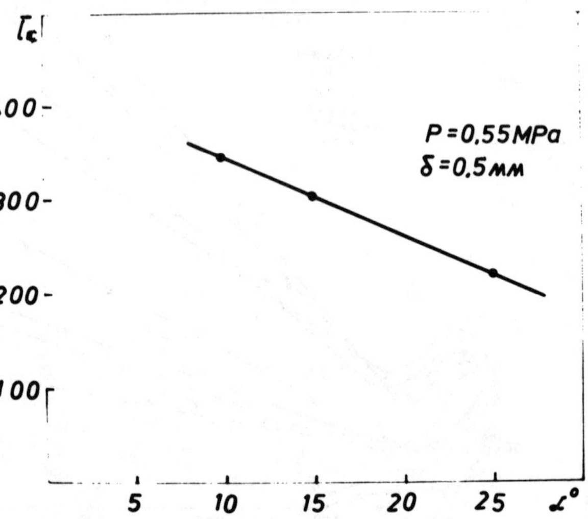

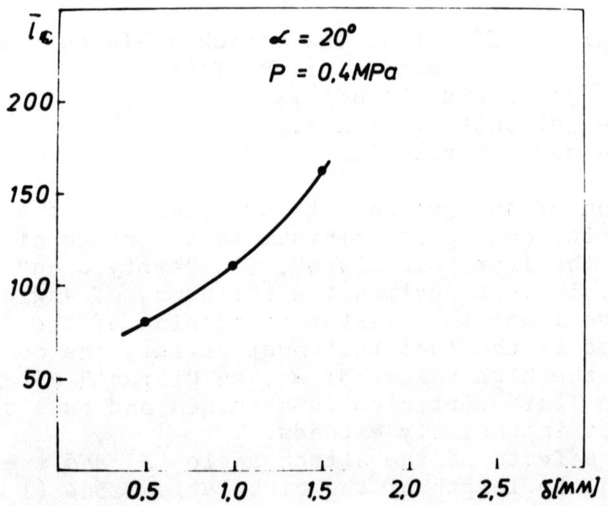

Figure 3. Effects of the attack angle (α) and the width of the nozzle slot (δ) on the length of the circulation zone (l_c).

Particularity of the investigated flow is the presence of a jet "torch" lying symmetrically around its axis. Of interest is the fact that the "torch" displays a tendency to approach the axis of the flow after the circulation zone has vanished.

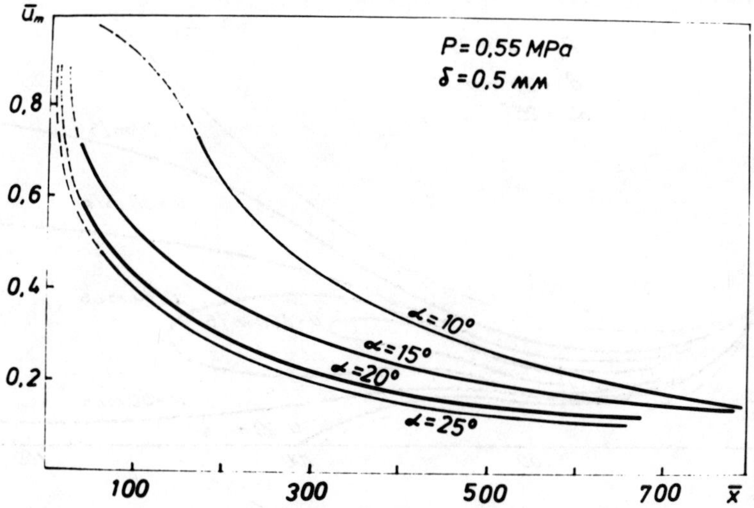

Figure 4. Effect of the attack angle (α) on the maximum value of the axis velocity component ($\bar{U}_m$).

Since the investigated flow is swirled considerable interest presents the variation of the static pressure. The distribution of the static pressure in the area of the flow may be illustrated by the constant pressure lines (isobars). In Fig. 6 the isobar types for three cases of outflowing are shown. Characteristic feature of isobars is that they follow the form of the jet expansion, the minimum values of pressure being in the circulation zone.

The minimum values of the difference between the static pressure in the jet and that of the surrounding environment are observed immediately after the initial cross-section. Along the jet length this difference decreases going to zero, whereupon a tendency of equilizing with the pressure of the surrounding environment is displayed (Figure 7). The increase of the angle between the two flows leads to an increase of the above said difference between pressures.

In order to present a clear quantitative analysis of the examined jet flow a graph integration of the velocity field and current picture of the flow was made. As a result lines for ψ = const which in this case close zones of constant flow rates were obtained. Moreover, $\psi_i = \frac{\psi_i}{\psi_o}$, and

$$\psi_i = 2\pi \int_0^{r_i} u_i r_i \, dr \qquad (2)$$

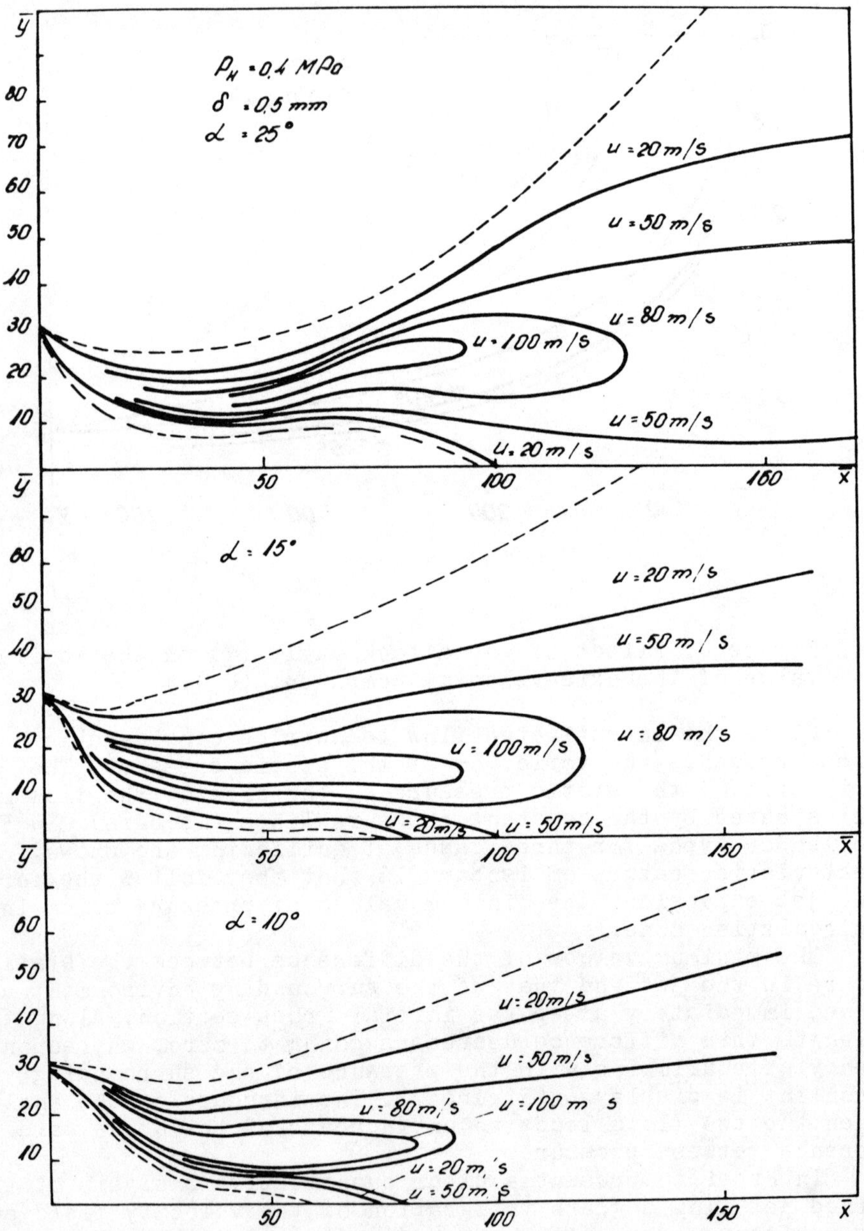

Figure 5. Development of the constant velocity lines along the flow length.

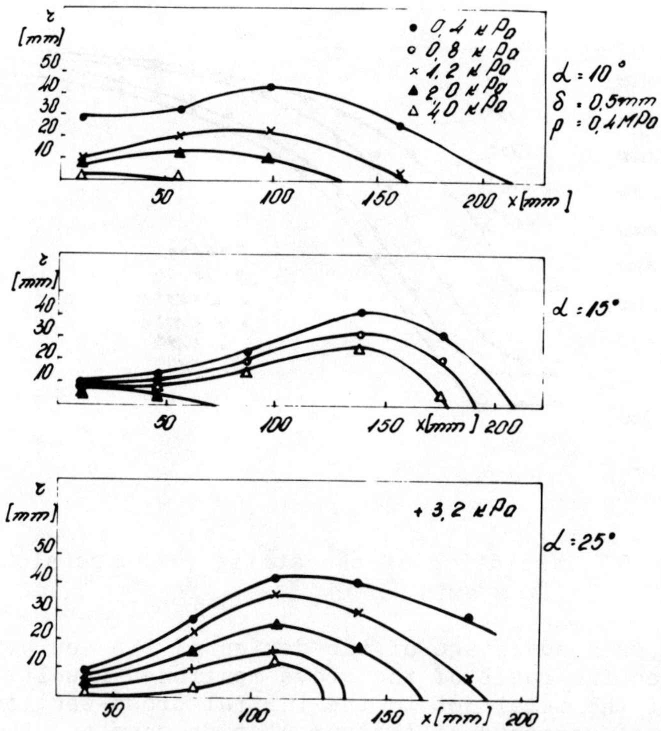

Figure 6. Variations of the static pressure in the section of flow

where: ψ_i is a flow rate through the cross-section of the ith radius, kg/sec.

ψ_0 - initial flow rate, kg/sec.

U_i - velocity of the flow through the ith radius, m/sec.

In Figure 8 the current picture of the flow for three regimes of outflow is shown. It is seen from the distribution of the flow rate shown that the ejection action of the jet appears and increases after the disappearing of the circulation zone in the section of intensive expansion of the flow.

From the studied cases of the outflowing with variation of the initial parameters it is seen that the increase of the attack angle leads to:

- constriction of the jet and its approach to the axis of symmetry with its minimum in the area of $110 \leq X \leq 210$
- rapid attenuation of $\overline{U}_m$
- rapid increase of the circulation zone
- displacement of the jet torch from its axis

- decrease of fluid ejection from the surrounding environment.

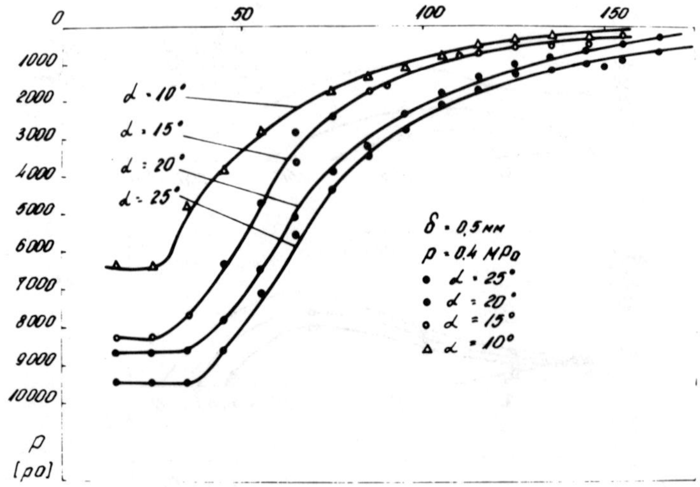

Figure 7. Variation of the static pressure along the flow axis.

The main advantage of the design of the new nozzle developed on the basis of the above mentioned results in the approach of the metal jet to the initial cross-section of the gas jet, i.e. decrease of the circulation zone and the possibility of returning back metal drops from the melt on the nozzle which leads to its blocking during atomization. Besides, the diameter of the annular slot is reduced aiming decrease of the relative consumption (flow rate) of gas. In order to keep the possibility to adjust the relation between the relative consumption of metal and gas in determined limits a version for reduction of the metal jet diameter is provided.

REFERENCES

1. Stoichev M., Production of Metal Powder by Atomization. Proceedings of PM Institute, 1980, 10 p.27 - 38.
2. Stoichev M., Antonov I. Investigation of Swirled Turbulent Jets Used in Two Fluid Two Jets Atomization. Scientific Session VNVU (Higher Marshal School), 13 - 15.05.1982, Paper 10.22.

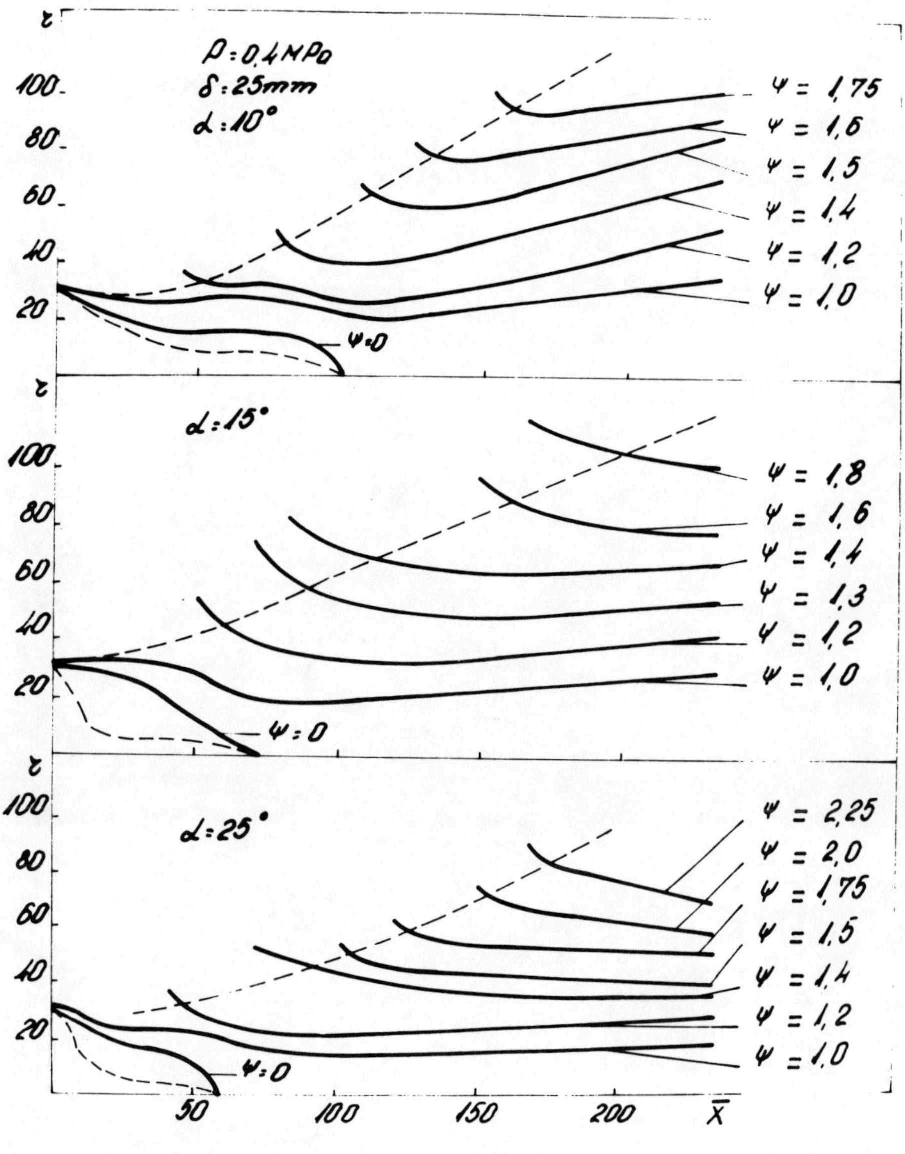

Figure 8. Development of the current lines along the flow length.

A MODERN POWDER METAL PRESS FOR UNIVERSAL APPLICATIONS

Winfried Beisel, President
Graebener Press Systems, Inc.

An unlimited range of applications

GRÄBENER Powder Compacting Presses of series GPS offer a vast range of application possibilities when mass-produced P/M components have to be manufactured at high precision and with high surface quality - particularly when the density must be uniformly distributed over the component cross-section. This is of special importance for precision components in the automotive and electrical industries as well as for the manufacturers of gearboxes and office machinery.

The GRÄBENER Powder Press GPS is the universal powder press for all crucial compacting applications:

▪ Compacting with the opposing-ram method

The standard GRÄBENER Powder Press compacts with the opposing-ram method, which is ideal to uniformly distribute the density over the whole component cross-section. This method is extremely reliable and economical due to simple tooling, short punch, and sturdy, compact adapter.

▪ Compacting with the withdrawal method

Conventional adapters can be applied with the GRÄBENER Powder Press GPS.

■ **Compacting with the horizontally split-die method.**

Complicated components and parts with undercuts and overlaps can be excellently compacted by the GRÄBENER Powder Press GPS.

■ **Compacting with the ejection method**

Simple compacted components are manufactured by using the ejection method, with static or floating die.

■ **Sizing - re-compacting**

The rigid design of the press and its extremely low deflection ensures precise sizing, high accuracy and excellent surface quality at high speed.

The advantages of GRÄBENER Powder Presses at a glance

The GRÄBENER Powder Press GPS operates fast, quiet, precise and reliable. It is a mechanical powder press which combines the knuckle-joint and cam drive (2 mechanical leading axes). Auxiliary and support functions are carried out hydraulically (up to 3 additional axes), thereby resulting in a combined hydro-mechanical press system (patent pending) to mutually benefit from the mechanical and hydraulic methods.

The knuckle-joint ensures a uniform density of the P/M components in the press and the cam drive precisely controls the lower ram. Due to its mechanical drive system, the machine functions reliably at high speed to achieve economical operation.

The filler shoe is synchronized to the press by means of cams.

The hydraulic system carries out the following auxiliary operations:

- Converting the system to various compacting methods
- Activating auxiliary punches
- Floating of the die plates

Brief summary

■ High speed
■ Excellent powder distribution
■ Uniform density of compacted components
■ Versatile applications
■ Reliable operation
■ Compaction within close tolerances with high surface quality

- Compact adapter
- Simple tooling system with short punches
- Virtually no effect of tool friction
- Low tooling costs
- Highly economical
- Non-polluting operation
- Simple to operate
- Reduced danger of relief or detentioning cracks in the compacted components

The universal powder press: A versatile modular system

GRÄBENER Powder Presses are mechanically driven. As they are of modular design, they can be converted simple and fast to perform a wide range of compacting requirements. All retro-fitting can be done at the job-site.

Basic machine for the opposing-ram method. A compacting method with future:

When compacting with the opposing-ram method, the tools carry out the shortest possible stroke. This enables only a minor displacement of the powder particles to achieve less friction, thereby producing the highest quality of compacted components. A static die during opposing-ram compacting also has a positive effect on the final accuracy.

A press with a bottom drive system is of extremely compact and rigid design to ensure smooth, vibration-free operation. The knuckle-joint forces the press frame and upper punch downwards towards the die, and the cam-driven lower ram forces the lower punch upwards.

The lower ram stroke can be infintely adjusted via a gear motor in the drive lever.

The opposing-ram cam can be individually adapted to the respective requirements. If the component dimensions differ only slightly, the machine will operate with the same double-action pressing cam, whereby the cam for complicated components can be individually adapted. The cams are equipped with change-over units and hydraulic clamping facilities to enable simple exchange.

Highlights of a modern powder press

The GRÄBENER Powder Press GPS is simple to operate and can be easily adjusted. All important functions are motorized and can be adjusted via push-buttons on the control panel:

- The upper punch
- The filling position
- The compacting movement
- The die retraction and upper compaction
- The lower punch stroke
- The withdrawal position
- The adapter connection

All necessary functions can be adjusted in small steps during the operation. Rough adjustments can be carried out much faster due to larger movements when in set-up mode.

The SPS control system ensures that the adjustments are safeguarded against each other, and all important parameters are displayed in writing on the control panel. This system can repeat a successfully tested programme.

The machine can be equipped with a force monitoring system which will automatically switch off the machine if it does not conform to preselected nominal values. A weighing device adjusts the filling height to ensure the correct component weight.

A fast tooling change is possible by exchanging the complete adapter which is either pneumatically or hydraulically clamped. The tool adapter exchange is made simple by the tool-change cart.

Retro-fitting of hydraulic elements and CNC controls to the mechanical system is possible to convert a GRÄBENER Powder Press into a multi-axis and highly sophisticated production unit.

Alternative compaction methods increase productivity and flexibility

Compacting with the withdrawal method

With an additional set of cams, the GRÄBENER Powder Press can be converted for the withdrawal method, thereby enabling the use of conventional adapters.

The opposing-ram or withdrawal method can be selected accordingly, and the change-over can be easily carried out by finger-tip control on the machine's control panel.

During the withdrawal method, the die is moved to the final compaction position against a precisely adjustable positive stop. The component is ejected by mechanically withdrawing the die via the cam-driven lower ram. The withdrawal stroke can be infinitely adjusted, thereby ensuring excellent ejection of the compacted component - even under high wall friction. The downward pressure for the withdrawal method can be precisely adjusted in the adapter upper section.

The split die method is ideal for complicated components

After installing a special adapter on the machine, complicated components or components with overlaps can be easily produced with a split die on a GRÄBENER Powder Press - even with the opposing-ram method. After compacting, the upper die segment is raised hydraulically and the finished component is ejected by the lower punch and discharged towards the front between the two die segments. Overfilling or underfilling is possible to ensure uniform material within a split die.

Absolute control of powder movement will prevent powder displacement to ensure that the component remains crack-free. A uniform powder density is also achieved.

DEVELOPMENT OF MULTIPHASE ALLOYS OF REFRACTORY METALS BY USING SINTERACTIVE COMPOSITE POWDERS

S. Haerdtle, R. Schmidberger

Dornier-System GmbH

SUMMARY

Powder production and sintering are being developed for a range of alloys based on the two refractory metals tungsten and molybdenum. Powder production is carried out by a reaction spray process (RSV). Sintering is done in various atmospheres where time and temperature have to be optimized for each alloy.
In cooperation with "Royal Ordnance Speciality Metals Ltd." (GB) and with EEC funding various alloys prepared by the RSV technique are being investigated concerning their suitability for different applications.

INTRODUCTION

Multiphase alloys consisting of tungsten and molybdenum as basemetals are usually been made by blending and subsequent pressing and sintering of different metal powders. Another way to produce such alloys is to sinter a skeleton of refractory metal and subsequent infiltration with liquid binder metal e.g. copper. Both of these conventional production routes produce coarse and sometimes inhomogeneous distribution of the different components.
The investigation described below starts out with prealloyed powders for sintering finegrained, homogeneous alloys.
The alloys investigated are being applied as:
- electrical contacts
- electrodes for spark erosion and resistance welding
- substrates for semiconductor devices.

One can expect to improve the following materials characteristics:

- mechanical resistance
- ductility
- plasticity
- burnoff resistance

POWDER PRODUCTION

Powder production is carried out by a reaction spray process. This process is based on thermal decomposition of salt solutions in a hot reaction chamber (Fig. 1). The solution contains all components in the exact ratio finally desired in the powder. Spraying of this solution into the hot reaction chamber by a nozzle leads to evaporation of the droplets within some microseconds. Fine powder particles containing all alloy components are obtained. The powder is collected in a high temperature filter unit. Waste gas is treated in a heat exchanger and a scrubber.

Particle size of such RSV powders ranges from 0.1 µm for primary particles up to some 100 µm for agglomerates. Agglomerates can easily be disintegrated to primary particles by a simple milling process. Each particle contains all components homogeneously distributed and in the accurate composition.
High sintering activity allows sintering temperatures far below temperatures used in conventional processing.
Fig. 2 shows a SEM micrograph of powder containing 80 % W and 20 % Cu. Since the powder production process is based on salt solutions there exists nearly no limitation concerning the ratio and number of alloying components.

POWDER PROCESSING

These powders are pressed either uniaxially or isostatically depending on sample geometry.
Sintering is done in various atmospheres preferably hydrogen and vacuum. Sintering atmosphere, temperature and duration depend on the composition of the alloy. In the following details are given for two different alloys - tungsten/copper and tungsten with yttriumoxide.

Tungsten/Copper:
The copper content varies from 10 % to 50 %.
Sintering is carried out in hydrogen at temperatures from 1000°C to 1100°C tungsten grainsize can be influenced by variation of duration of sintering time and temperature. The density obtained is nearly 100% of theoretical density. Homogeneity of the powder is transferred to the sintered speciment. Fig. 3a shows a micrograph of a tungsten/-copper sample with a copper content of 20 % by weight at a magnification 1000:1.

Tungsten containing 3 % Y_2O_3:
Sintering of this alloy is carried out in two steps. Up to a temperature of about 1000°C the atmosphere is hydrogen in order to remove oxygen. Above this temperature sintering takes place in vacuum. Optimum density is reached at about 1600°C, Fig. 3b shows a micrograph of such an alloy.

CONCLUSION

Homogeneous distribution of the various components in a powder is extraordinarily important for preparing fully densified finegrained sintered structures. When using RSV powder as sintering material the absolutely homogeneous distribution of alloy components in the powder can be maintained during sintering. RSV powder allows sintering at lower temperatures than in commonly used processes and no segregation occurs at liquid phase sintering.

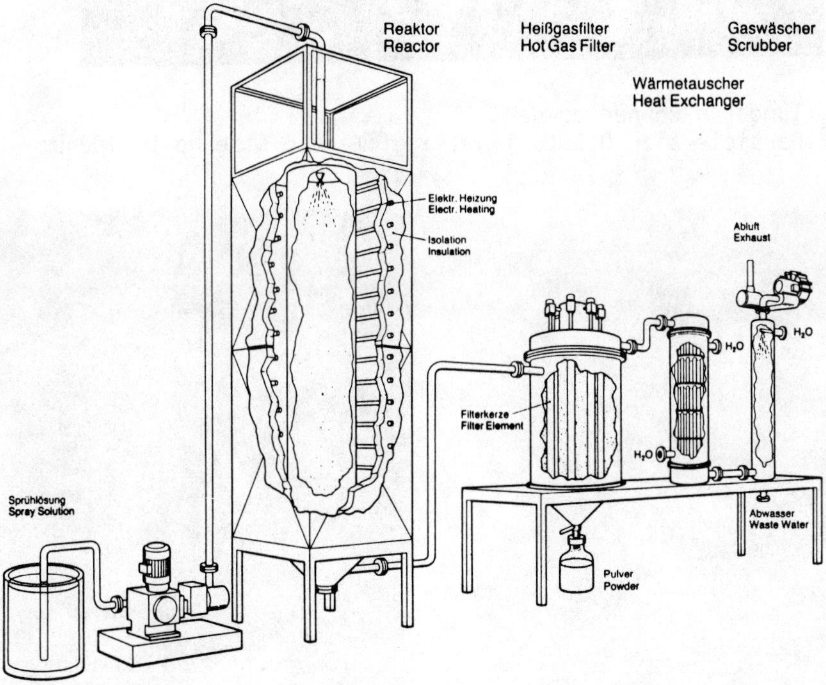

Fig. 1: Diagram of the reaction spraying process.

Fig. 2: Tungsten/Copper powder
Particle size 0.5 to 1 µm, agglomerate size up to 100 µm

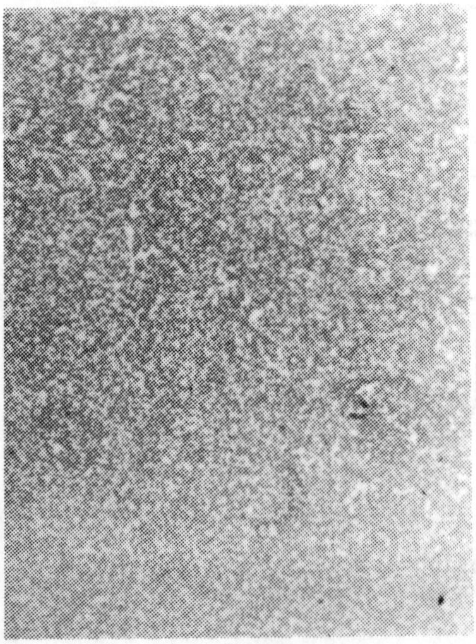

Fig.3a: Tungsten/Copper
Magnification 1000:1

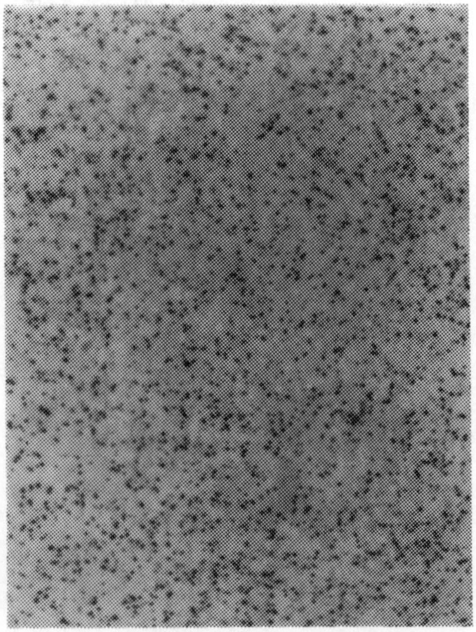

Fig.3b: Tungsten/3 % Y_2O_3
Magnification 1000:1

NOVEL AND ALTERNATIVE MATERIALS FOR PRODUCTION OF HARROW TINES

O. KRAEMER and L. ROESEN
Technological Institute,
DK-2600 Taastrup, Denmark

Summary

Tillage Tools Based on Wear Resistant PM-Materials.
Metal matrix composites manufactured by a full dense process were wear tested using a pin-on-disc abrasion tester. Measurement of tensile properties and metallurgical examination was also carried out. In the light of the laboratory work springtine point like specimens were tested at a typical Danish site. Test done so far looks promising. The wear resistance of the composites were up to 100% better than harden steel (45 HRC). The ranking order is as follows:

1. White cast iron-Fe composites
2. SiC-Fe composites
3. Al_2O_3-FE composites

Patent has been applied for.

INTRODUCTION

Through a litterature study of wear mechanism, and wear test in relation to agricultural tools, it is suggested that heterageneous materials with a dispersed hard phase have an optimal structure in a specific wear environment. The idea in this project is through a powder metallurgical proces route to control the structure of powder based composites by varying the volume ratio of iron powder and hard particles in the mixing state of manufacturing.

The wear resistance was measured on a pin-on-disc apparatus. Test results have proved to fit fairly good with wear resistance of metallic materials recorded in a working environment. Although it is important to keep in mind that the prediction of practical wear properties is empiric. The abrasive matter for this test was fixed siliconcarbide particles.

The following inexpensive material systems were chosen for further investigations:

- White cast iron particles in a steel matrix.
- Silicon carbide particles in a steel matrix.
- Alumina particles in a steel matrix.

The raw materials are all commercially available.

In an early stage of the experimental work it became clear that a full dense process is necessary, and it was decided to adapt powder forging.

PRODUCTION OF P/M COMPOSITES

Atomet 25 iron powder was mixed with "hard particles", lubrication and graphite powder. The micro structure were varied by changing the carbon content and volume content of hard particles (up to 40%).

Preforms of test pieces were compacted in a floating die. The compression ratio were held constant. Different pressures were obtained due to differences in compactability of the powder mixes. After dewaxing in hydrogen atmosphere the sintering process took place at approx. 1100°C depending on the type of hard particles. As a final process, powder forging was carried out.

RESULT AND DISCUSSION

White cast iron-Fe composites

It was found that the thermal decomposition of iron carbide particles (Höganäs WA60C) is sensitive to sintring parameters. Only materials containing large particles (>180 μ) of white cast iron have a structure of dispersed particles of partly decomposed cementite in a perlitic matrix with grain bondary cementite. No clear correlation between wear resistance and metallurgical structure was found. However, alloys with a high content of cast iron (20-40 vol. %) exhibit wear resistance which are up to 30% better than conventional harden steel for tillage purposes. Tensile strengths were found at all levels of carbide content. (Rm > 400 N/mm^2).

SiC-Fe COMPOSITES

SiC-bearing composites were produced and tested. Careful powder handling is needed in order to avoid segregation of the powder components. Some porosities were observed specially in case of 30 and 40 vol. % of SiC. The wear test showed that an increasing amount of SiC regardless of the particle size leeds to increasing wear resistance up to 30 vol. % SiC. The forging process had a destructive effect on the particles in case of 30 and 40 vol. %.

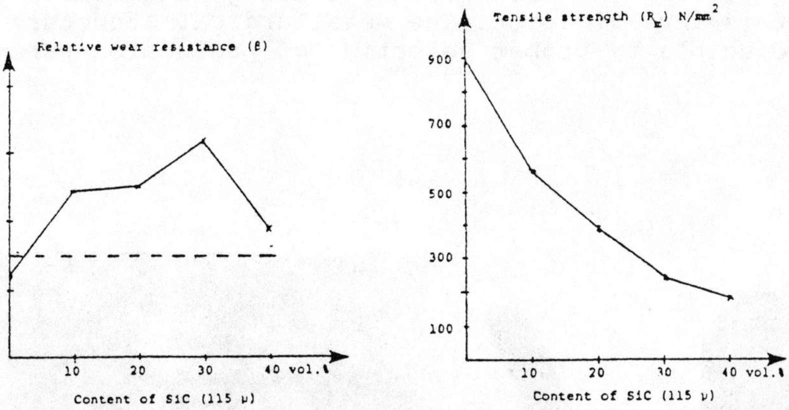

Fig. 1. Typical relative wear resistance and tensile strength of SiC-Fe composites. Wear data is recorded on an abrasion tester. The doted line indicates the wear resistance of hardened steel.

Only composites containing up to 20 vol. % SiC obtained satisfactory tensile strength with respect to tillage purposes. Among these materials a 100% increase in wear resistance is achived. A reaction zone between SiC and Fe was identified as FeSi-alloy and free graphite. It was concluded that the adhesion is weak and mainly physical.

Al_2O_3-Fe composites did not obtain good wear properties at SiC grinding paper.

PRACTICAL TESTS

In the light of the laboratory test results the most suitable composite materials have undergone field tests. Harrow tine like specimens have been tested in different field conditions. None of the tines failed by impact damage.

The tests carried out so far have been sucessful. The wear resistance is up to 100% better than reference tines made of harden steel (HRC ~ 45) which is normal for spring tine points.

A comparison of the wear resistances from the field tests and the abrasion tester shows that the ranking order changes. The most suitable material in the field tests was the iron carbide bearing tines. The empirical correlation seems only to applies to certain groups of materials.

Another pronounced difference is that the content of C in the matrix material influence on the wear resistance. This was not observed in the laboratory investigations.

Based on the field tests done so far, 500 powder based spring tine points in the proper geometry are during manufacture. The iron carbide bearing material is chosen for this investigation. The metallurgical structure is reproducable by proper selection of production parameters.

Fig. 2. A. Springtine point made of white cast iron powder and iron powder.(0.5x)
B. Metallurgical structure shows cementite rich areas (200x).

CONCLUSION

A comparison of the wear resistance from the field test and the abrasion tester shows that the correlation is relatively poor. The ranking order of changes with respect to wear properties. However, the laboratory test shows results of the same magnitute as measured in field tests.

It is believed that the particle bonding to matrix in a three body system such as that imposted by agricultural soil becomes important. That is the explanation why SiC-composites in spite of high particle hardness show poor wear properties in soil compared to white cast iron composites.

Composites containing cementite show the best overall performence in soil.

Patent has been applied for.

HIP - A WAY TO INCREASE LIFE TIME OF CAST INCONEL 713 C.

C. Persson, P. O. Persson, M. Ostensson
FFV Materials Technology
Box 13200
S-580 13 Linkoping
Sweden

OBJECTIVES

Present investigation includes creep testing of cast Inconel 713 C in the as-cast condition, heat treated condition and hot isostatic pressed (HIP) + heat treated condition and hot isostatic pressed (HIP) + heat treated condition. Nominal composition of Inconel 713 C is 12,5 CR, 4,2 Mo, 2,0 Nb, 6,1 Al, 0,8 Ti, 0,12 C, 0,012 B, 0,10 Zr and bal Ni.

Much attention has been paid by different investigators to understand and delay the creep phenomena in wrought material for turbine blades in jet engines. It is possible to determine the remaining creep life time of turbine blades by a non destructive testing method where the amount of voids is examined. The blades can be recovered by a heat treating operation.

A number of articles have been published about HIP-ing of wrought and cast nickel-based alloys (1-6). As a short summary of the literature can be mentioned that both creep and low cycle fatigue properties can be increased by HIP for cast as well as wrought materials. The properties can be increased for both as-produced components and for components that have been in service.

The HIP-temperature has to be in the region between γ'-solvus and the solidus temperature of the alloy.

To get the best microstructure of the HIP-ed material subsequent heat treatment has to be done.

The best effect of HIP-treatment is expected when it is done after that 50 to 85% of nominal creep life time is consumed.

The aim of this investigation is to find out:

if the creep life time of cast IN 713 C can be increased by heat treatment or HIP + heat treatment.

if heat treatment is as efficient as HIP combined with heat treatment.

at what time in the creep life cycle the treatment/s should be done.

Present investigation was planned to be performed in three steps.

1. Testing of cast samples. The samples are tested as-cast (reference), heat treated and HIP-ed + heat treated. Heat treatment and HIP + HT is done when approximately half of the nominal creep life time is consumed. This is performed by initial creep testing.

2. Creep testing of specimens taken from turbine blades that have been in service (app 600 h).

3. Creep testing of specimens from blades with even longer service times (1200 h). This step has not started as no blades with 1200 h running time are available.

The investigation is supported financially by Swedish Defense Material Administration.

CONCLUSIONS

Heat treating and HIP + HT of prestrained samples (270 h creep testing increases total creep life time of cast samples of Inconel 713 C.

HT increases total life time 75% compared to reference material.

HIP at 1230 C followed by HT increases total life time by 81%.

HIP at 1180 C + HT increases total life time by 107%.

HIP narrows the scatter in creep data.

Surface oxidation during creep testing results in absence of γ' in a surface layer. HIP + HT result in recovered microstructure.

METHODS

Test specimens for creep testing is shown in fig 1. The test was performed at 900°C and 190 MPa. Four groups of different treated specimens have been tested.

A. As-cast samples. Reference material. Tested until fracture.

B. Heat treated samples. Initially creep tested for 270 h and then heat treated at 1180°C/2h + ageing 925°C/12h.

C. HIP-ed + HT samples. Initially creep tested for 270 h and then HIP-ed 1230°C/105 MPa/2h followed by heat treatment at 1230°C/0,5h + ageing 925°C/12h.

D. Same as C but HIP at 1180°C/105 MPa/2h + ageing 925°C/12h.

Second part of the investigation, creep testing of samples from turbine blades, has just started. Due to the very complex shape and thin cross-section of the blade it is difficult to get samples. The samples have to be taken from the most stressed region.

The most stressed areas of the turbine blade are shown in fig 2. The accumulation of creep damage in the blade has been found to correlate to the degree of γ'-raftening in the microstructure.

No results are presently available from 2 and 3.

RESULTS

CREEP DATA

A summary of the creep data is shown in fig 3.

Group A, reference material, have a mean creep life time of 498 h.

Group B, heat treated, show a mean life time of 871 h. With the initial creep testing (270 h) before heat treatment it means an increase of total life time by 75%,

From the individual creep data it can be seen that the curves corresponding to heat treated samples have the same appearance as the curves for the reference material.

Group C, HIP 1230°C + HT, have a mean life time of 900 h. The stage three creep is delayed compared to reference material.

Group D, HIP 1180°C + HT, have a mean life time of 1029 h. Compared to reference material it means 107% increase.

METALLOGRAPHICAL INVESTIGATION

From the metallographical investigation it can be stated that as-cast samples contain porosity. After HIP no evidence of porosity is present.

After creep testing of HIP-ed + HT samples they do contain porosity. The total content of porosity is however much less compared to the as-cast condition. Mean size of the pores are approximately half of the size in the as-cast condition.

After initial creep testing for 270 h a surface zone, about 10 μm thick, depleted of γ' is present. Probably due to a too low chromium content caused by oxidation. After HIP the chromium is redistributed and γ' is present.

Normally cast material give rise to a rather wide scatter in creep life time. This is due to inhomogen structure and remaining porosity. In our investigation the widest scatter was found for group B, heat treated. It is quite obvious that both HIP - operations combined with heat treatment have narrowed the scatter band.

REFERENCES

1. HIP Processing - Potentials and Applications. W. J. van der Vet, Chromalloy Division Oklahoma Agard Conference Proceedings No 317 1981.

2. The Rejuvenation of Properties in Turbine Engine Hot Section Components by Hot Isostatic Pressing. P. H. Floyd, W. Wallace, J-P. A. Immarigeon. National Aeronautic Establishment Ottawa, Feb. 1981. (Aeronautical Report LR-605, NRCNO 19164).

3. HIP'ing Various Precision Cast Engine Components in Nickel-base Superalloys. M. Lamebrigts, E. Diderrich, D. Coutsouradis, Superalloys 1980. Proceedings of the 4th Int. Symposium.

4. Hot Isostatic Pressing of Alloy IN-718. M. Lamerigts, C. Herman, P. Louis, J. P. Wallemaca et al. Conference Proceedings: High Temperature Alloys for Gas Turbines, 1982 (Liege, Belgium).

5. Regeneration of the Creep Properties of a Cast Ni-Cr-base Alloy. H. R. Tipler, National Physical Laboratory, England. Agard Conference Proceedings No. 317, 1981.

6. "The Superalloys", Sims & Hagel, 1972.

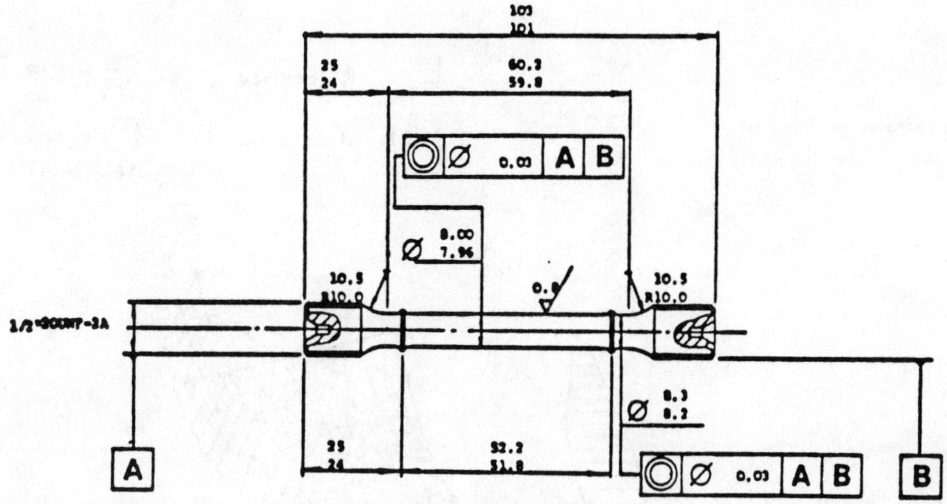

Figure 1. Creep testing specimen.

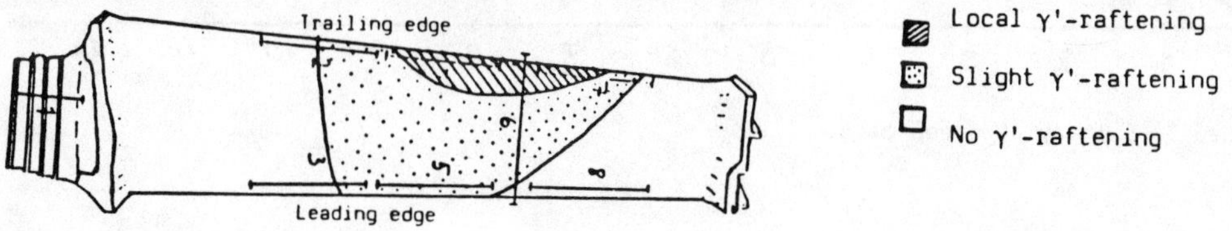

Figure 2. Second stage turbine blade from JT8D, modified version for the Swedish Airforce.

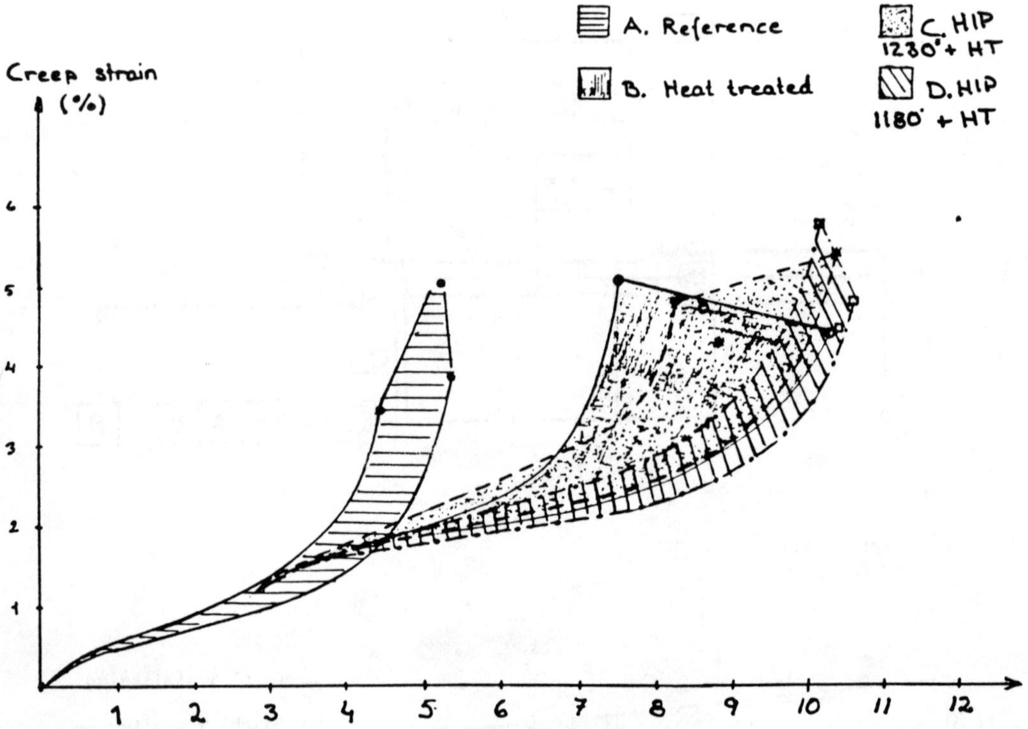

Figure 3. Summary of creep data.

MATERIAL DEFECTS IN A PM-NICKEL-BASE SUPERALLOY

B. Nowak, H. Lahodny*, E. Arzt*,
G. König and E. Affeldt

*Max Planck-Institut für Metallforschung
Stuttgart, FRG
MTU Motoren- und Turbinen-Union München GmbH
Munich, FRG

Abstract

The purpose of this work was the determination of the type, size and frequency of defects in a nickel-base superalloy (U700). In order to obtain a higher frequency of defects compared to the standard material the alloy was deliberately doped with organic defects (rubber particles). Eight different types of defects were documented and their influence on the surrounding material investigated by energy and wavelength dispersive X-ray analysis.

With the aid of a fully automated scanning device the intersection areas of the defects were determined (shape and size of the circumferential ellipses and the distribution of linear intercepts). The spatial defect size distributions were derived from the distributions of intercept areas and evaluated statistically. The resulting defect distributions were compared to those of unsieved powder.

1 Introduction

In modern aero-engines high strength nickel-base alloys are used for disc applications. In the production of these alloys the limitation of the number and size of defects is of fundamental importance. Defects such as pores as well as reactive and nonreactive inclusions are preferred sites for crack initiation. They are the reason for a considerable scatter in measured fatigue lives. Process control measures, especially the sieving of powders before hipping, are able to reduce the maximum size of material defects. The assessment of the influence of defects on the fatigue life requires knowledge of the type, size and frequency of defects.

2 Experimental procedure and statistical evaluation

U700 (HIP) has a nominal composition (in weight %) of 0.04 C, 16 Cr, 18 Co, 5.5 Mo, 3.7 Ti, 4.2 Al, 0.03 B, 0.06 Zr and balance Ni. The material was doped with organic defects (particles of fluorocarbon rubber; tradename VITON) with a concentration of 10 mg per kg powder and a size ≤ 150 micron. The chemical analysis of the rubber revealed the following metallic constituents (weight %): 2.0 Zn, 0.5 Ca, 0.05 Mg and 0.02 Fe.

Metallographic sections were taken and statistically evaluated with respect to material defects. The spatial size distribution was derived using the following approach: The intersect areas of the defects were described by ellipses and subsequently approximated by circles with diameters equal to the long axis of the ellipses. From the distribution of the intersect-circle diameters the distribution of spheres was calculated (tomato salad problem) according to a method described by Exner (1972); see Fig. 1.

3 Results

The following types of defects were established:

a) Pores

Pores were found at triple points of grains and within grains. Pores have the highest frequency and smallest size of all defects (500 defects per cm^2; defect size ≤ 34 micron; see Fig. 7).

b) Ellipsoid reactive inclusion

A number of defects were found which are characterized by an ellipsoid reaction zone larger than the particle size:

- γ' -agglomerations have the largest size of all the reactive inclusions (Figs. 2a and 3); Rings of Ti (C, N) and Mo-rich M_6C (Fig. 2b); Cr-inclusions (Fig. 2c); ZrO_2-inclusions (Fig. 2d); Nb/Ti-inclusions (Fig. 2e) and Al_2O_3-inclusions (Fig. 2f)

These defects have a lower frequency but a greater size than the pores (3.5 defects per cm^2; defect size ≤ 229 micron).

c) Viton defects

Organic nonmetallic inclusions are characterized by pronounced prior particle boundaries (Fig. 5). In the defect area and reaction zone the chemical compositions (Fig. 6) and hardness (Fig. 4) differ from the matrix. Viton defects have the lowest frequency and the greatest size of all defects (0.25 defects per cm^2; defect size ≤ 388 micron).

4 Summary and conclusions

In a model alloy (U700 doped with Viton) eight type of defects have been analyzed. The defects can be categorized in three classes: .

- pores
- ellipsoid reactive inclusions
- Viton defects

Pores and reactive inclusions were described by ellipsoids. Viton defects, the largest defects, show a rather irregular shape. Furthermore, they can be distinguished from the matrix by a much higher value of hardness. For that reason a marked influence of fatigue life may be expected. König (1984) has studied the effect of all these defects on the fatigue behaviour. The Viton defects reduced the fatigue life by more than a factor of 10 compared with undoped material.

Fig. 7 gives a summary of the frequency of defects as a function of size. For comparison, typical frequencies of defects in U700 of both sieved and unsieved powders are included. The diagram suggests that the number of reactive zones is about one decade higher than for defects in unsieved powder. However, if the fact is taken into account that the inclusions themselves are about one half of the reactive zone size, there is a reasonable correlation of the defect frequencies.

5 References

Betz, W. (1986), In: High Temperature Alloys for Gas Turbines and other Applications 1986. Proceedings of a Conference, Liege, Belgien, 6 - 9 Okt., 81

Exner, H.E. (1972), Int. Metall. Rev. 17, 25 - 42

König, G.W. (1984), In: PM Aerospace Materials, A Metal Powder Report Conference, MPIF, Bern, 12 - 14 Nov., 23

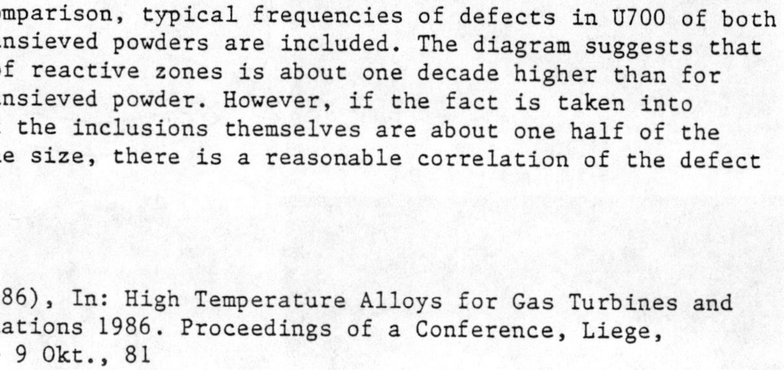

Fig. 1: Representative defect size distribution
a) two-dimensional
b) three-dimensional

Fig. 2: ELLIPSOID REACTIVE INCLUSIONS

Coarse (1-4um) γ'-precipitates are not observed on grain boundaries and in grains within the reaction zone.

Measurement section area: 100 cm^2

Measured defect count: 346

Lower measurement limit: 21 µm

Max. defect size: 229 µm

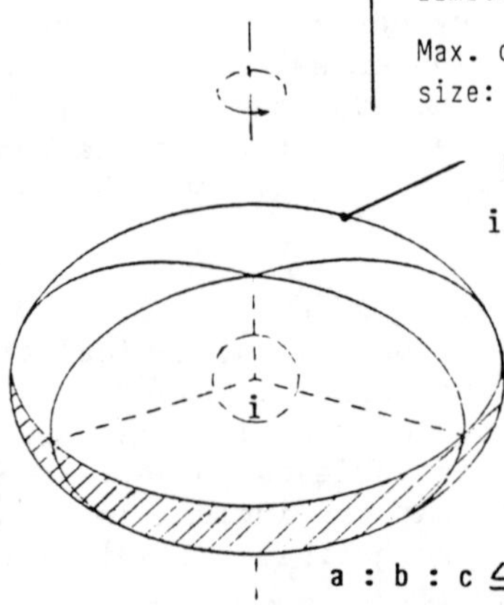

a : b : c $\leq$ 3 : 3 : 1

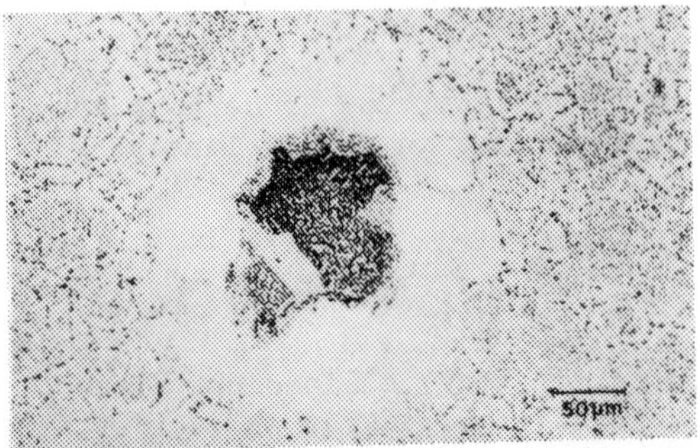

Fig. 2a: γ' - agglomeration

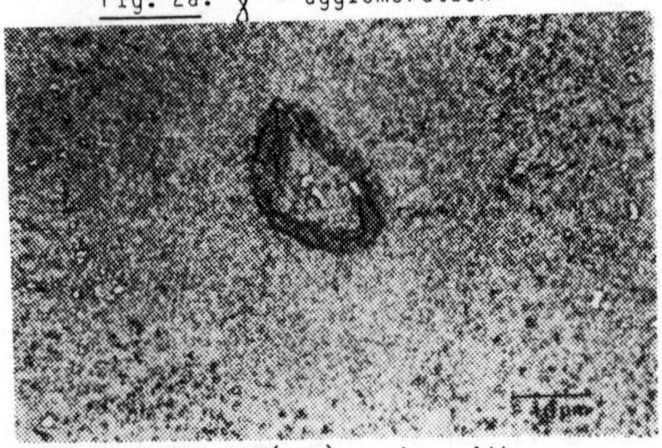

Fig. 2b: Ti(C,N) - ring with molybdenum rich M_6C

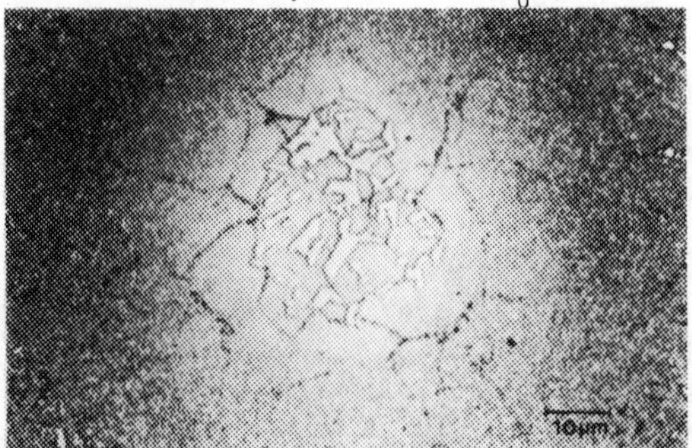

Fig. 2c: Cr - inclusion

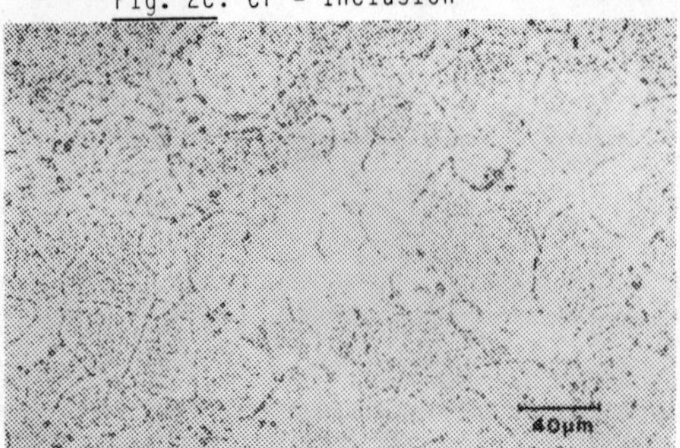

Fig. 2d: ZrO_2 - inclusion

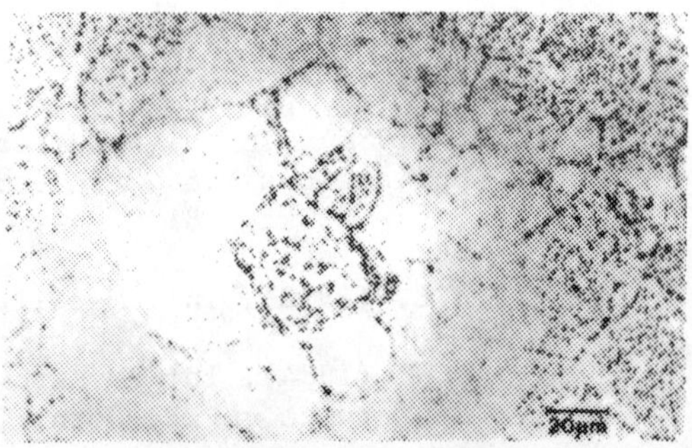

Fig. 2e: Nb/Ti - rich carbide

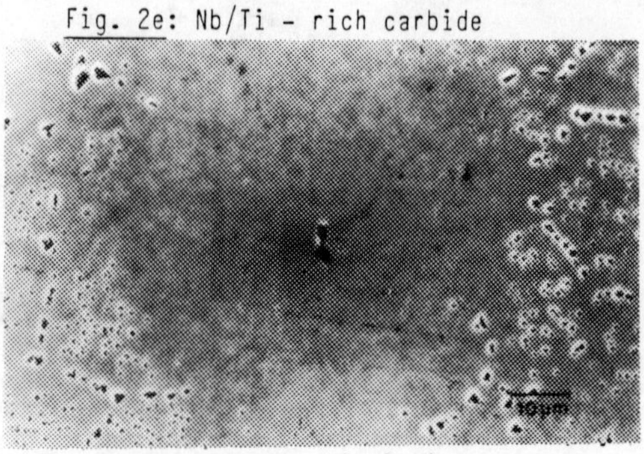

Fig. 2f: Al_2O_3 - inclusion

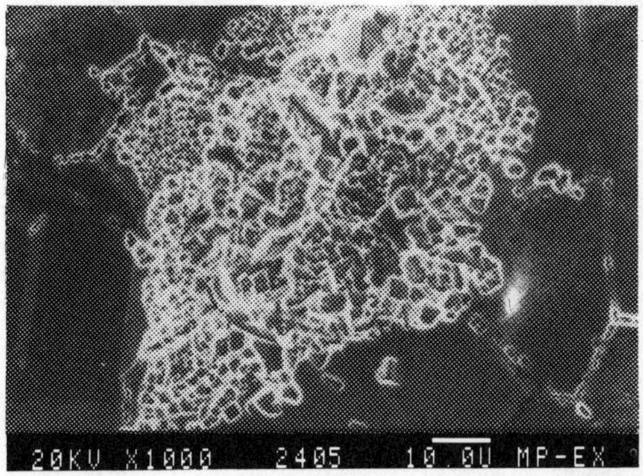

Fig. 3: γ' - AGGLOMERATIONS

Energy dispersive x-ray analysis (EDX) shows a higher concentration of γ' - formers (Ni, Al and Ti) in the defects than in the matrix. These types of defects can be observed after etching (Kallings - Adler 1:1; negative etch for γ' - precipitates). After a strong etch, cube shaped holes appear where the γ' - precipitate was dissolved.

Fig. 4: HARDNESS GRADIENT OVER THE CROSS SECTION

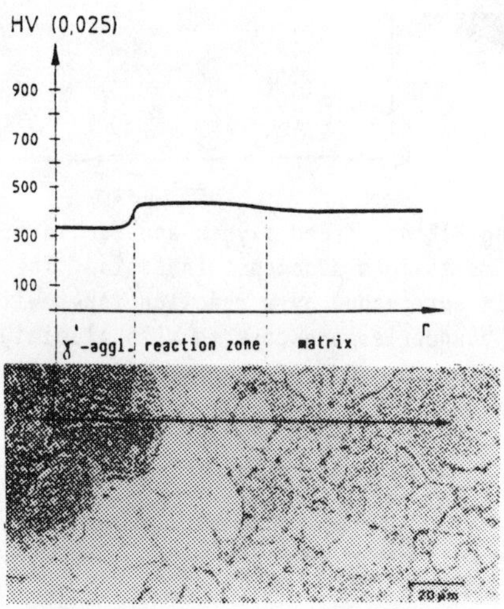

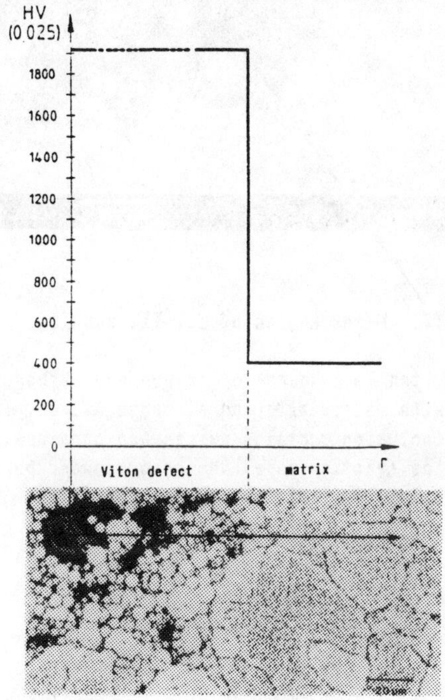

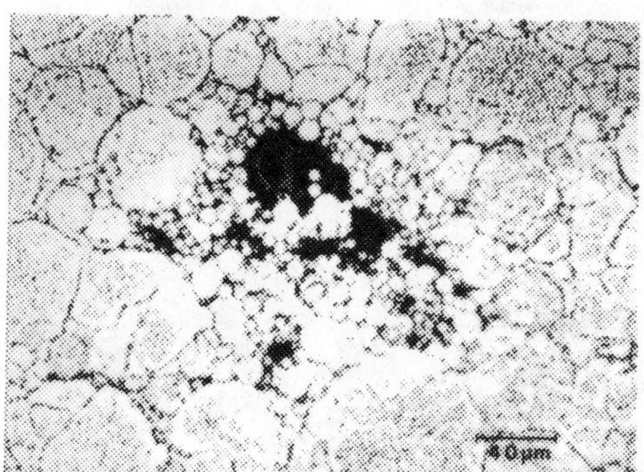

Fig. 5: VITON DEFECTS

Decoration of prior particle boundaries with carbides and/or oxides.

Measurement section area: 100 cm^2

Measured defect count: 25

Lower measurement limit: 39 µm

Max. defect size: 388 µm

Fig. 6: VITON DEFECT

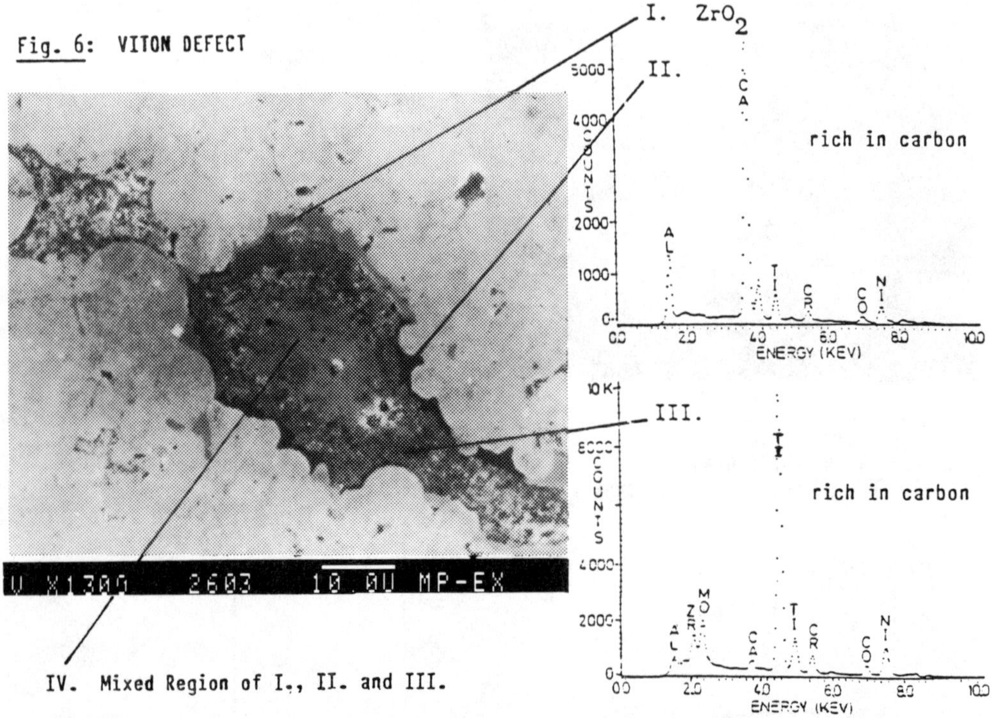

Viton is a source of oxygen and carbon. During HIPing, freed oxygen and carbon react with matrix alloying elements Al, Zr, Ti and Mo to form a compact inclusion. The inclusion contains oxides and carbides, and is surrounded by a reaction zone. Within the reaction zone, the prior powder particle boundaries are covered with aluminium oxide and titanium carbide particles.

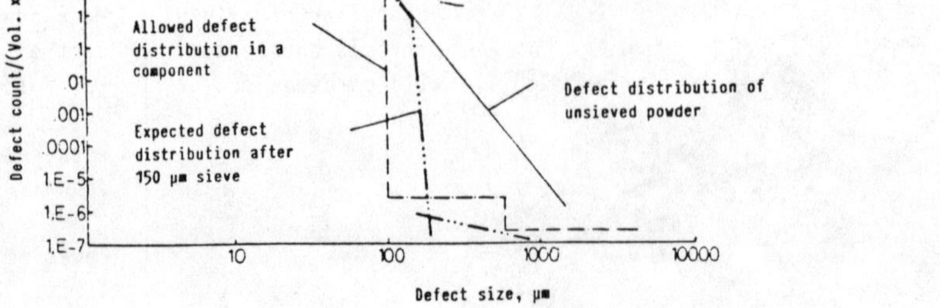

Fig. 7: Comparision of the distribution of sphere diameters with type defect distributions of UDIMET 700 according to Betz (1986).

PRESSURELESS SINTERING AND HIPPING OF β_{ss} - Si_3N_4 + SiC FOR CUTTING TOOLS

Albir A Layyous*, P Greil, G Petzow

Max Planck Institute, Heisenberg St 5, 7000 Stuttgart 80, West Germany
*ISCAR LTD, Box 34, Nahariya, Israel 22100

Abstract

Si_3N_4 / SiC-particulate composites were pressureless sintered at 1850°C and subsequently hot isostatically pressed with N_2 of 100 Mpa. Si_3N_4 - SiC chemical interaction was studied by XRD and Auger spectroscopy. The cutting tools were examined for cast iron and steel machining. The wear mechanism of the cutting tool edges and the mechanical properties of various composite compositions were analyzed.

Introduction

Management of any industry which deals with manufactured products is aware of the high cost of replacing worn tools. Any method, process or development which reduces wear, and hence lowers production costs, is met with great interest.

On increasing the machining speed to improve the production yield, the cutting-tool-edge temperatures increased up to 1000 - 1200°C(1) in the zone of contact with the fabricated metal. The mechanical and thermodynamic properties of the cutting tools at high temperatures are, therefore, important for successful high-speed machining.

Silicon-nitride based cutting tools have been found to be very effective in selected machining operations, especially when machining cast iron at high speed.

The rate of wear-cratering in particular of silicon nitride cutting tools is generally too high in steel machining. The fundamental reason is probably the solubility of Si and N in the hot steel chip flowing across the rake face.(1,2) Therefore, it will be of interest to examine the behavior of the Si_3N_4+SiC mixture as a cutting tool for steel machining.

While silicon ceramics exhibit better strength and thermal shock resistance, the silicon carbide ceramics are of higher hardness, creep and oxidation resistance. It will be worthwhile to design composite materials in order to make use of the improved properties of dispersed SiC particles in a Si_3N_4 matrix material.

The densification of the Si_3N_4/SiC composite is reduced in pressureless sintering while in hot pressing or HIPping, full densification is achieved. Pressureless sintering is an interesting process for producing complicate ceramic shapes such as cutting tools. Therefore, we concentrated on pressureless sintering and HIPping for producing the cutting tools.

Experimental

Different Si_3N_4+SiC particulate mixtures were prepared by pressureless sintering. The starting powders (Tables 1 and 2) were attritor milled with Al_2O_3 balls and subsequently dried with a rotation evaporator. After sieving, green bodies were prepared by isostatic pressing with 630 MPa. Sintering was done in nitrogen atmosphere (P=1.1 MPa). For all the Si_3N_4+SiC types, sintering conditions in boron nitride crucible were one hour heating, one hour holding at 1850°C and one hour cooling down to room temperature. All sintered specimens were subsequently HIPped for 15 minutes at 1850°C with an N_2-pressure of 100 MPa.

Fracture surfaces were observed by SEM, EDAX and AUGER spectroscopy for chemical analysis; XRD was used to analyze the phase compositions.

The cutting tools were shaped by diamond cutting and grinding the specimens with diamond wheels, using an 8mm, one-pass grinding to prevent crack formation on the ground surface.

The ISO designation of all the cutting tools was ISO CNGN 120408T. The tools were tested in machining on steel SAE 1045 and on grey cast iron. Machining conditions for steel and for cast iron were constant for all cutting tools (Table 3). The total wear of the cutting edge, designed as the crater width which formed during machining, was measured from graphs. The depth of the crater was designated as the wear amount in case of high wear rate, while the total wear (V_B) was designated as flank wear.

Results and Discussion

By adding SiC powder to Si_3N_4 the theoretical density of the pressureless sintered specimens decreased. For example, the sintered density of group A was 96% TD while the same mixture without the SiC addition was about 99.5% TD. Using HIP process after sintering did not improve the theoretical density. In the second group B, using HIP process after pressureless sintering increased the theoretical density from 93% TD 96% TD.

At 1850°C the Si_3N_4 matrix will sinter much faster than the SiC particles, and radial compressive and tensile hoop stresses were supposed to develop around SiC particles. Thus, the densificatin is related to the creep relaxation by viscous flow of the liquid phase. Creep relaxation rate of the matrix becomes slower with increasing volume fraction of SiC, resulting in reduced densification rate and residual porosity [3,5]. With an increase in the SiC content, the volumetric pore density seems to increase while the pore size decreases.

By HIPping after sintering, the mass transport, specially the liquid phase, was accelerated and close porosity was eliminated, while open porosity remained. In the HIP process, the high N_2 pressure stabilizes the Si_3N_4 matrix.

Thus, in pressureless sintered specimens free silicon was detected while after HIPping no free silicon remained in the bulk.

This reactive behavior of Si_3N_4 -SiC sintering additives in pressureless sintering was detected clearly in group B on the specimen surfaces where interceramic layers formed (Fig. 2), every layer with special structure, porosity and chemical composition. After HIPping, these layers remained on the surface with the deep open porosity.

These interceramic layers were not formed during pressureless sintering of the β_{SS}-Si_3N_4+10%SiC specimens, but free silicon was detected on the surface as well during sintering at the same conditions as for β_{SS}-Si_3N_4+40%SiC. The surface layers were analyzed by two different methods:

1) Auger spectroscopy

By increasing the SiC concentration in β_{SS}-Si_3N_4 , the conductivity was increased and charging problems during Auger spectroscopy were prevented. Thus, it was not possible to test any Si_3N_4 specimen because of the low electrical conductivity of the Si_3N_4 phase.

Elements were analyzed in the different layers (Fig. 3). The concentration of Si, C, N, Al and B changes from one layer to another:

A) Boron was concentrated only in the second porous layer. The boron diffused from the boron nitride crucible into the specimen surface. The boron concentration decreased as the distance increased inside the layer.The second layer was porous because of the reaction between boron and silicon nitride:

(1) $$Si_3N_4 + 4B \rightarrow 4BN + 3Si$$

the free silicon was removed intensively by diffusion from the second layer to the outer layer.

B) Carbon and silicon were enriched in the outer layer.

C) The surface layer was free from nitrogen; the highest nitrogen concentration was in the internal layers.

D) The third layer was enriched by aluminium more than the other layers.

2) X-Ray Diffraction Analysis

The relationship between different phases in the different layers was detected by X-ray diffraction. Every layer was subsequently removed after the X-ray analysis (Fig. 4).

A) Free silicon concentrated on the surface layer and decreased with increased distance from the surface.

B) SiC-6H polytype is stable in all layers at 1.1 MPa nitrogen pressure. It was detected at the inner boundary of the fourth layer, where the structure becomes more homogeneous.

C) Oxynitrides of aluminium, silicon and yttrium were formed in all layers.

Machining Tests:

The flank wear rate of group A in grey cast iron machining was 175μm/min, compared to pressureless sintering β_{SS}-Si_3N_4 with the same machining conditions (4 minutes), where the flank wear rate was 115mm/min.

The HIPping process did not improve the wear properties of the cutting tools; nor did increasing the SiC concentration to 40 wt% increase the wear rate to too high a level.

In steel machining all the cutting tool edges broke or showed deep crater formation after 30 seconds. Flank wear rate was 600mm/min and crater depth of 45mm after 30 seconds; while in pressureless sintering β_{SS}-Si_3N_4 , the flank wear rate with the same machining conditions was 480mm/min and crater depth of 18μm after 30 seconds (Fig. 5).

The chemical stability of Si_3N_4 at high temperatures against cast iron and steel is an important factor in explaining the differences in the machining results. During machining, cutting edges are in continuous contact with the machined material, and the wear forces develop high temperatures in the contact zone (~1200°C). This sytem acts as a diffusion couple. The chemical and mechanical properties in the small diffusion layer which formed in a short time, changed and removed mechanically from the silicon nitride surface. The continuous process increases the total wear. The diffusivity of Si and N (6,7) in the hot steel chip flowing across the rake face explains the cratering of the cutting edges; while in cast iron, the chips are smaller and not continous as are steel chips.

Silicon carbide addition to silicon nitride did not improve the wear properties, because of the diffusion of Si and carbon in steel and because of the undensified specimens.

Conclusions

1 When SiC content in β_{SS}-Si_3N_4 increased, the fractional density decreased. Hipping after pressureless sintering did not full densify the specimens.

2 A sintering reaction between SiC-Si_3N_4-sintering additives-sintering atmosphere and sintering crucible took place, and caused the formation of interceramic layers on the specimen surface layers. In group B (with 40% SiC content) Si_3N_4 decomposed on the surface, SiC remained stable, while free silicon concentrated on the specimen surfaces and in the bulk (more in group B than group A).

3 The addition of SiC to β_{SS}-Si_3N_4 increased the wear of the cutting tools in steel and cast iron machining.

References

1 B North, R D Baker, **R and HM**, March (1984), 46-51.
2 S Robb, **Ceramic Bulletin**, 62, No 2 (1983), 206-207.
3 P Greil, H Tanaka and G Petzow, **6th Cimtec**, Milan, Italy, June (1986).
4 F F Lange, **J Am Ceram Soc,** 56 (1973), 445.
5 R Raj and K Bordia, **ACTA MET**, 32 (1984), 1003.
6 J Gerald, Andre Ezis and D Franklin Runkle, **J Am Ceram Soc**, 68 (1985), 30-31.
7 Koichi Terao, Takao Susuki and Tadahisa Arahori, **J of Japanese Ceram Soc**, 94 (1986), 111.

List of Tables and Figures

Table 1: Producers' Characterizations of the Starting Powders in Nonoxide Ceramic Preparations

Table 2: Concentrations of Sintered Si_3N_4-SiC Mixtures

Table 3: Machining Conditions of Steel and Cast Iron by Different Cutting Tools

Fig. 1: X-ray Diffraction of pressureless sintering Si_3N_4-SiC Mixtures. Free Silicon Detected in Si_3N_4-40% SiC Substrate

Fig. 2: Surface Layers of Pressureless Sintering / Si_3N_4-40% SiC-9% Y_2O_3-6% Al_2O_3

Fig. 3: Auger Analysis of Surface Layers in Sintered Si_3N_4-40% SiC-9% Y_2O_3-6% Al_2O_3

Fig. 4: X-ray Diffraction Analysis of Surface Layers in Pressureless Sintering / Si_3N_4-40% SiC

Fig. 5: Cutting Tool Edges After Cast Iron and Steel Machining

Powder	Company	Composition (wt%)	BeT (m^2/g)
α Si_3N_4	LC10, H C Starck	N 38.2 O1.1-1.5 C 0.2 F 0.12 Fe 0.04 Al 0.15 Ca 0.03 Na 0.005 Free Si 0.1	14-17
Al_2O_3	A16, ALCOA	Al 52.9 MgO 0.1 O 47.0	
Y_2O_3	REO, VENTRON	99.99%	
α SiC	UF15-LONZA	Free Carbon 0.39 Free Silicon 0.12 O 0.81 Al 0.05 Fe 0.04 Ti 0.01	15±1

Table 1: Producers' Characterizations of the Starting Powders in Nonoxide Ceramic Preparations

Specimen	Si_3N_4	Al_2O_3	Y_2O_3	αSiC (wt%)
A	75	6	9	10
B	45	6	9	40

Table 2: Concentrations of Sintered Si_3N_4-SiC Mixtures

Machining Conditions	Steel (SAE1045)	Grey Cast Iron
Speed (m/min)	250	400
Feed (mm/min)	0.2	0.2
Depth of Cut (mm)	2	2
Time	30" (sec)	2', 4' and 12' (min)

Table 3: Machining Conditions of Steel and Cast Iron by Different Cutting Tools

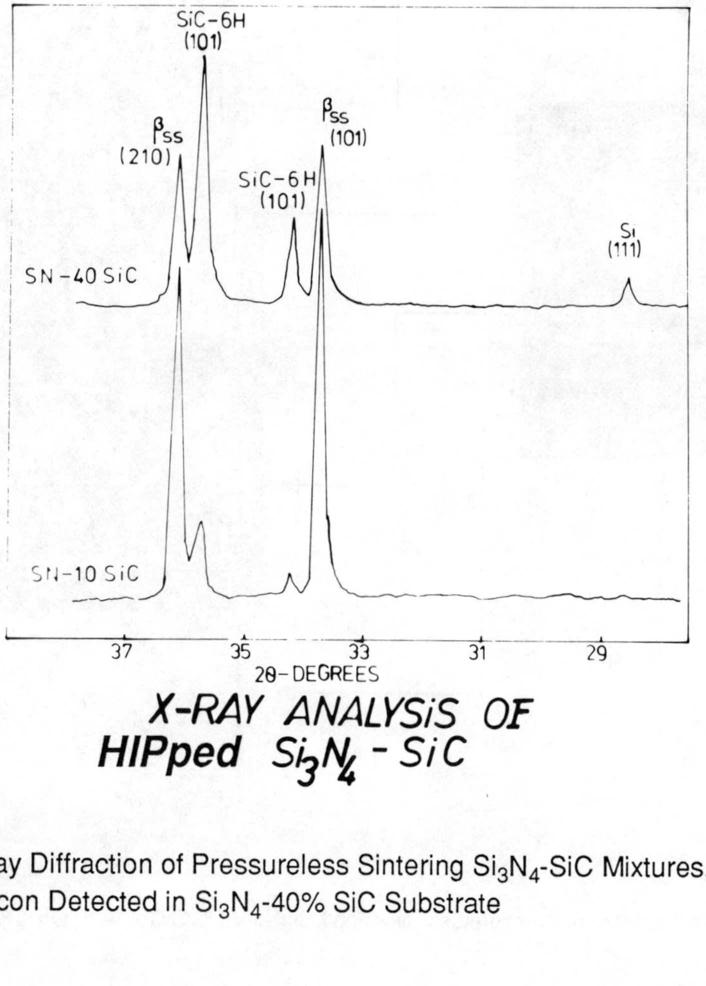

Fig. 1 : X-ray Diffraction of Pressureless Sintering Si_3N_4-SiC Mixtures. Free Silicon Detected in Si_3N_4-40% SiC Substrate

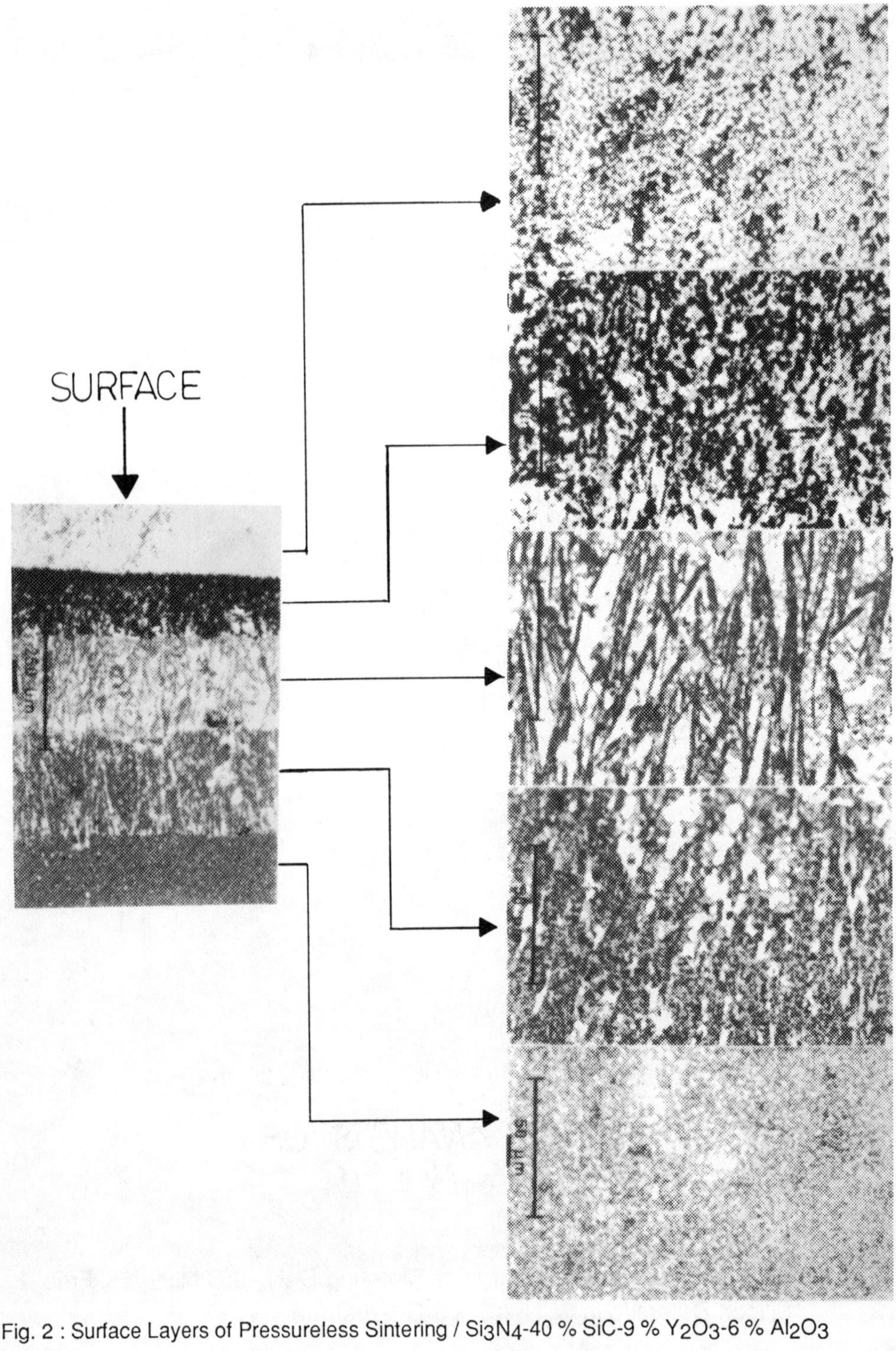

Fig. 2 : Surface Layers of Pressureless Sintering / Si_3N_4-40 % SiC-9 % Y_2O_3-6 % Al_2O_3

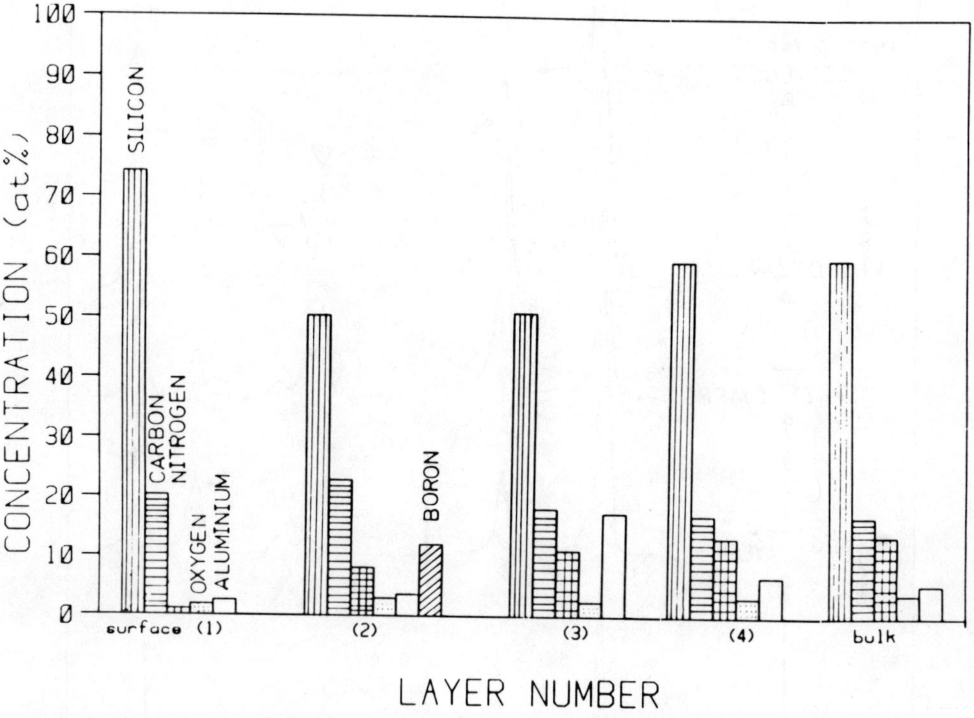

Fig. 3 : Auger Analysis of Surface Layers in Sintered Si_3N_4-40 % SiC-9 % Y_2O_3-6 % Al_2O_3

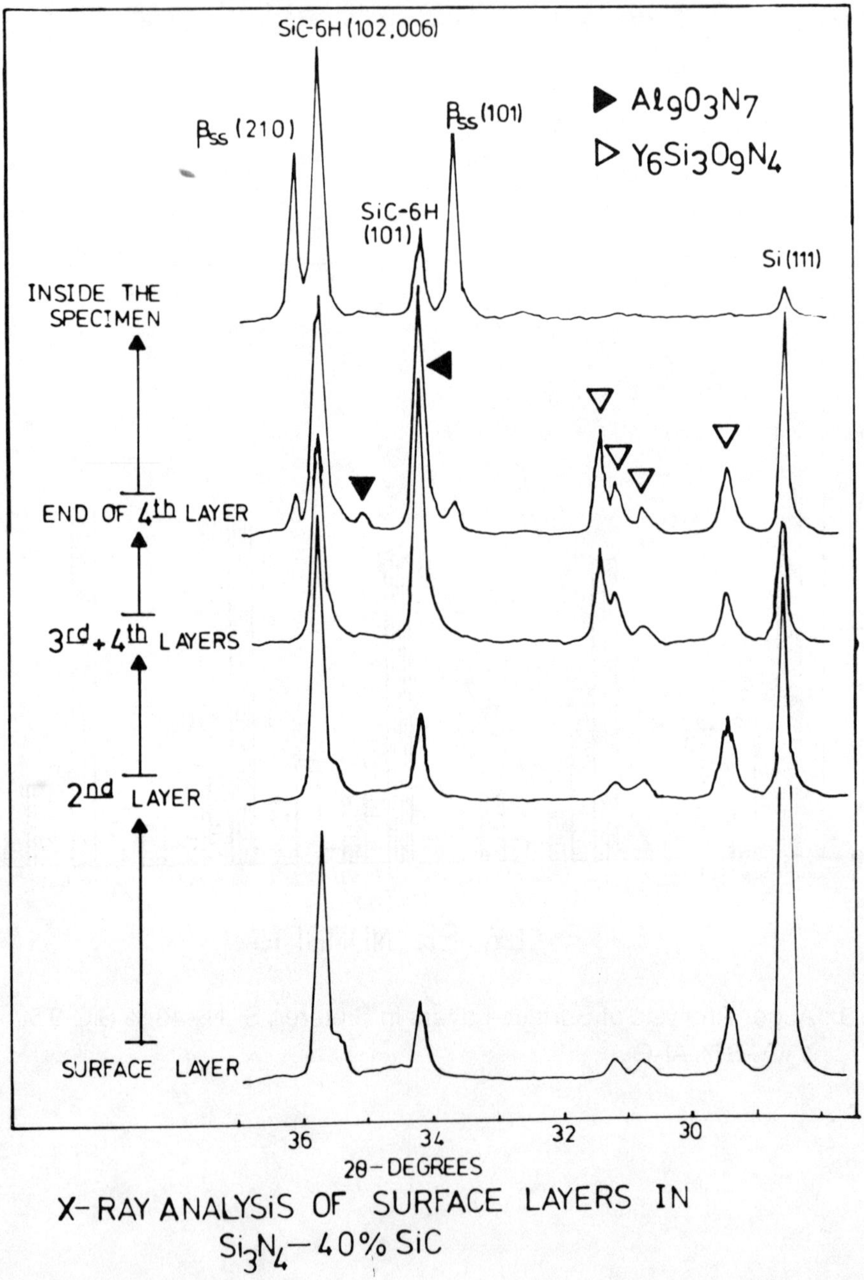

Fig. 4 : X-ray Diffraction Analysis of Surface Layers in Pressureless Sintering / Si_3N_4-40 % SiC

Fig. 5 : Cutting Tool Edges After Machining on Cast Iron (upper) and Steel (lower)

COMPRESSION MOLDING OF FINE CERAMIC POWDER BY USING WATER BINDER

BY

TAKEO NAKAGAWA, LIHONG ZHANG, HIROYUKI NOGUCHI
NOBUYUKI TAKAHASHI and KIYOSHI SUZUKI*

INSTITUTE OF INDUSTRIAL SCIENCE, UNIVERSITY OF TOKYO
7-22-1,Roppongi,Minatoku,Tokyo,Japan
*NIPPON INSTITUTE OF TECHNOLOGY

Abstract

Wet ceramic powder mixed with water was confirmed to have sufficient flowability in making complicated shapes, which leads to making the debinding process easier and shorter in powder injection molding process. Here a molding die is cooled beforehand under zero degree and the wet powder filled in a die cavity by compression molding starts to freeze and becomes solid. The frozen powder compact is ejected from the die cavity and is directly fed into a vacuum furnace and then sintered. Even if the water content is less than 35 volume percent, a simple cup shape was formed by using a forming tool like a cold forging tool. This flowability is improved remarkably by applying vibration to the tool. Thin and deep tubes were formed by this vibration forming of wet ceramic powder. In addition, a wire with a diameter of 0.1 ∿ 0.5mm was extruded by adding 1 ∿ 2wt.% PVA in water binder.

Introduction

In the injection molding of ceramic powder, more than 40 volume percent organic binder is added so as to yield sufficient flowability and shape fixability. Before sintering, the binder has to be extracted by heating the green compact in a furnace. The extracting speed is extraordinary low and a thick compact must be kept in the furnace for a long time.

The purpose of this research is to reduce the debinding time by substituting water binder for the organic binder. Here, only a small amount of organic binder is added to keep the shape after drying. This paper proposes several new compacting processes which utilize the flowability of the wet ceramic powder.

1. Freeze Compression Molding[1)]

First of all, the flowability of fine alumina powder mixed with water was experimentally investigated. The water was mixed with Al_2O_3 powder of 0.2 μm diameter as shown in Fig.1 and the wet powder was extruded from the container through a narrow nozzle. Fig.2 shows the extruding stage, in which a smooth wire was formed successfully and unexpected good flowability and shape fixability were observed without adding any organic binder.

The relation between the water content and the flowability was obtained by using a conventional spiral flowability test tool. The results are shown in Fig.3. The wet powder has sufficient flowability in case of the water content of 15 - 20wt.%, which makes it possible to fill the wet powder into a narrow cavity of a molding die.

Fig.1 Mixing Al_2O_3 powder with water

Fig.2 Extruding test using good flowability of wet Al_2O_3 powder

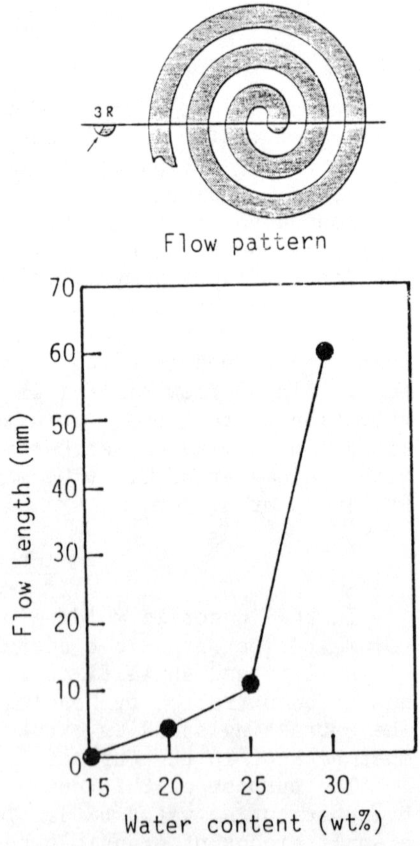

Fig.3 Result of flow test for Al_2O_3 powder

By utilizing this flowability, the freeze compression molding process as shown in Fig.4 is proposed. A wet powder billet including 45 - 50 vol.% water is prepared and is shaped to a final complicated shape by compression molding. The die cavity is cooled to -15 - -20°C and the wet powder filled in the cavity starts to freeze. After complete freezing, the frozen and solid powder compact is ejected from the cavity by ejecting pins. If the compact is simple and thick, it can be ejected even in the stage of partial freezing. The ejected frozen compact is put into a vacuum furnace. Here, the frozen compact is dried directly in the solid state in order to prevent the crack formation at the time of drying due to its dimensional shrinkage. Fig.5 shows the relation between the drying time and the temperature. The drying time can be reduced by increasing the drying temperature.

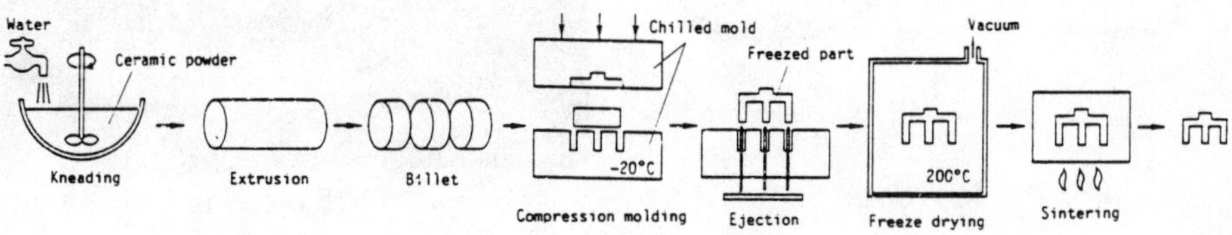

Fig.4 Process for freeze compression molding

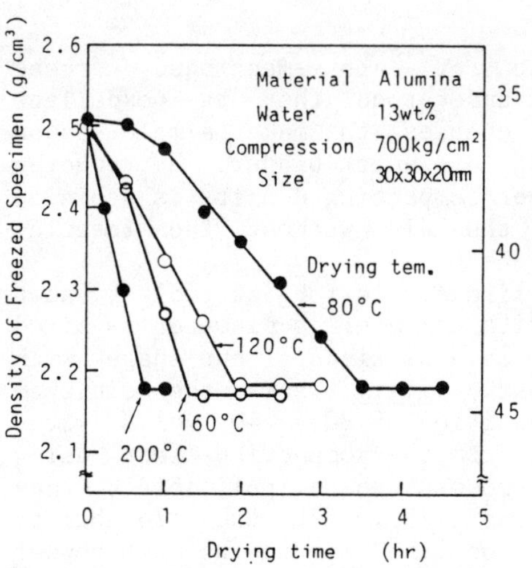

Fig.5 Relation between density of freeze drying samples and drying time

Fig.6 Simple molded specimen (Green alumina, 46 x 46 x 30mm, t=3mm)

Experimental molding was conducted. Fig.6 is a simple square cup molded by this process. Fig.7 is a little more complicated molded example. From these tests, the freezing time of 2 - 3 minutes in the mold is enough to be ejected without breakage. A small amount of organic binder should be added to keep the shape after drying and make the handling easier.

(a)Tool (b)Ejecting

Fig.7 Molded specimen with complicated shape

2. Compression Molding

From the experimental test of above-mentioned freeze compression molding, it can be understood that the excellent flowability with water binder enables to make simple shape compacts even in the condition of less water binder. By reducing the amount of water binder, higher compacting density is achieved and the compact is ejected from the die without the freezing procedure.

The compacting tool is just like a cold forging tool as shown in Fig.8. 99.8 % Al_2O_3 powder with 2μm average diameter is mixed with water including 2wt.% PVA. Several kinds of cup shape with bottom are formed by this process. Fig.9 shows the sintered example after compacting. The addition of 30 ∿ 40 vol.% water binder is found to be suitable. From the compacting test using different kinds of powder, this process can be applicable to any kind of ceramics, including SiC and Si_3N_4. But the flowability was confirmed to depend greatly on the scattering of powder diameter. Powder size distributions of typical two kinds of Al_2O_3 powder which have different flowability are shown in Fig.10. "A" powder has better flowability than "B". Uniformly sized powder with little scattering has not sufficient flowability and then has poor formability. On the contrary, largely scattered powder has better flowability and formability and accordingly the high density compact with a complicated shape can be obtained even in such less water binder condition.

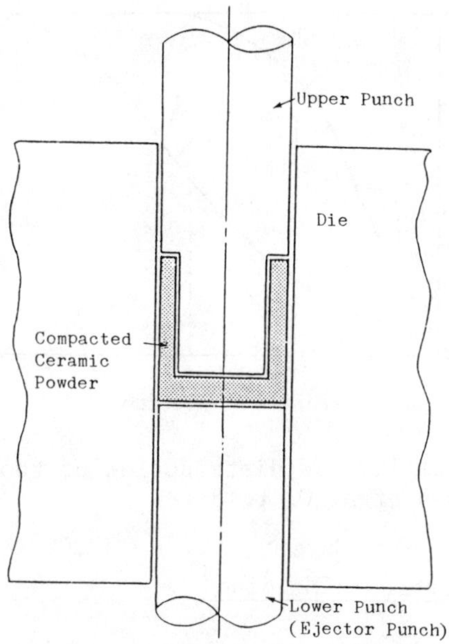

Fig.8 Compression molding tool

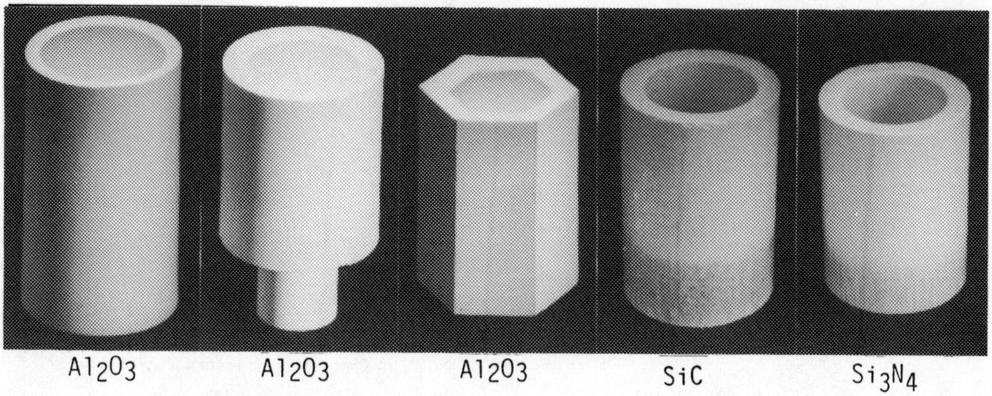

Al_2O_3 Al_2O_3 Al_2O_3 SiC Si_3N_4

Fig.9 Molded examples (sintered)

Alumina powder	Average particle size (μm)	Specibic surface area (m^2/g)	Purity (%)	True density (g/cm^3)
AL-170 "A"	2.0	2.5	99.8	3.93
AL-160SG "B"	0.6	7.0	99.9	3.91

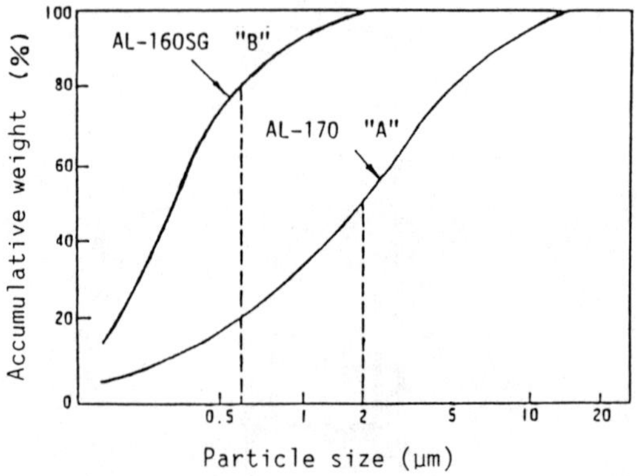

Fig.10 Particle size distribution of two different types of Al_2O_3 powder

3. Vibratory Compression Molding[3)]

Though the application of vibration is known not to be so effective in the dry compaction, the vibration was found to

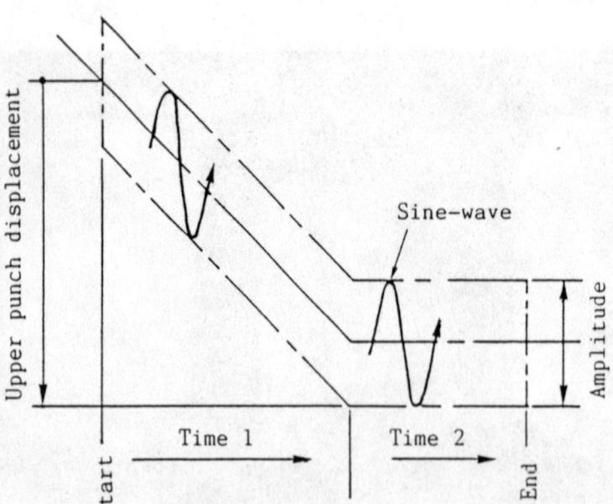

Fig.11 Vibration mode applied to punch

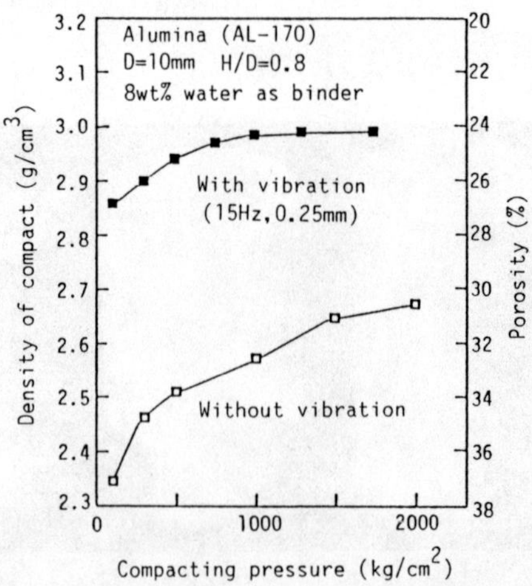

Fig.12 Effect of vibration on the increase of density of compact(after drying)

increase the flowability of wet powder to a great extent. In this experiment, the vibration mode shown in Fig.11 was applied to the punch. Fig.12 shows the effect of the vibration of 15Hz to the density of cylindrical compacts. In the experiments the frequency of 0 ∿ 60Hz and the amplitude of 0 ∿ 2.0mm were used. The effective range was more than 5Hz in frequency and 0.25mm in amplitude. Table 1 shows the recommended suitable water content and the compacting conditions for the two kinds of Al_2O_3 powder. By the application of vibration to the punch, water content can be reduced to 25 ∿ 28 vol.% in case of the "A" powder, while 28 ∿ 37 vol.% water is necessary in the compression molding without vibration. Fig.13 shows a little thinner cup made of Al_2O_3 "A" powder.

Table 1 Suitable conditions in vibratory compression molding

	Amplitude (mm)	Frequency (Hz)	Compacting pressure(kgf/cm²)	Water content	
				wt.%	vol.%
AL-170	2	25 - 30	600 - 700	8 - 9	25 - 28
AL-160SG-3	2	25 - 30	500 - 600	12 -13	35 - 37

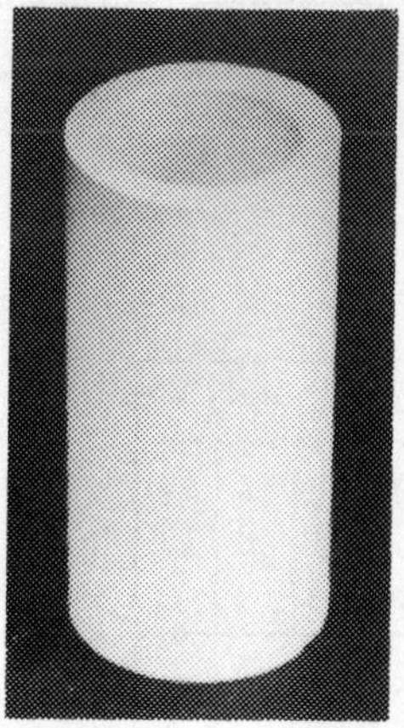

Fig. 13 Alumina Ceramic cup with thinner wall produced by vibratory compression molding with water binder

(Al_2O_3 powder: Al-170, D=10mm, t=1.5mm, H=25mm, bottom thickness=0.7mm)

4. Wire Extrusion[4)]

By changing the molding die shown previously in Fig.8 to a diamond nozzle with a small hole, it is possible to achieve wire extrusion of ceramics powder. Here, the wet ceramic powder including more than 40 vol.% water and a small amount of PVA was extruded without adding any vibration. Table 2 shows the characteristics of the ceramic powder used in the experiment and the results of the extrusion are shown in Table 3. Under the binding condition of 40 ∿ 50 vol.% water and 1 ∿ 2 wt.% PVA, the wet ceramic powder is formed to a long wire by extrusion. The several samples of the extruded wire are shown in Fig.14. Fig.15 shows SEM photographs of the green and sintered ceramic wires produced by this process.

Table 2 Characteristics of ceramic powder for extrusion test

Powder	Commercial No.	Particle size distribution	Specific surface area	True density
SiC 0.15μm	201A200204	-2.3 -1.0 -0.5 μm 100 87 62 wt%	23.9 (m^2/g)	3.22 (g/cm^3)
Al_2O_3	UA-5055	-3.0 -1.0 -0.5 μm 100 86 64 wt%	5.1 (m^2/g)	3.97 (g/cm^3)
Si_3N_4	SN-9S	-16 -5.0 -1.0μm 100 57 42vol%	7.0 (m^2/g)	3.18 (g/cm^3)

Table 3 Formability test results

Water binder		SiC					Al_2O_3			Si_3N_4		
vol.%		38	40	41	45	48	43	47	50	41	44	47
wt.%		16	17	18	20	22	16	18	20	18	20	22
(PVA:wt.%)		0.96	1.02	1.08	1.20	1.32	1.6	1.8	2.0	1.8	2.0	2.2
Diameter of nozzle	0.1mm	●	△	◎	○	○	●	△	○	●	△	○
	0.3	●	○	◎	○	○	●	○	○	●	△	○
	0.5	●	○	◎	○	○	●	○	○	●	△	○

● not formed △ poorly formed

○ formed ◎ perfectly formed

(a)Al_2O_3:Φ0.3mm (b)Si_3N_4:Φ0.1mm

(c)SiC:Φ0.5mm (d)SiC:Φ0.3mm

Fig.14 Extruded samples of wire(Green stage)

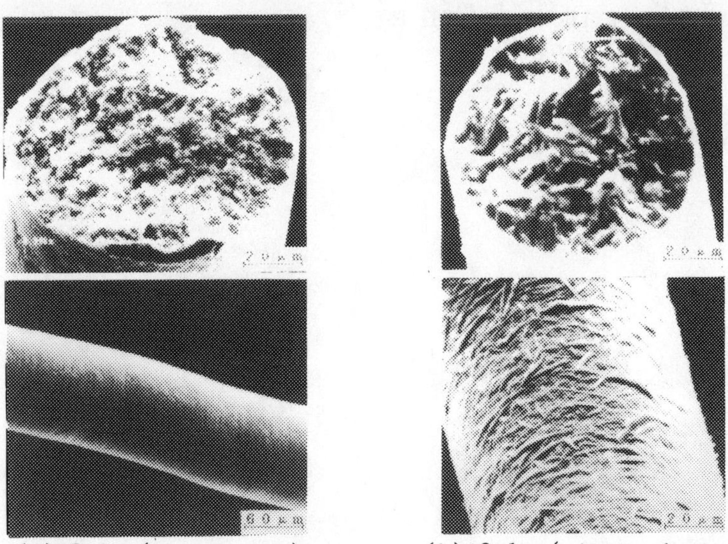

(a)ϕ0.1mm(green wire) (b)ϕ0.1mm(sintered wire)

Fig.15 SEM photos of green and sintered SiC wire

Concluding Remarks

Wet ceramic powder including a small amount of organic binder has sufficient flowability for being formed into a desired complicated shape. By making use of this flowability, special shapes of compact, which are difficult to be formed by conventional die molding, can be formed with a simple tool. Flowability is affected by the characteristics of the powder and the vibration of the tool. In this investigation, the vibration was applied only to the punch, but the vibration of the die is also expected to be effective to improve the flowability and formability.

Refferences

1)T.Nakagawa, N.Takahashi, H.Noguchi, L,Zhang : Freeze Injection and Compression Molding of Fine Ceramic Powder by Using Water Binder, Jour.Japan Soc. Powder and Powder Metallurgy 34(1987), 383

2)T.Nakagawa, L.Zhang, H.Noguchi : Compression Molding of Alumina Powder with Water Binder, Proc. of 1987 Meeting of Ceramic Soc. of Japan.

3)T.Nakagawa, L.Zang, H.Noguchi, K.Suzuki : Vibratory compression Molding of Ceramics Powder Using Water Binder, Proc. of 1988 Spring Conference, Japan Soc. Powder and Powder Metallurgy

4)T.Nakagawa, L.Zhang, H.Noguchi, K.Suzuki : Extrusion of Fine Ceramic Powder with Water Binder, Proc. of 1988 Meeting of Ceramic Soc.of Japan.

VARIATION OF THE MICROSTRUCTURE OF PM ASTROLOY LC AND ITS EFFECT ON STATIC STRENGTH AT ROOM TEMPERATURE

W. Track
H. Floege
H. Pillhoefer

MTU Motoren- und Turbinen-Union GmbH
Munich, FRG

The influence of different microstructures (grain size, grain size distribution, γ' characteristics) in PM Astroloy LC on static strength at RT was measured. A broad varity of structures was included in the tests.

Variations in microstructure:

- Grain sizes from ASTM 10 to ASTM 4-5 were obtained by using different HIP and solution treatment temperatures (Figure 1).
- The size distribution was on the one hand uniform according to ASTM-definition, on the other hand we gained a duplex structure with recristallized fine grains surrounding the prior coarse grains.

  This was achieved by combinations of different degrees of deformation and solution treatments (Figure 2).
- Particle sizes for γ' were set over a range from about 70 nm to 420 nm by age hardening and overaging. A size range from 15 nm to 300 nm was covered by different cooling rates after solutioning (Figure 3).
- The particle size distribution of γ' was defined by the volume ratio of the coarse, not solutioned γ'-particles to the fine γ'-particles newly precipitated after solution treatment. The ratio varied from about 18:20 to 0:38. It was controlled by solution temperatures between 1050°C and 1150°C (Figure 4).

It was possible to select the heat treatment parameters such that each structural feature could be varied independently. Therefore, the influence of each of the structural variables on the strength could be determined separately and quantitatively.

The influence of microstructure on yield strength can be described by the following statements (Figure 5):

- Different grain sizes, produce values of yield strength between 910 MPa and 1,100 MPa.

- Different grain size distributions, obtained by partial recrystallization coupled with the decreasing strain-hardening, gave rise to yield strengths between 1,080 MPa and 1,220 MPa.

- The various γ' distributions produced yield strengths from 1,060 MPa to 1,090 MPa.

- The coarsening of the γ' particles from 100 nm to 420 nm resulted in a decrease in the yield strength from 1,080 MPa to below 900 MPa.

  With a very small, finely dispersed γ' phase (dia. = 15 nm), a yield strength of 1,225 MPa was achieved.

Conclusions

Small particle sizes in the γ' phase (achieved by rapid cooling after solution heat treatment), and the degree of work hardening (set by low solution heat treatment temperatures) make a considerable and technically useful contribution to high RT strengths.

By combination of appropriate grain sizes, strain hardening and precipitation hardening, PM Astroloy LC was produced with yield strengths ranging from below 900 MPa up to 1,225 MPa and ultimate elongations between 36% and 18%.

It appears possible, however, that yield strengths from below 700 MPa up to 1350 MPa can be obtained.

FIGURE 1: Variation of Grain Size

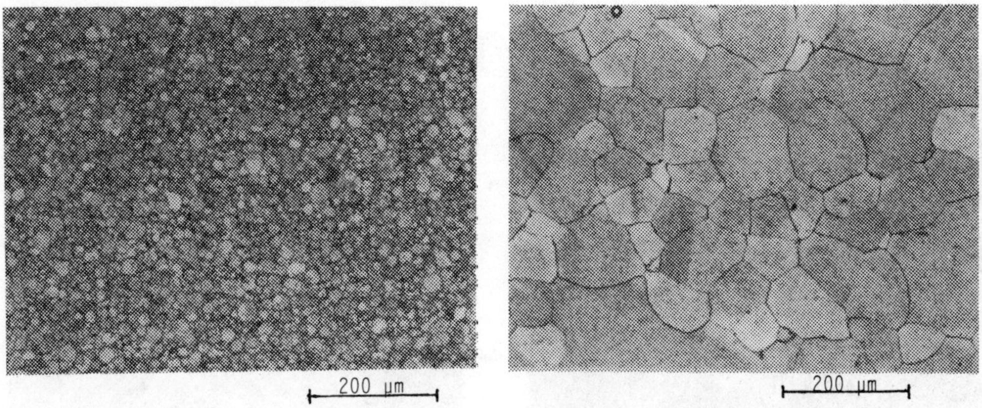

GRAIN SIZE: ASTM 9

HIP-treatment:
1100°C/1000bar/4h

solution-treatment:
1150°C/2h

GRAIN SIZE: ASTM 4-5

HIP-treatment:
1240°C/1000bar/24h

solution-treatment:
1150°C/2h

FIGURE 2: Grain Size Distribution

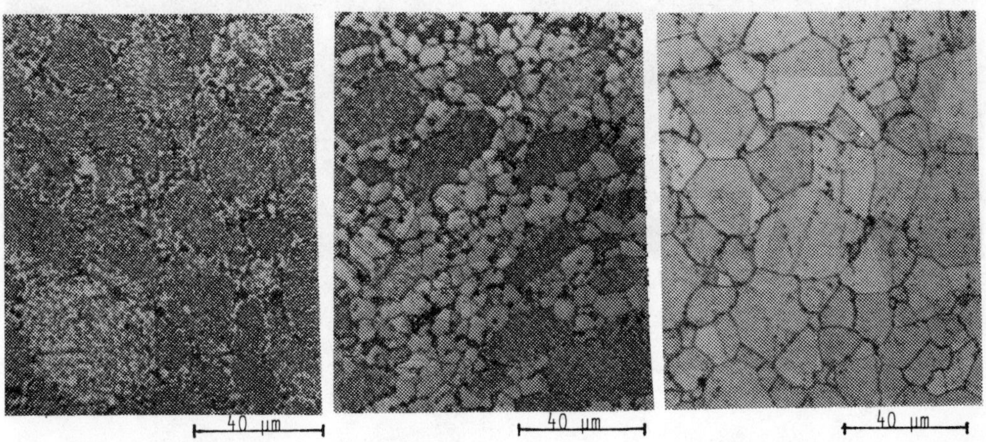

DEGREE OF RECRYSTALLISATION: 0 %

degree of working: 40%

solution-treatment: 1050°C/1h

DEGREE OF RECRYSTALLISATION: 70 %

degree of working: 40%

solution-treatment: 1115°C/1h

DEGREE OF RECRYSTALLISATION: 100 %

degree of working: 40%

solution-treatment: 1150°C/1h

FIGURE 3: Variation of γ'-Particle Size

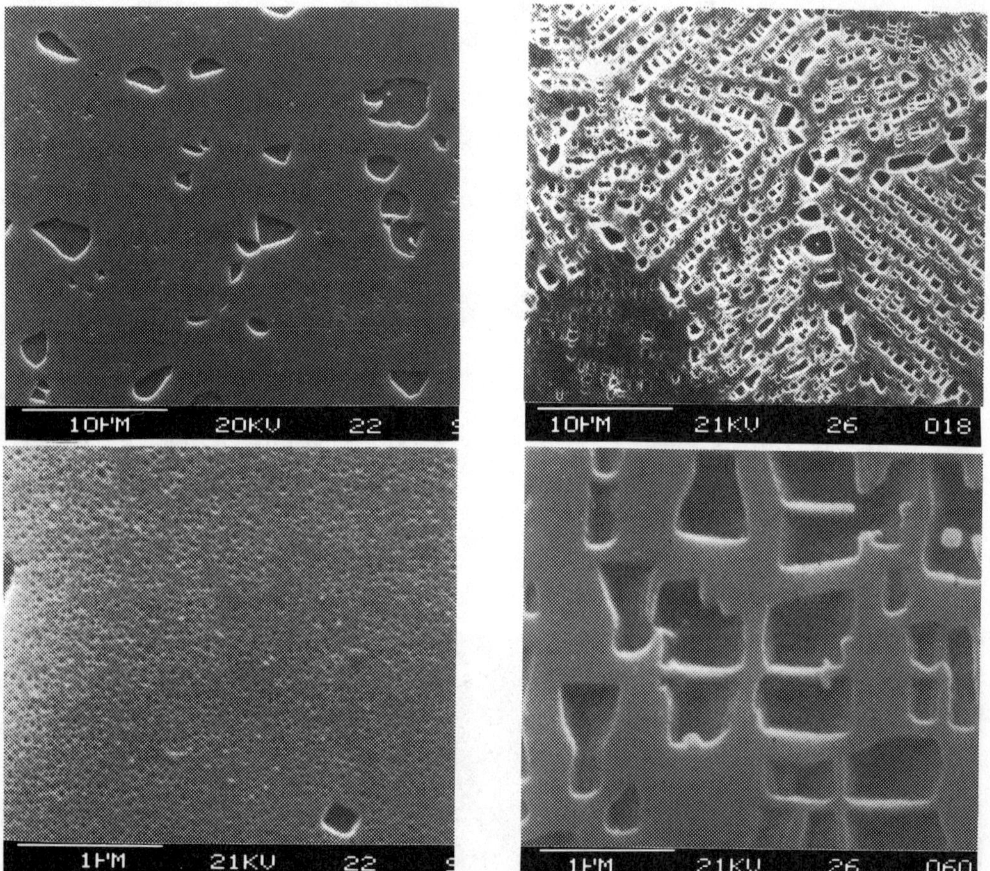

γ'-AVERAGE SIZE: 15 nm

solution-treatment:
1130°C/1h

cooling rate: 1800 °C/min

γ'-AVERAGE SIZE: 420 nm

solution-treatment:
1130°C/1h

cooling rate: 2 °C/min

FIGURE 4: Variation of γ'-Particle Size Distribution

18 VOL% γ'-PHASE,
NOT DISSOLVED

20 VOL% γ'-PHASE,
REPRECIPITATED

solution-treatment:
1090°C/2h

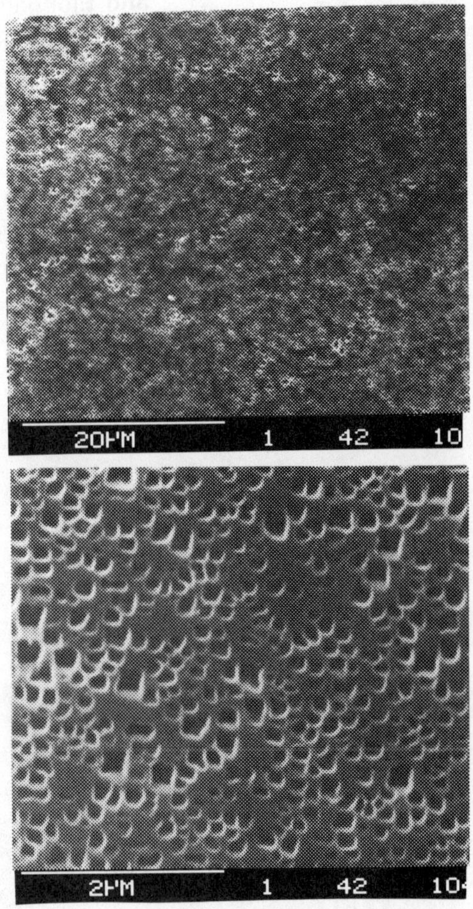

38 VOL% γ'-PHASE,
REPRECIPITATED

solution-treatment:
1150°C/2h

Figure 5 : Effects of Microstructural Variations on Yield Strength and Elongation at Room Temperature

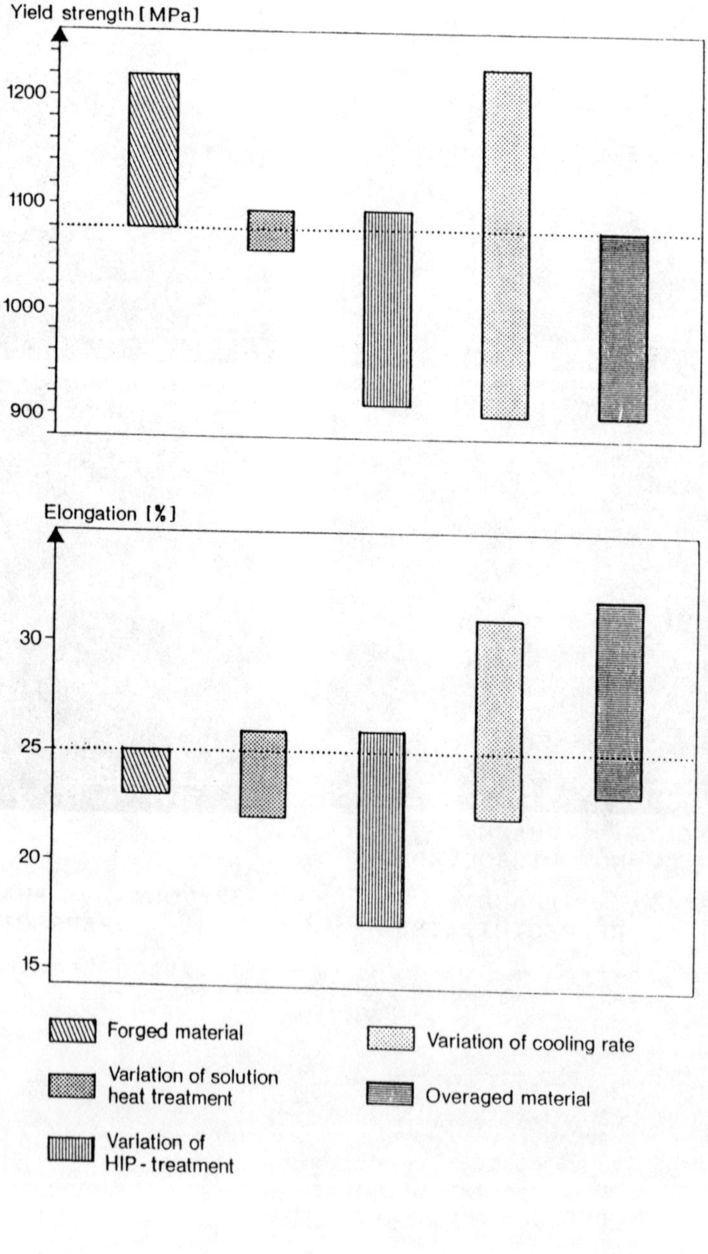

STUDIES ON THE SINTERING OF PARTS OF AUSTENITIC STAINLESS STEEL.

M. ROSSO*, G. SCAVINO*, E. GALETTO**, G. PESCE**

* Dipartimento di Scienza dei Materiali e
Ingegneria Chimica
Politecnico di Torino
** FIPS S.p.A.
TORINO - ITALY

Abstract

AISI austenitic stainless steel grades 304L and 316L samples were sintered at 1120 and 1190 °C in nitrogen-based atmosphere and in vacuum at 1200°C with nitrogen back-filling in the cooling stage. The effects of the sintering atmospheres and temperature were related to: dimensional change of sintered specimens in respect of dies size, rockwell hardness (HRB), radial crushing strength, unnotched impact strength, ultimate tensile strength and metallographic characteristics. The samples sintered in nitrogen-based atmosphere show the precipitation of Cr_2N, thus attaining higher hardness, crushing and tensile strengths, with the highest values reached by AISI grade 304L; conversely, the use of vacuum reduces the difference between AISI 304L and 316L steels; moreover, the ductility properties are increased.

Introduction

The powder metallurgy method for manufacturing stainless steel parts represents an important and growing segment of the P/M industry. These materials, also in the sintered state, have superior oxidation and corrosion resistance, good physical appearance, ductility and impact strength, wear resistance and nonmagnetic characteristics (austenitic steels) or hardenability (martensitic steels).

Both austenitic and martensitic stainless steel parts are used in a wide variety of applications in many different industries, thus they have various market segments.

The corrosion resistance of these materials depends strongly upon the sintering parameters and their degree of porosity. In fact, the corrosion resistance of a porous body, owing to the large specific surface, is inferior to a wrought material of the same composition /1/.

With regard to the processing parameters, the same basic considerations applicable to other P/M materials are of concern for stainless steels. However, as compared with the sintering of low alloyed ferrous parts, some precautions must be taken.

Low carbon and high nickel contents, in austenitic steels, contribute to good compressibility. However, the stainless steel powders require higher compacting pressures and yield lower green strength, being these also affected by the lubricant choice /2, 3/; i.e. lubricants which provide good compactibility generally cause lower green strength.

The steel powders with a high Cr content have poor sinterability /4/ and from the point of view of the reaction with atmospheres, nickel presents no more difficulties than iron, whereas chromium requires more attention. Then the atmosphere employed for sintering stainless steels must be low in oxygen and water to prevent oxidation; on the other hand, recent studies showed that it is impossible to utilize endothermic atmospheres for the correct sintering of chromium steels /5/.

The most important and most critical step in manufacturing stainless P/M parts is sintering. The pressing lubricant must be removed in a presintering stage, in order to avoid carbon pick up in austenitic steels and hence reduction of the corrosion resistance.

The mechanical properties of austenitic stainless steels depend upon the atmosphere in which they are sintered. The atmosphere generally used are constituted by dissociated ammonia or hydrogen; vacuum is also employed. Owing to the high cost of hydrogen atmospheres, these are scarcely adopted; the most widely used commercial atmosphere is dissociated ammonia (75 vol. % hydrogen and 25 vol. % nitrogen), but the increase of its price makes interesting the use of nitrogen-based atmospheres containing as little as 3% hydrogen.

Sintering in vacuum is the principal alternative to sintering in dissociated ammonia or nitrogen-based atmospheres. The vacuum sintering is increasingly adopted because of energy conservation and ecological considerations. Using vacuum there is the risk of evaporation at the sintering temperature of some elements, such as chromium, if the furnace pressure falls below the vapor pressure of these elements; thus the vacuum level must be properly controlled.

The nitrogen in the sintering atmosphere is definitely not neutral to stainless steel. In fact, sintering in dissociated ammonia introduces as much as 0.3% nitrogen in the composition, which increases the strength, but decreases the ductility of the material. Owing to these properties, desirable in some applications, recent material standards take into consideration also the production of austenitic stainless steel nitrogen alloyed powders /6/. For materials vacuum sintered, the properties are similar to those of materials sintered in hydrogen, unless the vacuum sintered material is allowed to pick up nitrogen during cooling, using nitrogen as back-fill gas /2, 3, 7, 8/.

It was observed that sintering in dissociated ammonia leads to the formation of chromium nitride with a loss of corrosion resistance /9/. The nitrogen content in the material depends upon the temperature range; the larger absorption of nitrogen occurs in the range 700-1100 °C and the maximum amount absorbed is lower than 9 wt% /1/; upon heating to higher temperatures, the nitrogen solubility decreases. While the rate of heating has no appreciable effect on the properties of sintered austenitic stainless steel parts, the rate of cooling from the sintering temperature has a significant influence /10/. Using commercial sintering times and temperatures, equilibrium nitrogen contents may range from 0.2 to 0.5 wt% /3/.

The sintering time and temperature also affect the austenitic stainless steel characteristics; i.e. at increasing temperatures corrosion resistance, tensile strength and shrinkage increase whereas yield strength decreases /3, 8/.

To contribute to the knowledge of properties of sintered austenitic stainless steels, samples of AISI grades 304L and 316L have been fabbricated from powders at different temperatures in vacuum or in atmospheres constituted by dissociated ammonia added of nitrogen. The mechanical and microstructural characteristics of sintered samples have been checked and related to the sintering parameters.

Experimental procedure

Samples, in the form of bar with a rectangular transverse section and of bushings, were produced by compacting BSA powders of AISI stainless steel grades 304L and 316L with added lubricant, 0.7 wt% litium stearate. The bar samples had green volume mass of $6.71 \cdot 10^{-3}$ kg/m^3 and green strength of 9,5 and 7,5 MPa for grades 304L and 316L respectively, their dimensions being ≈7x12x55 mm. The bushings had smaller values of volume mass, with dimensions approximately equal to 15 mm height, 17 mm inner diameter and 24 mm the outer. The sample composition is reported in table I.

The compacted samples were sintered in industrial plants in nitrogen-based atmosphere formed by dissociated ammonia with nitrogen additions (62.5 vol. % nitrogen and 37.5 vol. % hydrogen, -35 °C dew point) obtained by adding nitrogen to dissociate ammonia. The sintering was performed at 1120 and 1190 °C for 1 h, with a step at 500 °C to eliminate the lubricant. The total time of sintering cycle was 7.7 hrs approximately. For the sake of comparison, a series of specimens was also sintered in vacuum at 1200 °C for 1 h, with pre-heating steps. The total cycle time was of 6 hrs. The vacuum degree during the heating was of 133 Pa, whereas in the cooling was of 79.9 kPa using nitrogen as back-filling gas.

Table I: Chemical analysis (wt. %) of the AISI 304L and 316L steel samples.

AISI	C	Si	Ni	Cr	Mo	Mn	S	P	Fe
304L	0.015	0.67	10.28	18.65	-	0.11	0.007	0.002	bal
316L	0.012	0.86	12.90	17.85	2.22	0.17	0.006	0.015	bal

Following the sintering cycles, the dimensional changes from die size and the volume mass of the samples were measured. The bar specimens were subjected to tension tests, after machining a short calibrated length, and to unnotched impact tests. Impact was in the greater transversal dimension of the samples.

Crushing tests on the bushings were carried out determining the radial crushing strength. All the samples were subjected to Rockwell hardness tests and optical microscopy on their polished transverse section. Moreover, all the samples were controlled in order to verify the absence of ferromagnetic properties.

Results and discussion

The volume mass of sintered samples and the linear shrinkage due to the sintering process are listed in tables II for the bushings and III for bars. In table II the HRB hardness and the radial crushing strength are also reported, whereas in table III the ultimate tensile strength, the impact strength and the hardness are added.

The metallographic appearances of bar samples are reported in fig. 1; there is a great difference between the microstructure of specimens sintered in atmosphere and those sintered in vacuum. Only in the case of vacuum sintered samples the microstructure is of a correct austenitic type; whereas, in the specimens sintered in atmosphere, there is an evident precipitation of chromium nitride, Cr_2N. This precipitation appears higher in the AISI grade 304L samples.

Upon sintering in nitrogen-based atmospheres, in fact, the nitrogen amount absorbed at equilibrium, being proportional to the square root of the partial pressure, yields high quantities of nitrogen in solid solution; then, during the cooling below 1000 °C the Cr_2N can precipitate /3, 10/. This precipitation implies the existence of chromium-deficient regions, what can implicate a reduction of the corrosion resistance /1, 9/. In the vacuum sintering the presence of Cr_2N was not observed, on the other hand the pressure was very low and the cooling rate high, avoiding large nitrogen intake.

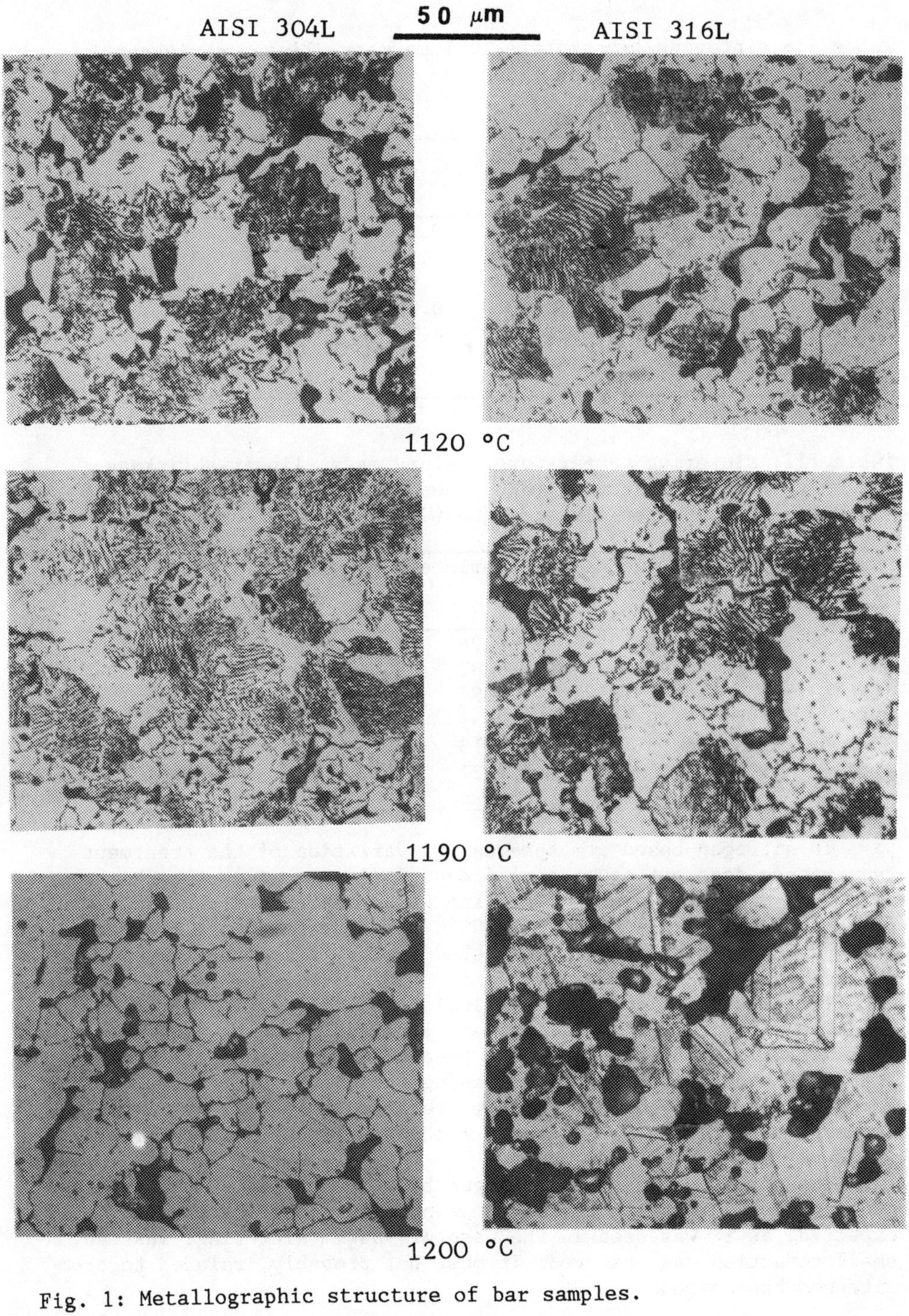

Fig. 1: Metallographic structure of bar samples.

Table II: sintering temperature, volume mass, linear shrinkage, rockwell hardness (HRB) and radial crushing strength (K), for bushing samples.

Sample AISI	T [°C]	Volume mass [kg/m^3]	Shrinkage %	HRB	K [MPa]
304L	1120	$6.39 \cdot 10^{-3}$	0.53	72	890
304L	1190	$6.43 \cdot 10^{-3}$	0.52	78	903
316L	1120	$6.34 \cdot 10^{-3}$	0.61	63	802
316L	1190	$6.36 \cdot 10^{-3}$	0.51	69	797
304L	1200	$6.52 \cdot 10^{-3}$	1.34	44.5	690
316L	1200	$6.43 \cdot 10^{-3}$	1.30	38	650

Table III: sintering temperature, volume mass, linear shrinkage, rockwell hardness (HRB), unnotched impact strength (UIS), ultimate tensile strength (UTS), for bar samples.

Sample AISI	T [°C]	Volume mass [kg/m^3]	Shrinkage %	HRB	UIS [J/cm^2]	UTS [MPa]
304L	1120	$6.79 \cdot 10^{-3}$	0.94	78	38	505
304L	1190	$6.73 \cdot 10^{-3}$	0.82	77	36	574
316L	1120	$6.77 \cdot 10^{-3}$	0.87	75	41	470
316L	1190	$6.72 \cdot 10^{-3}$	0.51	73	40	527
304L	1200	$6.78 \cdot 10^{-3}$	1.13	63	91	465
316L	1200	$6.81 \cdot 10^{-3}$	1.27	63	88	476

In nitrogen-based atmospheres the variation of the treatment temperature does not cause significant influence; i.e. the characteristics of materials are similar with the exception of the UTS. In fact, the samples sintered at 1190 °C have greater tensile strength in full agreement with what is indicated in the literature.

The highest values of mechanical properties, owing to the higher chromium nitride precipitation, were obtained with the AISI grade 304L specimens. On the other hand, the greater amount of nickel in the AISI stainless steel grade 316L, decreasing the quantity of nitrogen absorbed in the austenite /11/, reduces the Cr_2N precipitation, although the presence of Mo favours the nitrides formation /12/.

The shrinkage data for nitrogen base atmosphere sintering are peculiar, because an increase with sintering temperature would be expected, as it was seen in the case of vacuum sintering; instead, a small reduction was observed; it must be, probably, related to the nitrogen absorption.

The vacuum sintered pieces have, in respect of those sintered in atmosphere, higher values of unnotched impact strength and lower values of other mechanical characteristics, i. e. they are more ductile and tough. On the other hand, high absorbed nitrogen levels improve the strength of the steel /10/. It must be pinpointed that, generally, greater ductility properties of the sintered products are related to a greater degree of bonding and then better corrosion resistance. Moreover, the differences between the characteristics of the two type of stainless steel were reduced, if vacuum sintered, owing to the lack of chromium nitride precipitation.

Conclusions

The sintering of stainless steels constitutes the most critical step in the production of sintered parts of these alloys. In fact, the attainment of good results depends on the complet evaporation of the lubricant in the presintering stage and on the correct choice of the sintering parameters, i.e time, temperature and atmosphere.

The samples sintered in nitrogen-based atmosphere showed good hardness and strength characteristics, related to nitrogen absorption and chromium nitride precipitation, higher in the AISI grade 304L, but their ductility properties are lower than those of the vacuum sintered specimens, for which the precipitates are absent.

References

1) R.L. Sands, G.F. Bidmead and D.A. Oliver, Modern Development in Powder Metallurgy, Plenum Press, 2 (1966), 73-83.
2) R. W. Stevenson, P/M Stainless Steels, Metals Handbook, 7 (1984), 728-732.
3) D.L. Dyke and H.D. Ambs, Powder Metallurgy-Applications, Advantages and Limitations, ed. E. Klar, American Society for Metals, (1983), 123-144.
4) F. Eisenkolb, Stahl und Eisen, 78 (1958), 141-148.
5) G.F. Bocchini and C. De Cataldis, Metall. Ital., 78 (1986), 457-466.
6) F.V. Lenel, Powder Metallurgy-Principles and Applications, Metal Powder Industries Federation, (19), 429-430.
7) MPIF Standard No. 35 (1987/1988), Metal Powder Industries Federation.
8) Sintering of Stainless Steel, Metals Handbook, 7 (1984), 368-370.
9) H.S. Kalish and E.N. Mazza, Trans. AIME, 203 (1955), 304-310.
10) G. Lei, R.M. German and H.S. Nayar, Powder Metall. Int., 15 (1983), 70-76.
11) H.A. Wriedt, O.D. Gonzales, Trans. Met. Soc. AIME, 221 (1961), 532-535.
12) H. Schenck, M.G. Prohberg and F. Reinders, Stahl und Eisen, 83 (1963), 93-99.

HIGH RESOLUTION AUGER ANALYSIS OF 12CR-STEEL POWDER ANNEALED AT CONSOLIDATION TEMPERATURES

L. Nyborg and I. Olefjord

Department of Engineering Metals,
Chalmers University of Technology
S-421 96 Gothenburg, Sweden

INTRODUCTION

In order to improve the mechanical properties of PM compacts it is necessary to determine the compounds formed on the powder surfaces during atomization, handling and compaction. This can be accomplished by using ESCA (Electron Spectroscopy for Chemical Analysis) and Auger electron spectroscopy (1-11).

In previous studies of martensitic 12Cr-steel (1-5), it has been shown that the oxides formed during atomization are reduced as the temperature is increased up to the maximum consolidation temperature, while silica is co-precipitated with vanadium carbonitride on the prior particle boundaries (PPBs). Chromium carbide precipitates also on the PPBs.

The objective of this study is to further investigate the surface reactions which take place at high temperatures. The martensitic steel powder is annealed in a temperature gradient at temperatures between 800 and 1200°C and analysed using Auger electron spectroscopy.

EXPERIMENTAL PROCEDURE

Nitrogen-gas-atomized martensitic steel powder with an average particle size of 190 μm was provided by ABB-Powdermet in Sweden. The chemical composition of the steel is shown in Table 1.

TABLE 1. CHEMICAL COMPOSITION OF THE POWDER (wt%)

Fe	C	Si	Mn	S	P	Cr	Ni	Mo	V	Nb	N	O
84.9	0.2	0.3	0.6	0.02	0.01	12	0.5	1.0	0.3	0.1	0.06	0.0135

Particles ranging between 180 and 300 μm were collected by sieving and packed into a 500 x 10 mm quartz tube. The tube was connected via a valve to a vacuum pump and evacuated to about 10^{-4} mbar. The tube containing the powder was heated in a temperature gradient for 2 h with the tube ends held at 1200 and 20°C respectively.

Interparticle surfaces were prepared by fracturing the sintered powder into sections at the positions kept at 800, 900, 1000, 1100 and 1200°C. The specimens were fractured and handled in air. The chemical compositions of the surfaces were determined by utilizing Auger Multiprobe (PHI 600). The in-depth distributions of elements were obtained by successive ion etchings and analyses. For this purpose, the etch rate was calibrated on oxidized Ta and Fe plates. The oxygen content of the powder was measured by using chemical analysis.

RESULTS

Figure 1 shows a SEM micrograph of the interparticle fracture surface of the powder sintered at 900°C. The typical interparticle fracture shown in the figure was the same for all annealing temperatures.

Figure 2 shows that the oxygen content and the oxide thickness increase with decreasing temperature. The oxygen levels obtained at 800, 1000 and 1200°C are 330, 162 and 128 ppm respectively. The corresponding oxide thicknesses are 60, 9 and 3 nm. The oxide thickness before annealing is about 7 nm (2). Thus, the surface oxide is reduced at the higher temperatures and oxygen is transported from higher to lower temperatures. The increased oxygen activity causes oxide growth at temperatures of 1000°C and below.

Auger spectra recorded in typical cases from the surfaces of the annealed powder are shown in Fig. 3. Since the fracturing of the sintered powder was performed in air, part of the oxygen signals represent the air-formed oxide. A comparison of the spectra in Fig. 3 show that the Cr and Mn intensities decrease with increasing temperature, while the intensity of Si increases. Thus, the chromium and manganese oxides are reduced under the formation of silica as the temperature increases. The presence of the relatively strong C, N and V peaks in Fig. 3b-c shows that vanadium carbonitride precipitates on the powder surfaces at temperatures between 900 and 1000°C.

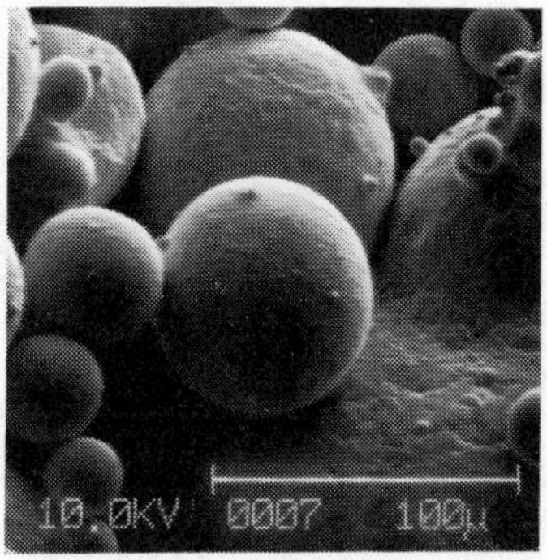

Figure 1. SEM micrograph of inter-particle surface of powder sintered at 900°C.

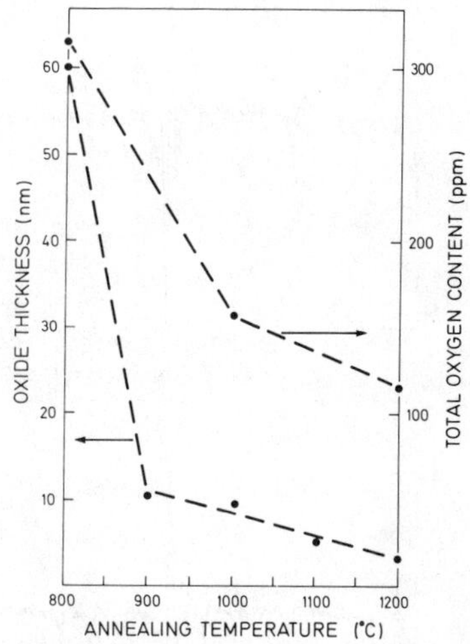

Figure 2. Oxide thickness and oxygen content versus the annealing temperature.

The spectra recorded from the powder annealed at 1100 and 1200°C (Fig. 3d-e) shows that the intensities of the N signal are very low. Furthermore, no peaks representing V or Mn are detectable. A relatively strong Si peak is recorded from the powder annealed at 1100°C indicating that the surface is partly covered with silica particles. High magnification SEM micrographs (Fig. 4a-b) show that discrete particles are present on the surfaces of the powder annealed at 1100 as well as at 1200°C. The particles in the former case are about 0.5 μm in size and cover a larger part of the surface than those in the latter case, where both smaller, 0.5 μm, and larger, 2.0 μm, particles appear on the surface. No such particles could be observed on the other samples.

Figure 5a shows an Auger spectrum recorded from the 1100°C-sample (ion-etched 3 nm). It appears that silica particles are present on the surface. Corresponding analyses of the 1200°C-sample (Fig. 5b-c) show that the larger, 2.0 μm, particles are manganese sulphides and that the smaller, 0.5 μm, particles contain Si, Nb, C, N, V, O and Mn. It is concluded that Si and Mn are bound in oxides co-precipitated with carbonitrides containing V and Nb.

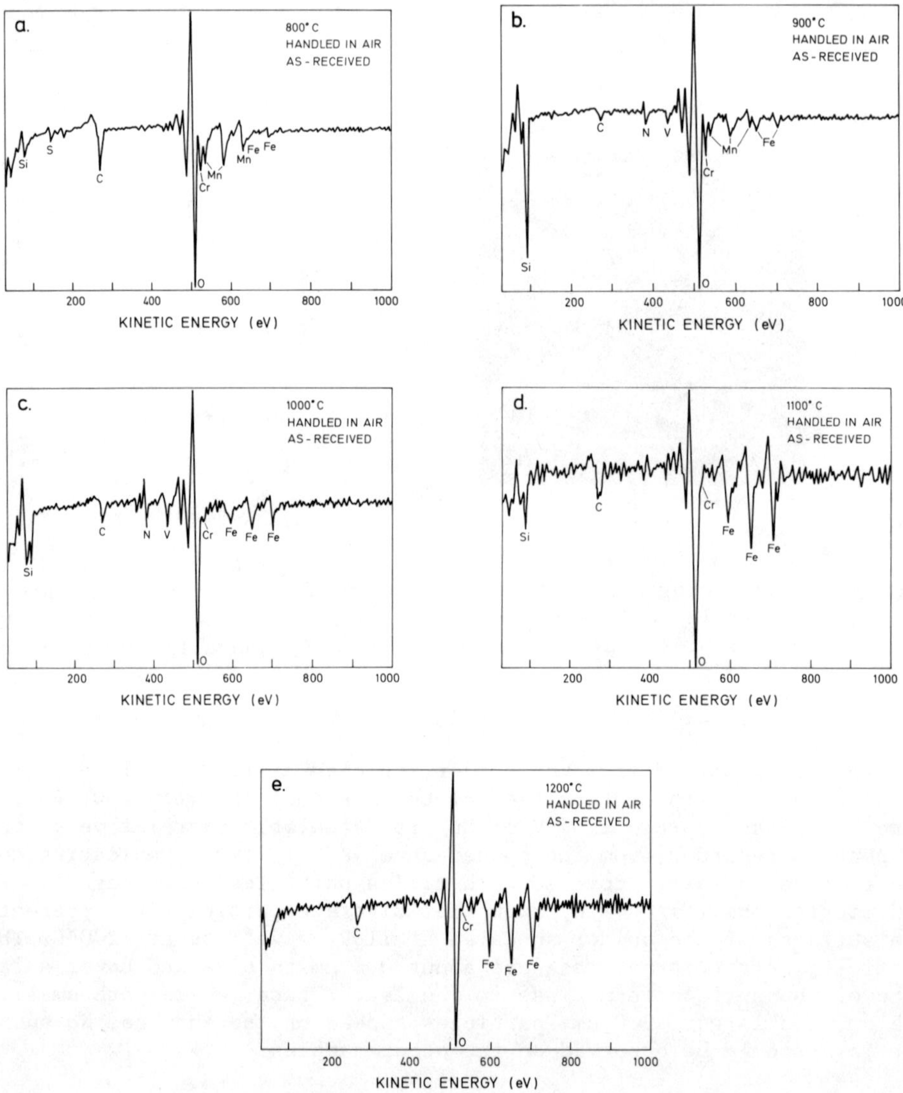

Figure 3. Auger spectra recorded from the powder annealed at: a) 800°C; b) 900°C; c) 1000°C; d) 1100°C; e) 1200°C.

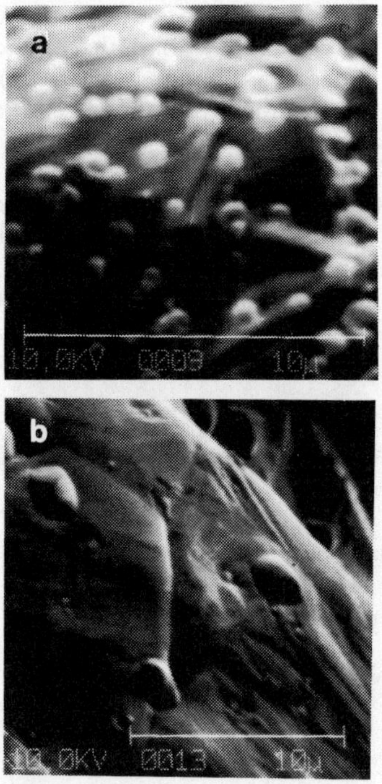

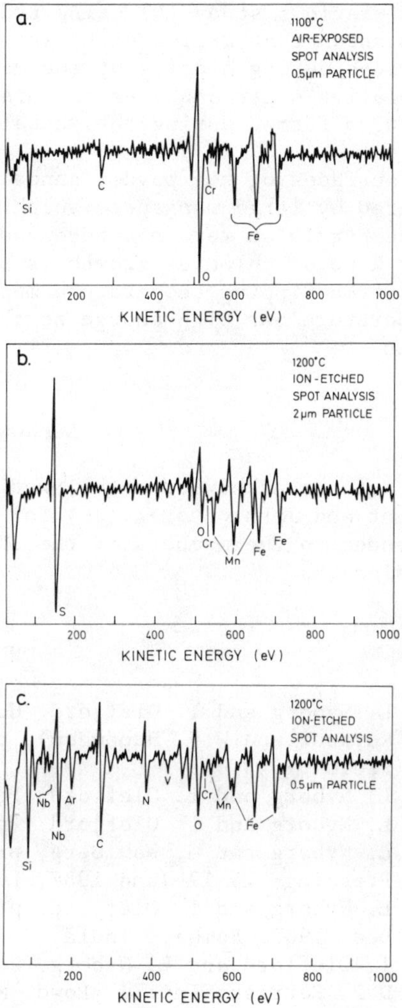

Figure 4. High magnification SEM micrographs of the powder annealed at: a) 1100°C; b) 1200°C.

Figure 5. High resolution Auger analyses of particles on the powder annealed at: a) 1100°C; b); c) 1200°C.

DISCUSSION

The annealing of powder in a temperature gradient permits the study of the surface reactions which take place at different temperatures. At high temperatures, the Fe-, Cr-, and Mn-oxides present on the surface of the martensitic steel powder are reduced, while silica particles are formed. Excessive oxygen is transported towards regions kept at lower temperatures, where the alloying elements are oxidized. The oxidation of Si and reduction of Fe-, Cr- and Mn-oxides at higher temperatures are controlled by the CO/CO_2 ratio obtained during annealing (11).

In a previous study (4) using transmission electron microscopy it has been shown that silica particles co-precipitate with vanadium carbonitrides during hipping of the martensitic stainless steel. The high resolution Auger analyses in this study show that such co-precipitates are also formed during the annealening of the powder.

The surfaces of the powder annealed at 1200°C are to a limited extent covered by large manganese sulphides about 2 μm in size. Such large sulphide particles were not observed on the other surfaces suggesting that this kind of sulphide growth is only obtained at very high temperatures. Thus, the hipping temperature must be lower than the sulphide formation temperature but high enough so that most of the surface oxide is reduced.

ACKNOWLEDGEMENTS

The authors acknowledge the Swedish National Board for Technical Development and Jernkontoret (JK) for financial support. Special thanks are extended to the members of the JK reference group for many helpful discussions.

REFERENCES

1. L. Nyborg and I. Olefjord, Horizons of Powder Metallurgy (eds. W.A. Kaysser and W.J. Huppmann), part 2, p. 755, 1986, Verlag Schmid, Freiburg.
2. L. Nyborg and I. Olefjord, Powder Metall. 31 (1), p. 33, 1988.
3. L. Nyborg and I. Olefjord, Powder Metall. Int., 2, p. 11, 1988.
4. L. Nyborg and G. Wahlberg, presented at Int. Conf. on Hot Isostatic Pressing, 15-17 June 1987, Luleå, Sweden, Centek Publ., Luleå.
5. L. Nyborg and I. Olefjord, presented at 1987 Int. P/M Conf., 7-10 Dec. 1987, Bombay, India.
6. I. Olefjord and L. Nyborg, Powder Metall., 28 (4), p. 234, 1985.
7. D.P. Ferriss, Int. J. Powd. Met. Powd. Technol., 19 (2), p. 11, 1983.
8. C. Aubin, J.H. Davidson and J.P. Trottier, Proc of the 4th Int. Symp. on Superalloys (ed. J.K. Tien), p. 345, 1980, American Society for Metals, Metals Park, Ohio.
9. P.N. Ross and B.H. Kear, Rapidly Quenched Metals III, vol. 3 (ed. B. Cantor), p. 102, 1978, Chameleon Press Ltd, London.
10. R.E. Waters, J.A. Charles and C. Lea, Metals Technol., 8, p. 194, 1981.
11. I. Olefjord and L. Nyborg, presented at this conference.

Author Index

This alphabetical author index covers Volumes 18-21.
Numbers indicate volume and title page of the article.

Achterath, W., 18-351
Ackerman, L., 19-425
Affeldt, E., 20-745
Agarwal, D. P., 18-597
Ahmed, M. N. H., 18-213, 20-391
Ahn, I. S., 19-235
Albiston, J. N., 19-75
Amafuji, K., 19-727
Amano, N., 19-533
Amaya, H. E., 18-323
Ameyama, K., 20-675
Anderson, I. E., 20-205
Andrievski, R. A., 20-13
Angers, R., 20-629
Antes, H. W., 18-197
Antonov, I. S., 20-713
Apelian, D., 19-397
Aren, B. G., 18-23
Aria, C. S., 18-403
Arnberg, L., 20-503
Arnhold, V., 21-55, 21-183
Arzt, E., 20-745
Ashdown, C. P., 20-169
Aslan, M., 18-11, 18-77
Aslund, C., 20-181
Backensto, A. B., 19-641
Backmark, U., 20-503
Backstrom, N., 20-503
Bagchi, T. P., 20-279
Barata Marques, M. J. M., 19-365
Barpanda, S., 20-279
Bas, J. A., 18-745
Bauer, R. E., 18-91
Baum, L. W., 18-363
Beddow, J. K., 20-293
Beisel, W., 20-723
Beiss, P., 19-281
Beltz, R. J., 20-409
Benjamin, J. S., 21-397
Benn, R. C., 21-479
Ber, A., 19-345
Berry, D. F., 21-123
Beuers, J., 18-351
Bewley, J. G., 20-169
Bhambri, A., 18-155
Bhattacharjee, G. S., 20-279

Bhave, P., 18-333
Bialo, D., 19-441
Bicudo, J. E. L., 18-433
Bocchini, G. F., 18-507, 18-691
Bonora, P. L., 21-313
Bose, A., 18-299, 19-139, 19-155, 20-511, 20-559
Botstein, O., 19-91
Bourguignon, L. L., 18-597
Brewin, P., 19-305
Brown, R. L., 18-57
Brown, S. B., 18-465
Brunel, G., 19-603
Burlingame, J. W., 18-197, 19-155
Buzolits, S. R., 18-197

Camus, G., 20-559
Carvalhinhos, H., 19-319
Castelli, J. M., 18-119
Castro, R. G., 19-743
Causton, R. J., 18-119
Chaiat, D., 18-645
Chandhok, V. K., 18-71
Chen, L-M., 21-619
Chen, X-L., 18-789
Chen, Y. T., 19-681
Cheney, R. F., 19-155
Chesney, P. F., 19-475
Choi, C. J., 20-587
Chopra, K. S., 21-361
Christian, K. D., 21-23
Ciloglu, C., 18-107
Clark, J. P., 21-723
Clement, T. P., 18-627
Coleman, D. S., 19-305
Collette, J. P., 21-275

Daga, R. L., 19-155
Dankoff, J. D., 20-409
Danninger, H., 21-171
de Haan, T. L. J., 19-441
de Rege, A, F., 21-327
Deo, B., 20-265
Diamond, S., 21-445
DiBari, G. A., 20-311
Dickson, J., 18-775
Dormon, W., 18-333
Dornisch, A. C., 21-699
Dornisch, J. T., 21-699
Dorweiler, H-J., 18-263
Druzhinin, L. K., 20-137
Du, D-M., 18-789
Dube, R. K., 20-265
Dumbleton, J. H., 20-361
Duquette, D. J., 20-559
Duszczyk, J., 19-441

Eck, R., 19-173
Eddington, D. J., 20-53
Eisenmann, M., 21-637
Ekbom, L. B., 19-63
Ekbom, R., 18-37
Eliasson, A., 19-63
Emmer, S., 19-697
Endoh, H., 21-157
Engdahl, P., 20-655
Estrada, J. L., 19-441

Fang, Z., 18-615
Farooq, S., 18-597
Faure, J. F., 19-425, 19-603
Fedrizzi, L., 21-313
Ferguson, B. L., 18-269, 21-685
Field, F. R., 21-723
Figliola, R. S., 20-205
Figueredo, A. M., 20-491
Fine, T. E., 20-125
Fischer, A., 21-637
Fischer, J. J., 21-461
Floege, H., 20-773
Forno, A. E. J., 20-451
Frazier, W. E., 21-495
French, C., 19-249
Frisch, A., 18-11, 18-77
Froes, F. H., 21-429
Fuchs, G. E., 20-531
Furuta, S., 20-575

Gagne, M., 18-107
Galetto, E., 20-779
Ge, C-C., 21-619
Gelinas, C., 20-629
Gennari, U., 19-201
German, R. M., 18-299, 18-391, 18-597, 19-139, 19-155, 20-511, 20-559, 21-23
Gorantiwar, V. S., 20-293

Gothard, D. O., 21-511
Gotman, I., 18-675
Goto, M., 19-379
Grazzini, H., 19-565
Greil, P., 20-751
Gustavson, L. J., 20-361
Gutierrez-Miravete, E., 19-457
Gutmanas, E. Y., 18-675,
20-421, 20-467

Haeberle, R. M., 21-461
Haerdtle, S., 20-729
Haerian, A., 19-43
Hale, T. E., 19-3
Hallen, A. M., 20-503, 21-755
Hallen, H., 20-187
Han, Y. C., 21-723
Hannigan, J. W., 21-415
Hashiguchi, D. H., 18-627,
20-109
Hashii, M., 20-675
Hauber, D., 18-661
Hausselt, J. H., 18-351
Havalda, A., 19-697
Hayashi, K., 18-287
Hayashi, T., 19-533
Heck, F. W., 20-311
Held, M., 20-63
Hens, K. F., 18-417
Hermes, T. V., 18-315
Hierath, L. L., 18-711
Hirai, Y., 21-575
Hofer, H., 19-201
Hohmann, M., 20-159
Homma, K., 21-535
Horiishi, N., 18-799
Horikoshi, E., 19-579
Hovis, Jr., V. M., 19-753
Hribernik, B., 20-681
Huang, P., 19-15
Huet, D., 20-669
Hulyal, R., 21-139

Igharo, M., 19-305
Iikawa, T., 19-579
Ikenoue, Y., 21-157
Isonishi, K., 20-675
Itabashi, M., 18-287
Iwai, K., 20-575

Jager, H., 20-681
James, W. B., 18-119
Japka, J. E., 20-339
Jin, J., 20-243
Johansson, R., 21-381
Johnson, P. K., 20-3
Jokinen, A., 19-547
Jonsson, S., 20-159

Kainer, K. U., 20-323
Kao, W. H., 20-531
Kapoor, D., 19-105, 19-121
Kawai, N., 20-575
Kawano, M., 21-605
Kaysser, W. A., 18-11, 18-77,
19-235, 20-79
Kenney, G. B., 20-169
Kim, D. K., 19-591
Kim, H-S., 19-271
Kim, S. U., 19-105
Kim, Y-W., 20-615
Kim, Y-H., 19-259
Kim, Y-L., 19-223
Kimura, T., 21-551
Klar, E., 21-123
Klausmann, R., 20-431
Kny, E., 19-201
Kobylanski, A., 20-669
Koczak, M. J., 18-213, 20-391
Konig, G., 20-745
Konig, W., 18-253, 18-263
Korevaar, B. M., 19-441
Kraemer, O., 20-733
Krishnamurthy, S., 20-615
Kubicki, R., 18-551
Kulkarni, K. M., 19-329
Kunert, P., 20-225
Kunze, H-D., 20-63, 20-707
Kuroishi, K., 20-599
Kuroishi, N., 19-533

Lahiri, D. K., 20-279
Lahodny, H., 20-745
Lai, H-Y., 19-23
Lal, S., 18-581
Lall, C., 18-363
Lasne, P., 18-449, 19-565
Laugee, C., 19-565
Lavernia, E., 19-457
Lawley, A., 19-397, 19-489,
20-467
Layyous, A. A., 20-691, 20-751

Leatham, A. G., 19-441, 19-475
Ledoux, L., 21-41
Lee, D., 18-417
Leismann, H., 21-637
Leister, T., 18-197
Lengyel, J. J., 20-643
Leroux, J., 20-669
L'Estrade, L., 20-187
Levaillant, C., 18-449, 19-565
Liao, C-M., 18-789
Lierke, E., 20-159
Lim, T. W., 18-287
Lin, I. J., 20-691
Lin, S. T., 18-597
Lin, Z., 20-443
Lindner, K-H., 18-263
Lindqvist, B., 21-67
Ling, X., 19-53
Liu, Y. X., 19-33
Ljunggren, R., 20-187
Lograsso, B., 20-559
Luna, J. M., 21-301
Lund, J. A., 18-569
Luo, X. J., 19-397

McGeever, J., 21-215
McGinn, K. S., 21-229
McIndoe, G. R., 18-71
Majima, K., 19-727
Manto, H., 21-535
Marcela, Z., 20-701
Marder, J. M., 18-627, 20-109
Marks, W. D., 19-671
Martins, I. M., 19-319
Masters, R. E., 20-409
Mathur, P., 19-397
Matsunaga, H., 21-563
Mayama, O., 21-563
Mazen, M. I., 20-391
Meinhardt, H. W., 20-225
Mellanby, I. J., 18-183
Mesquita, R. M. D., 19-365
Metelmann, O., 19-475
Michal, G. M., 20-491
Milkov, S. B., 20-481
Mirchandani, P. K., 21-479
Molinari, A., 21-313
Molins, C. B., 18-723, 18-745
Moll, J. H., 16-709
Moon, I-H., 19-223, 19-259, 19-271
Moon, J. R., 18-183
Moran, A. L., 19-503
Morandi, O., 20-35
Mordike, B. L., 20-323
Morioka, Y., 21-521
Mosca, E., 20-35
Moszumanski, R., 18-479, 20-701
Moyer, K. H., 18-535, 18-757
Muller, F., 21-55
Murphy, T. F., 18-119

Nadiv, S., 18-645, 20-691
Nagare, I., 21-591
Nagata, N., 20-575
Nakagawa, T., 20-763, 21-653
Nakai, T., 19-379
Nakatani, S., 19-379
Navara, E., 18-23
Nayar, H. S., 21-243
Neumann, W., 21-755
Neupaver, A. J., 19-655
Noguchi, H., 20-763
Northcutt, Jr., W. G., 19-753
Nowak, B., 20-745
Nurthen, P. D., 19-305
Nyborg, L., 20-93, 20-787

Obayashi, M., 21-197
Oberacker, R., 20-431, 21-85
Odani, Y., 19-533
Ogilvy, A. J. W., 19-441, 19-475
Ogura, K., 21-605
Oh, S-T., 19-223
Ohta, J., 21-605
Okazaki, K., 19-591
Okimoto, K., 18-799
Olefjord, I., 20-93, 20-787
Oliveira, M. M., 19-319
Ostenssen, M., 20-739

Pacher, O., 20-681
Pak, H-R., 19-591
Palko, W. A., 19-503
Park, J. C., 21-723
Park, W. W., 20-587
Park, Y. J., 20-587
Parten, C. W., 18-315
Patel, A. N., 21-445
Patel, K. B. M., 19-621

Patterson, B. R., 18-403, 18-615, 19-249
Paulson, A. S., 21-23
Payne, J. A., 19-3
Pease, D. A., 21-275
Pease III, L. F., 18-155, 20-643, 21-275, 21-327
Pellegrini, M. T., 18-433
Persson, C., 20-739
Persson, P. O., 20-739
Pesce, G., 20-779
Petersen, N., 20-377
Petzoldt, F., 20-63
Petzow, G., 18-11, 18-77, 20-79, 20-751
Pfitzinger, M. S., 21-229
Pillhoefer, H., 20-773
Planas, J., 18-723
Polanski, Z., 20-701
Porat, R., 19-345
Pregowski, J., 18-493
Prioul, C., 21-41
Prucher, T., 18-143
Puig, J., 18-745
Pulc, V., 19-697

Qi, X-W., 18-789
Qu, X., 18-569, 19-53
Queeney, R. A., 20-409, 20-443
Querbes, J. L., 19-565

Rabin, B. H., 20-511
Raghupathi, P. S., 18-231
Ramakrishnan, P., 20-293
Ranninger, C., 21-301
Raz, G., 18-645
Rees, E. L. LL., 20-311
Reinshagen, J. H., 19-655
Rhee, B. O., 18-417
Riblett, C. R., 18-711
Rober, G., 18-253, 18-263
Roebuck, B., 20-451
Roesen, L., 20-733
Rosso, M., 20-779
Rosso, M. L. P., 21-259
Roy, L. G., 21-327
Ruhkamp, J. D., 18-551
Ruiz, J. M., 21-301
Ryan, J. B., 18-535, 18-757
Rzesnitzek, K., 20-79

Saegusa, Y., 19-379
Sahoo, K. C., 20-279
St. Pierre, J., 18-323
Saito, T., 21-197
Sakamoto, Y., 21-575
Sakuma, H., 21-535
Sakurada, I., 21-605
Salak, A., 19-697
Sale, F. R., 19-75
Salje, H., 18-263
Santos, M., 19-319
Santos, M. B., 18-433
Sato, T., 19-579
Satoh, T., 18-799
Scavino, G., 20-779, 21-259
Schiefer, H., 18-263
Schiller, H., 18-263
Schlieper, G., 21-55
Schmidberger, R., 20-279
Scholz, B., 20-63, 20-707
Schwarz, R. B., 21-415
Semel F. J., 21-101
Senick, Jr., J. R., 18-231
Shaikh, Q., 19-305
Shaw, H. S., 18-119
Sheinberg, H., 21-415
Shimmura, T., 21-535
Shoji, K., 19-519, 19-727
Shpigler, B., 19-91
Sicken, R., 21-637
Sierra, C. M., 18-417
Sikka, V. K., 20-543
Simeonov, Y. T., 20-481
Singh, R. P., 19-489
Smith, D. W., 19-3
Smith, G., 21-215
Sonsino, C. M., 21-55
Stamberger, J., 20-681
Stanek, P. W., 19-743
Stickler, R., 21-3
Stoichev, M. S., 20-481, 20-713
Stoloff, N. S., 20-559
Stromgren, M., 18-253, 18-263
Suh, S. K., 20-587
Sulkowski, A., 20-701
Sundaresan, R., 21-429
Suzuki, K., 20-763, 21-157
Svilar, M., 21-123

Tabor, W., 18-493, 20-701

Takahashi, N., 20-763
Takajo, S., 21-521, 21-605
Takata, J., 21-535
Takeda, Y., 19-533, 20-599
Takigawa, H., 20-575
Tanase, T., 21-563
Teel, D., 18-333
Thompson, J. J., 21-495
Thornton, P. A., 18-231
Thummler, F., 20-431, 21-85
Tiainen, T., 21-415
Tingskog, T., 20-181
Tinzl, J., 19-173
Tiwari, A. N., 20-293
Tiziani, A., 21-313
Tokizane, M., 20-675
Toloui, B., 19-305
Tomasi, A., 21-313
Torralba, J. M., 21-301
Tracey, V. A., 21-745
Track, W., 20-773
Trapaga, G., 19-457
Trudel, Y., 18-107
Tu, L-L., 20-643

Uchino, M., 21-157
Uenishi, N., 20-599
Upadhyaya, G. S., 18-581, 21-139

Vedula, K., 20-491
Veltl, G., 20-707
Vossen, K., 18-253
Voyzelle, B., 20-629
Vu Chat, P., 20-669
Vujic, D., 19-121

Wahling, R., 19-281, 21-183
Wang, C., 19-53
Wang, J. C., 21-669
Wang, K. K., 20-361
Warbichler, P., 19-201
Warren, J., 18-391
Watwe, A. S., 21-479
Weber, G. G. A., 18-465
Weber, M., 18-263
Webster, J. C., 20-643
Weiss, B., 21-3
Wentland, W. L., 18-71
Whang, S. H., 19-105, 19-121
White, D. G., 18-1
Whitman, C. I., 20-27
Wojcik, C. C., 19-187
Wood, J. V., 19-305

Xia, Y-L., 21-619
Xiong, C., 19-53
Xu, R., 19-15, 19-53

Yang, H., 19-15
Yang, J., 19-23
Yokota, M., 19-519
Yolton, C. F., 16-709
Yoo, S-C., 19-271

Zak, D., 20-421
Zeng, D., 21-735
Zengin, O. Z., 21-171
Zhang, L., 20-763
Zhang, S. H., 19-33
Zhelazkov, Zh. H., 20-481
Zhou, W., 21-735

Subject Index

This alphabetical subject index was compiled with the key words supplied by the authors and covers Volumes 18-21. Numbers indicate volume and title page of the article.

Activated sintering
  bronze bearings, 19-621
  mechanical and physical properties, 19-259
  silicon, 19-753
  sintered properties, 19-259
  sintering behavior, 19-223
  tungsten-silver composites, 19-75
Agglomeration, 20-691
Aging heat treatment, 20-691
Alloy design, cemented carbides, 19-15
Alloy steel sintering, 20-669
Alloy steels, 18-535
Aluminides, 21-495
Aluminum alloys, 19-533, 21-495
  2xxx and 7xxx alloys, 19-547
  Al_3Ti intermetallic, 21-479
  aluminum-silicon, 19-441
  composite materials, 20-451
  rapid solidification, 19-519
Aluminum-hafnium, amorphous alloys, 21-415
Aluminum lithium, 20-691
Aluminum lithium hydride, 20-691
Amorphous alloy
  aluminum-hafnium, 21-415
  consolidation, 21-415
  hydrogen absorption, 21-415
Amorphous metals, 20-707
Amorphous powder, 20-63
Applications, 19-201, 20-701
  automotive, 18-155, 19-533, 20-669, 21-55, 21-157, 21-591
  gas and stream turbines, 18-37
  porous materials, 21-637
  split-die compaction, 20-35
Atmospheres, 21-259
  sintering atmospheres, 21-243
Atomic absorption spectroscopy (AA), 20-691

Atomization, 20-225, 20-691
  copper alloys, 20-125
  Fe-base alloys, 20-93
  Ni-base alloys, 20-93
  rapid solidification, 20-79
  two fluids/two jets, 20-713
Atomizing nozzle, 20-713
Attritor milling, 20-599, 20-707, 21-397
  carbides, 20-467
  elemental powders, 20-467
Auger spectroscopy, 20-751
Austenitic stainless steel, ternary addition, 18-581
Automotive industry, 21-591

Ball mill, 20-691
Bearing alloys
  copper base alloy, 19-681
  iron base alloy, 19-681
Bearings, 20-701
  bronze, 19-641
  contact wear resistance, 19-697
  fast sintering, 19-621
  microstructures, 19-621
Beryllium, 18-627
Billets
  hot extruded Al alloys, 19-533
  microstructure, 18-57
Binder layer thickness, 19-43
Binder phase, cobalt phase, 19-23
Bronze, elemental copper/tin, 19-641
Bronze microstructure, effects of oxygen, 19-671

CAD/CAM, computer control, 18-661
Carbides, 20-431
  cemented carbides, 19-3, 19-33
Carbon gasification reaction, at reduction processes, 20-243
Cast iron machining, 20-751
Catalytic mechanism, Oxygen Transfer Theory (OTT), 20-243
Cemented carbides
  tungsten heavy alloys, 19-43
  WC-Co hardmetal, 19-23
Ceramic powder, 20-763
Ceramic shapes, 20-751
Chemical application, heating bag, 20-279
Chemical compatibility, interfacial reactions, 20-531
Clean steel powders, 18-107
Cleanliness, 18-107
Co-reduction, alloy powder, 20-279
Cold isostatic compaction, 18-479
Cold repressing, 20-723, 21-157, 21-575
  billets, 18-493
Cold sintering, Co-base/Fe-base alloys and composites, 20-467
Cold welding, 20-691
Comminution, 19-3
Compacting, 20-763
  die lubrication, 18-723
  galling, 18-723
  green strength, 18-723
Compaction, 18-449, 18-465, 20-723
  explosive compaction, 18-645
  stress analysis in tooling, 18-433
Compaction tooling, 19-281
Composite, 19-441, 20-751, 21-653
Composite materials, 18-799, 19-475, 20-409, 20-421, 20-431, 20-443
  controlled porosity, 21-637
  copper tungsten microlaminates, 19-743
  deformation behavior, 20-451
  iron based, 20-377
  macrocomposite, 20-391

nickel aluminides, 20-531
nickel coated products, 20-311
property comparisions, 20-451
pseudoalloy cooper-carbon, 19-697
SiC/Al, 19-547
Compound materials, 18-263
Compressibility, 18-723
Compressive yield strength, 20-643
Computer aided design (CAD), 21-685
parts tooling, 18-231
Computer applications, 18-417
finite element modeling (FEM), 18-269
Computer projections, unique database, 21-699
Computer simulation, 18-479
Concentration profiles, 20-751
Consolidation
Fe-base alloys, 20-93
stainless steel, 20-787
Consolidation processes
cold isostatic pressing, 18-627
dynamic consolidation, 18-645
hot isostatic pressing, 18-627
process modeling, 18-23
vacuum sintering, 18-627
Consolidation under atmospheric pressure (CAP), nickel aluminide, 20-543
Constitutive behavior, 18-465
Continuous sintering furnaces, 21-243
pusher furnace, 18-315
Copper, lead, 20-125
Copper alloys, 18-645, 20-79
aging, 19-655
copper-zirconium alloys, 19-489
electrical conductivity, 20-503
processing, 19-655
spinodal decomposition, 19-655
Corrosion resistance, 19-727
high temperature materials, 21-215
sintering atmospheres, 21-215
Corrosion resistant stainless steel parts, market development, 20-27
Cost estimation, forging, 21-723
Cost reduction, 18-37
Cratering, 20-751
Customer/supplier communication, 18-691
Cutting materials
cemented carbides, 19-345
high speed steel, 19-271
tool wear, 19-365
Cutting tools, 20-751
Debinderizing, plastic injection molding, 18-315
Deformation behavior, 18-465, 18-507, 19-565
Densification, 21-669
Densification process, densification rate, 20-675
Density, 20-691
Density variation, 18-449, 18-465, 21-23
P/M copper, 18-287
Design, 21-685
tooling, 18-269
Design criteria, 18-23, 18-507, 20-205, 21-55
steam blackened materials, 21-275
Development/trends, 20-35
Diamond tool, 20-443
Die pressing, 18-465
Diffraction pattern, 20-691
Diffusion, 20-225
Dimensional change, copper/tin mixtures, 19-671
Dimensional control, 18-23, 19-641

Dispersion
iron base alloys, 21-461
mechanical alloying, 21-479, 21-511
nickel base alloys, 21-461
Dispersion strengthening, 20-503
cobalt base alloys, 20-361
Donor-Acceptor Catalyzing Theory, electron cycle, 20-243
Drillability indices, data analysis, 21-327
Dual-phase steel, 20-293
Duplex structures, 20-391
Dynamic properties, steam blackened materials, 21-275

Education problem, 20-13
Effective stress, hot pressing, 20-675
Electrical contact alloys, 18-597
Electrical materials, 20-339
contact materials, 19-75
Electro-discharge compaction (EDC)
consolidation process, 19-591
rapid solidification, 19-591
Embrittlement, steam blackened materials, 21-275
Enhanced alloying, 18-569
Enhanced sintering, P/M copper, 18-287
Experimental design, Taguchi orthogonal array, 18-551
Expert system, powder forging, 21-685
Explosive compaction, 20-63
Extrusion, 20-587, 21-575
aluminum alloys, 19-547
mechanical properties, 19-519

Fabrication, 18-691, 18-799, 21-723
net shape, 18-661
ring shaped components, 18-231
Fatigue, 21-67
Fatigue behavior, 21-41
constant and variable amplitudes, 21-55
critical defect sizes, 21-3
microstructure, 21-23
Fatigue properties, 18-155
crack propagation thresholds, 18-183
non-metallic inclusions, 18-143
rolling contact fatigue, 18-119
FeAl, 20-491
Ferrous alloys, 21-55
powder forging, 18-197
powder production, 20-53
Ferrous P/M, 18-535, 18-569, 20-339, 21-85, 21-563
heat treating, 21-301
high strength sintered steel, 21-197
nickel coated iron powder, 20-311
Fine grained structure, 20-691
Fine powder, 20-169, 20-225, 20-339
mechanical processing, 20-53
Finishing, 19-379
Finite element modeling (FEM), 18-269
Flame spraying, 18-789
Flank wear, 20-751
Fluidized bed
coating various substances, 20-311
jet milling process, 20-53
Flux-enhanced sintering, 18-569
Forging, 18-107, 18-119, 18-155, 18-213, 18-269, 19-565, 20-669, 21-591
ferrous alloys, 18-231
helical gears, 18-253
low alloy steels, 18-197
mechanical properties, 18-197

multiphase materials, 18-263
non-metallic inclusions, 18-143
Formability, forgeability, 19-533
Fracture, cemented carbides, 19-53
Fracture behavior, 19-281
Fracture toughness, 19-33, 19-43
mechanical properties, 19-345
Fracturing, 20-691
Full density, 21-183
Full density properties, 18-213
Full density sintering, high speed steels, 19-329
Fully dense materials and processing, 19-281

Gas atomization, 19-709, 20-169, 20-181, 20-187, 20-205, 20-225, 20-713
Gaseous reduction of magnetite, morphological changes during, 20-265
Grain growth, densification, 19-249
Grinding, 20-691

Hardness, 19-43, 20-691, 20-751, 21-361
Heat flow
compacts, 21-755
die chill, 18-269
Heat transfer, microstructural evolution, 19-457
Heat treating, 18-213, 20-773
nitocarburizing, 21-259
Heat treatment, transformation temperature, 19-23
Heavy metal alloys, 19-91
Hierarchical structures, 20-467
High density, 18-155, 19-235, 21-563, 21-575, 21-605
parts, 18-493
High speed steel, 19-271, 19-281, 20-421
cutting tools, 19-319
dimensional control, 19-305
effects of particle size and shape, 19-329
M-2 and T-15, 19-329
sintered properties, 19-305
wear of HSS, 19-365
wear resistant materials, 20-409
High strength steels, 21-67
High temperature compaction, 18-479
High temperature materials, 20-575, 21-157
hot isostatic pressing (HIPing), 20-739
HIP/peeling, strip, 18-57
Homogeneous alloy, 20-691
Homogenization, 21-171
Hot consolidation, 20-669
hot extrusion, 19-547
Hot isostatic pressing (HIPing), 19-91, 19-727, 20-225, 20-575, 20-599, 20-629, 20-745, 20-751, 20-773
alloy steels, 18-37
defect healing, 18-77
deformation behavior, 18-11
high speed steel, 19-319
near-net shapes, 19-709
nickel aluminide, 20-543
shape control, 18-23
steel, 19-697
tungsten powders, 19-249
Hot pressing, 20-691

Image analyzer, 20-187
Immiscible alloys, copper base, 20-125
Impact strength, infiltrated ferrous P/M, 21-123
Inclusions (metallic), 20-225
Inert gas atomization, 20-181
Infiltration, 20-481, 21-157
ferrous P/M, 21-123
iron copper parts, 20-279

Infiltration technology, market development, 20-27
Information flow, 20-13
Injection molding, 18-403, 18-417, 20-169, 20-763
  fine iron powder, 20-279
  intermetallic matrix composites, 18-299
  P/M copper, 18-287
  powder properties, 18-391
Ink jet, 18-661
Innovation cycle, P/M parts, 20-27
Intermetallic compounds, 20-559
  aluminides, 20-491
  liquid phase sintering, 20-511
  nickel aluminides, 20-531
Internal carburizing, 18-569
Iron, 21-41
Iron nucei, 20-265
Iron phosphorus alloys, 18-535, 18-757, 20-655
Iron/steel powder, 20-669
Isostatic pressing, 20-751

Liquid phase, 20-751
Liquid phase sintering, 18-535, 18-581, 18-597, 18-757, 19-155, 21-171
  Fe-P alloy, 18-551
  grain growth inhibition, 19-139
  high speed steels, 19-329
  microstructure, 18-615
  reactive sintering, 20-511
  supersolidus sintering, 21-139
  tungsten heavy alloys, 19-63
Lithium, 20-691
Low alloy ferrous powders, partially alloyed powders, 21-535
Low alloy steel, 21-85, 21-605
  fatigue behavior, 18-183
  mechanical properties, 21-171
Low alloy steel with chromium, 18-535
Low alloy steel powder, 21-521
  with Cr, 21-521
Low carbon steels, reactive alloys, 19-457
Low density high strength structural parts, sintered bush, 20-279

Machinability
  distaloy, 21-381
  heat treatment, 21-381
  high strength sintered steels, 21-381
  prealloyed manganese sulfide, 21-327
  sintered steel, 21-381
Machining, 19-379
Magnesium, 20-691
Magnesium sintered composite alloy
  composition and structure, 19-579
  mechanical properties, 19-579
  sintering condition, 19-579
Magnetic field compaction, 18-479
Magnetic materials, 18-745, 18-757, 20-701
  Fe-Nd-B, 18-775
  FeNiSO, 18-351
  magnetic properties, 18-799
Magnetic properties, 18-745, 18-757
  metal injection molding, 18-363
  testing methods, 18-363
Magnetite, 20-265
Manganese bearing inclusions, metallographical study, 21-327
Manganese sulfide
  compressibility, 21-361
  dimensional tolerance, 21-361
  machining of ferrous P/M parts, 21-361

sintered P/M parts, 21-361
tensile rupture strength, 21-361
Manganese sulfide additions, 18-155
Manufacturing, 18-107
tool steel, 20-681
Mass transport, 20-751
Master alloy, 20-691
Material selection criteria, 18-507
Mechanical alloying, 20-63, 20-691, 20-707, 21-397, 21-445
aluminum alloys, 21-511
aluminum-hafnium, 21-415
attritor milling, 20-431
commercial status, 21-461
intermetallic compounds, 21-479
process control, 21-461
production scale-up, 21-461
titanium alloys, 21-429
Mechanical behavior, 18-675, 20-421
Mechanical processes, jet milling with inert gas, 20-53
Mechanical properties, 18-37, 18-155, 18-333, 19-201, 19-709, 19-727, 20-751, 20-779, 21-41, 21-575
embrittlement, 20-655
Fe-VC/CO-WC, 20-467
Fe-P, 18-551
fracture behavior, 21-197
fracture toughness, 18-71
heat treatment, 21-605
iron nickel alloy, 20-643
nickel aluminide, 20-543
property comparision, 20-531
steam blackened materials, 21-275
strain aging, 19-173
tensile and fatigue, 20-361
tensile strength, 21-535
tungsten heavy alloys, 19-139
Melt spinning, 20-587, 21-495
Metal to ceramic joining, 18-675
Metal fiber, 21-653
Metal injection molding, 18-333, 18-351, 18-417
market development, 20-27
soft magnetic materials, 18-363
Metal matrix, 20-481
Metal matrix composites, 19-457
mechanical properties, 20-451
Metallization of ceramic powder, VC and WC, 20-467
Metal-polymer composites, 20-481
Microstructure, 19-33, 20-655, 20-707, 20-745, 20-773, 21-67, 21-495
bainite, 21-197
Co-base/Fe-base alloys and composites, 20-467
copper-zirconium alloy, 19-489
heat flow, 20-79
homogeneous, 20-125
microhardness variation, 20-323
RS cobalt base alloy, 20-361
spray deposited aluminum alloys and composites, 19-397
steam blackened materials, 21-275
toughness and hardness, 19-43
Microstructure control, liquid phase sintering, 19-139
Modeling, 18-269, 18-449
early-stage sintering, 21-669
sintering atmospheres, 21-215
Modulus of elasticity, 20-691
Molding, injection molding, 20-763

Molybdenum additions, 20-587
  tungsten heavy alloys, 19-139
Morphology, 21-41, 21-495

Near net shape, 21-183
Near net shape processing, nickel base alloy, 19-503
Net shape/near net-shape, 18-711
Ni-Cu alloys, 21-139
Ni_3Fe, 20-559
NiAl, 20-491
Nickel aluminide
  alumina fiber, 18-299
  intermetallic compound, 20-543
  nickel coated aluminum powder, 20-311
  synthesis, 20-511
Nickel base alloys, 18-675, 20-225, 20-575, 20-587, 20-599
  nickel aluminides, 21-479
  superalloys, 20-745, 20-773
Niobium alloy
  P/M manufacturing, 19-201
  physical properties, 19-187
Nitrogen, 20-751
Nitrogen/hydrogen atmospheres, metal injection molding, 18-315
Non-metallic inclusions, 18-107, 20-225, 20-745
  18-107, 18-119
Nondestructive evaluation, mechanical properties, 19-43

Order, 20-559
Ordnances, shaped charges, 20-629
Osprey process, 19-441, 19-475
Oxide dispersion strengthening
  aluminides, 20-491
  titanium oxide, 19-201
Oxygen content, 18-757, 20-225, 20-323
  hydrogen content, 20-187

P/M for electronics, 20-339
Palmqvist indentation test, 19-33
Partially alloyed steel powder, 21-521
Particle size, 20-691
  effect on sintering, 19-329
Particle size distribution, 20-181, 20-339, 20-745, 21-23
  log normal, 19-249
  steep curve, 20-53
Particulate composites, yttria particles, 18-581
Particulate reinforcement, titanium diboride, 20-451
Parts, 21-591
  fabrication, 19-753
  ferrous P/M, 20-35
  mechanical properties, 20-681
  physical properties, 20-681
  properties comparisions, 20-681
  wear resistant materials, 20-681
Parts applications
  automotive, 20-3
  biomedical, 20-3
  computers, 20-3
  injection molding, 20-3
  lock hardware, 20-3
  roll compacted strip, 20-3
  systems considerations, 20-27
Pb-Cu base bearings, 21-445
Phase redistribution processing, 21-445
Phenomena modeling, 18-479
Plasma processing, 19-155
Plasma spraying, 18-789
  beryllium, 19-753
  low pressure plasma spray, 19-743
Plastic deformation, tungsten heavy alloys, 19-139
Poisson's ratio, 20-643
Pore morphology, 21-67
Porosity, 20-751, 21-41
  hollow particles, 20-187

Porous material, 21-67,
21-637, 21-669
filter, 21-653
porous nickel, 21-745
Porous structure analysis,
21-745
Preform design, powder
forging, 21-685
Preforms, gears, 18-71
Press charge preweighing,
powder weighing,
18-711
Press systems
applications-automotive,
20-723
automation, 20-723
processing, 20-723
Pressure, 20-691
Pressureless sintering, 20-751
Process control, copper
strips, 21-735
Process modeling
billet cold rolling, 21-735
droplet consolidation at
substrate, 19-397
droplet-gas interaction,
19-397
process control, 20-205
Process variables,
manufacturing
conditions, 21-699
Processing, 18-213, 18-675,
18-691, 20-293,
20-629, 21-85
coining, 18-507
Cu-Ni-Sn alloys, 19-655
hot isostatic pressing
(HIPing), 18-71
sizing, 18-507
Production, 20-63, 20-169,
20-225, 20-713,
21-397
flame spraying, 19-519
fluidized bed process, 20-53
gaseous reduction processes,
19-75
nickel carbonyl
decomposition,
20-311
properties, 20-323
refractory alloy powder,
20-729
Production methods, 20-159
Properties, 20-225, 20-339,
20-773
Properties prediction,
mechanical
properties, 21-699
Property comparison
corrosion resistance, 21-511
mechanical properties,
21-511
nickel base alloys, 21-215
steam blackened materials,
21-275
Prototype, 18-661
Pulsatory compaction, 18-479,
20-701

Quality, dimensional
precision, 18-507

Rapid solidification, 19-155,
20-503, 20-587,
21-445, 21-495
gas atomization, 20-361
iron/steel powder, 20-323
melt spinning, 18-775
niobium alloys, 19-187
plasma spraying, 19-743
spray forming, 18-661
Rapid solidification
(RSM/RSP/RST),
20-629
Rapidly solidified powders,
18-597
Rate controlled sintering,
activated sintering,
19-235
Reactive sintering,
densification,
20-511
Recycling of steel swarf,
20-293
Refractory metals, tungsten
multi-phase
materials, 19-235
Refractory metal alloys,
19-91, 20-729
molybdenum, 19-173
niobium, 19-187
niobium aluminides, 21-479

Refractory metal powder, processing, 19-187
Repeated impact property, tungsten carbide-cobalt alloy, 19-53
Repressing, 18-213
  processing, 21-605
  rotary cold repressing, 18-493
Residual stresses, 20-391
Rotation evaporator, 20-751

Scanning electron microscopy (SEM), 20-691, 20-751
Selection criteria, high temperature materials, 21-215
Sieving, 20-751
Silicon, 20-751
Silicon carbide, 20-751
Silicon nitride, 20-751
  non-oxide additive, 21-619
  non-toxic non-oxide additive, 21-619
  thermo-mechanical properties, 21-619
Simulation, 18-449
Sintered alloy steel, 18-535, 20-431
Sintered high speed steel, 21-183
Sintered properties, 18-351
  relationship to machinability, 21-327
Sintering, 19-753, 20-691
  surface reactions, 20-787
Sintering additives, 20-751
Sintering atmospheres, 20-779
Sintering behavior, 19-641, 21-85
  simultaneous sintering, 21-669
Sintering furnaces, sintering equipment, 21-243
Sintering and HIPing, high speed steel, 19-319
Sintering mechanism, growth and coalescence, 19-63
Sintering process, 18-789
  carbides/cemented carbides, 18-91
  optimization, 18-551
Sintering productivity, increase during sintering, 21-243
Sintering quality, improvement during sintering, 21-243
Sintering under microgravity, space research, 19-63
Soft magnetic materials, 18-363
Soft magnetic properties, effect of sintering temperature, 18-363
Solution heat treatment, 20-691
Specific stiffness, 20-691
Spherical powders, 19-155, 20-169, 20-225
Spinel structure, 20-265
Spinodal decomposition, WC-TiC-ZrC-Co, 19-15
Sponge iron powder, 20-265
  Hoganas method, 20-243
  hydrogen reduced, 20-279
Spray atomization and deposition, liquid dynamic compaction, 19-457
Spray deposition/spray forming, 19-489
  Osprey process, 19-397
Spray forming
  piping, 19-503
  spray deposition, 19-441, 19-475
  strip casting, 19-457
Stainless steel, 18-757, 20-779, 21-313
  hardenable grade, 18-57
  porous structures, 21-637
Stainless steel 316L, 18-351
Statistical process control (SPC), 18-119, 21-23
Steam blackening of P/M parts, mechanical properties, 21-275

Steel machining, 20-751
Steel powder, oil-atomized powder, 21-197
Strain hardening, 20-691
Strength analysis, 21-745
Strength of bonds, 21-745
Stress analysis, 18-433
Sublimation of lithium, 20-691
Superalloys, 20-575, 20-599, 21-397
  mechanical properties, 20-739
Superconductivity, 18-789
Superplasticity, nickel aluminide, 20-543
Supersolidus sintering, 18-597
Surface analysis, 20-93
  stainless steel powder, 20-787
Surface hardening, by steam blackening, 21-275
Surface reactions, 20-93
Surface treatment, 21-259
  mechanical and thermal, 21-55
Swirled gas jet, 20-713

Theoretical density, 20-751
Thermal shock resistance, 20-751
Titanium alloys, 19-727
  titanium aluminides, 19-709, 21-479
  titanium carbide, 19-345
Tool steel, 19-271
  microstructure, 18-71
Tooling, 18-691
  stress analysis, 18-433
Tooling production, 18-691
Transient liquid phase, densification, 20-511
Transmission electron microscopy (TEM), 20-691
Tungsten heavy alloys, 18-645, 19-155
  liquid phase sintering, 19-139
Tungsten multiphase materials, 19-91, 20-279

Ultrasonic atomization process, 20-159
Upsetting, 18-213

Vacuum, 20-691
Vacuum melting, 20-181
Vacuum sintering, 21-183
  high speed steel, 19-319
  hot isostatic pressing (HIPing), 18-91
Vibratory mill, 20-691

Wear mechanism, 20-751
Wear of P/M, iron base alloy, 19-681
Wear prevention, solid lubrication, 20-125
Wear resistant coating, steam blackened ferrous materials, 21-275
Wear resistant materials, 19-305, 20-421, 20-431, 20-587, 21-157, 21-563
  composite materials, 20-377
  wear testing, 19-345
Wear testing, 19-305, 21-157
  composite materials, 20-377
  cutting tests, 19-365
Wustite, 20-265

Zn-22%Al alloy powder, superplasticity, 20-675

—NOTES—

—NOTES—

—NOTES—

—NOTES—

—NOTES—

—NOTES—

—NOTES—

—NOTES—

—NOTES—

—NOTES—

—NOTES—